AGRISCIENCE

Fundamentals & Applications

2nd Edition

In Memory of
Elmer L. Cooper

**Delmar Publishers
is proud to support
FFA activities**

AGRISCIENCE

Fundamentals & Applications

2nd Edition

ELMER L. COOPER

Delmar Publishers

I(T)P™ An International Thomson Publishing Company

Albany • Bonn • Boston • Cincinnati • Detroit • London • Madrid
Melbourne • Mexico City • New York • Pacific Grove • Paris • San Francisco
Singapore • Tokyo • Toronto • Washington

NOTICE TO THE READER

Cover photo courtesy of: Michael Dzaman
Cover design: Wendy A. Troeger

Delmar Staff:
Publisher: Tim O'Leary
Acquisitions Editor: Cathy L. Esperti
Developmental Editor: Cathy L. Esperti
Senior Project Editor: Andrea Edwards Myers
Production Manager: Wendy A. Troeger
Marketing Manager: Maura Theriault

COPYRIGHT ©1997
By Delmar Publishers
a division of International Thomson Publishing Inc.

The ITP logo is a trademark under license.

Printed in the United States of America

For more information, contact:

Delmar Publishers
3 Columbia Circle, Box 15015
Albany, New York 12212-5015

International Thomson Publishing Europe
Berkshire House 168-173
High Holborn
London, WC1V 7AA
England

Thomas Nelson Australia
102 Dodds Street
South Melbourne, 3205
Victoria, Australia

Nelson Canada
1120 Birchmount Road
Scarborough, Ontario
Canada M1K 5G4

International Thomson Editores
Campos Eliseos 385, Piso 7
Col Polanco
11560 Mexico D F Mexico

International Thomson Publishing Gmbh
Königswinterer Strasse 418
53227 Bonn
Germany

International Thomson Publishing Asia
221 Henderson Road #05-10
Henderson Building
Singapore 0315

International Thomson Publishing—Japan
Hirakawacho Kyowa Building, 3F
2-2-1 Hirakawacho
Chiyoda-ku, 102 Tokyo
Japan

     5 6 7 8 9 10 XXX 02 01 00 99 98 97

Library of Congress Cataloging-in-Publication Data

Cooper, Elmer L.
     Agriscience : fundamentals & applications / Elmer L. Cooper. —
2nd ed.
          p.   cm.
     Includes index.
     Summary: An agriscience textbook exploring such topics as
environmental technology, plant sciences, integrated pest
management, interior and exterior plantscape, animal sciences, food
science, and agribusiness.
     ISBN 0-8273-6278-1 (textbook)
     1. Agriculture—Juvenile literature.   [1. Agriculture.]
I. Title.
S495.C78   1995
630—dc20   95-15156
                                                  CIP
                                                  AC

CONTENTS

PREFACE

Agriculture education has undergone a drastic series of changes over the past decade, and it continues to change just as the industry and business of agriculture has experienced such changes. Today's agriculture education demands new approaches in teaching and new directions for learning. *Agriscience: Fundamentals and Applications,* 2nd Edition, is a textbook that addresses these new approaches and directions.

Written in a clear, easy-to-read style, *Agriscience: Fundamentals and Applications,* 2nd Edition, integrates basic biological and technological concepts with principles of agricultural production and business. The student is taken on a journey through all of today's agricultural areas including environmental technology, biotechnology, integrated pest management, plant sciences, interior and exterior plantscaping, animal sciences, food science, and agribusiness. The text is an introduction to the field of agriscience designed to acquaint the student with significant scientific and production concepts that are integrated throughout the text. Urban, suburban, exurban, and rural agriscience concerns are addressed.

Major additions to the second edition include a new unit on biotechnology, section openers to highlight the importance of the material in each section, and a Biotech Connection added to each unit to introduce students to current research in biotechnology applications that relate to each particular unit. Agri-Profiles provide students with insight into career areas. By highlighting aspects of various careers directly related to the information presented in each unit, students are motivated to investigate careers that excite and interest them. All data has been updated to the latest available, and every photograph has been replaced with new photographs in full color. The text is truly a state-of-the-art book.

Ten major sections of the text are further divided into thirty-six units. Each unit contains an agriscience career profile, a Biotech Connection, performance-based Objectives, Terms to Know, and a Materials List. Chapters are clearly subdivided, and the chapter-end reviews include objective questions, essay questions, and many student activities. Mathematical and verbal skills are emphasized in both the Student Activities and Self Evaluation questions. Scientific concepts and agricultural projects are addressed in the activities. A comprehensive glossary can be found at the end of the text. The career profiles along with several agriscience career chapters are designed to acquaint the student with the many technical and professional careers associated with agribusiness and agriscience.

Agriscience: Fundamentals and Applications, 2nd Edition, is a text designed for use in agriscience courses, agricultural survey courses, introductory agriculture courses, and general agriculture courses. It

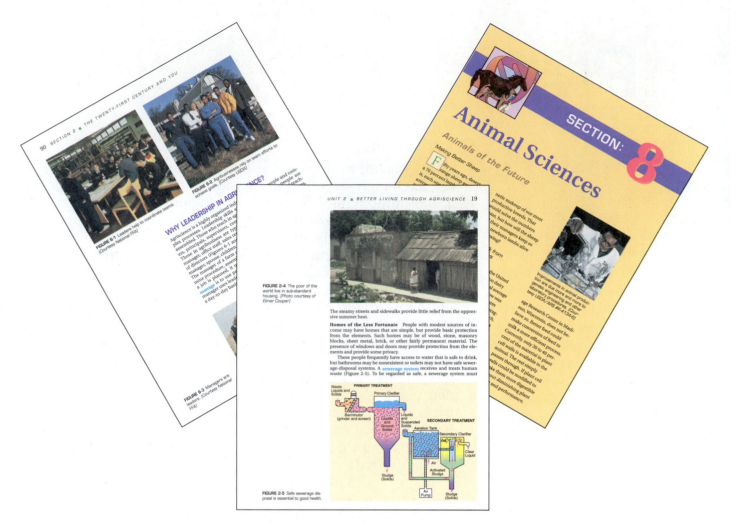

is an overview that introduces the student to many agriscience concepts, and readies them for a later in-depth treatment of each subject.

This text was prepared with the help of many individuals, associations, and corporations. Without their help and their commitment to agriculture and agriculture education, this text could not have been properly written.

The second edition of *Agriscience Fundamentals and Applications* has been carefully designed to enhance the study of agriscience. For best results, you may want to become familiar with the features incorporated into this text as well as with accompanying learning tools.

ADDED FEATURES

■ Full-color photos and illustrations to enhance learning.

■ Special section entitled Biotech Connection added to each unit. The Biotech Connections are descriptions of recent or current reseach in biotechnology related to the units under study. This feature provides the instructor and students with information, ideas, and references for further exploration, experimentation, and career considerations in the topical areas.

■ Career-focusing Agri-Profiles for each unit have been updated to include a wide variety of agribusiness, biotechnology, research, and agriculture careers.

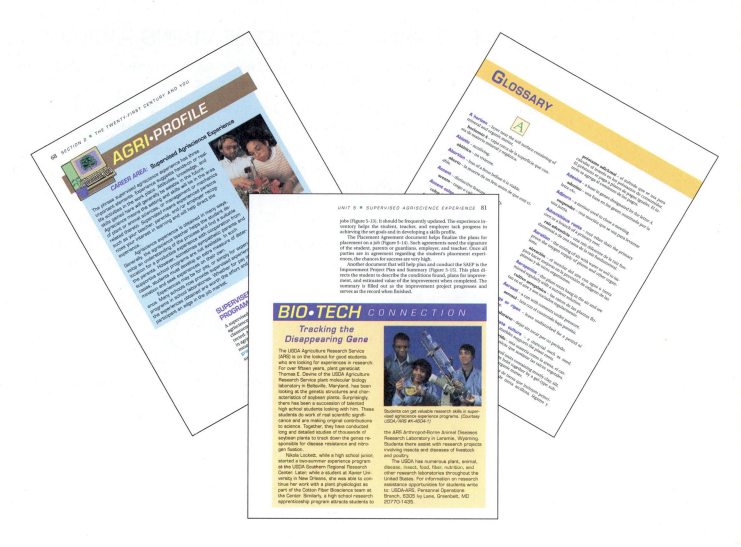

■ English-Spanish glossary with over 1,200 definitions provides an excellent study aid and references.

ENHANCED CONTENT

The updated and enhanced content addresses the evolving agriscience curriculum:

■ Broad applications to science, math, agriculture, natural resources, and the environment provide the appropriate balance for the agriscience curriculum.

■ New unit on biotechnology describes the field and its agriscience connection.

■ Expanded coverage of integrated pest management. This coverage provides modern concepts in pest control to minimize use of chemicals and emphasize biological and cultural control of pests.

■ Expanded coverage of food science and technology. Presents an integrated view of food science and the food industry from source to retail market, with emphasis on career possibilities.

■ New content on personal skills in career preparation, experiences, and leadership.

EXTENSIVE TEACHING/LEARNING PACKAGE

The complete supplement package was developed to achieve two goals:

1. To assist students in learning the essential information needed to continue exploration into the exciting field of agriscience.

2. To assist instructors in planning and implementing their instructional programs for the most efficient use of time and other resources.

Instructor's Guide to Text

The Instructor's Guide provides answers to the end-of-chapter questions and additional material to assist the instructor in the preparation of lesson plans.

Lab Manual

Order #0-8273-6941-7

This comprehensive lab manual reinforces the text content. It is recommended that students complete each lab to confirm understanding of essential science content. Great care has been taken to provide instructors with low cost, strongly science-focused labs.

Lab Manual Instructor's Guide

Order #0-8273-6942-5

The Instructor's Guide for the lab manual provides answers to lab manual exercises and additional guidance for the instructor.

Teacher's Resource Guide

Order #0-8273-6957-3

This new supplement provides the instructor with valuable resources to simplify the planning and implementation of the instructional program. It includes transparency masters; motivational questions and activities; answers to questions in the text; and lesson plans to provide the instructor with a cohesive plan for presenting each topic.

Computerized Testbank

Order #0-8273-7487-9

An all-new computerized testbank (DOS 3.5″and MAC compatible) with more than 500 new questions gives the instructor an expanded capability to create tests.

Agriscience Testmaster CD-ROM

Order #0-8273-7496-0

This testbank contains over 3,000 questions correlated to the following Delmar titles:

Cooper/*Agriscience: Fundamentals and Applications*, 2E
Cooper/*Agricultural Mechanics: Fundamentals and Applications*, 3E

Reiley/*Introductory Horticulture,* 5E
Herren/*The Science of Agriculture: A Biological Approach*
Ricketts/*Leadership: Personal Development and Career Success*
Camp/*Managing Our Natural Resources,* 3E
Gillespie/*Modern Livestock and Poultry Production,* 5E.

The CD-ROM also contains ESAgrade Electronic Gradebook and ESAtest III. These enable the instructor to create and administer on-line tests and keep track of grades.

Delmar's Agriscience Laserdisc Package

Order #0-8273-7300-7

Delmar's Agriscience Laserdisc Package offers instant accessibility to interactive video segments on four double-sided Level I laserdiscs directly corresponding to all the major content areas in agriculture. The discs can be obtained as a package or individually:

Disc 1: Animal Science
Order #0-8273-7301-5

Disc 2: Plant Science
Order #0-8273-7302-3

Disc 3: Business and Mechanical Technology
Order #0-8273-7303-1

Disc 4: Forestry and Natural Resources Management
Order #0-8273-7304-X

In addition, a correlation guide, complete with barcodes, correlates all barcodes contained on the laserdiscs with all of Delmar's top agriscience texts and lab manuals.

Color Transparency Package

Order #0-8273-6989-1

Forty color transparencies based on illustrations from the text provide the instructor with yet another means of promoting student understanding of agriscience concepts.

ACKNOWLEDGMENTS

The author and publisher wish to express their appreciation to the many individuals, FFA associations, and organizations that have supplied photographs and information necessary for the creation of this text. A very special thank you to all the folks at the National FFA organization and the USDA photo libraries, who provided many of the excellent photographs found in this textbook. Because of their efforts, this is a better book. Special thanks, also, to:

Chesapeake Bay Foundation
    Rodney M. Coggin, Public Information Director

National FFA Organization
    Bill Stagg, Photographer

United States Department of Agriculture
    John Kucharski—Chief, Photography Unit, ARS

Scott R. Bauer—Photographer, *Agricultural Research* magazine

Lloyd E. McLaughlin—Editor, *Agricultural Research* magazine

Linda R. McElreath—Associated Editor, *Agricultural Research* magazine

Agway, Inc.

FFA Chapter, Denmark, Wisconsin
    Ken Seering, Photographer

The faculty and students of Anderson Valley High School, Boonville, California

Michael T. Gill, Bedford, Texas

U.S. Fish and Wildlife Service
    Charles Cadieux, Photographer
    Bob Hines, Artist
    Lee Emery, Photographer
    Earl W. Craven, Photographer

Bureau of Sport Fisheries & Wildlife
    E. R. Kalmbach, Photographer
    Jack Dermid, Photographer
    E. P. Hadden, Photographer
    Gale Monson, Photographer

Rogers Vocational Center, Gardendale, Alabama
    Roy Holsomback, Photographer

John Deere Company

American Veterinary Medical Association

American Soybean Association

American Fisheries Society
    John Scarola, Photographer

American Shetland Pony Club, Peoria, Illinois

Omaha Livestock Market, Inc.

Riburn Industries, Inc.

SHOWCASE Magazine, copyright 1989

Calloway Nurseries, Hurst, Texas

Christian Children's Fund
Alison Cross, Public Relations Representative

FFA Chapter, Gilbert, Arizona
Joe Granio, Photographer

American Quarter Horse Association, Amarillo, Texas

Kevin Mathias, PhD, College Park, Maryland

The New Pesticide User's Guide, by Bert L. Bohmont

Vocational Agricultural Service, University of Illinois at Urbanaq-Champaign

The Fate of an Insecticide Discharged by Aircraft (Flint and van den Bosch, 1977, after von Rumker et al., 1974)

Science of Food and Agriculture, CAST

Texas Vocational Instructional Services

Introductory Horticulture, Fifth Edition, by Edward Reiley and Carroll Shry, copyright 1997 by Delmar Publishers

Ornamental Horticulture, Second Edition, by Jack E. Ingels, copyright 1994 by Delmar Publishers

Soil Science and Management, by Jack Ingels, copyright 1985 by Delmar Publishers

University of Maryland Cooperative Extension Service

The United Nations Fund for Population Activities

Instructional Materials Service, Texas A & M University

Flint, M. L. and R. van den Bosch, 1981. *Introduction to Integrated Pest Management.* Plenum Press, NY, page 178.

Penn State University, Entomology Department and Agricultural Extension Education Department

Maryland Magazine, Autumn 1988, "Choking Off Life of the Bay" by Johnstone Quinan, copyright the *Washington Post.*

Turfgrass Science and Management, Second Edition, by Robert Emmons, copyright 1995 by Delmar Publishers

Dairy Farm Management, by Thomas Quinn, copyright 1980 by Delmar Publishers

Modern Livestock and Poultry Production, Fifth Edition, by James Gillespie, copyright 1997 by Delmar Publishers

Agricultural Mechanics: Fundamentals and Applications, Third Edition, by Elmer Cooper, copyright 1997 by Delmar Publishers

Chicago Board of Trade

Managing Our Natural Resources, Third Edition, by William Camp, copyright 1997 by Delmar Publishers

Childworld Magazine

USDA, Soil Conservation Service

G. T. Moody High School, Agriculture Education Department, photo by Jim Jones

U.S. Department of Interior
Lavonda Walton, Audiovisual Office, FWS

Honey Bee Research, Weslaco, Texas

C. W. Hardeman, Scottsboro High School Agriculture Education Department

George Rogers Area Vocational-Technical School

ChemLawn Corporation

Brouwer Equipment, Ltd.

American Rabbit Breeders Association

National Livestock and Meat Board

Stebco Products

Pa. Game News
Bob Mitchell

The author and publisher gratefully acknowledge the unique expertise provided by the contributing authors to the text. Their work provides a special assurance that the content in each unit is at the cutting edge of the respective topic and is particularly appropriate for the level of the text. The contributing authors are:

Robert S. DeLauder, Agriscience Instructor, Damascus, Maryland;

Thomas S. Handwerker, PhD, Department of Agriculture, University of Maryland at Princess Anne;

Curtis F. Henry, Business Manager, College of Agriculture, University of Maryland at College Park;

Dr. David R. Hershey, Assistant Professor, Department of Horticulture, University of Maryland at College Park;

Robert G. Keenan, Agriscience Instructor, Landsdowne High School, Baltimore, Maryland;

J. Kevin Mathias, PhD, Institute of Applied Agriculture, University of Maryland at College Park;

Regina A. Smick, EdD, Academic Advisor and Instructor, College of Agriculture, Virginia Tech, Blacksburg; and

Gail P. Yeiser, Instructor, Institute of Applied Agriculture, University of Maryland at College Park.

Finally, the work of Dollye L. Cooper, wife of the author and technical problem solver, was especially helpful in standardizing the manuscripts of the contributing authors. Her help in word processing and other details of manuscript preparation was invaluable.

REVIEWERS FOR THIS EDITION

Ermanno Arizzi
Eureka Senior High School
Eureka, CA

Dale Layfield
Penn State University
University Park, PA

Paul Day
Northfield, MN

Glenn Linder
Pecatonica Area Schools
Blanchardville, WI

Eric Dyer
Davis, CA

Tom Paulsen
Carroll Community High School
Carroll, IA

J. E. Forgety
Clinton High School
Clinton, TN

Douglas Plant
Nordheim I.S.D.
Runge, TX

Greg Goodrum
Mexia High School
Mexia, TX

Mack Sanders
South-Doyle High School
Knoxville, TN

Roger King
Holmen, WI

Murdock Leroy Gillis
Ponce De Leon, FL

Joel Larson
Belle Plaine High School
Belle Plaine, MN

PAST REVIEWERS

Ferrell M. Bridwell
Paul M. Dorman High School
Spartanburg, SC

Robert A. Sills
Warren Hills Regional High School
Washington, NJ

Marlies Harris
University of California, Davis
Davis, CA

Glenn B. Sims
Windsor High School
Windsor, IL

Gerald McDonald
Spring Branch Education Center
Houston, TX

Jim Welles
Cherokee High School
Rogersville, TN

Steve McKay
Anderson Valley Agriculture Institute
Booneville, CA

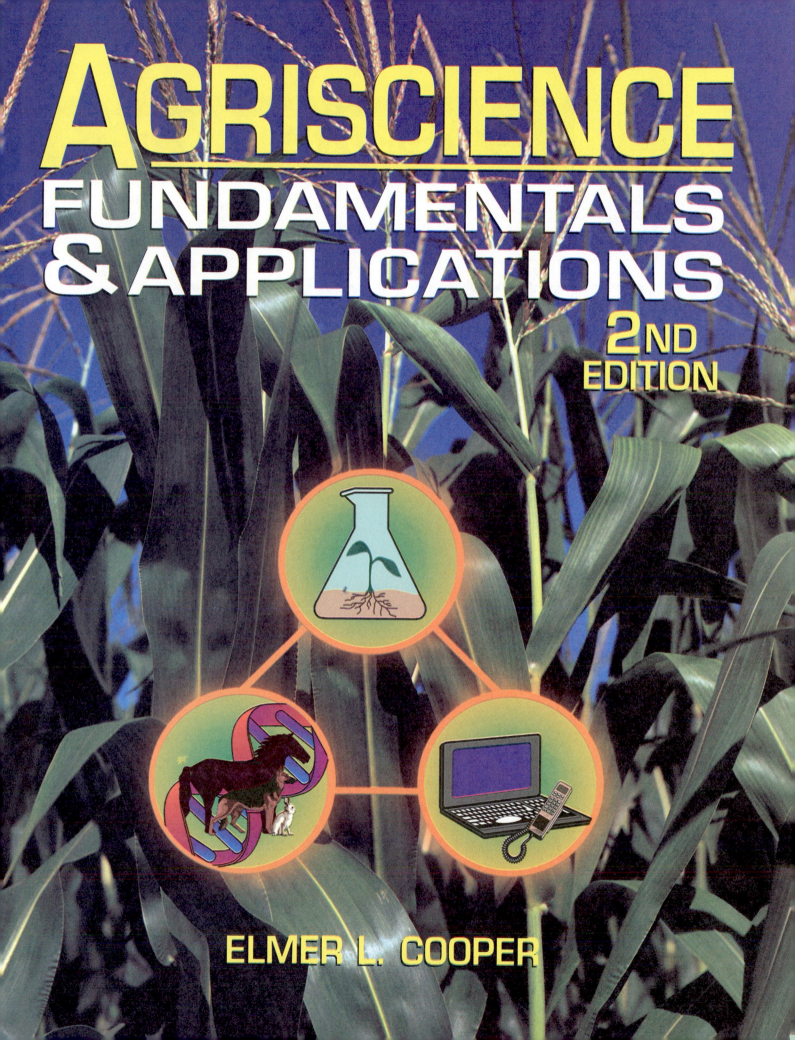

AGRISCIENCE

FUNDAMENTALS & APPLICATIONS

2ND EDITION

ELMER L. COOPER

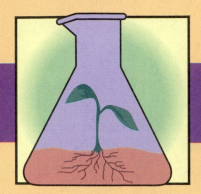

Agriscience in the Information Age

The Green Machine

Science and technology have opened up the door to areas of research never thought possible. Space station research, new frontiers to investigate, and our never ending quest for knowledge has lead to exciting new career areas.

You could be one of the people growing plants or animals in a space station miles above the earth. Or, you might be an engineer who designs the animal or plant-growing module of the space station, or a molecular geneticist or plant breeder tailor-making plants to grow well in low gravity, or a food scientist designing packaging for space-grown produce.

One fast growing career area is in plant science. As you will learn, plants are "green machines" which capture, package, and store energy from the sun through photosynthesis. They supply food and fiber for animals and humans to help sustain life. But, plants need human knowledge and energy to help them function in the overall "green machine" that constitutes our food, fiber, and natural resources system.

Whether you choose a job in plant or animal science, sales and marketing, mechanics, or processing, it is sure to be rewarding. By studying agriscience you open the door to exciting educational programs and careers that bring life and prosperity.

(Courtesy NASA)

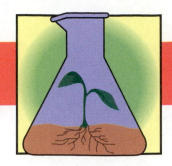

The Science of Living Things

OBJECTIVE

To recognize the major sciences that explain the development, existence, and improvement of living things.

MATERIALS LIST

✓ writing materials

✓ newspapers and magazines

✓ encyclopedias

COMPETENCIES TO BE DEVELOPED

After studying this unit, you should be able to:

■ define agriscience.

■ discover agriscience in the world around us.

■ relate agriscience to agriculture, agribusiness, and renewable natural resources.

■ state the major sciences that support agriscience.

■ describe basic and applied sciences that relate to agriscience.

Life in the United States and throughout the world is changing every moment of our lives. The space we occupy as well as the people we work and play with may be constant for a brief time. However, these are quick to change with time and circumstances. The things we need to know and the resources we have to use are constantly shifting as the world turns around us.

Humans have the gift of intelligence—the ability to learn and know (Figure 1-1). This permits us to compete successfully with the millions of other creatures that share the earth with us (Figure 1-2). In ages

FIGURE 1-1 Humans have the gift of intelligence and the ability to learn and know. *(Courtesy USDA/ARS #K-3414-1)*

FIGURE 1-2 The gift of intelligence has permitted humans to compete with animals even though most animals are superior to humans in other ways. *(Courtesy USDA/ARS #K-4703-12)*

FIGURE 1-3 Early humans had to rely on natural settings to shield them from danger and the elements. *(From Thode,* Technology, *Copyright 1994 by Delmar Publishers)*

past, humans have not always fared well in this competition. Wild animals had the advantages of speed, strength, numbers, hunting skills, and superior senses over humans. These superior senses of sight, smell, hearing, heat sensing, and reproduction all helped certain animals, plants, and microbes to exercise control over humans in order to meet their own needs.

The cave of the cave dweller, lake of the lake dweller, and cliff of the cliff dweller indicate human reliance on natural surroundings for basic needs (food, clothing, and shelter) (Figure 1-3). Those early homes gave humans some protection from animals and unfavorable weather. However, they were still exposed to the ravishes of disease, the pangs of hunger, the sting of cold, and the oppression of heat.

The world of agriscience has changed the comfort, convenience, and safety of people today. In the United States, we spend only 11 percent of our wages to feed ourselves (Figure 1-4). The agriscience, agribusiness, and renewable natural resources of the nation provide materials for clothing, housing, and industry at an equally attractive price.

AGRISCIENCE DEFINED

Agriscience is a relatively new term that you may not find in your dictionary. **Agriscience** is the application of scientific principles and new technologies to agriculture. **Agriculture** is defined as the activities concerned with the production of plants and animals, and related supplies, services, mechanics, products, processing, and marketing. Actually, modern agriculture covers so many activities that a simple definition is not possible. So, the U.S. Department of Education has used the phrase **"agriculture/agribusiness and renewable natural resources"** to refer to the broad range of activities in agriculture. Agriculture generally has some tie-in or tie-back to animals or plants. However, production agriculture, or farming and ranching, accounts for only one-fifth of the

FIGURE 1-4 Americans spend only 11 percent of their incomes on food. *(Courtesy USDA/ARS #CS-1081)*

FIGURE 1-5 Farming and ranching accounts for approximately one-fifth of the agricultural jobs in the United States. *(Courtesy USDA/ARS #K-4320-9)*

FIGURE 1-6 Agribusinesses are important to the life of most communities. *(Courtesy USDA/ARS #K-567-13)*

total jobs in agriculture (Figure 1-5). The other four-fifths of the jobs in agriculture are nonfarm and nonranch jobs. **Agribusiness** refers to commercial firms that have developed with or stemmed out of agriculture (Figure 1-6).

Renewable natural resources are the resources provided by nature that can replace or renew themselves. Examples of such resources are wildlife, trees, and fish. Some occupations in renewable natural resources are game trapper, forester, and waterman (someone who uses boats and specialized equipment to harvest fish, oysters, and other seafood).

Technology is defined as the application of science to an industrial or commercial objective. Hence, agriscience was coined to describe the application of high technology to agriculture. **High technology** refers to the use of electronics and state-of-the-art equipment to perform tasks and control machinery and processes (Figure 1-7). It plays an important role in the industry of agriculture.

Agriscience includes many endeavors. Some of these are aquaculture, agricultural engineering, animal science technology, crop science, soil science, biotechnology, integrated pest management, organic foods, water resources, and environment. **Aquaculture** means the growing and management of living things in water, such as fish. **Agricultural engineering** means the application of mechanical and other engineering principles in agricultural settings. **Animal science technology** refers to the use of modern principles and practice for animal growth and management (Figure 1-8). **Crop science** refers to use of modern principles in growing and managing crops. **Soil science** refers to the study of the properties and management of soil to grow plants. **Biotechnology** refers to the management of the genetic characteristics transmitted from one generation to another and its application to modern living. It may be defined as the use of cells or components of cells to produce products and processes (Figure 1-9).

The phrase **integrated pest management** refers to the many different methods used together to control insects, diseases, rodents, and

FIGURE 1-7 High technology is used extensively in agriscience. *(Courtesy USDA/ARS #K-3396-6)*

FIGURE 1-8 Veterinarians use animal sciences to help keep our pets and production animals healthy. *(Courtesy USDA/ARS #K-4807-1)*

FIGURE 1-9 Genetic engineering and other forms of biotechnology have become one of the most important priorities in research today. *(Courtesy USDA/ARS #K-5011-19)*

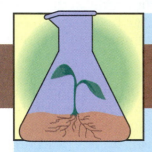

AGRI·PROFILE

CAREER AREA: Agriscientist

In recent years science has played an increasing role in the lives of plants and animals and the people around them. These living bodies include plants ranging in size from microscopic bacteria to the huge redwood and giant sequoia trees. They include animals from the one-celled amoeba to elephants and whales.

Only recently has science identified the nature of viruses and permitted humans to observe the submicroscopic world in which they exist. The electron microscope, radioactive tracers, computers, electronics, robots, and biotechnology are just a few of the developments that have revolutionized the world of living things. We call this world agriscience. Through agriscience, humans can control their destinies better than at any time in known history.

Students experience the wonder of living things. *(Courtesy USDA/ARS #CS-0146)*

Agriscience spans many of the major industries of the world today. Some examples are production, processing, transportation, selling, distribution, recreation, environmental management, and professional services. Studies and experiences in a wide array of basic and applied sciences are appropriate preparation for a career in agriscience.

other pests. **Organic food** is a term used for foods that have been grown without the use of chemical pesticides. **Water resources** cover all aspects of water conservation and management. Finally, **environment** refers to the space and mass around us. This generally means air, water, and soil.

AGRISCIENCE AROUND US

Whether you live in the city, town, or country, you are surrounded by the world of agriscience. Plants use water and nutrients from the soil and release water and oxygen into the air. Animals provide companionship as pets and assistance with work. Both plants and animals are sources of food. Many microscopic plants and animals are silent garbage disposals. They decay the unused plant and animal remains around us. This process returns nutrients to the soil and has many other benefits to our environment and well-being.

Agriscience encompasses the wildlife of our cities and country, and the fish and other life in streams, ponds, lakes, and oceans. Plants are used extensively to decorate homes, businesses, shopping malls, buildings, and grounds. When one crop is used less, another takes its place. This occurs even where land changes from farm use to suburban and urban uses.

Corn has long been referred to as king among crops in the United States. Yet, in one state, turf grass recently replaced corn as the number one crop. **Turf** is grass used for decorative as well as soil-holding purposes. This change has occurred as more land is being used for roads, housing, businesses, institutions, recreation, and other nonfarm uses (Figure 1-10).

FIGURE 1-10 Turf has become one of America's main crops. *(Courtesy USDA/ARS #CS-0311)*

BIO•TECH CONNECTION

The Agriscience Project

The scientific method is an excellent and widely used method for systematic inquiry and documentation of new findings. The agriscience student is encouraged to learn and use the scientific method for classroom, laboratory, and field studies. The following procedures will guide you in your quest for new knowledge in agriscience.

Step 1. Identify the Problem
Decide precisely and specifically what it is that you wish to find out. For example, "How much nitrogen fertilizer is needed to grow healthy bean plants?" Be careful to limit your topic to a single researchable objective. Your teacher can suggest other topics that could be researched.

Step 2. Review the Literature
This simply means read up on and become well informed about the topic. See what is already known about it. Magazines, newspapers, reference books, encyclopedias, trade journals, computer information systems, television, cooperative extension meetings, teleconferences, and personal interviews

The written report completes the research project and enables others to benefit from the new knowledge. *(Courtesy USDA/ARS 027)*

Many of the flowers used by florists in the United States come from Colombia, South America, and other foreign countries. Bulbs come from Holland and meat products are imported from Argentina. Lumber is shipped from the United States to Japan, only to return in the form of plywood and other processed lumber commodities. A drop in the price of sow bellies or an unexpected change in the price of grain futures in Chicago can affect business and investment around the world (Figure 1-11).

The great water-control projects on the Colorado River have permitted the transformation of the American Southwest from a desert to irrigated lands. This is now an area of intensive crop production that has stimulated national population shifts. Water management has transformed the great dust bowl of the American West into the "bread basket" of the world.

Agriscience helps create the magazines, newspapers, and journals of the nation. Similarly, radio and television rely on reporters, wildlife biologists, extension specialists, and others in agriscience. They help create radio and television programs about plants, animals, wildlife, market conditions, homemaking, gardening, lawn care, and dozens of related topics.

may be sources of appropriate information. Be sure to seek information on appropriate ways to conduct research on the type of problem you have chosen.

Step 3. Form a Hypothesis
After learning as much as you can about the topic, develop a hypothesis or statement to be proven or disproven, which will solve the problem. For example, "Trout grown in 60°F water will grow faster than trout grown in 75°F water."

Step 4. Prepare a Project Proposal
Prepare a proposal outlining how you think the project should be done. Include the timelines, facilities, and equipment required, as well as anticipated costs and a description of how you will do the project.

Step 5. Design the Experiment
Considering the information gathered in Step 2, develop a plan for carrying out the project so you can test the hypothesis. This is the most critical step in your research project. If this isn't done correctly, you may invest considerable time, work, and expense and end up with incorrect or invalid conclusions. The method or procedure should be carefully thought out and discussed with your teacher or other research authorities. This is to be sure that your design will actually measure what you are testing.

Step 6. Collect the Data
In this phase you actually conduct the experiment. Here you test and/or observe what takes place and record what you measure or observe.

Step 7. Draw Conclusions
Summarize the results. Make all appropriate calculations. Determine if the information allows you to accept or reject the hypothesis, or if the information is inconclusive.

Step 8. Prepare a Written Report
The written report provides you, your teacher, and other interested parties with a permanent record of your research. From this you can report to your peers, get course credit, and possibly apply for awards. Perhaps you can use a computer and hone your word-processing skills. For scientists, the written report becomes a permanent document that is kept by the research institution and becomes the basis for articles in research publications for all the world to see. The results become part of the "literature" on the topic.

FIGURE 1-11 Marketing in agriscience has become big business. *(Courtesy National FFA; FFA #18)*

AGRISCIENCE AND OTHER SCIENCES

Agriscience is really the application of many sciences. Colleges of agriculture and life sciences do research and teach students in these sciences. **Biology** is the basic science of plant and animal life. **Chemistry** is another basic science that deals with the characteristics of elements or simple substances. It includes the behavior of substances when combined with other substances. **Biochemistry** focuses on chemistry as it applies to living matter. These three are referred to as basic sciences. One must understand the basic sciences to work in the applied sciences.

Applied sciences utilize basic sciences in practical ways. For instance, **entomology** is the science of insect life. A knowledge of biology and chemistry is fundamental to entomology and the understanding of insects.

Agronomy is the science of soils and field crops, while **horticulture** is the science of fruits, vegetables, and ornamentals. **Ornamentals** are plants used for their appearance. Examples are flowers, shrubs, trees, and grasses. **Animal sciences** are applied sciences that involve animal growth, care, and management. They include veterinary medicine, animal nutrition, and animal production and care.

Agricultural economics addresses the management of agricultural resources, including farms and agribusinesses. Farm policy and international trade are important components of agricultural economics. **Agricultural education** covers teaching and program management in agricultural endeavors. Agricultural communications, journalism, extension service, and community development are also components of agricultural education. These and other disciplines are part of agriscience.

A PLACE FOR YOU IN AGRISCIENCE

What about career opportunities in agriscience? By 1990, it was reported that the nation's agricultural colleges were enjoying the strongest demand for graduates in a decade. A U.S. Department of Agriculture (USDA) study group forecasted a national shortage of 4,000 agricultural and life sciences graduates per year. The shortage became reality in the early 1990s. Employers are offering higher salaries and more job variety to agriscience college graduates than ever before. Consequently, graduates of high school agricultural, horticultural, or other agriscience programs can obtain good jobs and have rewarding careers (Figure 1-12). These are described in later units in this text. By studying agriscience, you open the door to exciting educational programs and careers that bring life and prosperity.

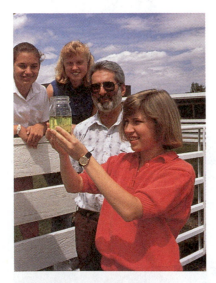

FIGURE 1-12 There are good jobs awaiting agriscience graduates. *(Courtesy USDA/ARS #K-4604-1)*

STUDENT ACTIVITIES

1. Write the Terms to Know and their meanings in your notebook.

2. List examples of animals that have better senses than do humans. Indicate the sense(s) along with the animals.

3. Write a paragraph or two on (1) cave, (2) lake, and (3) cliff dwellers. Explain how the types and locations of their homes provided protection from (1) animals and (2) unfavorable weather. An encyclopedia would be a good resource for this activity.

4. Ask your teacher to assign you to a small discussion group to talk about the responses to Activity 3.

5. Place a map of your school community on a bulletin board. Insert a colored map pin in every location of a farm, ranch, or agribusiness in your school community.

6. Talk to your County Extension Agent or another agricultural leader regarding the importance and role of agriscience, agribusiness, and renewable natural resources in your county.

7. Select one of the sciences mentioned in this unit. Prepare a written report on the meaning and nature of that science. Report to the class.

SELF EVALUATION

A. Multiple Choice

1. Humans have the ability to learn and know. This is known as
   a. achievement.
   b. intelligence.
   c. intuition.
   d. spontaneity.

2. The percentage of an average U.S. worker's pay used for food is
   a. 11 percent.
   b. 14 percent.
   c. 50 percent.
   d. 74 percent.

3. The best term to describe the application of scientific principles and new technologies to agriculture is
   a. agribusiness.
   b. renewable natural resources.
   c. farming.
   d. agriscience.

4. Harmful insects, rodents, and diseases are all referred to as
   a.   animals.
   b.   plants.
   c.   pests.
   d.   parasites.

5. Agriscience encompasses
   a.   wildlife and fish.
   b.   ornamental plants and trees.
   c.   farms and agribusinesses.
   d.   all of the above and more.

6. Irrigated lands are generally used for
   a.   intensive crop production.
   b.   wildlife refuges.
   c.   forests.
   d.   boating and fishing.

7. An example of a basic science is
   a.   agronomy.
   b.   aquaculture.
   c.   horticulture.
   d.   chemistry.

8. An example of an applied science is
   a.   animal science.
   b.   biochemistry.
   c.   biology.
   d.   chemistry.

9. One relationship of agriscience with many other sciences is that
   a.   agriscience is the application of many other sciences.
   b.   agriscience is entirely different from all other sciences.
   c.   agriscience is an old term and an old science.
   d.   agriscience is a very narrow science and easily defined.

10. The career and job outlook in agriscience is
   a.   a strong demand for college graduates.
   b.   a shortage of 4,000 trained workers per year.
   c.   higher salaries are being offered.
   d.   all of the above.

B. Matching

_____ **1.** Aquaculture   a. Commercial firms in agriculture
_____ **2.** Renewable resource   b. Electronics and ultramodern equipment
_____ **3.** Agribusiness   c. Growing in water

_____ **4.** Chemistry   d. Basic science of plants and animals

_____ **5.** High technology   e. Can replace itself

_____ **6.** Biology   f. Characteristics of elements

_____ **7.** Organic food   g. Space and mass around us

_____ **8.** Environment   h. Grown without chemical pesticides

C. Completion

1. Integrated pest management refers to the use of many different methods used together to _____ .

2. The transformation of the American Southwest from desert to irrigated lands was made possible, in part, by water-control projects on the _____ river.

3. By studying agriscience, you open the door to exciting educational programs that may lead to _____ .

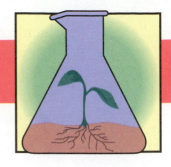

Better Living Through Agriscience

To determine important elements of a desirable environment and explore efforts made to improve the environment.

MATERIALS LIST

✔ paper

✔ pen or pencil

✔ current newspaper

✔ encyclopedias

✔ agriscience magazines

COMPETENCIES TO BE DEVELOPED

After studying this unit, you should be able to:

- describe the conditions of desirable living spaces.
- discuss the influence of climate on our environment.
- compare the influence of humans, animals, and plants on the environment.
- examine the problems of an inadequate environment.
- identify some significant world population trends.
- identify significant historical developments in agriscience.
- state practices used to increase productivity in agriscience.
- identify important research achievements in agriscience.
- describe future research priorities in agriscience.

TERMS TO KNOW

Sewerage system	Reaper	BelRus
Safe water	Combine	Aerosol
Polluted	Moldboard plow	Beltsville Small White
Condominium	Cotton gin	Green Revolution
Townhouse	Corn picker	Feedstuff
Famines	Barbed wire	Mastitis
Contaminating	Milking machine	Coccidiosis
Urine	Tractor	Impatiens
Feces	Legume	Hybrid
Parasites	Tofu	Deficiency
Immune	Katahdin	Lasers

Living conditions in the world vary extensively. In all countries, there are some very desirable places to live (Figure 2-1). Yet even in the highly developed countries, there are vast areas of poverty. How do you explain the differences in living conditions from one place to another? Why do living conditions vary from one community to another? From one neighborhood to another? From one house to another? The wealth and preferences of individuals explain some of the

FIGURE 2-1 In all countries there are some very desirable places to live. *(Courtesy National FFA; FFA #28)*

differences. Yet, the environment or the area around us has much to do with the quality of life. It also has much to do with the way we feel about ourselves and others.

VARIETY IN LIVING CONDITIONS

The population of the world is approaching 6 billion (Figure 2-2). Somewhere, a new child will have the distinction of being the 6 billionth human being living on the planet Earth (Figure 2-3). What will the home be like where "Baby 6 Billion" is born? Will there be adequate food? Will that child be warm, but not too warm or too cold? Will there be sufficient food? Will the child be kept free from serious illness? Will the family have a house or good living space they call home? Will they have clothing to permit them to live and work outside the home in relative comfort? What will be the quality of life of others around the 6 billionth human being? Will that child survive and will that person live a happy life? Positive answers to these questions would indicate a good environment for a person. These same questions should be asked for the rest of humankind.

The Homes We Live In

Homes of the Very Poor   Homes of the most destitute range in size and quality from nothing to a piece of cardboard or a scrap of wood on the ground. Many survive the freezing winters with only a tattered blanket on the warm sidewalk grates of our modern cities. For others, housing may take the form of a grass hut or a shack made of wood, cardboard, or scraps of sheet metal. The outdoors must provide water and washing areas, and serve as a receptacle for human waste. Large families generally share such homes with pets, poultry, or other livestock (Figure 2-4).

In cities, the poor frequently live in old buildings that are in very bad condition and with plumbing that does not work. Drugs, crime, poisonous lead paint, and disease are typical hazards for such people.

FIGURE 2-2 The population of the world is approaching 6 billion. (Courtesy of D.C. Committee to Promote Washington)

FIGURE 2-3 What kind of life for "Baby 6 Billion?" *(Courtesy NASA)*

FIGURE 2-4 The poor of the world live in sub-standard housing. *(Photo courtesy of Elmer Cooper)*

The steamy streets and sidewalks provide little relief from the oppressive summer heat.

Homes of the Less Fortunate People with modest sources of income may have homes that are simple, but provide basic protection from the elements. Such homes may be of wood, stone, masonry blocks, sheet metal, brick, or other fairly permanent material. The presence of windows and doors may provide protection from the elements and provide some privacy.

These people frequently have access to water that is safe to drink, but bathrooms may be nonexistent or toilets may not have safe sewerage-disposal systems. A sewerage system receives and treats human waste (Figure 2-5). To be regarded as safe, a sewerage system must

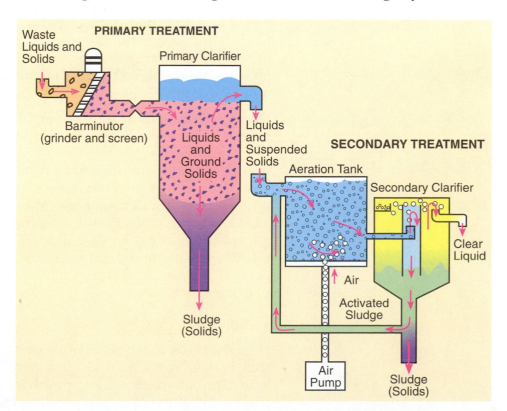

FIGURE 2-5 Safe sewerage disposal is essential to good health.

FIGURE 2-6 For much of the world's population, even a single community source of safe water is a luxury. *(Photo courtesy of Elmer Cooper)*

decompose human waste and release by-products that are free from harmful chemicals and disease-causing organisms. In most countries of the world, people rely on creeks or rivers to supply their drinking water, bathe the family, wash the clothes, and carry away the human waste.

People with low or modest income may live in housing with bathrooms and running water. However, maintenance of the systems may be poor and ignorance of the users may cause conditions that are health hazards.

In the undeveloped countries, the lower classes are fortunate if there is a source of **safe water** (water that is free of harmful chemicals and disease-causing organisms) at the village center (Figure 2-6). Modest and simple running-water facilities are generally the first evidence of community development in many such areas. Even a single faucet with unpolluted water is a major step forward for many communities. **Polluted** means containing harmful chemicals or organisms.

Homes of the Middle and Upper Classes   The middle- and upper-class people of the world can afford and enjoy housing that is clean, safe, and convenient. Such living spaces are often found as single houses. They are found in both rural and urban areas. In towns, villages, and urban areas, homes may also be in the form of townhouses, condominiums, or apartment buildings. A **condominium** is a building with many individually owned living areas or units. All living space of a single unit is generally on one floor. A **townhouse** is one of a row of houses connected by common side walls. Each unit is generally two or three stories high, giving the occupants more variety of living space.

Food

Until the 1970s, much of the world went to bed hungry. Only a few countries had sufficient food for their people. All countries had problems with distribution. So, not everyone had appropriate food for proper nutrition. Today, major **famines** (widespread starvation) are still a fact of life. In the current decade, serious famines have occurred in various parts of the world.

United Nations scientist John Tanner concluded that, in theory, the world could feed itself; but in practice it could not. It is estimated that nearly a billion people are not getting enough food for an active working life. While some countries enjoy an adequate food supply from their own production and imports, most have many individuals who do not get proper nutrition.

Family

Family may well be the dominant force that shapes the environment for most individuals. The family has control of the household activities and sets the priorities of its members. The family has considerable influence over the attractiveness of its surroundings and the warmth of relations among individuals.

For some, the family chooses the neighborhood and community where they live. A wise choice, however, is based on having the knowledge of better opportunities and the necessary resources to move to a better environment. For most of the world's population, the communities where individuals are born are the communities where they are raised and spend their lives.

Neighborhood and Community

Neighborhood and community have substantial influence on the environment in which we live. Some communities have tree-lined country roads with attractive fields, pastures, or woodlands to provide variety in the landscape. Other communities may have the advantage of attractive homes, businesses, or community centers (Figure 2-7). Urban areas may boast high-rise buildings for work and residence. These provide beautiful vistas of city lights or harbor scenes of commerce and recreation.

Neighborhoods and villages are parts of larger communities. These are influenced greatly by the families who live in the immediate area. If families work together toward common goals, they can shape the character, education, religious activities, social outlets, employment opportunities, and other broad aspects of their environment.

FIGURE 2-7 Good communities provide nice places to live and work. *(Courtesy National FFA; FFA #43)*

FIGURE 2-8 Topography is an important factor impacting the environment. *(Photo courtesy of Elmer Cooper)*

Climate and Topography

Climate and topography are also important factors affecting our environment. Average annual temperatures are very high near the equator. Yet people living near ocean waters, even in tropical areas, enjoy a moderate climate with cool breezes most of the time. Inland, the inhabitants are likely to experience hot, humid weather with high rates of rainfall. The high rainfall, in turn, stimulates heavy plant growth, resulting in jungle conditions. Similarly, sea-level elevations may create balmy 80°F temperatures, while a short trip to the top of a nearby mountain may reveal snow on its peak (Figure 2-8).

Northern areas, such as Alaska, may border on the Arctic Circle and have long, frigid winters. Yet those same latitudes enjoy summers suitable for short-season crops. People inhabit most areas of the earth, so the climate and topography where they find themselves create environmental conditions that influence their quality of life.

INFLUENCE OF HUMANS, ANIMALS, AND PLANTS

Humans have the option to enhance the environment. Or, we can work against the forces of nature to reduce the beauty, healthfulness, and safety of our surroundings. Humans and animals have body processes that use nutrients from the food we eat. We also have processes to remove waste products, poisons, and disease organisms from food, drink, and the air. However, the body is limited in its capacity to remove poisons and harmful organisms. Therefore, to remain healthy, humans and animals must limit their exposure to disease organisms and poisons.

Contamination by Humans and Animals

FIGURE 2-9 Food and water are easily contaminated.

A major problem for humans and animals is to avoid **contaminating** (adding material that will change the purity or usefulness of a substance) food and water with secretions from their own bodies. **Urine** and **feces** are liquid and solid body wastes, respectively. They are serious contaminants of food and water. Also, certain diseases can be spread by body contact or by breathing contaminated air. An example would be breathing air expelled by a sneeze of a person with a cold virus or sore throat (Figure 2-9).

There are serious animal diseases that spread from animal to animal by contact with body wastes. If animals have plenty of living space, this generally does not cause serious problems. But, as with humans, when animals are concentrated, health hazards increase. Fortunately, most diseases are spread among a given species of animal and not from one species to another. For instance, most diseases of dogs do not spread to cats. Similarly, most diseases of animals do not infect humans. However, there are some animal disorders that cause human sickness. Internal **parasites** (organisms that live on other organisms with no benefit to the hosts) and brucellosis are examples of animal disease organisms that may be transferred from animals to humans and create enormous health problems.

Contamination by Insects

Insects impact heavily on our environment. The cockroach is an unwelcome guest in many households of the world (Figure 2-10). Some cockroaches feed on human waste and then on the food of humans. They transmit disease from waste material to food and water. In poor housing conditions, they can move from household to household. In doing so, they leave disease organisms and possible illness in their wake.

Plants are not immune from disease and insect pests, nor damage from animals and humans. **Immune** means not harmed by an organism that is present. A plant relies on the nutrients and water it can extract from the soil and air. It cannot move, like animals and people, to sources of food and water. Therefore, it must depend on its tremendous capacity to reproduce in order to survive as a species. Reproduction is the plant's main function in life. However, plants accommodate both people and animals as they struggle to reproduce themselves. In doing so, they become an indispensable part of the environment.

Contamination by Chemicals

Certain chemicals are serious threats to our environment (Figure 2-11). Oil spills and industrial chemical discharges have caused

FIGURE 2-10 Cockroaches are serious household pests that may contaminate food or beverages. *(Courtesy USDA/ARS #K3272-3)*

FIGURE 2-11 Chemicals are needed in our modern society, but threaten our health if misused or abused. *(Courtesy USDA/ARS #K-4817-4)*

serious problems in rivers and streams. Chemical pesticides continue to threaten wildlife, fish, shellfish, beneficial insects, microscopic organisms, plants, animals, and humans. In the middle 1960s, American biologist Rachel Carson shocked the world with her book, *Silent Spring*. This was one of the first books to provide convincing evidence of environmental damage being done by pesticides.

In 1972, DDT was banned in the United States because of its damaging effects on the environment. This insecticide had been used to control mosquitoes, which carried the dreaded malaria organism. DDT was also a very effective chemical used against flies and it enjoyed widespread use in homes, on farms and ranches, and wherever flies were a problem. Yet, because of its damaging effect on the reproduction of birds, it had to be discontinued and safer substitutes were found.

It has been reported that over 10,000 different pesticides are registered for use in the states that surround one of our major coastal bays. Needless to say, careful management and control of so many different chemicals is absolutely essential. It requires the utmost care to avoid unacceptable damage to our environment.

PROBLEMS WITH AN INADEQUATE ENVIRONMENT

As we ponder the life of "Baby 6 Billion," mentioned earlier, we must wonder if we are doing our part to preserve and enhance the environment. Plants, animals, insects, soil, water, and air must be kept in reasonable balance or all will suffer. Excessive plant growth can infringe upon the space for humans and animals. Yet excessive populations of humans and animals can damage a plant species until it is unable to adequately reproduce itself. Too many animals can compete excessively with humans for food, water, and space. Some species of insects are regarded as harmful by people because they feed upon desirable crops or bother humans or livestock. However, many species of insects are beneficial to plants, animals, or humans.

Humans and animals tend to consume or remove plants, which hold soil in place and prevent erosion from wind and water (Figure 2-12). For instance, during the 1960s, most of the forests of China were cut and not replanted. Rapid and alarming soil erosion followed. The government then placed a high priority on reforestation and reversed the trend. Soil is needed to hold nutrients until plants need them. Similarly, we need the soil to filter and store clean water for plant growth and human and animal consumption. Plants take water from the soil and release water and oxygen to the air, which benefits humans and animals.

The number of human beings in the world is growing at the rate of 150 every minute; 220,000 a day; 80 million a year. At this rate, the earth's population will reach 6 billion by the year 2000; 7 billion by 2010; and 8 billion by 2022. Can the earth sustain such population growth? Will humans find enough to eat? Will we destroy our environment and, in doing so, destroy the system that supports life itself? Will we survive the competition of such population growth, but sacrifice our quality of life? Might we, in fact, improve our quality of life by using our intelligence to improve our environment?

FIGURE 2-12 Plants are necessary to conserve our soil. *(Courtesy USDA/ARS #K-5243-11)*

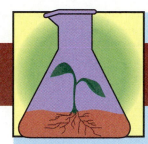

AGRI·PROFILE

CAREER AREA: Environmental Management

Management of the environment requires the attention of consumers as well as professionals. However, specialists in air and water quality, soils, wildlife, fire control, automotive emissions, and factory emissions all help maintain a clean environment against tremendous population pressures in many localities. Helicopter, airplane, and satellite crews gather important data for scientific analysis to help monitor the quality of our environment.

Individuals in environmental careers may work indoors or outdoors; in urban or rural settings; in boats, planes, factories, laboratories or parks; and in positions ranging from laborer to professional. Environmental concerns are high on global agendas today as nations attempt to head off global hunger and pollution.

Environmental management requires skills in observation, analysis, and interpretation. (Courtesy USDA/ARS #K-5184-1)

AGRISCIENCE IN OUR GROWING WORLD

The keys to a prosperous future, indeed the bottom line for survival of the world's population, can be found in agriscience. Agriscience is the science of food production, processing, and distribution. It is the system that supplies fiber for building materials, rope, silk, wool, cotton, and medicines. It provides the grasses and ornamental trees and shrubs that beautify our landscapes, protect the soil, filter out dust and sound, and supply oxygen to the air.

Agriscience accounts for 20 percent of the jobs in the United States. It is the mechanism that permits the United States and other developed countries of the world to enjoy high standards of living. It is the system that nondeveloped countries are using in their efforts to feed and clothe their bulging populations. They look to agriscience for the necessary technology to enter the twenty-first century on a par with other nations. As we look to the twenty-first century, we must look to agriscience to maintain and improve our quality of life.

The United States is a major world supplier of food. It is also a major supplier of fiber for clothing and trees for lumber, posts, piling, paper, and wood products. The use of ornamental plants and acreage devoted to recreation was never greater in the history of our country.

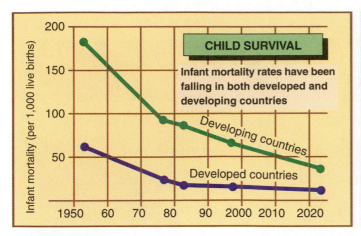
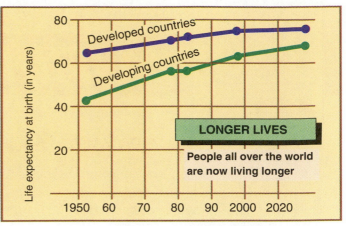

FIGURE 2-13 Worldwide child survival and life expectancy. *(Courtesy United Nation Fund for Population Activities)*

CHANGING POPULATION PATTERNS

The United Nations organization has reported that more children than ever before are surviving to adulthood. It also indicated that adults were living longer (Figure 2-13). Together these trends mean more population growth and more pressure on the environment. Advancements in medical science and services have made good health and longer lives a reality, but only for those who can afford good nutrition and modern health services. Similarly, through agriscience we have made substantial gains in providing food, fiber, and shelter for the world (Figure 2-14). At the same time, the environment has stayed reasonably clean, considering the impact of bulging populations.

In the past, persons under the age of 25 constituted the world's largest population group. This occurred because children were valued for the help they provided in making the family living. Further, the young were the backbone of a nation's labor force and provided the manpower for the nations' armies. In most countries, the young respected their elders and provided for the needs of the elderly within the family.

Honduras has the traditional population pattern, with its largest number of citizens younger than five years old. The number per age group then drops off to the smallest number occurring in the over 80-years-of-age group. When the Honduras population groups are displayed by sex in a bar graph, the graph takes the shape of a pyramid

#1. LABOR REQUIRED TO PRODUCE WHEAT, CORN, AND COTTON (IN HOURS)

	1800	1935-39	1955-59	1980-84	1990 or Later
Wheat (100 bu.)	373	67	17	7	7
Corn (100 bu.)	344	108	20	3	2.88
Cotton (1 bale)	601	209	74	5	5

#2. YIELDS PER ACRE OF WHEAT, CORN, AND COTTON

	1800	1940	1960	1985-86	1990 (Prelim)
Wheat (bu.)	15	15	20	34	39.5
Corn (bu.)	25	29	55	118	118.5
Cotton (lb.)	154	253	446	630	640.0

FIGURE 2-14 Changes in agriscience productivity.

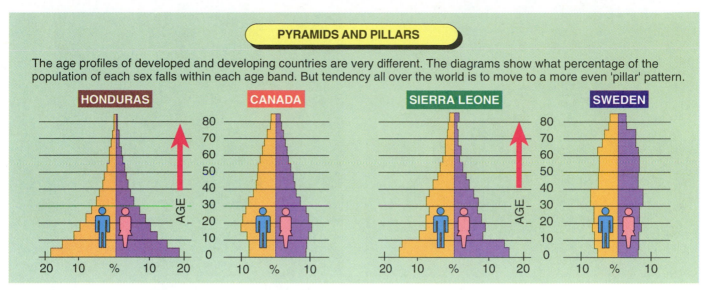

FIGURE 2-15 Age profiles and population patterns for developing and developed countries. *(Courtesy United Nations Fund for Population Activities)*

(Figure 2-15). Canada's pattern is slightly different. Its greatest population group is around the 20-year mark. Its graph reminds you of a Christmas tree, with its narrow bottom and cone appearance. Sweden, a country known for its excellent health services and high survival rate, has age brackets that are about equal. The population graph for that country resembles a column.

China has about one-fifth of the world's population, yet it has been reasonably successful at feeding its population by keeping about 70 percent of its work force on farms. In contrast, only about 2 percent of the work force in the United States is necessary to operate the nation's farms.

In the mid 1970s, China implemented a policy whereby each couple was limited to one child. They called it the 4-2-1 policy. This means that extended families consisted of four grandparents, two parents, and one child. What would be the outcome if such a policy were strictly enforced for several generations? You would expect the pyramidal shape of China's population graph to change to the shape of a Christmas tree, and, in time, to an upside-down pyramid! What would be the implications of feeding a nation with a population of mostly elderly people?

IMPACT OF AGRISCIENCE

History records little progress in agriculture for thousands of years. Then, starting in the early 1800s, the use of iron spurred inventions that revolutionized agriculture in the United States, British Isles, and northern Europe (Figure 2-16). However, for most of the world, the progress has been much slower.

Progress Through Engineering

Mechanization through inventive engineering was an important factor in America's agricultural development. The change from 90 percent to about 2 percent of the workers being farmers evolved over a 200-year

FIGURE 2-16 The inventions of the 1800s brought revolutionary changes in agriculture in the United States and Europe. *(Courtesy National FFA; FFA #19)*

FIGURE 2-17 Cyrus McCormick's reaper lead to the development of the modern combine. *(Photo courtesy of Elmer Cooper)*

period. Machines helped make this possible. The old saying that "necessity is the mother of invention" suggests the relationship between an inventor's problem and the use of previously acquired skills to solve that problem. The solution is frequently a new device, machine, or process.

American Inventors   America provided the inventors of many of the world's most important agricultural machines (Figure 2-17). In 1834, Cyrus McCormick invented the **reaper,** a machine to cut small grain. Later, a threshing device was added to the reaper and the new machine was called a combine. The **reaper** cut and threshed the grain in the field. Today, one modern combine operator can cut and thresh as much grain in one day as 100 persons could cut and bundle in the 1830s.

Thomas Jefferson's invention of an iron plow to replace the wooden plow of the time was of great significance. Later, in 1837, a blacksmith named John Deere experienced the frustration of prairie soil sticking to the cast-iron plows of the time. It became apparent that Jefferson's invention would not work in the rich prairie soils of the Midwest. Through numerous attempts at shaping and polishing a piece of steel cut from a saw blade, the steel **moldboard plow** evolved. That plow permitted plowing of the rich, deep prairie soils of the great American West for agricultural production, and launched the beginning of the John Deere Company.

In 1793, Eli Whitney invented the cotton gin. The **cotton gin** removes the cotton seed from cotton fiber. This paved the way for an expanded cotton and textile industry. In 1850, Edmund W. Quincy invented the mechanical **corn picker,** which removes ears of corn from the stalks. During the same era, Joseph Glidden developed **barbed wire,** with sharp points to discourage livestock from touching fences. This effective fencing permitted establishment of ranches with definite boundaries. In 1878, Anna Baldwin invented a **milking machine** to replace hand milking. In 1904, Benjamin Holt invented the **tractor,** which became the source of power for belt-driven machines as well as for pulling.

Formation of Machinery Companies   Many of the early inventors worked alone or with one or two partners. They were all workers in the area of agricultural mechanics and, as such, in agriscience. By the early 1900s, the inventors or other enterprising people had formed companies to produce agricultural machinery or process agricultural products. This made invention a continuing process. Successive invention was used to improve earlier inventions and develop new equipment and supplies to meet the needs of a changing agriculture.

The development of mechanical cotton pickers and corn harvesters greatly expanded the output per farm. Significant expansion of American agriculture also resulted from the development of irrigation technology. Since the end of World War II, the mechanization of American agriculture has moved at a breath-taking pace (Figure 2-18).

Mechanizing Undeveloped Countries   In the undeveloped countries of the world, many engineers, teachers, and technicians have sought simple, tough, reliable machines to improve agriculture. In such countries, America's highly developed, complex, computerized, and expensive machinery does not work for long. Most countries do not have people trained for the variety of agriculture mechanics jobs that are needed to support America's agriculture.

A machine with rubber tires is useless if a tire is damaged and repair services are not available. Similarly, failure of an electronic device may cause a $100,000 piece of machinery to become junk in the hands of an unskilled person in a country without appropriate repair facilities. This is the case in most undeveloped countries in Central and South America, Asia, and Africa. For the undeveloped nations of the world, other aspects of agriscience must become the vehicles for advancing agricultural productivity.

Improving Plant and Animal Performance

Humans have improved on nature's support of plant and animal growth since they discovered that the loosening of soil and planting of seeds could result in new and better plants. Even prior to that discov-

FIGURE 2-18 Since World War II, American agriscience has progressed at a breath-taking pace. *(Courtesy USDA/ARS #K-4911-19)*

ery, they probably aided growth by keeping animals away from plants until fruit or other plant parts edible to humans were yielded.

The human touch has permitted plants and animals to increase production and performance to the point where fewer and fewer are needed to produce the same amount of food.

Improving Life through Agriscience Research

Unlocking the Secrets of the Soybean   Americans have long appreciated the extensive research on the peanut done by the American scientist George Washington Carver. Carver is credited with finding over three hundred uses for the peanut. These include food for humans, feed for livestock, cooking fats and oils, cosmetics, wallboard, plastics, paints, and explosives.

Less known are the secrets of the soybean. The Chinese have known for centuries that the soybean is a versatile plant with many uses. Calling it the "yellow jewel," the Chinese are said to have grown the soybean 3,000 years ago. The strong flavor of the soybean itself is not appealing, but the bean is a legume and very nutritious. A **legume** is a plant that hosts nitrogen-fixing bacteria. These bacteria convert nitrogen from the air to a form that can be used by plants. Legume plants are excellent sources of protein for humans and animals.

A Chinese scholar is believed to have first made tofu from soybeans in 164 B.C. **Tofu** is a popular Chinese food made by boiling and crushing soybeans, coagulating the resulting soy milk, and pressing the curds into desired shapes. Today, tofu is a major food in the diet of China's huge population. It provides a reasonably healthful diet. Tofu can be fermented; marinated; smoked; steamed; deep-fried; sliced; shredded; made into candy; or shaped into loaves, cakes, or noodles.

Soy oil is the world's most plentiful vegetable oil. It is first extracted from the soybean, and the material that is left is processed into a

BIO•TECH CONNECTION

Agriscience and Biotechnology

Agriscience is heavily impacted by biotechnology which addresses the continuation of life. In ornamental horticulture one uses a myriad of plants to beautify the interiors of homes, businesses, and institutions. Such plants also consume the carbon dioxide gas and supply oxygen for humans, animals, insects, and other living matter. In outdoor settings, ornamental trees, shrubs, and turfgrass grace our yards, streets, parks, and other public areas. Our highways rely on

Biotechnology provides valuable mechanisms to improve life. [Courtesy USDA/ARS #K-5011-14]

protein-rich livestock feed known as soybean meal. The components of the soybean are used for hundreds of items. These range from dozens of food products, to lubricants, paper, chalk, paint, printing ink, and plastics (Figure 2-19). As early as 1940, Henry Ford evidently shocked journalists with an unusual demonstration. He slammed an ax into a Ford automobile-trunk door made from highly-resilient plastic. The new plastic was made from soybeans.

Baked Potatoes from the Northeast   Many improvements in our way of life can be traced to agriscience research. For instance, the U.S. Department of Agriculture developed many pest-resistant varieties of potatoes. A case in point is the work with the **Katahdin,** a popular potato variety of the 1930s. From the Katahdin, scientists developed the **BelRus,** a superior baking variety bred to grow well in the Northeast.

The Common Aerosol   Prior to World War II, death from malaria was commonplace in the tropics. The deaths of American soldiers from malaria triggered intensive research on the control of mosquitoes, the carrier of the malaria-causing organism. Development of the "bug bomb" resulted. Our present-day **aerosol** (a can with contents under pressure) resulted from that early research (Figure 2-20).

Turkey for the Small Family   In your grandparents' time, Thanksgiving was probably observed by having all the relatives in to consume the typical 30-lb turkey. As families became smaller and more scattered, the need for such large birds decreased; but even people with small families liked turkey. The 30-lb bird was too much, so the problem was to develop a breed of turkey that weighed 8 to 12 lb at maturity (Figure 2-21). A solution was the **Beltsville Small White** turkey, named after the Beltsville Agricultural Research Center in Maryland, where the breed was developed. Further research and development has yielded meat animals with higher yields of lean meat and less fat.

FIGURE 2-19 The soybean is the world's most important source of vegetable oil and provides the basic materials for hundreds of products. *(Courtesy of the American Soybean Association)*

plants to screen off on-coming traffic, provide living hedges, absorb sound, prevent soil erosion, and create a stimulating environment to keep motorists alert. Further, fruits, vegetables, grains, and forage crops of gardens, ranches, and farms provide the backbone of the world's food supply. And, of course, trees provide wood, paper, and other fiber products. Plants reproduce both sexually and asexually through many different processes. The productive capability of these and other plants have been greatly improved through the efforts of scientists, technicians, and growers.

Similarly, many animal species have been modified over the centuries by humans through domestication, selection, breeding, and care. Currently, biotechnology is improving the productivity of plants and animals and providing new foods and medicines to enrich our lives. Genetic engineering enables humans to modify and utilize microorganisms in our fight against harmful insects and other pests. Today, the earth is providing food, shelter, habitat, health care, and other essentials to more people than at any time in history. However, there is much malnutrition and starvation in the world. The causes tend to be rooted in deficiencies in government, national infrastructure, poverty, and lack of education. The technology of food production is believed to be adequate to meet the world food needs if it could be applied to maximum levels.

FIGURE 2-20 The common aerosol can resulted from agriscience research on insect control. *(Courtesy USDA/ARS #K-3760-13)*

FIGURE 2-21 The Beltsville Small White Turkey. *(Courtesy USDA/ARS)*

FIGURE 2-22 Attractants are valuable nonpolluting chemicals to lure insects to traps, bait, or systems of control through sterility. *(Courtesy USDA/ARS #K-4113-1)*

The Green Revolution   During the 1950s, starvation was rampant in many countries of the world. A major question was: could the world's agriculture sustain the new population growth? The solution was partly in the development of new, higher-yielding, disease- and insect-resistant varieties of small grain for developing countries. The result—the Green Revolution, a process whereby many countries became self-sufficient in food production in the 1960s by utilizing improved varieties and practices.

Cultivated Blueberries   Wild blueberries were enjoyed in earlier times when people had time to pick the tiny berries growing in the wild. But labor costs were too high to harvest such berries for sale. The solution—development of high-quality, large-fruited blueberry varieties from the wild. This started the new and valuable cultivated-blueberry industry of today.

Nutritional Values   Until rather recently, feeding of animals and human nutrition were based on poor methods of feed and food analysis. The problem—how can one recommend what to feed or what to eat if the content of food for humans and crops for livestock cannot be accurately determined? The solution—develop detergent chemical methods for determining the nutritional value of feedstuff (any edible material used for animals). The procedures are now used widely throughout the world in both human and animal nutrition.

Biological Attractants   The use of chemical pesticides provided short-term solutions to many insect-control problems. However, it soon became apparent that chemicals have disadvantages and other means of control must be found. A partial solution—discover chemicals that insects produce and give off to attract their mates (Figure 2-22). These were, in time, produced in the laboratory. Laboratory production of these chemicals permitted mass trapping of insects to survey insect populations for integrated pest-management programs.

Recent Breakthroughs in Agriscience

Mastitis Reduced The mastitis organism has always been a serious problem for dairy farmers. **Mastitis** is an infection of the milk-secreting glands of cattle, goats, and other milk-producing animals. The resulting loss of milk production adds millions of dollars yearly to the cost of milk in the United States. Recent research developed abraded plastic loops for insertion into cattle udders. The procedure resulted in a 75 percent reduction in clinical mastitis. The reduction in infections resulted in increased milk production, averaging nearly 4 lb of milk per cow per day.

Human Nutrition Recent studies in human nutrition have demonstrated the benefits of decreasing fat in the diet and increasing the proportion of fat from vegetable sources. This practice reduces high blood pressure and risk of heart attack. Much of the progress in human nutrition has grown out of research on animals and plants by agriscientists. Research on human nutrition has yielded new recommendations for healthful eating.

Fire-Ant Control Fire ants infest 230 million acres in the southern areas of the United States. Their presence in the warmer climates of the world is a constant threat to the well-being of humans and livestock. A new synthetic control for fire ants increases the ratio of nonproductive drone ants to worker ants. This ratio change gradually weakens the colony and causes it to die.

New Hope for Coccidiosis Control **Coccidiosis** is a disease that costs poultry growers nearly $300 million a year in the United States alone. Recently, the U.S. Department of Agriculture and research efforts by private industry genetically engineered a parasite constituent which stimulates birds to develop immunity to coccidiosis. Hopefully this will be the first step in the process of developing an effective vaccine against this persistent pest.

Exotic Flowers Horticulturists, gardeners, and hobbyists will be delighted with the new varieties of **Impatiens** (a popular, easy to grow, summer-flowering plant). Plant explorers introduced exotic new germ plasm and plant breeders developed a new technique called ovule-culture to develop hybrids and new kinds of Impatiens. A **hybrid** is the offspring of a plant or animal derived from the crossing of two different species or varieties.

Satellites and Nitrogen-Gas Lasers Nutrient deficiencies in growing corn and soybean crops are not easy to detect from the ground. **Deficiency** means something less available than needed for optimum growth. Yet new technology now permits the monitoring of deficiencies of iron, nitrogen, potassium, and other nutrients using nitrogen-gas **lasers** (devices used to determine wavelengths given off by the plants) from satellites. These wavelengths indicate the level of various nutrients in the plant (Figure 2-23).

Sugar Beet and Rice Hybrids The development of new varieties is a technique that has been used in agriscience for many decades to improve plant performance. A recent breakthrough has provided a sugar-beet hybrid with a high ratio of taproot weight to leaf weight. The hybrid yields about 15 percent more sugar per acre than previous

FIGURE 2-23 Infrared photos taken from satellites help diagnose problems such as ant infestations in fields and pastures. *(Courtesy USDA/ARS #K-3957-11)*

varieties. On the other side of the world, Chinese agronomists developed hybrid rice. Hybrid rice is capable of yielding up to 40 percent more rice per acre than traditional varieties.

A study of U.S. Department of Agriculture publications reveals great numbers of improved varieties, new products, and superior processes discovered or developed through agriscience research.

AGRISCIENCE AND THE FUTURE

By the mid 1990s, the average American farmer was said to be producing enough food and fiber for 128 people. Agriscience will become more important in the future. As the world's population increases, it will require a more sophisticated agriscience industry to provide the food, clothing, building materials, ornamental plants, recreation areas, and open-space needs for the world's billions. Americans will have to work more in the international arena, as more countries become highly competitive in agriscience and trade barriers are removed. Research and development will continue to play a dominant role as they lead the way in agriscience expansion in the future.

The USDA Agricultural Research Service has developed a strategic plan to guide the research efforts of the future. That strategic plan identifies and explains the main problems that confront the food and agriscience industry. It charts the minimum course of action that will provide the research needed for solutions. The six objectives of the strategic plan are to:

1. manage and conserve the nation's soil and water resources for a stable and productive agriculture;
2. maintain and increase the productivity and quality of crop plants;
3. increase the productivity of animals and the quality of animal products;
4. improve the delivery system and conversion of raw agricultural commodities into food and useful products for domestic consumption and export;
5. promote optimum human health and performance through improved nutrition; and
6. integrate scientific knowledge on agricultural production and processing into systems that optimize resource management and facilitate the transfer of technology to end-users.

In the early 1990s, a new and special research objective was developed to meet threats to the stability of American farms and other aspects of agriscience. American farm surpluses, low farm product prices, increased farmer reliance on chemical pesticides, lower government subsidies, and growing concerns about contamination of drinking water and food were cited as reasons for the new initiative. The new objective was to conduct research in ways to reduce the cost of production while maintaining production efficiency. The cost of fertilizers, chemicals, machinery, supplies, and equipment was thought to be too high for farmers to operate at a profit in the future. The new thrust was called "low input agricultural research and extension."

STUDENT ACTIVITIES

1. Write the Terms to Know and their meanings in your notebook.

2. Develop a bulletin board that illustrates the components of our environment.

3. Collect newspaper articles that describe environmental problems in your community.

4. Prepare a two- or three-page paper describing a good environment in which to live. Include factors such as home, community, air, water, cleanliness, wildlife, plants, and animals.

5. Ask your teacher to invite a public health official to your class to discuss health problems in the community and how they could be reduced by improving the environment.

6. Draw a chart that illustrates some relationships among plants, animals, trees, soil, water, air, and people.

7. Look up three prominent American inventors and describe the events that led to the inventions that made them famous.

8. Make a model of one of the machines that strongly influenced agriscience development.

9. Make a collage depicting some important discoveries, inventions, and developments in agriscience.

10. Assume that a bar graph depicting China's population for 1975 was pyramid-shaped like that of Honduras. Make a bar graph to represent what the population pattern will look like after two generations of 4-2-1 families.

11. Ask your teacher to arrange a field trip to study the variety of living environments in your community.

SELF EVALUATION

A. Multiple Choice

1. To be regarded as safe, a sewerage system must
   a. be connected to a city system.
   b. be constructed from concrete block.
   c. decompose human waste.
   d. discharge into a stream or river.

2. Safe water is
   a. any water pumped from wells.
   b. water collected from a roof.
   c. free of harmful chemicals and organisms.
   d. water taken from free-flowing rivers.

3. Apartments on one level in large buildings and owned by the residents are called
   a. condominiums.   c. townhouses.
   b. single houses.   d. villas.

4. The world's food capability as is indicates
   a. it could feed itself in theory, but not do so in practice.
   b. it could probably never keep up with population growth.
   c. food supplies outpace demand, and reduced production is recommended.
   d. widespread famine could not be helped by better distribution.

5. Contaminants of food and water include
   a. registered pesticides.
   b. contact by cockroaches.
   c. feces and urine.
   d. all of the above.

6. While plants cannot move to food and water, they survive because of
   a. their capacity to reproduce.
   b. their ability to survive without food and water.
   c. their roots, which extract water from any material.
   d. parasites that convert water to nutrients.

7. The population of the world is projected to increase to
   a. 6 billion by the year 2000.
   b. 7 billion by the year 2010.
   c. 8 billion by the year 2022.
   d. all of the above.

8. Agriscience in the United States
   a. accounts for 20 percent of the jobs.
   b. is likely to diminish in importance.
   c. reduces our standard of living.
   d. is being replaced by biotechnology.

9. Agricultural development would be described as
   a. rapid until about 1850.
   b. slow in most countries.
   c. very rapid in the United States in recent years.
   d. rapid in the past, but decreasing today.

10. The inventor of the iron plow was
    a. Cyrus McCormick.   c. Joseph Glidden.
    b. John Deere.   d. Thomas Jefferson.

11. The combine is a combination of the reaper and a
    a. corn picker.   c. cotton picker.
    b. cotton gin.   d. threshing device.

12. "Yellow jewel" is the name given by the Chinese to
    a. a special type of horse. c. garden peas.
    b. a very young emperor.   d. soybeans.

13. Beltsville Small White is a
    a. breed of rabbit.   c. type of building.
    b. breed of turkey.   d. variety of soybean.

14. The great advance in world food production in the 1960s was called the
    a. biological attractants.   c. Greening of America.
    b. Green Revolution.   d. Great Leap Forward.

B. Matching Group 1

_____ **1.** Neighborhood   a. Release oxygen into the air
_____ **2.** Starvation   b. A priority in China
_____ **3.** Parasite   c. Registered pesticides in one bay area
_____ **4.** Immune   d. Part of a community
_____ **5.** DDT   e. Not harmed by
_____ **6.** 10,000   f. Famine
_____ **7.** Cockroach   g. Lives on another organism
_____ **8.** Reforestation   h. Banned insecticide
_____ **9.** Plants   i. Feeds on human waste and food

Matching Group 2

_____ **1.** Aerosol   a. Sense nutrient deficiencies
_____ **2.** Barbed wire   b. "Bug bomb"
_____ **3.** Coccidiosis   c. Nitrogen fixation
_____ **4.** Cotton gin   d. Disease of poultry
_____ **5.** Impatiens   e. Sharp points to discourage livestock
_____ **6.** Laser   f. Colorful flower
_____ **7.** Legume   g. Eli Whitney
_____ **8.** Low-input   h. Curd from soybeans
              agriculture   j. New research objective
_____ **9.** Mastitis   k. Cyrus McCormick
_____ **10.** Milking machine   l. Infection of milk-secreting glands
_____ **11.** Steel moldboard   m. Anna Baldwin
              plow
_____ **12.** Reaper
_____ **13.** Tofu

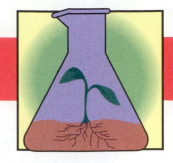

Biotechnology

OBJECTIVE

To examine elements of biotechnology

MATERIALS LIST

✓ paper

✓ pencil or pen

✓ encyclopedias

✓ agriscience magazines

COMPETENCIES TO BE DEVELOPED

After studying this unit, you should be able to:
- define biotechnology, DNA, and other related terms
- compare methods of plant and animal improvement
- discuss historic applications of biotechnology
- explain the concept of genetic engineering
- describe applications of biotechnology in agriscience
- state some safety concerns and safeguards in biotechnology

Bio	Clone	Gene splicing
Improvement by selection	Genetic engineering	Recombinant DNA technology
Selective breeding	Nucleic acid	Gene mapping
Genetics	Bases	Insulin
Heredity	Adenine (A)	Ice-minus
Genes	Guanine (G)	X-Gal
Generation	Cytosine (C)	Bovine somatotropin (BST)
Progeny	Thymine (T)	Porcine somatotropin (PST)
Deoxyribonucleic acid (DNA)	Mapping	

Biotechnology has become a very important tool in agriscience. It promises unprecedented advancements in plant and animal improvement, pest control, environmental preservation, and life enhancement. However, there are real dangers that this new power over life processes can lead to unmanageable consequences in careless, uninformed, or criminal hands. Therefore, governments, scientists, agencies, corporations, and individuals have moved cautiously in the pursuit of new benefits through biotechnology.

Bio means life or living, so biotechnology is the application of living processes to technology. While many definitions abound for biotechnology, one of the more popular definitions is the use of microorganisms, animal cells, plant cells, or components of cells to produce products or carry out processes.

FIGURE 3-1 Many foods owe their texture and taste to microorganisms that helped create them. *(Courtesy USDA/ARS #K-36072-2)*

HISTORIC APPLICATIONS OF BIOTECHNOLOGY

Living organisms have been used for centuries to alter and improve the quality and types of food for humans and animals. Examples include the use of yeast to make bread raise, bacteria to ferment sauerkraut, bacteria to produce dozens of types of cheeses and other dairy products, and microorganisms to transform fruit and grains into alcoholic beverages (Figure 3-1). Similarly, green grasses and grains have been stored in air-tight spaces and containers, such as silos, where bacteria converts sugars and starches into acids. The acids provide a desirable taste and protect the feed from spoilage by other microorganisms. The converted feed is called silage (Figure 3-2).

FIGURE 3-2 Silage is grains or green plant material preserved by the action of bacteria in an airtight environment. *[Courtesy National FFA; FFA #167]*

FIGURE 3-3 Improvement by selection means picking the best adults for breeding purposes. *[Courtesy National FFA; FFA #162]*

IMPROVING PLANT AND ANIMAL PERFORMANCE

Humans have improved on nature's support of plant and animal growth since they discovered that the loosening of soil and planting of seeds could result in new plants. Even prior to that discovery, they probably aided plant growth by keeping animals away from plants until they yielded fruit or other plant parts which were edible by humans.

Improvement by Selection

History documents the domestication of the dog, horse, sheep, goat, ox, and other animals thousands of years ago. Improvement by selection soon followed. **Improvement by selection** means picking the best plants or animals for producing the next generation (Figure 3-3). As people bought, sold, bartered, and traded, they were able to get animals which had desirable characteristics, such as speed, gentleness, strength, color, size, milk production, and the like. By mating animals with characteristics that humans preferred, the offspring of those animals would tend to imitate the characteristics of the parents and further intensify the desired characteristics. By accident, the owner was practicing **selective breeding** or the selection of parents to get desirable characteristics in the offspring.

The chariot armies of the Egyptians and Romans; the might of the Chinese emperors; the speed of the invading barbarians into northern Europe; the strength of mounts carrying armored knights into battle; and the evasive Arabians of the desert; all provide convincing testimony to early successes at breeding horses for specific purposes.

Improvement by Genetics

An Austrian monk named Gregor Johann Mendel is credited with discovering the effect of genetics on plant characteristics. **Genetics** is the biology of heredity. **Heredity** is the transmission of characteristics from an organism to its offspring through genes in reproductive cells.

Genes are components of cells which determine the individual characteristics of living things. Mendel experimented with garden peas. He observed there was definitely a pattern in the way different characteristics were passed down from one generation to another. **Generation** refers to the offspring, or **progeny,** of common parents.

In 1866, Mendel published a scientific paper reporting the results of his experiments. He had discovered that certain characteristics occurred in pairs, for example, short and tall in pea plants. Further, he observed that one of those characteristics seemed to be dominant over the other. If tall was the dominant characteristic, then tall plants crossed with tall or short plants produced mostly tall plants. But, some plants would be short. It was observed that the short characteristic could be hidden in tall plants in the form of a recessive gene. Such recessive genes could not express themselves in the form of a short plant unless both genes in the plant cells were the recessive genes for shortness. He also observed that short plants crossed with short plants always had short plants as offspring. This happened because there were no tall characteristics in either parent to dominate the characteristic of the offspring.

AGRI·PROFILE

CAREER AREA: Genetic Engineering

Genetic engineering cuts across many fields of endeavor. Procedures for genetic modification of organisms have been developing for over a decade. Biologists, microbiologists, plant breeders, and animal physiologists are some examples of specialists who might use genetic engineering in their work. The work settings for people include the field, laboratory, classroom, and commercial operations.

People involved in genetic engineering may have advanced degrees and be highly specialized in a narrow area of research, such as cellular biology. Others may work in applied research in areas, such as weevil control in small grains, nutrient requirements of small grains, or reproductive problems in dairy cattle. Others may be crop, animal, or pest control technicians who help manage the plants or animals that are the subjects of research. Still others may work in laboratories and devote most of their time to analysis and observation.

Because genetic engineering is a relatively new field and the applications are so numerous, the opportunities will expand as the field develops.

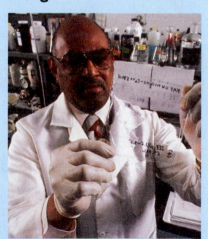

Using a DNA probe, an animal physiologist examines film showing gene patterns of various animals. *[Courtesy of USDA/ARS #K-1968-13]*

FIGURE 3-4 Mendel's extensive experimentation, observation, and record keeping provided the foundation for modern study of heredity. *(Courtesy National FFA; FFA #199)*

Mendel's work provides an excellent example of the power of the written word. His discoveries and conclusions would have been lost if they were not recorded. The usefulness of his discoveries was not recognized until long after his death. In 1900, other scientists reviewed his writings and built upon the observations and conclusions he had reported. Today, biologists credit his work as being the foundation for the scientific study of heredity (Figure 3-4). Principles of heredity apply to animals as well as plants. More information on the principles of heredity may be found in subsequent units of this text.

Improving Plants and Animals Through Biotechnology

More recently scientists have learned to improve plants, animals, and microbes by manipulating the genetic content of cells. This procedure permits more choices for the researcher and more rapid observation of results. With the manipulation of cellular material, the scientist can now alter the characteristics of microorganisms as well as of larger plants and animals. This new capability has some amazing implications for human efforts to improve the quality of life.

DNA—GENETIC CODE OF LIFE

Of the estimated 300,000 kinds of plants and over 1 million kinds of animals in the world, all are different in some ways. On the other hand, plants and animals have certain similar characteristics which lend themselves to classification and permit prediction of characteristics of offspring by viewing the parents. That is, the individual fertilized cell, called the embryo, contains coded information that determines what that cell and its successive cells will become. The coded material in a cell is call DNA. **DNA** is an acronym for **Deoxyribonucleic Acid.**

Scientists have a working knowledge of how genetic information is stored in a cell, duplicated, and passed on from cell to cell as they divide and the organism forms. Further, the process of transmitting genetic codes from parents to offspring and from parent to clone is common scientific knowledge. A **clone** is an exact duplicate of something.

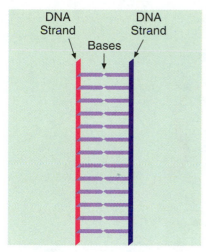

FIGURE 3-5 The components of DNA may be likened to a ladder with very close rungs.

A major breakthrough was made in the early 1980s as scientists developed the process of genetic engineering. **Genetic engineering** refers to the movement of genetic information in the form of genes from one cell to another. Genes are comprised of DNA.

It is believed that a universal chemical language unites all living things. It was observed in the early 1800s that all living organisms are composed of cells. And, that cells of microscopic organisms, as well as larger plants and animals, are basically the same. Then, in 1867, Friedrich Meischer observed that the nuclei of all cells contain a slightly acidic substance. He named the substance **nucleic acid.** Later the name was expanded to Deoxyribonucleic Acid or DNA. DNA in all living cells is similar in structure, function, and composition, and is the transmitter of hereditary information.

DNA occurs in pairs of strands intertwined with each other and connected by chemicals called **bases.** The pairs of DNA strands may be likened to the two sides of a wire ladder. And, the bases may be likened to the rungs of that wire ladder. The different bases are 1) **Adenine,** 2) **Guanine,** 3) **Cytosine** and 4) **Thymine.** The first letters of each name of the bases—**A, G, C,** and **T,**—have become known as the genetic alphabet of the language of life (Figure 3-5).

If one end of the wire ladder is held while the other end is twisted, the resulting shape would be called a double helix. This is the shape of DNA strands in a cell. Two strands of DNA and the bases between the strands compose a specific gene. The order or sequence of the bases between the DNA strands is the code by which a gene controls a specific trait. Therefore, each rung with its accompanying side pieces of DNA constitutes a gene containing the genetic code to a single trait (Figure 3-6). The genetic material in the cells of a given microbe, plant, animal, or human can be isolated and observed. Further, the trait or traits a given gene determines can be identified and/or the combination of genes that influence a single trait can be determined. The

FIGURE 3-6 The occurrence of DNA in a cell nucleus may be likened to a ladder twisted to form a double helix.

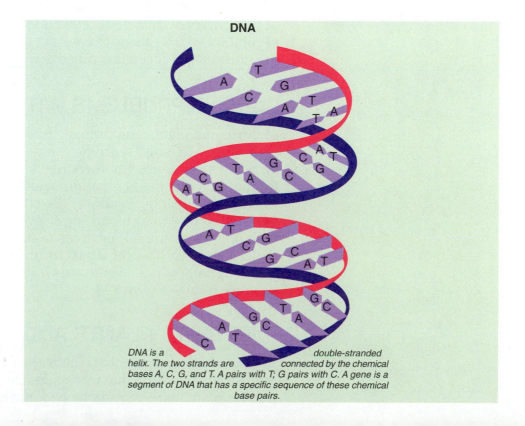

DNA is a double-stranded helix. The two strands are connected by the chemical bases A, C, G, and T. A pairs with T; G pairs with C. A gene is a segment of DNA that has a specific sequence of these chemical base pairs.

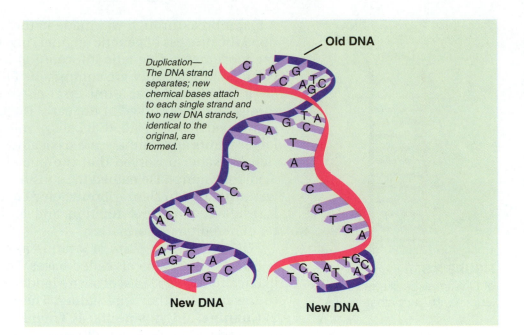

Duplication— The DNA strand separates; new chemical bases attach to each single strand and two new DNA strands, identical to the original, are formed.

Old DNA

New DNA

New DNA

FIGURE 3-7 DNA strands divide and bases attach themselves to the new strands to form identical genes for new cells.

matching of genes to traits is called **mapping.** Some examples of individual traits are hair color, tendency for baldness in humans, height of plants at maturity, and tendency of females to have twin offspring.

As cells divide, the DNA strands separate from each other and create duplicate strands to go to the new cells. Therefore, the genetic codes are duplicated and passed on from old cell to new cell as growth occurs and individuals reproduce (Figure 3-7).

Scientists can now remove individual genes carrying certain genetic information and replace them with genes containing other genetic instructions. By doing so a given characteristic or performance can be altered in the microorganism or other plant or animal. For instance, plants that are susceptible to being eaten by certain insects may be altered so they will have resistance to that insect. The process of removing and inserting genes into DNA is called **gene splicing,** or **recombinant DNA technology.** And, the process of finding and recording the location of genes is called **gene mapping.**

SOLVING PROBLEMS WITH MICROBES

Microscopic plants and animals lend themselves well to genetic engineering. Microbes reproduce quickly and can be genetically engineered to produce products needed by other plants, animals, and humans. One of the first commercial products made by genetic engineering was insulin. **Insulin** is the chemical used by people with diabetes to control their blood sugar levels. Previously, insulin was available only from animal pancreas tissue, was in short supply, and was very expensive. However, a bacterium called E. coli was genetically engineered to produce insulin much like cows produce milk and bees produce honey (Figure 3-8).

IMPROVING PLANTS AND ANIMALS

In 1988, California scientists made the first outdoor tests of ice-minus. **Ice-minus** is bacteria that was genetically altered to retard frost formation on plant leaves. Synthetic chemicals are now available to protect

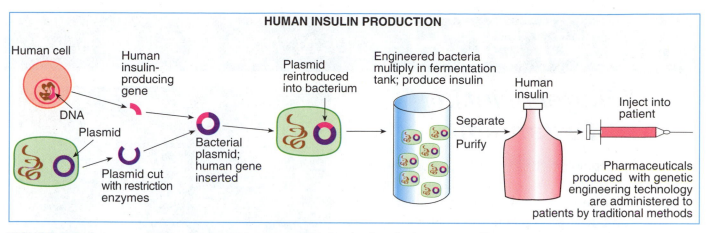

HUMAN INSULIN PRODUCTION

FIGURE 3-8 As a result of genetic engineering of bacteria, insulin is now readily available and relatively inexpensive.

fruit crops when temperatures fall 4 to 6 degrees below what would normally damage the fruiting process. Similarly in 1988, genetically altered bacteria was injected into elm trees in an effort to control the deadly Dutch Elm disease. Further, bacteria were genetically engineered so they turn a brilliant shade of blue in the presence of a compound called **X-Gal.** Such bacteria can be easily detected and traced in experimental or real-life situations. The ability to so mark microorganisms has made biotechnology a safer and more manageable enterprise.

In animal science, the hormone **bovine somatotropin (BST)** has long been known for its stimulation of increased milk production in cows. However, it was not available for commercial use until bacteria was altered to produce the hormone. Another example of hormone production by genetically altered bacteria is an animal hormone called **porcine somatotropin (PST)**, which increases meat production in swine.

It is now evident that genetic engineering and other forms of biotechnology hold great promise in controlling diseases, insects, weeds, and other pests. The plants and animals that we nurture for food, fiber, recreation, and preservation can benefit from biotechnology as well as humans. Further, the environment will be enhanced by less use of chemical pesticides and greater use of biological controls (Figure 3-9).

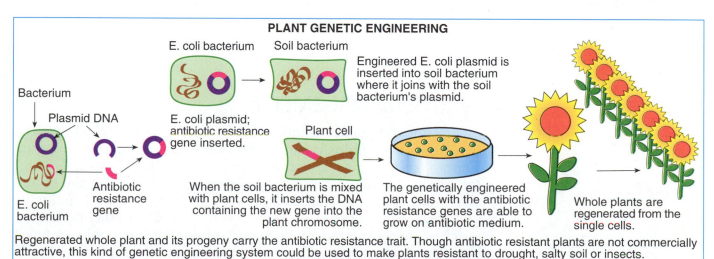

PLANT GENETIC ENGINEERING

Regenerated whole plant and its progeny carry the antibiotic resistance trait. Though antibiotic resistant plants are not commercially attractive, this kind of genetic engineering system could be used to make plants resistant to drought, salty soil or insects.

FIGURE 3-9 Plant genetic engineering is becoming commonplace in the battle against diseases, insects, weeds, and other pests.

BIO·TECH CONNECTION

"Fingerprinting" Organisms

USDA microbiologists at the National Animal Disease Center in Ames, Iowa were able to "crack" the mysteries of the three worst known cases of food poisoning in North America. DNA matching or "fingerprinting," was used to link 1) the persons who became ill, 2) the contaminated food that caused the poisoning, 3) the place where the food had been contaminated, and 4) the material where the food poisoning organisms had grown. Food poisoning is a life-threatening condition resulting from eating food containing toxic material produced by certain bacteria under unsanitary conditions.

Medical authorities wondered if the meningitis of a child in California was caused by food poisoning, possibly from a certain cheese she had eaten. Could the culprit be *Listeria monocytogenes*, bacteria that had caused deaths in southern California in 1985, New England in 1983, and Canada in 1981?

Researchers were using restriction enzyme analysis to check the isolates involving *Listeria monocytogene* from previous listeriosis outbreaks. By "fingerprinting" the isolates from each of the outbreaks, it was

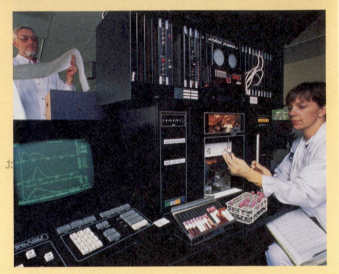

DNA analysis make "fingerprinting" or individual identification possible in all organisms. *(Courtesy USDA/ARS #K-4767-1)*

found that the isolates showed a characteristic DNA pattern for each particular episode.

The relatively simple method can be used by lab technicians and public health safety workers to track the spread of *Listeria* from the food-processing environment to the food product and on to the human patient. In analyzing the *Listeria* in the California case, the researchers found that the isolates recovered from the patient, the suspect cheese, and the cheese factory all matched!

WASTE MANAGEMENT

Environmental pollution and the elimination of waste products from home, business, industry, utility, government, the military, and other sources has become a major problem throughout the world. Landfills are becoming full and causing problems of leakage into the ground water, old dump sites are creating new problems, waste is piling up, and sewage and chemical disposal is a constant problem. The "tipping" or unloading fee for dumping solid waste material in landfills has risen from as low as $3.00 per ton in some communities in the early 1980s to over $146.00 per ton for New York City in the middle 1990s. Further, from 1993 levels, the environmental laws mandate a

FIGURE 3-10 Biotechnology is helping to solve the problems of waste disposal. *(Courtesy USDA #CS 336)*

reduction in solid waste disposal of 10 percent by 1995, 27 percent by 1997, and 40 percent by the year 2000.

Biotechnology is being used to help solve waste disposal problems. Already, genetically altered bacteria are used to feed on oil slicks and spills to transform this serious pollutant into less harmful products. Similarly, bacteria have been developed to decompose or deactivate dioxin, PCBs, insecticides, herbicides, and other chemicals in our rivers, lakes, and streams. And, bacteria strains are under development to convert solid waste from humans and livestock into sugars and fuel.

While great progress has been made in pollution reduction in some areas, pollution is still one of the world's greatest problems. Biotechnology has brought some spectacular breakthroughs in waste use and decomposition and promises much more in the future (Figure 3-10).

SAFETY IN BIOTECHNOLOGY

Federal and state governments monitor biotechnology research and development very closely. Much fear has been expressed about the dangers of genetically modified organisms. Therefore, appropriate policies, procedures, and laws have had to be developed as biotechnology has evolved. Research priorities and initiatives require discussion and interaction by scientists, government agencies, and other authorities. Products are tested in laboratories, greenhouses, and other enclosures before being approved for testing outdoors and in other less controlled environments. Even then, outdoor tests are first conducted on a small scale in remote places under careful observation. Under these conditions, the efficiency, safety, control, and environmental impact of new organisms are determined. If the new organism poses an unmanageable threat, it can be destroyed.

Biotechnology is rapidly gaining the public's confidence and has become an important part of our daily lives. Many of its potential benefits have already been realized and most believe that we have only scratched the surface. With proper safeguards, we can look confidently to a bright future in this emerging field. Many other applications of biotechnology are cited throughout this text.

STUDENT ACTIVITIES

1. Write the Terms to Know and their meanings in your notebook.

2. Read an article in an encyclopedia or other reference on the process of genetic engineering.

3. Report your findings from Activity 2 to the class.

4. Make a collage depicting some important discoveries, inventions, and developments in biotechnology.

5. Form a discussion group to explore the benefits and hazards of biotechnology.

6. Arrange for a resource person to speak on the ethical and moral issues surrounding developments in biotechnology.

7. Organize a class debate on the ethical and moral issues regarding research and the use of new discoveries in biotechnology.

SELF EVALUATION

A. Multiple Choice

1. Bio means
   a. a study of.
   b. life.
   c. three.
   d. science.

2. An example of a fermented food is
   a. applesauce.
   b. bologna.
   c. cheese.
   d. coffee.

3. The earliest method of livestock improvement was probably by
   a. biotechnology.
   b. cross breeding.
   c. gene splicing.
   d. selection.

4. The person providing the foundation for scientific study of heredity was
   a. Gregor Johann Mendel.
   b. George Washington Carver.
   c. Joseph Glidden.
   d. Thomas Jefferson.

5. The genetic code of life is
   a. clone.
   b. DNA.
   c. progeny.
   d. thymine.

6. Adenine, Guanine, Cytosine, and Thymine are all
   a. acids.
   b. bases.
   c. DNA.
   d. genes.

7. Recombinant DNA technology is also known as
   a. bovine somatotropin.
   b. gene splicing.
   c. porcine somatotropin.
   d. X-Gal.

8. Genetic engineering can be done to change
   a. animals.
   c. plants.
   b. microorganisms.
   d. all of the above.

9. An important contribution of biotechnology to waste management is
   a. bacteria that consume oil.
   c. ice-minus bacteria.
   b. disease resistant bacteria.
   d. human bacteria.

10. Chemical pollutants in water that may be decomposed or deactivated by bacteria include
    a. chlorine.
    c. iron.
    b. fluorides.
    d. PCBs.

B. Matching

_____ **1.** Fruits and grains   a. Makes "marked" bacteria turn blue

_____ **2.** Yeast   b. Controls blood sugar levels

_____ **3.** Silage   c. Used to make alcoholic beverages

_____ **4.** Genetics   d. Embryo

_____ **5.** Genes   e. Result of dominant gene

_____ **6.** Tall pea plants   f. Deoxyribonucleic Acid

_____ **7.** DNA   g. Causes bread to raise

_____ **8.** Fertilized cell   h. Fermented grains or grasses

_____ **9.** Insulin   i. Heredity

_____ **10.** X-Gal   j. DNA and bases

The Twenty-first Century and You

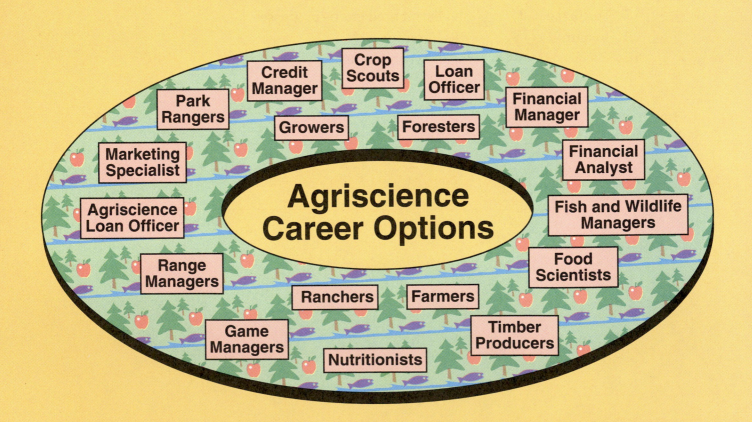

Agriscience
Career Options

Credit Manager
Crop Scouts
Loan Officer
Financial Manager
Park Rangers
Growers
Foresters
Financial Analyst
Marketing Specialist
Fish and Wildlife Managers
Agriscience Loan Officer
Food Scientists
Range Managers
Ranchers
Farmers
Timber Producers
Game Managers
Nutritionists

You and the "Green Machine"

The "green machine" can use your energy in science and engineering, business and financial management, production, renewable natural resources, communications, or other challenging and rewarding careers.

Futurists predict that during the next quarter century the most important discoveries in genetic engineering will be made in agriscience. Here you could help develop plants and animals that grow better and more efficiently, even in adverse situations. Nutritionists and food scientists study links between food, diet, and health. As one of them, you might work to ensure that new convenience foods are nutritious and healthy; or, you might design a new way of preserving, processing, or packaging food.

As a financial manager you might work for a bank or a credit agency as an agriscience loan officer, or for a company that sells supplies to producers and growers. As credit manager, loan officer, financial analyst, or marketing specialist you could utilize your knowledge of business and finance as well as production, processing, or distribution.

The farmers, ranchers, timber producers, and other growers are the foundation of the food and fiber system. To join their ranks is to play a most basic role in the future of our planet. For others loving the outdoors, a look at the work of foresters, range managers, game managers, fish and wildlife managers, park rangers, and crop scouts should be of interest.

Like all parts of our world touched by science and technology, the food, fiber, and renewable natural resources system is changing rapidly. Because it contributes to and is strongly influenced by trade and consumer lifestyles, the system must adjust continuously. If you get the appropriate training and experience, any of these careers can be yours.

Career Options in Agriscience

OBJECTIVE

To survey the variety of career opportunities in agriscience, observe how they are classified, and consider how you can prepare for careers in agriscience.

COMPETENCIES TO BE DEVELOPED

After studying this unit, you should be able to:

- define agriscience and its major divisions.
- describe the opportunities for careers in agriscience.
- compare the scope of job opportunities in farm and off-farm agriscience jobs.
- list activities in middle school, high school, and thereafter to help prepare for agriscience careers.
- identify resource people for obtaining career assistance in agriscience.

MATERIALS LIST

✓ paper

✓ pencil or pen

✓ bulletin board materials

✓ agriscience magazines and pictures

✓ *Occupational Outlook Handbook* or other agriscience career references

Production agriscience
Agriscience processing, products, and distribution
Horticulture

Forestry
Agriscience supplies and services
Agriscience mechanics

Profession
Agriscience professions

L ife is possible without many of our modern conveniences, but not without food. Basic to life is an adequate supply of suitable food and other products of the soil, air, and water. This includes food for nourishment, fiber for clothing, and trees for lumber. Less obvious are alcohols for fuel and solvents, oils for home and industry, and oxygen for life itself. The industry that provides these vital basic commodities is agriscience. American agriscience is the world's largest commercial industry, with assets of nearly $1 trillion (Figure 4-1).

DEFINITION

Agriscience is a term that includes all jobs relating in some way to plants, animals, and renewable natural resources. Such jobs occur indoors and outdoors. They include people in banking and finance; radio, television, and satellite communications; engineering and design; construction and maintenance; research and education; and environmental protection. All are in the field of agriscience if their products or services are related to plants, animals, and other renewable natural resources.

PLENTY OF OPPORTUNITIES

Approximately 21 million people are employed in agriscience careers. About 400,000 people are needed each year to fill positions in this field. Of those vacancies, only 100,000 are currently being filled by people trained in agriscience (Figure 4-2). That means there are many opportunities for you. About 20 percent of the careers in agriscience require college degrees, so you can use what you learn today in your current job, on your farm, or in agriscience classes to go directly to full time employment. However, if you choose to pursue a college degree in agriscience, many additional career opportunities will be open to you (Figure 4-3).

How Much is One Trillion Dollars?

You can count $1 trillion ($1,000,000,000,000) by using the following procedure:

- One bill every second
- Sixty bills per minute
- Thirty-six hundred bills per hour
- Eighty-six thousand per day
- Thirty-one million five hundred thirty-six thousand per year
- And continue counting for thirty-one thousand seven hundred and ten years!

FIGURE 4-1 The agriscience industry in the United States has assets of nearly $1 trillion ($1,000,000,000,000).

53

400,000 people are needed each year as replacements

100,000 trained people available

300,000 openings each year for additional people trained in agriscience

FIGURE 4-2 The employment outlook is good for people trained in agriscience.

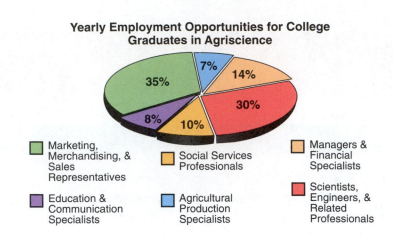

Yearly Employment Opportunities for College Graduates in Agriscience

35% | 7% | 14% | 30% | 8% | 10%

- Marketing, Merchandising, & Sales Representatives
- Social Services Professionals
- Managers & Financial Specialists
- Education & Communication Specialists
- Agricultural Production Specialists
- Scientists, Engineers, & Related Professionals

FIGURE 4-3 A college education opens additional doors in agriscience.

Careers That Help Others

Helping others is an extra bonus with a career in agriscience. Of the jobs in processing, marketing, production, natural resources, mechanics, banking, education, writing, and other areas, many are people-oriented jobs. This means you have the extra benefit of being a product or process specialist and receive the special appreciation of others. Considering the new medical procedures discovered through agriscience, more and more opportunities for helping others will develop.

Careers That Satisfy

You might ask, "How can an agriscience career benefit me? What's in it for me?" Of course, there's the money. In agriscience, salaries vary tremendously from job to job. Generally, the better qualified individuals will be able to earn more. Agriscience industries employ one-fifth of all workers in the United States. And, there are job openings for skilled individuals at various levels of expertise. You can be a sharp, well-paid agriscience mechanic right out of high school, or work for an advanced degree and be an agriscience engineer. You can do what you decide is best for you.

THE WHEEL OF FORTUNE

Agriscience is like a wheel with a large hub. The hub of that wheel is production agriscience, or farming and ranching. The rest of the wheel is the nonfarm and nonranch activities in agriscience. Since so many opportunities for rewarding careers exist in that wheel, it may be called a wheel of fortune.

Production Agriscience

Production agriscience is farming and ranching. It involves the growing and marketing of field crops and livestock. Careers in this area account for one-fifth of all jobs in agriscience. Some estimates indicate the average U.S. farmer produces enough food and fiber for 128 people (Figure 4-4). Large farm operators produce enough to feed over 200.

Most other agriscience careers are involved with goods and services that flow toward or away from production agriscience. Workers in

FIGURE 4-4 The average U.S. farmer produces enough food and fiber for 128 people. *(Courtesy USDA/ARS)*

FIGURE 4-5 Agriscience may be compared to a wheel of fortune.

those careers permit American farmers to supply goods so efficiently that American consumers spend only 14 percent of their income on food. This is the lowest percentage in the world. Out of five workers in agriscience, four have jobs that are not on farms. The nonfarm agriscience jobs may be in rural, suburban, or urban settings. The agriscience wheel of fortune contains a hub with one-fifth of the agriscience workers in production on farms and ranches. The rest of the wheel contains the four-fifths of the agriscience jobs in nonproduction-type careers that are off the farm (Figure 4-5).

Agriscience Processing, Products, and Distribution

Spin the wheel of fortune! What comes up for you? Agriscience processing, products, and distribution (Figure 4-6)!

Agriscience processing, products, and distribution are those parts of the industry that haul, grade, process, package, and market commodities from production sources. Pick any item of food, clothing, or other commodity. Trace it back to its source. Except for metals and stone, most objects can be traced back to a farm, ranch, forest, greenhouse, body of water, or other agriscience production facility. If you consider a deluxe hamburger, you can trace the beef, mayonnaise, tomato, lettuce, pickle, catsup, mustard, relish, bun, and sesame seeds back to farms where they were produced (Figure 4-7). The same is true of many ingredients in soda, coffee, chocolate, or any other beverage you choose.

Check the label in your coat. Is it made of cotton, nylon, polyester, leather, vinyl, rubber, or wool? Each can be traced to a farm, ranch, or plantation. Your search may take you to a Maryland farm, a California ranch, a Colombian coffee plantation, or a trapper's lodge.

People with careers and jobs in agriscience processing, products, and distribution make it all possible. From hauling to selling, processing to merchandising, inspection, and research—the commodity moves from its source to consumption. The USDA reports that the producer's share of the food dollar is as low as 15.4 cents for cereal and bakery products. The rest is for handling, processing, and distribution.

FIGURE 4-7 The components of a deluxe cheeseburger may have come from several states or even different countries. *(Courtesy National FFA; FFA #18)*

Ag Establishment Inspector
Butcher
Cattle Buyer
Christmas Tree Grader
Cotton Grader
Farm Stand Operator
Federal Grain Inspector
Food & Drug Inspector
Food Processing Supervisor
Fruit & Vegetable Grader
Fruit Distributor
Fruit Press Operator
Flower Grader
Grain Broker
Grain Buyer

Grain Elevator Operator
Hog Buyer
Livestock Commission Agent
Livestock Yard Supervisor
Meat Inspector
Meatcutter
Milk Plant Supervisor
Produce Buyer
Produce Commission Agent
Quality Control Supervisor
Tobacco Buyer
Weights & Measures Official
Winery Supervisor
Wood Buyer

FIGURE 4-6 Spin the wheel of fortune! What comes up for you? Agriscience processing, products, and distribution.

FIGURE 4-9 Horticulture provides many opportunities in urban as well as rural areas. *(Courtesy National FFA; FFA #242)*

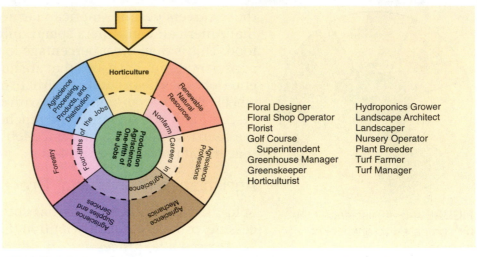

Floral Designer	Hydroponics Grower
Floral Shop Operator	Landscape Architect
Florist	Landscaper
Golf Course	Nursery Operator
Superintendent	Plant Breeder
Greenhouse Manager	Turf Farmer
Greenskeeper	Turf Manager
Horticulturist	

FIGURE 4-8 Spin the wheel of fortune! What comes up for you? Horticulture.

Horticulture

Spin the wheel of fortune! What comes up for you? Horticulture (Figures 4-8 and 4-9)!

Horticulture includes producing, processing, and marketing fruits, vegetables, and ornamental plants (turfgrass, flowers, shrubs and trees grown and used for their beauty). Horticultural production could be thought of as farming. But, since it is generally done on small plots, the production of horticultural crops is classified with horticulture rather than farming. Horticultural commodities are high-labor and high-income commodities.

The landscape designer, golf course superintendent, greenhouse supplier, greenhouse manager, flower wholesaler, floral market analyst, florist, strawberry grower, vegetable retailer—all are horticulturists. Recent data show more than 110,000 people employed by the floral industry alone.

Forestry

Spin the wheel of fortune! What comes up for you? Forestry (Figure 4-10)!

Forestry is the industry that grows, manages, and harvests trees for lumber, poles, posts, panels, pulpwood, and many other commodities. Americans have huge appetites for wood products.

Careers in forestry range from growing tree seedlings to marketing wood products. Many jobs in forestry are outdoors and require the use of large machines to cut trees, drag logs, and load trucks (Figure 4-11). Other jobs are service-oriented, such as the state or district forester whose job is to give advice and administer governmental programs. Many find enjoyable careers in forestry research, teaching, wood technology, and marketing.

Renewable Natural Resources

Spin the wheel of fortune! What comes up for you? Renewable natural resources (Figures 4-12 and 4-13)!

Renewable natural resources involve the management of wetlands, rangelands, water, fish, and wildlife. All require people with an appre-

FIGURE 4-11 The forestry industry provides many opportunities for work outdoors. *(Courtesy National FFA; FFA #187)*

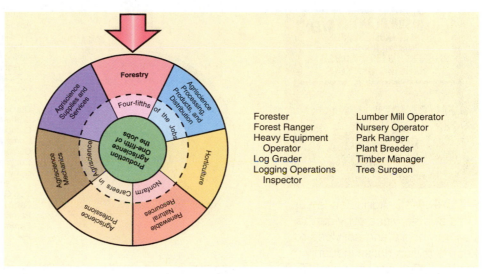

Forester
Forest Ranger
Heavy Equipment
  Operator
Log Grader
Logging Operations
  Inspector

Lumber Mill Operator
Nursery Operator
Park Ranger
Plant Breeder
Timber Manager
Tree Surgeon

FIGURE 4-10 Spin the wheel of fortune! What comes up for you? Forestry.

ciation for natural and scientific knowledge of plants and animals. This area of agriscience is attractive to those who enjoy working in parks, on game preserves, or with landowners to preserve and enhance natural habitat, plants, and wildlife. Water quality and soil conservation are state and regional concerns of high priority. New career opportunities in natural resource management are resulting from new efforts to save our rivers and bays.

Agriscience Supplies and Services

Spin the wheel of fortune! What comes up for you? Agriscience supplies and services (Figures 4-14 and 4-15)!

Agriscience supplies and services are businesses that sell supplies to agencies that provide services for people in agriscience. Examples of supplies are seed, feed, fertilizer, lawn equipment, farm machinery, hardware, pesticides, and building supplies. These businesses are

FIGURE 4-13 Effective resources management is critical for maintaining a balanced environment. *(Courtesy National FFA; FFA #226)*

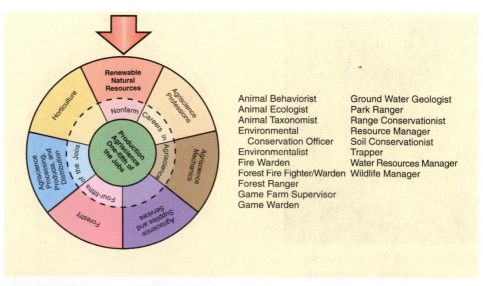

Animal Behaviorist
Animal Ecologist
Animal Taxonomist
Environmental
  Conservation Officer
Environmentalist
Fire Warden
Forest Fire Fighter/Warden
Forest Ranger
Game Farm Supervisor
Game Warden

Ground Water Geologist
Park Ranger
Range Conservationist
Resource Manager
Soil Conservationist
Trapper
Water Resources Manager
Wildlife Manager

FIGURE 4-12 Spin the wheel of fortune! What comes up for you? Renewable natural resources.

FIGURE 4-15 Agriscience supplies and services provide the vital materials and services to keep a trillion dollar industry moving. *(Courtesy National FFA; FFA #87)*

Animal Keeper	Field Sales
Animal Trainer	Representative,
Artificial Breeding	Agricultural Equipment
Distributor	Field Sales
Artificial Breeding	Representative,
Technician	Animal Health Products
Artificial Inseminator	Field Sales
Biostatistician	Representative,
Chemical Applicator	Crop Chemicals,
Chemical Distributor	Machinery
Computer Analyst	Harness Maker
Computer Operator	Harvest Contractor
Computer Programmer	Horse Trainer
Computer Salesperson	Insect & Disease Inspector
Custom Operator	Kennel Operator
Dairy Management	Lab Technician
Specialist	Meteorological Analyst
Dog Groomer	Pest Control Technician
Farm Appraiser	Pet Shop Operator
Farm Auctioneer	Poultry Field Service
Farrier	Technician
Feed Mill Operator	Poultry Hatchery Manager
Feed Ration Developer &	Poultry Inseminator
Analyst	Sales Manager
Fertilizer Plant Supervisor	Salesperson
Fiber Technologist	Service Technician
Field Inspector	Sheep Shearer

Aerial Crop Duster	Animal Groomer
Ag Aviator	Animal Health Products
Ag Chemical Dealer	Distributor
Ag Equipment Dealer	Animal Inspector

FIGURE 4-14 Spin the wheel of fortune! What comes up for you? Agriscience supplies and services.

operated by owners, managers, mill operators, truck drivers, sales personnel, bookkeepers, field representatives, clerks, and others.

People in these jobs provide the supplies for the industry. However, there are many in agriscience who seldom handle the commodities themselves. Instead they provide a service. Those who provide legal assistance, write agricultural publications, advise agriculturists on money matters, or provide advice on crops, livestock, pest control, or soil fertility are working in service occupations. Such jobs are for those who are more people-oriented than commodity-oriented.

Agriscience Mechanics

Spin the wheel of fortune! What comes up for you? Agriscience mechanics (Figures 4-16 and 4-17)!

FIGURE 4-17 Careers in agriscience mechanics are varied and challenging. *(Courtesy National FFA; FFA #290)*

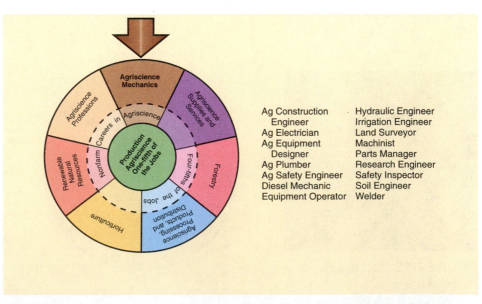

Ag Construction	Hydraulic Engineer
Engineer	Irrigation Engineer
Ag Electrician	Land Surveyor
Ag Equipment	Machinist
Designer	Parts Manager
Ag Plumber	Research Engineer
Ag Safety Engineer	Safety Inspector
Diesel Mechanic	Soil Engineer
Equipment Operator	Welder

FIGURE 4-16 Spin the wheel of fortune! What comes up for you? Agriscience mechanics.

AGRI·PROFILE

CAREER AREA: Agriscience Technician or Professional?

An assessment of career opportunities in agriscience by the United States Department of Agriculture revealed a bright outlook for college graduates. Other studies indicate a need for workers trained in agriscience for technical-level jobs.

The agriscience technician may be broadly trained in plant and animal sciences and employable in many fields. For the sharp individual with good work habits and a broad background in plant and animal sciences, the choices are extensive. Add agriscience mechanics skills to the package and the individual has access to dozens of career options.

Education for agriscience careers should begin at the high-school level or earlier. If one plans to work at the technician level, an early start will be especially helpful. The technician is expected to have first-hand experience and detailed knowledge of procedures and techniques. Such knowledge comes with experience in shops, laboratories, farms, greenhouses, fisheries, and on-the-job training situations, as well as in the classroom. Technicians generally have some special training beyond high school level, while professionals are required to obtain degrees at the bachelor, master, or doctorate levels.

Whether you prefer to be a technician or a professional, agriscience offers a broad array of career possibilities. *(Courtesy of USDA #036)*

Are you fascinated by tools and equipment? Are you challenged by something that does not work? Are you creative and do you like to build things? If so, a career in agriscience mechanics may be for you. **Agriscience mechanics** is the design, operation, maintenance, service, selling, and use of power units, machinery, equipment, structures, and utilities in agriscience.

Agriscience mechanics includes the use of hand and power tools, woodworking, metalworking, welding, electricity, plumbing, tractor and machinery mechanics, hydraulics, terracing, drainage, painting, and construction. Choose your level—indoors or out. Choose your role—employee, employer, or professional.

Agriscience Professions

Spin the wheel of fortune! What comes up for you? Agriscience professions (Figures 4-18 and 4-19)!

The word **profession** means an occupation requiring specialized education, especially in law, medicine, teaching, or the ministry.

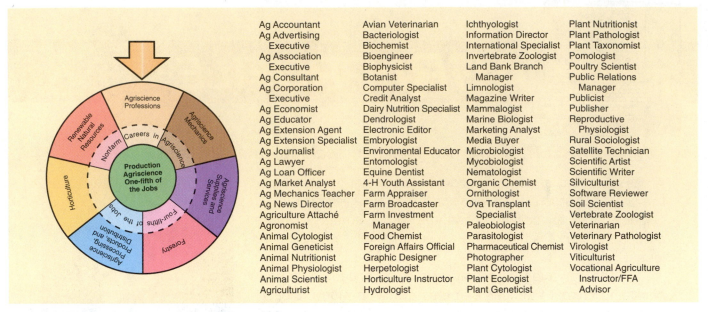

Ag Accountant	Avian Veterinarian	Ichthyologist	Plant Nutritionist
Ag Advertising	Bacteriologist	Information Director	Plant Pathologist
Executive	Biochemist	International Specialist	Plant Taxonomist
Ag Association	Bioengineer	Invertebrate Zoologist	Pomologist
Executive	Biophysicist	Land Bank Branch	Poultry Scientist
Ag Consultant	Botanist	Manager	Public Relations
Ag Corporation	Computer Specialist	Limnologist	Manager
Executive	Credit Analyst	Magazine Writer	Publicist
Ag Economist	Dairy Nutrition Specialist	Mammalogist	Publisher
Ag Educator	Dendrologist	Marine Biologist	Reproductive
Ag Extension Agent	Electronic Editor	Marketing Analyst	Physiologist
Ag Extension Specialist	Embryologist	Media Buyer	Rural Sociologist
Ag Journalist	Environmental Educator	Microbiologist	Satellite Technician
Ag Lawyer	Entomologist	Mycobiologist	Scientific Artist
Ag Loan Officer	Equine Dentist	Nematologist	Scientific Writer
Ag Market Analyst	4-H Youth Assistant	Organic Chemist	Silviculturist
Ag Mechanics Teacher	Farm Appraiser	Ornithologist	Software Reviewer
Ag News Director	Farm Broadcaster	Ova Transplant	Soil Scientist
Agriculture Attaché	Farm Investment	Specialist	Vertebrate Zoologist
Agronomist	Manager	Paleobiologist	Veterinarian
Animal Cytologist	Food Chemist	Parasitologist	Veterinary Pathologist
Animal Geneticist	Foreign Affairs Official	Pharmaceutical Chemist	Virologist
Animal Nutritionist	Graphic Designer	Photographer	Viticulturist
Animal Physiologist	Herpetologist	Plant Cytologist	Vocational Agriculture
Animal Scientist	Horticulture Instructor	Plant Ecologist	Instructor/FFA
Agriculturist	Hydrologist	Plant Geneticist	Advisor

FIGURE 4-18 Spin the wheel of fortune! What comes up for you? Agriscience professions.

FIGURE 4-19 Professional workers are becoming more important than ever as agriscience enters the information age. *(Courtesy USDA/ARS K-3401-10)*

Agriscience professions are those professional jobs that deal with agriscience situations. They cut across all divisions in the wheel of fortune.

Consider the agriscience teacher and the Cooperative Extension agent. Both must have a bachelor's or master's degree to be qualified. They may teach subjects in several or all divisions of agriscience.

Consider the veterinarian, agriscience attorney, research scientist, geneticist, or engineer—all these professions require advanced degrees and high levels of training. If you can meet the standard, then a career in an agriscience profession may be for you.

Computers in Agriscience

The use of computers is extensive in agriscience. This means there are many opportunities to combine computer skills with agriscience settings. These range from farms and ranches to horticulture and mechanics, from business and industry to research.

PREPARING FOR AN AGRISCIENCE CAREER

Career education is an important part of public education today. Since the early 1970s, the career education movement has spread from occupational education programs to the general curriculum. Many school systems now emphasize career education from kindergarten through adulthood.

As you consider a career in agriscience, it is important to consider how to meet the requirements to get started in that career. Some young people have an early start on careers in agriscience. They may have grown up on a farm or ranch. Their parents may have worked in one of the other areas of agriscience, such as horticulture, resource management, business, or teaching. They might have obtained jobs or worked for friends or neighbors who had agriscience businesses. The following are some suggestions for preparing for a career in agriscience.

BIO•TECH CONNECTION

When I'm Twenty Something!

Close your eyes and relax for a minute—then think about yourself in the year 2005. How old will you be then? Twenty something? You have finished your formal schooling and are into your career now. What kind of job do you have? Are you married or single? Is life enjoyable and your work challenging? Did you stay in your hometown, or are you in some other state or nation? Are you happy with your career? If you could start again, would you make the same choices? Now read the following questions and make mental notes of the answers that are most likely to be correct for you in the year 2005.

A DAY ON THE JOB IN THE YEAR 2005

DIRECTIONS: SELECT THE RESPONSE TO EACH ITEM THAT YOU THINK WILL BEST DESCRIBE YOU AND YOUR SITUATION IN THE YEAR 2005

1. What time do I generally wake up?
   ☐ Early morning   ☐ Afternoon   ☐ Evening

2. What is my work environment?
   ☐ Indoor   ☐ Outdoor   ☐ Both

3. Where am I working?
   ☐ In an office   ☐ In the community
   ☐ In a factory or shop   ☐ In my home

4. What type of clothes do I wear to work?
   ☐ Dress clothes (coat and tie or suit)
   ☐ Uniform
   ☐ Casual (open collar)
   ☐ Worn jeans

5. Am I married?
   ☐ Yes   ☐ No

6. If married, how long?
   ☐ Less than 3 years
   ☐ Three years or more

7. Number of children
   ☐ Zero   ☐ Three
   ☐ One   ☐ More than three
   ☐ Two

8. I will retire when I am:
   ☐ 54 years old or less
   ☐ 55 to 65
   ☐ Over 65

9. How much education do I have? My highest level of education:
   ☐ High school   ☐ 4-year college
   ☐ 2-year tech-   ☐ Master degree
      nical school   ☐ Doctor degree
      degree or   ☐ Other
      certificate

10. Do I like keeping up with technical trends and procedures related to my work?
    ☐ Yes   ☐ No

11. Where do I live?
    ☐ Home town   ☐ Not home state,
    ☐ Home state,   but in the U.S.
       but not home   ☐ Outside the U.S.
       town

12. I am working with or for:
    ☐ Government   ☐ A small company
    ☐ Education   (200 or less
    ☐ Self employed   employees)
    ☐ A large company
       (over 200 employees)

13. My income level makes me:
    ☐ Wealthy   ☐ Struggling financially
    ☐ Comfortable   ☐ Generally deprived

14. The amount of paid vacation is:
    ☐ Less than 5 days   ☐ 11 to 15 days
    ☐ 6 to 10 days   ☐ 16 days or more

15. The most important thing in my life is:
    ☐ Family   ☐ Time off from work
    ☐ Money and   ☐ The place where I live
       belongings
    ☐ Prestige and
       status

16. My work can best be described as:
    ☐ Producing information
    ☐ Producing or selling a product
    ☐ Distributing information
    ☐ Serving people

How well do your current career preparation activities mesh with your perception of yourself in the year 2005. The material in this and other units ahead should be helpful in determining your preferences and making plans and preparations for a successful and challenging career.

While in Middle School

- Make science projects with plants, animals, soil, water, energy, ecology, conservation, and wildlife.
- Research and report on the above.
- Join 4-H or Scouts and choose agricultural projects and merit badges.
- Volunteer to work on lawn, garden, greenhouse, farm, or conservation projects.
- Enroll in agriscience or other career education programs.

While in High School

- Enroll in agriscience classes, including plant science, animal science, agriscience mechanics, agribusiness, and farm management.
- Enroll in college-preparatory courses in English, math, and science.
- Join the FFA organization and participate in leadership, citizenship, and agriscience activities.
- Develop a broad, supervised, occupational agriscience experience program.
- Acquire hands-on, skill-development experiences.

After High School

- Obtain an agricultural job and plan ways to get additional training while on the job.
- Enter a community college and take courses that will transfer to the college of agriculture or life sciences at your state university.
- Enter a two-year program in technical agriculture.
- Enter a college of agriculture or life sciences and obtain a bachelor degree (B.S.), master degree (M.S.), and/or doctorate (Ph.D.).

You can obtain information on careers, schools, and colleges from many sources. The following suggestions may be helpful to you.

High School Agriscience Teachers

- Agriscience mechanics
- General agriscience
- Animal sciences
- Horticulture

Cooperative Extension Service

- Listed in your phone book under county or city government

State Department of Education

- Specialist in agriscience, agribusiness, and renewable natural resources
- FFA State Executive Secretary

Community Colleges and Other Post-secondary Institutions

■ Occupational Dean, Community College

■ Director or Dean, Institute or Technical School
Typical programs include: agriscience business management, farm production and management, ornamental horticulture, water resources, and wildlife.

■ Dean, College of Agriculture or Life Sciences
Typical programs include: agricultural education, agricultural and resource economics, agronomy (crops and soils), animal sciences, food science, forestry, horticulture, natural resources management, and poultry science.

■ Agricultural Education Coordinator, Department of Agricultural and Extension Education, Land Grant University

As new technologies and job opportunities emerge, so will the need for well-trained and educated new people. Agriscience is a diverse field with job opportunities available at all levels. Pick your area of interest, determine the level at which you wish to operate, obtain the appropriate education for the job, and follow through with a rewarding career into the twenty-first century!

STUDENT ACTIVITIES

1. Write the Terms to Know and their meanings in your notebook.

2. Using paper and pencil, calculate the amount of time needed to count the dollar value of assets in agriscience in the United States, as suggested in Figure 4-1.

3. Develop a bulletin board that illustrates the "Wheel of Fortune," with its listing of the broad categories of jobs or divisions in agriscience.

4. Develop a collage that illustrates the many jobs in agriscience.

5. Write the names of the divisions of agriscience, such as "Production Agriscience," "Agriscience Processing, Products, and Distribution," "Horticulture," etc., and list five jobs under each that look interesting to you.

6. Pick a job from the lists you developed for Activity 5 and write a one-page description of the job. Include the following sections in your description:

   a. job title;

   b. education/training required to get the job and advance in the field;

   c. working conditions on the job;

   d. advantages/benefits of the job;

   e. disadvantages of the job;

   f. salaries of beginning and advanced workers in the field; and

   g. aspects of the job that you like.

7. Using the job you researched for Activity 6, or another job or career area, list the things you should do during and after high school to prepare for a career in that area.

8. Determine the name and address of an appropriate official of a school or college and request information from that person about educational opportunities for you in his or her institution.

9. Develop a list of agriscience jobs in which computers are used.

SELF EVALUATION

A. Multiple Choice

1. The industry that provides commodities that are basic to life is
  a. aerospace.
  b. agriscience.
  c. biotechnology.
  d. transportation.

2. The number of workers in agriscience in the United States is approximately
  a. 21 million.
  b. 100 million.
  c. 100,000.
  d. 400,000.

3. The percentage of total jobs in agriscience that require a college education to enter the job is
  a. 10 percent.
  b. 20 percent.
  c. 41 percent.
  d. 60 percent.

4. The products and services that are provided in most of the areas in the agriscience wheel of fortune seem to flow to or originate from
  a. agriscience processing, products, and distribution.
  b. agriscience professions.
  c. horticulture.
  d. production agriscience.

5. The management of wetlands comes under the area of
  a. agriscience processing, products, and distribution.
  b. agriscience professions.
  c. horticulture.
  d. renewable natural resources.

6. Of all agriscience jobs, the percentage that are *not* on farms or ranches is
  a. 20 percent.
  b. 40 percent.
  c. 60 percent.
  d. 80 percent.

7. A producer's share of each dollar spent for bread and cereals in the United States is about
   a.   11 cents.
   b.   15 cents.
   c.   25 cents.
   d.   75 cents.

8. The number of floral industry workers in the United States is about
   a.   110,000.
   b.   220,000.
   c.   500,000.
   d.   1,000,000.

9. A student may join 4-H or Scout groups to learn agriscience concepts as early as
   a.   college.
   b.   high school.
   c.   middle school.
   d.   none of the above.

10. Agriscience classes in high school would logically include extensive instruction in
   a.   plants, animals, and agribusiness.
   b.   plants, animals, and social sciences.
   c.   plants, mechanics, and higher math.
   d.   food, fiber, and physics.

B. Matching

_____ **1.** Production
_____ **2.** Processing and distribution
_____ **3.** Horticulture
_____ **4.** Forestry
_____ **5.** Natural resources
_____ **6.** Supplies and services
_____ **7.** Mechanics
_____ **8.** Professions

a. Teacher or veterinarian
b. Hydraulics
c. Seed, feed, or lawn supply
d. Farming or ranching
e. Lumber
f. Ornamentals
g. Grading and packaging
h. Wildlife

Supervised Agriscience Experience

OBJECTIVE

To learn the rationale for and plan a supervised agricultural experience program.

COMPETENCIES TO BE DEVELOPED

After studying this unit, you should be able to:

- define supervised agriscience experience program (SAEP) terms.
- determine the place and purposes of SAEPs in agriscience programs.
- determine the types of supervised agriscience experience activities.
- explore the opportunities for supervised agriscience experience programs.
- set personal goals for an SAEP.
- plan your personal SAEP.

MATERIALS LIST

✓ paper

✓ pencil or pen

✓ bulletin board materials

✓ Student Interest Survey form (Figure 5-7)

✓ Resources Inventory form (Figure 5-8)

✓ Selecting a Supervised Agricultural Experience Program form (Figure 5-9)

✓ Experience Inventory form (Figure 5-13)

✓ Placement Agreement (Figure 5-14)

✓ Improvement Project Plan and Summary (Figure 5-15)

✓ Supplementary Agriscience Skills Plan and Record (Figure 5-16)

TERMS TO KNOW

Simulate
Real-world experience
On-the-job training
Supervised agriscience experience program (SAEP)
Supervised
Experience
Program
FFA
Project

Enterprise
Exploratory Supervised Agriscience Experience
Mentor
Agriscience literacy
Career exploration
Career
Entrepreneurship Supervised Agriscience Experience

Production or productive enterprise
Agribusiness
Placement Supervised Agriscience Experience
Improvement activities
Agriscience skills profile
Resources Inventory
Resumé

Agriscience programs in various schools teach basic principles and practices in plant and animal sciences, resources management, business management, agriscience mechanics, landscape design, leadership, and personal development. Such programs emphasize reading and math skill development and the application of scientific principles.

Classrooms are excellent places to learn fundamentals and theory through reading, study, discussion, and planning. School laboratories, such as greenhouses, agriscience mechanics shops, school land demonstration plots, farms, and animal production facilities, help provide experiences that simulate real-world experiences. **Simulate** means to look or act like. **Real-world experience** means conducting the activity in the daily routine of our society. Simulation is an excellent way to learn. It imitates the real world and may provide an ideal setting for the activity.

However, classroom experiences and simulation lack the thoroughness of real-world experiences. Also, the personal relationships, such as employer-employee, supervisor-subordinate, owner-worker, salesperson-customer, owner-government, owner-community, fellow workers, and employee competition, are rarely present in simulated activities. Therefore, a program is needed for the student to obtain real-world experiences and on-the-job training if agriscience education is to lead to a successful career. **On-the-job training** means experience obtained while working in an actual job setting. In agriscience, the method used for students to obtain real-world experiences is referred to as the **supervised agriscience experience program (SAEP).**

AGRI·PROFILE

CAREER AREA: Supervised Agriscience Experience

The phrase *supervised agriscience experience* has three important elements. Experience suggests hands-on or real-life activities in the work place. Attitudes, knowledge, and skills gained here will generally be salable in the future. *Agriscience* means the setting and skills will be in the area of plant or animal sciences, or management or mechanics related thereto. *Supervised* means experienced persons such as your teacher, parents, and/or employer recognizes your interest in learning and will help direct the experience.

Agriscience experience is obtained in many ways. Generally, the agriscience teacher helps the student develop an understanding of the process and find a suitable location for a productive experience with cooperation from the parents. However, sometimes sympathetic parents and supportive school programs are not available. In such cases, students must develop an extra measure of determination and seek experiences on their own.

Experiences may be for pay, or simply for experience. Many schools now provide supervised experience programs in school laboratories. Whether for pay or not, the experiences obtained are worth the effort and give the participant an edge in the job market.

Learning by doing is generally accepted as the best way to become proficient in complex procedures and psychomotor skills. *(Courtesy USDA/ARS #K-3395-1)*

SUPERVISED AGRISCIENCE EXPERIENCE PROGRAM DEFINED

A supervised agriscience experience program consists of all supervised agriscience experiences learned outside of the regularly scheduled classroom or laboratory. **Supervised** means to be looked after and directed. Agriscience in this phrase means business, employment, or trade in agriculture, agribusiness, or renewable natural resources. **Experience** means anything and everything observed, done, or lived through. **Program** means the total plans, activities, experiences, and records of the supervised agriscience activities.

PURPOSE OF SAEPs

Supervised agriscience experience programs provide opportunities for learning by doing. They provide the means for you to learn with the help of your teacher, parents, employer, and other adults experienced in the area of your interest. Student SAEPs are an essential part of effective agriscience programs.

Some important purposes and benefits of supervised agriscience experience programs are:

1. provide opportunities to creatively explore a variety of subjects about agriscience;

2. provide educational and practical experiences in a specialized area of agriscience;

3. provide the opportunity to become established in an agriscience occupation;

4. provide opportunities for earning while learning;

5. create opportunities for earning after graduation;

6. develop interests in additional areas of agriscience;

7. develop valuable work skills such as
   appreciate importance of honest work,
   improve personal habits,
   develop superior work habits,
   establish good relationships with others,
   keep effective records,
   fill out useful reports,
   follow instructions and regulations,
   contribute to your occupation, and
   contribute to your family, community, and nation;

8. permit individualization of instruction;

9. permit recognition for individual achievement; and

10. become established in an agriscience business.

SAEP AND THE TOTAL AGRISCIENCE PROGRAM

High school agriscience programs may be comprehensive and provide students with agricultural experience programs, leadership development programs, and laboratory experiences, along with classroom instruction. Leadership skills are developed through the FFA program, as discussed in Unit 6. These components are integrated so they complement each other (Figure 5-1). This integration should provide the most effective program for the student and make the best use of teacher time.

SAEP and Classroom Instruction

The SAEP is planned as part of the classroom instruction. Students use this instruction to learn how to plan an SAEP, decide what types are possible, choose activities for their own SAEPs, and make appropriate arrangements with parents, teachers, and employers. The student conducts the SAEP under the supervision of the teacher, who provides instruction on the necessary topics. The teacher also arranges for small group and individual instruction. This permits the student to use the classroom and laboratory to solve problems encountered with the SAEP.

A COMPREHENSIVE AGRISCIENCE PROGRAM

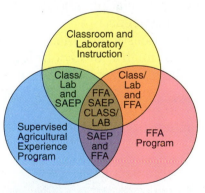

FIGURE 5-1 A comprehensive agriscience program should provide the student appropriate classroom/laboratory instruction, supervised agriscience experiences, and personal development through FFA.

FIGURE 5-2 Effective planning results in meaningful supervised agriscience experiences. *(Courtesy USDA/ARS #K-3405-11)*

SAEP and the FFA

The SAEP also overlaps with the FFA (an intracurricular youth organization for students enrolled in agriscience programs). The FFA has many activities that encourage students to do more in their SAEPs and provides contests that more fully develop useful skills for SAEPs. Further, the FFA provides awards and other recognition for achievements in SAEPs. These frequently lead to trips and other travel experiences greatly enrich the student's education.

FFA and Classroom Instruction

Finally, FFA is merged with classroom and laboratory instruction in a number of ways. The teacher has instructional units on the FFA in the classroom. Further, instruction is provided in public speaking, parliamentary procedure, and other leadership skills. The classroom setting can become the place where FFA teams polish their skills in preparation for upcoming contest or award activities. Therefore, the students who obtain the most benefit from their agriscience programs are those who take advantage of the opportunities provided through SAEPs and the FFA.

TYPES OF SAE PROGRAMS

Supervised agriscience experiences grow out of planned programs (Figure 5-2). The word **project** is used to describe a series of activities related to a single objective or enterprise, such as raising rabbits, building a porch, or improving wildlife habitat. The term **enterprise** generally refers to a type of animal or plant. Examples are dairy, beef, or rabbits, and corn, hay, turf, or poinsettia. Students may choose from various types of SAE programs (Figure 5-3).

Exploratory SAE

An **Exploratory Supervised Agriscience Experience** program is one where supervised activities are conducted to explore a variety of subjects about agriscience and careers in agriscience. Such experiences help students gain understanding in and appreciation about agriscience in order to satisfy personal interests and needs.

Exploratory supervised agriscience experiences might include investigations and experiences in small animal health, biotechnology, water rights, agriscience journalism, aquaculture, hydroponics, air pollution, crop science, tissue culture, agriscience engineering, and many other areas. The student's exploratory SAE program is planned by the student under the direction of the teacher in cooperation with

TYPES OF SUPERVISED AGRISCIENCE EXPERIENCE			
Exploratory SAEP	Entrepreneurship SAEP	Placement SAEP	Improvement Activities
	Production Enterprise Agribusiness Enterprise	Farm/Ranch Placement (for wages and experience) Agribusiness Placement (for wages and experience)	

FIGURE 5-3 Supervised Agriscience Experience Programs.

EXAMPLES OF ENTREPRENEURSHIP

A production enterprise is a crop, livestock, or agribusiness venture which the student has a degree of *ownership*. The student may own all or part of the enterprise.

Types of Crop Enterprises

- Corn Production
- Soybean Production
- Small Grain Production
- Greenhouse Production
- Nursery Production
- Hay Production
- Vegetable Production
- Fruit Production
- Forestry Production
- Christmas Tree Production

Types of Livestock Enterprises

- Commercial Cow-Calf Production
- Registered Breeding Stock Production
- Market Beef Production
- Dairy Production
- Feeder Pig Production
- Market Swine Production
- Sheep Production
- Poultry Production
- Horse Production
- Custom

Types of Agribusiness Enterprises

- Lawn Service
- Custom Farm Work
- Trapping and Pelt Sales
- Hunting Guide Service
- Tree Service
- Farm and Garden Supply Service
- Artificial Insemination Business
- Animal Care and Boarding
- Winery
- Fishing and Crabbing for Sales
- Custom

FIGURE 5-4 Agriscience students have many production projects from which to choose.

parent/guardian, mentor, and others who will help the student obtain the exploratory experiences. A **mentor** is a person whom you admire and who has skills that you would like to learn.

Exploratory SAEs are intended for students who wish to observe and experience a variety of areas in agriscience or to explore one or more areas not covered sufficiently in class to satisfy the student's interest. Such programs add an exciting dimension to courses at the elementary or middle school levels where agriscience literacy and career exploration are emphasized.

Agriscience literacy means education in or understanding about agriscience. Literacy does not require the student to become proficient in a given area, but requires general knowledge about it. **Career exploration** means learning about occupations and jobs that could possibly become one's future **career** or life's work.

Entrepreneurship SAE

Entrepreneurship Supervised Agriscience Experience refers to supervised activities conducted by students as owners or managers for profit. Emphasis is placed on developing the skills of the job or enterprise and operating in a profitable and professional manner. Students may develop and own plant or animal enterprises where they grow commodities such as flowers, fruits, vegetables, field crops, turfgrass, Christmas trees, nursery stock, forestry, small animals, wildlife, beef, sheep, swine, honey bees, earth worms, and others. An enterprise where the student owns the equipment, feed, seed, and other supplies for the enterprise is called a **production or productive enterprise.**

Entrepreneurship activities may also be conducted in agribusiness. An **agribusiness** entrepreneurship enterprise is one where the student buys and sells an agriscience commodity for profit, rather than raising or growing the commodity. Some examples are a pet business, florist shop, livestock sales business, game dressing service, crop scouting service, crop spraying service, feed sales, seed sales, flower vendor, auctioneer, agriscience mechanic, and trucker (Figure 5-4).

Placement SAE

Placement Supervised Agriscience Experience programs are those where the student is placed with an employer in a production unit such as a farm, ranch, green house, nursery, and aquaculture facility to produce commodities for wages. Or, the student is placed with an employer or mentor in an agency or agribusiness where commodities are bought and sold, or services are rendered. Some examples of agribusinesses are veterinary centers, kennels, feed or seed stores, pet shops, nursery outlets, florists, and garden centers. Some examples of agencies where students may be placed are Cooperative Extension Service (CES), Soil Conservation Service (SCS), Agricultural Stabilization and Conservation Service (ASCS), Forestry Service (FS), wildlife and environmental agencies, and school laboratories. Unfortunately, funds are generally not available for hiring students in most agencies. The emphasis in placement programs is learning real skills and becoming a professional in the activity.

Improvement Activities   **Improvement activities** are activities that improve the appearance, convenience, efficiency, safety, or value of a home, farm, ranch, agribusiness, or other agriscience facility. The student does not receive a wage or profit for conducting improvement activities. However, the student benefits by learning new skills and enjoying the benefits of the improvements. The owner of the facility should provide the materials and cover other expenses. Improvement activities provide the student with many opportunities to learn without the risks and commitment necessary for entrepreneurship activities (Figure 5-5).

Research Activities   Research is another possible area for students to obtain excellent supervised agriscience experiences. Research may be done in school laboratories, at home, on the job, or wherever suitable facilities may be found. Research is generally not regarded as a profit–making activity, but may be conducted to answer questions that can be profitable if applied in production settings. And, research projects may be part of community improvement activities or state research efforts such as stream monitoring, weather watch, forest fire watch, crop scouting, insect or weed monitoring, and crop reporting.

Research skills are valuable for solving problems and for working into interesting and good-paying careers. As a special incentive for teaching and learning research skills, the National FFA provides awards annually to outstanding agriscience students and teachers.

AGRISCIENCE SKILLS PLAN AND PROFILE

Students are encouraged to develop an agriscience skills profile. An **agriscience skills profile** is a record of skills developed and the level of competence in each. The better that competent people can document what they can do, the more likely they are to be hired in the jobs and careers they desire. Time and resources are limiting factors in preparing for a career, so every effort should be made to learn the most useful and interesting skills. Skills may be obtained in the classroom, laboratory, community, and SAE. Lists of appropriate skills for various areas in agriscience should be helpful for the student and teacher for planning curriculum and supervised agriscience experiences (Figure 5-6).

EXPLORING OPPORTUNITIES FOR SAEPs

Students should use great imagination when considering SAEPs. Some complain that they do not have opportunities for meaningful SAEPs. Yet other students in similar circumstances find or create opportunities for effective programs. Seek the advice of your teacher for ideas. Also, observe what successful students have done in the agriscience program and in your community. Then develop an SAEP that provides the most opportunity to learn and earn.

Personal Interest

Personal interest is an important factor in the success of an SAEP. Consider the kinds of activities you like to do and then build on those interests. A Student Interest Survey or Inventory should help you assess your natural interests and provide some guidance in developing an SAEP (Figure 5-7, page 76).

EXAMPLES OF IMPROVEMENT ACTIVITIES

Soil Improvement Programs

- Liming
- Fertilizing
- Drainage
- Erosion control
- Plow under green manure
- Introduce a cropping system
- Soil sampling

Building Improvement Programs

- Painting
- Window repair
- Roof repair
- Foundation repair
- Floor repair
- Siding repair
- Door repair
- Electric wiring
- Water systems installation
- Heating
- Lighting protection
- Feeding floor construction
- Remodeling
- Home sewage system installation

Fence Improvement

- Construction of new fence
- Fence replacement and repair
- Construction of flood gates
- Construction and repair of gates

Homestead Improvement Programs

- Plan and set out a windbreak
- Seed or reseed lawn
- Plant shrubs and trees
- Clean up homestead

Orchard, Small Fruits and Vegetables Improvement Programs

- Plan and set out a fruit tree orchard
- Plan and set out a small fruit garden
- Plan and grow a home vegetable garden
- Renovate an old orchard

Weed Control Programs

- Spray major weed areas on farm
- Mow or spray weeds in fence rows
- Pull weeds in corn
- Clip weeds in permanent pastures

Insect and Pest Control

- Rat control
- Corn borer control
- Japanese beetle control
- Livestock parasite control

Farm Management Programs

- Keep farm accounts
- Plan farm safety program
- Inventory farm equipment
- Keep checking account records

Agricultural Shop Programs

Agricultural Machine Repair and Reconditioning Programs

Livestock Improvement Programs

Crop Improvement Programs

Landscaping Improvement Programs

General Clean-up Program
- Remove dead trees or shrubs
- Remove and discard dead branches
- Remove unsightly junk, trash, and woodpiles
- Provide a specific storage area for all lawn and horticulture equipment
- Repair or remove broken lawn furniture
- Improve the grade if necessary
- Remove, replace or repair fences, sidewalks, step railings or porches
- Transplant trees, shrubs or flowers
- Make or improve the driveway
- Pick up nails and broken glass
- Remove unsightly rocks

Grounds Maintenance Practices
- Mow the lawn
- Trim or prune trees and shrubs
- Fertilize the lawn, trees, shrubs, and flowers
- Apply herbicides
- Apply insecticides
- Water the lawn, trees, shrubs, and flowers
- Repair trees
- Brace trees
- Prevent sunburn of plants
- Prevent insect damage and disease
- Set up, adjust, and move a sprinkler
- Edge the lawn
- Stake trees
- Set up rain gauge
- Record precipitation from rain gauge
- Mulch trees, shrubs, and flowers
- Rake the lawn

Improvement and Beautification Activities
- Plant new trees and shrubs
- Draw a landscape plan
- Seed, plug or sod lawn
- Plant flowers
- Relocate and replant trees and shrubs
- Renovate existing lawn
- Build a patio
- Make a window box
- Build a trellis
- Plant a windbreak

FIGURE 5-5 Improvement activities are available for agriscience students regardless of the home situation.

EXAMPLES OF AGRISCIENCE SKILLS

Agribusiness

- Operate cash register
- Display merchandise
- Keep inventory records
- Write sales tickets
- Deliver products
- Set up machinery
- Assemble equipment
- Greet customers
- Close sales
- Answer telephones
- Order merchandise
- Operate adding machine
- Wrap meat
- Cut carcass into wholesale cuts
- Cut wholesale cuts into retail cuts
- Compute sales tax

Agriculture Mechanics

- Arc weld metals
- Oxy-acetylene weld metals
- Cut details with oxy-acetylene
- Operate farm machinery
- Operate wood power tools
- Operate metal power and hand tools
- Recondition and sharpen tools
- Service air cleaner
- Service electric motor
- Store machinery
- Install rings or pistons
- Grind valves
- Wire electrical convenience outlet
- Change transmission fluid
- Fasten sheet metal with rivets
- Repair flat tire
- Calibrate field sprayer
- Pour concrete
- Lay reinforcement steel
- Use farm level

Corn

- Select seed
- Plant seed
- Prepare seedbed
- Calibrate corn planter
- Adjust planter for depth
- Apply dry fertilizer
- Apply anhydrous ammonia
- Conduct corn variety test
- Check harvest losses
- Apply herbicides and insecticides
- Operate combine
- Operate corn picker
- Identify weeds
- Identify insects
- Cultivate corn
- Dry corn artificially

Forages

- Innoculate legume seeds
- Rotate pastures
- Greenchop forages
- Bale hay
- Renovate permanent pastures
- Combine grasses and legumes
- Graze pastures properly
- Rotate pastures

Horticulture

- Plant vegetable garden
- Prepare garden plot
- Plant fruit trees
- Bud-graft
- Cleft-graft
- Whip-graft
- Prepare growing medium
- Sterilize soil
- Pot plants
- Water greenhouse plants
- Root cuttings
- Harvest crop
- Process vegetables
- Store produce
- Fertilize plants
- Determine plant diseases
- Treat deficiency symptoms
- Force blooming

Small Grains

- Calibrate grain drill
- Plant small grains
- Broadcast fertilizer
- Select adapted varieties
- Clean seeds
- Harvest small grains

Soil Management

- Test soil for lime requirements
- Lime soils
- Seed grass waterway
- Rotate crops
- Plant windbreak
- Test for fertilizer
- Fertilize soils
- Lay drainage tile
- Terrace fields
- Farm on contour
- Aerate soil
- Control erosion

Soybeans

- Test for germination
- Innoculate seed
- Take soil sample
- Control weeds
- Treat seed for storage
- Market beans
- Harvest soybeans
- Check harvest losses

Beef Cattle

- Assist cow in calving
- Dehorn
- Castrate bull calves
- Disinfect navel of baby calves
- Select herd sire
- Select replacement heifers
- Creep feed calves
- Cull poor producers
- Ear tag
- Tatoo
- Remove warts
- Drench
- Treat for bloat
- Treat for external parasites
- Select feeders
- Trim hooves
- Vaccinate for blackleg
- Vaccinate for IBR
- Vaccinate for brucellosis
- Formulate a balanced ration
- Ring bull
- Fit animal for show or fair
- Analyze production records
- Palpate to determine pregnancy

FIGURE 5-6 There are hundreds of agriscience skills that are useful for employment in agriscience.

EXAMPLES OF AGRISCIENCE SKILLS *(cont'd.)*

Dairy Cattle

- Select replacement heifers
- Dry cows
- Control external parasites
- Assist newborn calves in getting colostrum
- Vaccinate heifers for brucellosis
- Test cows for T.B.
- Cull low producers
- Dehorn
- Production test cows
- Participate in DHIA
- Treat mastitis
- Operate milking machines
- Clean facilities after milking
- Artificially inseminate cows
- Prevent milk fever
- Formulate balanced ration
- Wash udder with disinfectant
- Detect abnormal milk with strip-cup
- Feed according to production
- Clean and sterilize utensils

Poultry

- Clean and disinfect brooder
- Cull poor producers
- Select pullets
- Sex baby chick or poults
- Prevent cannibalism
- Keep production records
- Dress broilers
- Prevent breast blisters
- Stimulate egg production
- Disinfect laying house
- Grade eggs by candling
- Size eggs
- Debeak chicks
- Vaccinate for fowl pox
- Treat for external parasites
- Worm poultry
- Provide sanitary water
- Feed balanced rations
- Caponize cockerels
- Provide litter

Sheep

- Select ram
- Flush ewes
- Treat for external parasites
- Assist ewes at lambing
- Creep feed lambs
- Dock lambs
- Castrate lambs
- Artificially inseminate ewes
- Ear tag
- Shear sheep
- Tie fleece
- Cull farm flock
- Determine estrus in ewes
- Trim hooves
- Worm for internal parasites
- Keep production records

Swine

- Select herd boar
- Select replacement gilts
- Flush gilts
- Vaccinate sows for leptospirosis
- Provide farrowing stalls
- Clean facilities before farrowing
- Clean sow before farrowing
- Assist sow at farrowing
- Treat navels of baby pigs
- Clip needle teeth
- Creep feed pigs
- Ring hogs
- Treat for external parasites
- Inject iron in baby pigs
- Vaccinate for erysipelas
- Vaccinate for brucellosis
- Wean pigs at 4–6 weeks
- Weigh pigs at 56 days
- Formulate balanced ration
- Castrate boar pigs

FIGURE 5-6 *(Continued)*

Resources Inventory

A **Resources Inventory** is a summary of the resources that may be available for conducting SAEP activities. It includes information about your home, farm, work setting, and community that might be useful in considering your SAEP (Figure 5-8). Part of the inventory is a scale drawing of the property where you live or work. Making the scale drawing will help you realize what is available and help your teacher to suggest SAEP possibilities.

SELECTING AND IMPLEMENTING YOUR SAEP

After completing the Personal Interest Survey and the Resources Inventory, you should arrange a conference with your teacher. The conference should include discussions on your interests and a look at the possible production enterprises, improvement projects, and supplementary skills. After the conference you should record tentative plans for the SAEP (Figure 5-9). At this point you and your teacher should discuss the plan with your parents or guardians. If an employer is involved, he or she should become a partner in the planning process.

STUDENT INTEREST SURVEY		
Place an 'X' in the blank by the tasks that you like to do or would like to learn to do.		
Tasks Typical of Agribusiness	**Tasks Typical of Horticulture**	**Tasks Typical of Production**
_____ Delivering merchandise	_____ Applying pesticides	_____ Applying pesticides
_____ Displaying merchandise	_____ Arranging flowers	_____ Baling hay
_____ Driving trucks	_____ Balling and burlapping trees	_____ Building fences and buildings
_____ Keeping records	_____ Building patios	_____ Castrating animals
_____ Mowing lawns	_____ Edging flower beds	_____ Cleaning animals
_____ Operating cash registers	_____ Identifying plants	_____ Feeding animals
_____ Operating equipment	_____ Lifting heavy materials	_____ Getting up early
_____ Pricing merchandise	_____ Making Christmas decorations	_____ Handling manure
_____ Processing meat, milk, grains	_____ Making cuttings	_____ Harvesting crops
_____ Repairing equipment	_____ Mowing lawns	_____ Helping parents
_____ Selling merchandise	_____ Mulching beds	_____ Keeping records
_____ Stocking shelves	_____ Operating power machinery	_____ Lifting heavy materials
_____ Taking customer orders	_____ Planting bulbs	_____ Milking cows
_____ Taking inventory	_____ Planting grass	_____ Operating machinery
_____ Taking telephone orders	_____ Planting seeds	_____ Painting buildings
_____ Unloading trucks	_____ Planting trees and shrubs	_____ Planting crops
_____ Working outside	_____ Protecting plants from weather	_____ Plowing fields
_____ Working with people	_____ Pruning plants	_____ Repairing buildings
_____ Working with plants	_____ Raking leaves	_____ Repairing machinery
	_____ Selling plants	_____ Shearing sheep
	_____ Watering plants	_____ Showing animals
	_____ Weeding by hand	_____ Taking soil samples
	_____ Working in various weather	_____ Working in various weather
	_____ Working with people	_____ Working with animals

FIGURE 5-7 A Student Interest Survey should be helpful to you in choosing SAEP activities.

Securing a Job

If placement on a farm or in an agribusiness is part of the SAEP plan, you will need a brief **resumé** (a one-page summary of information about a job applicant (Figure 5-10). The prospective employer will be interested in your educational and occupational background, which will help determine your qualifications and experiences. Be sure to ask a minimum of three adults who know your character and qualifications if you may list them as references. Your prospective employer will probably want to contact them.

Resources available to the student for the Supervised Agricultural Experience Program (SAEP).

MAP OF HOME FARM AND/OR BUSINESS

Let each square represent any convenient acerage or square footage, i.e., 20, 40, 80, 100, etc.

One square = _____

LEGEND

—— Public Road	- - - Private Drive	☐ Farmstead	⌢ Terrace
➚ Terrace Outlet	➶ Natural Drain	⟋ Gullies	
--➤ Stream	⬭ Pond	✿ Trees	✳✳✳ Fence
▦▦▦ Railroad	= = Crossing		

Resources Inventory

1. Name _____ Age _____ Class _____

2. Address _____ Phone _____

3. Parent's Name _____ Occupation _____

4. Number in my family _____ Boys _____ Girls _____

5. I live: on farm _____ in town _____ on an acerage _____ .

6. Is land available for you to rent to produce crops? _____ yes _____ no

   a. If yes, how many acres? _____

   b. Which crops? _____

   c. Location of land? _____

7. Are facilities available for you to rent to produce livestock or livestock products? _____ If so,

   a. What type of livestock? _____

   b. Number _____

   c. Location of facilities _____

8. Do you have available space for a garden? _____ yes _____ no

9. Do you have facilities for mechanical work? _____ yes _____ no

10. Do you have a greenhouse available for your use? _____ yes _____ no

11. Would you be interested in producing livestock or crops on the school farm?

   _____ yes _____ no   If so, what type? _____

FIGURE 5-8 The Resources Inventory is especially helpful in planning production enterprises and improvement activities.

Personal Resumé

Name: John Smith

Address: 1234 Honeylocust Drive
         Frederick, Maryland 21701

Telephone: (301) 662-0000

Education: Junior at Frederick High School
           Majoring in Landscaping

Subjects Studied:
    Horticulture I
    Horticulture II
    Landscaping I
    Landscaping II
    Typing I

Student Activities:
    President: FFA
    Editor of school yearbook
    Tennis team and softball team

Special Skills:
    Can ball and burlap trees, use cash register, water
    plants, transplant plants, and operate a tractor.

Employment Experience:
    Hardees Fast Food: Worked as a cashier and cook
    Landscaped neighbor's yard

References:
    Mr. Ralph Rece, Principal, Frederick County Vo-Tech,
    Frederick, MD.
    Mrs. Holly Deane, Instructor, Frederick County Vo-Tech,
    Frederick, MD.
    Mr. Bernard Rose, Manager, Hardees Fast Food, Frederick,
    MD.

FIGURE 5-10 A personal resumé will help when seeking a job.

Selecting a Supervised Agriscience Experience Program for

_____
(Name of Student)

Instruction: Use this form to tentatively decide on a beginning agriscience SAEP. This information will be helpful in agriscience classes to develop detailed plans for obtaining experiences.

My interest areas in agriscience/horticulture are _____

Based upon my interest and opportunities available to me to get practical experience in agriscience, I plan to include the following in my SAEP.

1. Production or Productive enterprises (examples: beef, dairy, nursery production, Christmas trees)

_____
_____

2. Placement in an agribusiness (examples: supply store, florist shop, nursery, golf course, landscape contracting)

_____
_____

3. Improvement activities (examples: landscape your home, fertilize your lawn, plant trees)

_____
_____

4. Skills (examples: change spark plugs, weld, change the oil in small engines)

_____
_____

5. Other activities (example: projects on school facility)

_____
_____

FIGURE 5-9 Enterprises, improvement activities, and agriscience skills should be selected and recorded early in the school year.

PREPARING FOR THE JOB INTERVIEW

What Employers Look for in an Employee

- **Attitude**—The prospective employee should have a positive attitude about the job. He/she should show enthusiasm and a willingness to learn and work. Employers stress this as being one of the most important qualities they look for in prospective employees.
- **Experience**—Previous experience of the prospective employee is important. However, employers are usually willing to train the person with a positive attitude.
- **Appearance**—The prospective employee should be neat and clean, have hair combed and be well dressed. It is better to be overdressed than underdressed for an interview.
- **Posture**—It is important to stand and sit up straight. The employer will be observing the way you carry yourself and will make judgments accordingly.
- **Mannerisms**—Mannerisms are gestures that are made which may be annoying or could be welcomed. However, one should be aware of mannerisms. Do you:
  1. Yawn a lot? If so, others will think you're bored or worse—lazy.
  2. Fidget? Squirming may indicate lack of confidence or disinterest in the job.
  3. Daydream? Give your full attention to the interviewer.
- **Handshake**—Have a firm handshake; not bone crushing and not limp.

What Questions Should be Asked?

The following questions may be asked if the information is not provided by the interviewer.
- What type of jobs or tasks are to be done?
- What are the policies and procedures for workers?
- What are the working hours?
- What is the rate of pay?
- What arrangements are needed for time off?
- If you are uncertain about something that has been discussed in the interview you should ask the employer to clarify or explain.
- IN SUMMARY, BE POLITE AND ATTENTIVE DURING YOUR JOB INTERVIEW.

FIGURE 5-11 Preparation for the job interview will permit you to relax and give your total attention to the interviewer.

When your resumé is complete, be sure your teacher has approved the final version. Then you are ready to approach employers for a job that will help achieve the objectives of your SAEP plan. Your approach is critical, since first impressions are lasting impressions. It is important to dress in a businesslike manner, be courteous and confident, and conduct yourself according to standard interview procedures (Figure 5-11).

Refining the Plan

Plans should be worked out for each production enterprise. It is important to estimate the anticipated expenses and income, which will help you make financial arrangements to conduct the project. Also, some goals for the enterprise should be set, including the size of the project and some efficiency factor goals (Figure 5-12).

Employers are generally impressed with students who are eager to learn. In this regard, both the student and the employer can benefit from a carefully thought out statement of skills to be developed on the job. You should develop the list with the guidance of your teacher and prepare an Experience Inventory to record the completion of tasks or

SETTING GOALS FOR PRODUCTION IN THE SAEP

What goals should be set for the Supervised Agricultural Experience Program?

DEFINITION: A goal is the hoped for end result of hard work and should be challenging and realistic.

- Goals should be challenging!
- Goals should be reachable!
- SAEP goals should focus on scope, learning opportunities, and production efficiency factors.
- Parents, employers, agriculture teachers, and other qualified adults should help develop SAEP goals.
- Goals should be recorded.
- Goals should be analyzed and evaluated periodically and new goals developed.
- Goals provide direction and organization.
- Settings realistic goals should help increase profits.

What are efficiency factors?

DEFINITION: Efficiency factors are measures of production success and encourage enterprise improvement and profit. Examples of efficiency factors are as follows.

- **Size of Enterprise.** For animal weight or livestock products produced.
- **Rate of Gain and Production.**

$$\text{Beef: Percent of calf crop} = \frac{\text{Calves born alive}}{\text{Cows bred}}$$

$$\text{Poultry: Percent of egg production} = \frac{\text{Average eggs per hen}}{\text{Number days in production}}$$

$$\text{Sheep: Percent of lamb crop} = \frac{\text{Lambs born alive}}{\text{Ewes bred}}$$

$$\text{Swine: Pigs farrowed per litter} = \frac{\text{Live pigs farrowed}}{\text{Sows bred}}$$

$$\text{Weight produced per litter} = \frac{\text{Total production}}{\text{Number of litters}}$$

- **Returns and Feed Costs.** Round total income and value of feed fed to the nearest whole dollar.

$$\text{Returns per \$100 feed fed} = \frac{\text{Total Income}}{\text{Dollars worth of feed fed}} \times 100$$

$$\text{Returns per \$100 invested} = \frac{\text{Total Income}}{\text{Total Expenses}} \times 100$$

$$\text{Expense Per Cwt. of Production} = \frac{\text{Total Expenses}}{\text{Total Production}} \times 100$$

$$\text{Average weight sold} = \frac{\text{Total sales weight}}{\text{Animals sold}}$$

$$\text{Average price received} = \frac{\text{Total sales value}}{\text{Units sold}}$$

- **Feeding Efficiency**

  Note: Convert all corn to shelled corn basis (56 lb. per bu.) before figuring efficiency factors.
  Note: Poultry—1 unit is equal to 1 dozen eggs or 1.5 lb. of weight.
  Note: Dairy—1 unit is equal to 1000 lb. milk or 100 lb. weight.

Feed cost per Cwt. or per unit

$$\text{a. For swine or beef cattle} = \frac{\text{Total feed cost}}{\text{Lb. weight produced}} \times 100$$

$$\text{b. For sheep} = \frac{\text{Total feed cost}}{\text{Lb. wool + lb. weight}} \times 100$$

$$\text{c. For dairy or poultry} = \frac{\text{Total feed cost}}{\text{Units of production}}$$

Feed per Cwt. produced

$$\text{a. For hogs, beef cattle, or sheep} = \frac{\text{Lb. of feed fed}}{\text{Lb. weight produced}}$$

(For sheep, include wool with weight as in lb. above)

$$\text{b. For dairy or poultry} = \frac{\text{Lb. of each feed fed}}{\text{Units of production}}$$

- **Death Loss**

$$\text{\% death loss} = \frac{\text{Weight of dead animals}}{\text{Total animal weight produced}}$$

A low percent of death loss means a high enterprise rating for this item.

FIGURE 5-12 Goals should state the number, size, time lines, and efficiency factors you hope to achieve.

jobs (Figure 5-13). It should be frequently updated. The experience inventory helps the student, teacher, and employer tack progress in achieving the set goals and in developing a skills profile.

The Placement Agreement document helps finalize the plans for placement on a job (Figure 5-14). Such agreements need the signature of the student, parents or guardians, employer, and teacher. Once all parties are in agreement regarding the student's placement experiences, the chances for success are very high.

Another document that will help plan and conduct the SAEP is the Improvement Project Plan and Summary (Figure 5-15). This plan directs the student to describe the conditions found, plans for improvement, and estimated value of the improvement when completed. The summary is filled out as the improvement project progresses and serves as the record when finished.

BIO•TECH CONNECTION

Tracking the Disappearing Gene

The USDA Agriculture Research Service (ARS) is on the lookout for good students who are looking for experiences in research. For over fifteen years, plant geneticist Thomas E. Devine of the USDA Agriculture Research Service plant molecular biology laboratory in Beltsville, Maryland, has been looking at the genetic structures and characteristics of soybean plants. Surprisingly, there has been a succession of talented high school students looking with him. These students do work of real scientific significance and are making original contributions to science. Together, they have conducted long and detailed studies of thousands of soybean plants to track down the genes responsible for disease resistance and nitrogen fixation.

Nikola Lockett, while a high school junior, started a two-summer experience program at the USDA Southern Regional Research Center. Later, while a student at Xavier University in New Orleans, she was able to continue her work with a plant physiologist as part of the Cotton Fiber Bioscience team at the Center. Similarly, a high school research apprenticeship program attracts students to

Students can get valuable research skills in supervised agriscience experience programs. *[Courtesy USDA/ARS #K-4604-1]*

the ARS Arthropod-Borne Animal Diseases Research Laboratory in Laramie, Wyoming. Students there assist with research projects involving insects and diseases of livestock and poultry.

The USDA has numerous plant, animal, disease, insect, food, fiber, nutrition, and other research laboratories throughout the United States. For information on research assistance opportunities for students write to: USDA-ARS, Personnel Operations Branch, 6305 Ivy Lane, Greenbelt, MD 20770-1435.

EXPERIENCE INVENTORY

Directions: Complete the following information sheet by listing any experiences you have had or would like to gain in the field of agriculture.

Tasks or Jobs	Can perform without help	Can perform with help	Can help perform	Cannot or have not performed	Would like to learn how to perform	How or where to obtain experience
Examples						
• Drive tractor		X				Landscaping business
• Take cuttings					X	Vo. Ag. Class
• Keep records				X		My own & in class
1. _____						
2. _____						
3. _____						
4. _____						
5. _____						
6. _____						

FIGURE 5-13 The Experience Inventory is a device to plan and record experiences you hope to gain. It is also a mechanism for recording how well you have learned them.

Finally, a skills plan should be developed. These skills are chosen from lists that apply to the student's community and are recorded on a Agriscience Skills Plan and Record (Figure 5-16). The date completed should be added when the skill is acquired.

SUMMING UP

A carefully planned supervised agriscience experience program is both challenging and rewarding. The student learns by doing and earns while learning. Learning is at the highest level, because the laboratory is the real world. The teacher benefits from having highly motivated students, because the learning is interesting and rewarding. The employer benefits by having better-trained employees. The community benefits from the improvement projects and better-educated citizens.

PLACEMENT AGREEMENT

To provide a basis of understanding and to promote business-like relationships, this memorandum is established on _____, 19 ____. This work will start on _____, 19 ____, and will end on or about _____, 19 ____, unless the arrangement becomes unsatisfactory to either party before the ending date. Person (employer) responsible for training _____

The usual working hours will be as follows:

1. While attending school working hours shall be _____
   When not attending school working hours shall be _____
2. Provisions for overtime _____
3. Provisions for time off _____
4. Liability insurance coverage (type and amount) _____

Wages will be at the following rate(s):
Trial period _____
Remainder of the agreement period _____
And will be paid (when?) _____

A. IT IS UNDERSTOOD THAT THE EMPLOYER WILL (check the items that apply):
____ Provide the student with opportunities to learn how to do well as many jobs as possible, with particular reference to those contained in the planned program;
____ Coach the student in methods found desirable in implementing project activities and handling management problems;
____ Help the student and teacher make an honest appraisal of the student's performance;
____ Avoid subjecting the student to unnecessary hazards;
____ Notify the parents and the school immediately in case of accident or sickness and if any other serious problem arises;
____ Assign the student new responsibilities when he/she can handle them;
____ Cooperate with the teacher in arranging conferences with the student on supervisory visits; and/or
____ Other: _____

B. THE STUDENT AGREES TO (check the items that apply):
____ Do productive work, recognizing that the employer must profit from the student's labor in order to justify employment;
____ Keep the employer's interest in mind and be punctual, dependable, and loyal;
____ Follow instructions, avoid unsafe acts, and be alert to unsafe conditions;
____ Be courteous and considerate of the employer, the family, and others;
____ Keep such records of work experience and make such reports as the school may require;

PLACEMENT AGREEMENT (cont'd.)

____ Develop plans for management decisions with the employer and teacher; and/or
____ Other: _____

C. THE TEACHER, ON BEHALF OF THE SCHOOL, AGREES TO (check the items that apply):
____ Visit the student on the job at frequent intervals for the purpose of instruction and assurance that the student gets the most education out of the experience;
____ Show discretion in the time and circumstances of these visits, especially when the work is pressing; and/or
____ Provide appropriate job-related instruction at school; and/or
____ Other: _____

D. THE PARENTS AGREE TO (check the items that apply):
____ Assist in promoting the value of the student's experience by cooperating with the employer and the teacher of agriscience;
____ Satisfy themselves in regard to the living and working conditions made available to the student; and/or
____ Other: _____

E. ALL PARTIES AGREE TO:
____ An initial trial period of ____ working days to allow the student to adjust and prove himself/herself;
____ Discuss any issues concerning the job with the teacher before ending employment and/or
____ Other: _____

STUDENT's Signature _____
Address _____

Telephone Number _____
Social Security No. _____
PARENT's Signature _____
Address _____

Telephone Number _____

EMPLOYER's Signature _____
Address _____

Telephone Number _____

TEACHER's Signature _____
School Address _____

Telephone Number _____
School Telephone Number _____

FIGURE 5-14 A Placement Agreement states what every party is expected to do. It promotes good planning and reduces misunderstandings and conflicts.

IMPROVEMENT ACTIVITY PLAN AND SUMMARY

Improvement Project No. _____

A. Conditions found:

_____

_____

_____

_____

_____

_____

B. Plans for improvement (including costs):

_____

_____

_____

_____

_____

_____

C. Value of improvement when completed:

_____

_____

_____

_____

_____

_____

AGRICULTURAL IMPROVEMENT PROJECT SUMMARY

Started: _____, 19____   Completed: _____, 19____

Date	Jobs Done	Hours of Labor	Cost of Materials & Equipment

FIGURE 5-15 Improvement activities should be planned and records kept on the jobs, hours, and cost involved.

STUDENT ACTIVITIES

1. Write the Terms to Know and their meanings in your notebook.

2. Describe the relationships between 1) classroom instruction and supervised agriscience experience and 2) the FFA program and supervised agriscience experience.

3. Construct a bulletin board showing the relationships presented in Figure 5-1.

4. Study Figure 5-4 and write five ideas for production projects or enterprises to discuss with your teacher.

5. Discuss Figure 5-5 with your parents/guardians and select two or three improvement activities that you would like to conduct.

SUPPLEMENTARY AGRISCIENCE SKILLS PLAN AND RECORD			
Directions: Using the list of Supplementary Practices supplied by the instructor, complete the chart below by choosing skills you would like to include as part of your SAEP.			
Skills, Practices, Job, or Experience	Place to obtain skill	Date planned to obtain skill	Date completed
Examples:			
• Operate Cash Register	On-Job	Sept. 16	Sept. 16
• Bud-Graft	School Farm	February	Feb. 20
1._____			
2._____			
3._____			
4._____			
5._____			
6._____			
7._____			

FIGURE 5-16 Agriscience skills should be selected at the beginning of the year with plans for times and places to complete each item.

6. Review the examples presented in Figure 5-6 and choose twenty agriscience skills from enterprises other than your production projects and improvement activities.

7. Examine the tasks in the Student Interest Survey (Figure 5-7). Determine whether your interests are more in agribusiness, horticulture, production, or other areas of agriscience.

8. Using Figure 5-8 as a guide, draw a map to scale of your home, farm, or business where you can conduct an SAEP. Make an inventory of the resources that may be available for you to use in conducting an SAEP.

9. Talk with your teacher and parents/guardians. Write the names of the projects, activities, and skills that you definitely plan to do during the current year (Figure 5-9).

10. Prepare a personal resumé.

11. Apply and interview for a part-time job.

12. Develop an Experience Inventory (Figure 5-13).

13. Work out a Placement Agreement with an employer (Figure 5-14).

14. Set goals for production projects (Figure 5-12).

15. Make detailed plans for improvement activities (Figure 5-15).

16. Select definite agriscience skills. Write down where and when you plan to accomplish the skills (Figure 5-16).

SELF EVALUATION

A. Multiple Choice

1. Conducting an activity in the daily routine of our society is said to be
   a. laboratory experience.
   b. real-world experience.
   c. simulation.
   d. supervised occupational experience.

2. Which is not a purpose or benefit of SAEPs?
   a. Become established in an agriscience occupation.
   b. Permit early graduation.
   c. Permit individualized instruction.
   d. Provide educational and practical experiences.

3. Which is not a major component of a comprehensive agriscience program?
   a. Classroom/laboratory instruction.
   b. FFA.
   c. Memorization and recitation.
   d. Supervised occupational experience.

4. SAEPs should be planned
   a. at home with parents/guardians.
   b. in the classroom.
   c. on the job with employers.
   d. all of the above.

5. A student-drawn map of the home property is an important part of a
   a. job interview.
   b. Placement Agreement.
   c. Resources Inventory.
   d. resumé.

6. A production or productive project
   a. is the same as an improvement activity.
   b. is the same as a skill.
   c. may involve either ownership or placement for experience.
   d. must be done without pay or profit.

7. Skills should be selected
   a. as part of a Placement Agreement.
   b. early in the school year.
   c. from occupations outside of agriscience.
   d. from the same enterprise as the production project.

8. Improvement activities

a.   are always connected with employment.

b.   focus primarily on leadership development.

c.   must improve some part of the instructional program.

d.   should be without pay.

B. Matching

_____ **1.** Enterprises

_____ **2.** Experience

_____ **3.** Interest survey

_____ **4.** FFA

_____ **5.** Improvement project

_____ **6.** Agriscience

_____ **7.** Production/ Productive

_____ **8.** Program

_____ **9.** Project

_____ **10.** Resumé

_____ **11.** Supervised

_____ **12.** Skills

a. Series of related activities

b. Types of plants or animals

c. Total SAEP activities

d. Should result in wages or profit

e. Observed, done, or lived through

f. Assess interests

g. Leadership development

h. Beauty, safety, convenience, or value

i. Broad and flexible part of SAEP

j. Looked after or directed

k. Agriculture, agribusiness, and renewable natural resources

l. Information about a job applicant

Leadership Development in Agriscience

OBJECTIVE

To develop basic leadership skills.

COMPETENCIES TO BE DEVELOPED

After studying this unit, you should be able to:

- define *leader* and *leadership*.
- explain why effective leadership is needed in agriscience.
- state some major traits of good leaders.
- describe the opportunities for leadership development in FFA.
- demonstrate positive leadership skills.

MATERIALS LIST

✓ paper

✓ pencil or pen

✓ bulletin board materials

What is leadership in agriscience? Agriscience has been described as a broad and diverse field. It is not just horticulture or supplies and services; not just professions or products, processing, and distribution; not just mechanics or forestry; and not just renewable natural resources or production. Agriscience is all of these. Then what is leadership in agriscience?

LEADERSHIP DEFINED

Leadership is defined as the capacity or ability to lead. To lead is to show the way by going in advance or to guide the action or opinion of. To do this in agriscience, you must have knowledge of technical information and people. Similarly, you must know how to organize and manage activities. Most jobs are too big for one person. We can only do part of what needs to be done. Therefore, we need to use the help of others.

A leader uses the knowledge and skills of others to achieve a common goal. For instance, a quarterback on a football team will use leadership skills to coordinate the team players to achieve a touchdown. Similarly, the wise batter in baseball will hit the ball in a place that not only gets the batter on first base, but permits other runners to advance around the bases. That properly placed hit supports the common goal of achieving runs. A single base hit, where everyone advances and no one gets out, may be the best for the team. On the other hand, a line drive, which doesn't quite make the fence, may result in a third out with no chance for other players to score.

FIGURE 6-1 Leaders help to coordinate teams. *(Courtesy National FFA)*

FIGURE 6-2 Agribusinesses rely on team efforts to achieve goals. *(Courtesy USDA)*

WHY LEADERSHIP IN AGRISCIENCE?

Agriscience is a highly organized industry. It involves people and complex processes. Leadership skills are necessary whenever people are assembled. Those who teach in agriscience are part of a team of teachers, principals, supervisors, community advisory groups, and others. Those in agribusiness are typically parts of teams consisting of the manager, office staff, sales representatives, field personnel, and board of directors (Figures 6-1 and 6-2). Those on farms may be the owner, manager, spouse, children, hired help, or neighbors who assist at times. The manager of a farm or business must **plan** (think through, determine procedure, assemble materials, and train staff to do a job). Once a job is planned, it may be accomplished through management. To **manage** is to use people, resources, and processes to reach a goal. A manager uses leadership skills continuously in working with others on a day-to-day basis (Figure 6-3).

FIGURE 6-3 Managers are leaders. *(Courtesy National FFA)*

FIGURE 6-4 Class participation is a form of leadership. *(Courtesy National FFA; FFA #1)*

A **landscaper** is one who plans, plants, builds, or maintains outdoor ornamental plants and landscape structures. Sometimes working alone and sometimes working with others, the landscaper is a leader in many respects. When developing a plan for a customer, the landscaper exerts leadership. The landscaper knows the names, function, and performance of various plants. This information is used to develop an acceptable plan. Because the customer has personal ideas about landscaping, a professional must consider these in the plan. It may test the landscaper's leadership skills to lead the customer to an acceptable plan with plants that survive the weather and conform to acceptable landscape practices.

The landscaper and other agriscience personnel may be called upon as officers or members of professional organizations to give testimony before legislators or other public officials on the need for laws, regulations, or other actions that affect their work; or, it may be quite possible to end the day presiding over a meeting or taking minutes at a professional or civic meeting.

For the agriscience student, the need for leadership skills is also apparent. Having confidence to participate fully in class is important (Figure 6-4). To meet prospective employers and conduct a supervised agriscience experience program requires leadership skills. Functioning in the community, participating in group meetings, and making friends readily, all require acceptable leadership skills. Good leadership skills greatly improve individual marketability in the work world.

To be a good citizen, you must earn your way in life without infringing on the rights of others. Useful citizenship utilizes leadership to promote the common good in society. **Citizenship** means functioning in society in a positive way.

TRAITS OF GOOD LEADERS

Good leaders must have **integrity** (honesty). Without it, others cannot entrust an individual with the power to manage or control, even in minor things. A leader must have **knowledge,** which means familiarity, awareness, and understanding. Good leaders are dependable and have the **courage** (willingness to proceed under difficult conditions) and initiative to carry out personal and group decisions. To lead one must

AGRI·PROFILE

CAREER AREA: Leadership Development

Leadership development may be a career area or specialty for teachers, consultants, personnel managers, coaches, and others. However, many agriscience positions require good leadership capabilities as a tool for everyday use. Auctioneers, salespersons, managers, entrepreneurs, corporate executives, politicians, and anyone who directs others or routinely meets the public must have good leadership skills.

Leadership involves good planning, goal setting, and the ability to inspire others to work toward a common goal. Such skills as committee interaction, parliamentary procedure, and self-expression are important leadership techniques that are developed through study and practice. Such skills are used in church, civic, and community organizations as well as in work places. Group projects and club activities in agriscience provide excellent opportunities for leadership development.

The art of public speaking is one of the most powerful tools of a leader. *(Courtesy National FFA)*

initiative to carry out personal and group decisions. To lead one must communicate. This requires good speaking and listening skills.

In working with others, **tact,** or the skill of encouraging others in positive ways, is useful. Similarly, a sense of justice to ensure the rights of others is important. **Enthusiasm,** or energy to do a job and inspiration to encourage others, is useful. **Unselfishness** means placing the desires and welfare of others above yourself. It, too, is an important quality for good leadership. These traits permit **loyalty,** which means reliable support for an individual, group, or cause. These and other traits are achieved through effective leadership development.

LEADERSHIP DEVELOPMENT OPPORTUNITIES

Modern schools provide extensive opportunities for agriscience students to develop leadership skills. Students develop leadership in school organizations, athletics, and in classroom and shop situations. Some become leaders at home, on the job, or in community organizations.

4-H Clubs

The **Cooperative Extension Service** is an educational agency of the USDA and an arm of your state university. It provides educational programs for both youth and adults. Its programs include personal, home and family, community, and agriscience resources development. The Cooperative Extension Service also sponsors 4-H clubs. The **4-H** network of clubs is directed by Cooperative Extension Service personnel to enhance personal development and provide skill development in many areas, including agriscience (Figure 6-5). The four Hs in 4-H stand for head, heart, hands, and health. These provide the basis for the 4-H pledge, which is, "I pledge my head to clearer thinking, my heart to greater loyalty, my hands to larger service, and my health to better living for my Club, my community, my country, and the world."

Scout Organizations

Girl Scout and **Boy Scout** organizations provide opportunities for leadership development and skill development in agriscience and other areas. Scouts focus heavily on the out-of-doors and provide excellent leadership development and natural resources skills. They provide recognition through a system of merit badges, which are earned by learning skills and obtaining experiences in agriscience, as well as other areas (Figure 6-6).

FFA

The FFA is a youth organization that was developed specifically to expand the opportunities in leadership and agriscience skill development for students in public schools. Only students enrolled in agriscience courses are eligible for membership in FFA.

Aim and Purposes   The FFA is part of the agriscience curriculum in most schools where agriscience is offered. It is an important teaching tool. It serves as a laboratory for developing leadership and citizenship skills. These, in turn, are helpful in learning agriscience skills. The primary aim of the FFA is the development of agriscience leadership, cooperation, and citizenship. The specific purposes of the FFA may be paraphrased as follows.

1. To develop competent and assertive agriscience and leadership.

FIGURE 6-5 Leadership skills are developed through 4-H. *(Courtesy Rensselaer County 4-H Urban Summer Program)*

FIGURE 6-6 Scouting can provide valuable skills in leadership as well as agriscience. *(Courtesy Joni Conlon)*

2. To develop awareness of the global importance of agriscience and its contribution to our well-being.
3. To strengthen the confidence of agriscience students in themselves and their work.
4. To promote the intelligent choice and establishment of an agriscience career.
5. To stimulate development and encourage achievement in individual agriscience experience programs.
6. To improve the economic, environmental, recreational, and human resources of the community.
7. To develop competencies in communications, human relations, and social abilities.
8. To develop character, train for useful citizenship, and foster patriotism.
9. To built cooperative attitudes among agriscience students.
10. To encourage wise management of resources.
11. To encourage improvement in scholarship.
12. To provide organized recreational activities for agriscience students.

The Emblem The FFA emblem contains five major symbols that help demonstrate the structure of the organization (Figure 6-7). They are as follows.

1. **Eagle**—The emblem is topped by the eagle and other items of our national seal. The eagle was placed in the emblem to represent the national scope of the organization. It could also represent the natural resources in agriscience.
2. **Corn**—Corn is grown in every state in the United States. It reminds us of our common interest in agriscience, regardless of where we live.
3. **Owl**—The owl represents knowledge and wisdom. Use of this symbol in the emblem recognizes the fact that people in agriscience need a good education and that education must be tempered with experience to be of greatest usefulness.
4. **Plow**—The plow has been used to represent work-labor-effort. These qualities are needed to cause things to happen and to get results in agriscience.
5. **Rising Sun**—The rising sun is a symbol of the progressive nature of agriscience. It is symbolic of the need for workers in agriscience to cooperate and work toward common goals.

The FFA emblem may be constructed one symbol at a time. When assembled and dissembled in this manner, it is a good device to help others understand the FFA.

The Colors The official FFA colors are blue and gold. The shade of blue is national blue. The shade of gold is the yellow color of corn. Therefore, the colors are called national blue and corn gold.

Motto The FFA motto contains phrases that describe the philosophy of learning and development in agriscience. The motto is:

■ Learning to Do   ■ Earning to Live
■ Doing to Learn   ■ Living to Serve

FIGURE 6-7 The FFA emblem contains five important symbols. *[Courtesy National FFA; FFA #148]*

BIO•TECH CONNECTION

Be All You Can Be

On the day after Christmas in 1960, a twelve year-old boy with polio was delivered to a ranch for needy boys, along with his two brothers. Their broken home was without heat, and food was a scarce commodity. After five years of growing up on the ranch, this individual enrolled in a high school agriculture program and an FFA chapter.

While love of animals attracted him to the FFA, it was recognition for his successes on the parliamentary team that spurred him on. "It was the first time I had ever won anything," was his later observation. From the parliamentary team he advanced to area FFA president and on to state president and eventually delegate to the national FFA convention. It was there that he presented the historic motion to open FFA membership to females.

After graduation from college, this individual followed in the foot steps of his FFA advisor, teaching agriculture and inspiring young people to be all they could be. As FFA advisor he insisted that his students could prepare themselves for any career through leadership training, talking on their feet, developing responsibility, learning to manage money, experiencing team work, learning respect, setting goals, and planning ahead. These experiences help the individual win in life.

The experiences in FFA and teaching and advising helped develop the confidence and skills for a career in public life. After teaching for a while, he served eight years in the Texas Senate before going on to the United States House of Representatives. There his leadership skills were soon recognized, he was elected president of the Freshman Class of Congressman, and eventually became a valued member of the House Agriculture Committee.

What advice would this highly successful statesman give to young people pursuing careers in agriscience? The Honorable Congressman Bill Sarpalius suggests:

1. Be in the right frame of mind. Don't dwell on your handicaps or lack of ability like in public speaking or running. Forget, "I can't."
2. Avoid negatives.
3. Stay physically and mentally sharp. Don't let yourself get lazy.
4. Develop a religious background.
5. Concentrate on doing for others—not yourself.

In summary, he asserts, "To achieve all that is possible—we must attempt the impossible. To be as much as we can be—we must dream of being more!"

"Learning to Do" emphasizes the practical reasons for study and experience in agriscience. It also suggests ambition and willingness to productively use the hands as well as the mind. "Doing to Learn" describes procedures used in agriscience instruction at the doing level. Experiencing results from doing is the most permanent result of learning. "Earning to Live" suggests that FFA members intend to develop their skills and support themselves in life. And "Living to Serve" indicates an intention to help others through personal and community service.

Salute   The Pledge of Allegiance to the American flag is the official FFA salute. The words of the pledge are: I pledge allegiance to the flag of the United States of America and to the Republic for which it stands, one nation under God, indivisible, with liberty and justice for all."

Degree Requirements   The FFA has four degrees, which indicate the progress a member is making. These are Greenhand, Chapter, State, and American degrees. The Greenhand and Chapter FFA degrees are awarded by the FFA chapter in the agriscience department of the school. The State FFA degree is awarded by the State Association and the American FFA degree is awarded by the National FFA.

The Greenhand Degree is so named to indicate the member has learned beginning skills in FFA and is studying agriscience. To receive the Greenhand degree the member must meet the **requirements** spelled out in the current FFA *Official Manual*. In general, the requirements for the Greenhand degree are:

1. Be enrolled in an agricultural education course.
2. Have satisfactory plans for a supervised agricultural experience program.
3. Have learned the FFA creed, motto, and salute.
4. Describe the FFA emblem, colors, and symbols.
5. Explain the FFA Code of Ethics and proper use of the FFA jacket.
6. Have satisfactory knowledge of the history of the organization, and of the Chapter Constitution and Program of Activities.
7. Know the duties and responsibilities of members.
8. Own or have access to a copy of the *Official Manual* and FFA *Student Handbook*.
9. Submit a written application for the Greenhand degree.

The requirements for the other three degrees help the member to learn and grow professionally from one level to the next in the organization. The *Official Manual* contains the exact requirements for all degrees and other details of membership and chapter operation for the FFA.

Contests   The FFA sponsors numerous contests at various levels. The first level is in the local chapter in the school. The second level is the district or regional level within the state. The third is the State Association level, and the fourth is the national level. Local FFA advisors determine which contests are appropriate for students in their programs. The contests that are conducted at each level generally reflect the content of the instructional programs.

The purpose of FFA contests is to encourage agriscience students to develop technical and leadership skills, and practice these skills in friendly competition with other FFA members. Some types of contests that are typically for FFA groups are:

Agriscience business management	Horticulture bowl
Agriscience mechanization	Horticulture happenings
Agriscience production bowl	Land judging
Dairy cattle judging and evaluation	Livestock judging and evaluation
Dairy foods	Meats
Environmental and natural resources	Nursery/Landscape
Floriculture	Parliamentary procedure
Food science and technology	Poultry
Forestry	Public speaking
Horses	Small engine service
	Tractor operation
	Vegetables

Some FFA contests require contestants to know how to grade agriscience products, such as eggs, meat, poultry, fruits, and vegetables. Others require students to evaluate live animals, such as beef and dairy cattle, horses, poultry, rabbits, sheep, and swine. Some require mechanical abilities, such as welding, plumbing, electronics, irrigation, surveying, engine trouble-shooting and repair, painting, woodworking, and general tool use.

Some contests, such as Floriculture and Horticulture Happenings, require knowledge of the art and science of floral arrangement. Horticulture and Agriscience Production Bowls have questions about approved practices in agriscience, which are tossed out to teams of FFA members. They vie for points using game buzzers, and official judges determine team scores.

Contests, such as Forestry, Vegetables, and Nursery/Landscape, require students to identify plants, plant materials, insects, and diseases. Land Judging involves evaluating the soil and land and recommending appropriate management practices. Parliamentary procedure teams demonstrate their knowledge of correct parliamentary practices. These procedures are used to open meetings, conduct business, close meetings, and write minutes according to acceptable practice in the real world. Various types of speaking contests help individuals to sharpen their speaking skills.

Contests are organized so the participants are competing as individuals, teams of three or four participants, or chapters (Figure 6-8). An example of a chapter contest is the Chapter Award. This contest involves most or all of the students in the agriscience program, and provides valuable improvements to the program, school, and community.

Proficiency Awards The FFA has an extensive system of awards for individual members. These provide members with incentives and rewards for excellence in leadership and agriscience achievement. These awards change as the FFA expands to meet the changing nature of agriscience. Examples of proficiency awards that may be available at the chapter, state, and national levels are:

Agricultural Communications	Agricultural Processing
Agricultural Mechanical Technical Systems	Agricultural Sales and/or Service

FIGURE 6-8 Contests provide motivation, skill development, and recognition. *(Courtesy National FFA)*

Aquaculture
Beef Production
Cereal Grain Production
Dairy Production
Diversified Crop Production
Diversified Livestock
    Production
Emerging Agricultural
    Technology
Environmental Science
Equine Science
Feed Grain Production
Fiber Crop Production
Floriculture
Food Science and
    Technology
Forest Management
Forage Production
Fruit and/or Vegetable
    Production

Home and/or Community
    Development
Horticulture
Landscape Management
Nursery Operations
Oil Crop Production
Outdoor Recreation
Placement in Crop Production
Placement in Livestock
    Production
Poultry Production
Sheep Production
Small Animal Care and
    Management
Soil and Water Management
Specialty Animal Production
Specialty Crop Production
Swine Production
Turfgrass Management
Wildlife Management

FFA members may apply for the proficiency awards that fit their own situations. Medals and certificates are awarded to winners at the chapter level, while plaques and financial awards are given to those at the state and national levels. More information on proficiency and other award programs are described in the FFA *Student Handbook*.

PUBLIC SPEAKING

Oral communication skills are important for good leadership. Effective leaders must speak with individuals, committees, small groups, or in large forums. The ability to relax, speak clearly, and state what is pertinent to the subject at hand is very useful. These skills are developed by applying basic principles of speech preparation and organization.

Speeches may be prepared or extemporaneous. An **extemporaneous** speech is a speech delivered with little or no preparation. Most speaking is extemporaneous. The ability to do effective extemporaneous speaking is enhanced by delivering planned speeches. Therefore, the planned speech is used to teach public speaking skills in FFA.

The Plan

A speech should have at least three sections: the introduction, body, and conclusion. The plan should clearly identify the sections.

Introduction   The introduction indicates the need for and importance of the speech. It should be carefully planned and spoken with confidence. The introduction may be in the form of statements or questions. If the introduction is not to the point, does not fit the occasion, or is not delivered in a spirited manner, the audience may not listen to the rest of the speech. The introduction might be only a few lines long, but it must capture the audience's attention.

SPEECH PLAN

Introduction

Honorable judges, instructors, and fellow students
1. Rabbits, cows, plants, and plows—WE STILL NEED THEM!
2. When the family farm goes under, a piece of America goes under with it

Body

The 1986 census of Agriculture revealed . . .
1. The midsize farm is likely to be the true family farm
   - Owned and operated by the family
   - Family receives benefit from their work
2. Some believe the family farm is a relic of the past!
   a. Press fascination with bankruptcy sales
   b. Farms not in view from interstate highways
   c. Small population on farms
   d. Animal rights groups and unionizing efforts
3. The family farm endures in the United States
   a. Better managed farms buy the weaker
   b. Disease epidemics threaten specialized operations
   c. Farm retailing is on the rise
   d. Small farms are becoming legitimate

4. Farm bankruptcy must be minimized
   a. Farm failure affects general businesses
   b. We will miss fresh farm produce
   c. We will see more pollution
   d. We will depend more on imports

5. Are there remedies? Yes!
   a. Remove politics from marketing
   b. Recognize farmers as astute business people
   c. See the total industry of agriscience
   d. Tax farm land for farm use
   e. Increase agriculture land preservation programs

Conclusion

Indeed . . .
1. General George Washington nearly lost the continental army at Valley Forge from lack of food, shelter, and clothing.
2. It can happen to us if we don't maintain a healthy farm situation in the United States today
3. Don't give away our most valuable resource—the ability to feed, clothe, and shelter ourselves!

FIGURE 6-9 An outline of a winning speech on the importance of family farms.

Body The body of the speech contains the major information. It should consist of several major points which support one central theme or objective. Each major point is supported by additional information to explain, illustrate, or clarify the point. It is best to write the body of the speech in outline form (Figure 6-9).

An outline will help direct the thought and delivery of the speech. After outlining the speech, a carefully worded narrative may be written to help the speaker fully develop the content of the speech. However, it must be emphasized that a speech should be given from an outline to avoid the temptation to memorize the speech. Memorization of a speech has two serious pitfalls. One, there is the danger it will sound like someone else's words and lack authenticity. Two, if the line of thought is lost during delivery, the speaker may not be able to find the location in the narrative. This can greatly damage the quality of the speech.

Conclusion The conclusion should remind the audience of the major theme or central point of the speech and briefly restate the major points. Further, the conclusion should leave the audience feeling like they want to take action to implement or adopt what you have said. Some speeches call for action, while others call for changes in attitude or perception. The more powerful speeches move people to action. The words needed to do such a big job must be carefully planned.

FIGURE 6-10 Speaking can be enjoyable when the speaker has been properly trained. *(Courtesy National FFA; FFA #74)*

Giving the Speech

Giving the speech can be fun and provide much satisfaction (Figure 6-10). However, this fun and satisfaction does not come free. The speaker must have prepared the plan well and practiced the speech extensively. Practicing the speech until the content becomes very familiar helps speaking become nearly automatic. The speech should be given orally to yourself several times. Then practice in front of a mirror to observe facial expressions, posture, and gestures. Finally, give the speech in front of others and invite them to make suggestions to improve the delivery (Figure 6-11).

Books have been written on techniques to enhance speech delivery. However, for the beginner, a few basic and time-tested procedures should be helpful for effective speaking. Some suggested procedures for giving speeches are:

- Have your teacher read and make suggestions on the content of your written speech.
- Learn the content thoroughly through repeated thought and practice.
- Record the speech on a tape recorder and observe the sound, speed, power, and effectiveness of your voice. Make corrections to improve the delivery.
- Practice the speech in front of a mirror to observe posture, hand gestures, and facial expressions. Your posture should be erect and natural, with hands at your side or resting lightly on the edges of the podium. They should be used occasionally for gestures that emphasize a point, show direction, or indicate count.
- Ask your teacher for a score sheet for judging speeches. Deliver the speech in front of a trusted person who can check your delivery against the score sheet and provide suggestions for improvements. This may be a friend, relative, or teacher.
- Deliver your speech in front of your class for experience and suggestions.
- Ask your teacher to critique your speech for final approval.
- Deliver your speech in front of civic groups and/or in FFA public-speaking contests.
- Videotape your speech for later critique and review.

PARLIAMENTARY PROCEDURE

What is parliamentary procedure? Why are so many people familiar with it? Why is it important? Why should agriscience students be interested in learning parliamentary procedure? **Parliamentary procedure** is a system of guidelines or rules for conducting meetings. Most Americans who are influential in their communities are familiar with parliamentary procedure.

Parliamentary procedure is used to guide the meetings conducted by school groups, church groups, and civic organizations, such as Lions, Rotary, and Ruitan clubs. Agriscience students should be

Explanation of Score Sheet Points

Part I–For Scoring Content and Composition

1. *Content of the manuscript* includes:
   Importance and appropriateness of the subject
   Suitability of the material used
   Accuracy of statements included
   Evidence of purpose
   Completeness and accuracy of bibliography

2. *Composition of the manuscript* includes:
   Organization of the content
   Unity of thought
   Logical development
   Language used
   Sentence structure
   Accomplishment of purpose–conclusions

Part II–For Scoring Delivery of Production

1. *Voice* includes:
   Quality
   Pitch
   Articulation
   Pronunciation
   Force

2. *Stage Presence* includes:
   Personal appearance
   Poise and body posture
   Attitude
   Confidence
   Personality
   Ease before audience

3. *Power of expression* includes:
   Fluency
   Emphasis
   Directness
   Sincerity
   Communicative ability
   Conveyance of thought and meaning

4. *Response to questions* includes:
   *Ability to answer satisfactorily the questions on the speech which are asked by the judges indicating originality, familiarity with subject and ability to think quickly.

5. *General effect* includes:
   Extent to which the speech was interesting, understandable, convincing, pleasing, and held attention.

*NOTE: Judges should meet prior to the contest to prepare and clarify the questions asked.

FIGURE 6-11 A score sheet for evaluating speeches. *(Courtesy National FFA)*

National Public Speaking Contest   **Judge's Score Sheet**

PART I.   FOR SCORING CONTENT AND COMPOSITION

Items To Be Scored	Points Allowed	Points Awarded Contestant												
		1	2	3	4	5	6	7	8	9	10	11	12	13
Content of Manuscript	200													
Composition of Manuscript	100													
Score on Written Production	300													

PART II.   FOR SCORING DELIVERY OF THE PRODUCTION

Items To Be Scored	Points Allowed	Points Awarded Contestant												
		1	2	3	4	5	6	7	8	9	10	11	12	13
Voice	100													
Stage Presence	100													
Power of Expression	200													
Response to Questions	200													
General Effect	100													
Score on Delivery	700													

PART III.   FOR COMPUTING THE RESULTS OF THE CONTEST

Items To Be Scored	Points Allowed	Points Awarded Contestant												
		1	2	3	4	5	6	7	8	9	10	11	12	13
Score on Written Production	300													
Score on Delivery	700													
TOTALS	1000													
* Less Overtime Deductions, for each minute or major fraction thereof	20													
* Less Undertime Deductions, for each minute or major fraction thereof	20													
GRAND TOTALS														
Numerical or Final Placing														

* From the Timekeeper's record.

interested in learning this procedure so they can have their opinions heard and influence decisions that affect their lives. Parliamentary procedure is important because it permits a group to:

■ Discuss one thing at a time.
■ Hear everyone's opinion in a relaxed, courteous atmosphere.
■ Protect the rights of minorities.
■ Make decisions according to the wishes of the majority of the group.

Requirements for a Good Business Meeting

A good **business meeting** is a gathering of people working together to make wise decisions. Wrong decisions cause unhappiness, loss of income, inefficiency in business and social activities, injury, and other problems. The most common outcome of poorly run business meetings is the waste of time and lack of results. Meetings run by groups of individuals who know and use parliamentary procedure are smooth, efficient, orderly, brief, and get much done (Figure 6-12). Some requirements for a good business meeting follow.

A Good Presiding Officer   A **presiding officer** is a president, vice president, or chairperson who is designated to lead a business meeting. He/she should be committed to the goals of the organization and want to lead the group into good group decisions. A good presiding officer must know and use proper parliamentary procedure.

A Good Secretary   A **secretary** is a person elected or appointed to take notes and prepare minutes of the meeting. **Minutes** is the name of the official written record of a business meeting. Minutes should include the date, time, place, presiding officer, attendance, and motions discussed at the meeting. They should be written clearly, include all actions taken by the group, and be kept in a permanent secretary's book.

Informed Members   Informed members are members who are active in the organization and want to be part of the group. They give previous thought to issues to be discussed and gather useful information

FIGURE 6-12 An effective meeting in progress. *(Courtesy National FFA; FFA #141)*

about the issues. They share these thoughts with others in the meeting. This permits everyone to have the benefit of the best thinking in the group. It permits the best decisions to be made. Effective members know and use parliamentary procedure.

A Comfortable Meeting Room   The meeting place must be comfortable and free of distractions. A moderate temperature and good lighting are essential. Members should be seated so they can hear and see each other. For small groups, seating at a table or in a circle works well. Large groups must rely on a good sound system for the presiding officer to be heard, and members must speak clearly with good volume in order to be heard. Good public speaking skills and thorough knowledge of parliamentary procedure help the members considerably.

Conducting Meetings

The Order of Business   The **order of business** refers to the items and sequence of activities conducted at a meeting. The order of business is usually made up by the secretary. This generally grows out of an executive meeting. An **executive meeting** is a meeting of the officers to conduct the business of the organization between regular meetings. They also consider what needs to be discussed by the total membership at the regular meeting. The essential items in an order of business are:

- Call to order
- Reading and approval of minutes of the previous meeting
- Treasurer's report
- Reports of other officers and committees
- Old business
- New business
- Adjournment

Other items or activities that are frequently used or occur in orders of business are programs, speakers, or entertainment.

Parliamentary Practices

Use of the Gavel   The **gavel** is a small wooden hammerlike object used by the presiding officer to direct a meeting (Figure 6-13). It is used to call the meeting to order, announce the result of votes, and adjourn the meeting. It is also used to signal the members to stand, sit down, or reduce the noise level of the group. The gavel is a symbol of the authority of the office of president or chairperson, and should be respected by all attending the meeting.

In some organizations, like the FFA, a system of taps is used to signal the audience to do certain things. In FFA meetings, the gavel is used as follows:

- One tap—the outcome of or decision about the item under consideration has been announced by the presiding officer.
- Two taps—the meeting will come to order, members should sit down if standing, or members should be quiet except when recognized.
- Three taps—members should stand up.

FIGURE 6-13 A gavel is the symbol of authority of the presiding officer. *(Courtesy National FFA; FFA #138)*

Obtaining Recognition and Permission to Speak For a meeting to be orderly, members must speak one at a time and in some logical and fair sequence. The presiding officer is regarded as the "traffic controller," and calls on members as they request to be recognized or according to certain rules. To be recognized, the member should raise a hand to get the presiding officer's attention. The presiding officer should call the member by name; then the person should stand and address the presiding officer as Madame or Mr. Chairperson, or Madame or Mr. President. The individual should then proceed to speak. Both the presiding officer and the members should understand the correct classification of motions (Figures 6-14 and 6-15).

SUMMARY OF MOTIONS

Motion	Debatable	Amendable	Vote Required	Second	Reconsider
PRIVILEGED					
Fix time which to adjourn	No	Yes	Majority	Yes	Yes
Adjourn	No	No	Majority	Yes	No
Recess	No	Yes	Majority	Yes	No
Question of privilege	No	No	None	None	Yes
Call for orders of the day	No	No	None/2/3	None	No
INCIDENTAL					
Appeal	Yes	No	Majority	Yes	Yes
Point of order	No	No	None	No	No
Parliamentary inquiry	No	No	None	No	No
Suspend the rules	No	No	2/3	Yes	No
Withdraw a motion	No	No	Usually none	No	No
Object consideration of question Negative vote only	No	No	2/3	No	Yes
Division of the question	No	Yes	Majority	Yes	No
Division of the assembly	No	No	No	No	No
SUBSIDIARY					
Lay on table	No	No	Majority	Yes	No
Previous question before vote	No	No	2/3	Yes	Yes
Limit debate	No	Yes	2/3	Yes	Yes
Postpone definitely	Yes	Yes	Majority	Yes	Yes
Refer to committee	Yes	Yes	Majority	Yes	Yes
Amend	Yes	Yes	Majority	Yes	Yes
Postpone indefinitely	Yes	No	Majority	Yes	Yes vote only
Main motion	Yes	Yes	Majority	Yes	Yes
UNCLASSIFIED					
Take from table	No	No	Majority	Yes	No
Reconsider	No	No	Majority	Yes	Negative vote only

For more details on parliamentary procedure, see a parliamentary procedure book such as *Robert's Rules of Order*.

FIGURE 6-14 Parliamentary skills useful in FFA and other organizations. *(Courtesy National FFA)*

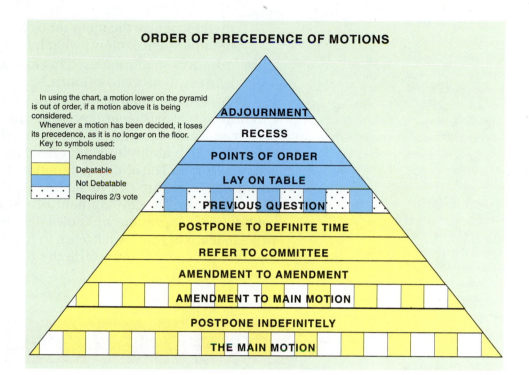

ORDER OF PRECEDENCE OF MOTIONS

In using the chart, a motion lower on the pyramid is out of order, if a motion above it is being considered.

Whenever a motion has been decided, it loses its precedence, as it is no longer on the floor.

Key to symbols used:

- ☐ Amendable
- ☐ Debatable
- ☐ Not Debatable
- ⋯ Requires 2/3 vote

ADJOURNMENT
RECESS
POINTS OF ORDER
LAY ON TABLE
PREVIOUS QUESTION
POSTPONE TO DEFINITE TIME
REFER TO COMMITTEE
AMENDMENT TO AMENDMENT
AMENDMENT TO MAIN MOTION
POSTPONE INDEFINITELY
THE MAIN MOTION

FIGURE 6-15 Correct order of precedence of motions. *(Courtesy National FFA)*

Presenting a Motion   A motion is a proposal, presented in a meeting, to be acted upon by the group. To present a motion, the member raises a hand and is recognized by the presiding officer. Then the member states, "Madame/Mr. President, I move that . . ." (and continues with the rest of the motion). The words "I move" are important to say when beginning the motion. Otherwise, you will be regarded as incorrect and not using good parliamentary procedure. In order for a motion to be discussed by the group, at least one other member must be willing to have the motion discussed. That second individual expresses this willingness by saying, "Madame/Mr. President, I second the motion."

Some Useful Motions   There are dozens of motions, but a few are basic ones that are generally known and widely used. Some of these are:

- **Main Motion**—a basic motion used to present a proposal for the first time. The way to state it is to get recognized and then say, "I move . . ."
- **Amend**—a type of motion used to add to, subtract from, or strike out words in a main motion. The way to present an amendment is to say, "I amend the motion by . . ."
- **Refer**—a motion used to refer to some other motion or to a committee or person for finding more information and/or taking action on the motion. The way to state a referral is to say, "I move to refer this motion to . . "
- **Lay on the Table**—a motion used to stop discussion on a motion until the next meeting. The way to table a motion is to say, "I move to table the motion."
- **Point of Order**—a procedure used to object to some item in or about the meeting that is not being done properly. The procedure to use is to stand up and say, "Madame/Mr. President, I rise to a point of order!" The presiding officer should

then recognize the member by saying, "State your point." The member then explains what has been done incorrectly.

■ **Adjourn**—a motion used to close a meeting. The procedure is to say, "I move to adjourn."

The FFA *Student Handbook* provides a listing of additional motions and how to use parliamentary procedure for more effective meetings.

Agriscience students are encouraged to develop effective leadership skills. The ability to work effectively as a member of a group is essential for all. The ability to function as a chairperson or officer creates more opportunities to serve and influence the environment in which we live. The development of self-confidence is essential. Self-confidence is a product of knowledge and skill. Therefore, all should strive to learn to speak well, learn to function in groups through parliamentary procedure, and utilize the opportunities in FFA for personal development.

STUDENT ACTIVITIES

1. Write the Terms to Know and their meanings in your notebook.

2. Make a list of the many ways that you exercise leadership in your family, school, and community.

3. Develop a bulletin board showing the symbols of the FFA emblem.

4. Develop a bulletin board illustrating the purposes of the FFA. Include the FFA colors.

5. Write down five contests and five proficiency awards for which you would like to try out. Discuss these with your classmates and teacher.

6. Prepare and present a three-minute speech on an agriscience topic to your class.

7. Ask your teacher to let you use a gavel and direct a mock class or FFA meeting to learn parliamentary skills.

8. Form a parliamentary procedure team of six or eight classmates and demonstrate various parliamentary skills to the class.

SELF EVALUATION

A. Multiple Choice

1. To lead is to
   - a. manage.
   - b. organize.
   - c. show the way.
   - d. all of the above.

2. Which is not a trait of a good leader?
   - a. Courage
   - b. Integrity
   - c. Selfishness
   - d. Tact

3. A leadership organization of the Cooperative Extension Service is
   - a. Boy Scouts.
   - b. FFA.
   - c. Girl Scouts.
   - d. 4-H.

4. Membership in FFA is limited to youth who
   a.   are in the country.
   b.   enroll in an agriscience program in school.
   c.   plan careers in agriscience.
   d.   seek leadership training.

5. Which is *not* a purpose of FFA?
   a.   Develop leadership
   b.   Intelligent choice of agriscience occupations
   c.   Promote scholarship
   d.   Promote self above others

6. The symbol which signifies that the FFA is a national organization is the
   a.   corn.
   b.   eagle.
   c.   owl.
   d.   rising sun.

7. The first line of the FFA motto is
   a.   doing to learn.
   b.   earning to live.
   c.   learning to do.
   d.   living to serve.

8. One requirement for the Greenhand Degree is
   a.   a plan for supervised agricultural experience.
   b. $70 earned from agriscience experience.
   c.   school grades of C or above.
   d.   unselfish attitude in FFA activities.

9. An FFA activity not generally organized as a contest is
   a.   dairy foods.
   b.   forestry.
   c.   land judging.
   d.   wildlife management.

10. One of the last items in an order of business is
   a.   new business.
   b.   officer reports.
   c.   reading of the minutes.
   d.   treasurer's report.

11. The largest part of a speech is the
   a.   body.
   b.   conclusion.
   c.   introduction.
   d.   summary.

12. The only acceptable way to start a motion is to say
   a.   "I believe . . .".
   b.   "I make a motion that . . .".
   c.   "I move . . .".
   d.   "I think . . .".

B. Matching

_____ **1.** Adjourn
_____ **2.** Amend
_____ **3.** Lay on the table
_____ **4.** Main motion
_____ **5.** Point of order
_____ **6.** Refer
_____ **7.** Three taps of gavel
_____ **8.** Secretary

a. Present a new proposal
b. Leave it to a committee
c. Correct some procedure
d. Close the meeting
e. Consider it at the next meeting
f. Change a motion
g. Prepare minutes
h. Members stand

Natural Resources Management

Stewards of the Land

Hikers, bikers, hunters, anglers, farmers, foresters, ranchers, caretakers, and occupants all must be good stewards of the land. All are dependent upon the land and rely upon the soil, air, water, wildlife, and other natural resources around us. Farmers, ranchers, and foresters rely on the land to grow the crops and animals of their businesses. Similarly, hunters and other recreational users of the land and water rely on the habitat to grow and sustain the plants and wildlife. All enjoy wildlife for sport and recreation.

Land for farming and ranching is typically owned by the families who occupy the land; and they have definite property rights to grant or deny access of others to their property for hunting, fishing, or other recreational pursuits. However, the good-citizen hunter or angler seeks permission of the owner to access private property and strives to protect or enhance the fish and wildlife population and habitat through legal and good stewardship practices. Both owners and good-citizen users share the love for animals and the respect for crops, pasture, and woodland. Both wish to conserve the quality of land, air, and water.

Farmers and ranchers plan or encourage growth of food and cover and habitat for wildlife. Further, they interact with wildlife biologists, game specialists, game officers, and other public authorities to nurture game populations and enforce game laws. Hunters and anglers, which includes farmers, ranchers, wildlife specialists, and the general public, all help keep wildlife populations in check by harvesting the excess game. Animals that can be found and killed or taken by humane and legal hunting and fishing practices are generally the excess mature males in the population who compete with the females and young for food and cover.

Game hunting limits are set by wildlife specialists via scientific methods in an effort to permit and encourage the removal of excess wild animals by hunting. Those animals that are not removed by hunting perish by disease, predation, and starvation. This is nature's way of keeping animal and plant populations in balance.

Public lands are owned by federal, state, or local governments and require employees in many specialty occupations to care for them. Foresters, biologists, fish and game managers, game officers, park rangers, horticulturists, scientists, technicians, and others all contribute to the upkeep and improvement of public lands.

The owners, managers, and users of both private and public lands must all cooperate to utilize and conserve soil, water, trees, crops, wild plants, livestock, and wildlife to be counted among the good stewards of the land.

Maintaining Air Quality

OBJECTIVE

To determine major sources of air pollution and identify procedures for maintaining and improving air quality.

MATERIALS LIST

- ✓ pencil and paper
- ✓ several specimen aerosol cans

COMPETENCIES TO BE DEVELOPED

After studying this unit, you should be able to:

- define air and identify its major components.
- analyze the importance of air to humans and other living organisms.
- determine the characteristics of clean air.
- describe common threats to air quality.
- describe important relationships between plant life and air quality.
- discuss the greenhouse effect and global warming.
- list practices that lead to improved air quality.

Air	Tetraethyl lead	Pesticide
Water	Carbon monoxide	Toxic
Soil	Radon	Asbestos
Habitat	Radioactive material	Greenhouse effect
Sulfur	Chlorofluorocarbons	Photosynthesis
Hydrocarbons	Ozone	
Nitrous oxides	Pest	

Life as we know it on our planet requires a certain balance of unpolluted air, water, and soil . **Air** is a colorless, odorless, tasteless mixture of gases. It occurs in the atmosphere around the earth and is composed of approximately 78 percent nitrogen, 21 percent oxygen, and 1 percent argon, carbon dioxide, neon, helium, and other gases. **Water** is a clear, colorless, tasteless, and near odorless liquid. Its chemical makeup is two parts hydrogen to one part oxygen. **Soil** is the top layer of the Earth's surface, suitable for the growth of plant life.

AIR QUALITY

Without a reasonable balance of air, water, and soil, most organisms would perish. Slight changes in the composition of air or water may favor some organisms, while causing others to diminish in number or health. Unfavorable soil conditions usually mean inadequate food, water, shelter, and other factors of habitat. **Habitat** is the area or type of environment in which an organism or biological population normally lives (Figure 7-1).

FIGURE 7-1 Clean air, clean water, and productive soil are necessary for good habitat for plants, animals, and humans. *(Courtesy Michael Dzaman)*

Threats to Air Quality

The mixture of gases we call air is absolutely essential for life. The air we breathe should be healthful and life supporting. Air must contain approximately 21 percent oxygen for human survival. If a human stops breathing and no life-supporting equipment or procedures are used, the brain will die in approximately 4 to 6 minutes. Further, air may contain poisonous materials or organisms that can decrease the body's efficiency, cause disease, or cause death through poisoning.

The circumference or distance around the Earth at its largest part is 24,000 miles. Yet abuse of the atmosphere in one area frequently damages the environment in distant parts of the world. Air currents flow in somewhat constant patterns, and air pollution will move according to those patterns. However, when warm and cold air meet, the exact air movement will be determined by the differences in temperature, the terrain, and other factors. Because we cannot control the wind, humans have an obligation to keep the air clean for their own benefit as well as that of society at large.

There are major worldwide threats to air quality. These include sulfur compounds, hydrocarbons, nitrous oxides, lead, carbon monoxide, radon gas, radioactive dust, chlorofluorocarbons, pesticide spray materials, and others (Figure 7-2). Most of these are poisonous to breathe and have other damaging effects.

Sulfur Sulfur is a pale-yellow element occurring widely in nature. It is present in coal and crude oil. It also combines with oxygen to form harmful gases, such as sulfur dioxide, when these and other fuels are burned. Therefore, most smoke and exhaust from homes, factories, or motor vehicles contains some harmful compounds unless special equipment is used to remove them. Once these invisible gases are in the air, they combine with moisture to form sulfuric acid, which falls as acid rain. Acid rain has been found to damage and kill trees and other plants. It also has a corrosive effect on metals.

Hydrocarbons Hydrocarbons have become serious problems as factories and motor vehicles have become so prevalent during the twentieth century. In the United States, hydrocarbon emissions (by-products of combustion or burning) are held in check by special emission-control equipment on automobiles and special equipment called stack scrubbers in large industrial plants. Hydrocarbon output is controlled on automobiles by crankcase ventilation, exhaust gas recirculation, air injection, and engine refinements, such as four valves per cylinder. Without this equipment, it would be impossible to breathe in our major cities and in heavy stop-and-go traffic on major highways.

Nitrous Oxides and Lead Nitrous oxides are compounds containing nitrogen and oxygen. They make up about 5 percent of the pollutants in automobile exhaust. Yet they are damaging to the atmosphere and must be removed from exhaust gases. This group of chemicals was the most difficult and perhaps the most costly group to remove from automobile emissions. Scientists and engineers solved the problem partly by installing catalytic converters in automotive exhaust systems.

Hot exhaust gases from the engine flow through a honeycomb of platinum in the catalytic converter. The reaction converts the nitrous oxides into harmless gases. Until recently, all gasoline contained **tetraethyl lead,** a colorless, poisonous, oily liquid. This lead improved the

FIGURE 7-2 Automobiles, trucks, homes, and factories burn gasoline, oil, coal, and wood, which release products of combustion that pollute the air. (Courtesy Michael Dzaman)

burning qualities of gasoline and improved the control of engine knocking. However, tetraethyl lead ruined catalytic converters. So catalytic converters could not be used until a substitute for tetraethyl lead was found in the 1970s.

Tetraethyl lead is a poisonous product. Unfortunately, it is still used in some gasoline for trucks and tractors where catalytic converters are not used. Lead and nitrous oxides are still major pollutants of the atmosphere.

Carbon Monoxide Carbon monoxide is one of the automotive gases that cannot be removed with our present technology. It is the colorless, odorless, and poisonous gas that kills people in automobiles with leaking exhaust systems. A similar hazard exists when engines are run in closed areas without adequate ventilation. Victims fall asleep and die. Carbon monoxide emissions may be reduced by keeping engines in good repair and properly tuned.

Radon Radon has become a hazard in homes in many parts of the United States. Radon is a colorless, radioactive gas formed by disintegration of radium. It moves up through the soil and flows into the atmosphere at low and usually harmless rates.

However, if a house or other building is constructed over an area where radon gas is being emitted, a hazardous condition can result. Radon gas can accumulate in buildings with cracks in the basement floor or walls or enter through sump holes. The problem can be prevented by tightly sealing all cracks and/or providing continuous ventilation either below the basement floor or throughout the building (Figure 7-3).

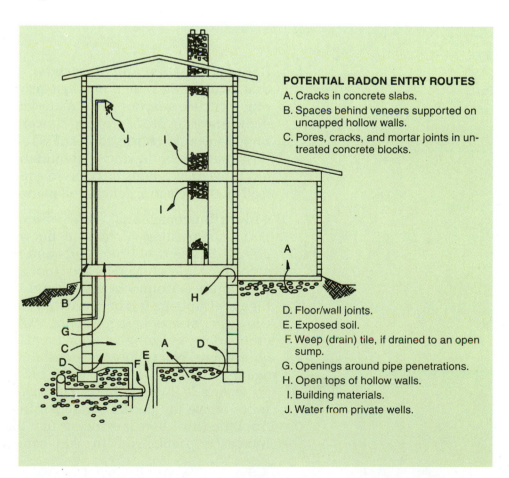

POTENTIAL RADON ENTRY ROUTES
A. Cracks in concrete slabs.
B. Spaces behind veneers supported on uncapped hollow walls.
C. Pores, cracks, and mortar joints in untreated concrete blocks.
D. Floor/wall joints.
E. Exposed soil.
F. Weep (drain) tile, if drained to an open sump.
G. Openings around pipe penetrations.
H. Open tops of hollow walls.
I. Building materials.
J. Water from private wells.

FIGURE 7-3 Ventilation systems must be correctly designed and carefully maintained to keep interior areas free of pollution.

AGRI·PROFILE

CAREER AREA: Air Quality Control

Careers in air quality are available with the weather services of local, state, and national agencies. Local and network radio and television stations all utilize weather reporters and meteorology forecasters. Of equal importance are those who monitor and help to improve air quality. Technicians collect and chemists analyze samples of air taken from various places in the atmosphere, buildings, and homes. Employees of environmental protection agencies and environmental advocacy groups are important links in our efforts to maintain a healthful environment.

Air quality specialists advise and assist industry in reducing harmful emissions from motor vehicles and industrial smokestacks. Not to be overlooked are the entomologists, who monitor the winds for signs of invading insects, and plant pathologists, who watch for airborne disease organisms. Career opportunities in air quality maintenance and improvement will undoubtedly increase in the future.

Weather scientists at work.
(Courtesy USDA/ARS #K-2228-11)

Radioactive Dust and Materials Radioactive material is material that is emitting radiation. Dust resulting from atomic explosion or other nuclear reaction, and materials contaminated by atomic accidents or wastes are of growing concern. The damage from radioactivity ranges from skin burns to sickness to hereditary damage to death. The possibility of worldwide contamination and other hazards from serious nuclear accidents has caused a reduction in construction of atomically fueled electric generation plants.

Chlorofluorocarbons Chlorofluorocarbons are a group of compounds consisting of chlorine, fluorine, carbon, and hydrogen. They are used as aerosol propellants and refrigeration gas These materials are very stable. Once released from an aerosol can or cooling system, they bounce around in the air and eventually float upward into the high atmosphere. It is believed that chlorofluorocarbons will survive in the upper atmosphere for about 100 years. Meanwhile, their chlorine atoms destroy ozone without themselves being destroyed. In newer equipment, chlorofluorocarbons are being replaced by less polluting agents.

Ozone is a compound that exists in limited quantities about 15 miles above the Earth's surface. It filters out harmful ultraviolet rays from the Sun. There is evidence that the ozone layer is being damaged. Increased problems with skin cancer and damage to the body's immune system is occurring. Most living organisms will be exposed to the damaging effect of ultraviolet rays.

In 1987, there was an historic international conference where at least thirty-seven nations agreed to schedule cutbacks in the production of chlorofluorocarbons. However, given the possible damage by this pollutant, it may make sense to stop all production immediately.

Pesticide Spray Materials   A **pest** is a living organism that acts as a nuisance. A **pesticide** is a material used to control pests. Many pesticides are chemicals mixed with water so they can be sprayed on plants, animals, soil, or water to kill or otherwise control diseases, insects, weeds, rodents, and other pests. Spray materials are pollutants if they carry **toxic** (poisonous) materials or are harmful to more than the target organism. Spray materials used to control pests, such as diseases, insects, weeds, rodents, and others, are generally harmful to the air if they are not used exactly as specified by the government and the manufacturer. Poisons may be thinned out or diluted by air movement, but excessive toxic materials can overburden the atmosphere's ability to cleanse itself. The continued use and abuse of chemicals to control pests is an area of growing concern in maintaining air quality.

Asbestos   **Asbestos** is a heat and friction resistant material. It was used extensively in the past for brake and clutch linings of vehicles, shingles for house siding, steam and hot-water pipe insulation, ceiling panels, and other products.

Unfortunately, asbestos fibers are very damaging to the lungs and cause disease and death. Therefore, there are now state and national laws and codes requiring the removal of asbestos from public buildings, industrial settings, and general use.

The Greenhouse Effect and Global Warming

Light sources from the sky seem to be changing. Bright sunshine and the suntan that many sought in the past now seem to be a threat, especially if over done. The excessive heat and drought of summer is frequently accompanied by parched and withered crops. Authorities also have found a definite link between extensive exposure to the Sun and the occurrence of the deadly melanoma skin cancer. Further, there is scientific evidence that skin-damaging and life-threatening ultraviolet rays from the Sun now reach the Earth's surface in greater intensity than in the past.

The Greenhouse Effect   When sunlight passes through a clear object, such as a glass window, it heats the air on the opposite side of that object. If the warmed air is not cooled or flushed out, heat builds up under the glass. Examples of the phenomenon may be observed under sky lights, auto windshields, and in greenhouses. The glass also absorbs some of the energy from the light and gives it off or radiates it as heat to the interior area. The result is a buildup of heat causing the interior to become warmer than the outside. This heat buildup from the rays passing through the clear object and the resulting heat being trapped inside is known as the **greenhouse effect.**

The Sun's rays include many different colors of light and types of rays. Some of the more familiar rays are ultraviolet and infrared. The ultraviolet rays are known for their extensive skin damage and other life-threatening effects from overexposure to the Sun. Infrared rays are

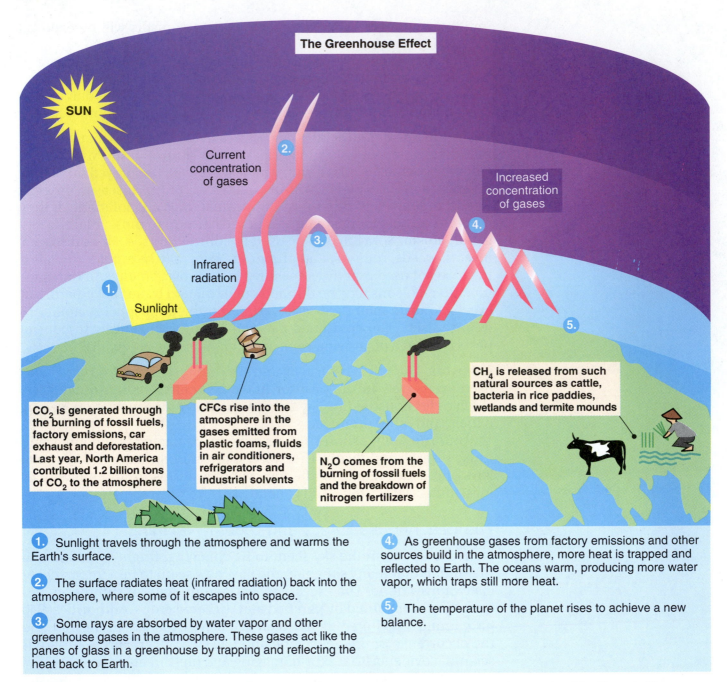

The Greenhouse Effect

SUN

Sunlight

Current
concentration
of gases

Infrared
radiation

Increased
concentration
of gases

CO_2 is generated through the burning of fossil fuels, factory emissions, car exhaust and deforestation. Last year, North America contributed 1.2 billion tons of CO_2 to the atmosphere

CFCs rise into the atmosphere in the gases emitted from plastic foams, fluids in air conditioners, refrigerators and industrial solvents

N_2O comes from the burning of fossil fuels and the breakdown of nitrogen fertilizers

CH_4 is released from such natural sources as cattle, bacteria in rice paddies, wetlands and termite mounds

1. Sunlight travels through the atmosphere and warms the Earth's surface.

2. The surface radiates heat (infrared radiation) back into the atmosphere, where some of it escapes into space.

3. Some rays are absorbed by water vapor and other greenhouse gases in the atmosphere. These gases act like the panes of glass in a greenhouse by trapping and reflecting the heat back to Earth.

4. As greenhouse gases from factory emissions and other sources build in the atmosphere, more heat is trapped and reflected to Earth. The oceans warm, producing more water vapor, which traps still more heat.

5. The temperature of the planet rises to achieve a new balance.

FIGURE 7-4 The greenhouse effect on land, sea, and air. *(Adapted from material provided by Electric Power Research Institute)*

emitted from any warm object, such as a hot stove, glass warmed by the passage of ultraviolet rays, or the presence of an open fire.

The gases in the Earth's atmosphere serve as a relatively clear object through which sunlight passes. The crust of the Earth absorbs and radiates, as well as reflects, heat back into the air above it. The atmosphere encircles the Earth and creates the greenhouse effect (Figure 7-4).

Scientists report that the buildup of heat within this global "greenhouse" is increasing and is creating changes in average temperatures and climate. This increased global warming is believed to be a serious threat to the environment and life itself. The greenhouse effect must be stabilized or reduced if we are to restore air quality to its former state (Figure 7-5).

Predicted Change	Global Average	Regional Average
Temperature	+4° to +9° F	−5 to +18° F**
Sea Level	+4 to +40 in	
Precipitation	+7% to +15%	−20 to +20%
Direct Solar Radiation	−10 to +10%	−30 to +30%
Evaporation/Transpiration	+5 to +10%	−10 to +10%
Soil Moisture	?	−50 to +50%
Runoff	increase	−50 to +50%

Source: Schieder, S. (1990). Prudent planning for a warmer planet. *New Scientist,* 128(1743).
**To interpret, if the greenhouse effect produces the results that some scientists predict, these kinds of changes could occur. The average global temperature could increase between 4°and 9° F. Regional changes could range from a drop of 5° to an increase of 18° F in different parts of the world. Other categories of possible changes are given.

FIGURE 7-5 Global warming is believed to be caused by the greenhouse effect.

Hole in the Ozone Layer   The ozone layer over the south pole has been found to be less dense, or thinner, than in the past. The thinner ozone layer lets more of the damaging ultraviolet rays through. This thinner area is referred to as a "hole in the ozone layer," which is enlarging at an alarming rate. Scientists have identified the buildup of chlorofluorocarbons as the cause of this hole. Of equal concern is the buildup of air pollutants that contribute to global warming even where the ozone layer has been damaged less than over the south pole. Scientists and governments of the world are trying to find ways to make continued improvements in living conditions without experiencing additional reductions in air quality and other environmental factors.

Global Warming—Some Causes and Remedies   Scientists believe they have identified some of the important factors that contribute to the intensity of the greenhouse effect and to the resultant global warming. Further, they predict increases in global warming if current trends continue (Figure 7-6). Fortunately, the greatest factors are caused by humans, and, therefore, can be remedied if we agree to take collective action.

Carbon Dioxide (CO_2)   Carbon dioxide is a major product of combustion. Our robust appetites for food, clothing, consumer goods, heated and air conditioned spaces, and transportation has led to a lifestyle that requires huge volumes of fuels to be burned to raise food, manufacture goods, and move vehicles. Our highways in and around large cities are clogged with autos, interstate highways are loaded with trucks, rivers and other waterways carry heavy shipping traffic, homes and commercial buildings use extensive volumes of electricity for light and temperature control, and farms and factories have huge machines and heating devices—all consuming fuel that expels carbon dioxide and other pollutants into the air.

Green plants can utilize some carbon dioxide from the air and convert it into plant food, oxygen, and water. However, in most parts of the world, the mass of green plants is being reduced, while the expulsion of pollutants increases dramatically. The practice of slash and burn in the jungle regions of the world threatens to remove vast areas and volumes of plant growth with little hope of replacement. Increased

Carbon Dioxide (CO₂) The relative contribution of the greenhouse gas CO₂ to the global warming trend is expected to be about 50 percent by the year 2020

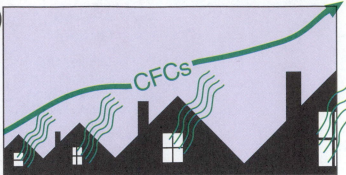

Chlorofluorocarbons (CFCs) about 25 percent of greenhouse effect by 2020

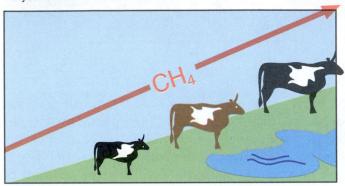

Methane (CH₄) About 15 percent of greenhouse effect by 2020

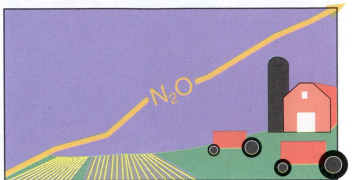

Nitrous Oxide (N₂O) About 10 percent of greenhouse effect by 2020

FIGURE 7-6 Increases in the concentrations of four major pollutants and their predicted effects on global warming by the year 2020.

carbon dioxide in the atmosphere is projected to account for 50 percent of the increase in global warming by the year 2020.

Chlorofluorocarbons (CFCs)   Since discovery of the damage that CFCs do to the ozone layer, other propellants are being used in pressurized spray containers and CFCs are being recaptured from refrigeration units before they are discarded. Still, the escape of CFCs into the air and their effect on the ozone layer is projected to account for 25 percent of the increase in global warming by the year 2020.

Methane (CH₄)   Methane gas is a product of decaying organic matter, such as human and animal waste and dead plant and animal materials. Methane rises from piles, pits, and other accumulations of animal manure, peat bogs, sewerage, and other decaying waste. Some large farms are now using carefully engineered systems to capture the gas from large manure holding areas. The methane gas is then used as fuel for engines driving generators to provide electricity for the farm and for sale. Methane is projected to account for 15 percent of the increase in global warming by the year 2020.

Nitrous Oxide (N₂O)   Nitrous oxide has long been a troublesome pollutant from gasoline engines. The ever increasing numbers of autos, trucks, tractors, heavy equipment, aircraft, chain saws, lawn motors, boats, and other engine-driven applications has off-set the tremendous improvements made in emission reduction from individual engines. Because nitrous oxide emissions are still increasing, it is projected that they will account for 10 percent of the increase in global warming by the year 2020.

Controlling or Reversing Global Warming   Reversing the trend toward continued global warming and the problems it will bring will require real changes in the way we do some things. With good research, wise government, and environmentally sensitive business, we can slow down or stop the decline of our air quality brought by human-caused pollution. Even the small engines and lawn and garden equipment are now seen as serious contributors to pollution, and changes are needed to address this problem. For instance, leaf blowers create pollution through engine exhaust. Pollution is created by the generation of electricity consumed by the electric motor when expelling dust and dirt into the air. Experts now observe that a gasoline lawn mower can cause fifty times as much pollution per horsepower as a modern

BIO·TECH CONNECTION

The Battle to Breathe

U.S. Department of Agriculture scientists are tackling what could be the toughest battle of the century—the battle to breathe. While ozone depletion is a serious problem in the upper atmosphere, ozone as a product of combustion hovers just above the Earth's surface as a pollutant which decreases air quality.

It has been estimated that American farmers experience at least a billion dollars a year in lost crop yields due to air pollution. Research indicates that cutting the level of ozone in the air by 40 percent would mean an extra $2.78 billion for agricultural producers.

USDA and University of Maryland scientists discovered that treatment of plants with the growth hormone Ethylenediura (EDU) can reduce damage by ozone. One soil drench was found to effectively protect some plants from damage and reduce the sensitivity of others to excessive ozone in the air. Similarly, injection of EUD in the stems of shade trees could protect them from damage if located in highly polluted areas. EUD alters enzyme and membrane activity within the leaf cells where photosynthesis takes place.

A biological aide uses a porometer to measure the impact of CO_2 enrichment of the air on the transpiration rate and stomata activity of a soybean leaf. (Courtesy USDA/ARS #K-3204-1)

In addition to ozone, other major air pollutants damaging to crops include peroxyacetetyl nitrate, oxides of nitrogen, sulfur dioxide, fluorides, agricultural chemicals, and ethylene. The challenge of reducing the amounts of pollutants in the air and finding ways to reduce the effect of pollutants and living organisms will continue to challenge future generations.

FIGURE 7-7 A lawn mower may create fifty times more pollution per horsepower than a modern truck engine. *[Courtesy National FFA; FFA #28]*

FIGURE 7-8 A chain saw emits large amounts of air pollutants. *[Courtesy National FFA; FFA #186]*

truck engine (Figure 7-7). Similarly, a lawn mower running one hour may create as much pollution as an automobile traveling 240 miles. Further, a chain saw running two hours emits as much pollution from hydrocarbons as a new car running from coast to coast across the United States (Figure 7-8).

Air and Living Organisms

Oxygen in the air is consumed by plants and animals when they breathe. Animals, including humans, use oxygen to convert food into energy and nutrients for the body. Further, animals breathe out or exhale carbon dioxide gas. It seems we would eventually run out of oxygen. Fortunately, plants give off oxygen during the day. They create oxygen through the process of **photosynthesis** (a process where chlorophyll in green plants enables those plants to utilize light, carbon dioxide, and water to make food and release oxygen).

Maintaining and Improving Air Quality

Air quality can be improved by reducing or avoiding the release of pollutants into the air and removing the pollutants that are there. Some specific practices to reduce air pollution are:

1. Stop using aerosol products containing chlorofluorocarbons.
2. Provide adequate ventilation in tightly constructed and heavily insulated buildings.
3. Have buildings checked for the presence of radon gas.
4. Use exhaust fans to remove cooking oils, odors, solvents, and sprays from interior areas.
5. Clean and service furnaces, air conditioners, and ventilation systems regularly.
6. Maintain all sawdust, wood chip, paint spray, welding fume, dust removal, and general ventilation systems to function at their most efficient levels.

7. Keep gasoline and diesel engines properly tuned and serviced.
8. Keep all emissions systems in place and properly serviced on motor vehicles.
9. Observe all codes and laws regarding outdoor burning.
10. Report any suspicious toxic materials or conditions to the police or other proper authorities.
11. Avoid the use of pesticide sprays as much as possible.
12. Use pesticide spray materials strictly according to label directions.

STUDENT ACTIVITIES

1. Write the Terms to Know and their meanings in your notebook.

2. Make a pie chart illustrating the components of air.

3. Stand in a safe location about 5 ft behind and to the side of a parked truck or bus with its gasoline engine running. Notice the smell of the exhaust. Stand in the same relative position from an automobile that is less than three years old or has less than 40,000 miles on it. What differences do you observe in the truck and automobile exhausts? Why? Which do you think causes more pollution of the air?

4. Examine the written material on all of the gas pumps at a filling station or store. Which types of gas contain lead? Which types of vehicles can use gas containing lead? Which should not use leaded gasoline?

5. Examine three different aerosol cans. Which products use chlorofluorocarbons for the propellant? Why is it unwise to use such products?

6. Talk with an automobile tune-up specialist about the effect of engine adjustments on the content of exhaust gases.

7. Ask your teacher to invite an air-pollution specialist to your class to discuss the problems of air pollution in your town, county, or state.

8. Do a research project on the greenhouse effect and its relationship to global warming.

9. Obtain a radon test kit and do a test for radon in your home.

SELF EVALUATION

A. Multiple Choice

1. Air is
   a. 78 percent argon.
   b. 21 percent nitrogen.
   c. 21 percent oxygen.
   d. 10 percent carbon dioxide.

2. Pure water is
   a. a mixture of gases.
   b. colorless.
   c. one part hydrogen to two parts oxygen.
   d. odorless.

3. Without proper air to breathe, a human can survive only about
   a. 6 minutes.
   b. 12 minutes.
   c. 2 hours.
   d. 12 hours.

4. Radon gas is a widespread threat to air quality
   a. on the highway.
   b. in factories.
   c. in homes.
   d. in wooded areas.

5. Radioactive dust is likely to be caused by
   a. improperly adjusted furnaces.
   b. cracks in basement floors.
   c. a damaged ozone layer.
   d. nuclear reactions.

6. Chemicals used to kill insects are always called
   a. pests.
   b. pesticides.
   c. pollutants.
   d. toxic materials.

7. One ingredient *not* associated with photosynthesis is
   a. carbon dioxide.
   b. oxygen.
   c. radon.
   d. water.

8. Chlorofluorocarbons have been found to damage
   a. aerosol sprays.
   b. the ozone layer.
   c. refrigeration units.
   d. water pumps and equipment.

9. Poisonous gas we cannot remove from auto exhaust is
   a. carbon monoxide.
   b. hydrocarbons.
   c. nitrous oxides.
   d. radon.

10. The most reliable source of information on the use of a pesticide is

    a. experienced applicators.

    b. an extension service.

    c. personal experience.

    d. the product label.

11. The buildup of heat resulting from sunlight passing through glass and heating trapped air in the interior area is called

    a. the greenhouse effect.

    b. infrared energy.

    c. radiation.

    d. ultraviolet.

12. A product of decaying plant or animal matter is

    a. chlorofluorocarbons.

    b. methane.

    c. nitrous oxide.

    d. ozone.

B. Matching

_____ **1.** Carbon monoxide	a.	Tetraethyl
_____ **2.** Chlorofluorcarbons	b.	Pale yellow
_____ **3.** Hydrocarbons	c.	5 percent of auto exhaust
_____ **4.** Lead	d.	Damages ozone layer
_____ **5.** Nitrous oxides	e.	Filters ultraviolet rays
_____ **6.** Ozone	f.	Diseases, insects, weeds
_____ **7.** Radon	g.	Chlorophyll, light, carbon dioxide
_____ **8.** Pests	h.	Causes death from auto exhaust
_____ **9.** Photosynthesis	i.	Leaks into houses
_____ **10.** Sulfur	j.	Pollutant from autos and factories

Water and Soil Conservation

OBJECTIVE

To determine the relationships between water and soil in our environment and the recommended practices for conserving these resources.

COMPETENCIES TO BE DEVELOPED

After studying this unit, you should be able to:

- define *water*, *soil*, and related terms.
- cite important relationships between land characteristics and water quality.
- discuss some major threats to water quality.
- describe types of soil water and their relationships to plant growth.
- cite examples of enormous erosion problems worldwide.
- describe key factors affecting soil erosion by wind and water.
- list important soil and water conservation practices.

MATERIALS LIST

✓ three growing plants in 4- to 6-inch pots for Student Activity

✓ kitchen scale or laboratory balance scale

✓ plant watering containers

✓ plant growing area

Potable

Fresh water

Domestic

Tidewater

Food chain

Universal solvent

Water cycle

Deserts

Irrigation

Precipitation

Evaporation

Water table

Fertility

Saturated

Free water

Gravitational water

Capillary water

Hygroscopic water

Purify

No-till

Contour

Cover crop

Erosion

Port

Aquifer

Sheet erosion

Gully erosion

Mulch

Conservation tillage

Plant residue

Contour practice

Strip cropping

Crop rotation

Organic matter

Aggregates

Lime

Fertilizer

Grass waterway

Terrace

Overgrazing

Conservation plan

As you fly across the vast continents of the Americas, Europe, Asia, and Africa, there is a feeling that land is an endless resource (Figure 8-1). The great oceans of the world combine to provide an even larger area than that covered by the land masses of the world.

FIGURE 8-1 The vastness of land resources on the Earth. *(Courtesy USDA #K-5051-05)*

125

Yet both resources have become limited and there is real concern that we are rapidly depleting them. In developed countries, a safe and adequate water supply is generally threatened only by temporary shortages. Such shortages usually create only mild inconveniences, such as restrictions on the use of water for nonessential tasks like washing cars and watering lawns. But in Third World countries, a safe water supply is a luxury; and sufficient volume for household use is unusual. While there seems to be a sufficient volume of water in most areas of the world, supplies of usable water are generally insufficient due to misuse, poor management, waste, and pollution.

Similarly, productive land is becoming a scarce commodity and ownership is very expensive. Good land is sought by individuals for homes, farms, and recreation; and by businesses for banks, stores, warehouses, car lots, and other things. Industry needs land for factories and storage areas. Government needs land for roads, bridges, buildings, parks, recreation, and military facilities. Most parties look for the best land—land that is level, with deep and productive soil. Such land is in great demand because it provides a firm foundation for roads and buildings, grows crops, supports trees and shrubs, and is easy to modify according to the desired use (Figure 8-2).

FIGURE 8-2 Good land is in high demand. (A) Housing developments, (B) Apartments, condominiums, business, and industry, (C) Roads and bridges, (D) Farm land. *(Courtesy of Elmer Cooper)*

FIGURE 8-4 Clean, fresh water serves many life-supporting functions including nutrient carrier, waste transporter, coolant, home for aquatic plants and animals, oxygen for fish, and cleanser for humans and animals. *(Courtesy Wendy Troeger)*

FIGURE 8-3 Most of the Earth's surface is covered by water. *(Courtesy Chesapeake Bay Foundation)*

THE NATURE OF WATER AND SOIL

Water

We live on the water planet! Most of the Earth's surface is covered with water (Figure 8-3). The oceans and lakes are vast. Most Americans live near an ocean, river, lake, or stream. And those who do not, have access to water from deep wells, rivers, or lakes. Our bodies and the bodies of plants and animals are about 90 percent water. Therefore, we can survive only a few short days if our supply of **potable** (drinkable, that is, free from harmful chemicals and organisms) water is cut off.

Water is an essential nutrient for all plant and animal life. Water transports nutrients to living cells and carries away waste products. Water cools the body. It serves so many useful functions that a sufficient supply of potable water is one of the first considerations for a healthy community (Figure 8-4).

Fresh Versus Salt Water   Most of the water on the earth is not fresh water and is not suitable for humans to drink or use except for transportation. **Fresh water** refers to water that flows from the land to oceans and contains little or no salt. The water in our oceans contains heavy concentrations of salt. Similarly, our bays and tidewater rivers contain too much salt for **domestic** (of or pertaining to the household) use. **Tidewater** refers to the water that flows up the mouth of a river as the ocean tide rises or comes in. Salt water is not fit for animal consumption, nor for plant irrigation.

Food Chains   The bays and oceans provide excellent support for bacteria, algae, and certain water plants. These, in turn, are consumed by insects and small finfish and shellfish. Insects and small fish are then consumed by the larger fish. Large fish, finfish, and shellfish, in turn, are eaten by people, animals, and birds. Similar relationships exist among plants and animals on land and in streams and rivers. The interdependence of plants and animals for food constitutes what is known as a **food chain** (Figure 8-5).

Energy relationships are often represented by a diagram like the one below of a food or energy chain.

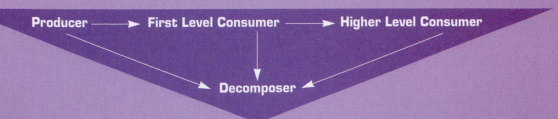

The arrows show the direction of energy and nutrient flow. No arrows lead from the decomposers because they do not pass on energy, only nutrients. Because decomposers work at every level, and are commonly the same species at every level, they are usually omitted from depictions of the food chain. Students should be reminded that although decomposers are not shown, they are acting at each level. Listed below are examples of food chains.

1. Algae ⟶ Oyster ⟶ Oyster Borer (Sea Snail)

2. Marsh Grass ⟶ Duck ⟶ Hawk

3. Algae ⟶ Copepod ⟶ Fish ⟶ Hawk

If arrows were added to show all the relationships, e.g., the duck in number 2 eating the algae in numbers 1 and 3, the food chains would form a food web showing all the possible feeding relationships in an ecosystem, as in the example below.

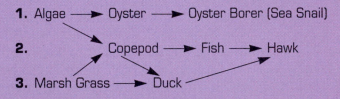

The chart that follows identifies the producers and the levels of the consumers, extracted from the preceding food web chart.

Consumer Level	Example 1	Example 2	Example 3
Third		Hawk	
Second	Oyster Borer	Fish	Hawk
First	Oyster	Copepod	Duck
Producers	Algae	Algae	Marsh Grass

FIGURE 8-5 Food chains demonstrate the interdependence of plants and animals on each other.

The Universal Solvent   Water has been described as the **universal solvent** (a material that dissolves or otherwise changes most other materials). Nearly every substance will rust, corrode, decompose, dissolve, or otherwise yield to the presence of water. Therefore, water is seldom seen in its pure form. It generally has something in it.

Some minerals in water are healthful and give the water a desirable flavor. However, water frequently carries toxic or undesirable chemicals or minerals. Further, water may contain decayed plant or animal remains, disease-causing organisms, or poisons. Ocean water may be described as a thin soup. It is like our blood. It gathers and transports nutrients, is the habitat for microorganisms, carries life-supporting oxygen, cleanses and purifies, and neutralizes and removes wastes.

Scientists tell us that our rivers, bays, and seas have reached their breaking point. People are now polluting air, water, and land faster than nature can cleanse and purify these resources (Figure 8-6). The sea contains all of the dissolved elements carried by the rivers through all time to the low places of the Earth's crust. The water itself may evaporate and be carried back to the land as moisture in the clouds. Eventually it may fall to the land again as rain or snow. However, the minerals remain in the ocean. Many of those substances may be toxic or otherwise threatening to the living organisms of the sea. The water that falls as rain or snow would be pure if it were not contaminated by pollutants when falling through the air.

The Water Cycle   Moisture evaporates from the earth, plant leaves, freshwater sources, and the seas to form clouds. Clouds remain in the air until warm air masses meet cold air masses. This causes the water vapor to change to a liquid and fall as rain, sleet, or snow. The cycling of water between the water sources, atmosphere, and surface areas is called the **water cycle.** (Figure 8-7).

Land

Land provides us solid foundations for buildings, nutrition and support for plants, and space for work and play. It provides storage for water. Further, it serves as a heat and compression chamber that has converted organic material into coal and oil throughout the existence of the Earth.

Soil is an important component of land. To be productive, soil must be made up of the correct proportions of soil particles, have the correct balance of nutrients, contain some organic matter, and have adequate moisture. Much of the Earth's crust is too rocky or has an incorrect balance of nutrients for crop production. Similarly, much land is covered by only a thin layer of productive soil or is too steep to permit cultivation. Where there is some useful soil, however, trees may survive. Forest lands provide lumber, poles, paper, and other products. Here too, humankind can benefit from the pleasures of wildlife and recreation.

Large areas of the Earth have soil with a usable balance of soil particles and minerals for plant growth, but insufficient water. Areas with continuous, severe water shortages are called **deserts.** Fortunately, some desert areas have been made productive through the use of modern irrigation practices. **Irrigation** is the addition of water to plants to supplement that provided by rain or snow. Many nations have desert areas, but do not have the money, technical knowledge, or water to make their deserts productive. Most nations of the world have such limited land resources that all land must be used to its greatest capacity.

During the 1930s, a large area of western Oklahoma and neighboring states became known as the "dust bowl" when lack of rainfall and increased wind shifted the dry soil and ruined the productivity of the land. Later, the productivity returned as weather patterns changed and soil and water conservation practices were implemented.

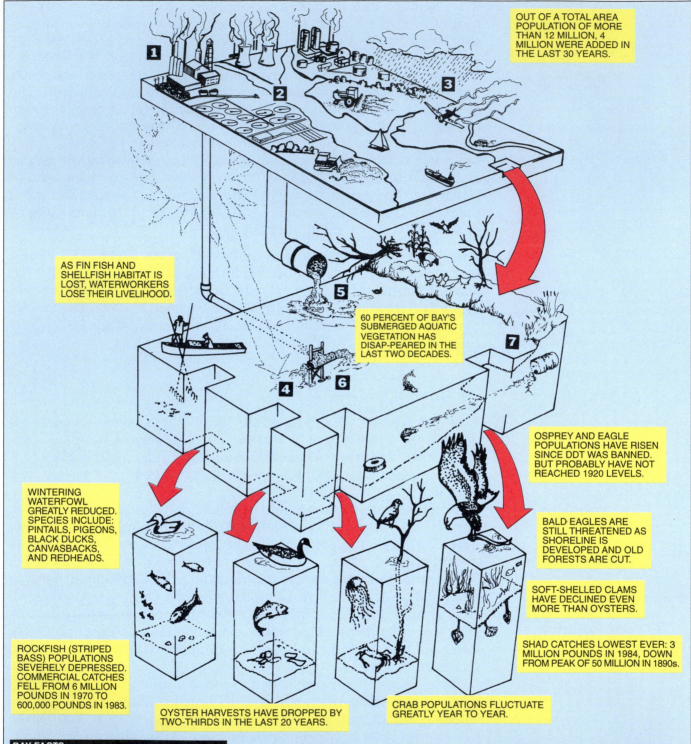

OUT OF A TOTAL AREA POPULATION OF MORE THAN 12 MILLION, 4 MILLION WERE ADDED IN THE LAST 30 YEARS.

AS FIN FISH AND SHELLFISH HABITAT IS LOST, WATERWORKERS LOSE THEIR LIVELIHOOD.

60 PERCENT OF BAY'S SUBMERGED AQUATIC VEGETATION HAS DISAP-PEARED IN THE LAST TWO DECADES.

OSPREY AND EAGLE POPULATIONS HAVE RISEN SINCE DDT WAS BANNED. BUT PROBABLY HAVE NOT REACHED 1920 LEVELS.

WINTERING WATERFOWL GREATLY REDUCED. SPECIES INCLUDE: PINTAILS, PIGEONS, BLACK DUCKS, CANVASBACKS, AND REDHEADS.

BALD EAGLES ARE STILL THREATENED AS SHORELINE IS DEVELOPED AND OLD FORESTS ARE CUT.

SOFT-SHELLED CLAMS HAVE DECLINED EVEN MORE THAN OYSTERS.

ROCKFISH (STRIPED BASS) POPULATIONS SEVERELY DEPRESSED. COMMERCIAL CATCHES FELL FROM 6 MILLION POUNDS IN 1970 TO 600,000 POUNDS IN 1983.

SHAD CATCHES LOWEST EVER: 3 MILLION POUNDS IN 1984, DOWN FROM PEAK OF 50 MILLION IN 1890s.

OYSTER HARVESTS HAVE DROPPED BY TWO-THIRDS IN THE LAST 20 YEARS.

CRAB POPULATIONS FLUCTUATE GREATLY YEAR TO YEAR.

BAY FACTS

SIZE: 195 miles long, 3.5 to 30 miles wide. Average depth: 24 feet.
DRAINAGE BASIN: 64,000 square miles, 50 major tributaries feed it.
WATER: Fresh water from tributaries mixes with salt water from Atlantic. Bay acts like sink, trapping pollutants. Only 1 percent are flushed out to sea.
PLANTS, ANIMALS: More than 2,000 species in bay shoreline.
1 INDUSTRIAL WASTES: Thousands of commercial, industrial facilities discharge water containing toxic chemicals, metals, nitrogen, phosphorous into bay.

Also: cooling needs can lead to corrosion, chlorine contamination.
2 MUNICIPAL SEWAGE: Water discharged from treatment plant contains nitrogen, phosphorous, toxic chemicals.
3 RAIN RUNOFF: Sediment from farms, forests, urban areas carries fertilizers, pesticides, herbicides. Over past 30 years, farmers have doubled the amount of fertilizers they use. Since 1960, herbicide use has tripled.
4 NUTRIENTS: High nutrient levels are most severe on northern, middle bay areas and tributaries, leading to excessive algae growth and less light filtering down to allow grasses to grow below water surface. This

gives waterfowl less food. Algae dies quickly, using up oxygen in the water and making it unsuitable for aquatic life. This leads to fewer fish, shellfish, oysters.
5 TOXIC CHEMICALS: Rain washes sediment into bay. This often includes chemicals, herbicides, pesticides, toxic wastes from industry.
6 CHLORINE: It's used in bay for disinfecting drinking water, sewage, industrial processes. It's suspected of hindering spawning runs of migrating fish.
7 HEAVY METALS: Cadmium, mercury, copper from industrial wastewater, sewage treatment plants can be toxic. Lead from auto exhaust, iron and zinc from industrial discharge, shore erosion.

FIGURE 8-6 Water resources of the world are losing their productivity. *(Adapted from, and used with permission from The Washington Post and "MD Magazine," Autumn 1988)*

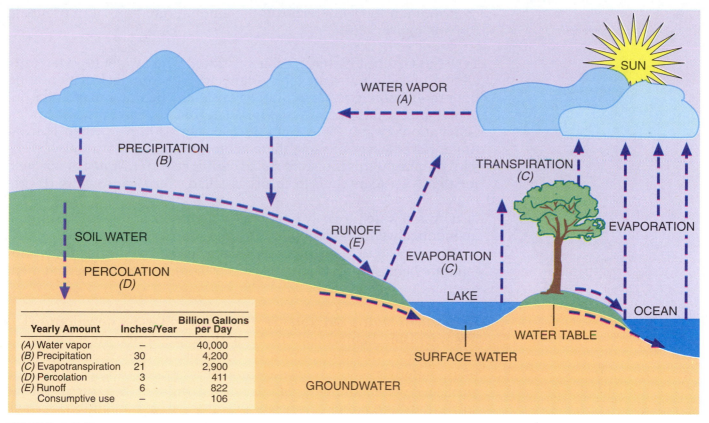

Yearly Amount	Inches/Year	Billion Gallons per Day
(A) Water vapor	–	40,000
(B) Precipitation	30	4,200
(C) Evapotranspiration	21	2,900
(D) Percolation	3	411
(E) Runoff	6	822
Consumptive use	–	106

FIGURE 8-7 The water cycle.

Relationships of Land and Water

Precipitation   Land and water are related to each other in many ways. Land in cold regions or high altitudes holds moisture on its surface in the form of snow. This moisture is then released gradually to feed the streams and rivers after the **precipitation** (the formation of rain and snow) has stopped. Precipitation is caused by the change of water in the air from a gaseous state to a liquid state. It then falls to the land or bodies of water. Moisture-laden, warm air clouds contact cold air masses in the atmosphere and precipitation results. The clouds are formed by water changing from a liquid to a gas when it is evaporated by air movement over land and water. **Evaporation** means changing from a liquid to a vapor or gas.

Land as a Reservoir   Land serves as a container or reservoir for water (Figure 8-8). Water soaks down into the soil and forms a **water table** (the level below which soil is saturated or filled with water). Water held below the water table may run out onto the surface in the form of springs. These create or feed streams, which, in turn, form rivers and flow into bays and oceans. Since ancient times, people have known to dig wells below the water table so water could be extracted for human needs.

Water moves upward in the soil from the water table to provide water for plant roots. From the roots, water travels throughout the plant; and much of it evaporates from the leaves to contribute to the moisture supply in the atmosphere. Water helps soil by improving its physical structure. It is essential for microorganisms that live in the soil and

FIGURE 8-8 Land is a reservoir for water. [*Courtesy USDA/ARS #K-4521-9*]

contribute to the soil's **fertility** (the amount and type of nutrients in the soil) and structure.

Types of Ground Water   Soil is **saturated** when water is added until all the spaces or pores are filled. Plants will die from lack of air around their roots if soil remains saturated very long. The water that drains out of soil after it has been wetted is called **free,** or **gravitational, water.** Gravitational water feeds wells and springs. However, when gravitational water leaves the soil, there is moisture left. Plant roots can absorb or take up this moisture called **capillary water.** Water that is held too tightly for plant roots to absorb is called **hygroscopic water** (Figure 8-9).

Ground water is easily polluted by chemicals and manure through abandoned wells and other depressions. Also, the constant use of irrigation can lead to increased salt concentration. These and other problems are being addressed by soil conservation and water management districts.

Benefits of Living Organisms   Both soil and water benefit from living organisms. Plants break the fall of raindrops and reduce damage to the soil from the impact of water. When plants drop their leaves and plant materials accumulate, they provide a rain-absorbing layer on the surface. When plant materials decay, they improve the structure of the soil. Plants, therefore, contribute to water absorption and reduce soil erosion. Further, decayed plant and animal matter contributes substantially to the nutrient content of the soil. Plants assist in moving water around by taking it up through the roots and releasing it in the atmosphere. Worms, insects, bacteria, and other small and microscopic plants and animals contribute by decomposing plant and animal matter. They also contribute in other important ways.

FIGURE 8-9 Plants use only capillary water. However, hygroscopic water contributes to soil structure, and gravitational water is held in reserve for future use of plants and animals.

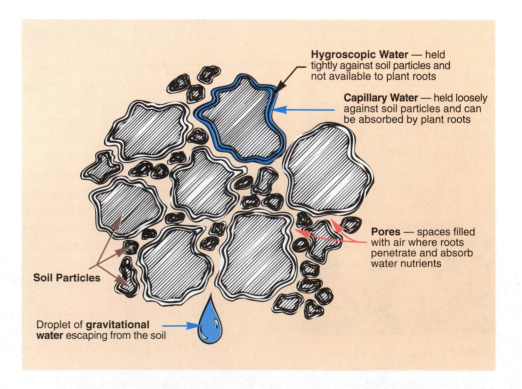

Hygroscopic Water — held tightly against soil particles and not available to plant roots

Capillary Water — held loosely against soil particles and can be absorbed by plant roots

Pores — spaces filled with air where roots penetrate and absorb water nutrients

Soil Particles

Droplet of **gravitational water** escaping from the soil

AGRI·PROFILE

CAREER AREA: Soil Conservation Hydrology

Some major challenges in soil and water conservation are to keep rain water on the land where it falls, keep soil in place and stabilized with plants, minimize pollution of fresh water, and wisely manage land and water to maintain ecological balance. For instance, soil and water scientists at the University of California are developing ways to reuse salt-laden irrigation water. It is estimated that California alone may be generating upwards of 1.6 trillion gallons of sub-surface drainage water annually.

Soil and water conservation career options provide extensive opportunities for inside and outside work. Work may be found in Soil Conservation Service (SCS) or Agricultural Stabilization and Conservation Service (ASCS) offices. Or, one may work in the field served by such offices in most counties of the United States. Typical job titles are conservation technician, farm planner, soil scientist, and soil mapper.

The United States Department of Interior, state departments of natural resources, city and county governments, industry, and private agencies hire people with soil and water conservation expertise. Such workers manage water resources for recreation, conservation, and consumption. They may work as consultants, law enforcement officers, technicians, administrators, heavy equipment operators, and the like.

Salt-laden waste water flows from irrigated land in California's Imperial Valley into the Sulton Sea. *(Courtesy USDA/ARS #K-1918-10)*

CONSERVING WATER AND IMPROVING WATER QUALITY

How can we reduce water pollution? How can soil erosion be reduced? What is the most productive use of water and soil without polluting or losing these essential resources? These are important questions deserving correct answers now! Farmers have long appreciated the value of these resources and generally used them wisely. However, economic conditions, governmental policies, production costs, farm income, personal knowledge, and other factors influence the use of conservation practices by farmers and other land users. Every citizen, business, agency, and industry impacts on air and water quality. Similarly, all have some influence on the use of our land and water resources (Figure 8-10).

THE COLORADO RIVER BASIN ... AND ITS PLUMBING

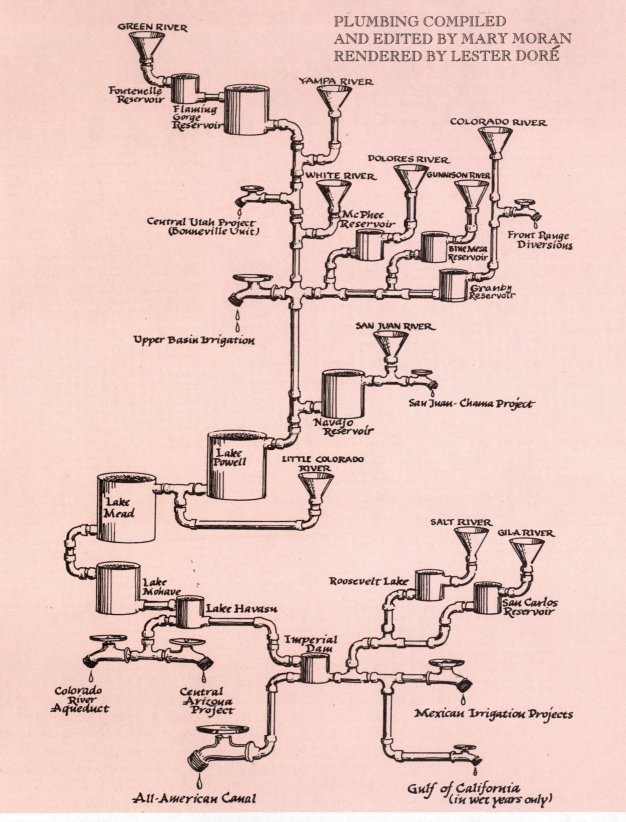

PLUMBING COMPILED
AND EDITED BY MARY MORAN
RENDERED BY LESTER DORÉ

FIGURE 8-10 Water sources and uses from the Green, Colorado, and other rivers to the Gulf of California. *(Courtesy Dinosaur National Monument, National Park Service)*

The improvement of water quality can be achieved by proper land management, careful water storage and handling, and appropriate use of water. Some practices that help reduce water pollution are as follows:

1. Save clean water. Whenever we permit a faucet to drip, leave water running while we brush our teeth, or flush toilets excessively, we waste water. Further, taking long showers, leaving water running while we wash automobiles or livestock, and using excessive water for pesticide washup wastes clean water. Once clean water is mixed with contaminants, it is not safe for use for other purposes until it is cleaned or repurified. **Purify** means removal of all foreign material.

2. Dispose of household products carefully. Many products under the kitchen sink, in the basement, or in the garage are threats to clean water (Figure 8-11). Never pour paints, wood preservatives, brush cleaners, or solvents down the drain. They will eventually enter the water supply, rivers, or oceans. Use all products sparingly and completely. Put solvents into closed containers and permit the dissolved materials to settle out. Then reuse the solvent. Stuff empty containers with newspaper to absorb all liquids before discarding. Then send the containers to approved disposal sites.

3. Care for lawns, gardens, and farmland carefully. Improve soil by adding organic matter. Mulch lawn and garden plants. Use proper amounts and types of lime and fertilizer. Only till soil that will not erode excessively. Cover exposed soil with a new crop immediately. Water soil only when it is excessively dry, and then water until the soil is soaked to a depth of 4 to 6 inches.

4. Practice sensible pest control. Many insecticides kill all insects—both harmful and beneficial ones. Insecticides also pollute water. Therefore, use cultural practices, such as crop rotations and resistant varieties, instead of insecticides whenever possible. Encourage beneficial insects and insect-eating birds by improving habitat around lawn, garden, and fields. Eliminate pools of stagnant water to eliminate mosquitoes. Keep untreated wood away from soil to avoid termite, other insect, and rot damage. Follow all pesticide label instructions exactly.

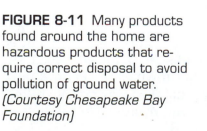

FIGURE 8-11 Many products found around the home are hazardous products that require correct disposal to avoid pollution of ground water. (Courtesy Chesapeake Bay Foundation)

5. Control water run-off from lawns, gardens, feedlots, and fields. Keep soil covered with plants. Construct livestock facilities so manure can be collected and spread on fields. Use no-till cropping. **No-till** means planting crops without plowing or disking the soil. Alternate strips of close-growing crops (like small grains or hay) with row crops (like corn or soybeans). Farm on the **contour** (following the level of the land around a hill). Plant cover crops where regular crops do not protect the soil. A **cover crop** is a close-growing crop planted to temporarily protect the soil.

6. Control soil erosion. Reduce the volume of rain water carried away by minimizing the amount of blacktop or concrete surface constructed. Use grass waterways in low areas of fields. Add manure and other organic matter to soil to increase water-holding capacity. Construct terraces on long or steep slopes. Leave steep areas in trees and sloping areas in close-growing crops.

7. Avoid spillage or dumping of gasoline, fuel, or oil on the ground or in storm drains. Turn over used petroleum products to truck and automobile service centers for disposal.

8. Keep chemical spills from running or seeping away. Do not flush chemicals away. The chemicals will damage lawns, trees, gardens, fields, and ground water. Sprinkle spills with an absorbent material such as soil, kitty litter, or sawdust. Then place the chemical-laden absorbent material in a strong plastic bag and discard it according to local recommendations.

9. Properly maintain your septic system (if your home has one). Avoid using excessive water. Do not flush inappropriate materials down the toilet. Do not plant trees where their roots might interfere with sewer lines, septic tanks, or field drains. Don't run tractors and other heavy equipment over field drains.

These recommendations are some practices that should help increase water supplies and decrease water usage and pollution.

LAND EROSION AND SOIL CONSERVATION

The Problem

Land Erosion—A Worldwide Problem Land **erosion** (wearing away) is a serious problem worldwide (Figure 8-12). Both wind and water are capable of wearing away soil. The food and fiber production capabilities of large nations are being compromised due to extensive damage from soil erosion. Numerous cases can be cited from history where soil erosion has caused enormous problems. Yet, as we move into the 21st century, the world is making the same mistakes in many areas that caused extensive losses in the past.

Consider, for example, an important colonial **port** (a town having a harbor for ships to take on cargo) along the eastern coast of what is now the United States. The colony was established in the year 1634. By 1706, the port had been replaced by another town as the central port for the colony. The harbor area of that original port was rapidly filling with soil that was washed in from the surrounding area. The water was no longer deep enough for the ships to move.

FIGURE 8-12 Erosion of soil means loss of productive soil, damage to machinery, additional costs of production, and pollution of streams, rivers, bays, and oceans. *(Courtesy USDA/ARS #K-1021-15)*

Consider the massive size of the Mississippi River Delta (land created by the soil that water deposits at the mouth of a river). The Mississippi River Delta is over 15,000 square miles in area. In order for the Mississippi River Delta to be formed, the river had to deposit soil on the bottom of the Gulf of Mexico and continue until it reached the surface of the water. The soil gets into the river from smaller rivers and streams receiving run-off water from fields in the heartland of America. Therefore, one must conclude that the Mississippi River Delta was built from the bottom to the surface of the Gulf with America's best topsoil. Yet, in these modern times, it is estimated that the amount of soil being dumped by the river into the Gulf every day would fill a freight train 150 miles long. Further, it has been estimated that the amount deposited at the mouth of the Mississippi in one year is sufficient to cover the entire state of Connecticut with soil 1 in. deep, or over 8 ft in 100 years! Similar deltas exist at the mouth of the Nile and most other great rivers of the world.

During the late sixties and early seventies in the People's Republic of China, much of the forested land was cleared. It soon became apparent that much of that land depended on the trees to prevent soil erosion. Seeing the extent of damage from soil erosion, the country has responded with a national policy of rapid reforestation. There are over 1 billion people in China and all are expected to contribute to the tree planting effort. However, it will take many years for newly planted trees to grow sufficiently to protect the soil again. Unfortunately, the soil washed away during the absence of trees can never be returned to the land. Soil scientists report that it takes 300 to 500 years for nature to develop one inch of topsoil from bedrock.

Another serious threat to large areas of the world is the farming system known as "slash and burn." This is used in the tropical rain forests of South America, Africa, and Indonesia. In these areas, impoverished farmers cut or slash the jungle growth and burn the plant residues. The extensive burning causes serious air pollution and destroys useful fuel for cooking. Then they raise crops for several years until the soil is drained of its meager fertility and soil erosion takes its toll. They then move on to uncleared land and repeat the cycle. Sadly, once the land is cleared, it becomes difficult for plant growth to become re-established and the land is left permanently ruined. Reportedly, there are 50 acres of tropical rain forest lost every minute through this process. It is said that, in some regions, 80 percent of the rainfall is attributed to transpiration from trees. When the trees are removed, such areas are subject to drought.

National Problems Each year, about 1.6 billion tons of soil are worn away from 417 million acres of U.S. farmland and deposited into lakes, rivers, and reservoirs. One ton equals 2,000 lb. While some soils are deep and can tolerate a certain amount of erosion, many fragile soils cannot. The USDA's National Resources Inventory indicates 41 million acres, or about 10 percent, of our nation's cropland is highly erodible at rates of 50 or more tons per acre per year.

Of growing concern is the contamination of ground water under large areas of the United States. Ground-water pollution emerged as a public issue in the late 1970s. The first reports documented sources of contamination associated with the disposal of manufacturing wastes. By the early 1980s, several incidences of ground water contamination by pesticides used on field crops were confirmed. Ground water contamination can threaten the health of large populations. For instance, in New York State, the vast **aquifer** (a water-bearing rock formation) that underlies Long Island represents the only supply of drinking water for more than 3 million people. In the United States, major aquifers underlie areas that contain thousands of square miles of land, which often includes several states. Contamination of the aquifer in one place means contamination of the water for large areas.

Soil Conservation

Preventing and Reducing Soil Erosion Soil erosion can be reduced or stopped by good land management. Fortunately, management practices that reduce soil erosion increase water absorption and retention (Figure 8-13). This, in turn, increases the supply of fresh water available for crops, livestock, wildlife, and people. Further, it reduces dryness and dustiness, which are threats to good air quality. Most soil conservation methods are based on either (1) reducing raindrop impact, (2) reducing or slowing the speed of wind or water moving across the land, (3) securing the soil with plant roots, (4) increasing absorption of water, or (5) carrying run-off water safely away. If water is free to run down a hill, it increases in volume and speed, picks up and pushes soil particles, and carries them off the land. This results in sheet and gully erosion and deposits soil, nutrients, and chemicals into streams. **Sheet erosion** is removal of layers of soil from the land while **gully erosion** is soil removal that leaves trenches.

The following are some recommended practices to reduce or prevent wind and water erosion.

1. Keep soil covered with growing plants. Plants reduce the impact of raindrops, reduce the speed of wind and water movement across the land, and hold soil in place when threatened by moving wind or water.

FIGURE 8-13 Plant growth holds the key to soil and water conservation. (A) Surface mulch protects the soil around cotton seedlings, (B) Extensive root systems of soybeans, and (C) Revegetation of sandbars with willow trees. *(Courtesy USDA/ARS #K-3226-6, #K-3213-1, #K-5Z10-18)*

2. Cover the soil with a mulch. **Mulch** is a material placed on soil to break the fall of raindrops, prevent weeds from growing, and/or improve the appearance of the area.

3. Utilize conservation tillage methods. **Conservation tillage** means using techniques of soil preparation, planting, and cultivation that disturb the soil the least and leave the maximum amount of plant residue on the surface. **Plant residue** is the plant material left when a plant dies or is harvested.

4. Use contour practices in farming, nursery production, and gardening. **Contour practice** means conducting all operations, such as plowing, disking, planting, cultivating, and harvesting, across the slope and on the level. Any grooves or ridges created by machinery go around the slope or hill. When water tries to run down the hill, it encounters the grooves and ridges. Because these are level, the water tends to soak into the soil. This holds the water for future use.

5. Use strip cropping on hilly land. **Strip cropping** means alternating strips of row crops with strips of close-growing crops. Examples of close-growing crops are hay, pasture, and small grains, such as wheat, barley, oats, and rye. Examples of row crops are corn, soybeans, and most vegetables. The strips of close-growing crops capture run-off water from the row crops and prevent it from entering streams.

6. Rotate crops. **Crop rotation** is the planting of different crops in a given field every year or every several years. Crop rotation permits close-growing crops to retain water and soil, and tends to rebuild the soil after losses incurred while row crops occupied the land.

7. Increase organic matter in the soil. **Organic matter** is dead plant and animal tissue. Nonliving plant leaves, stalks, branches, bark, and roots decay and become organic matter. Similarly, animal manure and dead insects, worms, and animal carcasses decompose to make organic matter. The decomposed organic matter forms a gel-like substance that holds soil particles in absorbent granules called **aggregates.** An aggregated soil is a water-absorbing and nutrient-holding soil. Organic matter also releases nutrients that improve plant growth.

8. Provide the correct balance of lime and fertilizer. **Lime** is a material that reduces the acid content of soil. It also supplies nutrients such as calcium and magnesium to improve plant growth. **Fertilizer** is any material that supplies nutrients for plants.

9. Establish permanent grass waterways. A **grass waterway** is a strip of grass growing in the low area of a field where water can gather and cause erosion.

10. Construct terraces. A **terrace** is a soil or wall structure built across the slope to capture water and move it safely to areas where it will not cause erosion.

11. Avoid overgrazing. **Overgrazing** refers to damage to plants or soil due to animals eating too much of the plants at one time. This reduces the plant's ability to hold soil or recover after grazing.

12. Use land according to a conservation plan. A **conservation plan** is a plan developed by soil and water conservation specialists to use land for its maximum production and water conservation without unacceptable damage to the land.

The USDA's Soil Conservation Service (SCS) and Agriculture Stabilization and Conservation Service (ASCS) provide advice, technical assistance, and funds to assist land owners with soil and water conservation practices. Further, the Environmental Protection Agency (EPA) monitors ground water quality, and enforces point and nonpoint pollution laws to help protect our land and water.

There are many careers in the field of water and soil conservation. The material in this unit hints at the problems created by humans as they use the natural resources around them to provide food, water, shelter, recreation, and other resources for living. The health, wealth, peace, and general welfare of humanity depends upon our skill in conserving these most basic resources.

BIO•TECH CONNECTION

Safe-guarding Ground Water

Authorities estimate that nearly 330,000 tons of pesticides are applied yearly on U.S. crops. Added to this are the chemicals used to control fleas, flies, ticks, termites, roaches, and other pests of livestock and humans. Agricultural chemicals have enabled us to feed and clothe ourselves and many others in the world; but pollution from these same chemicals have become a problem and a threat to our air, water, and land.

Applying these pesticides where they are intended and keeping them there until they change or biodegrade into harmless products is a major concern in protecting our water supplies. To help safe-guard our surface and ground water from pesticide pollution, the USDA Agricultural Research Service has developed a specific strategy. Research priorities of the 1990s are shaped by this strategy.

The ultimate goals of the plan are to (1) provide the American farmer with cost-effective best management practices to ensure ample supplies of food and fiber at a reasonable cost while reducing pesticide movement into the ground water, (2) identify the factors that accelerate or retard pesticide movement, and (3) provide computer models that will quickly and accurately predict contamination.

Further, USDA research is being concentrated in six areas to prevent contamination of ground water by crop pesticides. These are:

- Conservation tillage practices
- Integrated pest management
- Improved pesticide application technology
- Improved water/pesticide management practices

Areas of Potential Groundwater Contamination from Agricultural Chemicals

Nitrates Only
Pesticides Only
Nitrates and Pesticides

Ground water contamination by agricultural chemicals is considered a risk in many areas. *(Courtesy USDA/ARS August '89 p. 2 -USDA/ERS, AER 576. Neilson & Lee, "The Magnitude and Cost of Groundwater Contamination from Agricultural Chemicals: A National Perspective," 1987)*

- New methods of pesticide analysis and decontamination
- Improved computer models

The computer models under development are decision-enhancing tools that use a database and computer program to aid the selection of management practices. Models integrate data from many sources that are frequently not available to the decision makers.

Computer models help identify pesticide processes related to movement in various soils, climates, and other environmental conditions. New technologies, such as advances in slow-release formulations, improved pesticide application scheduling, and selective placement, are under study. Today ground water contamination by agricultural chemicals is considered a risk in many of the major crop and livestock production areas of the country.

STUDENT ACTIVITIES

1. Write the Terms to Know and their meanings in your notebook.

2. Collect a sample of drinking water from each of five sources as follows: a safe spring or well, bottled pure water, and faucets attached to (a) galvanized pipe, (b) plastic pipe, and (c) copper pipe. (Do not run off the water before obtaining the samples from the faucets.) Taste a small amount of each sample and describe the taste. Do the samples have different tastes? If so, what causes the differences?

3. Study the eating habits of one species of birds in your community. Describe the food chain that accounts for the survival of that species. What effect did the use of DDT as an insecticide have on that species of birds before DDT was banned from use?

4. Obtain a rain gauge and record the precipitation on a daily basis for several months. What variations did you observe from week to week? How do you explain the variations? What effect did these variations have on the agricultural activities of the community?

5. Conduct an experiment to determine the effect of soil water on plant growth. Obtain three inexpensive pots of healthy flowers or other plants. Each pot must be the same size and type, have the same amount and type of soil, and contain the same size and number of plants. Use the following procedure:

a. Mark the pots "1," "2," and "3."

b. Plug the holes in the bottoms of pots 2 and 3 so water cannot drain from the pots. Leave pot 1 unplugged so it has good drainage.

c. Add water to pot 1 as needed to keep the plant healthy for several weeks. Use it as a comparison specimen (called the "control").

d. Add water slowly to pot 2 until water has filled the soil and the water is just level with the surface of the soil. All pores of the soil are now filled and the soil is "saturated." Weigh the pot and record the weight as "A."

e. Remove the plug from the bottom of pot 2 and permit the water to drain out. After water has stopped flowing from the drain hole, immediately weigh the pot again. Record the weight as "B." The water that flowed from pot 2 when the plug was removed is the free or gravitational water. Weight of the gravitational water, "D," should be calculated and recorded using the formula $A - B = D$. Do not add any more water to pot 2 for 2 weeks.

f. With the hole plugged in pot 3, add water slowly until the water is just level with the top of the soil. Do not remove the plug, and keep pot 3 filled with water to the saturation point for 2 weeks.

g. Keep all three pots in a good growing environment for 2 weeks and record all observations.

h. When the plant in pot 2 wilts badly due to lack of water, weigh the pot and record the weight as "C."

   Make the following calculations:

   Weight of the gravitational water $(D) = A - B$

   Weight of the capillary water $(E) = B - C$

   The weight of the hygroscopic water can only be determined by driving the remaining water from the soil by heating the soil in an oven.

6. Ask your teacher to help you design and conduct a project that demonstrates some or all of the following:

a. Effect of the force of raindrops on soil.

b. Effect of soil aggregation on absorption.

c. Effect of slope on erosion.

d. Effect of living grass on erosion control.

e. Effect of plant residue on erosion control.

SELF EVALUATION

A. Multiple Choice

1. Most of the earth's surface is covered with
   a. crops.
   b. farms.
   c. trees.
   d. water.

2. The bodies of plants, animals and humans consist of about what percent water?
   a. 10 percent
   b. 40 percent
   c. 70 percent
   d. 90 percent

3. The universal solvent is
   a. gasoline.
   b. paint thinner.
   c. varsol.
   d. water.

4. The content of ocean water may be likened to
   a. fresh water.
   b. pure water.
   c. thin soup.
   d. varsol.

5. The interdependence of plants and animals on each other for food is known as
   a. domestic.
   b. food chain.
   c. symbiosis.
   d. universal relationship.

6. The land serving as a heat and compression chamber gives us
   a. building foundations.
   b. coal and oil.
   c. crops.
   d. wildlife habitat.

7. Ground water which is available for plant root absorption is called
   a. capillary.
   b. free.
   c. gravitational.
   d. hygroscopic.

8. Improvement of water quality can be achieved by
   a. appropriate use of water.
   b. careful water storage and handling.
   c. proper land management.
   d. all of the above.

9. The problem of land erosion is
   a. characteristic of developed countries only.
   b. found world-wide.
   c. mostly found in poor countries.
   d. not a substantial problem in view of food surpluses.

10. A conservation plan
   a. conserves soil and water.
   b. is prepared by professionals.
   c. maximizes productivity of land.
   d. all of the above.

B. Matching

_____ **1.** Contour
_____ **2.** Delta
_____ **3.** Evaporation
_____ **4.** Fertility
_____ **5.** Irrigation
_____ **6.** No-till
_____ **7.** Port
_____ **8.** Precipitation
_____ **9.** Purify
_____**10.** Water table

   a. Amount and type of nutrients
   b. Without plowing or disking
   c. Town with a harbor
   d. Removal of foreign material
   e. Soil deposited by water
   f. Top of saturated soil
   g. Rain and snow
   h. Supplemental water
   i. On the level
   j. Change from liquid and gas

C. Completion

1. An _____ is a water-bearing rock formation.

2. The process of creating narrow and deep trenches eroded in soil is known as _____ _____.

3. Material placed on soil to break the fall of raindrops is called _____.

4. Alternating row crops with close-growing crops is known as _____ _____.

5. _____ is used to reduce acid in soil.

6. One inch of topsoil may be formed from bedrock in about _____ years.

7. _____ _____ is dead plant and animal material in soil.

8. A _____ _____ may be planted to temporarily protect soil from erosion.

9. In the United States about _____ tons of soil are worn away from farmland each year.

10. About _____ acres or 10 percent of the U.S. cropland is highly erodible.

Soils and Hydroponics Management

OBJECTIVE

To determine the origin and classification of soils, and identify effective procedures for soils and hydroponics management.

MATERIALS LIST

✓ writing materials

✓ newspapers

✓ topography map

COMPETENCIES TO BE DEVELOPED

After studying this unit, you should be able to:

- define terms in soils, hydroponics, and other plant-growing media management.
- identify types of plant-growing media.
- describe the origin and composition of soils.
- discuss the principles of soil classification.
- determine appropriate amendments for soil and hydroponics media.
- discuss fundamentals of fertilizing and liming materials.
- identify requirements for hydroponics plant production.
- describe types of hydroponics systems.

TERMS TO KNOW

Media	Permeable	Alkalinity
Medium	Capability classes	Neutral
Hydroponics	Capability subclasses	Primary nutrients
Decomposed	Capability unit	Complete fertilizer
Leaf mold	Organic matter	Fertilizer grade
Compost	Mineral matter	Active ingredients
Sphagnum	Horizon A	Broadcasting
Peat moss	Clay	Incorporated
Bogs	Silt	Band application
Water-logged	Sand	Side-dressing
Perlite	Tillable	Top-dressing
Vermiculite	Topsoil	Starter solutions
Leached	B horizon	Foliar sprays
Microbes	C horizon	Knife application
Parent material	Bedrock	Legumes
Horizon	Coarse-textured (sandy) soil	Nitrates
Profile	Medium-textured (loamy) soil	Nitrogen fixation
Residual soils	Fine-textured (clay) soil	Aggregate culture
Alluvial deposits	Structure	Water culture
Lacustrine deposits	Aggregates	Solution culture
Loess deposits	Crumbs	Nutriculture
Colluvial deposits	Amendment	Aeroponics
Glacial deposits	pH	Continuous-flow systems
Percolation	Acidity	Inert

T he role of plants in our environment and their importance in our lives has been discussed in previous units. Plants are necessary to nourish the animals of the world and maintain the balance of oxygen in our atmosphere. However, they depend upon soil, water, and air as their media for support. **Media** is plural for medium. **Medium** is a surrounding environment in which something functions and thrives.

PLANT-GROWING MEDIA

For discussion in this unit, the word media will be used to mean the material that provides plants with nourishment and support though their root systems.

Types of Media

Media comes in many forms. The oceans, rivers, land, and man-made mixtures of various materials are the principal types of media for plant growth. Seaweed, kelp, plankton, and many other plants depend upon water for their nutrients and support. It is only recently that we have come to understand the tremendous amount of plant life in the sea. The plant life in oceans and rivers are important for feeding the animal life of the seas and bodies of fresh water. Humankind has used fish as a staple food since the beginning of time.

It has long been known that water could be used to promote new root formation on the stems of certain green plants and completely support plant growth on a limited basis. Recently, however, it has been found that food crops can be grown efficiently without soil in structures where plant roots are submerged in or sprayed with solutions of water and nutrients. These solutions feed the plants, while mechanical apparatus provide physical support. The practice of growing plants without soil is called **hydroponics.** Hydroponics has become an important commercial method of growing green plants. This will be discussed later in this unit.

Soil

Soil is defined as the top layer of the Earth's surface, suitable for the growth of plant life. It has long been the predominant medium for cultivated plants (Figure 9-1). In early years, humans accepted the soil as it existed. They planted seeds using primitive tools and did not know how to modify or enhance the soil to improve its plant-supporting performance.

Ancient civilizations discovered that plant-growing conditions were improved on some land where deposits were left after river waters flowed over the land during flood season. Similarly, other land was ruined by flood waters. Therefore, early efforts to improve plant-growing media was a matter of moving to better soil. Obviously, good soil was a valuable asset and, therefore, was the cause of intense personal disputes and wars among nations.

Other Media

In addition to water and soil, certain other materials will hold water and support plant growth. Fortunately, some of the best nonsoil and nonwater plant-growing media are partially **decomposed** (decayed) plant materials. Hence, plants tend to improve the environment where they grow. One common material available around most homes is leaf mold and compost. **Leaf mold** is partially decomposed plant leaves. **Compost** is a mixture of partially decayed organic matter such as leaves, manure, and household plant wastes. Decaying plant matter should be mixed with lime and fertilizer in correct proportions to support plant growth (Figure 9-2).

There is a group of pale or ashy mosses called **sphagnum.** These are used extensively in horticulture as a medium for encouraging root growth and growing plants under certain conditions. **Peat moss** consists of partially decomposed mosses which have accumulated in water-logged areas called **bogs. Water-logged** means soaked or saturated with water. Both substances have excellent air- and water-holding qualities.

FIGURE 9-1 Soil has long been the predominant medium for plant growth. *(Courtesy USDA/ARS)*

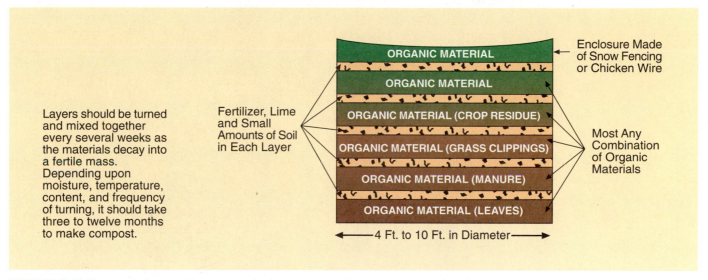

Layers should be turned and mixed together every several weeks as the materials decay into a fertile mass. Depending upon moisture, temperature, content, and frequency of turning, it should take three to twelve months to make compost.

Fertilizer, Lime and Small Amounts of Soil in Each Layer

Enclosure Made of Snow Fencing or Chicken Wire

ORGANIC MATERIAL

ORGANIC MATERIAL

ORGANIC MATERIAL (CROP RESIDUE)

ORGANIC MATERIAL (GRASS CLIPPINGS)

ORGANIC MATERIAL (MANURE)

ORGANIC MATERIAL (LEAVES)

Most Any Combination of Organic Materials

◄─── 4 Ft. to 10 Ft. in Diameter ───►

FIGURE 9-2 Compost is excellent organic matter.

Many other sources of plant and animal residues may become plant-growing media. For instance, a fence post may rot on the top and hold moisture from rainfall. In time, a plant seed may be deposited in the rotted wood medium, germinate, root in the wood, and start to grow. Even horse manure mixed with straw is used extensively as a medium for growing mushrooms. In this instance, both animal residue (manure) and plant residue (straw) combine to make an effective medium.

Some mineral matter can also become plant-growing media. For instance, volcanic lava and ash remain black, gray, and barren only briefly. Soon the layer cools and cracks. Seeds settle into the cracks and moisture causes the seeds to germinate. Roots then penetrate and break up the volcanic residue. Soon the area is covered with plant life.

Horticulturists use certain mineral materials in plant-growing areas, too. **Perlite** is a natural volcanic glass material having water-holding capabilities. Perlite is used extensively for starting new plants. **Vermiculite,** a mineral matter from a group of mica-type materials, is also used for starting plant seeds and cuttings.

ORIGIN AND COMPOSITION OF SOILS

Factors Affecting Soil Formation

Productive soils develop on the Earth's surface as the atmosphere, sunlight, water, and living things meet and interact with the mineral world (Figure 9-3). If soil is suitable for plant growth to a depth of 36 in. or more, that soil is regarded as "deep." Many soils of the Earth are much more shallow than this. Plants attach themselves to the soil by their roots, grow, manufacture food, and give off oxygen. Plants and animals of various sizes live on and in the soil, utilizing carbon dioxide, oxygen, water, mineral matter, other plants and animals, and products of decomposition.

Soils vary in temperature, organic matter, and the amount of air and water they contain. The kinds of soils formed at a specific site are determined by the forces of climate, living organisms, parent soil material, topography, and time. Each factor varies by local conditions.

FIGURE 9-3 A soil ecosystem.

Climate and Location

Climatic factors, such as temperature and rainfall, greatly affect the rate of weathering. When temperature rises, the rate of chemical reactions increases, and growth of fungi, organisms (such as bacteria), and plants increases. The rate and amount of rainfall in a locality greatly affect the soil. In areas of high rainfall, the soils usually are leached and somewhat acidic. **Leached** means having certain contents removed by water. If the land is covered by trees, the action of high temperatures and moisture on leaf residues generally creates an acidic soil. Rainfall during cold weather has less effect on the soil than during warm or hot weather.

Slope and location of a field also affect erosion and drainage and thus influence soil formation. Moreover, free water in the soil carries fine particles to the deeper layers and tends to produce "layering." Too much water prevents or retards microbial growth and may exclude air by water-logging. Water and temperature also have the effect of swelling and contracting soil particles.

Living Organisms

Living organisms, such as microbes, plants, insects, animals, and humans, exert considerable influence on the formation of a soil. Certain types of soil bacteria and fungi aid in soil formation by causing decay or breakdown of the plant and animal residues in the soil. Carbon dioxide and other compounds essential to soil formation are released by microbic activity. **Microbes** are microscopic plants and animals. Without soil microbes, organic materials would not decay.

Numerous insects, worms, and animals contribute to the formation of soil by mixing the various soil materials. Earthworms consume and digest certain soil substances and discharge body wastes. This aids decomposition and soil mixing. All such dead organisms add to soil organic matter.

FIGURE 9-4 Human activities greatly influence soil development when land is cleared or otherwise disturbed. *(Courtesy National FFA)*

Human activity also influences soil formation as cultivation, bulldozing, and construction projects disturb the surface layer. The clearing of land removes native plant life and greatly modifies soil-forming activities (Figure 9-4).

Parent Material

Parent material is the horizon of unconsolidated material from which a soil develops. **Horizon** means layer. Parent materials compose the C horizon of a typical soil profile. **Profile** means a cross-sectional view of soil (Figure 9-5).

Parent materials formed in place are called **residual soils.** Other soils are transported and deposited by water, wind, gravity, or ice. **Alluvial deposits** are transported by streams, and **lacustrine deposits** are left by lakes. **Loess deposits** are left by wind, **colluvial deposits** by gravity, and **glacial deposits** by ice.

The kind of parent material from which soil is formed influences the many characteristics of that soil. Natural fertility and texture are influenced greatly by the parent material of a profile.

Topography

Slope and drainage affect soil formation both directly and indirectly. On a steep slope, loose material is moved downward by run-off water, gravity, and movement of humans and animals. This movement not only breaks up soil materials and adds them to the lower levels, but it exposes subsoil materials along the upper slopes. The movement of soil materials has a pulverizing effect on the material being moved as well as on the material left behind.

Slope affects the distribution of water that falls on the Earth's surface. On level areas, the water soaks in and moves through the soil in a process called **percolation.** On nonlevel land, the water tends to run off and moves some surface soil with it. Soils that develop on level land tend to be poorly drained, while soils on level to gently rolling slopes are better drained and more productive (Figure 9-6).

Drainage (or lack of it) affects the water table in a particular field or area. The water table has a direct bearing on soil formation, especially

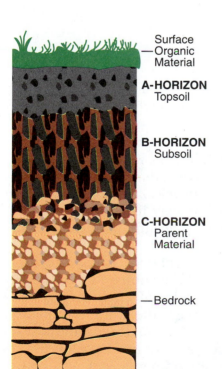

Surface
—Organic
Material

A-HORIZON
Topsoil

B-HORIZON
Subsoil

C-HORIZON
Parent
Material

—Bedrock

FIGURE 9-5 A soil profile.

FIGURE 9-6 Topography greatly influences soil formation. *(Courtesy Michael Dzaman)*

if it is near the surface. When a soil is saturated with water, little or no air can penetrate it. The lack of air reduces the action of fungi, bacteria, and other soil-forming activities in the soil.

A wet soil, therefore, is a slow-forming soil and usually is low in productivity. Because of the lack of air, undecomposed organic matter will accumulate in a wet soil. This organic matter generally causes the soil to be a blackish color. Poor drainage, accompanied by free water in the soil, reduces or retards plant growth and affects soil formation.

Time

Soils are formed by the chemical and physical weathering of parent material over time, as affected by climate, living organisms, and topography. Therefore, time itself is regarded as a factor in soil formation. Chemical weathering is the result of the chemical reactions of water, oxygen, carbon dioxide, and other substances that act upon the rocks, minerals, organic matter, and life that compose the soil. The leaching action of water hastens the weathering process by removing soluble materials, and chemicals react with each other to form new chemicals in the soil.

Weathering

Weathering refers to mechanical forces caused by temperature change such as heating, cooling, freezing, and thawing. As these processes occur, rocks, minerals, organic matter, and other soil-forming materials are broken into smaller and smaller particles until soil is formed. Soils at different stages of weathering will differ widely. Weathering causes soils to develop, mature, and age much as people do.

Soils develop rapidly, mature, and then develop certain characteristics of age. Plant nutrients are released quickly from the minerals, plant growth increases, and organic matter accumulates. Soils age more slowly as they continue to weather.

Eventually, nutrients in the soil are depleted. Water moving through the soil leaches away many soluble portions. At this stage many soils are acidic because the limestone originally in them is gone. As the supply of nutrients in the soil decreases, the amount of plant growth is reduced to the point where the organic matter decomposes faster than it is produced. When soils become acidic and have lost their native fertility, they require expensive amendments to keep them productive.

BIO•TECH CONNECTION

The Anatomy of a Sinkhole

Soil and water scientists go underground to check water quality! Doug Boyer squeezes, winds, crawls, stoops, glides, jumps, splashes, plunges, soaks, wades, and slithers his way through the muddy, narrow passage. Beneath farms of West Virginia's Greenbrier County lies "The Hole." The Hole is part of a six-square-mile hole basin. Here underground water flowing through limestone bedrock has dissolved the limestone and left caverns, sinking streams, and sinkholes. Such sinkholes are found in many parts of the nation.

Sinkholes are funnel-shaped depressions in the Earth's surface formed when soil settled into underground cavities and caves. These can provide perfect collection basins for surface water and pollutants enroute to underground water reserves unless special management practices around each such sinkhole are followed.

The Hole is one of thirty-seven drainage basins selected for study under the United States Clean Water Act. The purpose of these studies is to ensure that surface and underground water supplies are protected through the safe use of fertilizers, manure, and pesticides.

As part as the water quality study, easily accessible water samples are taken at least weekly and cave samples are taken occasionally. Samples are analyzed for nitrates, selected bacteria, herbicides, and other indicators of pollution. Bill Balfour, a volunteer with the West Virginia Association for Cave Studies, has mapped much of The Hole and much of the 400 miles of other caverns mapped in Greenbrier County. He introduced Boyer and other soil and water scientists and technicians to the cave system and continues to push back the frontiers of knowledge in the field.

Water quality experts take water-droplet samples and examine a well casing beneath a sink hole to check for water pollution. A sinkhole is a depression in the surface of the Earth caused by surface soil caving into cavities below left after limestone or other rock materials have dissolved. *(Courtesy USDA/ARS: Agricultural Research, August 1993, p. 7)*

In **permeable** (permitting movement) soils, the fine clay particles tend to move downward from the surface soil into the subsoil during the weathering process. This movement, together with further breakdown of the rock material, accounts for the fact that many soil types have a higher percentage of clay in the subsoil than in the surface soil.

SOIL CLASSIFICATION

Soil scientists have developed a system for mapping soils according to the physical, chemical, and topographical aspects of the land. Such maps have lines showing the outline of soil types and provide numerical codes keyed to large amounts of information about the land. The experienced soil technician can obtain a wealth of information about the land by consulting a soils map and accompanying material.

Land Capability Maps

Soil mapping and land classification have been a priority of the U.S. Soil Conservation Service (SCS) throughout most of the twentieth century. Now the SCS can provide maps and classification information for almost any area in the United States, down to small areas on individual farms. Soil Conservation personnel work with local farmers to develop farm plans. These provide recommendations for land use, cropping systems, crop production practices, and livestock systems. These plans are designed to maximize production while controlling soil erosion and enhancing productivity.

Land capability maps indicate (1) capability class, (2) capability subclass, and (3) capability unit.

Capability Class   **Capability classes** are the broadest groups on a soils map and are designated by Roman numerals I through VIII (Figure 9-7). The numerals indicate progressively greater limitations and narrower choices for practical use of the land as follows.

Class I—soils have few limitations that restrict their use.

Class II—soils have moderate limitations that reduce the choice of plants or require moderate conservation practices.

Class III—soils have severe limitations that reduce the choice of plants, require special conservation practices, or both.

Class IV—soils have very severe limitations that reduce the choice of plants, require very careful management, or both.

Class V—soils are not likely to erode but have other limitations, impractical to remove, that limit their use.

Class VI—soils have severe limitations that make them generally unsuitable for cultivation.

Class VII—soils have very severe limitations that make them unsuitable for cultivation.

Class VIII—soils and landforms have limitations that nearly prevent their use for commercial plants.

Land capability maps are usually color-coded for ease in differentiating capability classes.

Capability Subclasses   **Capability subclasses** are soil groups within one class. They are designated by adding a small letter *e, w, s,* or *c* to the

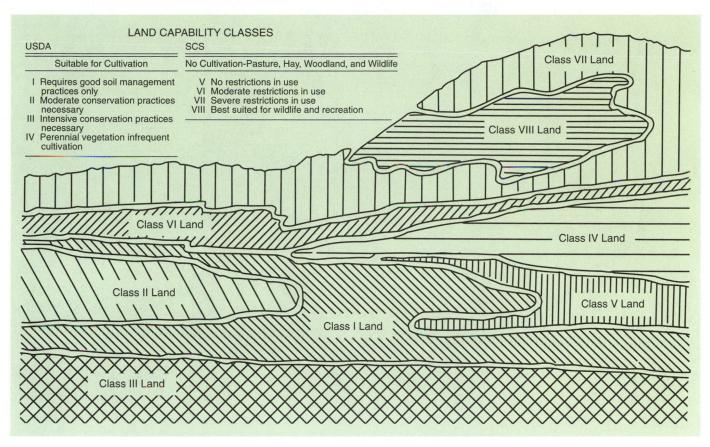

FIGURE 9-7 The eight classes of land in the United States.

class numeral (for example, IIe). The letter *e* indicates the main limitation is risk of erosion unless close-growing plant cover is maintained; *w* indicates weather in or on the soil interferes with plant growth or cultivation. The letter *s* indicates the soil is limited mainly because it is shallow, droughty, or stony; and *c,* used in only some parts of the United States, indicates the chief limitation is climate—too cold or too dry (Figure 9-8).

In Class I there are no subclasses because the soils of this class have few limitations. On the other hand, Class V contains only the subclasses indicated by *w, s,* or *c,* because the soils in Class V are subject to little or no erosion. However, they have other limitations that restrict their use to pasture, range, woodland, wildlife habitat, or recreation.

Capability Units The soils in one **capability unit** (soil groups within the subclasses) are enough alike to be suited to the same crops and pasture plants, and require similar management. They have similar productivity capabilities and other responses to management. The capability unit is a convenient grouping for making many statements about the management of a soil.

Capability units are generally designated by adding numbers (0 to 9) to the subclass symbol (for example, IIIe4 or IIw2). Thus, in one symbol, the Roman numeral designates the capability class or degree of limitation; the small letter indicates the subclass or kind of limitation; and the Arabic numeral specifically identifies the capability unit within each subclass. A map legend for each soil grouping is included with the soil and land capability map. This type of map is also an

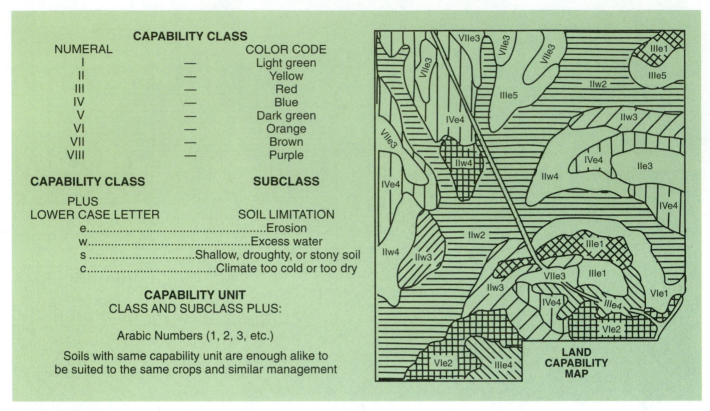

FIGURE 9-8 Land capability map.

example of how symbols can be effectively used. They make it possible to place much information in a small space on the map by use of a code.

Use of Maps

When the landowner and the soil conservationist start planning for the most intensive use of a farm, they need a soil and land capability map. This will help them prepare a conservation plan or proposed land-use map. The soil conservationist and the landowner, through the use of the soil and land capability map, discuss the kinds of soil on the farm. The current and original land uses are also discussed and noted on the map. In developing a proposed land-use map, the soil conservationist must know the personal goals or objectives of the landowner and plans for developing the land.

Many things can be involved in reaching land use decisions, field by field. Field boundaries may need to be changed so all the soil in each field is suited for the same purpose and management practices. The desired balance between cropland, pasture, woodland, and other land uses will need to be considered. Appropriate livestock enterprises should also be considered to match the land's characteristics and potential. If there is a good potential for income-producing recreation enterprises in the community, an area may be used for hunting, camp-sites, or fishing.

The landowner and soil conservationists must consider how to treat each field to get the desired results. The SCS conservationist can give many good suggestions, but the landowner must decide what to do and when and how to do it. As planning decisions are made, the conservationist will record them in narrative form and make them part of the plan map. This becomes the farm conservation plan. It is the guide for the farming operation in the years ahead. A conservation plan is required to obtain federal support under the numerous plans used to encourage good farming practices and price support.

Workers from the Soil Conservation Service are also available to give on-site technical assistance in applying and maintaining the farm conservation plan. They provide management advice and services to nonfarm landowners, developers, strip-mining operations, and other activities where soil is involved (Figure 9-9).

PHYSICAL, CHEMICAL, AND BIOLOGICAL CHARACTERISTICS

Soil Profile

Undisturbed soil will have four or more horizons in its profile. These are designated by the capital letters *O, A, B,* and *C* (Figure 9-10). The O horizon is on the surface and is comprised of organic matter and a small amount of mineral matter. **Organic matter** originates from living sources such as plants, animals, insects, and microbes. **Mineral matter** is derived from nonliving sources such as rock materials.

Horizon A is located near the surface and is mineral matter and organic matter. It contains desirable proportions of organic matter, very fine mineral particles called **clay,** medium-sized mineral particles called **silt,** and larger mineral particles called **sand.** The appropriate proportion of these creates soil that is **tillable,** or workable with tools and equipment. With the presence of desirable plant nutrients, chemicals, and living organisms, the A horizon generally supports good plant growth. The A horizon is frequently called **topsoil.**

The **B horizon** is below the A horizon and is generally referred to as subsoil. The mineral content is similar to horizon A, but the particle sizes and properties will differ. Because organic matter comes from decayed plant and animal materials, the amount naturally decreases as distance from the surface increases.

The **C horizon** is below the B horizon and is composed mostly of parent material. Horizon C is important for storing and releasing water

FIGURE 9-9 Soil Conservation Service personnel must advise on the many uses of land. Here high altitude photography helps document changing conditions. *(Courtesy USDA #K-5218-03)*

Horizon	Name	Colors	Structure	Processes Occurring
O	Organic	Black, dark brown	Loose, crumbly, well broken up	Decomposition
A	Topsoil	Dark brown to yellow	Generally loose, crumbly, well broken up	Zone of leaching
B	Subsoil	Brown, red, yellow, or gray	Generally larger chunks, may be dense or crumbly, can be cement-like	Zone of accumulation
C	Parent material (slightly weathered material)	Variable — depending on parent material	Loose to dense	Weathering, disintegration of parent material or rock

FIGURE 9-10 Characteristics of soil horizons.

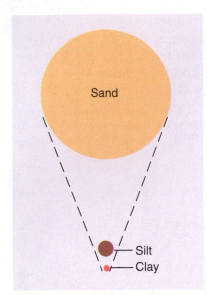

FIGURE 9-11 Relative sizes of sand, silt, and clay particles.

SIZE OF SOIL PARTICLES	
Name	Size, Diameter in Millimeters
Fine gravel	2–1
Coarse sand	1.00–0.50
Medium sand	0.50–0.25
Fine sand	0.25–0.10
Very fine sand	0.10–0.05
Silt	0.05–0.002
Clay	less than 0.002

FIGURE 9-12 Range of sizes of soil particles.

to the upper layers of soil but does not contribute much to plant nutrition. It is likely to contain larger soil particles and may have substantial amounts of gravel and large rocks. The area below horizon C is called **bedrock.**

Texture

Texture refers to the proportion and size of soil particles (Figures 9-11 and 9-12). Texture can be determined very accurately in the laboratory. However, it can also be determined by mechanical analysis or by the feel of the soil as follows (Figure 9-13).

1. Make a stiff mud ball.
2. Rub the mud ball between the thumb and forefinger.
3. Note the degree of coarseness and grittiness due to the sand particles.
4. Squeeze the mud between the fingers and then pull your thumb and fingers apart.
5. Note the degree of stickiness due to the clay particles.
6. Make the soil slightly more moist and note that the clay leaves a "slick" surface on the thumb and fingers.

The outstanding physical characteristics of the important textural grades, as determined by the "feel" of the soil, are as follows (Figure 9-14).

Coarse-textured (sandy) soil is loose and single grained. The individual grains can be seen readily or felt. Squeezed in the hand when dry, it will fall apart when the pressure is released. Squeezed when moist, it will form a cast, but will crumble when touched.

Medium-textured (loamy) soil has a relatively even mixture of sand, silt, and clay. However, the clay content is less than 20 percent. (The characteristic properties of clay are more pronounced than those of sand.) A loam is mellow with a somewhat gritty feel, yet fairly smooth and highly plastic. Squeezed when moist, it will form a cast that can be handled quite freely without breaking.

Fine-textured (clay) soil usually forms very hard lumps or clods when dry. It is usually very sticky when wet and is quite plastic. When the moist soil is pinched between the thumb and fingers, it will form a long, flexible "ribbon." A clay soil leaves a "slick" surface on the thumb and fingers when rubbed together with a long stroke and a firm pressure. The clay tends to hold the thumb and fingers together due to its stickiness.

Structure

Soil **structure** refers to the tendency of soil particles to cluster together and function as soil units called aggregates (Figure 9-15). **Aggregates,** or **crumbs,** contain mostly clay, silt, and sand particles held together by a gel-type substance formed by organic matter.

Aggregates absorb and hold water better than individual particles. They also hold plant nutrients and influence chemical reactions in the soil.

Another major benefit of a well-aggregated soil, or a soil with good structure, is its resistance to damage by falling raindrops. When hit by falling rain, the aggregate stays together as a water-absorbing unit,

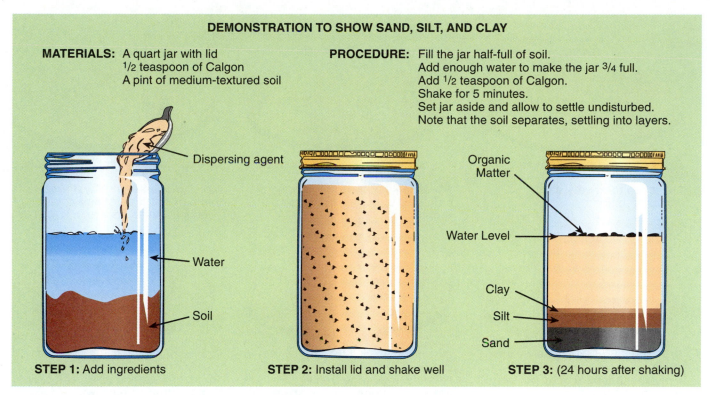

DEMONSTRATION TO SHOW SAND, SILT, AND CLAY

MATERIALS: A quart jar with lid
1/2 teaspoon of Calgon
A pint of medium-textured soil

PROCEDURE: Fill the jar half-full of soil.
Add enough water to make the jar 3/4 full.
Add 1/2 teaspoon of Calgon.
Shake for 5 minutes.
Set jar aside and allow to settle undisturbed.
Note that the soil separates, settling into layers.

Dispersing agent

Water

Soil

STEP 1: Add ingredients

STEP 2: Install lid and shake well

Organic
Matter

Water Level

Clay

Silt

Sand

STEP 3: (24 hours after shaking)

FIGURE 9-13 Mechanical analysis of soil to determine texture.

rather than separating into individual particles. When aggregates on the surface of soil dry out, they remain in a crumbly form and permit good air movement. However, dispersed soil particles run together when dry and form a crust on the soil surface. The crust prevents air exchange between the soil and atmosphere and decreases plant growth. The process and benefits of aggregation are applicable mostly to fine- and medium-textured soils.

Organic Matter

As seen from the previous discussion, organic matter plays an important role in soil structure. Soil is a living medium with a great variety of living organisms. Some groups of organisms of the plant kingdom are:

A. Roots of higher plants.

B. Algae: green, blue-green, diatoms.

C. Fungi: mushroom fungi, yeasts, molds.

D. Actinomycetes of many kinds: aerobic, anaerobic, autotrophic, heterotrophic.

Some examples of groups of organisms from the animal kingdom that are prevalent in soils are:

A. Those that subsist largely on plant material: small mammals, insects, millipedes, sow bugs (wood lice), mites, slugs, snails, earthworms.

B. Those that are largely predators: snakes, moles, insects, mites, centipedes, spiders.

C. Microanimals that are predatory, parasitic, and live on plant tissues: nematodes, protozoans, rotifers.

SOIL TEXTURAL CLASSES

SAND

0-10% Clay
0-15% Silt
85-100% Sand

Dry: loose and single grained; feels gritty.
Moist: will form very easily crumbled ball.

LOAMY SAND

0-15% Clay
0-30% Silt
70-90% Sand

Dry: silt and clay may mask sand; feels loose, gritty.
Moist: feels gritty; forms easily crumbled ball; stains fingers slightly.

SANDY LOAM

0-20% Clay
0-50% Silt
43-85% Sand

Dry: clods easily broken; sand can be seen and felt.
Moist: moderately gritty; forms ball that can stand careful handling; definitely stains fingers.

LOAM

7-27% Clay
23-52% Sand
28-50% Silt

Dry: clods moderately difficult to break; mellow, somewhat gritty.
Moist: neither very gritty nor very smooth; forms a firm ball; stains fingers.

SILT LOAM

0-27% Clay
0-50% Sand
50-88% Silt

Dry: clods difficult to break; when pulverized feels smooth, soft, and floury; shows fingerprints.
Moist: has smooth or slick "buttery" or "velvety" feel; stains fingers.

CLAY LOAM

27-40% Clay
15-53% Silt
20-45% Sand

Dry: clods very difficult to break with fingers.
Moist: has slightly gritty feel; stains fingers; ribbons fairly well.

SILTY CLAY LOAM

27-40% Clay
0-20% Sand
40-73% Silt

same as **CLAY LOAM** but very smooth.

SANDY CLAY LOAM

20-35% Clay
0-28% Silt
45-80% Sand

same as **CLAY LOAM.**

CLAY

0-45% Sand
0-40% Silt
40-100% Clay

Dry: clods cannot be broken with fingers without extreme pressure.
Moist: quite plastic and usually sticky when wet; stains fingers. (A silty clay feels smooth, a sandy clay feels gritty.)

FIGURE 9-14 Major soil textural classes.

FIGURE 9-15 Good soil structure is very important in medium and heavy soils. *(Courtesy National FFA)*

Living Organisms

Living organisms excrete cell or body wastes that become part of the organic content of soil. Further, the microbes of the soil and the remains of larger plants and animals decompose or decay into soil-building materials and nutrients.

People who grow indoor plants at home, garden, farm, produce greenhouse crops, or grow nursery stock generally find it useful to add organic materials to the soil. Popular sources of organic matter for soil amendments are peat moss, leaf mold, compost, livestock manure, and sawdust.

Some important benefits or functions of organic matter in soil are:

1. makes the soil porous.
2. supplies nitrogen and other nutrients to the growing plant.
3. holds water for future plant use.
4. aids in managing soil moisture content.
5. furnishes food for soil organisms.
6. serves as a store house for nutrients.
7. minimizes leaching.
8. serves as a source of nitrogen and growth-promoting substance.
9. stabilizes soil structure.

Other Properties of Soils

Soil scientists, managers, technicians, and operators must be aware of numerous other properties of soils in their work. Some of these are external factors such as land position, slope, and stoniness. Further, soil color, depth, drainage, permeability, and erosion are important considerations.

MAKING AMENDMENTS TO PLANT-GROWING MEDIA

The term **amendment** is used here to mean addition to or change in. Most soil amendments are made to add organic matter, add specific nutrients, or modify soil pH. The **pH** is a measure of the degree of acidity or alkalinity. **Acidity** is sometimes referred to as sourness and **alkalinity** is referred to as sweetness. The pH scale ranges from 0 (maximum acidity) to 14 (maximum alkalinity). The midpoint of the scale is 7, which is **neutral,** meaning neither acid nor alkaline (Figure 9-16).

Crops grow best in media with a narrow pH range unique to that plant species (Figure 9-17). Most plants require a pH somewhere between 5.0 and 7.5. Some crops, such as potatoes and blueberries, prefer a soil pH around 5.5. Alfalfa requires a pH of 6.5 to 7.0.

Liming

Areas that were historically covered by trees, such as the northeastern, western, and northwestern parts of the United States, develop acidic soils. When cleared of trees, such soils require additional lime to raise the pH for the efficient production of most farm crops.

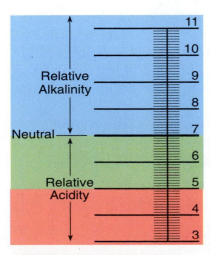

FIGURE 9-16 The pH scale is neutral in the middle. It gets progressively more acidic from 7 down to 3 and progressively more alkaline from 7 up to 11.

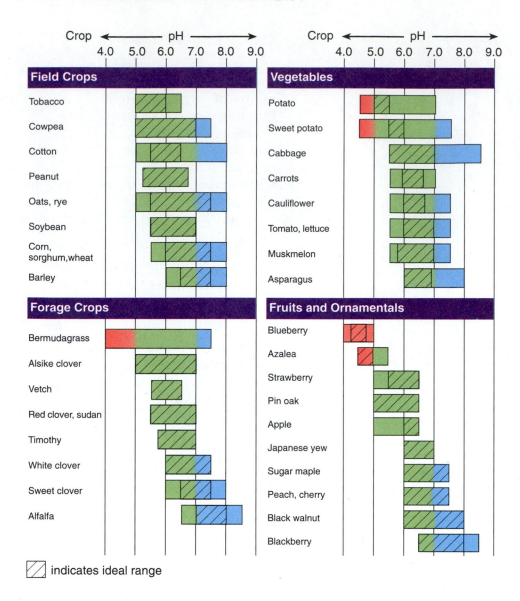

indicates ideal range

FIGURE 9-17 Plants grow in pH ranges from approximately 4 (very acid) to 8.5 (very alkaline).

A pH test can be performed using a pH test kit on-site, or soil samples can be sent to a university or commercial laboratory for analysis. Laboratory analyses generally include an analysis of phosphorus, potash, and magnesium, as well as pH. Liming and fertilizing recommendations may also be provided by testing laboratories and universities (Figure 9-18).

How to Take a Soil Sample in Your Lawn or Garden

1. Select an appropriate sampling tool (spade, auger, or soil tube).
2. Make a sketch dividing the area into sampling areas—for example, front lawn, garden, flower bed, slope, back lawn. Appropriately label each area (see Figure 9-18, view #1).
3. When taking samples, avoid wet or bare spots. Soils that are substantially different in plant growth or past treatment should be sampled separately, provided their size and nature make it feasible to fertilize or lime each area separately.

View 1   *Random Samples

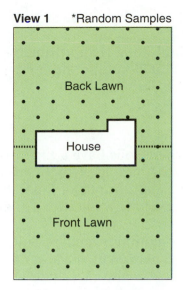

View 2   *Random Samples

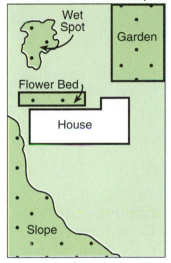

View 3

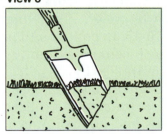

View 4

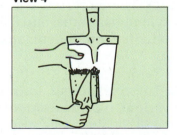

When using a spade, first make a V-shaped cut in the ground. Then, remove a 1-inch slice from one side of the cut (View 3). Take a 1-inch strip from the middle of the slice for the soil sample (View 4).

4. After removing surface litter, take a sample from the correct depth. This is 2 in. for established lawns, and about 6 in. for gardens, flower beds, farm crop land, and other areas to be tilled.

5. Submit a separate composite sample for each significantly different area—for example, front lawn, back lawn, flower bed. Your composite sample for each area should include a small amount of soil taken from each of ten to twenty randomly selected locations in the area represented by each sample.

6. When using a spade, first make a "V" shaped cut. Then remove a 1-in. slice from one side of the cut (see Figure 9-18, view #3). Now take a 1-in. strip from the middle of this slice (see Figure 9-18, view 4). This represents the soil from one spot in the sample.

7. **Air-dry the soil; do not use heat.** Mix the soil from a composite in a clean bucket. Place about 1 pint of this mixture into the sample box. Use a separate box for each composite. Fill in the blanks on the box or information sheet for each box.

8. Send soil sample(s) and information sheet(s) to the soil test laboratory.

Note: The Cooperative Extension Service in most counties and/or cities can arrange for soil testing.

The pH Test   The amount of agricultural lime required to raise soil to the desired pH level is indicated by a pH test. The same pH value may require different amounts of agricultural lime. This is due to the fact that soils contain varying amounts of organic matter, clay, silt, and sand. The greater the organic matter and clay content, the greater the amount of lime required to correct the acidity.

Even though all the lime required by your soil is applied, do not expect the pH to move up quickly. It will generally require two to three years for all of the lime to be used and the desired pH to be reached.

Mix Well   The soil on one acre 6 in. deep weighs about 2 million pounds. Therefore, it takes a lot of mixing to distribute a relatively small amount of lime with the soil. Liming recommendations are based on a specific plowing depth, such as 6 or 7 in. Application rates may need to be adjusted for deeper or more shallow tilling.

Standard Ground Limestone   Lime recommendations are based on standard ground limestone, which should contain a minimum of 50 percent lime oxides (calcium oxide plus magnesium oxide). About 98 percent should pass through a 20-mesh sieve, with a minimum of 40 percent passing through a 100-mesh sieve. The higher the percentage of limestone passing through a 100-mesh screen, the faster this limestone will correct soil acidity. The best limestone will have the greatest calcium and/or magnesium content and will be ground to a small particle size. It is more important to finely grind a high-magnesium stone than a high-calcium stone. A high-magnesium or dolomitic limestone

FIGURE 9-18 Procedure for taking a soil sample to conduct tests for pH and nutrient levels in soil. Such tests are necessary for matching lime and fertilizer applications to plant needs for optimum yields.

should always be used when a magnesium deficiency is indicated by a soil test.

One ton [2,000 lb] of standard ground limestone is approximately equivalent to 1,500 lb of hydrated lime or 1,100 lb of ground burned limestone.

Correcting Excessive Alkalinity

A reduction of soil alkalinity is desirable where soils have a high pH and the alkaline condition causes unsatisfactory crop growth. In most cases, this condition will exist where heavy applications of lime are made at one time or where lime has been applied to a soil with a high pH.

To lower the pH value of soil, use sulfur or aluminum sulfate. Sulfur at 1½ lb per 100 ft² or aluminum sulfate at 5 lb per 100 ft² will lower the alkalinity, under most conditions, by ½ pH. Aluminum sulfate acts rapidly, producing acidification in 10 to 14 days. Sulfur requires 3 to 6 months. Broadcast the material over the surface and thoroughly work it into the soil. For full benefit, sulfur should be applied in the fall, after garden crops are harvested. Aluminum sulfate may be applied in early spring.

Use chemicals cautiously to lower alkalinity. Before these measures are taken, you should consider other factors that may be responsible for poor growth (such as drought, insect and disease injury, and fertilizer burning).

Fertilizers and Fertilizing

Essential plant nutrients are discussed in Unit 16, "Plant Physiology." However, it should be noted that nitrogen, phosphorus, and potassium are known as the three **primary nutrients.** These three ingredients must be present for a fertilizer to be called a **complete fertilizer.**

The proportions of nitrogen, phosphorus, and potassium are known as the **fertilizer grade,** expressed on a fertilizer container as percentages of the contents of the container by weight. Therefore, a 100-lb container of fertilizer with a grade of 10-10-10 contains 10 percent nitrogen, 10 percent phosphorus, and 10 percent potassium.

If the total amount of fertilizer (100 lb) is multiplied by the percentage of each ingredient, the pounds of each ingredient may be calculated. Therefore, 0.10 (percent of nitrogen) x 100 (total lb of fertilizer) = 10 lb of nitrogen. The amount of phosphorus and potassium is also 10 lb each.

Fertilizer is frequently shipped in 80-lb bags. Therefore, one bag of 10-10-10 fertilizer would have 8 lb nitrogen, 8 lb phosphorus, and 8 lb potassium for a total of 24 lb of **active ingredients** (components that achieve one or more purposes of the mixture).

Some popular grades of fertilizer are 5-10-5, 5-10-10, 10-10-10, 6-10-4, 0-15-30, 0-20-20, 8-16-8, and 8-24-8. These grades are formulated to meet the needs of various crops on various soils. The amount and grade of fertilizer to apply are determined by (1) the specific crop to be grown, (2) the desired yield or performance of the crop, (3) fertility of the soil, (4) physical properties of the soil, (5) previous crop, and (6) type and amount of manure applied. Therefore, decisions on rate of application must be made on a local basis. Soil tests, tissue tests, and

HEALTHY leaves shine with a rich, dark green color when adequately fed.

PHOSPHOROUS (phosphate) shortage marks leaves with red-dish-purple, particularly on young plants.

POTASSIUM (potash) deficiency appears as a firing or drying along the tips and edges of lowest leaves.

NITROGEN hunger sign is yellowing that starts at tip and moves along middle of leaf.

MAGNESIUM deficiency causes whitish stripes along the veins and often a purplish color on the underside of the lower leaves.

DROUGHT causes corn plants to have a gray-ish-green color: leaves may roll up to the size of a pencil.

DISEASE, helminthosporium blight, starts in small spots, gradually spreads across leaf.

CHEMICALS may sometimes burn tips, edges of leaves and at other contacts. Tissue dies, leaf becomes whitecap.

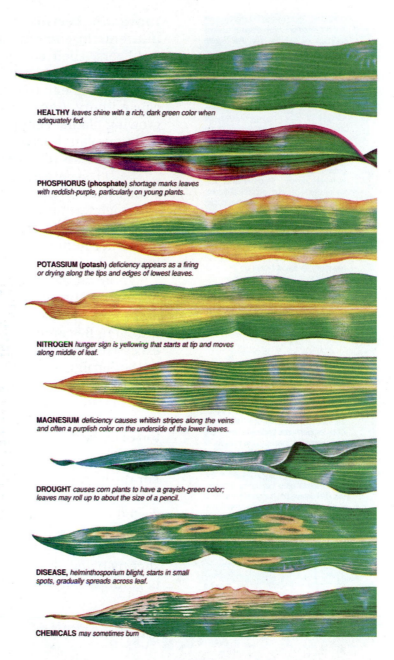

HEALTHY *leaves shine with a rich, dark green color when adequately fed.*

PHOSPHORUS (phosphate) *shortage marks leaves with reddish-purple, particularly on young plants.*

POTASSIUM (potash) *deficiency appears as a firing or drying along the tips and edges of lowest leaves.*

NITROGEN *hunger sign is yellowing that starts at tip and moves along middle of leaf.*

MAGNESIUM *deficiency causes whitish stripes along the veins and often a purplish color on the underside of the lower leaves.*

DROUGHT *causes corn plants to have a grayish-green color; leaves may roll up to about the size of a pencil.*

DISEASE, *helminthosporium blight, starts in small spots, gradually spreads across leaf.*

CHEMICALS *may sometimes burn*

FIGURE 9-19 Corn plants are good indicators of nutrient deficiencies and other factors that impact plant health and productivity. *(Courtesy Potash and Phosphate Institute, Norcross, GA)*

plant observations are useful techniques for determining fertility needs (Figure 9-19).

Organic Fertilizers   Organic fertilizers include animal manures and compost made with plant or animal products. Organic commercial fertilizers include dried and pulverized manures, bone meal, slaughterhouse tankage, blood meal, dried and ground sewage sludge, cottonseed meal, and soybean meal.

Organic fertilizers have certain definite characteristics: (1) Nitrogen is usually the predominating nutrient, with lesser quantities of phosphorus and potassium. One exception to this is bone meal, in which phosphorus predominates, and nitrogen is a minor ingredient. (2) The nutrients are only made available to plants as the material decays in the soil, so they are slow acting and long lasting. (3) Organic materials alone are not balanced sources of plant nutrients, and their analysis in terms of the three major nutrients is generally low.

Inorganic Fertilizers   Various mineral salts, which contain plant nutrients in combination with other elements, are called inorganic fertilizers. Their characteristics are different from organic fertilizers. (1) The nutrients are in soluble form and are quickly available to plants. (2) The soluble nutrients make them caustic to growing plants and can cause injury. Care must be used in applying inorganic fertilizers to growing plants. They should not come in contact with the roots or remain on plant foliage for any length of time. (3) The analysis of chemical fertilizers is relatively high in terms of the nutrients they contain.

Fertilizers are needed to replenish mineral nutrients depleted from a soil by crop removal or by such natural means as leaching. Some soils with high fertility may need only nitrogen or manure. Use of fertilizers that also contain small amounts of copper, zinc, manganese, boron, and other minor elements is not considered necessary for most soils, but may be needed in certain soils and for certain crops.

There are also unmixed fertilizers that carry only one element (Figure 9-20). Most important of these unmixed materials are the nitrogen and phosphate carriers. Nitrogen carriers vary from 16 to 45 percent nitrogen. Be careful in using nitrogen materials. Too much may cause excessive and soft growth.

Superphosphate fertilizers carry only phosphorus. Phosphorus promotes flower, fruit, and seed development. It also firms up stem growth and stimulates root growth. Superphosphate may be added to manure to give a better balance of nutrients for plants (100 lb to each ton of horse or cow manure and 100 lb to each half-ton of sheep manure).

Fertilizer Applications   There are many ways to apply fertilizers. For the home lawn, the most likely method is **broadcasting** (spreading evenly over the entire surface). In the case of gardens or fields, broadcasted fertilizer may be **incorporated** or mixed into the soil by spading, tilling, plowing, or disking.

Band application places fertilizers about 2 in. to one side of and slightly below the seed. This method is used extensively for row crops in gardens and fields.

	PLANT NUTRIENTS IN FERTILIZER MATERIALS					
Materials	Nitrogen N%	Phosphorus P₂O₅%	Potassium K₂O%	Calcium %	Magnesium %	Sulphur %
Ammonium Nitrate	33.5	—	—	—	—	—
Ammonium Sulphate	21	—	—	—	0.3	23.7
Urea	45	—	—	1.5	0.7	.02
Sodium Nitrate	16	—	0.2	0.1	.05	.07
Calcium Nitrate	15	—	—	19.4	1.5	.02
Calcium Cyanamide	21	—	—	38.5	.06	0.3
Anhydrous Ammonia	82	—	—	—	—	—
Superphosphate	—	20	0.2	20.4	0.2	11.9
Liquid Phosphoric Acid	—	58	—	—	—	—
Ammonium Phosphate	11	48	0.2	1.1	0.3	2.2
Potassium Chloride	—	—	62	—	0.1	0.1
Potassium Sulphate	—	—	53	0.5	0.7	17.6

FIGURE 9-20 Plant nutrients in fertilizer materials.

Side-dressing is done by placing fertilizer in bands about 8 in. from the row of growing plants. This method is popular for row crops such as corn and soybeans.

Top-dressing is a procedure where fertilizer is broadcast lightly over close-growing plants. Top-dressing is used for adding nitrogen to small grain, hay, and turf crops after they are established.

Starter solutions are diluted mixtures of fertilizer used when plants are transplanted. Their purpose is to provide small amounts of nutrients that will not burn the tender roots of young plants.

Other methods of applying fertilizers include the application of **foliar sprays** directly onto the leaves of plants, and **knife application** of anhydrous ammonia gas into the soil. The practice of adding liquid fertilizer to irrigation water is also used extensively in the United States.

Using Manure

Animal manure is a valuable product when handled properly. Its content of plant nutrients makes it a valuable fertilizer material. Additionally, the organic matter aids in developing and maintaining structure in soils (Figure 9-21).

To obtain the most nutrient value from manures, these practices should be followed.

1. Use adequate bedding to absorb all of the liquids.
2. Balance the phosphorus in cow manure by adding superphosphate at the rate of 2 lb per cow per day. Apply in gutter or before bedding in sheds or loafing pens.
3. Spread manure evenly over fields. An 8- to 12-ton application per acre is recommended. (Fifty bushels of manure and litter weigh about one ton.)
4. When possible, incorporate manure into the soil immediately after spreading.
5. Do not spread on steep slopes when the ground is frozen.
6. When storing manure, keep it compact and under cover.
7. Prevent liquid run-off from manure holding areas into water shed areas.
8. Apply to crops that will give best response, such as corn, sorghums, potatoes, and tobacco.

When applying manure, the amount of commercial fertilizer should be reduced accordingly (Figure 9-22).

FIGURE 9-21 Animal manure is valuable for plant feeding as well as soil building.

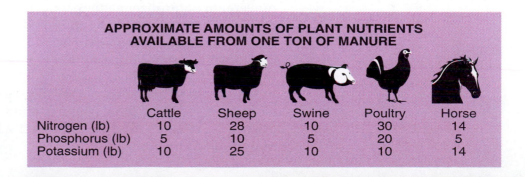

APPROXIMATE AMOUNTS OF PLANT NUTRIENTS AVAILABLE FROM ONE TON OF MANURE

	Cattle	Sheep	Swine	Poultry	Horse
Nitrogen (lb)	10	28	10	30	14
Phosphorus (lb)	5	10	5	20	5
Potassium (lb)	10	25	10	10	14

	N NITROGEN	P PHOSPHORUS	K POTASSIUM
Suppose the fertilizer recommendations <u>per acre</u> for a certain crop are:	150 lb	100 lb	100 lb
If well-managed cow manure is applied at the rate of 10 tons per acre, it can be determined from Figure 9-21 that the 10-ton application would provide:	100 lb	50 lb	100 lb
Therefore, the amount of nutrients that must be provided per acre through commercial fertilizer is:	50 lb	50 lb	0 lb

The shortfall of ingredients listed above could be made up by applying 500 lb. of 10-10-0 commercial fertilizer per acre.

NOTE: Caution must be exercised when estimating the nutrient values of manure due to variations in liquid and solid content captured from the animal, amount and type of bedding, and procedures used to handle and store the manure.

FIGURE 9-22 When manure is applied to land, the amount of commercial fertilizer used should be adjusted to allow for the nutrients in the manure.

Legume Crops

Legumes are plants in which certain bacteria utilize nitrogen gas from the air and convert it to **nitrates** (the form of nitrogen used by plants). Some examples of legumes are beans, peas, clovers, and alfalfa. The process of converting nitrogen gas to nitrates by bacteria in the roots of legumes is called **nitrogen fixation.**

Nitrogen fixation reduces or eliminates the need for adding expensive nitrogen fertilizer to legume crops. Further, when the roots of legume plants decay, large amounts of nitrates are left behind for the next crop. Hence corn, which requires large amounts of nitrogen, should follow alfalfa or clover in a field, because the legumes leave unused nitrogen behind.

Rotation Fertilization

Many crop rotations start with a small-grain crop. The preparation of the seedbed for small grains provides an excellent opportunity to put lime and fertilizer into the feeding zone for roots of the sod crop that normally follows. Lime and phosphorus move very slowly in the soil, and unless the roots contact these elements, they cannot provide the seedling's high requirements for phosphorus in the critical first year of growth.

A properly limed and fertilized sod crop is the backbone of a successful crop rotation. The growth of nutrient-enriched grass and legume roots throughout the soil provides a most favorable medium for natural soil-building processes. The addition of organic matter, the movement of fertilizer elements by the roots into the soil, and the production of root channels by sod roots produce soils that absorb water better, erode less, are easier to work, and grow better crops.

Phosphorus and potassium added to sod will benefit the sod and, in addition, will be placed in the best location and be in the best form to supply these elements to the long-season row crops which follow. Nitrogen produced by the legumes or added to grass will be present in organic form and will be released in the best possible way for the row and small-grain crops.

The composition of soils is complex and depends on many factors. Soil is dynamic and changing all the time. Nature has provided many cycles to help provide for soil renewal. However, the scientific management of soils makes them more productive and helps to ensure productive soils for efficient production for future generations.

HYDROPONICS

The term hydroponics refers to a number of types of systems used for growing plants without soil (Figure 9-23). Some major systems are:

1. **Aggregate culture,** in which a material such as sand, gravel, or marbles supports the plant roots.

2. **Water culture, solution culture,** or **nutriculture,** in which the plant roots are immersed in water containing dissolved nutrients.

3. **Aeroponics,** in which the plant roots hang in the air and are misted regularly with a nutrient solution.

4. **Continuous-flow systems,** in which the nutrient solution flows constantly over the plant roots. This system is the one most commonly used for commercial production.

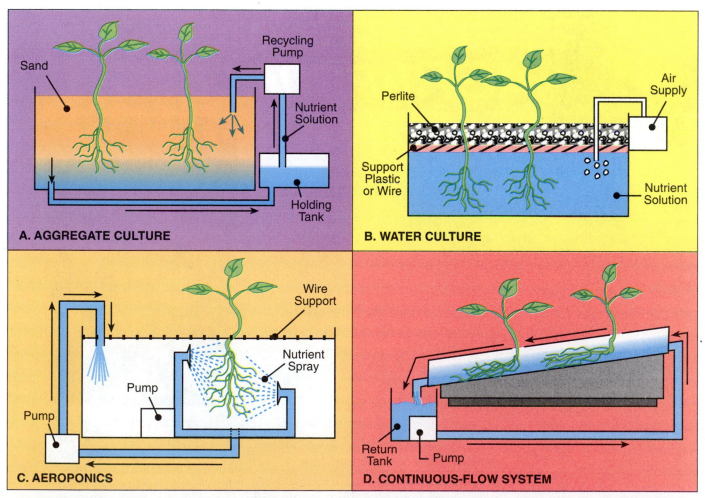

FIGURE 9-23 There are many types of hydroponic systems including (A) Aggregate culture; (B) Water culture; (C) Aeroponics; and (D) Continuous-flow systems. *(Adapted from material by Keith Staley, Middletown High School, Middletown, MD)*

Hydroponics is growing in importance as a means of producing vegetables and other high-income plants. In areas where soil is lacking or unsuitable for growth, hydroponics offers an alternative production system. Equally good crops can generally be produced in a greenhouse in conventional soil or bench systems.

When plants are grown hydroponically, their roots are either immersed in or coated with a carefully controlled nutrient solution. The nutrients and water are supplied by the solution alone and not by aggregates or other inert materials that support the roots. **Inert** means inactive. Without the presence of soil to absorb and release nutrients, nutrients must be carefully controlled in a hydroponic system.

Plant Growth Requirements

Hydroponically grown plants have the same general requirements for good growth as soil-grown plants. The major difference is the method by which the plants are supported and the nutrients are supplied for growth and development.

Water   Providing plants with an adequate amount of water is not difficult in a water culture system. During the hot summer months a large tomato plant may use one-half gallon of water per day. However, quality can be a problem. Water with excessive alkalinity or salt content can

AGRI·PROFILE

CAREER AREAS: Agronomy/Hydroponics

Electronic devices have become tools of the trade for soil and water research and management. Career options in soil and hydroponics management overlap those in soil and water conservation to a certain extent, and additional opportunities occur in management such as producers on farms and in hydroponics greenhouses. The practice of hydroponics is not new, but hydroponics for commercial production has captured the imagination of the world. The recent popularity of hydroponics operations provides new career opportunities, especially in urban areas.

Soil and water management specialists are in demand on the global scene. Progressive Third World countries need help in policy development, education, and project management. They hope to leap from their primitive agriculture of the past to the agriscience of the present in a few short years.

Soil scientist Ronald Schnabel (left) and hydrologic technician Earl Jacoby study natural riparian zone processes that lessen the impact of upstream agriculture on water quality. *(Courtesy USDA/ARS #K-5050-3)*

result in a nutrient imbalance and poor plant growth. Softened water may contain harmful amounts of sodium. Water that tests above 320 parts per million of salts is likely to cause an imbalance of nutrients.

Oxygen　Plants require oxygen for respiration to carry out their functions. Under field and normal greenhouse conditions, oxygen is usually adequate as provided by the soil. When plant roots grow in water, however, the supply of dissolved oxygen is soon depleted, and damage or death soon follow unless supplemental oxygen is provided. Where supplemental oxygen is needed, a common method of supplying oxygen is to bubble air through the water. It is not usually necessary to provide supplementary oxygen in aeroponic or continuous-flow systems.

Mineral Nutrients　Green plants must absorb certain minerals through their roots to survive. These minerals are supplied by soil, organic matter, or soil solutions. The elements needed in large quantities are nitrogen, phosphorus, potassium, calcium, magnesium, and sulfur. The nutrients needed in tiny amounts are iron, manganese, boron, zinc, copper, molybdenum, and chlorine. An over-supply of any nutrient is toxic or detrimental to plants. In hydroponic systems, all nutrients normally supplied by soil must be included in the water to form the solution or media.

Light　All vegetable plants and many flowers require large amounts of sunlight. Hydroponically grown vegetables, like those grown in a garden, need at least 8 to 10 hours of direct sunlight each day to produce well. Electrical lighting is a poor substitute for sunshine because most indoor lights do not provide enough intensity to produce a crop. Incandescent lamps supplemented with sunshine or special plant-growth lamps can be used to grow transplants but are not adequate to grow the crop to maturity. High-intensity lamps, such as high-pressure sodium lamps, can provide more than 1,000 foot-candles of light. They may be used successfully in small areas where sunlight is inadequate. However, these lights are too expensive for most commercial operations.

Spacing　Adequate spacing between plants will permit each plant to receive sufficient light in the greenhouse. Tomato plants, pruned to a single stem, should be allowed 4 ft^2 per plant. European seedless cucumbers should be allowed 7 to 9 ft^2, and seeded cucumbers need about 7 ft^2. Leaf lettuce plants should be spaced 7 to 9 in. apart within the row and 9 in. between rows. Most other vegetables and flowers should be grown at the same spacing as recommended for a garden.

Greenhouse vegetables will not do as well during the winter as in the summer. Shorter days and cloudy weather reduce the light intensity and thus limit production.

Temperature　Plants grow well only within a limited temperature range. Temperatures that are too high or too low will result in abnormal development and reduced production. Warm-season vegetables and most flowers grow best between 60° and 80°F. Cool-season vegetables, such as lettuce and spinach, should be grown between 50° and 70°F.

Support　In the garden or field, plants are supported by roots anchored in soil. A hydroponically grown plant must be artificially supported with string, stakes, or other means.

Hydroponics in the Classroom and Laboratory

Hydroponics has become an important teaching area in Agriscience programs. The use of common plastic bottles, photographic film cans, and low-cost aquarium supplies permit students and teachers to set up and perform numerous research and demonstration projects in the classroom or laboratory. The material that follows was provided by Dr. David R. Hershey, Assistant Professor in Horticulture at the University of Maryland.

A solution culture system (Figure 9-24) may be constructed as follows:

1. Fill two, dark 2-liter plastic soda bottles with hot water to loosen the glued label and base. Remove the labels from both bottles and the base from one bottle—keep the base for future use.

2. Using a fine-point felt-tip marker, draw a line around the bottle 9 in. up from the bottom.

3. Using short-bladed scissors, cut on the line made in Step 2 and remove the upper portion of the bottle. This will be the reservoir.

4. Using a cork-borer, drill, or scissors, make a hole in the center of the bottle base approximately ½ in. in diameter to accommodate a plant stem. Close all except one of the existing (pre-punched) holes in the base with black vinyl electrician's tape to prevent light passage. The base is then placed on the reservoir as a dome-shaped lid.

5. Insert about 8 in. of a 2-foot length of aquarium tubing into the open pre-punched hole. To keep the tubing in the reservoir rigid, place a plastic drinking straw over the tubing. Attach the free end of the aquarium tubing to an aquarium air pump with an air control valve.

Nutrient Solutions   There are many nutrient-solution recipes, but Hoagland Solution No. 1 is used widely and can be modified to create

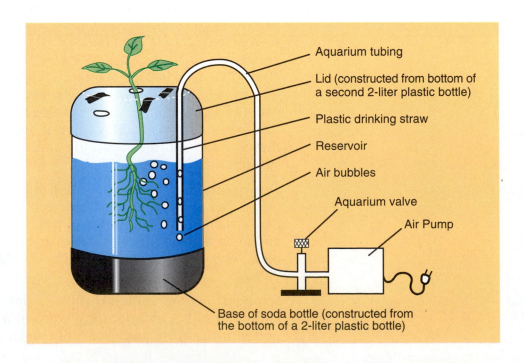

FIGURE 9-24 A static solution culture system built from two plastic soda bottles. *[Adapted from material by Dr. David R. Hershey, University of Maryland]*

Aquarium tubing

Lid (constructed from bottom of a second 2-liter plastic bottle)

Plastic drinking straw

Reservoir

Air bubbles

Aquarium valve

Air Pump

Base of soda bottle (constructed from the bottom of a 2-liter plastic bottle)

Stock Solution			milliliters of stock solution per liter of nutrient solution[x]								
Formula	Name	grams /liter	Complete	–N	–P	–K	–Ca	–Mg	–S	–Fe
$Ca(NO_3)_2 \cdot 4H_2O$	Calcium nitrate, 4-hydrate	236	5	–	4	5	–	4	4	5
KNO_3	Potassium nitrate	101	5	–	6	–	5	6	6	5
KH_2PO_4	Monopotassium phosphate	136	1	–	–	–	1	1	1	1
$MgSO_4 \cdot 7H_2O$	Magnesium sulfate, 7-hydrate	246	2	2	2	2	2	–	–	2
FeNa EDTA[z]	Iron EDTA	18.4	1	1	1	1	1	1	1	–
Micronutrients[y]		–	1	1	1	1	1	1	1	1
K_2SO_4	Potassium sulfate	87	–	5	–	–	–	3	–	–
$CaCl_2 \cdot 2H_2O$	Calcium chloride, dihydrate	147	–	2	–	–	–	–	–	–
$Ca(H_2PO_4)_2 \cdot H_2O$	Calcium phosphate, monobasic	12.6	–	10	–	10	–	–	–	–
$Mg(NO_3)_2 \cdot 6H_2O$	Magnesium nitrate, 6-hydrate	256	–	–	–	–	–	–	2	–

[z]Ferric-sodium salt of ethylene-diamine tetraacetic acid. Differs from original Hoagland recipe, which used Fe tartrate.

[y]Contains the following in grams/liter:

2.86 H_3BO_3, (boric acid) 0.08 $CuSO_4 \cdot 5H_2O$, (copper sulfate, 5-hydrate)

1.81 $MnCl_2 \cdot 4H_2O$, (manganese chloride, 4-hydrate) 0.02 $H_2MoO_4 \cdot H_2O$, (85% molybdic acid).

0.22 $ZnSO_4 \cdot 7H_2O$, (zinc sulfate, 7-hydrate)

[x]For Mn-, B-, Cu-, Zn-, and Mo-deficient solutions, substitute micronutrient stock solutions missing one of the five salts in the regular micronutrient stock solution. For Cl-deficient micronutrient stock solution, substitute 1.55 $MnSO_4 \cdot H_2O$ (manganese sulfate, monohydrate) for 1.81 $MnCl_2 \cdot 2H_2O$.

FIGURE 9-25 Preparation of modified Hoagland nutrient solutions for nutrient deficiency system development. *(Courtesy D. R. Hoagland and D. I. Arnon, "The Water-Culture Method for Growing Plants Without Soil," California Agricultural Experiment Station Circular 347, revised 1950)*

nutrient deficiencies (Figure 9-25). Often Hoagland solution is used at less than full strength. Nutrient stock solutions are prepared using a balance to weigh out the salts and a graduated cylinder or volumetric flask to measure the water. Plastic soda bottles can be used to store stock solutions. When stored in a dark place at room temperature, stock solutions should last for many years. To prepare the nutrient solution, add measured volumes of stock solutions to a measured volume of water. Stock solutions can be dispensed using pipets or graduated cylinders. Complete hydroponic salt mixtures can be purchased and greatly simplify nutrient solution preparation.

Aeration Solution cultures are typically aerated using an aquarium air pump and aquarium air tubing. In a soda bottle system, the aquarium air tubing is inserted into the reservoir through one of the prepunched holes, and the tubing in the reservoir is made rigid by slipping a plastic soda straw onto the tubing. To prevent clogging, a piece of cotton or aquarium filter floss is placed in the aquarium tubing as it exits the pump. The filter is necessary to either trap dirt from the air or pieces of the pump diaphragm. Adjust the aeration to a gentle rate of one to three bubbles per second. Some plants do not benefit from aeration but most do. An alternative to pump aeration is to let the top third of the root system remain uncovered by solution in the humid air in the top of the reservoir.

Maintenance Water must be added to static solution cultures every few days or so to replace water lost by transpiration and evaporation.

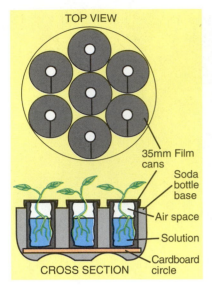

FIGURE 9-26 Film cans with a ¹/₄-inch hole punched in the center of each lid and a slit cut from the edge of each lid to the center hole make excellent static culture containers. Numerous film cans may be contained in the base taken from a two-liter soda bottle. *(Adapted from material by Dr. David R. Hershey, University of Maryland)*

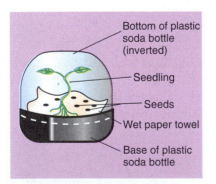

FIGURE 9-27 A germination chamber may be constructed with the bottom cut from a two-liter soda bottle inverted over a soda bottle base. The dome section is lined with a wet paper towel, and seeds are pushed against the towel to absorb moisture and germinate prior to transplanting. *(Adapted from material by Dr. David R. Hershey, University of Maryland)*

Nutrient solutions are typically replaced with fresh solutions on a schedule, such as every one or two weeks. Frequency of solution replenishment depends on the rate of plant growth relative to the volume of nutrient solution. An electrical conductivity (EC) meter is very useful for determining when a nutrient solution has been depleted. To prevent interruption of growth, replace the solution when the solution EC is about half the original value.

Plants   Bean, corn, sunflower, and tomato are often used for teaching hydroponics because they grow rapidly from seed; but they are often difficult to handle because of their large size, high light requirements, and need for staking. Wisconsin Fast Plants, which are becoming a standard plant for teaching use, are excellent for solution culture in 35-mm film cans (Figure 9-26). Fast plants go from seed to flower in 2 weeks under a bank of six, 4-foot, 40-watt fluorescent lamps and remain under a foot in height.

Houseplants, such as piggyback (*Tolmiea menziesii*), wandering jew (*Tradescantia* species), evergreen euonymus (*Euonymus japonica*), coleus, common philodendron, and pothos (*Epipremnum aureum*), also are excellent in teaching hydroponics. They thrive with low light levels, are readily available, and root quickly in solution. Radish and lettuce are easily grown from seed. A carrot placed in tap water in the 2-liter soda bottle system described earlier will sprout lateral roots, produce a shoot, and flower. Pineapple fruit tops easily root in solution.

Germination   Seeds for hydroponics are conveniently germinated in containers of perlite, which is inert and easily removed from roots prior to transferring the plants to solution. Small seeds, such as fast plants or coleus, can be germinated on a paper towel in a seed germinator built from a 2-liter soda bottle. The base is removed from a bottle and a four-inch bottom section of the bottle is inverted over the base to complete the germinator. A wet paper towel is used to line the domed section, and small seeds are pressed into the towel. The seeds remain stuck until the seed germinates and the seedling is transplanted (Figure 9-27).

Individual and Class Projects

The potential of hydroponics in the classroom appears very great. Many solution culture research papers could probably be adapted for teaching use. Students have the ability to do new solution culture experiments at little cost by using plants that have never been grown in solution culture before, such as most house plants and bedding plants. Experiments could be done by students in the areas of nutrient deficiencies, toxicities, carbon dioxide and oxygen deficiencies, pH, fertilizer testing, growth regulators, nitrogen fixation, shoot-to-root ration, bulb forcing, and others. The possibilities for hydroponics projects are nearly endless.

Future of Hydroponics

Hydroponics is increasing in use commercially, and will undoubtedly become increasingly important in the future. Research is expanding and new techniques are being developed. The use of nutrient solutions as media for growing plants will be an important part of agriscience in the future.

1. Write the Terms to Know and their meanings in your notebook.

2. Observe soils and other media used in flower pots, trays of vegetable seedlings, greenhouses, gardens, road banks, construction sites, and other places.

3. Examine the living organisms in a shovel full of soil taken from an outdoor area that is damp, moist, and high in organic matter.

4. Invite a Soil Conservation Service professional in to discuss soil mapping and land-use planning.

5. Obtain a land-use map from SCS for your home, farm, or other area. Study the material and report your findings to the class.

6. Observe a profile at a cut in a road bank, stream bank, or hole. Identify the O, A, B, and C horizons.

7. Do a mechanical analysis of a sample of soil using a fruit jar, water, and a dispersing agent such as calgon (Figure 9-13).

8. Observe the feel of some of the soil used in Activity 7.

9. Obtain a pH test kit and test samples of soil from home.

10. Do research on models and procedures for a home or school hydroponics unit. Plan, build, and use the unit to experiment with hydroponic production of various plants.

11. Build and use a composting structure.

SELF EVALUATION

A. Multiple Choice

1. An example of plant-growing media is
   a. soil.
   b. water.
   c. perlite.
   d. all of these.

2. Which is *not* organic matter?
   a. leaf mold
   b. peat moss
   c. sphagnum moss
   d. vermiculite

3. Decay of organic matter is caused by
   a. large animals.
   b. microbes.
   c. rodents.
   d. water.

4. Which is *not* a factor affecting soil formation?
   a. hydroponics
   b. gravity
   c. ice
   d. water

5. Humans affect soil formation by
   a. acid.
   b. alkaline.
   c. bulldozing.
   d. weathering.

6. The land class with the fewest limitations is
   a. Class I.
   b. Class III.
   c. Class VI.
   d. Class VIII.

7. The land classes suitable for field crop production are
    a.   I, II, IV, VI.
    b.   I, II, III, IV.
    c.   I, IV, V, VIII.

8. The horizon that is most supportive of plant growth is
    a.   horizon O.   c.   horizon B.
    b.   horizon A.   d.   horizon C.

9. The smallest soil particle is
    a.   clay.   c.   sand.
    b.   gravel.   d.   silt.

10. Good soil structure means
    a.   aggregated.   c.   raindrop endurance.
    b.   crumbly.   d.   all of the above.

11. Soil pH is generally raised by adding
    a.   sulfur.   c.   lime.
    b.   nitrogen.   d.   complete fertilizer.

12. Hydroponics refers to
    a.   aggregate culture.   c.   nutriculture.
    b.   aeroponics.   d.   all of the above.

13. The importance and use of hydroponics is
    a.   increasing.   c.   about the same.
    b.   decreasing.   d.   a new field.

14. Hydroponically grown plants differ principally from soil-grown plants by
    a.   light requirements.   c.   oxygen requirements.
    b.   nutrient requirements.   d.   support mechanisms.

15. The best and most economical light source for growing plants hydroponically is
    a.   incandescent.   c.   sodium.
    b.   fluorescent.   d.   sunlight.

B. Matching (Group I)

_____	**1.** Colluvial	a.	Deposited by ice
_____	**2.** Alluvial	b.	Deposited by lakes
_____	**3.** Glacial	c.	Formed in place
_____	**4.** Lacustrine	d.	Deposited by gravity
_____	**5.** Loess	e.	Class II land with erosion problem
_____	**6.** Leaching	f.	Cross section of soil
_____	**7.** Organic matter	g.	Makes soil dark in color
_____	**8.** Residual	h.	Removal of soluble materials
_____	**9.** IIe	i.	Deposited by streams
_____	**10.** Profile	j.	Deposited by wind

B. Matching (Group II)

_____ **1.** Horizon A	a.	Mostly organic matter
_____ **2.** Horizon B	b.	Parent material
_____ **3.** Horizon C	c.	Topsoil
_____ **4.** Horizon O	d.	Subsoil
_____ **5.** Coarse	e.	5-10-5 problem
_____ **6.** Medium texture	f.	Equal parts of nitrogen, phosphorus, and potassium
_____ **7.** Fine texture	g.	Sulfur
_____ **8.** 10-10-10	h.	Loamy soils
_____ **9.** Complete fertilizer	i.	Clay soils
_____ **10.** Lowers pH	j.	Sandy coils

C. Completion

1. Soil is defined as the top layer of the Earth's surface suitable for the growth of _____.

2. Three groups of plants found in soil are _____, _____, and _____.

3. Three groups or types of animals that live in the soil are _____, _____, and _____.

4. Five important benefits of organic matter in soil are _____, _____, _____, _____, and _____.

5. In solution culture, about one-third of the roots must be exposed to air, or air must be bubbled through the solution to prevent damage or death from lack of the element _____.

6. The six mineral elements taken up by roots and needed in large quantities by plants are _____, _____, _____, _____, _____, and _____.

7. The seven mineral elements taken up by roots and needed in tiny quantities by plants are _____, _____, _____, _____, _____, _____, and _____.

Forestry Management

MATERIALS LIST

- ✓ bulletin board materials
- ✓ samples of forest and wood products
- ✓ tree leaf collection
- ✓ forestry reference materials

COMPETENCIES TO BE DEVELOPED

After studying this unit, you should be able to:

- define forest terms.
- describe the forest regions of the United States.
- discuss important relationships among forests, wildlife, and water resources.
- identify important types and species of trees.
- describe how a tree grows.
- discuss important properties of wood.
- apply principles of good wood-lot management.
- describe procedures for seasoning lumber.

Forest land	Hardwood	Hardness
Timberland	Pulpwood	Shrinkage
Forest	Clear cut	Warp
Trees	Plywood	Ease of working
Shrubs	Veneer	Grain
Lumber	Cambium	Wood lot
Board foot	Annual rings	Silviculture
Forestry	Xylem layer	Arboriculture
Evergreen	Sapwood	Seedlings
Conifers	Phloem	Forester
Softwood	Inner bark	Virgin
Deciduous	Heartwood	

In the United States there are 483 million acres of timberland and 248 million acres of other forest land for a total of 731 million acres. This represents about one-third of the total land in the United States. **Forest land** is defined by the USDA as land at least 10 percent stocked by forest trees of any size. **Timberland** is defined as forest land that is capable of producing in excess of 20 cubic feet per acre per year of industrial wood and that has not been withdrawn from timber utilization by statute or administrative regulation.

A **forest** is a complex association of trees, shrubs, and plants, which all contribute to the life of the community (Figure 10-1). **Trees**

FIGURE 10-1 A forest contains trees, shrubs, and other plants, as well as animal life. *(Courtesy of USDA)*

are woody perennial plants, with a single stem that develops many branches. Trees usually grow to more than 10 ft high. **Shrubs** are bushy woody plants with multiple stems. A productive forest is one that is growing trees for lumber or other wood products on a continuous basis. **Lumber** is boards that are sawed from trees. It is bought and sold by the board foot. A **board foot** is a unit of measurement for lumber that is equal to 1" × 12" × 12" (Figure 10-2).

Forest land may include parks, wilderness land, game refuges, and other areas where harvesting of trees is not permitted. When you consider that there are 860 species of trees in the United States, it is evident that forestry is an important part of the economy. **Forestry** is the management of forests. An unbelievable variety of products come from trees (Figure 10-3).

Trees in the forests of the United States are divided into two general classifications: evergreen and deciduous. **Evergreen** trees do not shed their leaves on a yearly basis. Evergreen trees of commercial importance are mostly conifers. **Conifers** are evergreen trees that have needle-like leaves and produce lumber called **softwood**. **Deciduous** trees shed their needles or leaves every year and produce lumber called **hardwood.**

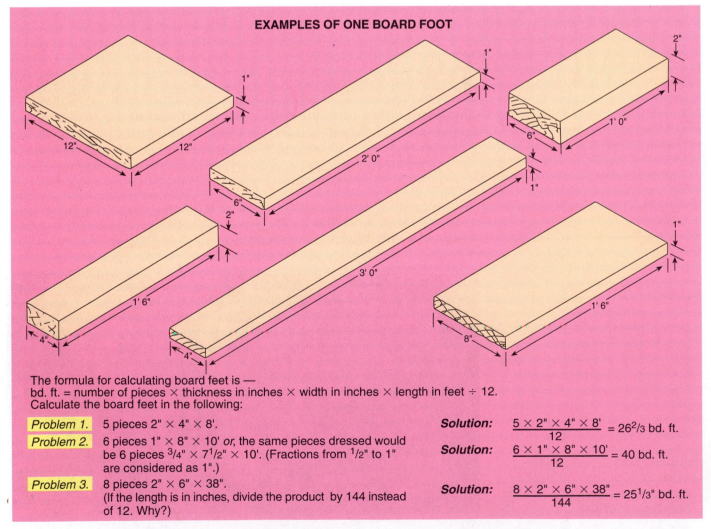

EXAMPLES OF ONE BOARD FOOT

The formula for calculating board feet is —
bd. ft. = number of pieces × thickness in inches × width in inches × length in feet ÷ 12.
Calculate the board feet in the following:

Problem 1. 5 pieces 2" × 4" × 8'.

Solution: $\dfrac{5 \times 2" \times 4" \times 8'}{12} = 26^{2}/_3$ bd. ft.

Problem 2. 6 pieces 1" × 8" × 10' *or,* the same pieces dressed would be 6 pieces $^3/_4$" × $7^1/_2$" × 10'. (Fractions from $^1/_2$" to 1" are considered as 1".)

Solution: $\dfrac{6 \times 1" \times 8" \times 10'}{12} = 40$ bd. ft.

Problem 3. 8 pieces 2" × 6" × 38".
(If the length is in inches, divide the product by 144 instead of 12. Why?)

Solution: $\dfrac{8 \times 2" \times 6" \times 38"}{144} = 25^{1}/_3$" bd. ft.

FIGURE 10-2 Lumber is bought and sold by the board foot. One board foot has a volume of 144 cubic inches. Any combination of dimensions that equals 144 cubic inches is one board foot.

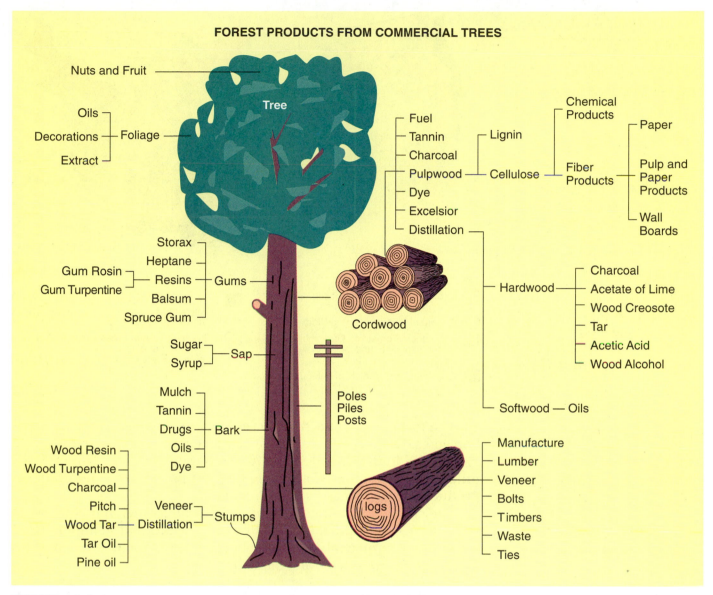

FIGURE 10-3 Forest products derived from commercial tree species.

FOREST REGIONS OF THE UNITED STATES

There are six major forest regions in the continental United States. They are the Northern Forest, Central Forest, Southern Forest, Bottomland Forest, West Coast or Pacific Coast Forest, and Western Interior Forest (Figure 10-4). There are also small areas of tropical or subtropical forest in southern Florida and southeastern Texas.

There are also two forest regions in Alaska and two in Hawaii. The Alaskan forest regions include Coastal Forest and Interior Forest. Hawaiian forests include the Wet Forest and Dry Forest (Figure 10-5).

Northern Forest

This region is located in the northeastern United States. It extends from Maine to southern Ohio and Indiana, and west to Wisconsin and Minnesota. The forests are mostly mixtures of hardwoods and softwoods, although there are some pure stands of softwoods.

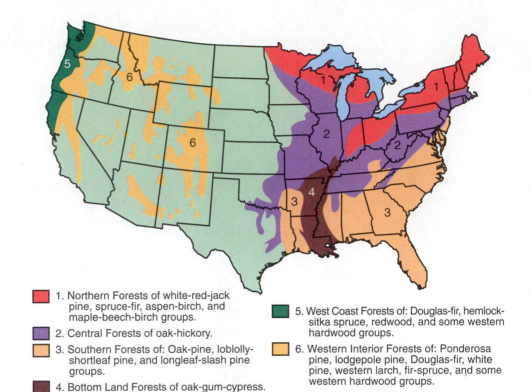

FIGURE 10-4 Primary forest regions of the forty-eight contiguous states. *[Adapted from material provided by USDA Forest Service]*

1. Northern Forests of white-red-jack pine, spruce-fir, aspen-birch, and maple-beech-birch groups.
2. Central Forests of oak-hickory.
3. Southern Forests of: Oak-pine, loblolly-shortleaf pine, and longleaf-slash pine groups.
4. Bottom Land Forests of oak-gum-cypress.
5. West Coast Forests of: Douglas-fir, hemlock-sitka spruce, redwood, and some western hardwood groups.
6. Western Interior Forests of: Ponderosa pine, lodgepole pine, Douglas-fir, white pine, western larch, fir-spruce, and some western hardwood groups.

The forests of the northern or eastern region cover about 115 million acres. They account for about 15 percent of the total yearly lumber production in the United States. About 26 percent of the United States **pulpwood** (wood used for making fiber for paper and other products) production also comes from this region.

The conifers that are important in this forest region include white pine, spruce, jack pine, balsam fir, white cedar, and hemlock. Hardwoods of commercial value are beech, oak, aspen, birch, walnut, maple, gum, cherry, ash, and basswood.

Central Forest

FIGURE 10-5 Forest regions of Alaska and Hawaii. *[Adapted from material provided by USDA Forest Service]*

This forest region is located mostly east of the Mississippi River and south of the Northern Forest region. It contains more varieties and

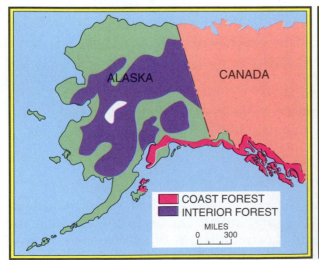

species of trees than any other forest region. It is composed mostly of hardwood trees. Hardwoods of commercial importance in the central forest region include oak, hickory, beech, maple, poplar, gum, walnut, cherry, ash, cottonwood, and sycamore. The conifers that are of economic value in this region include Virginia pine, pitch pine, shortleaf pine, red cedar, and some hemlock.

Southern Forest

This region is located in the southeastern part of the United States. It extends south from Delaware to Florida, and west to Texas and Oklahoma. It is the forest region with the most potential for meeting the future lumber and pulpwood needs of the United States.

The most important trees in the southern forest are conifers. They include Virginia, longleaf, loblolly, shortleaf, and slash pines. Oak, poplar, maple, and walnut are hardwood trees of economic importance.

Bottomland Forest

These forests occur mostly along the Mississippi River. They contain mostly hardwood trees and are often among the most productive of the U.S. forest regions. This is due to the high fertility of the soils in this area. Oak, gum, tupelo, and cypress are the major hardwood species found here.

West Coast, or Pacific Coast, Forest

This region is located in northern California, Oregon, and Washington. It is the most productive of the forest regions in the United States and has some of the largest trees in the world. The approximately 48 million acres of West Coast Forest provide more than 25 percent of the annual lumber production in the United States. About 19 percent of the pulpwood and 75 percent of the plywood produced in the United States come from trees grown in the West Coast Forest region.

Trees in the West Coast forests include 300-ft-tall redwoods and giant Sequoias that may be as much as 30 ft in diameter. Douglas fir, ponderosa pine, hemlock, western red cedar, Sitka spruce, sugar pine, lodgepole pine, noble fir, and white fir are conifers that are important in this region. Important hardwood species include oak, cottonwood, maple, and alder.

Western Interior Forest

The forests of the Western Interior Forest region are much less productive than those of the West Coast region. This region is divided up into many small areas and extends from Canada to Mexico. About 27 percent of the lumber produced in the United States comes from the 73 million acres in this forest region.

Most of the trees of commercial value in the Western Interior forests are the western pines. They are western white pine, ponderosa pine, and lodgepole pine. Spruce, fir, larch, western red cedar, and hemlock also grow there in small quantities. Aspen is the only hardwood of commercial importance in the Western Interior Forest region.

Tropical Forest

The tropical or subtropical forests of the continental United States are located in southern Florida and in southeastern Texas. These compose the smallest forest region in the United States. The major trees in this region are mahogany, mangrove, and bay, which are unimportant commercially. However, they are very important ecologically.

Alaskan and Hawaiian Forests

The Alaskan Coastal Forest region extends from southeastern Alaska to Kodiak Island. Trees of importance are Sitka spruce and western hemlock. There are about 6 million acres included in the Alaskan Coastal Forest region.

The Alaskan Interior Forest consists of about 105 million acres of forest land. Most of the wood products harvested in this region are used locally because of the difficulty in transporting them to other areas. Trees of commercial importance are spruce, aspen, and birch.

The Wet Forest region of Hawaii produces ohia, boa, tree fern, kukui, tropical ash, mamani, and eucalyptus. Most of these woods are used in the production of furniture and novelties.

The Dry Forest region of Hawaii produces koa, haole, algaroba, monkey pod, and wiliwili. None of these are of commercial value.

RELATIONSHIPS BETWEEN FORESTS AND OTHER NATIONAL RESOURCES

The relationships between forests and other natural resources, such as water and wildlife, are important to the overall well-being of the ecological system.

Forests filter rain as it falls and help reduce erosion of the soil. They trap soil sediment and help maintain water quality. Trees and shrubs of the forest are also instrumental in removing much of the pollutant materials from the air and from water run-off. Similarly, the roots of trees filter excess nutrients from water run-off. They also help reduce the harmful effects of excess fertilizer nutrients that enter streams and underground water systems. Algae, fungi, mosses, and numerous other plant and animal forms make their homes in forests.

The relationships between forests and many types of wildlife are numerous. Forests provide food, shelter, protection, and nesting sites for many species of birds and animals. They also help maintain the quality of streams so that fish and other types of aquatic life can live and thrive. The shade provided by forests helps to maintain proper water temperatures for the growth and reproduction of aquatic life.

The wildlife found in the different types of forests varies considerably (Figure 10-6). Some species must have the open areas that occur naturally or that have been clear cut by people. A forest that has been clear cut has had all of the marketable trees removed from it. Other wild birds and animals require mature forests to provide their needs in life. As forests change naturally or as a result of harvesting, the types of wildlife that inhabit it also change.

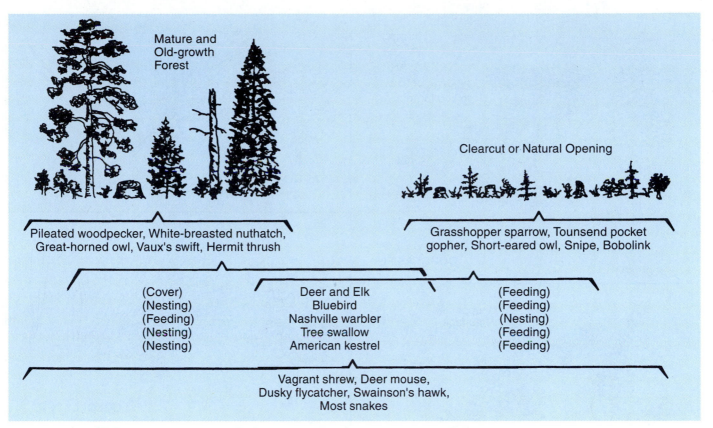

FIGURE 10-6 Preference of twenty wildlife species for different forest habitats.

SOME IMPORTANT TYPES AND SPECIES OF TREES IN THE UNITED STATES

Trees may be described in terms of the lumber they produce. Characteristics such as hardness, weight, tendency to shrink and warp, nail- and paint-holding ability, decay resistance, strength, and surface qualities are utilized to evaluate the usefulness of lumber.

Softwoods

The softwoods, or needle-type evergreens, that are important commercially in the United States include Douglas fir, balsam fir, hemlock, white pine, cedar, southern pine, ponderosa pine, and Sitka spruce (Figure 10-7).

Douglas Fir   Douglas fir is probably the most important species of tree in the United States today. It grows to a height of more than 300 ft and a diameter of more than 10 ft. About 20 percent of the timber harvested in the United States each year is Douglas fir. One-hundred-year-old stands of Douglas fir can produce 170,000 board feet of lumber per acre. This is five to six times the production of most other softwood species. Douglas fir is very popular as construction lumber and for the manufacture of plywood. **Plywood** is a construction material made of thin layers of wood glued together.

Balsam Fir   Found in the forests of the Northeast, the lumber from balsam fir is used mostly for framing buildings. Balsam fir trees have

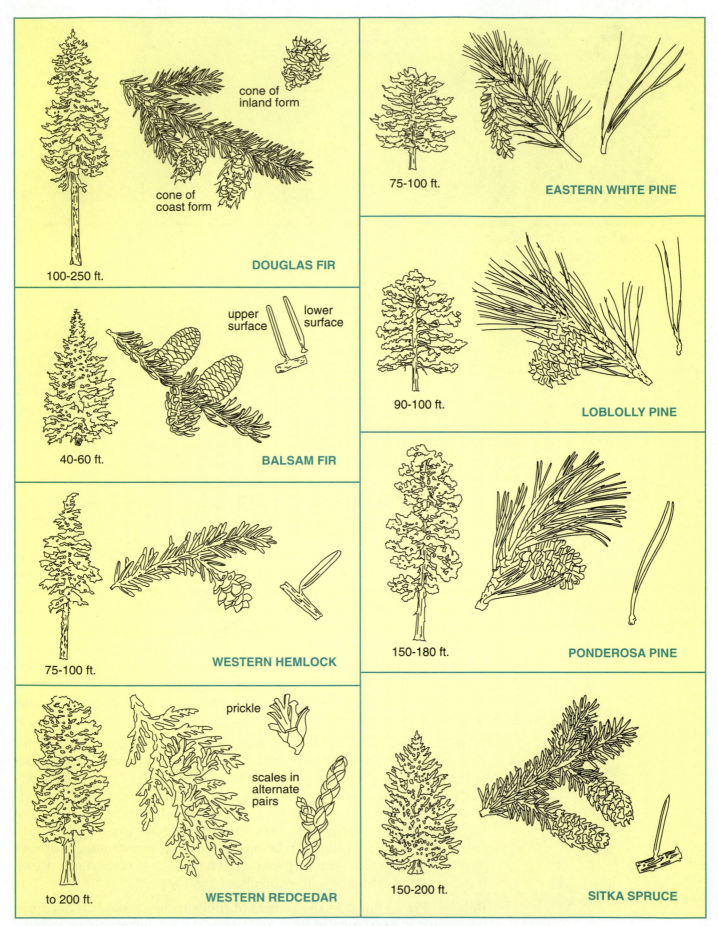

FIGURE 10-7 Some species of softwood trees commercially important for wood products.

soft, dark-green needles and a classic triangular shape when grown at low densities. They are often used as Christmas trees.

Hemlock: Eastern and Western   Eastern hemlock is strong and is often used for building material. At times, it can be brittle and difficult to work. Eastern hemlock grows over most of the Northern Forest range.

Western hemlock grows in the Pacific Coast Forest region, where yearly rainfall averages 70 in. Western hemlock lumber is very strong and is one of the most important sources of construction-grade lumber. It is also important for pulpwood.

Cedar: Eastern Red, Eastern White, and Western Red   Eastern red cedar is used for fence posts because it is very resistant to decay. It is also used for the lining of chests and closets because its odor repels many insects. White cedar is a swamp tree with decay-resistant wood that is often used for shingles and log homes. Western red cedar resembles redwood in appearance. It is used where decay resistance, rather than strength, is important.

White Pine   White pine lumber is soft, light, and straight-grained. It has less strength than spruce or hemlock, and is more popular as a wood for cabinetmaking. White pine grows from Maine to Georgia.

Southern Pine   Included in the category of southern pine is longleaf pine, shortleaf pine, loblolly pine, and slash pine. The southern pines grow in the southern and south Atlantic states. Lumber from southern pine is used for construction. It is also used for pulpwood and plywood.

Ponderosa Pine   Growing from California to Alaska, Sitka spruce trees attain a height of 300 ft and a diameter of 18 ft. Lumber from Sitka spruce is of very high quality. It is very strong, straight, and even-grained. Sitka spruce is also used in large quantities for pulpwood.

Hardwoods

Hardwoods come from deciduous trees. Commercially important species of hardwoods in the United States include birch, maple, poplar, sweetgum, oak, ash, beech, cherry, hickory, sycamore, walnut, and willow (Figure 10-8).

Birch   Easily recognized by their white bark, birch grow in areas where summer temperatures seldom exceed 70°F. Birch lumber is dense and fine-textured. It is used for furniture, plywood, paneling, boxes, baskets, and veneer, as well as for many small novelty items. **Veneer** is a very thin sheet of wood glued to a cheaper species of wood and used in paneling and furniture making.

Maple   Maple lumber is classified as both hard and soft. Hard maple lumber is heavy, strong, and very hard. It is used for butcher blocks, workbench tops, flooring, veneer, and furniture. Soft maple is only about 60 percent as strong as hard maple, and is used in the same applications. Some species produce sweet sap that is made into maple syrup.

Poplar   Poplar grows over most of the eastern United States. Poplar is classified as a hardwood due to its deciduous structure. However, lumber from poplar trees is soft, light, and usually knot-free. Poplar

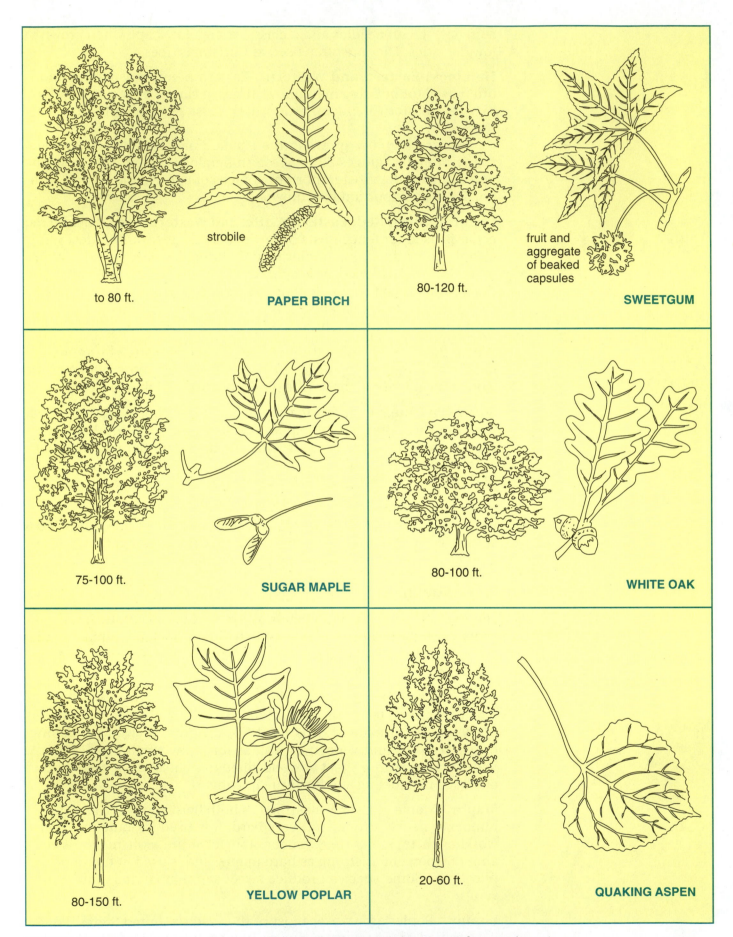

strobile

to 80 ft.

PAPER BIRCH

80-120 ft.

fruit and
aggregate
of beaked
capsules

SWEETGUM

75-100 ft.

SUGAR MAPLE

80-100 ft.

WHITE OAK

80-150 ft.

YELLOW POPLAR

20-60 ft.

QUAKING ASPEN

FIGURE 10-8 Some species of hardwood trees commercially important for wood products.

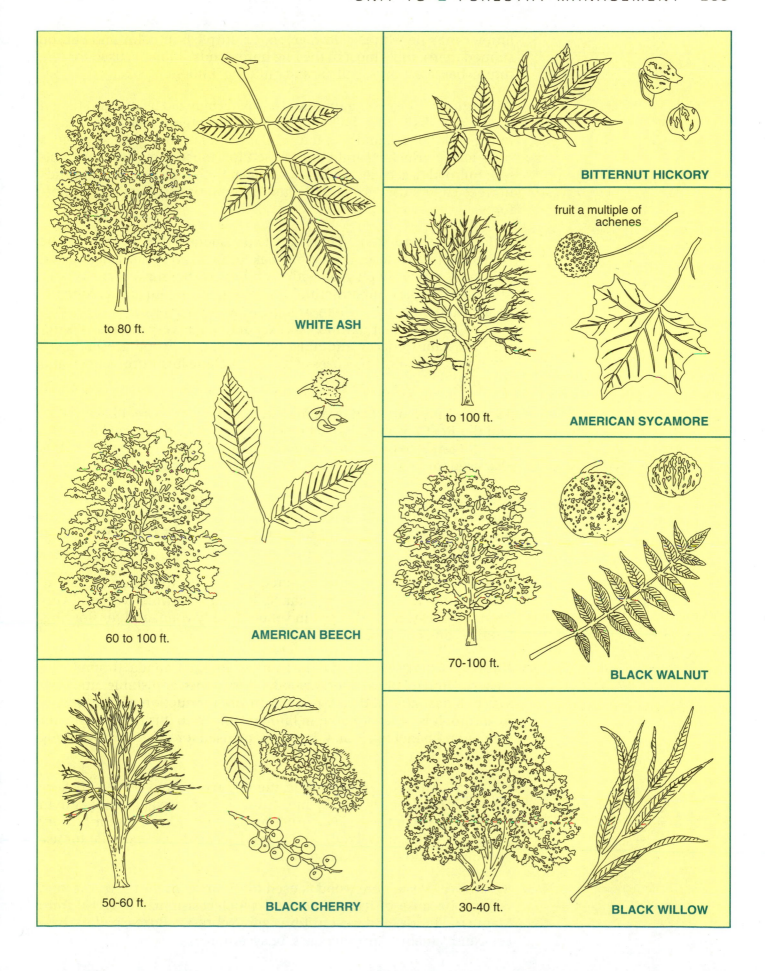

BITTERNUT HICKORY

to 80 ft.   **WHITE ASH**

fruit a multiple of achenes

to 100 ft.   **AMERICAN SYCAMORE**

60 to 100 ft.   **AMERICAN BEECH**

70-100 ft.   **BLACK WALNUT**

50-60 ft.   **BLACK CHERRY**

30-40 ft.   **BLACK WILLOW**

lumber may be white, yellow, green, or purplish in color and can be stained to resemble most of the fine hardwoods. Poplar is used for furniture, baskets, boxes, pallets, and building timbers.

Sweetgum   The sweetgum tree is easily recognized by its star-shaped leaves and distinctive ball-shaped fruit. Sweetgum trees grow to as much as 120 ft tall and 3 to 5 ft in diameter. Lumber from sweetgum trees has interlocking grain and is used for house trim, furniture, pallets, railroad ties, boxes, and crates. The gum that comes from wounds in trees can be used as natural chewing gum or as a flavoring or perfume.

Oak: White and Red   There are two general types of oak in the United States: white and red. White oak lumber is very hard, heavy, and strong. Its pores are plugged with membranes that make it nearly waterproof. It is used for structural timbers, flooring, furniture, fencing, pallets, and other uses where strength of wood is a necessity.

Red oak is similar to white oak except that it is very porous. It is not very resistant to decay and must be treated with wood preservatives when used outside. Chief uses of red oak include furniture, veneer, and flooring.

Aspen   This hardwood tree grows in the Northeast, Great Lake states, and the Rocky Mountains. Aspen is very rapid-growing, making lumber that tends to be weaker than most construction-grade timber. It is also used for pulpwood.

Ash   Ash lumber is heavy, hard, stiff, and has a high resistance to shock. It also has excellent bending qualities. It is popular for use in handles, baseball bats, boat oars, and furniture. It resembles oak in appearance.

Beech   Grown in the eastern United States, beech is heavy and hard, and is noted for its shock resistance. It is hard to work with and very prone to decay. Beech is used in veneer for plywood, and for flooring, handles, and containers.

Cherry   Cherry can be found from southern Canada through the eastern United States. Cherry wood is very dense and stable after drying. It is desirable and very popular in the production of fine-quality furniture. It is expensive and in limited supply, so it is used mostly for veneer and paneling; it may, however, be used for other woodworking purposes.

Hickory   HIckory grows best in the eastern United States. Hickory lumber is hard, heavy, tough, and strong. It is somewhat stronger than Douglas fir when used as construction lumber. Other uses for hickory include handles, dowel rods, and poles. Hickory is also popular for use as firewood and for smoking meat.

Sycamore   Sycamore wood is used for flooring in barns, trucks, and wagons because of its strength and shock resistance. It grows from Maine to Florida, and west to Texas and Nebraska. Boxes, pallets, baskets, and paneling are other uses for sycamore.

Black Walnut A premier wood for the manufacture of fine furniture, black walnut grows from Vermont to Texas. The wood has straight grain and is easily machined with woodworking tools. Because walnut is slow-growing and very desirable, it is often made into veneer to get more use from its chocolate-brown heartwood. It is also the source of black walnut nut meats.

Black Willow Most of the black willow of commercial value is grown in the Mississippi River Valley. It is soft, light, and has a uniform texture. Willow is used mostly in construction for subflooring, sheathing, and studs. Some willow is also used for pallets and for interior components of furniture. Black willow is sometimes a low-cost substitute for black walnut, because it has a similar brown appearance when finished.

There are many other domestic softwoods and hardwoods that grow in the United States. Many of these are important in local areas. The types discussed here are but a sampling, rather than a definitive list, of the commercially important trees in the United States.

AGRI·PROFILE

CAREER AREAS: Dendrology/Silviculture/Forestry

Forestry is known as a career area for rugged individuals who prefer the outdoors and working in relative isolation. However, many jobs in forestry are in urban areas and involve much indoor work. The United States Forestry Service hires large numbers of forestry technicians and managers. Many forestry jobs do involve an extensive amount of outdoor work, but most jobs provide a desirable mix of outdoor and indoor work.

Forestry includes the work of the dendrologist engaged in the study of trees, the silviculturist specializing in the care of trees, the forestry consultant who advises private forest land owners, lumber industry workers, government foresters, loggers, national and state forest rangers, and firefighters. A relatively new position is that of the urban forester, responsible for the health and well-being of the millions of trees found in parks, along streets, and in other areas of our cities.

The arborist is an urban forester whose work may include planting, transplanting, pruning, fertilizing, or tree removal.

There are many challenging careers in the forestry industry. *(Courtesy FFA; FFA #186)*

TREE GROWTH AND PHYSIOLOGY

Trees use carbon dioxide (carbon and oxygen) from the air, and water (hydrogen and oxygen) from the soil to manufacture simple sugars in their leaves. The leaves then use additional carbon, hydrogen, and oxygen to convert simple sugars into complex sugars and starches. Nitrogen and minerals from the soil are then used to manufacture proteins, the building blocks for growth and reproduction.

A tree typically starts from a seed. For instance, oak trees grow from seeds called acorns, pine trees start from seeds in pine cones, and peach trees grow from peach seeds. Trees may also sprout and grow from stumps or other tree parts. When a seed germinates, a shoot grows upward to form the top growth, and roots grow downward and outward to form the root system.

Both roots and shoots extend themselves by growth at the tips through cell division and elongation. At the same time, tree roots, stems, and trunks grow in diameter by adding cell layers near their outer surfaces. This growth layer in a tree root, trunk, or limb is called the **cambium.** The outward growth of the cambium in one year creates **annual rings** as seen in the cross section of a root, trunk, or limb.

Water and minerals are taken in by the roots and transported up to the leaves through a layer of cells called the **xylem layer,** or **sapwood.** The xylem is located just inside of the cambium layer. Just outside the cambium is another layer of cells called the **phloem,** or **inner bark,** which carries food manufactured in the leaves to the stems, trunk, and roots. Each year the tree grows new cambium, xylem, and phloem tissues, and the older sapwood becomes heartwood. **Heartwood** is the inactive core that gives a tree strength and rigidity (Figure 10-9).

PROPERTIES OF WOOD

The various properties of wood determine the uses for which it is best adapted. It should be noted that there are wide variations within a specific species of tree and that the properties discussed here are general in nature.

Hardness

The property of **hardness** refers to a wood's resistance to compression. Hardness determines how well it wears. It is also a factor in determining the ease of working the wood with tools. Splitting and difficulty in nailing are problems when wood is too hard.

Weight

The weight of wood is a good indication of its strength. In general, heavy wood is stronger than light wood. The moisture content affects the weight of wood, so comparisons should always be made between woods with similar moisture contents.

Shrinkage

The change in dimensions of a piece of wood as it reacts to changes in humidity or temperature is referred to as **shrinkage.** The amount of

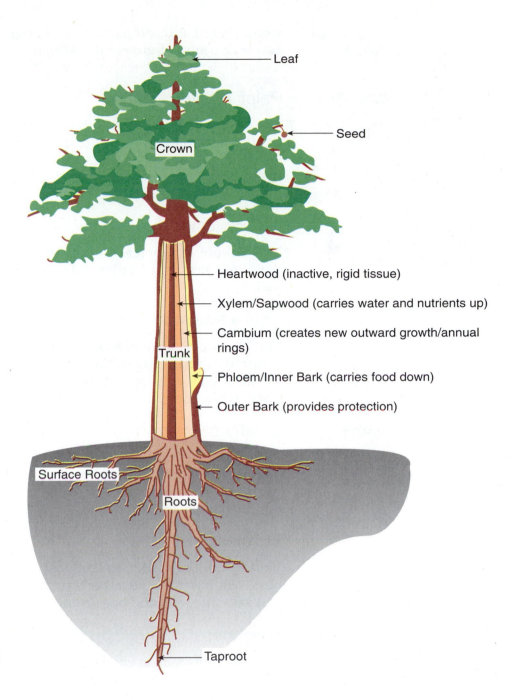

Leaf

Seed

Crown

Heartwood (inactive, rigid tissue)

Xylem/Sapwood (carries water and nutrients up)

Cambium (creates new outward growth/annual rings)

Trunk

Phloem/Inner Bark (carries food down)

Outer Bark (provides protection)

Surface Roots

Roots

Taproot

FIGURE 10-9 Major parts of a tree.

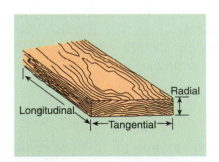

Radial

Longitudinal

Tangential

FIGURE 10-10 Dimensions of wood shrinkage.

expansion and contraction of wood greatly affects construction techniques used in building items from wood (Figure 10-10).

Warp

Warp refers to the tendency of wood to bend permanently due to moisture change. The tendency to warp is more of a problem in some types of wood than others. Warping is caused by uneven drying of wood across its three dimensions.

Ease of Working

In wood, **ease of working** refers to the level of difficulty in cutting, shaping, nailing, and finishing the wood. It is influenced by the hardness of the wood and by the characteristics of the grain. The **grain** of

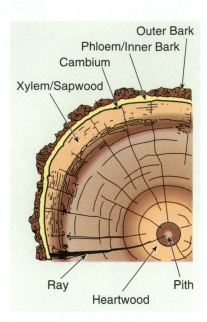

FIGURE 10-11 The annual rings seen in the cross section of a log or stump reflect the conditions of growth experienced by the tree on a year-to-year basis and determine the appearance of the grain.

wood is hard and soft patterns caused by growth of the tree as it adds successive annual-growth rings (Figure 10-11). In general, soft wood is easier to work than hard, dense wood.

Paint Holding

The ability of wood to hold paint may be determined by the type of paint, surface conditions, and methods of application, as well as characteristics of the wood itself. Moisture content, amount of pitch in the wood, and the presence of knots will all affect how well wood holds paint.

Nail Holding

Nail-holding ability refers to the resistance of wood to the removal of nails. In general, the harder and denser the wood, the better it will hold nails.

Decay Resistance

The resistance of wood to microorganisms that cause decay is a chief factor in selecting wood. Natural resistance to insects that live in, tunnel through, or devour wood also affects the choice of species for a given application.

Bending Strength

The ability of wood to carry a load without breaking is a measure of bending strength. Bending strength is important when determining types and sizes of lumber to use for rafters, beams, joists, and other building applications.

Stiffness

The resistance of wood to bending under a load is referred to as stiffness. When wood used in construction is not sufficiently stiff, ceilings and walls may flex and buckle and the wallboards crack under load.

Toughness

The ability of wood to withstand blows is called toughness. Hardwoods are much more resistant to shock and blows than softwoods. This characteristic is particularly important in woods used for tool handles.

Surface Characteristics

The appearance of the wood's surface is sometimes the determining factor in its use. The pattern of the grain greatly influences the staining characteristics and beauty of the finished product. The number and type of knots and pitch pockets are surface characteristics that need to be considered when selecting lumber for a project.

WOOD LOT MANAGEMENT

The proper management of a wooded area or wood lot involves more than just the harvest of trees and the removal of unwanted species. A **wood lot** is a small, privately owned forest. The production of trees for

BIO•TECH CONNECTION

Conserving Our Biodiversity

Our biodiversity is at risk! Not far from the shores of Florida and Texas, the tropical rain forests of Central and South America, as well as parts of Africa and southeast Asia, are evaporating from the face of the Earth. Commonly called a jungle, a tropical rain forest is a hot, wet, green place characterized by an enormous diversity of life and a huge mass of living matter. Here, biologists estimate that more than one-half of the plant and animal species of the world is found. Unfortunately, many of these species are not found anywhere except in the tropical rain forests. However, rain forests occupy only 7 percent of the land area of the Earth. Moreover, their total area is diminishing at an alarming rate!

Abundant moisture and warm temperatures encourage tremendous plant growth year-round. Plants and moisture then provide an excellent environment for animals and micro-organisms to grow, develop, and diversify. Rain forests are covered by a dense canopy formed by the crowns of trees up to 150 ft high. Under the shade of this canopy are smaller trees and shrubs creating a second layer called the understory. The two layers are typically woven together by strong woody vines, which may exceed 700 ft in length. The third layer, or forest floor, is typically dark and gloomy, but easy to walk through.

Palms, herbs, and ferns dot the forest floor. The jungle floor, with its shrubs and trees, is home to a myriad of insects and animals. Annual rainfall may reach 400 in., and the daily rate exceeds the rate of evaporation. Life is plentiful. However, the soil beneath the jungle is poor, shallow, and very erodible. As human populations expand into the rain forests, the foliage is cut and

Rain forests are our richest source of plant and animal species and biodiversity. *(Courtesy of Elmer Cooper)*

burned to make way for logging, grazing, and cropping.

Sadly, the land will generally support grazing or crop production for only a few years until it is worn out, depleted of minerals, and eroded. The occupants then typically move on to virgin jungle land to slash and burn a new tract and repeat the cycle. Once the jungle is destroyed, crops extract the scant nutrients, the soil is soon eroded, and the land will not support the resurgence of jungle. Therefore, once cleared, the jungle area is lost forever. It is estimated that one-half of the world's original tropical rain forests is gone. Further, at the current rate of destruction, only one-quarter will remain beyond the year 2000.

Because the rain forests are the storehouse for more than one-half of the genetic diversity in the entire world, species are being lost at an alarming rate. Many species are becoming extinct before they are discovered and recorded. In fact, it is estimated that scientists have discovered and named only one-sixth of all the plant and animal species in rain forests. Earth's biodiversity is indeed threatened by human hands.

harvest is a long-term investment, and mistakes in management take a long time to correct.

Some of the factors that need to be considered in the management of wood lots include soil, water, light, type of trees, condition of trees, markets available, methods of harvesting, and replanting. Using scientific methods in the management of forests is called **silviculture.** Similarly, the care and management of trees for ornamental purposes is called **arboriculture.**

Restocking a Wood Lot

The least expensive method of replacing trees harvested from a wood lot is natural seeding. Sources of seed for the desired species must be available in the forest, and conditions must be right for seed germination in order for natural seeding to take place. If seed from natural sources is not available, seeds from other sources may be planted on the forest site.

A surer method of restocking a wood lot is to plant trees of the desired species rather than rely on seeds to do the reforestation. In most cases, **seedlings** (young trees started from seeds) are planted during late winter and early spring, before the new season's growth begins. Wood lots can be planted with one species of tree or a mixture of several compatible species (Figure 10-12).

Management of a Growing Wood Lot

Management of a wood lot is much more involved than just sitting back and watching it grow. Proper care and management is important if the forestry enterprise is to be successful.

Trees that are of no commercial value should be removed as soon as possible to eliminate competition for light, moisture, and nutrients. Because these "weed" trees are removed when they are small, there is seldom a market for them and they are left on the wood-lot floor to decay. If weed trees are of sufficient size, however, they may be used for firewood.

When all the trees of a wood lot are nearly the same age (typically 15 to 30 years old), they often need to be thinned. Trees should be thinned any time that the crowns or branches occupy less than one-third of the height of the trees. Usually, about one-fourth to one-third of the trees in a wood lot are removed during thinning.

Trees that are being grown for lumber are often pruned of side branches to produce a better quality log. Prompt removal of side branches helps keep logs free of knots. Only rapidly growing trees should be pruned. Branches should be pruned flush with the trunk of the tree. Pruning is usually done during the fall and winter, when the trees are dormant.

Planning a Harvest Cutting

A wood lot should never be harvested without a plan. A harvesting plan can maximize the income from the wood lot over many years. The use of a forester's services is usually wise in developing a harvesting plan. A **forester** is a person who studies and manages forests.

There are several systems of harvesting a wood lot. They include clear cutting, seed-tree cutting, shelterwood cutting, diameter limit cutting, and selection cutting.

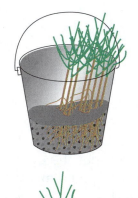

FIGURE 10-12 Reforestation may be done by seed trees, mechanical seed distribution, or seedling planting.

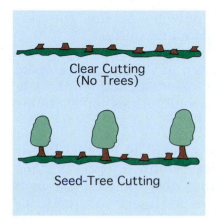

FIGURE 10-13 Clear cutting is generally done in small patches, but seed-tree cutting is an economical way of harvesting and reseeding large areas.

Clear cutting   Clear cutting is a system of cutting timber where all of the trees in an area are removed. It was used extensively in cutting the **virgin** forests (those that have never been harvested) of the United States when there was little thought of reforestation of the cut area.

Clear cutting is usually done in small patches, ranging in size from one-half acre to 50 acres. Prompt reforestation with desirable species of trees on the clear cut areas is essential. This will prevent erosion and the loss of fragile forest soil.

Seed-Tree Cutting   Harvesting trees by the seed-tree cutting method is very similar to clear cutting. The primary difference is that enough seed-bearing trees are left uncut to provide seeds for reforestation (Figure 10-13). The trees left to produce seeds should be very representative of the desired species and free from insect, disease, or mechanical damage.

The removal of the seed trees usually takes place after the cut area is repopulated with desirable seedlings.

Shelterwood Cutting   In shelterwood cutting, enough trees are left standing after harvesting to provide for reseeding of the wood lot. They will also protect the area until the young trees are well established. After the young trees become established, the residual trees are harvested.

Diameter Limit Cutting   When harvesting trees by the diameter limit method, all trees above a certain diameter are cut. Slow-growing or diseased trees are often left standing when harvesting is done using this method. Rapid-growing trees of desirable species may be cut before they reach their production potential. Often this method leaves a wood lot with only slow growing trees that may be undesirable. Diameter limit cutting is used advantageously to remove trees left from previous harvesting, which reduces competition with younger trees.

Selection Cutting   This is the usual method of harvest used when the trees in a wood lot are of different ages. Selecting the trees of harvest can be a difficult task for the average person and obtaining the services of an experienced forester is usually wise. This system of harvesting allows for fairly frequent income and maintains the wood lot's esthetic value.

Protecting a Wood Lot

A wood lot must be protected from fire, pests, and domestic animals if it is to yield up to its potential. It is estimated that more than 13 billion board feet of timber are destroyed each year by pests. This represents nearly 25 percent of the estimated net growth of forests and wood lots each year.

Fire   Millions of dollars worth of timber are destroyed by fire each year. In addition to actually killing trees, fire slows the growth of others, and damages some so that insects and diseases may destroy them. Fire also burns the organic matter on the forest floor, which takes nutrients away from the trees and exposes the soil to erosion.

To help prevent fires, debris should be removed from around trees. Weeds, brush, and other trash around the edges of a wood lot should be removed. Not allowing human use, such as hunting and fishing, during dry periods may also reduce the potential for fire. The construction of permanent firebreaks is useful for fire control.

A plan of how to deal with fire is important in minimizing damage should a fire occur. State and county foresters can assist in developing fire prevention and control plans. These include planning for water storage, and procedures for notifying proper authorities and obtaining appropriate equipment and help.

Pests   Insects and diseases cause more damage to existing forests than fires. Pests cause trees to be weak and deformed. They consume leaves, damage bark, and retard the growth of trees. They kill billions of trees each year. However, control of diseases and insects in forest lands is difficult and expensive.

Removal of dead, damaged, or weak trees may help reduce disease and insect problems. Not only are weak trees attacked, but healthy ones are sometimes favorite targets of pests. Prompt action in dealing with outbreaks of insects and diseases will do much to ensure a profit at harvest time.

Domestic Animals   The grazing of wood lots by cattle and sheep usually results in the destruction of all small seedlings in the forest. It also eliminates most of the wood-lot floor coverage. Livestock may also strip the bark from trees, causing them to die. Wood lots do not provide much food for grazing animals, so it is still wise to exclude livestock from them.

Harvesting a Wood Lot

Before timber is harvested, a market for it must be found. Various methods of marketing the timber should be explored to determine the most profitable alternatives. There are many uses for forest products, and trees should be marketed to maximize their value.

The actual harvest of a wood lot can be done by the owner or by a contractor skilled in timber harvest. Another alternative is to sell the standing timber to a company that later harvests and markets it.

Regardless of how the timber is harvested, it is extremely important that contracts covering all pertinent information be drawn up and signed by all parties involved.

SEASONING LUMBER

The proper seasoning of lumber is essential to protect it from damage. As soon as a tree is cut, it starts to lose moisture. If the wood dries too slowly, it may be subject to rot, stain, or insect damage. If it is allowed to dry too quickly, lumber may twist and warp or split. This may make it unsuitable for most uses.

Wood that is sawed for lumber should be stacked immediately after sawing to allow for even drying (Figure 10-14). The stacking should take place on a level, well-drained location. It should be off the ground and stacked so that air can circulate freely around each board. The stacked lumber should also be protected from weather. The amount of time required for lumber to dry depends on the thickness of the lumber and the species of tree from which it was cut. Times range from 30 to 200 days.

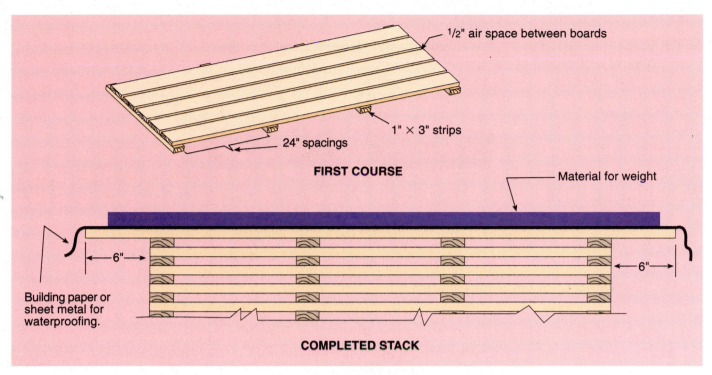

½" air space between boards

1" × 3" strips

24" spacings

FIRST COURSE

Material for weight

6"

6"

Building paper or
sheet metal for
waterproofing.

COMPLETED STACK

FIGURE 10-14 Freshly sawed lumber must be properly stacked to prevent warping and end-splitting as the lumber dries.

The forestry industry in the United States produces about 16 billion cubic feet of wood products each year. The demand for wood products, such as lumber, pulp wood, posts, and pilings, is almost 19 billion cubic feet per year. Only by carefully managing our forestry resources and replacing our trees can the United States fulfill its need for wood and wood products in the future.

STUDENT ACTIVITIES

1. Write the Terms to Know and their meanings in your notebook.

2. Make a collection of forest products.

3. Make a collection of tree leaves, bark, or twigs that are of economic importance in your area.

4. Make a bulletin board showing the many uses of forests and forest products.

5. Write a report on a species of trees or a type of forest product of interest to you.

6. Visit a local forest and identify the types of trees growing there.

7. Have a forester visit the class and speak about forest management.

8. Visit a wood-processing plant operation to learn how trees are processed into other products.

9. Write a report on a career in the forestry or wood products industry.

10. Study Figure 10-2 and learn how to calculate board feet (BF). Determine the number of board feet in the following:

    a. A board that is 1" × 12" × 12'. BF = _____   d. Six 2" × 4" × 8' boards.   BF = _____

    b. A plank that is 2" × 8" × 16'. BF = _____   e. Twenty 2" × 10" × 16' boards. BF = _____

    c. A 4" × 4" × 8' board.   BF = _____

11. What is the total board feet in Activity 10?

12. If the items in Activity 10 were rough-sawed lumber and selling at $.20 per board foot, what would be the cost of the total order? $_____

SELF EVALUATION

A. Multiple Choice

1. Trees that are used for making paper are called
a. timber.   c. pulpwood.
b. lumber.   d. veneer.

2. The scientific management of forests is
a. silviculture.   c. pomology.
b. aboriculture.   d. olericulture.

3. There are about _____ acres of productive forests in the continental United States.
a. 105 million   c. 500 million
b. 235 million   d. 735 million

4. The most important commercial species of trees in the United States is
a. oak.   c. redwood.
b. Douglas fir.   d. walnut.

5. About 75 percent of the wood for plywood is harvested from the _____ Forest region.
a. Northern   c. Alaskan Interior
b. Central   d. Pacific Coast

6. A forest that has never been harvested is called
a. virgin.   c. clear cut.
b. hardwood.   d. seedling.

7. The seed-tree method of harvesting
a. cuts all trees over a certain diameter.
b. cuts all trees under a certain diameter.
c. cuts about one-third of the trees in a wood lot.
d. cuts all but a few trees left for seed.

8. Which of the following was not stated as a property of wood?
a. nail-holding ability   c. color
b. bending strength   d. surface characteristics

9. The yearly demand for wood products in the United States is about _____ cubic feet.
a. 19 billion   c. 190 billion
b. 23 billion   d. 253 million

10. In a wood lot with even-aged trees, the lot usually needs to be thinned when the trees are _____ years old.
a. 1 to 5   c. 10 to 15
b. 5 to 10   d. 15 to 30

B. Matching

_____	**1.** Tree	a.	Thin layers of wood glued together
_____	**2.** Shrub	b.	Wood from conifers
_____	**3.** Wood lot	c.	Small, multistemmed plant
_____	**4.** Veneer	d.	Management of wooded land
_____	**5.** Plywood	e.	Wood from deciduous trees
_____	**6.** Lumber	f.	Thin slices of wood
_____	**7.** Forestry	g.	Small forest
_____	**8.** Softwood	h.	Tree with needle-like leaves
_____	**9.** Hardwood	i.	Woody, single-stem plant
_____	**10.** Conifer	j.	Boards sawed from trees

C. Completion

1. The air drying of lumber is referred to as _____.

2. Nearly _____ percent of the estimated net growth of forests each year is lost to fire and pests.

3. _____ is a type of harvesting where every tree over a certain size is cut.

4. Oak is a type of _____ or _____ tree.

5. Douglas fir is an example of an _____ or _____ tree.

6. Cutting all of the trees in an area is called _____.

7. The _____ Forest region is located along the Mississippi river.

8. The forest region with the most potential for meeting future needs for forest products is the _____ Forest region.

Wildlife Management

OBJECTIVE

To determine the relationship between wildlife and the environment and approved practices in managing wildlife enterprises.

MATERIALS LIST

✓ bulletin board materials

✓ reference materials on wildlife, wildlife management, and pollution

COMPETENCIES TO BE DEVELOPED

After studying this unit, you should be able to:
- define wildlife terms.
- identify characteristics of wildlife.
- describe relationships between types of wildlife.
- understand the relationships between wildlife and humans.
- describe classifications of wildlife management.
- identify approved practices in wildlife management.
- discuss the future of wildlife in the United States.

TERMS TO KNOW

Wildlife	Prey	Predation
Habitat	Parasitism	Commensalism
Vertebrate	Warm-blooded animals	Competition
Predators	Mutualism	Wetlands

Wildlife has been part of the life of humans since the beginning of time. **Wildlife** are animals that are adapted to live in a natural environment without the help of humans. Early humans followed herds of wild animals and killed what they needed in order to live and survive. They observed what the animals did and what they ate to determine what was safe for human consumption. Early humans also used wildlife as models for their artwork and in many of their ceremonial rites.

As settlers came to the new world, wildlife often provided the bulk of the food available until food production systems could be developed. Supplies of wildlife seemed to be inexhaustible as the skies were blackened with the flight of millions of passenger pigeons, and herds of bison created dust storms as they migrated on the vast prairie (Figure 11-1).

Unfortunately, supplies of wildlife were not and are not unlimited. Humans have destroyed wildlife **habitat** (the area where a plant or animal normally lives and grows). Humans have polluted the air and water supplies, killed wildlife in tremendous numbers, and generally disregarded the needs of wildlife. As a result, many species of wildlife in the United States can benefit greatly from proper management.

CHARACTERISTICS OF WILDLIFE

All **vertebrate** animals (animals with backbones) are included in the classification of wildlife. They have many of the same characteristics as humans. Growth processes, laws of heredity, and general cell structure are common to both humans and animals. When populations become too dense, disease outbreaks occur, populations suffer from starvation, and disposal of waste becomes a problem.

The wildness of the animal itself is a characteristic that allows the animal to survive without interference or help from humans. The animal's wildness often contributes to the interest that humans have in wildlife. Characteristics identified as wildness are what attract hunters to hunting and fishermen to fishing. Bird watching and wildlife

FIGURE 11-1 In the early years of our country, the supply of wildlife appeared to be limitless. *(Courtesy U.S. Fish and Wildlife Service)*

FIGURE 11-2 The wildness of animals is fascinating to most people. *(Courtesy Chesapeake Bay Foundation)*

photography would be far less fascinating if wildlife were less wild and wary of humans (Figure 11-2).

With few exceptions, wildlife lives in an environment over which it has no control. Wildlife must be able to adapt to whatever it is presented in terms of food and environment, or it will perish. It must also possess natural senses that allow it to avoid predators and other dangers. **Predators** are animals that feed on other animals. The animal being eaten by the predator is the **prey.**

The ability to avoid over-population is a characteristic of many groups of wildlife. Establishing and defending territories is one way that wildlife may naturally avoid overpopulation. The stress of overpopulation causes some animals to slow or stop reproducing altogether.

WILDLIFE RELATIONSHIPS

Every type of wildlife is part of a community of plants and animals where all individuals are dependent on others. Any attempt to manage wildlife must take into account the relationships that exist naturally. Because relationships within the wildlife community are constantly changing, it is very difficult to set standard procedures for their management.

The balance of nature is actually a myth because wildlife communities are seldom in a state of equilibrium. The numbers of various species of wildlife are constantly increasing and decreasing in response to each other and to many external factors such as natural disasters. These include fires, droughts, and disease outbreaks. The interference of humans also often upsets the natural balance of nature.

Some of the natural relationships that exist in the wildlife community include parasitism, mutualism, predation, commensalism, and competition.

Parasitism

The relationship between two organisms, either plants or animals, in which one feeds on the other without killing it is called **parasitism.** Parasites may be either internal or external. An example of a parasitic relationship is the wood tick, which may live on almost any warm-blooded animal. **Warm-blooded animals** have the ability to regulate their body temperatures.

Mutualism

Mutualism refers to two types of animals that live together for mutual benefit. There are many examples of mutualism in the wildlife community. Tick pickers are birds that remove and eat ticks from many of the wild animals in Africa, to the mutual benefit of both. The wild animals have parasites removed from them, and the birds receive nourishment from the ticks. A moth that lives only on a certain plant is also the only pollinator of that plant in several relationships. Some plant seeds will germinate only after having passed through the digestive tract of a specific bird or animal.

FIGURE 11-3 Predation is a natural occurrence among wildlife. *(Courtesy U.S. Fish and Wildlife Service; Photo by Pedro Ramirez, Jr.)*

FIGURE 11-4 Competition among wildlife species helps to keep the animal populations in balance. *(Courtesy U.S. Fish and Wildlife Service; Photo by John D. Wendler)*

Predation

When one animal eats another animal, the relationship is called **predation** (Figure 11-3). Predators are often very important in controlling populations of wildlife. Foxes are necessary to keep populations of rodents and other small animals under control. Populations of predators and prey tend to fluctuate widely. When predators are in abundance, prey becomes scarce because of overfeeding. When prey becomes scarce, predators may starve or move to other areas. This then permits increase in the population of the prey.

Commensalism

Commensalism refers to a plant or animal that lives in, on, or with another, sharing its food, but not helping or harming it. One species is helped, but the other is neither helped nor harmed. Vultures waiting to feed on the leftovers from a cougar's kill is an example of commensalism.

Competition

When different species of wildlife compete for the same food supply, cover, nesting sites, or breeding sites, **competition** exists. When competition exists, one species may increase in numbers while the other declines. Often the numbers of both species decrease as a result of competition. Owls and foxes both competing for the available supply of rodents and other small animals is an example of competition (Figure 11-4).

The various relationships that exist between species of wildlife make it necessary to consider more than just one species any time that management is contemplated. Understanding the relationships that exist in the entire wildlife community is essential if wildlife management programs are to be successful.

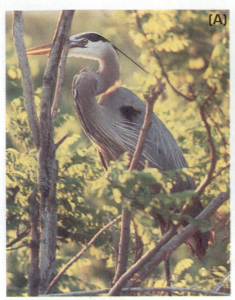

FIGURE 11-5 Some species just seem to belong in the ecosystem. Shown in their natural habitats are the (A) egret *(Courtesy Chesapeake Bay Foundation)*, (B) eastern cottontail *(Courtesy U.S. Fish and Wildlife Service; Photo by William Janus)*, and (C) arctic hare *(Courtesy U.S. Fish and Wildlife Service)*.

RELATIONSHIPS BETWEEN HUMANS AND WILDLIFE

Relationships between humans and wildlife may be biological, ecological, or economic. Biological relationships exist because humans are animals that are very similar to wildlife. Relationships may be ecological because humans are but one species among nearly 1 million species of animals that inhabit the planet Earth (Figure 11-5).

The economic relationship that exists between humans and wildlife are important. Originally humans were dependent on wildlife for food, clothing, and shelter. Today, there are six positive values of wildlife's relationship with humans—commercial, recreational, biological, aesthetic, scientific, and social.

The harvesting and sale of wildlife and/or wildlife products is an example of the commercial relationship between humans and wildlife. Raising wild animals for use in hunting, fishing, or other purposes also falls into this category.

Hunting and fishing, as well as watching and photographing wildlife, are included in the category of recreational relationship (Figure 11-6). Although it is estimated that more than $2 billion is spent each year on hunting and fishing and at least another $2 billion on other recreational uses of wildlife, many of the recreational values of wildlife are intangible.

The value of the biological relationship between humans and wildlife is difficult to measure. Examples of the biological relationship include pollination of crops, soil improvement, water conservation, and control of harmful diseases and parasites (Figure 11-7).

Aesthetic value refers to beauty. Watching a butterfly sipping nectar from a flower, a fawn grazing beside its mother, or a trout rising to a hatch of mayflies are all examples of the aesthetic value of wildlife. Wildlife also provides the inspiration for much artwork. Even though the aesthetic value of wildlife is not measurable in economic terms,

FIGURE 11-6 Hunting and fishing are popular recreational uses of wildlife. *(Courtesy U.S. Fish and Wildlife Service; Photo by Richard Baldes)*

FIGURE 11-7 Biological values of wildlife and human relationships include the pollination of crops by honeybees. *(Courtesy USDA/ARS #K-4716-1)*

wildlife can contribute much to the mental well-being of the human race.

Using wildlife for scientific studies often benefits humans. The scientific relationship between humans and wildlife has existed from the beginning of time, when early humans watched wild animals to determine which plants and berries were safe to eat.

The value of the social relationship between humans and wildlife is also difficult to measure. However, wildlife has the ability to enhance the value of their surroundings simply by their presence. They provide humans the opportunity for variety in outdoor recreation, hobbies, and adventure. They also make leisure time much more enjoyable for humans.

CLASSIFICATIONS OF WILDLIFE MANAGEMENT

Wildlife management can be divided into several classifications for ease in developing management plans. Techniques for management vary tremendously according to classification. Some wildlife management classifications are farm, forest, wetlands, stream, and lakes and ponds.

The management of farm wildlife is probably the most visible wildlife management classification. The development of fence rows, minimum tillage practices, improvement of wood lots, and controlled hunting are all techniques that have long been used to manage farm wildlife. Rabbits, quail, pheasants, doves, and deer are the types of wildlife most normally managed in this category.

Forest wildlife is more difficult to manage. Plans must be developed so that timber and wildlife can exist at desired populations and possibly be harvested. Management of forest wildlife may include population controls to prevent destruction of habitat. Deer, grouse, squirrels, and rabbits are usually wildlife species included in forestry wildlife management programs (Figure 11-8).

FIGURE 11-8 Habitat for forest wildlife is difficult to manage. *(Courtesy National FFA)*

The most productive wildlife management areas are wetlands. **Wetlands** include all areas between dry upland and open water. Marshes, swamps, and bogs are all wetland areas. Because these areas are very sensitive to changes in environmental conditions, careful management of them for wildlife is essential. The wetlands are home to ducks, geese, muskrats, raccoons, deer, pheasants, grouse, woodcock, fish, frogs, and thousands of other species of wildlife.

The management of running water or streams is often a difficult task. Water pollution and the need for clean water for a growing human population continue to increase at a rapid pace. Potential damage to the wildlife in streams from chemical pollution, the building of dams, roads, home construction, and the drainage of swampland are critical considerations for the stream wildlife manager.

Management of wildlife in lakes and ponds is normally somewhat easier than it is in streams because water is standing rather than running. Population levels of pond wildlife, oxygen levels, pollutants, and the availability of food resources are all concerns of the pond and lakes wildlife manager.

AGRI·PROFILE

CAREER AREAS: Wildlife Biologist/Manager/Officer

Wildlife biologists work with fish and game species of land, fresh water streams and lakes, tidal marshes, bays, seas, and oceans. Wildlife biologists generally have master or doctorate level degrees in biology. They tend to utilize the basic sciences in their work.

Wildlife managers typically have associate or bachelor level degrees. They work in government agencies, advising land owners and managing game populations on public lands. Their work frequently requires the use of helicopters, small planes, snowmobiles, all terrain vehicles, horses, and land rovers, as well as time in the wild on foot.

Wildlife officers interact continuously with the hunting and fishing community. They advise governments in establishing fish and game laws and programs for habitat improvement. Wildlife officers have the backing of strict laws and stiff penalties for offenders. However, much of their time is spent on educating the public and obtaining private assistance in improving habitat and maintaining game populations.

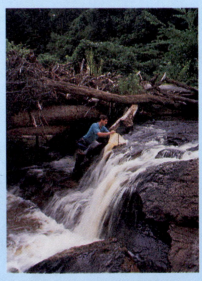

Conducting game counts and doing habitat analyses provide information for wildlife managers. *[Courtesy USDA/ARS; USDA #K-5213-3]*

APPROVED PRACTICES IN WILDLIFE MANAGEMENT

Farm Wildlife

Management of wildlife on most farms is usually a by-product of farming or ranching. It is often given little attention by the farmer or rancher except when wildlife causes crop damage and financial loss.

Much of the management of farm wildlife involves providing suitable habitat for their living, growth, and reproduction. This may involve leaving some unharvested areas in the corners of fields, planting fence rows with shrubs and grasses that provide winter feed and cover, or leaving brush piles when harvesting wood lots.

The timing of various farming operations is also important in a farm wildlife management program. When possible, mowing should be delayed until after pheasant eggs have hatched in the spring. Crop residues should be left standing over the winter to provide food and cover. Planting crops attractive to wildlife on areas that are less desirable as cropland is an excellent farm wildlife management practice. Providing water supplies for wildlife during dry periods is often necessary to maximize the numbers of farm wildlife on the area being managed (Figure 11-9).

Harvesting farm and ranch wildlife by hunting has been shown, by extensive research, to have little impact on spring breeding populations. Excess populations of farm and ranch wildlife that are not harvested by humans usually die during the winter. Even heavy hunting pressures seldom result in severe damage to wildlife populations. The sale of hunting rights to hunters is a way to increase the income of many farms and ranches. In addition, it often means the difference between profit and loss in the farming enterprise.

Management of wildlife on game preserves or farms set up specifically for hunting often differs drastically from other wildlife management programs. Often species of animals and birds that are not native to the area are raised and released on the preserve. Native wildlife species may also be raised in pens and released to the farm or preserve expressly for harvest by hunters.

FIGURE 11-9 Providing water for wildlife is an important wildlife management practice. *(Courtesy Wendy Troeger)*

Forest Wildlife

FIGURE 11-10 Habitat improvement is an important part of the wildlife manager's job. *(Courtesy of National FFA #255)*

The types and numbers of forest wildlife in any specific woodland are dependent on many factors. These include type and age of the trees in the forest, density of the trees, natural forest openings, types of vegetation on the forest floor, and the presence of natural predators.

Management of forest wildlife is usually geared toward increasing numbers of desired species of wildlife. If desired populations of wildlife are present, the management goal is usually to maintain those populations. Sometimes, numbers of certain species of forest wildlife increase to the point where destruction of habitat occurs. When this happens, control measures may have to be instituted to restore proper balance.

The steps in developing a forest wildlife management plan should include taking an inventory of the types and numbers of wildlife living in the forest area to be managed. Goals for the use of the forest and the wildlife living in it need to be developed. The third step in the development of a forest wildlife management plan is determining the types and populations of wildlife that the forest area can support and how best to manage the forest so that required habitat is provided.

The requirements for forest wildlife include food, water, and cover. These necessities must be readily available to the desired species of forest wildlife at all times. Management practices that meet these requirements include making clearings in the forest so that new growth will make twigs available for deer to feed on. Another practice is selective harvesting so that trees of various ages exist in the forest to make a more suitable habitat for squirrels and many other species of forest wildlife. Leaving piles of brush for food and cover is also a management practice that leads to increased production of forest wildlife. Care in managing harvests of forest products so that existing supplies of water are not contaminated is also important in good wildlife management.

Deer, grouse, squirrels, and rabbits are the forest wildlife species that are usually targeted for management because they are valuable for recreational purposes, especially hunting (Figure 11-10). They may also be managed to prevent the destruction of valuable forest trees and other products.

It should be noted that during times of over-population of forest animals, especially deer, it is seldom a good idea to provide supplemental food. Natural losses should nearly always be allowed to occur, including starvation of excess animals or allowing heavier-than-normal hunting pressures. Artificial feeding of wildlife populations usually results in further population increases and an extension of the problem.

Wetlands Wildlife

No area of American land is more important to wildlife than the wetlands. Wetlands include any land that is poorly drained—swamps, bogs, marshes, and even shallow areas of standing water (Figure 11-11). The wetlands are constantly changing as wet areas fill in with mud and decaying vegetation. They eventually become dry land that contains forests.

Wetlands provide food, nesting sites, and cover for many species of wildlife. Ducks and geese are probably the most economically impor-

FIGURE 11-11 No area of American land is more important to wildlife than the wetlands. *(Courtesy of Chesapeake Bay Foundation)*

FIGURE 11-12 Wetlands are important nesting areas for millions of ducks and geese in the United States. *(Courtesy U.S. Fish and Wildlife Service)*

tant type of wildlife that needs the wetlands for survival (Figure 11-12). Other types of wildlife found in the wetlands include woodcock, pheasants, deer, bears, mink, muskrats, raccoons, and thousands of other lesser known species.

Management of wetlands for wildlife may include impounding water. Open water areas should occupy about one-third of the wetlands for optimum use by wildlife. The depth of the standing water should not be more than about 18 in.

The management of the plant life in the wetlands is also important. This may include cutting trees to open up the wetland area. Many species of wildlife require large open areas in order to thrive. Care must be taken not to remove hollow trees that are used as nesting sites for some species of wildlife. Wetland areas can also be opened up by killing excess trees rather than cutting them. This provides resting areas for many types of wetlands wildlife.

Establishing open, grassy areas around wetlands also helps to attract many types of wildlife to the area. Planting millet, wild rice, and other aquatic plants in the wetlands also helps to attract wildlife.

A serious hazard to wetlands wildlife is pollution. Pollution of water flowing into the wetlands area may come from agriculture, industry, or the disposal of domestic wastes. Because pollutants are trapped in the mud and silt of the wetlands, the effects of pollutants are often long-lasting.

In areas lacking natural nesting sites, populations of some wildlife species can be greatly increased by providing artificial nesting sites. Wood duck boxes, old tires, and islands surrounded by open water provide safe nesting sites for many species of wetland wildlife.

Raising certain species of ducks in captivity and later releasing them in wetlands areas has been important in maintaining viable duck populations. This has been important as more and more natural duck nesting areas have been destroyed to meet the needs of the human race.

Stream Wildlife

Stream wildlife can be divided into two general categories—warm-water and cold-water. These categories are based on the water temperatures at which the wildlife, primarily fish, can best grow and thrive. There is little or no difference in the management practices between

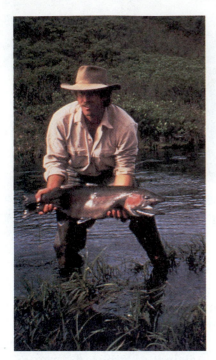

FIGURE 11-13 Cool, rapid-flowing brooks and mountain streams provide appropriate habitat for trout. *(Courtesy U.S. Fish and Wildlife Service)*

the two categories of wildlife. In general, fish are the type of stream wildlife that is managed, although many other types of wildlife depend on streams for their existence (Figure 11-13).

As land is developed, forests are harvested, and civilization is expanded, streams and their wildlife populations come under increasing pressure. Because we cannot build new streams, it is essential that existing ones be managed properly.

Management practices for streams include preventing stream banks from being overgrazed by livestock. Fencing the stream to limit access by livestock is also wise to reduce pollution and the destruction of stream banks.

Good erosion control practices on lands surrounding streams is important to help maintain clear, clean water. It is also important to prevent chemical pollutants from entering streams (Figure 11-14).

Maintaining stream-side forestation is important in regulating stream temperatures during the warm summer months. Some species of fish stop feeding and may even die when stream temperatures become too high. The amount of dissolved oxygen in warm water is also much lower than it is in cold water. Without adequate oxygen, aquatic wildlife dies.

Anything that impedes the flow of a stream also serves to change it, to the detriment of many species of stream wildlife and the benefit of some others. Trout must have swiftly moving, cool water in which to thrive, while catfish are perfectly happy in sluggish streams (Figure 11-15).

Care must be taken to maintain desirable species of wildlife in the stream. Introducing new species of wildlife in the stream may result in reducing native wildlife already in the stream.

The maintenance of population levels of stream wildlife that are in balance with the available food supply is important. Too many fish for the available food supply normally results in stunted fish of no value to fishermen. This situation does provide an increased food supply for some types of birds and animals that use streams for their food supply. Overfishing of predatory species of fish, such as bass or northern pike, may allow perch or sunfish to overpopulate the stream and become stunted. Often, the only way to restore streams to a desired mix of fish species is to remove the unwanted species. This is accomplished by netting, poisoning, or electric shocking. These techniques are gener-

FIGURE 11-14 Soil erosion causes siltation of streams, which destroys the habitat for desirable species. *(Courtesy Chesapeake Bay Foundation)*

FIGURE 11-15 Catfish are perfectly happy in slow-moving streams. *(Courtesy of U.S. Fish and Wildlife Service)*

BIO•TECH CONNECTION

Alaskan Adventure

Sierra Stoneberg of Hinsdale, Montana had developed extensive knowledge and experience of the outdoors by the time she graduated from high school. As early as the seventh grade, she became interested in botany through the Montana Range Days program, where students experience first-hand elements of botany, biology, and other range sciences. Her high school summers were filled with FFA activities and experiences with the Soil Conservation Service.

The summer following high school graduation brought an Alaskan adventure studying moose, mountain sheep, and other major wildlife species. As a volunteer at the Kenai Lake Work Center, she, along with others at the Center, worked forty hours a week for lodging, board, and plane fare.

Incredible, huge green trees and mosses created vivid memories of the high alpine country. Similarly, the tiny plants and lichens on the trees and rocks contributed a unique hue to the surroundings.

Scientific observations and data collection were important facets of her many treks into the wilderness. These included

High school and college summers in the field provide valuable experiences for future wildlife biologists or range scientists. *(Courtesy National FFA)*

monitoring of fertilizing results on rangeland for sheep herds and range grass response to management practices. Scouting for eagle nests and moose habitat created an interesting and stimulating blend of mental and physical activities. Even a close encounter with a black bear provided valuable experience in becoming a national FFA Proficiency Award winner and for a future in wildlife biology and range science.

netting, poisoning, or electric shocking. These techniques are generally legal only for authorized and specially trained personnel.

The artificial rearing and stocking of desired species of stream wildlife is a management practice that is important in many streams. Typically, game species of fish are stocked in streams for fishermen to catch and remove. Often few or no fish survive to reproduce, and stocking must take place each year.

The regulations of sport fishing is often necessary to maintain desirable populations of game fish. This may include closed seasons, minimum size limits, creel limits, and restricted methods of catching fish.

Lakes and Ponds Wildlife

The management practices for wildlife in lakes and ponds are usually very similar to those for managing stream wildlife. Pollution must be controlled. Wildlife populations must be managed to maintain desired

FIGURE 11-16 Sunfish provide food for bass and sport for fishing in ponds and lakes. *(Courtesy Chesapeake Bay Foundation)*

mixes of species. Harvest and use must also be controlled to ensure wildlife for the future.

However, there are some differences between management of wildlife in streams and management of wildlife in ponds and lakes. Because the water in lakes and ponds is normally standing, the amount of oxygen available for aquatic life sometimes becomes critical in the hot summer months. In small ponds, artificial means of incorporating oxygen into the water may be used to prevent fish kills.

Water temperatures in lakes and ponds are more variable than they are in streams. This means that different species of fish are usually dominant in ponds and lakes. In many ponds and lakes, fish populations are predominantly largemouth bass and sunfish (Figure 11-16).

When it is necessary or desirable to rid a pond or lake of unwanted species of fish such as carp, it is often much easier to do so because the water is contained. Sometimes it is possible to drain the body of water to remove the unwanted species of wildlife. More often, the body of water is simply poisoned so that all fish and other species of pond wildlife are killed. The pond is then restocked with desirable wildlife species.

The management of wildlife is a very imprecise business. Often, specific species of wildlife are managed to the detriment of others. It is reasonably clear that any human interference in the wildlife community results in changes that are not always to the benefit of much of that community.

THE FUTURE OF WILDLIFE IN THE UNITED STATES

A bright future for all of the wildlife species is not ensured in the United States because the needs of the human population continue to compete with those of the wildlife community. The outlook for wildlife in the future is not all bleak, however. Humans have recognized that it is possible to satisfy the needs of wildlife and the demands of humans if careful management practices are instituted (Figure 11-17).

FIGURE 11-17 Ponds and lakes can accommodate fish and wildlife as well as provide recreation for the family. *(Courtesy Chesapeake Bay Foundation)*

Careful studies of how humans and wildlife can peacefully co-exist are being made. There are sincere attempts being made to reduce the pollution of the environment. Polluted areas are being cleaned up. More extensive testing of new chemicals and other pesticides and their environmental effects is being conducted. The effects of new construction on wildlife habitat is also studied before that construction begins.

Establishing large acreages in national parks and wildlife refuges is also important for the future well-being of the wildlife in the United States. More emphasis on management of wildlife resources, rather than simply exploitation, also bodes well for the future of the country's wildlife. Fish and game laws are important elements in preserving and enhancing wildlife.

Realistically, some species of wildlife will decline and cease to exist in the future, while other species will proliferate. This has been the case in the past and will continue to be so in the future, with or without the interference or help of the human population.

STUDENT ACTIVITIES

1. Write the Terms to Know and their meanings in your notebook.

2. Visit a wildlife area and identify the species of wildlife that you see there.

3. Construct a bulletin board showing the species of wildlife that are important to hunters and fishermen in your area.

4. Have a wildlife manager visit your class and explain management practices that he or she uses in the area that he or she manages.

5. Write a report on the effects of pollution on wildlife in some area of the United States.

6. Participate in a New Year's Day bird count.

7. Plant a feed patch area for wildlife.

8. Participate in a stream or other wildlife area cleanup program.

9. Visit a zoo and list the species of wildlife that are endangered.

10. Develop a list of endangered species of wildlife in your area and what is being done to prevent their extinction.

11. Write a report on a species of wildlife that interests you.

12. Have a local farmer or rancher speak to the class on what measures he or she is taking to prevent polluting the environment of wildlife.

13. Invite a birdwatcher or wildlife photographer to discuss his or her given hobby or profession with the class.

A. Multiple Choice

1. _____ is a species of fish adapted to cold, running water in streams.
   a.   Carp
   b.   Catfish
   c.   Trout
   d.   Sunfish

2. Forest wildlife survive best in forests that are
   a.   of mixed-age trees.
   b.   deciduous.
   c.   evergreen.
   d.   of even-age trees.

3. Trees growing along streams help to
   a.   regulate water flow.
   b.   provide food for aquatic wildlife.
   c.   provide oxygen for fish.
   d.   regulate stream temperatures.

4. The most important wildlife management area is the
   a.   farm.
   b.   stream.
   c.   wetlands.
   d.   forest.

5. When two species of wildlife live together for the benefit of both, the relationship is called
   a.   mutualism.
   b.   predation.
   c.   commensalism.
   d.   competition.

6. The raising and sale of wildlife for hunting is an example of what type of relationship between humans and wildlife?
   a.   biological
   b.   recreational
   c.   social
   d.   commercial

7. One species of fish that is often undesirable in lakes and ponds is the
   a.   bass.
   b.   carp.
   c.   sunfish.
   d.   trout.

8. Rabbits, quail, pheasants, doves, and deer are wildlife often targeted for management in

    a.   farm areas.

    b.   forests.

    c.   wetlands.

    d.   none of the above.

9. Management of deer is a concern of _____ wildlife managers.

    a.   forest

    b.   farm

    c.   wetlands

    d.   all of the above

10. Wetlands should be made up of about _____ shallow, standing water for optimum wildlife use.

    a.   one-fourth

    b.   one-third

    c.   one-half

    d.   two-thirds

B. Matching

_____ **1.** Predation

_____ **2.** Parasitism

_____ **3.** Mutualism

_____ **4.** Commensalism

_____ **5.** Competition

    a.   One type of wildlife living and feeding on another type

    b.   Two types of wildlife eating the same food

    c.   One type of wildlife eating another type of wildlife

    d.   One type of wildlife living in, on, or with another type but without helping or harming it

    e.   Two types of wildlife living together for the benefit of both

C. Completion

1. Inspiration for artwork is a/an _____ value of wildlife.

2. Fishing is a/an _____ value of wildlife.

3. Pollination of crops is a/an _____ value of wildlife.

4. The ability of wildlife to increase the value of their surroundings simply by their presence is a/an _____ value of wildlife.

5. The observation of wildlife by early humans to determine what was safe to eat is a/an _____ value of wildlife.

Aquaculture

COMPETENCIES TO BE DEVELOPED

After studying this unit, you should be able to:

- explain the food chain in a freshwater pond.
- discuss water quality and list eight measurable factors.
- describe three major aquaculture production systems.

MATERIALS LIST

- ✓ bulletin board materials
- ✓ small freshwater aquarium (optional)
- ✓ 30 g of NaCl with 1,000 ml flask

Water covers three-quarters of the Earth's surface. This resource produces both plants and animals that are used to feed the world. **Aquaculture** is the management of this aquatic environment to increase the harvest of usable plant and animal products. The **aquaculturist** is a person trained in aquaculture who must understand where and how organisms live, eat, grow, and reproduce in water. The manipulation of these factors determines the success of the aquaculture system, because natural fisheries have reached their capacities to produce fish (Figure 12-1).

FIGURE 12-1 Natural fisheries have reached their capacities to produce more fish. *(Courtesy Chesapeake Bay Foundation)*

FIGURE 12-2 The oceans represent our largest water resource. *(Courtesy of Elmer Cooper)*

These production systems are part of an integrated industry that requires specialized products and services. Aquaculturists include nutritionists, feed mill operators, pathologists, fish hatchery managers, processing managers, researchers, and growers. We use these services to produce fresh and processed seafoods, shellfish, and ornamental fish and plants.

Natural fisheries are fish production areas that occur in nature without human intervention. They include the rivers, oceans, continental shelves, reefs, bays, lakes, and rivers. These fisheries are currently fished by sophisticated fishing fleets so efficient that yearly catches (yields) have leveled off or are decreasing due to insufficient supply of fish. The population of the world continues to increase; therefore, aquaculture must be used to produce more aquatic plants and animals for food. Understanding the aquatic environment, the biology of the organisms, and how to control the production of aquatic plants and animals is essential.

THE AQUATIC ENVIRONMENT

The oceans represent the largest expanse of water resources in the world (Figure 12-2). They represent a reservoir of water containing soluble nutrients and materials washed from the land. Over time, the evaporation of water into the atmosphere has increased the concentration of these nutrients until the **salinity,** or mineral content, of the water is high. We call this seawater. The concentrations of these nutrients or salts are so high that land plants and animals are unable to survive if we use this water for irrigation or drinking (Figure 12-3).

Rain contains only small amounts of salts. Therefore, accumulation of water on land and its flow into the seas generates a **gradient,** or measurable change over time or distance, in salinity. This affects the types of organisms that can flourish.

The Salinity Gradient

Freshwater **wetlands,** such as marshes, ponds, and streams, are generated by rainwater. Here, natural rainfall accumulates and provides

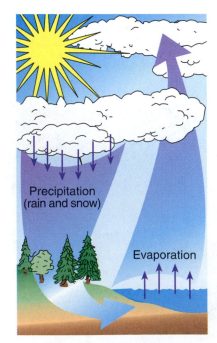

Precipitation
(rain and snow)

Evaporation

FIGURE 12-3 The natural water cycle.

FIGURE 12-4 Freshwater wetlands serve as transition areas between aquatic and terrestrial plants and animals

aquatic environments that represent the transition between aquatic and **terrestrial** (land) plants and animals. **Amphibians,** such as frogs, turtles, and reptiles, are animals that live part of their lives in these fresh waters and the remaining period on land. Several plant species, such as cattails *(Typha spp.)*, watercress *(Nasturtium officinale)*, water spinach *(Ipomoea reptans)*, and rice, also require a transitional period of flooding and drainage to flourish (Figure 12-4).

As water flows from fields and urban areas into large streams and lakes, this run-off accumulates more soluble nutrients, and the salinity increases. The profile of plants and animals change as certain other organisms are more adapted to this changed environment. Flows accumulate into rivers that empty into **bays, esturaries,** and **saltwater marshes.** Bays are open waters along coastlines where fresh water and salt water mix. Estuaries are ecological systems influenced by brackish or salty water. The saltwater marshes are lowlands influenced by tidal waters. The fresh water mixes with the seawater to create unique growing conditions for various fish, shellfish, and aquatic plants. This **brackish water** is a mixture of fresh water and salt water that fluctuates with the tide, flow of the rivers, and weather.

Several types of seaweeds are commercially produced in the Orient. These include types of red, green, and brown algae. Similar plant growth of algae must be maintained in intensive fish systems to feed fry. **Fry** are small, newly hatched fish.

The migration of saltwater salmon upstream to **spawn** (lay eggs) in freshwater streams illustrates the gradient effect on the life cycle of aquatic organisms. Although adult salmon cannot survive in fresh water, the young eggs must hatch in low-salinity water and then migrate back to the ocean.

The Aquatic Food Chain

The aquatic environment constantly changes to maintain a balance of organisms that function in the food chain or system. A simple illustration is the makeup of a freshwater pond. The food chain is fueled by sunlight. Green plants and algae use this energy to grow and utilize nutrients they absorb from the water. These plants are eaten by fish and

FIGURE 12-5 The aquatic food chain in a freshwater pond.

other animals, which are preyed upon by larger animals (fish, reptiles, etc.) and sometimes caught by humans. These large animals return nutrients to the water as waste or carrion that are reabsorbed by plants for growth. The maintenance of this food chain supports life (Figure 12-5).

The aquaculturist must understand the effect of any management on this cycle and compensate with technology or design in an aquaculture production system.

General Biology

The biology of aquatic plants and animals is very similar to that of terrestrial plants and animals. Both eat and sleep, reproduce, and interact with the environment. Green plants harvest energy from the Sun through photosynthesis and absorb nutrients from the water to manufacture carbohydrates, proteins, and cellulose. They serve as the waste recyclers in the aquatic environment by constantly absorbing waste products (nutrients) and contributing to the food chain. Like any land plant, they respond to fertilization, shading, competition, insects, disease, and weather. Aquatic plants are composed of many parts, as discussed in unit 15, Plant Structures and Taxonomy. Certain parts help them compete within the aquatic environment. Green plants absorb carbon dioxide from the water and release oxygen during photosynthesis. During the night, plants reabsorb a smaller amount of oxygen and release carbon dioxide.

Aquatic animals, particularly fish, complement the relationship with plants by generating carbon dioxide during respiration and releasing soluble nutrients through waste products and decay. Figure 12-6 illustrates the anatomy of a fish and shows the specialized **gills,** or lungs, that absorb oxygen and release carbon dioxide into the water.

In this competitive environment of predator and prey, the plants provide shelter and food that maintain the animals' life cycle.

Shellfish are aquatic animals with a shell or shell-like extensions. Nonmotile adult shellfish, including clams, mussels, oysters, and others, occupy a unique niche in the aquatic environment. Located on their bottoms, these organisms have developed an efficient pumping

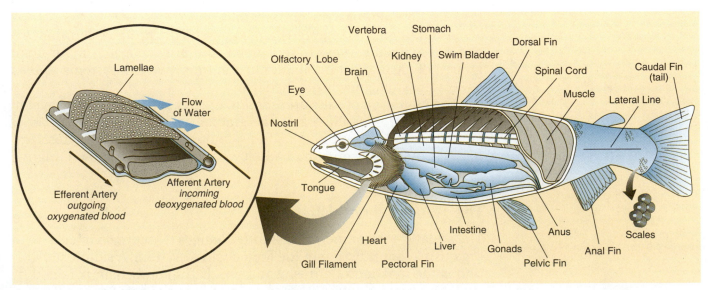

FIGURE 12-6 The anatomy of a fish.

mechanism that filters great quantities of water, filtering out edible microscopic plants and animals.

Crustaceans are a group of aquatic organisms with exoskeletons. These organisms **molt,** or replace their outer shells as they grow. Salt-water lobsters (*Homarous spp.*), crawfish (*Procambarus spp.*), and the various crabs, shrimps, and prawns are important crustaceans. These mobile organisms are characterized by hard exoskeletons that must soften and split as the animals molt and secrete larger shells.

Aquaculture Production

The aquaculturist, in an attempt to increase the production of any aquatic organism, must monitor and maintain the optimum **water quality.** Water quality has several chemical and physical characteristics that interact within the water and must be measured and maintained within a narrow range to satisfy growth and development (Figure 12-7).

There are several different types of water-quality test kits available to test the following characteristics.

Salinity is the measurement of total mineral solids in water. It is measured by either electrical conductivity (ohms/cm) or is calculated against known standards and converted to **ppt** (parts per thousand). The higher the concentration of salts, the greater the conductivity reading. A seawater standard can be made with artificial sea salts or by dissolving 29.674 g of NaCl in one liter of H_2O. Measurements of salinity usually range from 0 to 32 ppt (fresh water to salt water).

Dissolved oxygen (oxygen in water) is a function of temperature, pressure, and atmospheric oxygen. The cooler the water, the higher the pressure. The more contact with atmospheric oxygen, the higher and faster oxygen can be solubilized in water. Measured by oxygen probes or chemical tests, the results are reported as 0–10 **ppm** (parts per million). Water at 85°F (30°C) is saturated at about 8 ppm. Most fish can survive at levels as low as 3 ppm but quickly become stressed and succumb to other problems. Rainbow trout must have excellent, or high, levels of dissolved oxygen and can only be cultured in oxygen-saturated water (Figure 12-8).

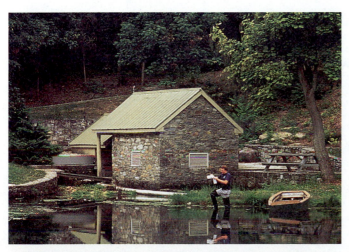

FIGURE 12-7 Water quality measurements must be taken by instruments or by chemical tests. *(Courtesy of USDA/ARS #K-4246-12)*

FIGURE 12-8 Oxygen content of water is increased by churning and mixing water through air. *(Courtesy of Elmer Cooper)*

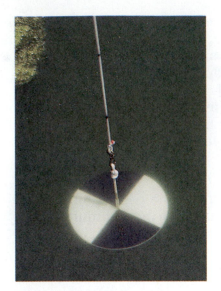

FIGURE 12-9 Measuring turbidity with a Secchi disc. *(Courtesy of Elmer Cooper)*

The measurement of acidity or alkalinity in water is the pH. This factor affects the toxicity of soluble nutrients in the water. Using a pH meter or paper tape, this measurement is recorded as a number from 1 to 14. Readings below 7 are acidic solutions, 7 is neutral, and numbers above 7 are alkaline. Most aquatic plants and animals grow well in water with a pH between 7 and 8.

Water hardness is measured by chemical analysis and is expressed as ppm calcium. This element is essential in the development of the exoskeletons of shellfish and crustaceans. It also serves as a chemical **buffer** that stabilizes rapid shifts in pH.

Turbidity in water is caused by the presence of suspended matter. High turbidity limits photosynthesis and visibility. A simple method for estimating pond turbidity uses a white Secchi disc that is lowered into the water. When visibility is impaired, the depth is recorded. The greater the depth, the less turbid the water (Figure 12-9).

Temperature limits the adaptive range of almost all aquatic organisms. Sunlight warms the upper surface of open water but does not penetrate it. Deep waters, cool-region currents, and melting winter covers can affect water temperature.

Ammonia/nitrite/nitrate is a group of nitrogen compounds generated by aquatic animals, first as urea and ammonia. These waste products are converted first to nitrite by microscopic organisms in the

AGRI·PROFILE

CAREER AREA: Aquaculture Researcher

Catfish farming and other aquaculture enterprises are big business in many nations of the world. Similarly, catfish farming is one of the fastest-growing food production enterprises in the United States. Cultured seafood production is rapidly approaching the volume of that taken from natural waters.

The rapid growth in aquaculture has spurred research-and-development activities. These in turn have stimulated career opportunities in animal science, nutrition, genetics, physiology, aquaculture construction, facility maintenance, pollution control, fish management, harvesting, marketing, and other areas.

As productive land becomes more scarce among the global resources, aquaculture will play an increasing role in providing high-quality food at affordable prices.

Animal physiologist Cheryl Goudie of the Catfish Genetic Research Unit at Stoneville, Mississippi applies gentle pressure to cause a female catfish to release her eggs for artificial fertilization. *(Courtesy of USDA/ARS #K-5325)*

water, and then to nitrate. They are ultimately converted to nitrogen gas or are absorbed by plants. The accumulation of both ammonia and nitrite is toxic to fish and often limits commercial production. Total ammonial nitrogen, or **TAN,** is recorded by chemical assay in ppm. This does not reflect the toxicity of the measured amount, because the toxicity of ammonia is dependent on the pH. Generally, levels of TAN are maintained below 1 ppm.

Toxins represent a host of materials that adversely affect the growth and development of aquatic plants and animals. These include agricultural chemicals, pesticides, municipal wastes, and industrial sludges. Chemical analyses are difficult and often inconclusive.

Selection of Aquaculture Crops

The actual selection of aquatic crops that may be grown is dependent on the resources and experience of the aquaculturist. Like terrestrial crops, each species of aquatic crop has a particular set of water-quality needs to ensure survival and reproduction. A discussion of a few of the well-known aquatic crops should indicate the diversity of this commercial industry. Characteristics will also vary between species. Ideal conditions should be requested from your local county extension office and local aquaculturists.

Trout and **salmonids** are high-quality fish products in high demand. The trout flourishes in high-quality water. Dissolved oxygen must be kept above 5 ppm. Salmonids also require low salinity, cool temperature 60°F (15°C), and low turbidity. The TAN must be maintained at less than 0.1 ppm.

Catfish (*Ictalurus spp.*) represent one of the fastest-growing aquaculture industries in the United States. Current figures project that over 130,000 acres of ponds are producing catfish annually. Catfish thrive at 75–79°F (24–26°C), a pH between 6.6 and 7.5, a water hardness of 10 ppm, and dissolved oxygen above 4 ppm.

Crawfish (*Procambarus spp.*) thrive in freshwater lakes and streams. Good growth occurs at 70–84°F (21–29°C), a pH close to neutral, a water hardness of 50–200 ppm, salinity up to 6 ppt, and dissolved oxygen above 3 ppm.

Clams, crabs, and oysters are cultivated in bays and estuaries subject to tidal flows. Good growing conditions vary greatly with species, but include approximately 6–20 ppt salinity, 15–30°C, dissolved oxygen above 1 ppm, and adequate amounts of microscopic organisms (low turbidity). Production should improve on cultivated sites when improved **spat,** or young oysters, are developed for stocking (Figure 12-10).

Shrimp and prawn are being cultured in brackish-water ponds and estuaries. Conditions for good growth are temperatures above 25°C, a salinity of 20 percent, high levels of micro-organisms, and dissolved oxygen above 4 ppm. Hatchery techniques are complex and involve several distinct growing stages.

Production Systems

The cultivation of any aquatic organism by the aquaculturist integrates the necessary cultural requirements with existing resources. This blend has developed three general production programs.

Open ponds, rivers, and bays are stocked with natural or cultured young and are maintained with densities that are balanced with the

FIGURE 12-10 The highly prized blue crab of the Chesapeake Bay. *(Courtesy of USDA/ARS #K-3692-3)*

FIGURE 12-11 Commercial fish ponds are subject to weather conditions. *(Courtesy of Rick Parker)*

existing ecosystem. Competing species are controlled, and natural recycling techniques are encouraged. This form of aquaculture can use both natural and constructed ponds (Figure 12-11). Some care must be taken to prevent any peaks in TAN. Overfertilization stimulates rapid algae blooms, with high levels of dissolved oxygen during photosynthesis but very low levels during early morning hours. Stress leads to high levels of mortality. Even lower dissolved-oxygen levels and fish kills can occur. The aquaculturist must be careful to monitor incoming water for toxins and other suspended materials. Harvesting must involve draining or **seining** the entire production area. Seining is the removal of fish with nets.

Caged Culture

Caged culture represents a more capital-intensive program. Aquatic animals or plants are contained in a small area, and waste products are removed by the flushing action of the natural waters (Figure 12-12). Confining growing fish in floating-pen cages or shellfish on suspended float tables are techniques for managing increasing densities of aquatic organisms. As is true on a cattle feedlot, the aquaculturist confines the

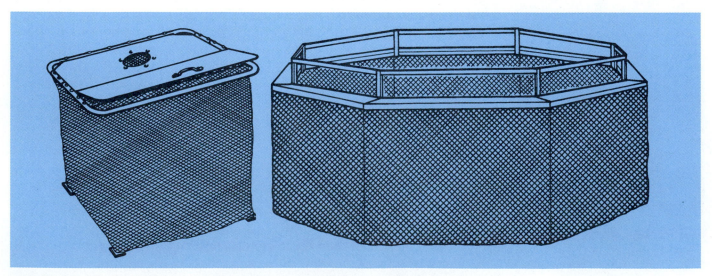

FIGURE 12-12 Fish can be raised in pens or cages sitting or suspended in water.

crops in a limited area and provides the necessary feed and cultural management. The natural ebb and flow of the water removes waste products and replenishes dissolved oxygen.

Cage culture can be designed for both natural waters and newly constructed ponds. Aquaculturists have a better idea of growth rates and can adjust feeding ratios more economically. Some growers have reported problems when a pond **rolls over** (changes water quality suddenly during cetain weather conditions and brings the less-oxygenated water to the surface). Fish in cages are unable to move and can be stressed or killed. In this intensive production system, the aquaculturist must ensure adequate nutrition, disease control, predatory control, and physical maintenance like any terrestrial animal producer. The young stock must be caught from natural waters or produced in controlled hatcheries. Successful operations include the production of Atlantic salmon off the coast of Norway, Nova Scotia, and Maine. Hybrids of striped bass have been cultured in cages in Maryland and California. Trout have been cultured in net pens suspended in mountain streams and ponds.

Shellfish growers in Japan and the United States have demonstrated the production of oysters, clams, and mussels on suspended float tables or ropes. Production is increased by higher water quality and ease of harvest.

Recirculating Tanks

Many areas of the world lack sufficient water resources to maintain a viable aquaculture industry. Recirculating systems must circulate the waste water through a biological purifier and return it to the growing tank (Figure 12-13). This complicated process is similar to that in a city waste-treatment plant. The system must remove the solid fish wastes, soluble ammonia/nitrite/nitrates, and carbon dioxide, and replace dissolved oxygen. The pH must be maintained and integrated into the biological needs of the bacteria that populate the biological filters. As in any pond, if any single parameter is ignored, the organisms become stressed and production is decreased.

Extensive research in aquaculture has been done at The University of Wisconsin with yellow perch; the University of Maryland with striped bass, trout, and tilapia; the Walt Disney World EPCOT Center with numerous species; Mississippi State University with catfish; and many other institutions in the United States.

Hatcheries

The development of more intensive aquaculture systems will depend on a constant supply of high-quality young organisms. The natural fisheries are threatened by overfishing, pollution, and habitat destruction. Hatcheries are investigating the parameters that affect fish breeding habits and attempt to induce spawning, fry, or larvae production. **Larvae** are mobile aquatic organisms that grow into nonmobile adults.

These advances in fish and shellfish management allow the industry to develop improved breeds and hybrids to support improved production. Improved genetic lines of trout, catfish, and salmon are already commercially available. The increase of existing ornamental fish within a country also prevents the introduction of imported fish diseases and parasites.

FIGURE 12-13 New technology is making fish production profitable in tanks in sheds, warehouses, greenhouses, and other enclosures. *(Courtesy of National FFA)*

Aquaculture and Resource Management

The demand for aquaculture products will continue to increase. This demand will stimulate a tremendous growth in commercial aquaculture. During the expansion of the commercial aquaculture industry, the conflicts for scarce natural resources and clean water, the impact on recreational areas, and the potential pollution effects will need to be resolved by trained specialists.

BIO·TECH CONNECTION

Aquaculture Research Is Big Time!

The Aquaculture Research Center in Baltimore, Maryland was established to provide combined research facilities for the University of Maryland Center of Marine Biotechnology, the $160 million Columbus Center for Marine Sciences, and the National Aquarium in Baltimore. The Center was established by converting a harbor-side shipbuilding structure into a state-of-the-art aquaculture research facility. Further, the Columbus Center will eventually occupy two large piers in the Inner Harbor area to complete the aquaculture research complex.

Scientists are growing high densities of valuable fish, such as striped bass, in closed systems using biological filters for cleaning the water to make the facility environmentally friendly. Tools of biotechnology are being used to acquire knowledge about native species and develop novel solutions to problems associated with commercial finfish and shellfish culture. Studies in the environmental, hormonal, and molecular regulation of spawning will enable growers to have access to seedstocks whenever they need them, rather than being dependent upon the normal spawning cycles.

The study of fish genes and hormones that control fish growth and the study of fish nutrition will decrease the time and cost of raising fish to market weight. Of primary im-

Acidity (left) and solids (right) tests are among the many tests needed periodically to keep fish growing in this 7,400 gallon recycled-water aquaculture system. (Courtesy USDA/ARS #K-4248-2)

portance is the study of devastating diseases that have nearly eliminated the commercial oyster population and certain other species in many natural waters. The worldwide demand for seafood is projected to increase by 70 percent by the year 2025. Yet the world's natural fisheries are already stressed beyond sustainable limits. To meet the increased demand for seafood with decreasing harvests from natural waters, it is believed that aquaculture will have to increase its output by seven times the 1993 levels.

With its working participatory science laboratory and public spaces for observation, the Aquaculture Research Center is an international focal point for scientists, a major research center, and an educational center.

STUDENT ACTIVITIES

1. Write the Terms to Know and their meanings in your notebook.

2. Visit a local supermarket or seafood market and list the seafood products. Classify them as fish, shellfish, or crustaceans; freshwater or saltwater products; or imported products.

3. Describe why some of the above products might not be produced in your area.

4. Locate three local aquaculturists and discuss their production systems. How do they maintain water quality?

5. Visit a local pet store. Describe the various parts of a freshwater aquarium. How is the water quality maintained in this recirculating system?

6. Make a bulletin board illustrating the food chain of a freshwater pond.

7. Set up a class aquarium and discuss the balance between plants and animals in the system.

8. Make salt water.

SELF EVALUATION

A. Multiple Choice

1. The aquaculturist must understand how aquatic organisms
  a. eat.
  b. reproduce.
  c. live.
  d. all of the above.

2. The yearly catch of fish from natural waters is
  a. increasing.
  b. holding constant or decreasing.
  c. hard to determine.
  d. mostly catfish.

3. The highest salinity is measured in
  a. pond water.
  b. irrigation water.
  c. creeks.
  d. ocean water.

4. The accumulation of salts in water occurs most often when
  a. water runs across agricultural land.
  b. water settles in a pond.
  c. water collects in a drainage ditch.
  d. water is lost through evaporation.

5. Brackish water is
   a. colored black.
   b. located in tidal areas.
   c. collected from small creeks and branches.
   d. mostly high in salinity (20–34 ppt).

6. Water quality is less affected by
   a. fish density.   c. chemical run-off.
   b. weather.   d. fish species.

7. The greater the density of fish in a system, the
   a. smaller the tank.
   b. larger the fish.
   c. the more difficult the management.
   d. the higher the temperature.

8. A fish kill can occur when a pond "rolls over"
   a. because of the temperature shock.
   b. because the cages sink to the bottom.
   c. because of low levels of dissolved oxygen.
   d. because the fish turn upside down.

B. Matching

____ **1.** Salinity   a.   Measured in ppm calcium
____ **2.** Dissolved oxygen   b.   Recorded as degrees F or C
____ **3.** Turbidity   c.   Measured as ppm or percent
____ **4.** Temperature   d.   Depth of visual penetration
____ **5.** Ammonia   e.   TAN
____ **6.** Hardness   f.   Conductivity or ppt

C. Completion

1. Both plants and animals are part of the food _____.
2. The _____ _____ are where fresh water and seawater mix.
3. The _____ of a fish absorb oxygen from the water.
4. Between 32 and 24 ppt is the salinity of _____.
5. Trout need a _____ dissolved oxygen concentration than clams to grow and mature.
6. Crawfish must _____, or break out of their exoskeletons to grow.
7. Salmon must return to _____ water to spawn and complete their life cycles.
8. Pond culture relies mostly on _____ recycling of fish waste products.
9. The recirculating production systems must treat the fish wastes with _____ filters.

Integrated Pest Management

A Balancing Act!

Honeybees are a major pillar in the food, fiber, flower, and ornamentals production arena in the United States. We rely on them as the only method of pollinating certain species and count on them to do some of the pollination of nearly all species of plants. Bees enter flowers of plants to gather nectar and pollen for their own food and nourishment of their young. Their service to humans and animals in helping plants to get pollinated, and thereby produce seed and fruit, is a service we cannot do without. Most plants could not reproduce and survive without producing seed. Their safety from pesticides used to control harmful insects is always at the top of the agenda for entomologists.

Honeybees are friends we cannot do without. However, the Africanized hybrids can be dangerous and threaten to decrease bee productivity in the United States. *(Courtesy USDA/ARS #K-3652-12)*

Honeybees are numbered among the many insects that are beneficial to us. While they can sting if threatened, honeybees in the United States are of the European type and are predictable. Except for the inexperienced person approaching a bee-hive, honeybee stings are generally a single sting by a single bee. Except for the relatively few individuals who have life-threatening allergic reactions to bee stings, honeybees pose little threat, because they act individually when they are aggravated enough to sting.

Enter the Africanized honeybee—a hybrid cousin of the domestic pollinator, famous for its aggressiveness and dubbed "killer bee!" On October 15, 1990, the first Africanized honeybee swarm to migrate naturally to the United States was identified by entomologists near Brownsville, Texas. The swarm was promptly destroyed, according to standard procedure.

Africanized honeybees resulted when in 1957 bees were imported from Africa to Brazil by a Brazilian scientist. The plan was to experiment by crossbreeding them with the domestic European bees prevalent in the Americas to develop a better strain of honeybees for the tropics. Unfortunately, some African bees were inadvertently released in the countryside and promptly interbred with the domestic bees. The new hybrids and their descendents are known as Africanized honeybees. They have migrated as far south as Argentina and as far north as the United States.

Unfortunately, Africanized honeybees have different dispositions then the domestic bees of the United States. They tend to defend their colonies more vigorously, stinging in greater numbers and with less provocation. One bee is likely to inflict many stings. Therefore, there is greater danger in an encounter with the Africanized bees. Further, U.S. agriculture and the beekeeping industry fear that domestic bees interbred with Africanized bees may become harder to manage as pollinators of crops and may not be as efficient as honey producers.

The challenge of observing, detecting, and stopping the northward migration of Africanized honeybees will be a top priority of government inspectors, entomologists, beekeepers, farmers, and the citizens at large. At the same time, animal behaviorists will study the bee's habits and look for ways to manage them. Geneticists will study the bee's genes and look for ways to genetically engineer future bees so as to decrease their objectionable habits and enhance their abilities as pollinators and honey makers.

Biological, Cultural, and Chemical Control of Pests

OBJECTIVE

To develop an understanding of the major pest groups and some elements of effective pest-management programs.

COMPETENCIES TO BE DEVELOPED

After studying this unit, you should be able to:

- define *pest, disease, insect, weed, biological, cultural, chemical,* and other terms associated with integrated pest management.
- know how the major pest groups adversely affect agriscience activities.
- describe weeds based on their life cycles.
- describe both the beneficial and detrimental roles that insects play.
- recognize the major components and the causal agents of disease.
- explain and understand the concept of integrated pest management.

MATERIALS LIST

✓ insect net

✓ killing jar

✓ insect mounting pins

✓ insect specimen labels

✓ pictures of pests

TERMS TO KNOW

Diseases
Vector
Insect
Arachnid
Pest
Defoliates
Pesticide
Weeds
Pathogens
Annual weed
Biennial weed
Perennial weed
Rhizome
Node
Stolon
Meristematic tissue
Entomophagous
Exoskeleton

Metamorphosis
Instars
Plant disease
Causal agent
Disease triangle
Abiotic (nonliving) diseases
Biotic diseases
Fungi
Hyphae
Mycelium
Bacteria
Appressorium
Viruses
Symptom
Mosaic
Nematodes
Integrated pest management (IPM)

Key pest
Pest population equilibrium
Economic threshold level
Monitoring
Quarantine
Targeted pest
Eradication
Pheromone
Cultivar
Biological control
Cultural control
Clean culture
Trap crop
Chemical control
Pesticide resistance
Pest resurgence

The ability to control pests by either chemical, cultural, or natural control methods has afforded the American people an unprecedented standard of living. We often take for granted an unlimited food supply, good health, a stable economy, and an aesthetically pleasing environment. Without effective pest control strategies, our standard of living would decrease.

Good pest management practices have resulted in dramatic yield increases for every major crop. A single U.S. farmer in 1850 could only support himself and four people, while in the 1990s a farmer can provide food and fiber for more than 130 people. The ability to control plant and animal **diseases** or disorders vectored by insects and arachnids has reduced the incidence of malaria, typhus, and Rocky Mountain spotted fever. A **vector** is a living organism that transmits or carries a disease organism. An **insect** is a six-legged animal with three body segments. An **arachnid** is an eight-legged animal, such as a spider or a mite.

The impact of pest management in maintaining a stable economy can be seen on a regional and national basis. The regional economy suffered shortly after the cotton boll weevil's introduction into the

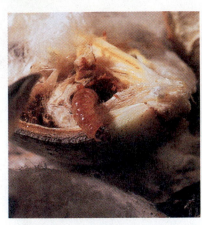

FIGURE 13-1 A sliced open cotton boll showing a pink boll-worm and the damage it has done. *(Courtesy of USDA/ARS #K-2886-13)*

FIGURE 13-2 Pear fruits that yellowed and shrivelled when the fire blight disease cut off the flow of nutrients from the tree to the fruit. *(Courtesy of USDA/ARS #K-5300-1)*

United States in 1892. The weevil devastated much of the cotton crop in the early 1900s (Figure 13-1). Similarly, the potato blight disease in Ireland caused famine and mass migration of Irish people to other parts of the world in 1845. Today many blights are still serious threats to our crops (Figure 13-2).

TYPES OF PESTS

The word **pest** is a general name for any organism that may adversely affect human activities. We may think of an agricultural pest as one that competes with crops for nutrients and water, **defoliates** plants (strips a plant of its leaves), or transmits plant or animal diseases. The major agricultural pests are weeds, insects, and plant diseases. However, other types of pests exist. Some examples, and the classes of **pesticides** or chemicals used for killing them, are listed below.

Type of Pest	Class of Pesticide
mites, ticks	acaricide
birds	avicide
fungi	fungicide
weeds	herbicide
insects	insecticide
nematodes	nematacide
rodents	rodenticide

Damage by pests to agricultural crops in the United States has been estimated to be one-third of the total crop production potential. Therefore, an understanding of the major pest groups and their biology is required to ensure success in reducing crop losses to pests.

Weeds

Weeds are plants that are undesirable and are often considered out of place (Figure 13-3). The definition of a weed is therefore a relative term. Corn growing in a soybean field or white clover growing in a turf-grass are examples of weeds.

Weeds can be considered undesirable for any of the following reasons:

1. They compete for water, nutrients, light, and space, resulting in reduced crop yields.
2. They decrease crop quality.
3. They reduce aesthetic value.
4. They interfere with maintenance along rights-of-way.
5. They harbor insects and disease **pathogens** (organisms that cause disease).

Weeds can be divided into three categories based on their life spans and their periods of vegetative and reproductive growth.

Annual Weeds An **annual weed** is a plant that completes its life cycle within one year. Two types of annual weeds occur, depending upon the time of year in which they germinate. A winter annual germinates in the fall and will actively grow until late spring. It will then produce

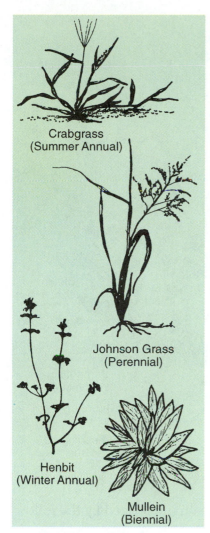

Crabgrass
(Summer Annual)

Johnson Grass
(Perennial)

Henbit
(Winter Annual)

Mullein
(Biennial)

FIGURE 13-3 Different types of plants considered to be weeds and their life cycles. *(Courtesy of Maryland Cooperative Extension Service)*

seed and die during periods of heat and drought stress. Examples of winter annuals are chickweed, henbit, and yellow rocket.

A summer annual germinates in the late spring, with vigorous growth during the summer months. Seed are produced by late summer and the plant will die during periods of low temperatures and frost. Examples of summer annuals are crabgrass, spotted spurge, and fall panicum.

Biennial Weeds   A **biennial weed** is a plant that will live for two years. In the first year the plant produces only vegetative growth, such as leaf, stem, and root tissue. By the end of the second year, the plant will produce flowers and seeds. This is referred to as reproductive growth. After the seed is produced, the plant will die. There are only a few plants that are considered biennials. Some examples are bull thistle, burdock, and wild carrot.

Perennial Weeds   A **perennial weed** can live for more than two years and may reproduce by seed and/or vegetative growth. The production of rhizomes, stolons, and an extensive rootstock are ways in which perennial plants reproduce vegetatively. A **rhizome** is a stem that runs underground and gives rise to new plants at each joint, or **node.** A **stolon** is a stem that runs on the surface and gives rise to new plants at each node. These plant parts have **meristematic tissue** (tissue capable of starting new plant growth). Examples of perennial weeds are dandelion, Bermuda grass, Canada thistle, and nutsedge.

Insects

Beneficial Insects   Scientists estimate that there are more than 1 million species of insects that inhabit the world. A majority of them are beneficial, or helpful, to humans. For example, insects are necessary for plant pollination. In the United States it is estimated that bees pollinate more than $1 billion worth of fruit, vegetable, and legume crops per year. Honey, beeswax, shellac, silk, and dyes are just a few of the commercial products produced by insects. Many insects are entomophagous and help in natural control of their insect species. **Entomophagous** insects feed on other insects. Insects that inhabit the soil, act as scavengers, or feed on undesirable plants all play important roles. These insects increase soil tilth, contribute to nutrient recycling, and act as biological weed control agents.

Finally, insects are at the lower levels of the food chain. Thus, they support higher life forms, such as fish, birds, animals, and humans.

Insect Pests   When compared to the total number of insect species, there are relatively few species that cause economic loss. However, it is estimated that crop losses due to such insects total more than $4 billion annually.

Insects can cause economic loss by feeding on cultivated crops and stored products. They can also vector plant and animal diseases, inflict painful stings or bites, or act as nuisance pests.

Insect Anatomy   Insects are considered to be one of the most successful groups of animals present on Earth. Their success in numbers and species is attributed to several characteristics, which include their anatomy, reproductive potential, and developmental diversity.

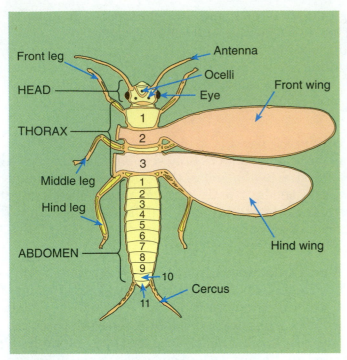

FIGURE 13-4 A diagram of an adult insect.

FIGURE 13-5 The different types of insect damage.
(Courtesy Maryland Cooperative Extension Service)

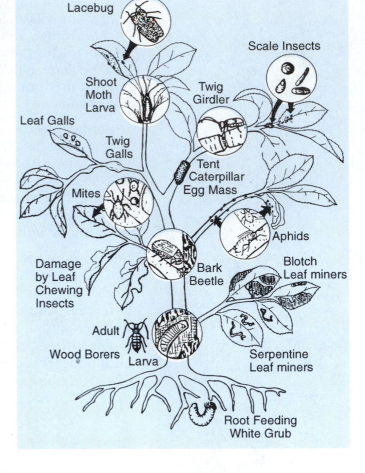

Insects are in the class Insecta and are characterized by the following similarities (Figure 13-4):

1. Each has an **exoskeleton,** which is the body wall of the insect. It provides protection and support for the insect.
2. The exoskeleton is divided into three regions: head, thorax, and abdomen.
3. There are segmented appendages on the head called antennae, which act as sensory organs.
4. Three pairs of legs are attached to the thorax of the body.
5. Wings are present (one or two pairs) in a majority of species. This permits mobility and greater utilization of habitat.

Feeding Damage Insects have either chewing or sucking mouthparts. Damage symptoms caused by chewing insects are leaf defoliation, leaf mining, stem boring, and root feeding. Insects with sucking mouthparts produce distorted plant growth, leaf stippling, and leaf burn (Figure 13-5).

Development As an insect grows from an egg to an adult, it passes through several growth stages. This growth process is known as **metamorphosis.** The two types of metamorphosis are gradual and complete.

Gradual metamorphosis consists of three life stages: egg, nymph, and adult (Figure 13-6). As a nymph, the insect will grow and pass through several **instars** (the stage of the insect between molts). Each time the insect sheds its exoskeleton, or molts, it passes into the next

FIGURE 13-6 Gradual metamorphosis of the chinch bug. (A) egg, (B) to (F) first to fifth instars, (G) adult.

instar. For example, chinch bugs have five instars before they reach adult form, but will vary in size, color, wing formation, and reproductive ability. When the insect reaches the adult stage, no further growth will occur.

Complete metamorphosis consists of four life stages: egg, larva, pupa, and adult (Figure 13-7). The larval stage is the period when the insect grows. As larvae molt, they pass to the next larval instar. A Japanese beetle will have three larval instars before developing to the pupa stage. The pupa is a resting period. It is also a transitional stage of dramatic morphological change from the larva to adult.

Plant Diseases A **plant disease** is any abnormal plant growth. The occurrence and severity of plant disease is based on these three factors:

1. A susceptible plant or host must be present.
2. The disease organism, or causal agent, must be present. A **causal agent** is an organism that produces a disease.
3. Environmental conditions conducive to the causal agent must occur.

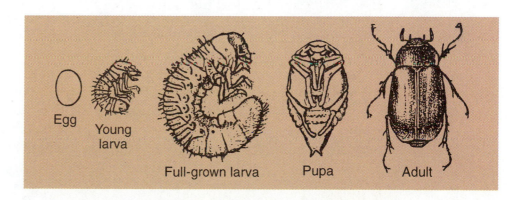

Egg   Young larva   Full-grown larva   Pupa   Adult

FIGURE 13-7 Complete metamorphosis of the June beetle.

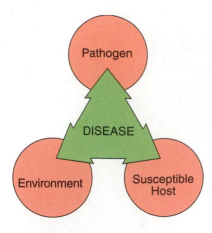

FIGURE 13-8 Components of the disease triangle.

The relationship of these three factors is known as the **disease triangle** (Figure 13-8). Disease-control programs are designed to affect each or all of these three factors. For example, if crop irrigation is decreased, a less favorable environment will exist for the disease organism. Breeding programs have introduced disease resistance into new plant lines for many different crops. Pesticides may also be used to suppress and control the disease organism.

Causal Agents for Plant Disease

Diseases may be incited by either abiotic factors or biotic agents. **Abiotic (nonliving) diseases** are caused by environmental or manmade stress. Examples of abiotic diseases are nutrient deficiencies, salt damage, air pollution, and temperature and moisture extremes.

AGRI·PROFILE

CAREER AREA: Entomology/Plant Pathology

The work of entomologists and plant pathologists is never done. They do battle in the laboratory, in buildings, or in the field against insects and diseases that consume or ruin much of what we produce. The advice and service of these specialists is sought to control diseases and damaging insects, as well as to encourage beneficial ones.

Entomologists and plant pathologists attempt to control or reduce the buildup of damaging insect and disease populations. Such work may include assessing damage; attracting, trapping, counting, and observing insects; and advising, directing, and assisting those who attempt to control insects and plant diseases. The mysteries of some insects are so great that scientists must specialize on just a few insects to be truly knowledgeable about them. Gone are the days when chemicals were our chief means of controlling insects and diseases. Rather, we now utilize a variety of techniques collectively known as integrated pest management.

Plant pathologists observe tree damaged by fire blight disease (foreground) and healthy, fire blight-resistant trees (background). *(Courtesy USDA/ARS #K-5310-1)*

Career opportunities exist for field and laboratory technicians, as well as for degree-holding specialists. Neighborhood jobs may include termite control, scouting, spraying, crop dusting, inspecting, monitoring, selling, and managing field research projects. Honey bee specialists may manage hives to pollinate crops for improved seed and fruit production.

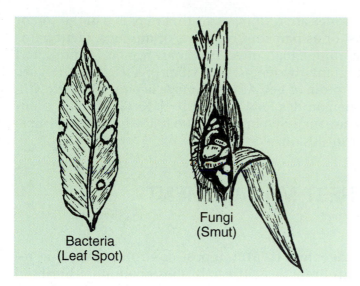

FIGURE 13-9 Examples of biotic diseases.

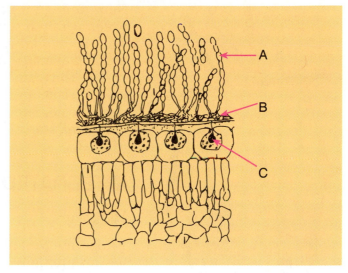

FIGURE 13-10 Powdery mildew of rose. (A) conidia (a sexual spore), (B) mycelium of fungus, and (C) fungal haustoria.

Biotic means living. **Biotic diseases** are caused by living organisms (Figure 13-9). Examples of causal agents or organisms are fungi, bacteria, viruses, nematodes, and parasitic seed plants. Organisms are parasites if they derive their nutrients from other living organisms. Examples and a discussion of causal agents for plant diseases follow.

Fungi   Fungi (plural for fungus) are the principal causes of plant disease. **Fungi** are plants that lack chlorophyll. Their bodies consist of threadlike vegetative structures known as **hyphae** (Figure 13-10). When hyphae are grouped together, they are called **mycelium.** Fungi can reproduce and cause disease by producing spores or mycelia. Spores can be produced asexually or sexually by the fungus. For example, a mushroom produces millions of sexual spores under its cap. These spores can be dispersed by wind, water, insects, and humans.

Bacteria   **Bacteria** are one-celled or unicellular microscopic plants. Relatively few bacteria are considered plant pathogens. Being unicellular, bacteria are among the smallest living organisms. Bacteria can enter a plant only through wounds or natural openings, such as a stem lenticel. Bacteria can be disseminated in ways similar to fungi, but they do not produce appressoria. An **appressorium** is the swollen tip of a hypha by which the fungus attaches itself to the plant. Several important bacterial diseases are fire blight of apples and pears and bacterial soft rot of vegetables.

Viruses   Plant viruses are pathogenic, or disease-causing, organisms. **Viruses** are composed of nucleic acids surrounded by protein sheaths. They are capable of altering a plant's metabolism by affecting protein synthesis. Plant viruses are transmitted by seeds, insects, nematodes, fungi, grafting, and mechanical means, including sap contact. Viral diseases produce several well-known symptoms. A **symptom** is the visible change to the host caused by a disease. These symptoms are ring spots, stunting, malformations, and mosaics. A **mosaic** symptom is a leaf pattern of light- and dark-green color.

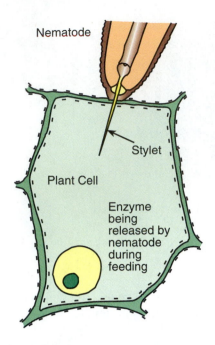

FIGURE 13-11 A nematode feeding on a plant cell. The stylet is a tiny spear-like feeding structure.

Nematodes Nematodes are roundworms that may live in the soil or water, within insects, or as parasites of plants or animals. Plant parasitic nematodes are quite small, often less than ¼ inch (4 mm), and produce damage to plants by feeding on stem or leaf tissue (Figure 13-11). The main symptom of nematode damage is poor plant growth, which results from nematodes feeding on the roots. The major plant parasitic nematodes are included in one of the following three groups: root-knot, stunt, or root-lesion.

INTEGRATED PEST MANAGEMENT

History

Integrated pest management (IPM) is a pest-control strategy that relies on multiple control practices. It establishes economic threshold levels in determining control actions. The concept of integrated control is not new. Entomologists had developed an array of cultural and natural controls for the boll weevil and other insect pests by the early 1900s.

However, our approach to pest management during the period 1940–1969 moved to a major reliance on chemical pesticides. Alternate control strategies were deemphasized, because chemical control gave excellent results at a low cost.

It was not until 1972 that a major change in policy occurred in the United States to encourage other pest-control strategies. Natural, biological, and cultural control programs began to be introduced as alternatives to chemical pest control.

The more recent trend toward reduced use of chemicals for pest control was triggered by a book published in 1962 entitled *Silent Spring*, by biologist Rachel Carson. After 1962, heavy reliance on chemical pest management began to be questioned. Carson's book created a public awareness of the environmental pollution that results from the overuse of pesticides. Adverse effects from pesticide misuse and/or overuse were beginning to occur as well. These included pest resurgence, resistance to pesticides, and concern over human health from exposure to pesticides.

Since the 1970s, great strides have been made in the development and implementation of IPM programs. The end result has been to lessen dependency on chemical use, while still achieving acceptable pest control.

Principles and Concepts of IPM

The following concepts or principles are important in understanding how IPM programs should operate.

Key Pests A key pest is one that occurs on a regular basis for a given crop (Figure 13-12). It is important to be able to identify key pests and to know their biological characteristics. Further, the weak link in each pest's biology must be found if management of the pest is to be successful.

Crop and Biology Ecosystem The integrated pest manager must learn the biology of the crop and its ecosystem. The ecosystem of the crop consists of the biotic and abiotic components that are present

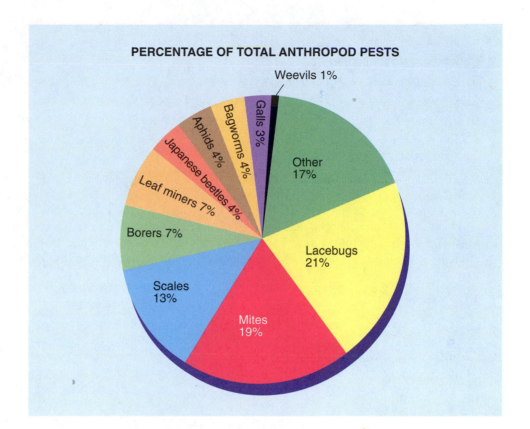

PERCENTAGE OF TOTAL ANTHROPOD PESTS

Weevils 1%
Galls 3%
Bagworms 4%
Aphids 4%
Japanese beetles 4%
Leaf miners 7%
Borers 7%
Scales 13%
Mites 19%
Lacebugs 21%
Other 17%

FIGURE 13-12 Arthropod pests and their percentages in six Maryland communities. *(Adapted from material from Entomology Department, University of Maryland)*

within the crop. The biotic components of the ecosystem are the living plants and animals. The abiotic components are nonliving factors, such as soil and water. Examples of human-managed ecosystems are a field of soybeans, a turfgrass area, or a poultry production operation.

Ecosystem Manipulation   With IPM, an attempt is made to understand the influence of ecosystem manipulation on lowering pest populations (Figure 13-13). To illustrate this concept, the manager must ask, What would happen to the pest population equilibrium if a disease-resistant plant were introduced? **Pest population equilibrium** occurs when the number of pests stabilizes or remains steady. The introduction of disease-resistant plants should lower the pest population below the economic threshold level. The **economic threshold level** is the point where pest damage is great enough to justify the cost of additional pest control measures.

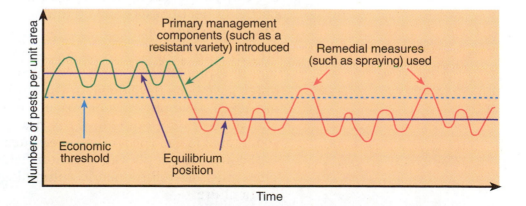

FIGURE 13-13 The effect of lowering the equilibrium position of a pest. *(Adapted from material from University of Maryland, Entomology Department)*

Numbers of pests per unit area

Primary management components (such as a resistant variety) introduced

Remedial measures (such as spraying) used

Economic threshold

Equilibrium position

Time

Economic Injury Threshold for Alfalfa Weevil; Number of Larvae From 30-Stem Sample

How to use table below:

1. Use plant height category that fits the field.
2. Estimate the value of crop in dollars per ton of hay equivalent and the cost to spray an acre.
3. From monitoring the field, find the number of alfalfa weevil larvae from a sample of 30 stems.
4. The number in each small box indicates the number of larvae per 30-stem sample that is required before a spray application would be profitable under these conditions.

EXAMPLE:

Plants in the field are 20 inches high (use Category II), hay is valued at $80 per ton, cost to spray is $8.00 per acre, and you collected 40 larvae from the sample of 30 stems. The number in the box common to $80 and $8 is 75. This means that under these conditions, 75 larvae are needed before a spray would be profitable. Since you collected only 40 larvae, a spray at this time will not be profitable.

Value of hay per ton	Category I plant height 12 to 18 inches						Category II plant height 18 to 24 inches						Category III plant height 24 to 30 inches					
$ 60	91	114	137	160	183	225	99	124	149	174	199	240	104	130	156	182	209	260
$ 80	68	85	102	119	136	171	75	94	113	131	150	186	78	97	117	137	157	195
$100	54	68	81	95	108	137	62	75	90	105	120	149	63	78	94	110	126	156
$120	45	57	68	79	91	114	50	62	75	87	100	124	52	65	78	91	105	130
$140	39	49	59	68	77	99	43	54	64	75	86	107	45	56	67	78	90	112
$160	34	43	51	60	68	86	37	47	56	65	75	93	39	49	58	68	79	98
	$8	10	12	14	16	20	$8	10	12	14	16	20	$8	10	12	14	16	20

Cost of insecticide application per acre

FIGURE 13-14 Chart to determine economic threshold level for the alfalfa weevil. *(Adapted from material from Pennsylvania State University)*

Threshold Levels The level of a pest population is important. For instance, the mere presence of a pest may not warrant any control measures. But at some point, the damage created by insects will be great enough to warrant control measures. Various threshold levels are developed to determine if and when a control measure should be implemented. This prevents excessive economic loss of plants to pest damage (Figure 13-14).

Economic threshold levels have been determined by first developing a pest-damage index (Figure 13-15). It is crucial in the decision-making process to know the level of pest infestation that will cause a given yield reduction. Pest populations are measured in several different ways. They can be counted in number of pests per plant or plant part, number of pests per crop row, or number of pests per sweep with a net above the crop.

Monitoring For IPM to be successful, a **monitoring** (checking) or scouting procedure must be performed. Different sampling procedures have been developed for various crops and pest problems (Figure 13-16). The presence or absence of the pest, amount of damage, and stage of development of the pest are several visual estimates a scout must make. The method used must be speedy and accurate.

Scouts are people who monitor fields to determine pest activity. They must be well trained in entomology, pathology, agronomy, and horticulture.

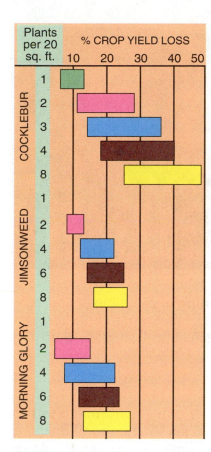

FIGURE 13-15 The pest-damage index for several weeds in soy beans.

PEST-CONTROL STRATEGIES

Pest-control programs can be grouped into several broad categories. These include regulatory; biological; cultural; physical and mechanical; and chemical.

Regulatory Control

Federal or state governments have created laws that prevent the entry or spread of known pests into uninfested areas. Regulatory agencies also attempt to contain or eradicate certain types of pest infestations. The Plant Quarantine Act of 1912 provides for inspection at ports of entry. Plant or animal quarantines are implemented if shipments are infested with targeted pests. A **quarantine** is the isolation of pest-infested material. A **targeted pest** is a pest that, if introduced, poses a major economic threat.

If a targeted pest becomes established, an eradication program will be started. **Eradication** means total removal or destruction of a pest. This type of pest control is extremely difficult and expensive to administer. In California, the Mediterranean fruit fly was eradicated at a cost of $100 million in 1982. This program relied on chemical spraying, sanitation, sterile male releases, and pheromone traps to ensure complete eradication. A **pheromone** is a chemical secreted by an organism to cause a specific reaction by another organism of the same species.

Host Resistance

The development of plants having pest resistance is an extremely effective control practice. The advantages of resistant varieties are as follows:

1. Low cost
2. No adverse effect to the environment
3. A significant reduction in pest damage
4. Ability to fit into any IPM program

Breeding programs attempt to identify and select plants with pest resistance. Currently, new plant cultivars with improved resistance to pests are released annually. A **cultivar** is a plant developed by humans, as distinguished from a natural variety.

Biological Control

Biological control means control by natural agents. Such agents may be predators, parasites, and pathogens. A predator is an animal that feeds on a smaller or weaker organism. An example of a predator is the lady beetle. Aphids are the lady beetle's principal prey. Parasites are organisms that live in or on another organism. The Braconid wasp is parasitic on the caterpillars of many moths and butterflies. Pathogens are organisms that will produce disease within their hosts. For example, the bacteria *Bacillus popilliae* is a pathogen, because it causes the milky spore disease in Japanese beetle grubs.

Successful biological control programs reduce pest populations below economic thresholds and keep the pests in check. Such

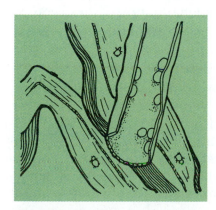

FIGURE 13-16 Random sampling of a plant stem and leaf to determine pest populations and damage. *(Courtesy Maryland Cooperative Extension Service)*

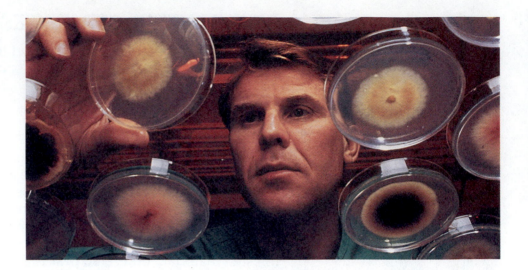

FIGURE 13-17 Plant pathologist Rick Bennent examines fungi that may be used for biological control of weeds. *(Courtesy of USDA/ARS #K-4652-1)*

programs require a thorough understanding of the biology and ecology of the beneficial organism, as well as of the pest. Careful research can even fit desirable plant pathogens against undesirable weeds (Figure 13-17).

Cultural Control

Cultural control is the attempt to alter the crop environment to prevent or reduce pest damage. It may include such agricultural practices as soil tillage, crop rotation, adjustment of harvest or planting dates, and irrigation schemes. Other practices that are considered cultural control are clean culture and trap crops. **Clean culture** refers to any practice that removes breeding or overwintering sites of a pest. This may include removal of crop leaves and stems, destruction of alternate hosts, or pruning of infested parts. A **trap crop** is a susceptible crop planted to attract a pest to a localized area. The trap crop is then either destroyed or treated with a pesticide.

Physical and Mechanical Control

Physical and mechanical control programs use direct measures to destroy pests. Examples of such practices are steam sterilization, hand removal, cold storage, and light traps. Implementation of these control practices are costly and provide varying pest-control results.

Chemical Control

Chemical control is the use of pesticides to reduce pest populations. Chemical control programs have been very cost effective. However, various problems occur if this practice is misused or overused. Problems that can develop are environmental pollution, pesticide resistance, and pest resurgence. **Pesticide resistance** is the ability of an organism to tolerate a lethal level of a pesticide. **Pest resurgence** refers to a pest's ability to repopulate after control measures have been eliminated or reduced.

Integrated pest management seems to be our best defense against pests. Biological and cultural controls are favored when they are effective. However, we cannot control certain pests without the use of chemical pesticides. Under such circumstances, it is very important to use chemical pesticides safely.

BIO•TECH CONNECTION

Bioengineering a Better Mosquito!

The common mosquito is a scourge from the Alaskan tundra to the equatorial marshes. The biological need of the female to obtain a blood meal to provide protein before laying her eggs causes the mosquito to be an annoyance and a threat to humans and animals alike. The small amount of blood extracted during a "bite" is not the problem. First, there is discomfort of the bite. However, the biggest problem lies in the life-threatening yellowfever or malaria-causing microorganisms that certain mosquitos transfer to humans during their blood meals. The search for methods to rid the environment of mosquitos is continuous.

Is there any such thing as a "good" mosquito? Yes, if the mosquito is a nonmalaria mosquito that replaces a malaria carrier. Geneticist Andrew Cockburn has developed a genetic engineering technique that may be useful in creating a malaria-resistant mosquito. Such mosquitos could theoretically thin out or replace their protozoan-carrying cousins and reduce or eliminate the malaria disease in humans.

A promising technique is the use of tiny needles, new genes, and insect eggs spun in an ordinary lab vortex in saline solution. The needles gently pierce the eggs and allow new material to enter. Cockburn has transferred either a test gene or dye into eggs of house flies, fruit flies, and stable flies. So far, the mosquito eggs have proven to be too tough for penetration. However, Cockburn believes he will eventually find a combination of needle number and vortex speed that will permit him to slip material into mosquito eggs.

When the technique for egg-penetration is worked out, perhaps a gene for malaria resistance will be found and a malaria-resistant population of mosquitos grown. Using modern insecticides, the mosquito population in a given area could then be killed off and the malaria-resistant mosquitos released to re-populate the area. You would still have biting mosquitos, but you wouldn't get malaria!

[A]

[B]

Scientists have tried for decades to control the dreaded mosquito. (A) mosquito in blood-sucking position *(Courtesy USDA/ARS #K-4158-7)* and (B) an arm treated with a modern mosquito repellent *(Courtesy of USDA/ARS #K-4158-19)*

STUDENT ACTIVITIES

1. Write the Terms to Know and their meanings in your notebook.

2. Use reference materials to find and list ten examples of beneficial insects and ten examples of insect pests.

3. Name five insects that are beneficial some of the time and pests some of the time.

4. Make an insect collection, with the insects properly identified and named.

5. Make a drawing of an insect and label the various body parts, appendages, and mouthparts.

6. Research a major crop in your community and discuss the key pests and measures recommended to control those key pests.

7. Develop a collage showing pests of plants in your community.

8. Make a weed collection and identify each sample. Your instructor may have a weed identification key for this purpose.

SELF EVALUATION

A. Multiple Choice

1. A biennial weed will live for
   a. 1 year.
   b. 2 years.
   c. 3 years.
   d. more than 3 years.

2. The major causal agent of plant disease is
   a. nematodes.
   b. bacteria.
   c. viruses.
   d. fungi.

3. The number of insect species in the world is estimated to be
   a. 100,000.
   b. 500,000.
   c. 1 million.
   d. none of the above.

4. The term *instar* refers to the development stage of
   a. plants.
   b. fungi.
   c. bacteria.
   d. insects.

5. Plant diseases are vectored by
   a. wind.
   b. rain.
   c. insects.
   d. none of the above.

6. A nematode is a type of
   a. fungus.
   b. roundworm.
   c. annual plant.
   d. insect.

7. A type of regulatory control is
   a. plant quarantine.
   b. sanitation.
   c. crop rotation.
   d. soil tillage.

8. The control practice that relies on the introduction of parasites and predators is
a. cultural.
c. biological.
b. chemical.
d. host resistance.

9. A threshold level is also known as the
a. pesticide residues.
c. degree of pest control.
b. control program.
d. pest concentration.

10. A pesticide used to control diseases is a/an
a. fungicide.
c. insecticide.
b. nematacide.
d. acaricide.

B. Matching

_____ **1.** Biotic
_____ **2.** Pest
_____ **3.** Summer annual
_____ **4.** Acaricide
_____ **5.** Instar
_____ **6.** Appressorium
_____ **7.** Abiotic

_____ **8.** Entomophagous
_____ **9.** Symptom
_____ **10.** IPM

a. The insect stage between molts
b. A chemical used to control mites and spiders
c. Integrated pest management
d. A visible change to the host caused by pests
e. Adversely affects human activities
f. Diseases caused by living organisms
g. A plant that germinates in the summer and lives for only one year
h. Structure by which a fungus attaches itself to a plant
i. Nonliving factors
j. Insects on which other insects feed

C. Completion

1. Soil tillage is an example of _____ control.
2. An _____ will control weeds.
3. A _____ lives in or on another organism.
4. Insects have _____ pairs of legs.
5. Complete metamorphosis consists of the following stages: egg, _____, _____, and adult.
6. _____ resistance is when a plant has developed its own defensive response to pests.
7. _____ control will utilize quarantine practices.
8. Monitoring is essential for _____ programs to work successfully.
9. _____ pests occur on a regular basis for a given crop.
10. Causal agents of disease are _____, _____, _____, and _____.

UNIT: 14

Safe Use of Pesticides

The development and use of pesticides has provided many benefits for both the producer of agricultural commodities and the consumer. However, there are also risks associated with pesticide use. These risks will be reduced with proper pesticide application, storage, and disposal. Improper use of pesticides will increase the risk of environmental contamination and any adverse effects on human health.

Pesticide use is a very controversial issue in America today. It is important to balance objectively the benefits and the risks associated with pesticide use. The Environmental Protection Agency (EPA) is conducting benefit-risk assessments on each pesticide currently registered. When a pesticide is registered for use and is applied according to label directions, the benefits greatly outweigh the risks.

Some benefits of pesticide use are summarized as follows:

1. Increased yields of food and fiber
2. Reduced loss of stored products
3. Increased crop quality
4. Economic stability
5. Better health
6. Environmental protection and conservation.

The increase in crop and animal yields and improved quality of crops and livestock products resulting from the use of pesticides have been adequately documented (Figure 14-1). It has been estimated that the average total income spent on food in the United States would increase from the current 11 percent to 30 percent without the protection

Crop	Estimated Percent Increase	Major Pests Controlled
Corn	25	Weeds, rootworms, corn borers, seedling blights
Cotton	100	Pink bollworms, boll weevils, nematodes, boll rots
Alfalfa seed	160	Weeds, alfalfa weevils
Potatoes	35	Tuber rots, blackleg, soft rots, blights
Onions	140	Botrytis blights, neck rot, smut, onion maggots

FIGURE 14-1 Yield increases from pesticide applications. *[Adapted from material from "The New Pesticide User's Guide"]*

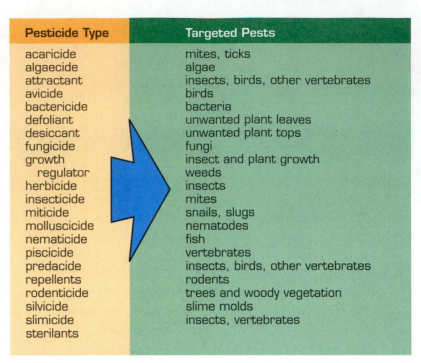

Pesticide Type	Targeted Pests
acaricide	mites, ticks
algaecide	algae
attractant	insects, birds, other vertebrates
avicide	birds
bactericide	bacteria
defoliant	unwanted plant leaves
desiccant	unwanted plant tops
fungicide	fungi
growth regulator	insect and plant growth
herbicide	weeds
insecticide	insects
miticide	mites
molluscicide	snails, slugs
nematicide	nematodes
piscicide	fish
predacide	vertebrates
repellents	insects, birds, other vertebrates
rodenticide	rodents
silvicide	trees and woody vegetation
slimicide	slime molds
sterilants	insects, vertebrates

FIGURE 14-2 Classification of pesticides based on the target pests. *[Adapted from material from "The New Pesticide User's Guide"]*

offered by pesticides. Benefits to human health are best illustrated where insecticides are used to control insects that carry and spread malaria and typhus diseases. Minimum-tillage or no-tillage practices have reduced soil erosion. These practices would not be possible without the use of pesticides. Many other substantial benefits to the quality of life resulting from the proper use of pesticides could be cited.

HISTORY OF PESTICIDE USE

The use of chemicals to control pests is not new. Elements such as sulfur and arsenic were among the first chemicals used for this purpose. An **element** is a uniform substance that cannot be further decomposed by ordinary chemical means. Homer, about 1000 B.C., wrote that sulfur had pest-control ability. The Chinese, about A.D. 900, discovered the insecticidal properties of arsenic sulfide, a chemical **compound** (a chemical substance that is composed of more than one element).

Until the late 1930s, pest-control chemicals or pesticides were mainly limited to **inorganic compounds** (any compounds that do not contain carbon). Examples of other inorganic pesticides are mercury and Bordeaux mixture. Bordeaux mixture is a combination of copper sulfate and lime. It is used for plant disease control.

A majority of today's pesticides are synthetically produced **organic compounds** (compounds that contain carbon). The organic chemistry involved in pesticide production is often complex and extremely diverse. However, a classification system for pesticides that is based on the type of pest being controlled is useful (Figure 14-2). The major pesticide groups are herbicides, insecticides, and fungicides.

In the United States, pesticide use totaled more than 820 million pounds and cost $6 billion in 1988. Currently, EPA has registered more than 600 chemicals that are formulated into some 30,000 products for pest control. Of the three major pesticide categories, the largest volume was for herbicides, followed by insecticides and fungicides.

HERBICIDES

Herbicides can be grouped into several major categories based on application method, type of control, and chemical structure. The terminology for herbicide use, type of control, and chemical family follows.

Selective Herbicides

A **selective herbicide** kills or affects a certain type or group of plants. The selectivity of a herbicide can be caused by many different factors. Some of these factors include:

1. differences in herbicide chemistry, formulation, and concentration.
2. differences in plant age, morphology, growth rate, and plant physiology.
3. environmental differences such as temperature, rainfall, and soil type.

Nonselective Herbicides

A **nonselective herbicide** controls or kills all plants. These herbicides are used for many different purposes. Examples of their use are for railroad rights-of-way, industrial areas, fence rows, irrigation and drainage banks, and renovation programs.

Contact Herbicides

Contact herbicides will not move or translocate within the plant. They only affect that part of the plant with which they come in contact. They are often used for controlling annual weeds.

Systemic Herbicides

A **systemic herbicide** is absorbed by the plant and is then translocated in either xylem or phloem tissue to other parts of the plant. **Xylem tissue** is where water and minerals are transported within the plant. **Phloem tissue** is responsible for transporting carbohydrates within the plant.

Preemergence and Postemergence Herbicides

A **preemergence herbicide** is applied prior to weed or crop germination. A **postemergence herbicide** is applied after the weed or crop is present.

CHEMICAL FAMILIES OF HERBICIDES

Herbicides are chemicals that are used to control weeds. These compounds can affect plant growth in many different ways. Currently, more than twenty-three chemical families of herbicides have been developed. Each one has a unique chemical structure and a different site or mode of action. **Mode of action** is a term used to describe the way in which herbicides adversely affect plant growth. Several of the more important chemical herbicide families and their characteristics are described in this section.

Acetanilides Acetanilides interfere with cell division and protein synthesis. They are applied as either preemergence or preplant applications for control of annual grasses and some annual broadleaf weeds. Two popular herbicides within this category are Lasso and Dual. They are used extensively for weed control in corn and soybeans.

Dinitroanilines   These act on root tissue, preventing root development in seedling plants. They are preemergence herbicides applied to prevent weed germination and should be incorporated into the soil. The dinitroanilines are deactivated very quickly by **photodecomposition** (chemical breakdown caused by exposure to light), **volatilization** (changing to gases), and chemical processes. Examples of these herbicides are Balan and Prowl. They are used for control of annual grasses and broadleaf weeds in many different crops.

AGRI·PROFILE

CAREER AREAS: Pesticide Applicator/Pesticide Specialist/Chemist

Tens of thousands of chemical formulations have been developed in recent years to control insects, diseases, weeds, nematodes, rodents, and other pests. We must use chemicals to help control pests. However, these materials are likely to be hazardous to the operator, plants, animals, or the environment if not properly used.

Pesticide applicators are employed by farmers and ranchers, lawn service companies, farm and garden supply firms, termite control companies, highway departments, and railroads, as well as self-employed persons. Special training and licensing is required for handling most pesticides.

Pesticide applicators may work in homes, buildings, fields, forests, and even the holds of ships. They may dust, spray, bait, or fumigate, depending on the setting and the pest. Pesticide applicators must always use protective clothing and other devices to protect against accidental poisoning. Their tools may be as simple as aerosol cans or as complex as orchard spray rigs or specially equipped helicopters.

Airplane crews are important members of the teams that apply chemical and biological pesticides to large crop, range, and forest areas. *(Courtesy USDA/ARS #K-3663-15)*

Phenoxys Phenoxy herbicides affect plants by causing overstimulation of growth. They perform best when applied as postemergence foliar sprays. They are selective herbicides that affect broadleaf weeds in grass crops. The herbicide 2, 4-D was first used in 1942, and is still widely used.

Triazines These are photosynthetic inhibitors that interfere with the process of photosynthesis. They are preemergence herbicides, used to control both annual and broadleaf weeds in grass crops. Aatrex and Princep are two examples of these herbicides. They are primarily used in corn weed-control programs.

INSECTICIDES

Insecticides are chemicals used to control insects. They can affect the insect in many different ways. The classification system of insecticides may be based on chemical structure and/or mode of action, as shown in Figure 14-3.

The botanical, inorganic, and oil insecticides are some of the original chemicals used for insect control. Sulfur, for example, may be used to control mites. Its effectiveness as an insecticide was discovered thousands of years ago.

Rotenone is a chemical present in the roots of certain species of legume plants. It affects insects by inhibiting respiratory metabolism and nerve transmission. Rotenone, first used in 1848, is a botanical insecticide still used today. It is a contact and stomach-poison insecticide. Rotenone is principally used in vegetables for controlling fleas, beetles, loopers, Japanese beetles, and many other insects.

Superior oils are highly refined oils and are applied to ornamentals and citrus crops to control scale insects, mites, and other soft-bodied insects. Oils act by excluding oxygen from the insect, thus causing suffocation. It is also believed that oils may destroy cell membrane

Chemical Group	Mode of Action	Common/Trade Names
Inorganics	Protoplasmic Poisons Physical Poisons	Sulfur Silica Aerogel
Oils	Physical Poisons	Superior Oils
Botanicals	Metabolic Inhibitors	Pyrethrum Rotenone
Synthetic Organics 　Chlorinated Hydrocarbons	Nerve Poison	Lindane Endosulfan-Thiodan
Organophosphates	Nerve Poison	Diazinon-Spectracide Parathion-Thiophos
Carbamates	Nerve Poison	Carbofuran-Furadan Carbaryl-Sevin
Pyrethroids	Nerve Poison	Permethrin-Ambush Fluvalinate-Mavrik
Insect Growth Regulators (IGRs)	Alter Insect Growth	Methoprene-Altosid Kinoprene-Enstar
Biorational/Microbial Insecticides	Wide Range of Activity	*Bacillus thuringiensis*

FIGURE 14-3 Insecticide classification system.

function. Oils were first used in the early 1900s to control San Jose scale in fruit orchards.

The synthetic organic group of insecticides was principally developed after 1940. This group includes the chlorinated hydrocarbons, organophosphates, carbamates, pyrethroids, insect growth regulators, and other minor classes. A majority of these insecticides adversely affect nerve transmission. The chlorinated hydrocarbons were the first to be synthesized (manmade). Released in 1940, DDT was the first and most popular chlorinated hydrocarbon. This chemical had excellent insecticidal properties and was used extensively. However, environmental and health problems were linked to DDT and other insecticides within this group. Because of these risks, many of the chlorinated hydrocarbons (such as DDT, aldrin, dieldrin, and chlordane) are banned from use in the United States.

The organophosphate and carbamate insecticides are the principal insecticides used to control insects today. Approximately 60 percent of all insecticides produced in the United States are from these two groups. They control insects by affecting the nervous system. The potential for pesticide poisoning of people and livestock is high for these insecticides. They will affect the nervous systems of humans and animals in a manner similar to the way they affect the insects they are designed to kill. The dose, or the amount, of insecticide applied is the discriminating factor with respect to insect control or human poisoning.

FUNGICIDES

Fungicides are chemicals used to control plant diseases caused by fungi. Chemical control of plant diseases is more difficult than it is for weed and insect control. Fungi are plants without chlorophyll. They are parasites of other plants. The fungicide must be selective enough to control the fungus but not adversely affect the host. Also, fungi have many generations each growing season. Therefore, reapplication of the fungicide will be required to provide effective control.

Protectant Fungicide   A protectant fungicide is applied prior to disease infection. This will provide a chemical barrier between the host and the germinating spores. However, this barrier will be broken down by environmental weathering of the fungicide. Rainfall, sunlight, temperature, and plant growth are the major causes of this breakdown. The fungicide will have to be reapplied if adequate disease control is expected.

Eradicant Fungicide   An eradicant fungicide can be applied after disease infection has occurred. These fungicides act systemically and are translocated by the plant to the site of infection. They offer a longer control period than do protectant fungicides, because they are not so prone to environmental weathering.

Chemical Structure

Fungicides, like insecticides and herbicides, can also be classified into different chemical families. Figure 14-4 lists some of the major groups of fungicides.

Chemical Family	Fungicide Activity	Common/Trade Name
Benzimidazoles	Eradicant	Benomyl-Tersan 1991
Dicarboximides	Protectant	Captan-Captane
Dithiocarbamates	Protectant	Mancozeb-Dithane M45
Oxathiins	Eradicant	Carboxin-Vitavax

FIGURE 14-4 Fungicide classification system.

Inorganic Fungicides   The elements sulfur, copper, mercury, and cadmium, or mixtures of them, are some of the oldest pesticides used by humans. They protect many ornamentals and turfgrasses from diseases and are examples of inorganic pesticides.

Organic Fungicides   Organic fungicides are a newer group of fungicides. These are used both as protectants and as eradicants.

PESTICIDE LABELS

The information on a pesticide-container label instructs the user on the correct procedures for application, storage, and disposal of the pesticide. The pesticide label is a summary of information gathered by the pesticide manufacturer and is required for product registration. The registration process is estimated to take 8 to 10 years and costs up to $20 million per pesticide. Several of the studies required to meet federal standards are summarized as follows:

1. Chemical and Physical Properties: water solubility, volatility, movement in soils, stability to heat and light, and other factors affecting environmental stability.
2. Toxicology studies: determine acute oral, dermal, and inhalation toxicity to various animals; evaluate chronic toxicity, including any effect on reproduction or the ability of the pesticide to be a carcinogen.
3. Residue analysis: determine the amount of pesticide residues at the time of harvest, develop safe tolerance levels for any pesticide residue.
4. Metabolism studies: determine application exposure, determine consumer exposure to pesticide residues, establish safety practices that minimize exposure.

The label is a legal document indicating proper and safe use of the product. Pesticide use that differs from that specified on the label is a misuse. Such use is illegal, and the offender can be charged with civil or criminal penalties. When improperly applied, a pesticide can pose danger to the applicator, the environment, and other people. Therefore, it is important to read, understand, and follow the information on the pesticide label.

The pesticide label and other labeling information must meet federal standards. The label consists of a front panel and a back panel on the product (Figures 14-5 and 14-6). If there is insufficient room on the

FIGURE 14-5 Front panel of a sample pesticide label. (*Adapted from material from Vo-Ag Services, University of Illinois*)

panels, additional labeling information will be attached, in booklet form, to the product. An outline of a label follows.

Use Classification

Pesticides are classified as either general use or restricted use. A **general-use pesticide** poses minimal risk when applied according to label directions.

Restricted-use pesticides pose a greater risk to humans and the environment. Therefore, anyone applying a restricted-use pesticide must be properly trained and certified. Applicator certification is administered by each state. An applicator must meet a minimum set of standards and is often evaluated by tests in order to be certified.

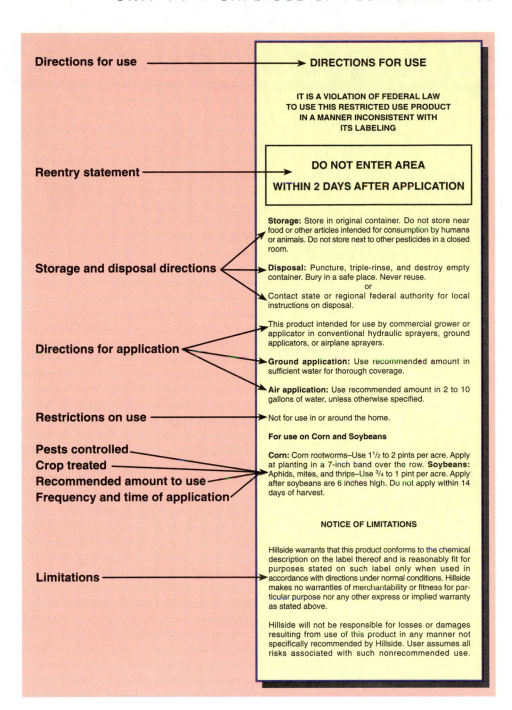

FIGURE 14-6 Back panel of a sample pesticide label. *(Adapted from material from Vo-Ag Services, University of Illinois)*

Trade Name

A **trade name** is the manufacturer's name for its product. It appears on the label as the most conspicuous item. The manufacturer will use the name in all of its promotional campaigns. The same chemical may have several different trade names, depending on the type of formulation and patent rights.

Formulation

Formulation refers to the physical properties of the pesticide. The pesticide chemical or active ingredient will often have to be modified to allow for field use. These modifications may include adding inert ingredients, such as solvents, wetting agents, powders, or granules, to

BIO•TECH CONNECTION

Substituting Biological for Chemical Pesticides

The sound of a roaring cyclone sprayer or the sight of an aircraft fogging a field or tree tops creates an image of chemical pesticide application in the minds of most people. However, more and more such sprays are the new, user-friendly, environmentally safe, biological-control pesticides. These materials generally contain bacteria, fungi, viruses, or other microbes that attack the targeted host, but are harmless to other living organisms and the environment.

The newer, safer pesticides increase the chances of eradicating or permanently subduing some of our most troubling and costly pests. For instance, the damage caused by

New technology permits pesticide application equipment to use less water and less pesticide to achieve equal or better control of pests.

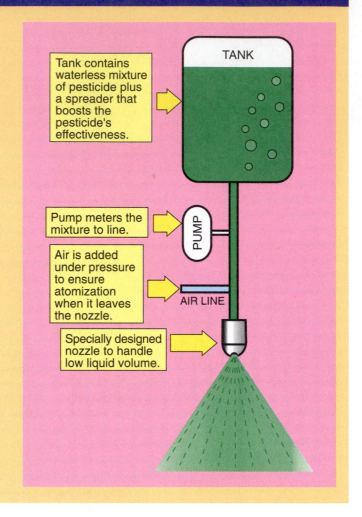

Tank contains waterless mixture of pesticide plus a spreader that boosts the pesticide's effectiveness.

TANK

Pump meters the mixture to line.

PUMP

Air is added under pressure to ensure atomization when it leaves the nozzle.

AIR LINE

Specially designed nozzle to handle low liquid volume.

---

the pesticide. This will result in different formulations. Some examples of the different types of formulations and their abbreviated label names follow.

Granules—G
Solutions—S
Wettable Powders—WP or W
Soluble Powders—SP
Dry Flowables—DF
Emulsifiable Concentrates—EC or E
Dusts—D

Common Name

The **common name** is given to a pesticide by a recognized authority on pesticide nomenclature. A pesticide is identified by a trade name, common name, and chemical name. It may have several trade names, but will have only one common and one chemical name. The common name, or generic name, identifies the active pesticide ingredient and can be used for comparison shopping.

and efforts made to control cotton bollworms and tobacco budworms on the Mississippi Delta costs an estimated $50 million per year using chemical pesticides. Marion Bell, an Agriculture Research Service entomologist at Stoneville, Mississippi believes he has a better way—the *Heliothis* nuclear polyhedrosis virus. *Heliothis* has been found to be very specific to the bollworm and tobacco budworm because it works only on insect larvae that have an alkaline midgut.

Bell specializes in microbial control of insects. He and other appropriate authorities have determined that treating the entire 4.7 million acres of the Mississippi Delta with *Heliothis* would cost about $7 million. However, the team is aware of the natural fears of the populace with such a widespread, general spraying program. Farmers are concerned about the effect of any general spraying program on their crops, aquaculturalists are concerned that such material could contaminate their catfish ponds, and the general public are concerned about any spray that may be damaging to the environment.

Therefore, they opted for an extensive educational program coupled with a spray program on small areas until the system is proven and better accepted by the public. The educational program included informational releases and face-to-face contacts with farmers, congressional representatives, extension personnel, environmental and health officials, physicians, civic clubs, and private interest groups. The message was, "The approach is solid and the virus is harmless to every living thing except the target pests."

Before a test sight was sprayed, brochures were given to the people affected, and catfish farmers were contacted to get permission to spray the virus around their ponds. The first year, spray planes were used to apply the virus mixture at the rate of 2 ounces of virus mix in 2 gallons of water per acre. Spray trucks were used to spray certain areas not accessible to aircraft. The researchers found the control does work, and they will fine-tune the procedure for use in future years.

Ingredients

The percentages by weight of both the active and inert ingredients are stated on the label. The active ingredient is identified by its chemical name. This is the name of the chemical structure of that pesticide. The inert, or inactive, ingredients do not have to be listed by their chemical names. The label must state only the total percentage of inert material.

Net Contents

The label will give the amount of product in the container. This is referred to as **net contents.** This quantity can be expressed in gallons, quarts, pints, and pounds.

Signal Words and Symbols

The **signal words** and **symbols** describe acute toxicity of the pesticide. The different categories are based on LD_{50} and LC_{50} values and on skin and eye irritation. The signal words used on pesticide labels are (1) danger—poison, (2) warning, and (3) caution. These words are used to alert the person handling or using the pesticide to the poisoning effect of contact with the chemical.

Pesticides with high toxicity, where only a few drops to one teaspoon will kill a 150-lb person, are labeled "DANGER—POISON." Pesticides with moderate toxicity, where one to two tablespoons will kill a 150-lb person, are labeled "WARNING." Those restricted-use pesticides requiring more than two tablespoons of the chemical to kill a 150-lb person are labeled "CAUTION." Obviously, even the pesticides with "CAUTION" as the signal word must be handled with extreme care.

Acute toxicity is measured by determining LD (lethal dose)$_{50}$ values when the pesticide is absorbed through the skin or is ingested orally. These values are determined by inhalation studies. **LD$_{50}$** is the amount or dose of the pesticide that is required to kill 50 percent of test populations. It is expressed in milligrams (mg) of pesticide per kilogram (kg) of body weight. The lower the LD$_{50}$ value, the more toxic a pesticide.

LC$_{50}$ is the lethal concentration of the pesticide in the air that is required to kill 50 percent of test populations. It is expressed in micrograms per liter (μg/l) or in parts per million (ppm). The lower the LC$_{50}$ value, the more toxic the pesticide.

All labels must bear the statement, "KEEP OUT OF REACH OF CHILDREN," regardless of pesticide toxicity. Figure 14-7 shows the toxicity ratings for the various signal words.

Precautionary Statements

These statements on the label will list any known hazards to humans, animals, and the environment. They will advise the user how to minimize any adverse effect that the pesticide may have. The categories that are normally listed are as follows:

1. Hazards to Humans and Domestic Animals
2. Statement of Practical Treatment
3. Environmental Hazards
4. Physical and Chemical Hazards

Establishment and EPA Registration Numbers

The establishment number establishes the manufacturer. The EPA registration number indicates that the product has passed the review process imposed by the EPA.

Toxicity Rating	Label Signal Words	Lethal Oral Dose, 150-pound Man	Oral LD$_{50}$ (mg/kg)	Dermal LD$_{50}$ (mg/kg)	Inhalation LC$_{50}$ (μg/l or ppm)
High	Danger-Poison	few drops to 1 teaspoon	0-50	0-200	0-2,000
Moderate	Warning	1 teaspoon to 1 ounce (2 tablespoons)	50-500	200-2,000	2,000-20,000
Low	Caution	1 ounce to 1 pint+ or 2 pounds	500-5,000	2,000-20,000	200,000+
Very low	Caution	1 pint+ or 2 pounds+	5,000+	20,000+	....

FIGURE 14-7 The toxicity ratings and signal words for pesticides. *(Adapted from material from College of Agriculture, University of Illinois)*

Name and Address of Manufacturer

All pesticide labels must contain the name and address of the company that manufactures and distributes the pesticide.

Directions for Use

The correct amount, timing, and mixing of the pesticide is given under the directions for use section of the label. The label will also list the different pests that are controlled, the application technique, and any other specific directions for optimum control.

Misuse Statement

The **misuse statement** appears on the label to remind the user to apply the pesticide according to label directions. Problems associated with pesticides, whether they involve environmental pollution or human poisoning, often occur due to pesticide misuse.

Reentry Information

Specific directions on reentering a treated area will appear under this heading. Only a few pesticides require reentry times of more than a day after application. However, even if the pesticide label does not contain specific restrictions, no one should ever be allowed to enter a treated area until the pesticide has dried.

Storage and Disposal Directions

This section will describe the proper storage and disposal of the pesticide. It is recommended that you purchase only the amount of pesticides needed for the current season. Stockpiling them will only increase storage risks and, ultimately, the problem of pesticide disposal, if they can no longer be used.

Notice of Limitations

The manufacturer guarantees that the product will perform as the label states. The company also conveys inherent risks to the user if the pesticide is applied in a manner inconsistent with the label. The manufacturer will limit its liability in case of lawsuits stemming from misuse of the pesticide.

RISK ASSESSMENT AND MANAGEMENT

Risk Measurement

An experienced pesticide applicator who understands the hazards of pesticides will take steps to reduce risks. The risk, or hazard, of pesticide applications has been expressed by the following:

$$\text{Risk (Hazard)} = \text{Toxicity} \times \text{Exposure}$$

The hazard, or risk, is the relationship between the toxicity of the pesticide and the exposure or use of the pesticide. **Toxicity** is a measure of how poisonous a chemical is. These data may be expressed in several ways. **Acute toxicity** describes the immediate effects (within 24 hours) of a single exposure to a chemical. Acute toxicity data based on dermal (skin), oral (by mouth), and inhalation (breathing) exposure routes have been determined. Signal words on pesticide labels indicate acute toxicity values. **Chronic toxicity** measures the effect of a chemical over a long period of time. To determine this information, the chemical is administered at low levels, with repeated exposures to the test animals. The effect of the chemical on reproduction or as a potential carcinogen, and any other adverse effects are evaluated.

Limiting Exposure

A pesticide's toxicity cannot be changed, but risk can be managed by addressing the exposure component of the formula. Many things can be done to reduce exposure. Examples of practices that can reduce exposure follow.

1. Select a pesticide formulation with a lower exposure potential. For example, granule formulations have a much lower exposure potential than do emulsifiable concentrates.
2. Use protective clothing and other safety equipment during the time of pesticide mixing, application, and disposal.
3. Apply pesticides during weather conditions that will not cause pesticide drift and those that provide for the most effective control.
4. Check all application equipment for proper working condition before applying pesticides.
5. Store pesticides and application equipment properly.

The pesticide label will provide guidance concerning acceptable protective clothing or gear for the application of the pesticide. Recommended protective clothing and gear will vary according to the toxicity of the pesticide. Even if no special gear is required, it is best to minimize your exposure to all pesticides by selecting appropriate clothing.

Appropriate clothing includes long pants, a long-sleeved shirt, nonabsorbent shoes, and socks. Avoid the use of any leather clothing, particularly shoes, because leather will absorb pesticides. The use of heavy denim clothing will provide good repellency to any pesticide and can be washed to remove any pesticide residue.

Special Gear

Gloves and Boots   Unlined rubber or neoprene gloves and boots will significantly reduce pesticide exposure (Figure 14-8). Any type of cloth-lined or leather boot or glove will only increase exposure if the lining becomes contaminated. Gloves should be tucked inside sleeves if you are working below the waist. They should be left outside your sleeves if you are working above the waist. Pants legs should be placed over the boots. By following these rules, you can prevent material from entering the inside of protective clothing or gear.

Hat and Coveralls   Absorption of pesticides through the skin and into the body is highest in the scrotal area, ear canal, forehead, and scalp (Figure 14-9). The use of an appropriate hat and coveralls will minimize the exposure to pesticides in these areas. Lightweight,

FIGURE 14-8 Proper clothing and safety equipment is essential in the safe use of pesticides. *(Courtesy of USDA)*

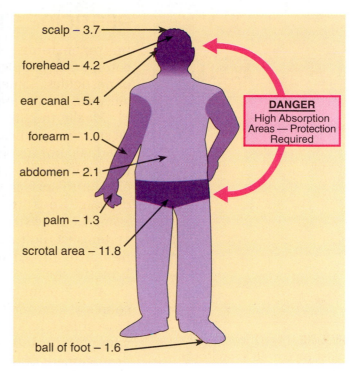

FIGURE 14-9 Dermal absorption sites and rates of pesticide absorption into the body. Comparisons are based on forearm absorption rates.

one-piece, repellent coveralls with hoods are available, and they provide excellent protection.

Apron   During the mixing and loading operations, the applicator is exposed to pesticide concentrates. The use of a rubber or neoprene apron at this time will prevent pesticide concentrates from splashing onto the chest, waist, and legs of the applicator.

Goggles and Face Shield   Eyes are extremely sensitive to many pesticides. The use of goggles and face shields is recommended when mixing pesticide concentrates or working in a spray, dust, or fog.

Respirators   Respirators reduce the inhalation of pesticide fumes and/or dust. A recommended respirator for pesticide applications is a cartridge respirator that will absorb toxic fumes and vapors, and filter any dust particles in the air. These respirators are often used during the mixing and application of a pesticide.

Personal Hygiene

Dermal exposure is the principal method of pesticide entry for the applicator. Personal hygiene can drastically reduce this type of exposure. Washing or showering at the end of the work day will remove any pesticide residue on the body. In case of a pesticide spill or splash at the work site, water can be used to immediately remove the material from the skin. After pesticide use, washing your hands prior to eating or using the restroom will further decrease pesticide exposure. Cleaning protective gear and clothing should also be done to prevent any residual exposure to pesticides on these objects.

FIGURE 14-10 Pesticides must be stored in approved storage units.

PESTICIDE STORAGE

Improper storage of pesticides can pose as much danger to the applicator, other people, animals, and the environment as misapplication of pesticides. Some important considerations in selecting a storage facility are the site location and building specifications.

Ideally, the site should be separate from other equipment or material storage facilities. This will reduce risk by decreasing exposure of individuals not involved in pesticide applications. The building should not be located on a floodplain, where flooding will introduce pesticides into surface water. It should be built to prevent any run-off or drainage from the site onto sensitive areas. Spill and drainage containment for large storage facilities is highly recommended. Containment systems would trap the pesticides and aid in emergency situations to minimize any environmental damage if the pesticides were to move from the site.

A well-planned storage building should be well ventilated, have a source of heat and water, be fireproof, have a secure locking system, and have sufficient storage area. The storage area should be well marked with placards indicating the presence of pesticides (Figure 14-10). The arrangement of the pesticides within the storage area should allow for ease in handling and safety. Tips to provide good storage conditions are:

1. Separate each pesticide class for storage on its own shelf.
2. Keep products off the floor.
3. Store containers so the labels remain in good condition and the containers remain orderly.
4. Practice good housekeeping.

HEALTH AND ENVIRONMENTAL CONCERNS

The current use pattern of pesticides has caused a heightened awareness of their risks. Human health and environmental quality are the major issues in assessing the hazards of pesticide use. The Environmental Protection Agency must conduct a benefit-risk assessment for each pesticide that is registered or re-registered for use. This is an extremely controversial issue. The EPA presently defines acceptable risk to the public at one death per million due to pesticide exposure.

Human Health

The number of lethal pesticide-related poisonings in the United States has decreased over the years. In one year, the total number of accidental deaths from all causes in the United States was 92,000, while deaths attributed to pesticide poisoning numbered only 27. However, more than 100,000 nonfatal pesticide poisoning cases per year have been estimated. A majority of the reported deaths were children involved in accidental ingestion of pesticides in the home. The causes have been traced to improper handling and storage practices by homeowners and other consumers.

The residues of a few pesticides used on food crops can pose potential health problems as carcinogens. A **carcinogen** is a material capable of producing a cancerous tumor. In 1987, the National Academy of Science estimated that certain types of pesticide residues in food crops "may" cause up to 20,000 cancers per year. This estimate was based on a "worst-case" scenario. It assumes that exposure to the pesticide would be continuous over 70 years, with the maximum allowable pesticide residue on the food when eaten. Obviously, careful handling and preparation of food eliminates most of this risk.

Environmental Concerns

After a pesticide is applied, not all of the pesticide reaches or remains in the target area (Figure 14-11). When this happens, the pesticide is often considered an environmental pollutant.

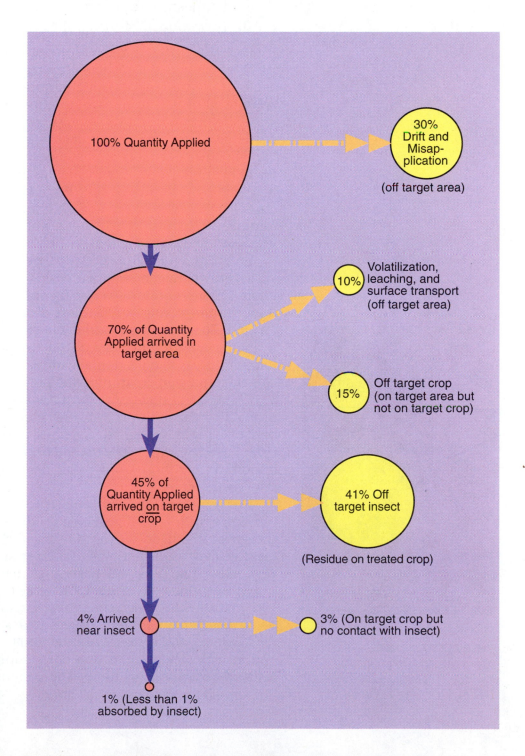

FIGURE 14-11 The movement of a pesticide after discharge from an aerial spray plane. *(Adapted with permission of Flint, M. L. and Van den Bosch, R. From* Introduction to Integrated Pest Management, *1981. Plenum Press, New York. 240 pp.)*

The movement of a pesticide from the designated area may occur in several ways. Drift, soil leaching, run-off, improper disposal and storage, and improper application are some of the major causes of a pesticide's becoming an environmental pollutant. Natural resources that can be contaminated are ground and surface water, soil, air, fish, and wildlife.

Surveys have shown that more than 50 percent of the counties in the United States have potential ground water contamination from agrichemicals. The three main factors affecting ground water contamination by agrichemicals are:

1. Soil type and other geological characteristics
2. The pesticide's persistence and mobility within the soil
3. The production and application methods of pesticide users

Pesticide drift is a major cause of soil and air contamination. **Drift** is the movement of a pesticide through the air to nontarget sites. It will occur at the time of pesticide application when small spray particles are moved by air currents to nontargeted areas. Also, vapor drift of a pesticide may occur after an application. **Vapor drift** is movement of pesticide vapors due to chemical volatilization of the product.

The adverse effects of pesticides on fish and wildlife may directly result in animal mortality. Pesticides may also indirectly influence animal feeding or reproduction. Pesticide labeling will indicate any potential harm to wildlife, and this information should be heeded to minimize risk. Fish, birds, bees, and other animals will be affected when pesticides reach them or their habitats.

Environmental contamination by agrichemicals can be decreased via several management practices. The use of integrated pest management (IPM) programs is reducing pesticide use. If pesticides are used, then proper mixing, application, storage, and disposal must be performed. These practices will decrease any adverse effect on the environment. An attempt must also be made to minimize any effect that temperature, soil type, rainfall, and wind patterns may have on a pesticide becoming an environmental pollutant.

Chemical pesticides are an important part of our food- and fiber-production capability. They are necessary to maintain our current standard of living in the United States and the world. However, they do create risks if used by improperly trained or careless individuals. Therefore, it is important that those using pesticides be properly informed and follow label instructions.

It is important that consumers wash and handle food products to minimize the intake of pesticide residues on food (Figure 14-12). While the government does extensive research to arrive at safe pesticide-residue levels, it is in the best interest of the consumer to wash fruits, vegetables, and all plant parts that may contain pesticide residues. This precaution will further reduce the hazard of pesticides.

Similarly, the wise person will become familiar with the uses of pesticides. Such uses include pesticides to control roaches, termites, flies, insects, and diseases of lawn and garden plants; insect repellants and insecticides used for outdoor camping and recreation; and other everyday common practices. The benefits of pesticides far outweigh the risks, if sensible use is observed.

FIGURE 14-12 The government does extensive research to arrive at safe pesticide-residue levels. (*Courtesy USDA/ARS #K-1992-9*)

STUDENT ACTIVITIES

1. Write the Terms To Know and their meanings in your notebook.

2. Ask your instructor to show examples of an approved pesticide applicator's respirator, goggles, gloves, boots, clothing, and other protective items.

3. Prepare and present a class demonstration on the proper use of one or more items of protective clothing and devices for safe handling of pesticides.

4. Do research and prepare to defend the position that pesticide residues on food in the United States is or is not a serious problem. Arrange for at least three classmates to prepare for a debate presenting various issues regarding the use of pesticides.

5. Work with the classmates arranged for in Activity 4 and present a debate in class. Have two people take the position that pesticide residues on food are dangerous and must be reduced. Have the other two persons defend the position that current levels of pesticide residues are safe and the benefits of using pesticides outweigh the risks to consumers.

6. Ask a professional pesticide applicator with a special interest in pesticide safety to demonstrate safe pesticide-application principles to the class.

7. Collect newspaper and magazine articles on accidents involving pesticides. Study the articles and determine how each accident could have been avoided. Report your conclusions to the class. *Note:* Many pesticide accidents occur in or on the home, lawn, and garden—do not overlook such cases.

8. Conduct a survey of pesticide storage-and-use practices in your home, farm, and place of employment. Correct all safety violations.

SELF EVALUATION

A. Multiple Choice

1. An example of an inorganic pesticide is
   a. pyrethrum.
   b. rotenone.
   c. Bordeaux mixture.
   d. organophosphate.

2. The total amount of pesticides used in the United States is
   a. 820 million pounds.
   b. 420 million pounds.
   c. 620 million pounds.
   d. 1.2 billion pounds.

3. A preemergence herbicide is applied
   a. after the weed or crop is present.
   b. before the weed or crop is present.
   c. at the time of planting.
   d. none of the above.

4. Which herbicide family inhibits photosynthesis?

a. acetanilines

b. dinitroanilines

c. phenoxys

d. triazines

5. An example of botanical insecticide is

a. 2, 4-D.

b. rotenone.

c. diazinon.

d. sulfur.

6. Pesticide registration often takes

a. between 1 and 2 years.

b. between 3 and 4 years.

c. between 8 and 10 years.

d. 15 years.

7. Pesticide risk can be decreased by

a. proper pesticide exposure.

b. reading the label.

c. minimizing pesticide exposure.

d. all of the above.

8. The signal word for a highly toxic pesticide is

a. CAUTION.

b. WARNING.

c. DANGER-POISON.

d. none of the above.

9. An example of protective clothing or gear that will minimize inhalation of a pesticide is

a. respirator.

b. boots.

c. gloves.

d. coveralls.

10. The number of lethal pesticide poisoning cases per year is

a. less than 20.

b. between 20 and 30.

c. between 40 and 60.

d. more than 100.

B. Matching

_____ **1.** Carcinogen

_____ **2.** Chronic toxicity

_____ **3.** Water

_____ **4.** DF

_____ **5.** Signal word

_____ **6.** Acute toxicity

_____ **7.** SP

_____ **8.** Drift

_____ **9.** Protectant

_____ **10.** Eradicant

a. The effect of a single exposure to a pesticide

b. Movement of a pesticide through the air to nontarget sites

c. Repeated exposures to low doses of a pesticide

d. A soluble powder pesticide formulation

e. A material capable of producing a tumor

f. A fungicide used after disease infection

g. Used to remove pesticides from the body

h. A dry, flowable pesticide formulation

i. Describes the acute toxicity of a pesticide

j. A fungicide used prior to disease infection

C. Completion

1. An _____ will control insects.

2. A _____ herbicide will control all types of plants.

3. A systemic herbicide will be translocated in the _____ and _____ tissue.

4. The amount of money spent on pesticides in 1988 was _____ billion dollars.

5. A pesticide will have only one _____ and one _____ name.

6. To reduce exposure to pesticides, the use of _____ is recommended for clothing and boots.

7. Environmental and health hazards of a pesticide are listed under the _____ _____ of the label.

8. The _____ number identifies where the pesticide was manufactured.

9. LD refers to the _____ _____ of a pesticide.

10. The _____ name is the manufacturer's name for its product.

Plant Sciences

The Quest for More and Better Plants!

Scientists, growers, propagators, breeders—everyone gets into the act. How can we get plants that are different, plants to provide materials for new medicines and industrial products; make our flowers and ornamentals more beautiful; make our present plants more disease, insect, drought, and cold or heat resistant? Where can we find trees that grow and produce wood faster? These questions all drive our quests in plant science.

During World War II, the USDA sent botanist Richard Schultes to South America in search of rubber trees resistant to diseases. Such trees were discovered, which opened the way for rubber plantations in the Western Hemisphere. These and other plants, which USDA explorers collected, changed agriculture around the world. For example:

- Curare vines that provide a muscle relaxant used in surgery
- Ucu uba tree in Columbia with bark that may provide a highly desired suppressant and skin medicine
- Rootstock from China for growing peaches
- Navel oranges from Brazil that created a new California industry
- Durum wheats from Russia that set the standard for United States varieties for years
- A peanut from Peru with genes resistant to two major diseases in the United States
- Wild oats that resulted in one of the most disease-resistant oat varieties ever developed
- California's avocado industry was started with Germplasm from Mexico
- Sorghum, dates, tung oil, and numerous forage grasses from around the world now grown in the United States.

There are over 1,600 species of plants used for medicinal purposes by people of the Columbian Amazon. However, only a few of these have been looked at by scientists. The USDA Chief of the Agriculture Research Service coordinates about ten trips a year in search of new plants and plant materials. The need for plant collecting, cataloging, and preserving is more urgent each year. The genetic base for many of the crops we take for granted is very narrow. Further, the encroachment of modern civilization on plants in remote places of the world is wiping out some important sources of new genes. Such plants could be the genetic material needed to help current important species survive or introduce new species with new uses.

Small samples of some of the diversity of plant materials are brought back to the United States. The following guidelines are used by modern plant hunters and collectors when exploring for or making new discoveries.

- Trips are to be organized as collaborations between the U.S. and the host country.
- All collected material is divided at least equally with the host country.
- All germplasm collected with the support of the USDA is deposited in the National Plant Germplasm System and is available to all valid users.
- All collection must "be done with a conservation ethic in mind."
- Collection must not endanger natural plant populations, with enough left behind so that the plant population can regenerate naturally.

Scientists believe that genetic diversity can be better preserved when new plants are cataloged, and most of them left and protected where they are found.

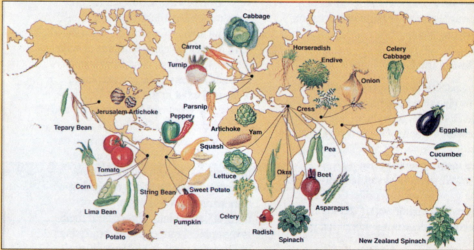

Centers of origin for some major fruits, vegetables, grains, and oil crops.
(Courtesy USDA/ARS)

Plant Structures and Taxonomy

MATERIALS LIST

✓ plant collection materials

✓ bulletin board materials

COMPETENCIES TO BE DEVELOPED

After studying this unit, you should be able to:

■ draw and label the major parts of plants.

■ describe the major functions of roots, stems, fruits, and leaves.

■ draw and label the parts of a typical root, stem, flower, fruit, and leaf.

■ explain some of the variations found in the structures of root systems, stems, flowers, fruits, and leaves.

■ describe the relationship of plant parts to fruits, nuts, vegetables, and crops.

Adventitious roots

Taproot

Ornamental

Fibrous roots

Root cap

Area of cell division

Area of cell elongation

Xylem

Phloem

Area of cell maturation

Root hairs

Stems

Woody

Herbaceous

Bulbs

Corms

Rhizomes

Tubers

Vascular bundles

Node

Internode

Axillary bud

Axil

Lenticels

Terminal bud

Vegetative bud

Flowering bud

Leaf

Margins

Simple leaf

Compound leaves

Leaf blade

Petiole

Photosynthesis

Cuticle

Epidermis

Palisade cells

Spongy layer

Chloroplasts

Mesophyll

Stoma

Guard cells

Flower

Bract

Stamen

Filament

Anther

Pollen

Pistil

Stigma

Style

Ovary

Ovules

Perfect flower

Imperfect flower

Pollination

Petals

Corolla

Sepals

Caylx

Fruit

Vegetable

Nut

Taxonomy

Genus

Species

Binomial

Variety

Plants are a basic part of the food chain. Without plants, the web of life cannot exist, and most animals and humans will die. A knowledge of plant growth is essential. To have a better understanding of plants, it is necessary to identify the parts that make up plants. The casual observer sees stems, branches, leaves, and possibly flowers and some nuts or fruits. The agriscience technician, however, will see a series of interconnected tissues and organs that depend on each other to function. The technician knows that all of the organs do not need to be present at one time, but is aware that each cell or organ (composed of cells) has an important role in the successful growth of the plant. Plant technicians and scientists are concerned with the efficient growth of plants. A plant may become stressed if one or more parts are absent or not functioning properly.

FIGURE 15-1 Seeds, nuts, and fruits are plant parts commonly used for food. *(Courtesy USDA/ARS #K-3839)*

The basic industry of agriculture is dependent on the proper functioning of plants. The animal grower needs many kinds of plants to feed livestock and poultry. The plant industry needs plants for seeds to sell to feed mills and to farmers for growing crops and other plants. The horticulture industry needs plants for seeds and cuttings, and plants for food such as fruits, vegetables, and nuts. Plants are also needed for landscaping the inside and outside of homes and office buildings.

To be successful with plants, one must have a knowledge of plant parts and how they function. Such knowledge is essential, whether you are growing, selling, or using plants.

THE PLANT

Plants are composed of many parts. Each part is important in the overall life and function of a plant. The root system is normally under the ground and is responsible for anchoring the plant and supplying water and nutrients. The stem, or trunk, is normally above the ground and functions as a support system for the rest of the above-ground parts. Leaves constitute the food-manufacturing parts of plants. Flowers come in many sizes, colors, and shapes, and function as the seed-producing parts of the plant. Healthy plants produce seeds, nuts, fruits, and vegetables. These parts are popular foods for animals and humans. They are also used for reproduction of the plant (Figure 15-1).

ROOTS

Root Systems

The largest part of the plant is often the root system. Roots take up more space in the soil than does the top part of the plant seen in the air above the ground. In fact, some roots will go down into the soil 6, 8, or even 10 ft. Some plants, such as squash, are said to have miles of roots.

While many roots are in the soil, there are other types that we see above the ground and may not consider as roots. Some plants, such as poison ivy (*Rhus radicans*) and English ivy (*Hedera helix*), have roots that help them climb trees, walls, and sides of buildings. These are called adventitious roots. **Adventitious roots** appear where roots are not normally expected.

The mistletoe, a popular Christmas plant, has roots that penetrate the bark of trees in the upper branches, or crown. These roots grow into the xylem and phloem tissues of the host plant and extract nutrients that originate in the soil. The dodder (*Cuscuta campestris*) has soil roots that die off as the plant gains a foothold in a plant. The dodder plant forms rootlike attachments that penetrate the stem of the host plant and extract nutrients from that plant.

Root systems are generally either taproot or fibrous roots (Figure 15-2). Knowledge of these two types of root systems can be of value in caring for and handling plants. The **taproot** is the main root of a plant and generally grows straight down from the stem. It is a heavy, thick root that does not have many side, or lateral, branches. Taproots are often used for human and livestock consumption, because they are food-storage organs. Carrots (*Daucus carota*) and sugar beets (*Beta*

FIGURE 15-2 Corn plants showing extensive fibrous root systems at a very early age. *(Courtesy USDA/ARS #K-5169-5)*

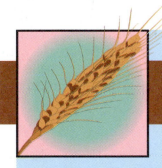

AGRI·PROFILE

CAREER AREAS: Botany/Biology/Taxonomy

Everyone relies heavily on our system of plant classification. Without it, we could not order, buy, sell, or instruct others about a plant unless we had the plant in the presence of both parties at the same time. Taxonomists devote their careers to identifying, classifying, and teaching others about plants. An important and exciting part of their work is that of discovering, studying, and naming new plants in their appropriate places in the classification system. Consider the excitement of discovering plants in far away lands or even at home that have not been observed or recognized even by the most knowledgeable specialists to date.

Botany is the study of plants, and biology is the study of both plants and animals. A knowledge of plant structures is critical to both. Similarly, consumers of plants for food, ornamentation, medicine, shade, wood, and other uses should have some knowledge of plant structures. Plant structures affect plant nutrition, functions, disease susceptibility, adaptation, and use.

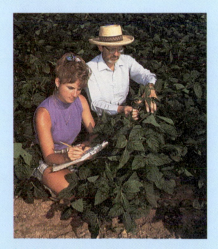

In the last two centuries, plant scientists have studied individual plant specimens and devised a system of classification of plants based on characteristics of plant parts. *(Courtesy USDA/ARS)*

FIGURE 15-3 Taproots of carrots, beets, and radishes are excellent sources of food.

vulgaris) are examples (Figure 15-3). Some plants with taproots are used for ornamental purposes. An **ornamental** plant is used to improve the appearance of a structure or area.

Plants that have taproot systems have the ability to survive periods of drought. Because they grow deep into the soil and have few fine secondary roots, taproots do not stabilize the soil very well. **Fibrous roots** are generally thin, somewhat hairlike, and numerous. The fibrous root system is normally very shallow. Grasses, corn, and many ornamentals, such as *Begonia semperflorens,* are good examples of plants with fibrous root systems. There are many small, thin-branched roots in this type of system. The result is that they are able to hold soil much better than taproot systems. However, fibrous root systems dry out more easily. Therefore, they cannot tolerate drought conditions.

Root Tissues

While there are different systems, all roots look similar when they are examined on the inside (Figure 15-4). The parts of the root have very specific functions in the plant. A knowledge of these parts is helpful in diagnosing diseases and other dysfunctions of plants.

The Root Cap The **root cap** is the outermost part of the root. It protects the tender growing tip as the root penetrates the soil. The root cap is a tough set of cells that is able to withstand the coarse conditions that the root encounters as it pushes its way through soil with rock and small sand particles. As the root cap wears away, the cells are replaced by more cells that develop at the root tip. This portion of the root is known as the area of cell division.

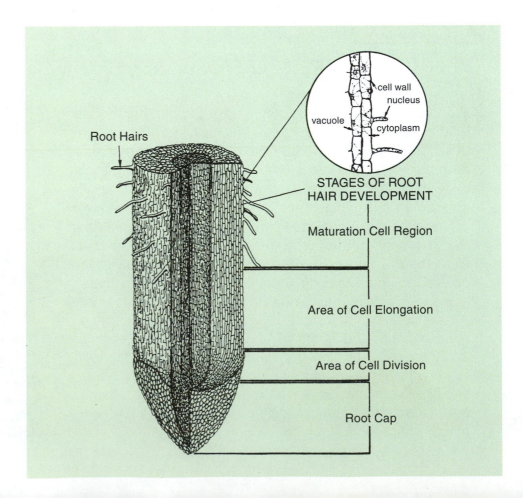

FIGURE 15-4 Cross section of a root showing the major internal parts.

Area of Cell Division and Elongation   The **area of cell division** provides new cells that allow the root to grow longer. The cells in this area multiply in two directions. The small and tougher cells are produced on the front edge of this region. They replace cells of the root cap that are worn off or destroyed as the tip pushes its way through the soil. Small and more tender cells are produced on the back of this area. They are used as the root tip grows longer. The area of cell division is actually very thin, maybe as thin as a strand of hair.

The next area, as you move toward the base of the plant, is the **area of cell elongation.** In this area the cells start to become longer and specialized. They also begin to look like the older cells and will start to do their specific jobs.

Xylem and Phloem   There are many types of cells in the root. Perhaps the most important ones are the xylem and the phloem cells. The **xylem** cells are responsible for carrying the water and nutrients that are in the soil to the upper portion of the plant. The **phloem** cells function as the pipeline to carry the manufactured food down from the leaves to other plant parts, including the roots, where it is used or stored. There are other cells in the root, and some will be discussed later.

Area of Cell Maturation   The **area of cell maturation** is where cells mature. This is also where the root hairs emerge. **Root hairs** are small microscopic roots. They will rise from existing cells located on the surface of the root (Figure 15-4). It is the job of root hairs to take in water and nutrients. Water and nutrients move into the root hairs, enter the xylem, and move to the upper portions of the plant. Root hairs are small and very tender. They will break off very easily. This means plants must be handled very carefully when transplanting. Once the root hairs are broken off, they cannot regrow or be replaced.

Although roots are normally hidden from view, they are still important parts of the plant. Roots need the same care and consideration as the other parts in order for plants to grow well.

STEMS

Stems are among the first things seen by the casual observer when looking at plants (Figure 15-5). Stems and branches are noticeable in the winter when the leaves are gone. They are easily seen as the plant grows. **Stems** support the leaves, flowers, and fruit.

Types of Stems

Aboveground stems are of two types: woody and herbaceous. **Woody** stems are tough and winter hardy. They often have bark around them. **Herbaceous** stems are succulent, often green, and will not survive winter in cold climates.

Not all stems are erect, aboveground structures. Some grow along the ground or even underground. Some stems have specialized jobs to perform. Such stems are referred to as modified stems. Examples of modified stems are bulbs, corms, rhizomes, and tubers. **Bulbs** are shortened stems that are surrounded by modified leaves called scales. Some

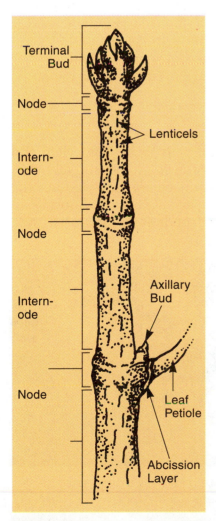

Terminal Bud

Node

Lenticels

Intern-ode

Node

Axillary Bud

Intern-ode

Leaf Petiole

Node

Abcission Layer

FIGURE 15-5 Important parts of stems.

FIGURE 15-6 Bulbs, such as Easter lilies and onions, are shortened stems surrounded by modified leaves called scales. *(Courtesy USDA/ARS)*

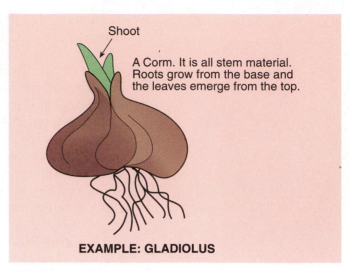

Shoot

A Corm. It is all stem material. Roots grow from the base and the leaves emerge from the top.

EXAMPLE: GLADIOLUS

FIGURE 15-7 Corms, such as the gladiola, are stems that are thick, compact, and fleshy.

examples of bulbs are Easter lilies (*Lillium longiflorum*) and onions (*Allium sp.*) (Figure 15-6). **Corms** are thickened, compact, fleshy stems. An example of a corm is the gladiola (*Gladiola sp.*) (Figure 15-7). **Rhizomes** are thick stems that run below the ground. Johnson grass and the iris (*Iris germanica*) are examples of plants with rhizomes (Figure 15-8). **Tubers** are thickened, underground stems that store carbohydrates. We often eat an example of this type of stem, the Irish potato (*Solanum tuberosum*) (Figure 15-9).

Parts of Stems

Stems have some of the same internal parts as roots. The xylem and the phloem continue to run the length of the stem and into all of the branches of the plant. In a subclass of plants called dicotyledons, the xylem and phloem occur together in tissue called **vascular bundles.** In another important subclass called monocotyledons, the xylem and

FIGURE 15-8 Rhizomes are thick stems that grow underground near the surface and give rise to new plants at each node. Johnson grass and iris are examples of plants with rhizome stems. *(Photo courtesy of Progressive Farmer Van Cleveland)*

FIGURE 15-9 The potato is really a specialized stem called a tuber. *(Courtesy USDA/ARS #K-4016-5)*

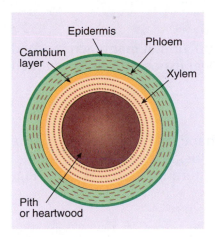

FIGURE 15-10 This cross section of a stem shows the xylem and phloem cells that make up the vascular system of plants.

another important subclass called monocotyledons, the xylem and phloem occur in separate areas (Figure 15-10).

Some important external parts of plant stems are the node, internode, axillary bud, lenticels, and terminal bud. The **node** is the portion of the stem that is swollen or slightly enlarged where buds and leaves originate. The **internode** is the area between the nodes. The **axillary bud** grows out of the axil. The **axil** is the angle above a leaf stem or flower stem and the stalk. The function of the axillary bud is to develop into a leaf or branch. The **lenticels** are pores in the stem that allow the passage of gases in and out of the plant. The **terminal bud** is located on the tip or top of the stem or its branches. It may be either a vegetative or flowering bud. The **vegetative bud** will produce the stem and leaf growth of the plant. The **flowering bud** will produce flowers.

BIO·TECH CONNECTION

Inside a Cell: Plant Biotechnology at Work

This electron micrograph shows some of the components in a soybean seed cell. For better visibility, computer processing has colored seed storage proteins purple and stored oil yellow. Red areas are cellular compartments where synthesis of the stored protein and oil occurs. The electron micrograph was taken by Eliot Herman of the Agriculture Research Service, and the color enhancement was performed by Terry Yoo of the Science and Technology Center for Computer Graphics and Scientific Visualization at the University of North Carolina.

Through the use of standard and electron microscopes, along with other modern technology, the scientist can peer into the most minute parts of plant structures. Further, they can improve the images seen through various instruments via computer enhancement. The fields of molecular biology, cellular biology, and genetics search the mysteries of the cell and its components, and seek to manipulate its biology to bring about desired changes. Larger parts of

A computer-enhanced view of the interior of a soybean cell. *(Courtesy USDA/ARS)*

plants are observed with hand glasses or the naked eye. Qualities of hardness may be indicated by touch and other characteristics determined by taste and smell.

Plant structures permit the plant to function as a whole. If one part becomes diseased, it will limit the performance of the rest of the plant and may lead to plant death. If the roots cannot obtain sufficient water and nutrients, the leaves cannot manufacture food for the plant. If the stem grows too fast without developing strength, the plant may topple and die. If seeds do not form and mature properly, the plant may not be able to reproduce and perpetuate itself. The whole is dependent upon the health of its parts.

LEAVES

The **leaf** of a plant has a very important function. It manufactures food for the plant by using light energy. The leaf exposes itself to as much sunlight as possible.

Leaf Margins

Plants may be identified by the edges, shape, and arrangement of the leaves. The leaf edges are known as **margins.** Leaf margins are named or described according to the toothed pattern on each leaf edge (Figure 15-11).

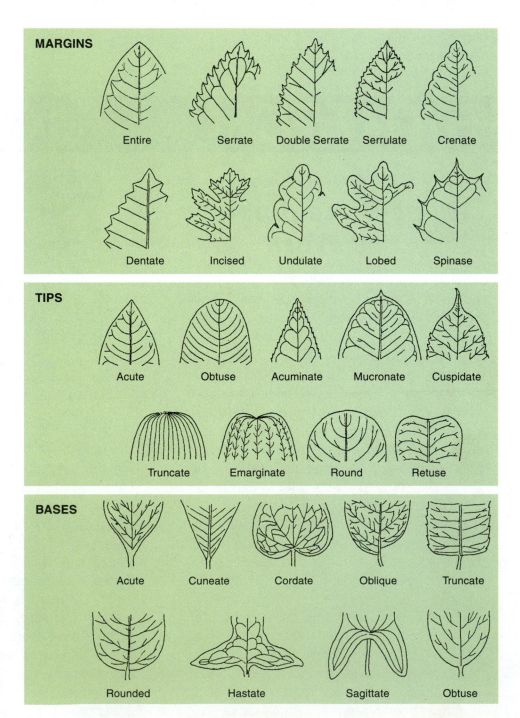

FIGURE 15-11 Leaf margins are helpful in identifying plants. *(From Bridwell,* Landscape Plants: Their Identification, Culture, and Use. *Copyright 1994 by Delmar Publishers)*

Leaf Shape and Form

Leaves vary in shape and form according to their species. Therefore, knowledge of the name given to each leaf shape and form is useful in identifying the plant (Figure 15-12).

Types of Leaves

Leaf types vary according to the species. Therefore, leaf type is also used to identify plant species. A single leaf arising from a stem is called a

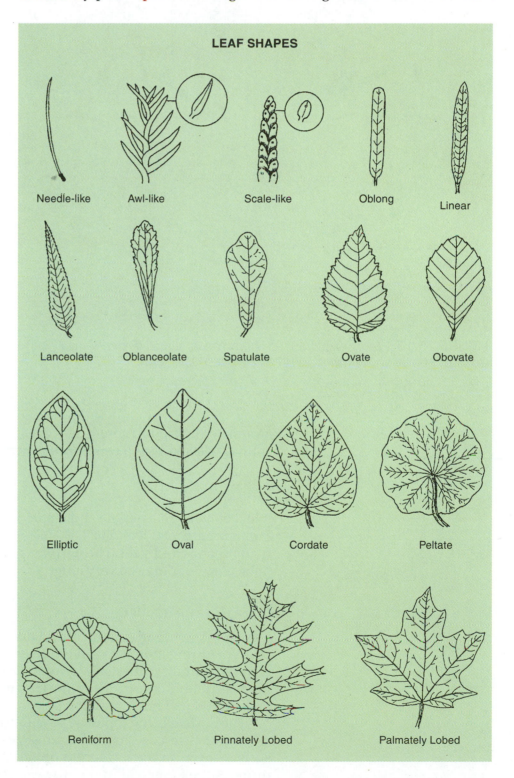

LEAF SHAPES

Needle-like Awl-like Scale-like Oblong Linear

Lanceolate Oblanceolate Spatulate Ovate Obovate

Elliptic Oval Cordate Peltate

Reniform Pinnately Lobed Palmately Lobed

FIGURE 15-12 Names given to various leaf shapes. *(From Bridwell,* Landscape Plants: Their Identification, Culture, and Use. *Copyright 1994 by Delmar Publishers)*

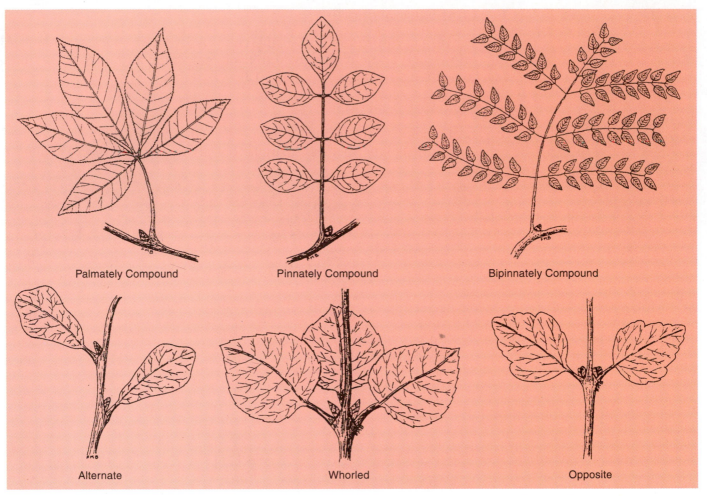

Palmately Compound

Pinnately Compound

Bipinnately Compound

Alternate

Whorled

Opposite

FIGURE 15-13 Examples of various leaf arrangements. *(From Bridwell,* Landscape Plants: Their Identification, Culture, and Use. *Copyright 1994 by Delmar Publishers)*

simple leaf. Two or more leaves arising from a common point on the stem is referred to as a **compound leaves** (Figure 15-13).

Leaf Parts

A leaf consists of a petiole and blade (Figure 15-14). These are the most familiar parts of a leaf. The **leaf blade** is the wide portion. It may be of many shapes and sizes. The **petiole** is the stem of the leaf. It may be almost absent or may be very long.

Internal Structure

The leaf is the food-manufacturing unit for the plant. The process of manufacturing food is called **photosynthesis.** The food that is created through photosynthesis in leaves enables the plant to grow. The process of photosynthesis is illustrated in Figure 15-15 and discussed in greater detail in a later unit.

The **cuticle** is the topmost layer of the leaf. It is waxy and functions as a protective covering for the rest of the leaf. The **epidermis** is the surface layer on the lower and upper sides of the leaf. The epidermis protects the inner leaf in many ways. The elongated, vertical **palisade**

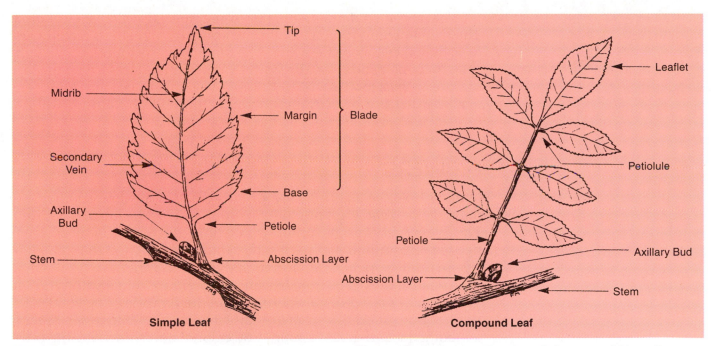

FIGURE 15-14 Parts of a leaf. *(From Bridwell,* Landscape Plants: Their Identification, Culture, and Use. *Copyright 1994 by Delmar Publishers)*

cells give the leaf strength and are the sites for the food-manufacturing process. These cells, as well as the lower **spongy layer,** contain chloroplasts. **Chloroplasts** are the parts of the cells that contain chlorophyll and are necessary for photosynthesis to be carried out. The lower layer is irregular and allows the veins, or vascular bundle, to extend into the leaf. The layer of palisades and spongy tissue is often referred to as the **mesophyll.**

The vascular bundle contains the xylem and the phloem. These are an extension of the same tissue that is located in the root and runs through the stem to the leaves. The xylem brings the water and minerals from the root. The phloem carries the manufactured food from the leaf to the various parts of the plant to nourish plant tissue or be stored.

FIGURE 15-15 The leaf is the plant-food-manufacturing part of plants. *(From Reiley and Shry,* Introductory Horticulture, *5E. Copyright 1997 by Delmar Publishers)*

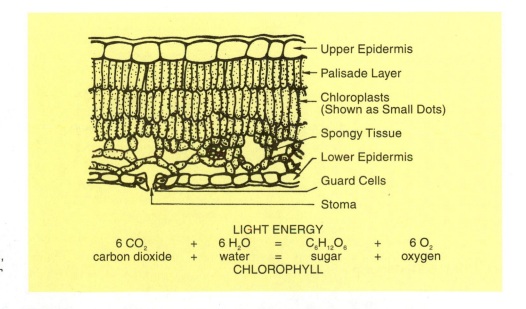

The lower epidermis contains some special cells called **stoma.** They are openings that allow for the exchange of carbon dioxide and oxygen, as well as some water. The stoma are surrounded by **guard cells** that open and close the stoma. If the plant is stressed by the lack of water or by a low light level, the guard cells will close the stoma. The result is that the plant cannot manufacture food because it will not have all the necessary ingredients.

FLOWERS

Many people see plants mostly for the beauty of the flower. Others see only a fruit to eat. Fruit production is only part of the job of the flower. The **flower** has, as its primary function, the production of seeds needed to continue the species. It is with this structure that the plant scientist will work to produce new and different varieties.

Not all of the beauty that is seen as flowers are actually flowers. The poinsettia (*Euphorbia pulcherrima*) and the flowering dogwood (*Cornus florida*), for example, have modified leaves called bracts. A **bract** is a modified leaf that is often brightly colored and showy. People often see the red or white bracts and call them flowers. Their function is to protect the flower parts, as well as attract insects for pollination.

Flower Structure

Flowers are composed of many parts. These include the filament, anther, pollen, stigma, style, ovary, petals, and sepals (Figure 15-16). These are the most important parts of the flowers.

The male part of the flower is the **stamen.** It consists of the filament, anther, and pollen. The **filament** supports the anther. The **anther** manufactures the pollen. The **pollen** is the male sexual reproductive cell. The female part of the flower, the **pistil,** is made up of the stigma, style, and ovary. The **stigma** receives the pollen. The pollen travels down the **style** and into the **ovary.** The ovary contains **ovules.** These are the eggs, which are the female reproductive cells. When the eggs are fertilized by the pollen, they will ripen into seeds.

If a flower contains all of the parts mentioned above, it is a **perfect flower.** If one or more of the parts are missing, it is considered an

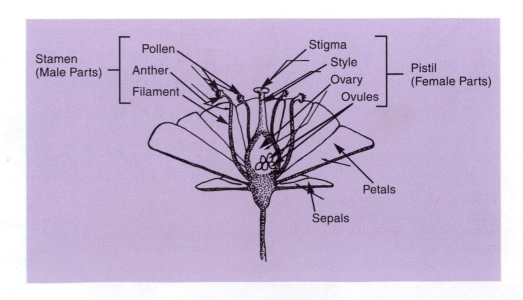

FIGURE 15-16 Major parts of flowers.

CROSS SECTION OF AN APPLE

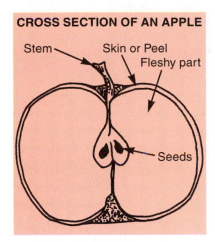

FIGURE 15-17 Seeds develop from the fertilized egg, or ovum, in the ovary to become the fruit of the plant.

imperfect flower. In some plants, particularly the flowering plants used in horticulture, it is desirable to remove the anther sacs before the pollen ripens. This prevents pollination and stains from the pollen on the petals of the flower. **Pollination** means the union of the pollen with the stigma. In some cases, orchids for example, the unfertilized flower will last for many months.

In plant breeding, the anther sac is removed from the plant to prevent natural pollination. It may be destroyed, or it may be used to pollinate another flower to create a new variety. Many hybrids are created in this way.

The colored **petals** attract insects or other natural pollinators. The flower petals are collectively called the **corolla.** The **sepals** together function as a protective device for the developing flower. Collectively, sepals are called the **caylx.**

Fruits, Nuts, and Vegetables

After fertilization, the ripening seed develops in the pistil. The pistil then enlarges and becomes the **fruit** (Figure 15-17). The fruit may be of many different shapes and sizes (Figure 15-18). The true fruit are the seeds that carry the male and female genetic characteristics of the plant. However, the fleshy material surrounding the mature seed, and the seed itself, is commonly called the fruit of the plant. The purpose of the fleshy part of the fruit is to attract animals and humans to the seed to help spread it over wide areas. This helps in the reproduction of the plant. Entire fruit or just the seed may be moved by wind, water, animals, and humans. Often the fruit is eaten by animals and humans as a source of food. Then the seeds may be discarded, where they can take

FIGURE 15-18 Fruits from different plants vary in size, shape, and taste. (A) Pineapple growing in Hawaii and (B) Peaches in Georgia. *(Courtesy USDA/ARS (A) #K-4281-1 (B) #K-4964-13)*

root and grow into new plants. People assist greatly in spreading seeds and starting new plants when they plant seeds for crops.

There are many kinds of fruits and vegetables. The two terms are sometimes used incorrectly. A **vegetable** can be any part of a plant that is grown for its edible parts. This can be a root, stem, leaf, or ripened flower. However, fruit, a ripened or mature ovary, is a specific plant part. A **nut** is also a type of fruit.

PLANT TAXONOMY

Importance of Classifying Plants

Taxonomy is the science, laws, and principles of classification. In biology, taxonomy provides the means for classifying organisms into established categories according to characteristics. Such classification makes it easier to understand and remember plants and animals by the similarities and differences found in their structures and parts. Living organisms are given Latin names to help scientists and technicians around the world communicate better. Latin is regarded as the universal language for those in professions dealing with the biological sciences. Agriscience has its origin in the biological sciences.

There are about 300,000 species of plants that have been identified and classified. The plant names are based on Latin descriptions and must be approved by a special committee of plant scientists. Carl Linnaeus, a Swedish botanist, developed the present system of plant classification in 1753.

If there were no botanical classification to identify and classify plants, many different species would carry one common name. For example, all clovers would be identified as clover, even though crimson clover is a winter annual, sweet clover is a biennial, and white clover is a perennial.

Complete classifications of field corn and petunia are shown in Figure 15-19.

Binomial System Used in Classifying Plants

When identifying a plant by its scientific name, it is not necessary to give its entire classification. Rather, a specific plant can be identified by using the genus and species only, because the genus and species name is not used in combination for any other plant or animal. For example, grain sorghum is *Sorghum* (genus) *vulgare* (species). **Genus** is

FIGURE 15-19 Example of a field crop (Yellow corn) and an ornamental plant (Petunia) showing their complete botanical classifications.

	Yellow Corn	Petunia
Common Name:	Corn	Petunia
Kingdom:	Plant	Plant
Phylum:	Spermatophyta (seed plants)	Embryophyta
Subphylum:	Angiosperm (seed in fruit)	Angiosperm
Class:	Monocotyledonae (single leaf seed)	Dicotyledonae (two-seed leaf)
Order:	Graminales (grasslike families)	Tubiflorea
Family:	Gramineae (grass family)	Solanaceae
Genus:	Zea (the corns)	Petunia
Species:	Mays (dent corns)	Hybridea
Variety:	Reid's yellow dent	Blue Moon

the taxonomic category between family and species. It is customarily capitalized when written along with a species name. **Species** is the subgroup under genus. A species name is generally not capitalized when written in combination with its genus. To compare a plant or animal name with that of a human's, the species corresponds to the person's first name and the genus to the person's last name. The system of using genus and species in combination is referred to as a binomial system of classification. **Binomial** means consisting of two names.

Some species are broken down into varieties. A **variety** is a subgroup of plants developed by people, as opposed to species that originate in the wild. Variety is a rank within a species. When writing a plant variety name, it is generally capitalized. An example is Triumph wheat.

STUDENT ACTIVITIES

1. Write the Terms to Know and their meanings in your notebook.

2. Observe the plants that are commonly grown in your area. Classify them by the type of root system that they have.

3. Make a chart of the plants listed in Activity 2. Indicate their common and scientific names, and discuss their responses to drought or their water maintenance requirements.

4. Make a collection of different kinds of leaves. Classify each according to its shape and type of margin.

5. Sketch the parts of roots, stems, and leaves. Label all items.

6. Make a bulletin board showing the major parts of plants.

7. Ask your teacher to provide a microscope and slides of plant tissue. Diagram the plant parts and label the cells and other structures that you see.

8. Record the common and scientific names of plants listed in this unit and learn the correct spelling of each.

SELF EVALUATION

A. Multiple Choice

1. The primary function of the root is to
   a.   make sure that the plant will grow.
   b.   anchor the plant and supply water and nutrients.
   c.   ensure that the plant can be propagated.
   d.   hold up the stem of the plant and provide propagation material.

2. The portion of the root that takes in the water and plant nutrients is the
    a.   root cap.
    b.   area of root division.
    c.   root hair.
    d.   area of cell maturation.

3. The major types of root systems are
    a.   area of cell division and fibrous.
    b.   fibrous and root cap.
    c.   cuttings and root hairs.
    d.   fibrous and taproot.

4. The area of cell division is
    a.   responsible for the production of new cells on the tip of the root.
    b.   where the cells will start to specialize.
    c.   located in the area where the root hairs start to erupt from the wall of the epidermal cell.
    d.   where the roots drop off on special plants like the dodder.

5. The phloem
    a.   is the pipeline that carries the water and nutrients from the soil to the leaves.
    b.   is the part of the stem that gives support to the node.
    c.   is the part of the leaf that holds it to the stem.
    d.   carries the manufactured food from the leaves to the roots.

6. Herbaceous stems
    a.   are tough and have bark around them.
    b.   come from herbs.
    c.   are green and are not winter hardy.
    d.   are part of the bulb.

7. The node
    a.   is the part of the stem that supports the flower.
    b.   is the part of the stem where the leaf is attached.
    c.   is the part of the stem that carries the nutrients.
    d.   will become detached when dry weather sets in.

B. Matching

_____ **1.** Root cap   a.   Located on the tip of the stem
_____ **2.** Terminal bud   b.   The wide portion of the leaf
_____ **3.** Leaves   c.   Protects the root tip as it moves in soil
_____ **4.** Cuticle   d.   Manufacture food for the plant
_____ **5.** Blade   e.   The topmost layer on the leaf
_____ **6.** Guard cells   f.   Surround the stoma

C. Completion

1. The roots are responsible for _____ the plant.
2. An _____ plant is a plant used to improve the appearance of an area.
3. The area of _____ _____ is where the cells start to become specialized.
4. _____ are thick stems that run below the ground.
5. _____ are pores in the stem that allow the gases to pass through the stem.
6. _____ _____ are cells that give the leaf strength.
7. The _____ is the male part of the flower.
8. When a flower contains the stamen, pistil, petals, and sepals, it is considered a _____ flower.
9. The _____ is an enlargement that results after fertilization.
10. A _____ can be any part of a plant that is grown for its edible parts.

D. True or False

_____ 1. Plants with many thin, hairlike roots have fibrous root systems.
_____ 2. Plants with taproot systems are more likely to survive in a dry period.
_____ 3. The root cap protects the young growing tip of the root.
_____ 4. The xylem carries the nutrients and water down the stem of the plant.
_____ 5. Herbaceous stems are tough and winter hardy.
_____ 6. Bulbs are stems that are thick and compact.
_____ 7. The vascular bundle is only in the leaf of the plant.
_____ 8. The leaf consists of the petiole and the blade.
_____ 9. The stomates allow for the passage of gases through only the leaf surface.
_____ 10. The bract is the colored petal located in the flower.

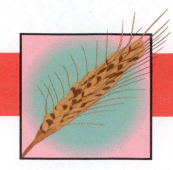

Plant Physiology

OBJECTIVE

To determine how plants make food and to describe the relationships among air, soil, water, and essential plant nutrients for good plant growth.

MATERIALS LIST

✓ writing materials

✓ encyclopedias

✓ corn or bean plant

COMPETENCIES TO BE DEVELOPED

After studying this unit, you should be able to:

- explain how plants make food.
- describe the roles of air, water, light, and media in relation to plant growth.
- trace the movement of minerals, water, and nutrients in plants.
- describe the ways that various plants store food for future use.
- compare the activity in a plant during exposure to light and periods of darkness.
- explain how plants protect themselves from disease, insects, and predators.

TERMS TO KNOW

Physiology	Semi-permeable	Deficiency
Photosynthesis	membrane	pH
Chlorophyll	Pores	Precipitate
Chloroplasts	Plant nutrition	Acid
Glucose	Plant fertilization	Alkaline
Light intensity	Macronutrients	Leached
Respiration	Micronutrients	Chlorosis
Transpiration	Ion	Marginal burn
Turgor	Anions	Scorch
Osmosis	Cations	

The life of a plant from its beginning to its maturity is a complex process. Many factors influence and directly control how a plant grows and what it produces. Growth, as in all living organisms, occurs by the division of cells and their enlargement as the plant increases in size (Figure 16-1). As the plant grows to maturity, the cells

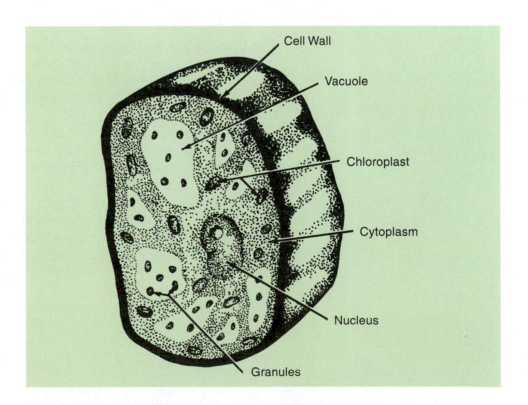

FIGURE 16-1 Major parts of a plant cell.

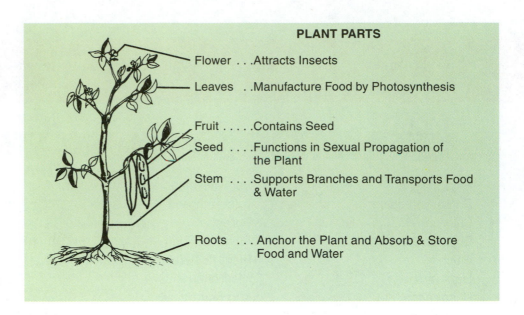

PLANT PARTS

Flower . . .Attracts Insects

Leaves . .Manufacture Food by Photosynthesis

Fruit . . . . .Contains Seed

Seed . . . .Functions in Sexual Propagation of
the Plant

Stem . . . .Supports Branches and Transports Food
& Water

Roots . . . Anchor the Plant and Absorb & Store
Food and Water

FIGURE 16-2 Major parts of a typical plant.

are produced, divide, grow, and become specialized organs. These specialized organs are stems, leaves, roots, flowers, fruits, and seeds (Figure 16-2). The study of how these organs function and the complex chemical processes that permit the plant to live, grow, and reproduce is **physiology.** An understanding of the processes of germination, photosynthesis, respiration, absorption of water and nutrients, translocation, and transpiration will enable the agriscience technician to maximize production of plants. The technician who works with interior and other ornamental plants must especially understand the environment of the plant, because it is not in its native habitat.

PHOTOSYNTHESIS

The most important chemical process in the atmosphere is photosynthesis. Without this chemical process, maintenance of life on this planet would not exist. Plants need carbon dioxide to manufacture food. Animals need oxygen to live. The complex chemical process of photosynthesis permits both to live and to support each other. Although the process is complex, it is not difficult to understand.

Photosynthesis is a series of processes in which light energy is converted to chemical energy in the form of a simple sugar. Chlorophyll and chloroplasts are also essential in this process. **Chlorophyll** is the green material inside the chloroplast. It is the substance that gives the green color to plant leaves. **Chloroplasts** are small membrane-bound bodies inside cells that contain the green chlorophyll pigments. The chloroplasts are located in the mesophyll of the leaf. They are the sites of the actual conversion of solar energy (light) into stored energy (simple sugars).

Photosynthesis is the conversion of carbon dioxide and water in the presence of light and chlorophyll into glucose, oxygen, and water. **Glucose** is a simple sugar and contains the building blocks for other nutrients. A simple chemical definition of photosynthesis is:

$$6CO_2 + 12H_2O \xrightarrow[\text{chlorophyll}]{\text{light energy}} C_6H_{12}O_6 + 6O_2 + 6H_2O$$

The rate at which the foodmaking process occurs depends on and varies with the light intensity, temperature, and concentration of carbon dioxide in the atmosphere. **Light intensity** is also known as the quality of light, or the brightness of light. Light must be present with sufficient brightness for the process to be successful. Plants have been able to adapt to various levels of light brightness. A knowledge of the level of light required for plants to grow well is essential, particularly for indoor plant production.

Temperature is also an important factor in the process of food manufacturing in the leaf. Photosynthesis occurs best in a temperature range of 65° to 85°F (18° to 27°C). Extremes of temperatures slow down or completely stop the process of photosynthesis. A lack of carbon dioxide will affect photosynthesis, too. Carbon dioxide is especially important in the beginning of the process. Under normal outdoor conditions, its availability is not a problem. However, in enclosed conditions such as those found in a greenhouse, a carbon dioxide shortage could be a limiting factor. To correct this problem, a carbon dioxide generator might be used (Figure 16-3).

RESPIRATION

All living cells carry on the process of respiration. **Respiration** is a process by which living cells (plant or animal) take in oxygen and give off carbon dioxide. Unlike photosynthesis, which occurs only in the light, respiration occurs both day and night. It is not easily measured during the day, because the presence of photosynthesis will mask or obscure the occurrence of respiration. Respiration is a breaking-down process. It uses the sugars and starches produced by photosynthesis and converts them into energy. The chemical equation for respiration is:

$$C_6H_{12}O_6 + 6O_2 \longrightarrow 6CO_2 + 6H_2O + \text{heat (energy)}$$

A comparison of the activities that occur during photosynthesis and respiration may be helpful in understanding the two processes (Figure 16-4).

FIGURE 16-3 A technician measures the impact of CO_2 enrichment on transpiration rate and stomata activity. *(Courtesy USDA/ARS #K-3750-7)*

Photosynthesis	Respiration
1. Food is produced.	1. Food is used for plant energy.
2. Energy is stored.	2. Energy is released.
3. It occurs in cells that contain chloroplasts.	3. It occurs in all cells.
4. Oxygen is released.	4. Oxygen is used.
5. Water is used.	5. Water is produced.
6. Carbon dioxide is used.	6. Carbon dioxide is produced.
7. It occurs in sunlight.	7. It occurs in dark as well as light.

FIGURE 16-4 Comparison of the activities that occur during photosynthesis and respiration.

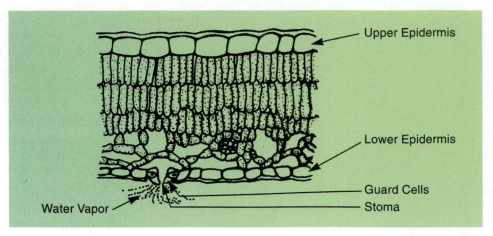

FIGURE 16-5 The process of transpiration. *[From Reiley and Shry*, Introductory Horticulture, 5E. *Copyright 1997 by Delmar Publishers]*

TRANSPIRATION

Transpiration is the process by which a plant loses water vapor (Figure 16-5). The loss takes place primarily through the leaf stoma. The plant transpires about 90 percent of the water that enters through the roots. Water saturates all of the spaces between the cells throughout the plant. About 10 percent of the water that enters from the roots is used in chemical processes and in the plant tissues. Functions of this water include transporting minerals throughout the plant, cooling the plant, moving sugars and plant chemicals, and maintaining turgor pressure. **Turgor** means a swollen or stiffened condition as a result of being filled with liquid.

Transpiration is greatly influenced by humidity, wind and other air movement, and temperature (Figure 16-6). As humidity in the air around the plant increases, the rate of transpiration decreases. Conversely, as humidity decreases, the rate of transpiration increases. Increased air movement around the plant increases the rate of transpiration, due to the evaporation caused by air movement. Similarly, as temperature increases, the rate of transpiration increases.

Often during dry weather or when plants are not watered, transpiration causes the plant to lose water faster than it can be replaced by the root system. When this occurs, the guard cells will close the stoma in the leaves, thus slowing down the rate of transpiration. This mechanism enables the plant to preserve the water it contains. If there is water in the soil, the plant may wilt slightly, but it will recover. However, if there is insufficient moisture in the soil, the plant may not be able to recover.

SOIL

Productive soil provides a natural environment for the root zone. There are air, water, and nutrients for the plant. Root hairs penetrate the pore spaces in the soil and absorb plant nutrients (Figure 16-7). A process called osmosis is used to get nutrients into root cells so the nutrients can be transported to the remainder of the plant. **Osmosis** is a process whereby the flow of a fluid through a semipermeable membrane separating two solutions permits the passage of the solvent but not the dis-

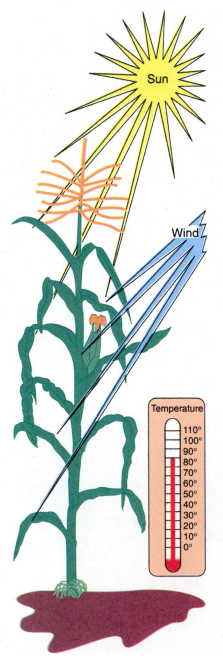

FIGURE 16-6 Factors that influence transpiration.

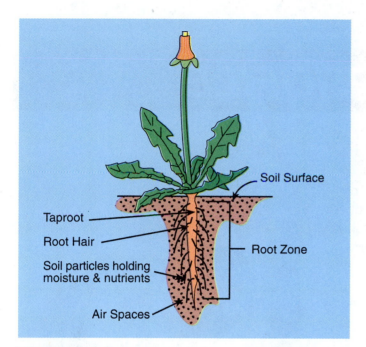

FIGURE 16-7 Root hairs in soil pores.

Taproot

Root Hair

Soil particles holding
moisture & nutrients

Air Spaces

Soil Surface

Root Zone

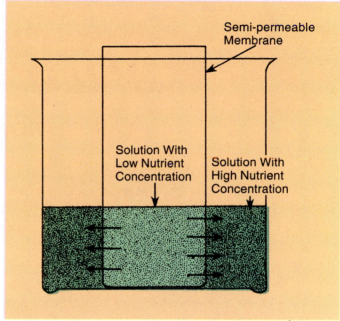

FIGURE 16-8 The process of osmosis.

Semi-permeable
Membrane

Solution With
Low Nutrient
Concentration

Solution With
High Nutrient
Concentration

solved substance. The liquid will flow from a weaker to a stronger solution, thus tending to equalize concentrations. (Figure 16-8). A **semi-permeable membrane** is a structure that permits a solution to move through it in a direction controlled by the concentration of solutions. The epidermis of a root hair is a semi-permeable membrane. When fertilizer is added to the soil, a high concentration of nutrients occurs in that soil. When the soil moisture dissolves nutrients from the fertilizer, they can move through the semi-permeable membrane of the root hair. They enter the cells of the plant, where nutrient concentrations are lower. Once inside the root-hair cell, nutrients can be transported to other parts of the plant as needed.

To allow the root hairs to move through the soil, the soil must have spaces between the particles of sand, silt, and clay. Such spaces are called **pores.** Their job is to store air, water, and nutrients, as well as permit root penetration.

AIR

The air or atmosphere that surrounds the aboveground portion of the plant must supply carbon dioxide as well as oxygen. Generally this is no problem when the plants grow outdoors in fields. When plants are transplanted into artificial or unnatural environments, consideration must be given to the quality of the air surrounding the plant. In a greenhouse or other enclosed system, the quality of the atmosphere needs monitoring. The presence and levels of carbon dioxide and pollutants must be understood for maximum production. With certain crops in greenhouses, such as carnations and roses, the addition of some carbon dioxide might be desirable to increase crop production. In areas where crops are growing near industrial plants or cities, or along major highways, the technician must be aware of the many types of pollutants that might cut production or severely damage the plants.

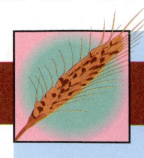

AGRI·PROFILE

CAREER AREA: Plant Physiology

Physiology refers to the many functions that occur inside of plants. These include familiar activities such as osmosis, nutrient uptake, translocation, respiration, photosynthesis, food movement, and food storage. Other complex functions are known only to specialists in the field.

Plant physiologists work closely with technicians and scientists in other fields of plant science. They may be consultants to or collaborators with specialists in agronomy and horticulture. The work of plant physiologists is typically done as college or university faculty, employees of state or national research institutes, or specialists with agriscience corporations developing and selling seeds and plant materials.

Plant physiologists Marcia Holden and Douglas Luster inspect tomato plants for iron deficiency. *(Courtesy USDA/ARS #K-4191-5)*

WATER

The most essential ingredient for all life is said to be water. Nutrients in the soil must first be dissolved in water before they can be absorbed via the roots. When inside the plant, water carries the nutrients to the leaves, where these nutrients chemically combine with water in the photosynthetic process. Sugars and other plant foods manufactured in the leaves are then transported throughout the plant by water. Water helps to control the temperatures in and around plants through transpiration. Finally, water gives the plant support by keeping the cells in a turgor state. It is important, therefore, that water used for plant production be of good quality and in adequate supply.

PLANT NUTRITION

This is often confused with plant fertilization. There is a difference. **Plant nutrition** refers to availability and type of basic chemical elements in the plant. **Plant fertilization** is the process of adding nutrients to the soil or leaves so these chemicals are added to the growing environment of the plant. Before chemicals supplied as fertilizer can be taken up and used by plants, they generally undergo various changes.

Essential Nutrients

There are sixteen elements that are essential for normal plant growth. These are required in various amounts by plants and must be available in the relative proportions needed if the plants are to produce well.

FIGURE 16-9 Soils and nutrient-management specialists do cooperative research to determine the best rates of fertilizer to optimize corn growth. *(Courtesy USDA/ARS #K-3694-5)*

Three are used in huge amounts and are obtained from the atmosphere and water around the plant. They are carbon (C), hydrogen (H), and oxygen (O). There are six elements that are used in relatively large amounts and are called **macronutrients.** The macronutrients are nitrogen (N), phosphorus (P), potassium (K), calcium (Ca), magnesium (Mg), and sulfur (S). These are all obtained from the soil.

An additional seven elements are used in small quantities and are called **micronutrients** (trace elements). The micronutrients are also obtained from the soil. They are boron (B), copper (Cu), Chlorine (Cl), iron (Fe), manganese (Mn), molybdenum (Mo), and zinc (Zn). For a plant to grow at maximum efficiency, it must have all essential plant nutrients (Figure 16-9). The absence of any one of these nutrients will cause the plant to grow poorly or show some signs of poor health.

Remembering the Sixteen Plant Nutrients

Various schemes have been devised to help you remember the names of the sixteen plant nutrients. One technique is to first learn the chemical symbols. Use the symbols to make a logical string of words that are easy to remember. One such string of words that uses the symbols of most of the nutrients is, "C. Hopkin's cafe, mighty good." By remembering this phrase, you can recall the symbols of ten of the sixteen nutrients as follows: C HOPKNS CaFe Mg (carbon, hydrogen, oxygen, phosphorus, potassium, nitrogen, sulfur, calcium, iron, and magnesium). The remaining ones are boron, copper, chlorine, manganese, molybdenum, and zinc. Can you devise a string of words to help you remember the symbols of these micronutrients?

Ions

Plant nutrients are absorbed from the soil-water solution that surrounds the root hairs of the plant. In fact, 98 percent of the nutrients obtained from the soil is absorbed in solution, while the other 2 percent is extracted by the root directly from soil particles. Most of the nutrients are absorbed as charged ions. An **ion** is an atom that has an electrical charge.

Ions are either negatively charged and called **anions,** or positively charged and called **cations.** The electrical charges in the soil are paired

1. **HEALTHY** *stalk has normal size. Cut-away stalk section below ear shows health, white pith area.*
2. **POTASSIUM** *is needed when the cut-away shows dark brown discoloration at the nodes.*
3. **PHOSPHORUS** *shortage leads to weak, spindly stalks, often barren of ears. Note purple coloration of lower leaves.*
4. **SUCKERS** *may form when corn gets too much nitrogen early in the season. Cut-away section also shows corn borer damage.*
5. **DISEASE** *symptoms found in stalks include the black bundles in the upper cut-away, and the darker pith in the lower cut-away. Stalk rot works inside first, causing early dying and breakage of the stalk, and shrivelled ears with chaffy, low-test-weight grain.*

FIGURE 16-10 Nutrient deficiencies decrease plant health, vigor, and growth. Deficiency of many nutrients can be determined by observing the plant. *(Courtesy Potash and Phosphate Institute, Norcross, GA)*

so that the overall effect in the soil is not changed. These ions compete and interact with each other according to their relative charges. For example, nitrogen, in its nitrate form, has a negative charge and chemical formula NO_3-. Therefore, nitrates are anions with negative charges.

On the other hand, potassium has a positive charge and is an example of a cation (K+). Potassium nitrate, $K+NO_3-$, is a combination of potassium and nitrate consisting of one nitrate ion and one potassium ion. Calcium nitrate, $Ca++(NO_3-)_2$, has two nitrate ions and one calcium ion. The reason is that the calcium cation has two positive charges. As you might guess, this could be confusing, but it serves to illustrate the need to understand chemistry to manipulate plant fertility if conditions are not ideal in the natural environment.

The balance of ions is important and needs to be carefully monitored for good plant growth. Opposite charges attract each other, but ions with similar charges compete for chemical reactions and interactions in the soil-water environment. Some ions are more active than others and might be able to compete better in the soil. Further study would be needed to thoroughly understand why soil tests may indicate the presence of a certain element in sufficient amounts for plant growth, yet the plants may show deficiency symptoms (Figure 16-10).

Deficiency means a shortage of a given nutrient available for plant use. A good example of a nutrient deficiency symptom is blossom-end rot of tomato. This is common in gardens and occurs when there is not enough water to dissolve and carry calcium to the plant in sufficient quantities. The end opposite the stem is called the blossom end. The calcium deficiency produces a tomato that looks good from the top but, when picked, the bottom end is rotten.

Soil Acidity and Alkalinity

The chemistry of plant elements in the soil can be affected by pH. Soil **pH** is a measurement of acidity (sourness) and alkalinity (sweetness) (Figure 16-11). Many of the nutrients in soil form very complex combinations and are capable of precipitating out of solution, where they are unavailable to the plant. **Precipitate** occurs when a solid is dropped out of solution. If the soil pH is **acid,** or extremely low, some micronutrients become too soluble and occur in concentrations high enough to harm the plants (Figure 16-12).

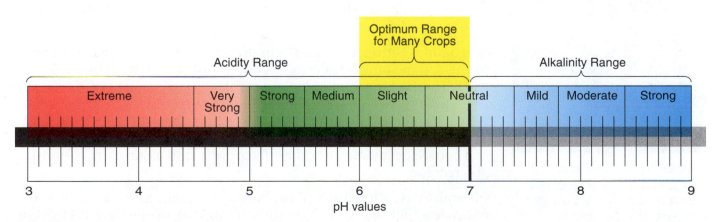

FIGURE 16-11 The pH scale.

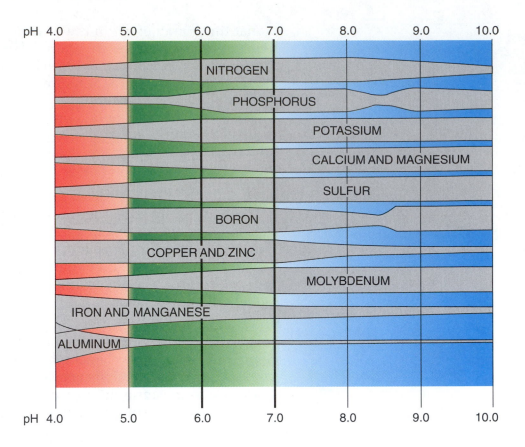

FIGURE 16-12 The effect of soil pH on nutrient availability.

On the other hand, if soil pH is very high, in the **alkaline** range, many of the nutrients can be precipitated out and not be available to the plants. The pH of soils can be determined with low-cost test kits. Fortunately, soil pH can be corrected by adding lime if the pH needs to be raised and sulfur if it needs to be lowered. Such practices are common, because soil pH is seldom perfect for the crop being grown (Figure 16-13).

Plant Nutrient Functions

The importance of carbon, hydrogen, and oxygen has already been discussed under the topic of photosynthesis. The other nutrients have very specific functions and must be available in the appropriate amounts and form. The effects of plant nutrients may be likened to a chain—the weakest link will determine how much the chain will pull. Similarly, the nutrient in shortest supply will determine the maximum growth achievable by the plant.

Nitrogen   Nitrogen is present in the atmosphere as a gas. It is added to the soil in some fertilizers. Because it exists in nature as a gas, it is easily **leached** (washed out of the soil). Nitrogen is responsible for the vegetative growth of the plant and its dark-green color. When nitrogen is lacking, some deficiency signs are reduced growth and yellowing of the leaves. This yellowing is referred to as **chlorosis.** Excess nitrogen can cause succulent, weak, spindly growth, and a darker-than-normal green color.

Phosphorus   In nature, phosphorus is present as a rock and is not easily leached out of the soil. It is important in seedling and young plant growth. It will help the plant develop a good root system. Some

FIGURE 16-13 Soil scientist Charles Foy compares barley plants grown in soils having different pHs. *(Courtesy USDA/ARS #K-3212-1)*

symptoms of a deficiency of phosphorus are reduced growth, poor root systems, and reduced flowering. Thin stems and browning or purpling of the foliage are also signs of poor phorphorus availability.

Potassium Potassium is mined as a rock and made into fertilizer, but it can be leached from the soil. If too much potassium is present, it can cause a nitrogen deficiency. A lack of potassium will show up as reduced growth or shortened internodes, and sometimes as **marginal burn** or **scorch** (brown leaf edges). Dead spots in the leaf, and plants that wilt easily, can also be an indication of a potassium deficiency.

Calcium Calcium is often supplied by adding lime to the soil. It can be leached out and does not move very easily throughout the plant. Too much calcium can cause a high pH and reduce the availability of some elements to the plant. A lack of this element can stop bud growth and result in death of root tips, cupping of mature leaves, and blossom-end rot of many fruits. Pits on root vegetables can also be a sign of calcium deficiency.

Magnesium Magnesium can be added by using high-magnesium lime. It can also be leached from soil. If magnesium is lacking, some reduction of growth and marginal chlorosis can be noticed. In some plants, even interveinal chlorosis can be seen. Cupped leaves and a reduction in seed production can also be symptoms of magnesium deficiency. Foliage plants are commonly lacking this nutrient.

Sulfur Present in the atmosphere as a result of combustion, sulfur is often an impurity in fertilizers carrying other nutrients. As a result, it is rarely deficient. However, if sulfur is deficient, a yellowing of the entire plant may result.

Deficiencies and exesses of nutrients cause typical symptoms in plants (Figure 16-14).

Nutrient	Excess	Deficiency
Iron	Rare	Interveinal chlorosis, especially on young growth
Zinc	Might appear as an iron deficiency	Interveinal chlorosis, reduction in leaf size, short internodes
Molybdenum	Not known	Interveinal chlorosis on older leaves; may also affect leaves in the middle of the plant
Boron	A blackening or death of tissue between veins	Failure to set seed; death of tip buds
Copper	Might occur in low pH; will appear as an iron deficiency	New growth small and misshapen, wilted
Manganese	Brown spotting on leaves; reduced growth	Interveinal chlorosis of the leaves and brown spotting; checkered effect possible

FIGURE 16-14 Both excess and deficiency of most nutrients will cause predictable symptoms in plants.

BIO·TECH CONNECTION

Diagnosing an Eminent Killer

A subtle and mysterious killer, citrus blight, has eluded plant pathologists and other scientists for over a century. Recognized by citrus growers and the federal government as a deadly disease of citrus trees, the U.S. Horticultural Research Laboratory was established in Eustis, Florida in 1892 to research this problem. Citrus blight is a mysterious disorder that renders a tree worthless for fruit production. It is most likely to attack young trees that are just beginning to produce fruit, but it also kills trees 20, 30, or even 50 years old. No cure or prevention is known for the citrus blight. First diagnosed in Florida in 1874, it causes physiological changes in the tree and leads first to a yellowing of leaves and eventually to leaf wilt. With no cure and no chance of recovery, diseased trees are destroyed as soon as the disorder is observed. Authorities estimate that one-half million trees valued at $60 million are lost each year to the blight.

USDA plant pathologist Michael Bausher believes he has a found a way that may lead to early diagnosis and reduce losses from the disorder. At the U.S. Horticultural Research Lab in Orlando, Bausher discovered unique proteins in the leaves of diseased trees that are not present in healthy trees, nor in trees with other diseases or stress problems. Bausher prepared, freeze-dried, and partially purified an antigen derived from ground-up leaves of blighted trees. When the antigen was injected into rabbits, he discovered the rabbits produced a unique antisera. The rabbit-produced antisera was then found to react positively with the unique proteins from the blighted trees. However, the antisera reacted negatively when exposed to

Pale green, chloritic leaves and reduced leaf size typical of citrus blight are displayed by plant pathologist Michael Bausher. *(Courtesy USDA/ARS #K-3323-3)*

proteins derived from tissue of healthy trees and trees with other disorders or diseases.

Growers usually remove trees at the first indication of citrus blight to cut the losses from the disease. Unfortunately, other disorders may cause yellowing or mottling of leaves, which are the first observable symptoms of the blight. Therefore, it is hard to tell how many trees with correctable disorders are sacrificed. It is hoped that a process can be developed wherein proteins can be used as biological markers to help identify trees with citrus blight before the visual symptoms appear. This early detection would permit growers to remove only trees that really have the citrus blight and spare those that have other correctable disorders. Further, the biological markers could possibly help researchers develop trees that are resistant to the blight.

FOOD STORAGE

When the plant makes its food through photosynthesis, it often manufactures more than it needs to maintain itself. This excess is often stored in the plant for future use. Such food may be stored in roots, stems, seeds, or fruits.

Roots

The most common type of root that serves as a storage organ is the taproot. Some common examples of plants with extensive storage capacity are sugar beets, carrots, radishes, and turnips. The sugars and carbohydrates are transported down the phloem and into the root cells. They are held here as the root enlarges. Most of the time, this type of plant is a short-term crop that does not take long to mature. This type of root system is easy to dig or harvest.

Stems

The stems of plants usually contain cells that are necessary for plant support. Some specialized stems, however, are excellent food-storage organs (Figure 16-15). Some are used for propagation and some for food. The most common specialized stem used for food is the tuber. It is an enlarged portion of a stem containing all of the parts of a normal stem. Nodes, internodes, and buds can be identified in tubers. The Irish potato is an example of a tuber.

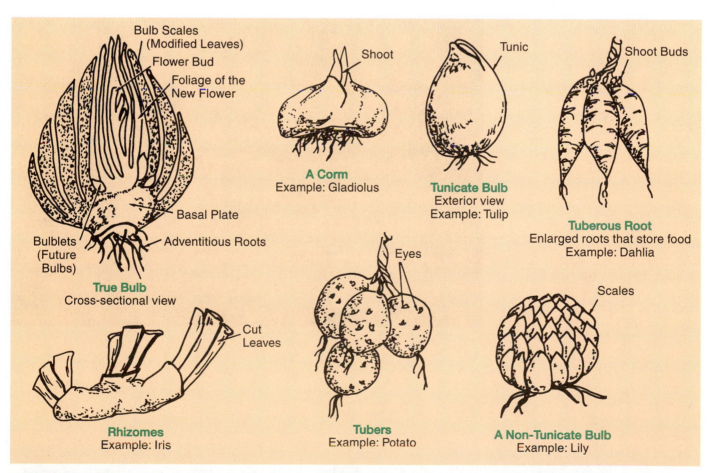

FIGURE 16-15 Examples of stems that are major food storage organs for the plant. *(Modified from Ingels, Ornamental Horticulture. Copyright 1994 by Delmar Publishers)*

Corms and bulbs are other examples of specialized stems that contain large amounts of food made by photosynthesis. A rhizome is yet another. Often these are used for propagation purposes.

Seeds

As the ovule of a plant matures, it stores food for the young embryo to start its growth when it germinates. Both animals and humans utilize seeds as major food sources. They help the plant by spreading the seed to new locations, increasing the plant's chances for survival.

SUMMARY

This unit has covered some basic principles of plant physiology. Physiology is complex and must be studied in great depth for proficiency. Plant physiologists typically have master or doctorate degrees. However, most technicians and scientists in the plant sciences will have some training in plant physiology. A basic knowledge of soils and how plants use nutrients and function in general will help greatly in the successful production and management of plants. Farmers, ranchers, and growers generally have access to sophisticated technology (Figure 16-16).

FIGURE 16-16 Missouri farmer Bill Holmes and specialists from the Space Remote Sensing Center examine soil fertility variations in the Holmes' cropland. *(Courtesy USDA/ARS #K-4914-2)*

STUDENT ACTIVITIES

1. Write the Terms to Know and their meanings in your notebook.

2. Make a bulletin board showing the cross-section of a leaf. Label the various cells and leaf parts. Include the formula for photosynthesis.

3. Write a letter to a person new to plant production and explain the importance of photosynthesis.

4. Collect plants or pictures to make a display that could be used by others to help identify nutrient deficiencies.

5. Select a crop that interests you and conduct research to determine the optimum nutrient requirements.

6. Set up a demonstration to explain osmosis.

7. Make a list of all the chemical symbols of plant nutrients. Write a sentence or story to help you remember them.

SELF EVALUATION

A. Multiple Choice

1. The study of functions and the complex chemical processes that allow plants to grow is known as
   a. plant taxonomy.
   b. plant physiology.
   c. plant nutrition.
   d. photosynthesis.

2. Chlorophyll is important in plants because it
   a. creates an atmosphere where it can determine the osmotic pressure.
   b. allows the plant to make good xylem tissue.
   c. gives the green color to plants.
   d. is also known as the chloroplasts.

3. The rate at which photosynthesis is carried out depends on
   a.   the amount of fertilizer in the water.
   b.   the amount of oxygen in the atmosphere.
   c.   the amount of respiration carried on during the daylight hours.
   d.   the light intensity, temperature, and concentration of carbon dioxide.

4. Photosynthesis will work best in which temperature range?
   a.   50° to 60°F
   b.   60° to 70°F
   c.   65° to 85°F
   d.   85° to 95°F

5. Respiration
   a.   uses food for plant energy.
   b.   stores energy.
   c.   occurs in cells that contain chlorophyll.
   d.   uses carbon dioxide.

6. Plant nutrition is
   a.   plant food added to the plant pot.
   b.   use of basic chemical elements in the plant.
   c.   chemical processes providing plants with elements for growth.
   d.   the measurement of acidity (sourness) and alkalinity (sweetness).

B. Matching

_____ **1.** pH   a.   Movement through a semi-permeable membrane
_____ **2.** Osmosis   b.   Addition of nutrients to the plant growing environment
_____ **3.** Corm   c.   Storage organ for excess plant food
_____ **4.** Root   d.   Site of photosynthesis
_____ **5.** Sulfur   e.   Measurement of acidity and alkalinity
_____ **6.** Leaves   f.   Macronutrient
_____ **7.** Fertilization   g.   Specialized stem

C. Completion

1. Extremes of temperature will slow down or completely stop _____.

2. Respiration will occur only in the _____.

3. When the temperature increases, the rate of transpiration _____.

4. Soil provides a natural environment for the _____ _____.

5. The spaces in between the soil particles, where the soil water is found, is called the _____.

D. True or False

_____ **1.** Respiration is a building process that uses sunlight to work.

_____ **2.** The process by which the plant loses water is perspiration.

_____ **3.** All plant elements perform the same function in the plant.

_____ **4.** Humidity refers to the amount of water in the atmosphere.

Plant Reproduction

TERMS TO KNOW

Propagation
Reproduction
Sexual reproduction
Asexual reproduction
Clone
Vegetative
Hybrid
Hybrid vigor
Germinate
Dormant
Imbibition
Scarify
Viable
Germination rate

Cuttings
Fungicide
Rooting hormone
Stem tip cuttings
Stem section
    cuttings
Cane cuttings
Heel cuttings
Single-eye cuttings
Double-eye cutting
Leaf cutting
Leaf petiole cuttings
Leaf section cuttings
Split-vein cuttings

Root cuttings
Layering
Simple layering
Tip layering
Air layering
Grafting
Scion
Rootstock
Stock
Graft union
Bud grafting
T-budding
Budding rubber
Tissue culture

Plant **propagation,** or **reproduction,** is simply the process of increasing the numbers of a species, or perpetuating a species. The two types of plant propagation are sexual and asexual. **Sexual reproduction** is the union of an egg (ovule) and sperm (pollen), resulting in a seed. Two parents creating a third individual is referred to as sexual propagation. In plants, it involves the floral parts. It may involve one or two plants. **Asexual reproduction** utilizes a part or parts of only one parent plant. The purpose is to cause the parent plant to make a duplicate of itself. The new plant is a **clone** (exact duplication) of its parent. Because this type of reproduction uses the vegetative parts of the plant, namely the stems, roots, or leaf, it is often referred to as **vegetative** propagation.

Some advantages of sexual propagation are that it is often less expensive and quicker than some other methods. It is the only way to obtain new varieties and also capture hybrid vigor. A **hybrid** is a plant obtained by crossbreeding. **Hybrid vigor** refers to the tendency of hybrid plants to be stronger and survive better than purebred plants. Sexual propagation is a good way to avoid passing on some diseases. In some plants, it is the only way to propagate them.

Asexual propagation has many advantages as well. In some cases, it is easier and less expensive to obtain plants this way. In some species or cultivars, it is the only way to propagate them.

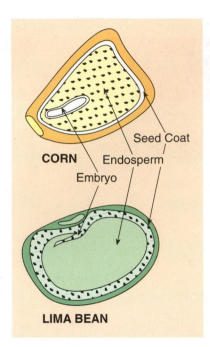

FIGURE 17-1 Parts of a seed.

SEXUAL PROPAGATION

A seed is made up of the seed coat, endosperm, and embryo (Figure 17-1). The seed coat functions as a protector for the seed. Sometimes it can be very thin and soft, or it may be very hard and impervious to water or moisture. The endosperm functions as a food reserve. It will supply the new plant with nourishment for the first few days of life. The embryo is the young plant itself. When a seed is fertilized and matures, it will be dormant. When it is given favorable conditions, it will **germinate,** which means that the seed will start to sprout, and the plant will grow.

Seed propagation starts with quality seed. Crop production by sexual reproduction means consideration can be given to the type of plant needed or the variety that is best adapted to a particular area or purpose. Hybrid plants are developed by cross-pollinating two different varieties. Many varieties on the market are the result of hybridization or crossbreeding. Seeds of hybrid plants cost more than open-pollinated varieties. However, the increased quality of the plants generally offsets the increased cost of the seed. New varieties are being developed for disease and insect resistance. It is natural to expect the seeds of such improved varieties to cost more than standard or regular seeds. Some varieties have unusual cultural or product characteristics (Figure 17-2).

Seeds collected from plants used for production other than seed production generally will not save money in the long run. Seeds from such plants are often small. They are often poorly managed and improperly handled and stored. It is recommended that seed saved from season to season be stored in a sealed jar at 40°F (4.4°C) and low humidity.

Germination

When seed is harvested, or collected, it is normally mature and in a **dormant** or resting state. To germinate, or start to grow, it must be placed in certain favorable conditions. The four environmental factors that must be right for effective germination are water, air, light, and temperature.

Water **Imbibition** (the absorption of water) is the first step in the germination process. The seed, in its dormant stage, contains very little water. The imbibition process allows the seed to fill all its cells with water. If other conditions are favorable, the seed then breaks its dormant stage and germinates.

A good germination medium is important. The medium must not be too wet or too dry. An adequate and continuous supply of water must be available. This is often difficult to control with crops directly seeded in the field. It is much easier to control in crops planted for transplanting. A dry period during the germination process will result in the death of the young embryo. Too much water will result in the young seed rotting. In some species, the seed coat is very hard and water cannot penetrate to the endosperm. In cases like this it is necessary to scarify the seed.

A common way to **scarify** seed is to nick the seed coat with a knife or a file. Another method is to soak the seeds in concentrated sulfuric acid. This requires special care and experience, because sulfuric acid is a dangerous material. Another technique is to place seeds in hot water, that is 180° to 212°F (82.2° to 100°C), and allow them to soak as the

FIGURE 17-2 At the National Seed Storage Laboratory in Fort Collins, Colorado, seeds are packaged in flexible, moisture-proof bags. *(Courtesy USDA/ARS #K-1657-16)*

water cools. This process takes 12 to 24 hours. A warm, moist scarification process may be utilized by simply placing the seeds in warm, damp containers and letting the seed coat decay over a period of time.

Air Respiration takes place in all viable seed. **Viable** seed is alive and capable of germinating. Oxygen is required. Even in nongerminating seeds, a small amount of oxygen is required even though respiration is low. As germination starts, the respiration rate increases. It is important that the seed be placed in good soil or media that is loose and well drained. If the oxygen supply is limited or reduced during the germination process, germination will be reduced or inhibited.

Light Some seeds are stimulated to grow by light. Some are inhibited by the presence of light. It will be necessary to have some knowledge of the presence or absence of special light requirements. Many of the agronomic crops do not require light for germination. In fact, light will inhibit germination. Ornamental bedding plants are more likely to require light for germination. Some crops requiring light for germination are ageratum, begonia, impatiens, and petunia. Lettuce also requires light for successful germination. Seeds of these plants are often deposited on the surface of soil by nature, and the grower should follow the same procedure for successful germination.

Temperature Heat is another important requirement for germination. The **germination rate,** or percentage of seed that germinates, is affected by the availability of heat. Some seeds will germinate over a wide range of temperatures, while others have more narrow limits. In the agronomic crops that are directly seeded in the field, the only way to control heat is to plant when the ground is warm. In horticultural crops, particularly bedding plants and perennials, knowledge of a plant's specific heat requirements for germination should result in more efficient production. The germination requirements are often listed in the seed catalogs (Figure 17-3).

Plant	Time to Seed Before Last Frost (Weeks)	Germination Time (Days)	Germination Temperature Requirements (Degrees F)	Germination Light Requirements Light (L) Dark (D)
Ageratum	8	5-10	70	L
Aster	6	5-10	70	–
Begonia	12+	10-15	70	L
Coleus	8	5-10	65	–
Cucumber	4	5-10	85	–
Eggplant	8	5-10	80	–
Marigold	6	5-10	70	–
Pepper	8	5-10	80	–
Portulaca	10	5-10	70	D
Snapdragon	10	5-10	65	L
Tomato	6	5-10	80	–
Watermelon	4	5-10	85	–
Zinnia	6	5-10	70	–

FIGURE 17-3 Time, temperature, and light requirements for some common flower and vegetable plants.

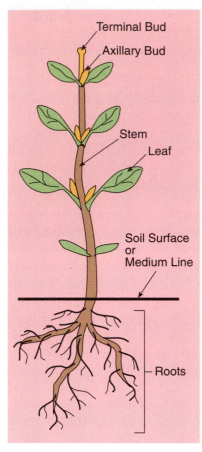

FIGURE 17-4 Vegetative parts of plants.

ASEXUAL PROPAGATION

As stated previously, asexual propagation is using the vegetative parts of the plant to increase the number of plants (Figure 17-4). The primary advantages are economy, time, and plants that are identical to the parents. The primary methods of asexual propagation are cuttings, layering, divison, grafting, and tissue culture.

Stem Cuttings

Herbaceous and woody plants are often propagated by **cuttings** (vegetative parts that the parent plant used to regenerate itself). Types of cuttings are named for the parts of the plant from which they come. There are stem tip cuttings, stem cuttings, cane cuttings, leaf cuttings, leaf petiole cuttings, and root cuttings.

The procedure for taking cuttings is relatively simple. The equipment needed is a sharp knife or a single-edge razor blade. Sharp equipment will make the job easier and will reduce injury to the parent plant. To prevent the possibility of diseases spreading, it is best to dip the cutting tool in bleach water made with one part bleach to nine parts water. The tool can also be dipped in rubbing alcohol.

The flowers and flower buds should be removed from all cuttings. This allows the cutting to use its energy and food storage for root formation instead of flower and fruit development. A rooting hormone containing a fungicide is used to stimulate root development. A **fungicide** is a pesticide that helps prevent diseases.

Rooting hormone is a chemical that will react with the newly formed cells and encourage the plant to make roots faster. The proper way to use a rooting hormone is to put a small amount in a separate container and work from that container. This procedure will ensure that the rooting hormone does not become contaminated with disease organisms. Do not put the unused hormone back in the original container.

Cuttings are normally placed in a medium consisting of coarse sand, perlite, soil, a mixture of peat and perlite, or vermiculite. It is best to use the correct medium for a specific plant in order to obtain the most efficient production in the shortest possible time. The rooting medium should always be sterile and well drained, with moisture retention ability to prevent the medium from drying out. The medium should be moistened before inserting the cuttings. It should then be kept continuously and evenly moist while the cuttings are forming roots and new shoots.

Stem and leaf cuttings do best in bright, but indirect light. However, root cuttings are often kept in the dark until new shoots are formed and start to grow. For most plants, the most popular method of making cuttings is by stem cutting. On herbaceous plants, stem cuttings may be made almost any time of the year. However, stem cuttings of many woody plants are normally taken in the fall and/or the dormant season.

Stem Tip Cuttings   Taken from the end of the stem or branch, **stem tip cuttings** normally include the terminal bud. A piece of stem between 2 and 4 in. long is selected, and the cut is made just below the node. The lower leaves that would be in contact with the medium are removed. The stem is dipped in the rooting hormone and is gently

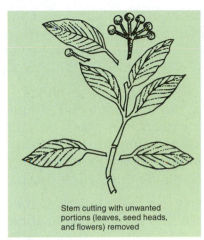

Stem cutting with unwanted portions (leaves, seed heads, and flowers) removed

FIGURE 17-5 Stem tip cuttings.

tapped to remove the excess rooting hormone. The cutting is then inserted into the rooting medium. The cutting should be inserted deep enough so the plant material will support itself. It is important that at least one node be below the surface of the medium (Figure 17-5).

Stem Section Cuttings Stem section cuttings are prepared by selecting a section of the stem located in the middle or behind the tip cutting. This type of cutting is often used after the tip cuttings are removed from the plant. The cutting should be between 2 and 4 in. long, and the lower leaves should be removed. The cutting should be made just above a node on both ends. It is then handled as a tip cutting. Make sure that the cutting is positioned with the right end up. The axial buds are always on the tops of the leaves.

Cane Cuttings Some plants, such as the dumbcane (*Diffenbachia sp.*), have canelike stems. These stems are cut into sections that have one or two eyes, or nodes, to make cane cuttings. The ends are dusted with activated charcoal or a fungicide. It is best to allow the cane to dry in the open air for one or two hours. The cutting is then placed in a

AGRI·PROFILE

CAREER AREAS: Plant Breeding/ Plant Propagation/ Crop Improvement/Tissue Culture

Plant breeders' objectives might include making plants faster growing, disease resistant, drought tolerant, insect resistant, wind resistant, frost tolerant, more beautiful, or better flavored, depending on the uses of the plants. Much plant breeding occurs in greenhouses, with plants having small flowers that require the use of magnifying glasses and tweezers to transfer pollen or remove reproductive parts. Generally, such research is followed by field trials and seed production, which gives the plant breeder variety in the settings where work is done.

Asexual reproduction involves rooting, budding, grafting, layering, and other procedures other than pollination. Tissue culture, a procedure developed in biotechnology, permits the production of thousands of new plants identical to a superior plant. The procedure is relatively cheap and easy, and is used extensively to reproduce ornamental plants. Many jobs are available in the area of plant reproduction.

The plant breeder or plant geneticist searches for plants, transfers genes, or crossbreeds plants from various sources in an effort to develop new varieties with desired characteristics. *(Courtesy USDA/ARS #K-5146-16)*

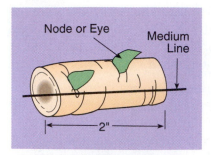

FIGURE 17-6 Cane cutting.

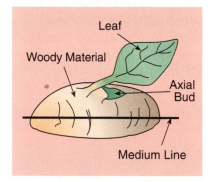

FIGURE 17-7 Heel cutting.

horizontal position with half of the cane above the surface of the medium. The eyes, or nodes, should be facing upward. This type of cutting is usually potted when the roots and new shoots appear (Figure 17-6).

Heel Cuttings   **Heel cuttings** are used with woody-stem plants. A shield-shaped cut is made about halfway through the wood around the leaf and the axial bud. Rooting hormone may be used as it is in the other types of cuttings. The cutting is inserted horizontally into the medium (Figure 17-7). If space or stock material is in short supply, propagators might make the single-eye and the double-eye cuttings.

Single-eye Cuttings   When the plant has alternate leaves, **single-eye cuttings** are used. The eye refers to the node. The stem is cut about ½ in. above and below the same node (Figure 17-8). The cutting may be dipped in rooting hormone and then placed either vertically or horizontally in the medium.

Double-eye Cuttings   When plants have opposite leaves, the **double-eye cutting** is the preferred type. It is prepared the same way as the single-eye cutting (Figure 17-9).

Leaf-type Cuttings

For many of the indoor herbaceous plants, a leaf-type cutting will produce plants quickly and efficiently. This type of cutting will not normally work for woody plants, however.

Leaf Cuttings   A cutting made from a leaf without a petiole is referred to as a **leaf cutting.** To prepare a leaf cutting, detach the leaf from the plant with a clean cut and dip the leaf into the rooting hormone. Place the leaf cutting vertically into the medium. New plants will form at the base of the leaf and may be removed when they have formed their own roots (Figure 17-10).

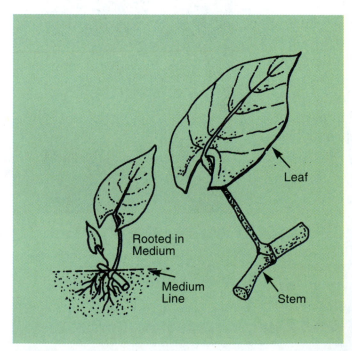

FIGURE 17-8 Single-eye cutting.

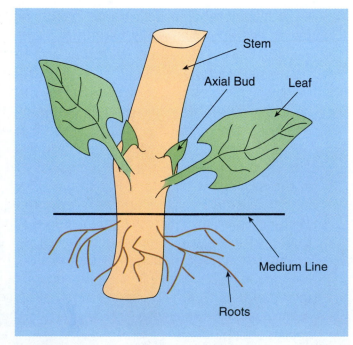

FIGURE 17-9 Double-eye cutting.

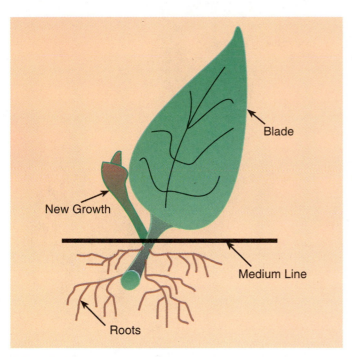

FIGURE 17-10 Leaf cutting.

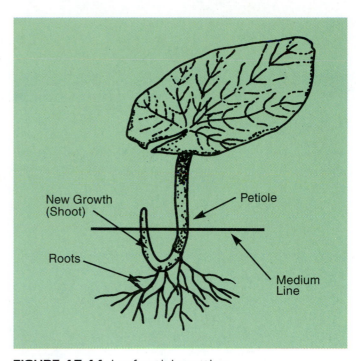

FIGURE 17-11 Leaf petiole cutting.

Leaf Petiole Cuttings For **leaf petiole cuttings,** a leaf with a petiole about ½ to 1½ in. is detached from the plant. The lower end of the petiole is dipped into the rooting medium and is then placed into the medium. Several plants will form at the base of the petiole (Figure 17-11). These plants may be removed when they have developed their own roots. The cutting may be left in the medium to form new plants.

Leaf Section Cuttings Fibrous-rooted begonias are frequently propagated using **leaf section cuttings.** The begonia leaves are cut into wedges, each containing at least one vein (Figure 17-12). The sections are then placed into the medium.

New plants will form at the vein that is in contact with the medium. A section-type leaf cutting is made with the snake plant *(Sanseveria sp.).* The leaf is cut into sections 2 to 3 in. long. It is a good practice to

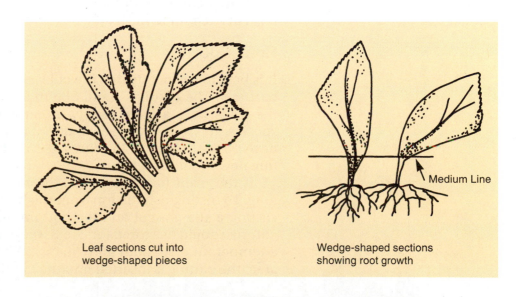

FIGURE 17-12 Leaf section cutting.

Leaf sections cut into wedge-shaped pieces

Wedge-shaped sections showing root growth

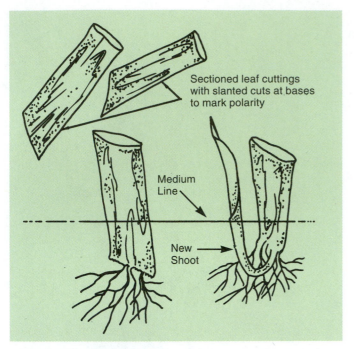

FIGURE 17-13 A leaf section of a snake plant.

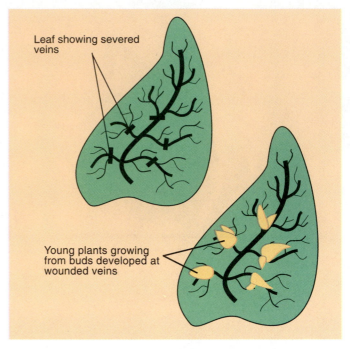

FIGURE 17-14 A split-vein cutting.

make the bottom of the cutting on a slant and the top straight. This is done so you can tell the top from the bottom (Figure 17-13). The sections are placed in the medium vertically. Roots will form reasonably soon and new plants will start to appear. These are to be cut off from the cutting as they develop root systems. The original cutting may be left in the medium for more plants to develop.

Split-vein Cuttings   Split-vein cuttings are often used with large leaf types, such as begonias and other large-leaf plants. With split-vein cuttings, the leaf is removed from the stock plant and the veins are slit on the lower surface of the leaf (Figure 17-14). The cutting is then placed on the rooting medium with the lower side down. It might be necessary to secure the leaf to make it lie flat on the surface. A good method is to use small pieces of wire, bending them like hair pins and pushing them through the leaf to hold it in place. The new plants will form at each slit in the leaf.

Root Cuttings

It is best to use plants that are at least 2 to 3 years old for making **root cuttings.** The cuttings should be made in the dormant season when the roots have a large supply of carbohydrates in reserve. In some species, the root cuttings will develop new shoots, which in turn will develop root systems. In others, the root system will be produced before new shoots develop.

If the plant has large roots, the root section should be 4 to 6 in. long. To distinguish the top from the bottom of the root, make the top cutting a straight cut and the bottom one a slanted cut. This type of cutting should be stored for 2 to 3 weeks in moist peat moss or sand at a temperature of about 40°F (4.4°C). When removed from the storage area, the cutting is inserted into the medium in a vertical position. The

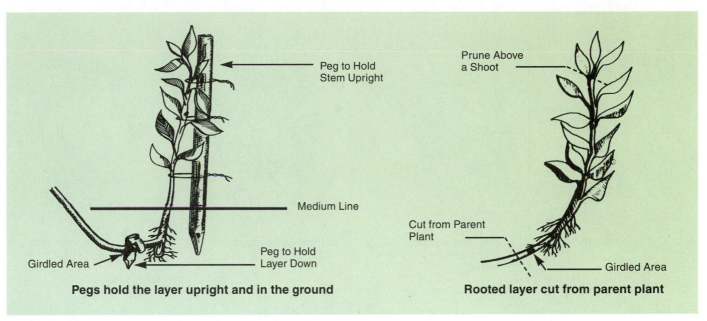

FIGURE 17-15 Simple layering. *(From Reiley and Shry,* Introductory Horticulture, *5E. Copyright 1997 by Delmar Publishers)*

slanted cut should be down and the top straight cut just level with the top of the medium. If the plant typically has small roots, a section 1 to 2 in. long is used. The cutting is placed horizontally ½ in. below the surface of the medium.

LAYERING

In many plants, stems will develop roots in any area that is in contact with the media while still attached to the parent plant. After roots form, shoots develop at the same point. An advantage of this type of vegetative propagation is that the plant does not experience water stress, and sufficient carbohydrates are supplied to the new plant that is forming. A discussion of some of the more common methods of **layering** follows.

Simple Layering

Simple layering is a very easy method that can be used on azaleas, rhododendrons, and other plants. A stem is bent to the ground and is covered with medium. It is advantageous to wound the lower side of the stem to the cambium layer. The last 6 to 10 in. of the stem is left exposed (Figure 17-15).

Tip Layering

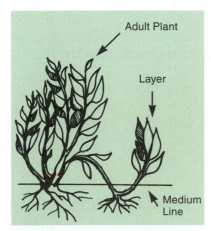

FIGURE 17-16 Tip layering. *(From Reiley and Shry,* Introductory Horticulture, *5E. Copyright 1997 by Delmar Publishers)*

Raspberries and blackberries are propagated using **tip layering.** With this method, a hole is made in the medium, and the tip of a shoot is placed in the hole and covered. The tip will start to grown downward and will then turn to grow upward. Roots will form at the bend. When the new tip appears above the medium, a new plant is ready to be transplanted. It will be necessary to separate the new plant from the parent by cutting the stem just before it enters the medium (Figure 17-16).

1. Prepare stem.

2. Soak Sphagnum moss and squeeze out excess water.

3. Pack damp moss over girdled area and tie.

4. Wrap with plastic and tape ends tightly.

FIGURE 17-17 Air layering.

Air Layering

Many foliage plants are propagated using **air layering.** Some ornamental trees, such as dogwood, can also be reproduced by this type of layering. The stem is girdled with two cuts about 1 in. apart. The bark is removed. The wound is dusted with a rooting hormone and surrounded with damp sphagnum moss (Figure 17-17). Plastic is wrapped around the moss-packed wound and tied at both ends. In a few weeks, depending on the plant, roots will appear throughout the moss. The stem is cut just below the newly formed root ball, and the ball is planted into a well-drained potting medium.

Division

Some plants are easily propagated by dividing or separating the main part into smaller parts. If the plant has rooted crowns, they are separated by cutting or pulling them apart. The resulting clumps are planted separately. If the stems are not attached to each other, they are pulled apart. If the crowns are joined together by horizontal stems, they are cut apart with a knife (Figure 17-18). It is a good practice to dust the divided plants with a fungicide.

Some plants that grow from bulbs or corms form little bulblets or cormels at their bases. To produce more plants from this type, simply separate the newly formed plant part and place it in a good medium (Figure 17-19).

Grafting

Grafting is a procedure for joining two plant parts together so they grow as one. This method of asexual propagation is used when plants do not root well as cuttings or when the root system is inadequate to support the plant for good growth. Grafting will allow the production of some unusual combinations of plants. For instance, several varieties of apples can be grown on one tree. Some nut trees can be made to grow varieties other than their own. Some unusual foliage plants can

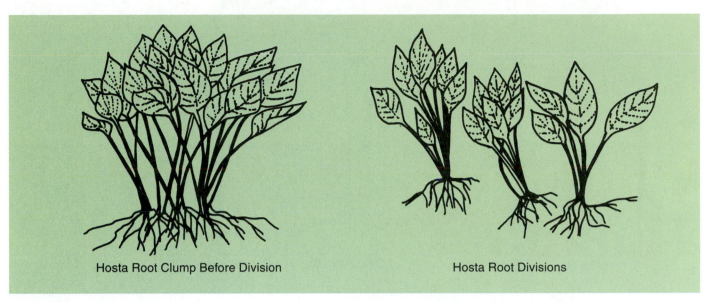

Hosta Root Clump Before Division

Hosta Root Divisions

FIGURE 17-18 Division.

also be made by grafting. Finally, dwarf fruit trees are created by grafting regular varieties on dwarfing root stock.

The top part of the plant that is to be propagated is called the **scion.** The **rootstock,** or **stock,** will be the new plant's root system and will supply the nutrients and water. The **graft union** is where the two parts meet (Figure 17-20).

To ensure successful grafting, the following conditions are necessary: (1) the scion and the rootstock must be compatible, (2) each must be at the right stage of growth, (3) the cambium layer of each section must meet, and (4) the graft union must be protected from drying out until the wound has healed.

There are many types of grafts. Some common grafts are whip, or tongue, graft; bark graft; cleft graft; bridge graft; and bud graft. Each type is used for a special purpose. The most commonly used and the easiest is the bud graft.

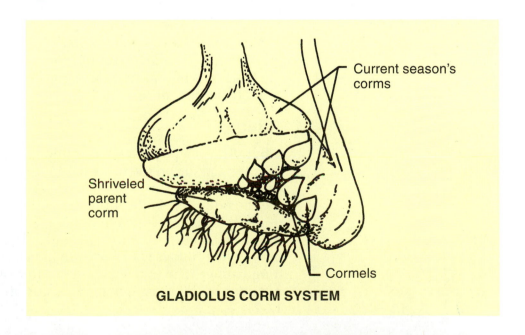

Current season's corms

Shriveled parent corm

Cormels

GLADIOLUS CORM SYSTEM

FIGURE 17-19 Separation.

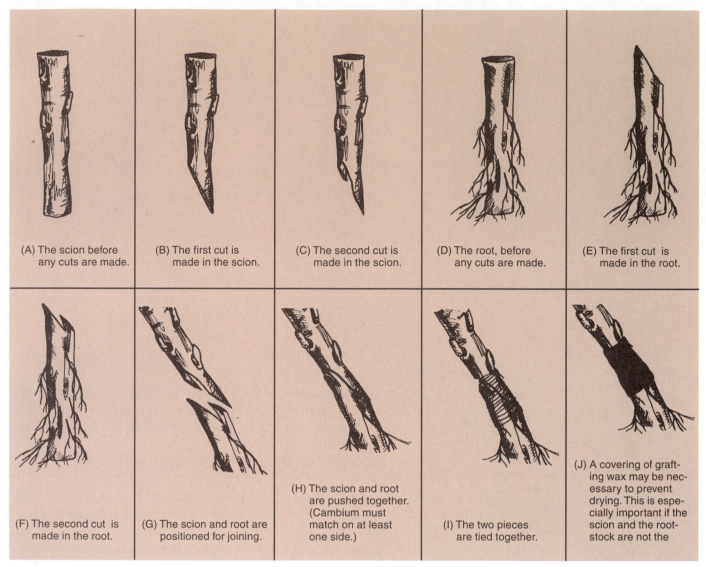

(A) The scion before any cuts are made.

(B) The first cut is made in the scion.

(C) The second cut is made in the scion.

(D) The root, before any cuts are made.

(E) The first cut is made in the root.

(F) The second cut is made in the root.

(G) The scion and root are positioned for joining.

(H) The scion and root are pushed together. (Cambium must match on at least one side.)

(I) The two pieces are tied together.

(J) A covering of grafting wax may be necessary to prevent drying. This is especially important if the scion and the rootstock are not the

FIGURE 17-20 Typical grafts. *(From Reiley and Shry,* Introductory Horticulture, 5E. *Copyright 1997 by Delmar Publishers)*

Bud Grafting

The union of a small piece of bark with a bud and a rootstock is called **bud grafting.** It is most useful when the scion material is in short supply. This type of grafting is faster and will make a stronger union than other types of grafting.

T-Budding

T-budding is a popular type of bud graft, where a vertical cut about ½ in. long is made on the rootstock. A horizontal cut is made at the top of the vertical cut. The result is a T-shape. The bark is loosened by twisting the point of a knife at the top of the T. A small shield-shaped piece of the scion, including a bud, bark, and a thin section of the wood, is prepared. The bud is pushed under the loosened bark of the stock plant. The union is wrapped with a piece of rubber band called a **budding rubber.** The bud is exposed. After the bud starts to grow, the remainder of the rootstock plant is cut off above the bud graft (Figure 17-21).

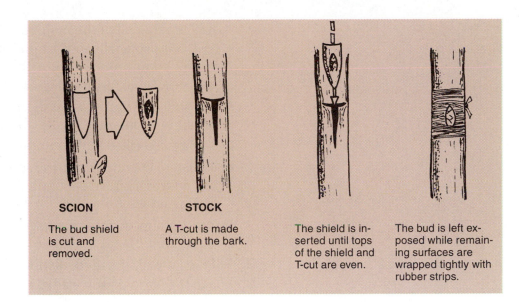

SCION
The bud shield is cut and removed.

STOCK
A T-cut is made through the bark.

The shield is inserted until tops of the shield and T-cut are even.

The bud is left exposed while remaining surfaces are wrapped tightly with rubber strips.

FIGURE 17-21 Making a T-bud. *(From Ingels,* Ornamental Horticulture, 2E. *Copyright 1994 by Delmar Publishers)*

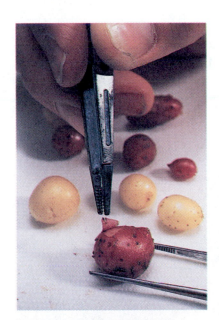

FIGURE 17-22 Only small amounts of tissue are needed to produce an exact clone of a plant when propagating via tissue culture. *(Courtesy USDA/ARS #K-4825-1)*

TISSUE CULTURE

A relatively new method of plant propagation is micropropagation, or **tissue culture.** Instead of using a large part of the plant as in other types of vegetative, or asexual, propagation, a very small and actively growing part of the plant is used (Figure 17-22). The result is many new plantlets from a section of a leaf. The process must be done in a clean atmosphere, and is not successful in the greenhouse or other traditional propagation areas. Tissue culture requires the use of very sanitary conditions. There are many commercial tissue-culture laboratories producing a large variety of plants today.

Advantages over Traditional Methods

The greatest advantage of this type of propagation is that numerous plants can be propagated from a single disease-free plant. Plants can be propagated more efficiently and economically than with traditional methods of asexual reproduction. The main disadvantage is that the work area must be clean. All of the equipment must be sterile. In commercial production by tissue culture, there is more expense in equipment and facilities than there is with traditional methods of propagation.

The materials necessary for tissue culture are (1) a clean, sterile area in which to work, (2) clean plant tissue, (3) a multiplication medium, (4) a transplanting medium, (5) sterile glassware, (6) sterile tools, (7) a scalpel, razor, or an X-acto®knife, and (8) tweezers.

Preparing Sterile Media

The first step in preparing for tissue culture is to prepare the medium in which the tissue will grow. The medium is commercially available from many of the scientific supply houses. The Virginia Cooperative Extension Service offers the following formula and procedure for preparing media for experimentation in tissue culture on a small basis.

1. Use a quart jar to mix the following materials:
- ⅛ cup of sugar
- 1 tsp of soluble, all-purpose fertilizer. The label will

indicate that all of the major and minor elements are present. It is especially important that the soluble fertilizer contain ammonium nitrate. If it is lacking, add ⅓ tsp of 35–0–0 soluble fertilizer.

- 1 tablet (100 mg) of inositol (myo-inositol). This can be obtained from most health-food stores.
- ¼ of a pulverized tablet containing 1 to 2 mg of thiamine
- 4 Tbsp of coconut milk, the source of cytokinin. This is obtained from a fresh coconut. Freeze the remainder for later use.
- 3 to 4 grains of a rooting hormone containing 0.1 active ingredient IBA

2. Fill the jar with purified, distilled, or deionized water.
3. Shake the jar to dissolve all materials.

After the medium is dissolved, prepare the culture tubes using either test tubes with lids or other suitable glass containers. Fill the culture tubes one-quarter of the way with sterile cotton balls. Use one or two per tube. They do not need to be packed tightly. Pour the prepared medium into the culture tubes to just below the top levels of the cotton. Place the lids on loosely.

After all medium is placed in culture tubes, it is ready to be sterilized. Sterilization may be done in two ways: (1) heat in a pressure cooker for 30 minutes or (2) heat in an oven for 4½ hours at 320°F. After they are removed, place the culture tubes in a clean area and allow them to cool (Figure 17-23). If several days will go by before using all of the tubes, wrap them in small groups in plastic wrap or foil before sterilizing.

Sterilizing Equipment and Work Areas

The tools and equipment used for tissue culture must also be sterilized. This can be done as the medium is sterilized by placing the tweezers, razor blade scalpel, or knife in the pressure cooker or oven. After the initial sterilization, they may be cleaned by dipping them in alcohol before and after each use.

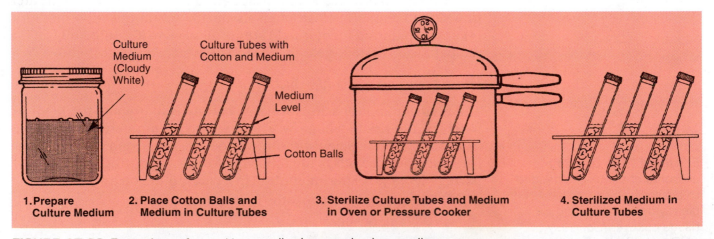

FIGURE 17-23 Procedures for making sterile tissue culturing medium.

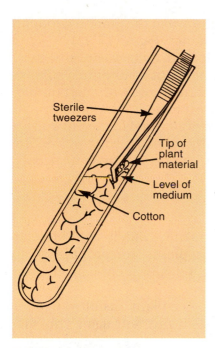

FIGURE 17-24 Plant material in culture tube.

The work area must be thoroughly cleaned and sterilized. Wash the area with a disinfectant. Keep a mist bottle filled with a mixture of 50 percent alcohol and sterilized water to spray work areas and tools, as well as the hands and arms of the propagator.

Preparing Plant Tissue and Placing in the Culture Tube

After the growing medium is properly prepared and cooled and the work area properly cleaned, the next step is to prepare the plant tissue. Various parts of the growing plant may be tissue cultured. For the production of vigorous plantlets, use only actively growing portions. With some species of plants, only a small, ¼-inch-square section of the leaf is used, while for others, ½ in. of the shoot tip is used. With ferns, ¼ in. of the tip of the rhizome is used.

Remove the part of the plant to be used and discard the excess plant material. Submerge the plant part into a solution of one part commercial bleach and nine parts water for about 10 minutes. This will disinfect the plant tissue. Remove the tissue with sterile tweezers and rinse the material in sterile water. When the plant part has been disinfected in the bleach solution, it can be handled only with sterile tweezers and must not touch any nonsterile surface.

When the plant material has been disinfected and rinsed, remove any damaged tissue with a sterilized scalpel or razor blade. Remove the lid from a properly prepared culture tube or jar and place the plant material on the agar. Take care that the plant material is not completely submerged (Figure 17-24). Recap quickly to avoid contamination from the air. It is best if the material is placed in front of the propagator so that work is not done over the uncapped culture tube.

Reminder!

IT IS IMPORTANT THAT ALL WORK AND THE TRANSFERRING OF MATERIALS BE DONE QUICKLY AND IN A CLEAN ENVIRONMENT. Scrub all areas with disinfectant and clean all tools with a disinfectant solution. Any contamination may lead to unsuccessful work. Bacteria and fungus will grow in unclean culture tubes and will overtake the new plant growth.

Storing Tissue Cultures

After all plant material has been cultured, put the cultures in an evenly warmed (70° to 75°F, or 21.1° to 23.9°C) and well-lighted area. The plant tissue will NOT do well in direct sunlight. If any contamination has occurred, it will be evident in 48 to 96 hours as mold or rotting on the medium. If contamination occurs, remove the contaminated tubes and wash them for reuse.

When the plantlets have grown to a satisfactory size, take them out of the culture tubes and transplant them into a good growing medium. Remember, the plantlets are very fragile, so handle them carefully. As

FIGURE 17-25 A young peach tree after being moved from the original culture tube and allowed to grow on sterile medium in a sterile container. *(Courtesy USDA/ARS #K-3279-11)*

each plant is removed from its culture tube, wash the plant thoroughly and transplant it into its own culture container (Figure 17-25).

When the plant is well established in its own culture container with viable roots and top growth, transplant the plant again into a suitable potting mix. Place the plant inside a protected area with high humidity. The plants are coming out of a well-protected environment with plenty of humidity and light. After they have adapted to the pot and are growing well, they may be treated like any other growing plant. This process will take about 3 to 6 weeks from the beginning to a successfully growing plant.

Lab: Cloning African Violets

Plant cloning by tissue culture is one of the most widely used biotechnologies. Most potatoes and many houseplants are propagated by cloning. **Cloning** generates multiple, genetically identical offspring from the nonsexual tissues of a parent plant.

In theory, cloning is simple: Cut a leaf off a plant, disinfect it, cut it into fragments, then plant the fragments in nutrient agar. This may take 30 minutes.

In practice, contaminants from the air, hands, and tools quickly take over. Instead of healthy clones, you get colorful molds and bacteria. You can minimize contamination using a simple hood and aseptic handling techniques.

Cloning African violets in the classroom is a long-term project, but can be done within class periods (Figure 17-26). The first stage takes about 30 minutes. Six to eight weeks later, the violets can be transplanted to a mini-greenhouse. In another few weeks, the plant will be ready for repotting.

Aseptic Technique

Aseptic handling is a must for successful plant tissue culture. The culture vessel is a battleground between rapidly growing microbes and slowly regenerating plant fragments. Five simple techniques of aseptic handling will minimize transfer of contaminants and, thus, growth of molds and bacteria. These are:

FIGURE 17-26 African violets can be readily propagated via tissue culture. *(Courtesy Michael Dzaman)*

1. Wash hands thoroughly and scrub nails using regular soap and paper towels. Do not touch your face or other objects or put your hands in your pockets. Such practices put contaminants back on your hands and can recontaminate your plants, equipment, work area, or medium.
2. Keep your hands from passing over open vessels.
3. Touch vessels far from the rim, neck, etc. Keep the caps on when not in use.
4. Grasp tools as far from the working ends as possible.
5. Use sterile materials.

Practice Activities

1. Make thumbprints on bacteriologic nutrient agar plates before and after handwashing. Incubate overnight in a warm (not above 99°F, or 37°C) place. Bacterial colonies will appear on both plates. These bacteria are normal for us but will hinder cloning.
2. Study violet leaf fragments using a hand lens or dissecting microscope. Note the hairs protruding from the leaf's upper surface. Now look at your fingers. Note the many ridges. Where do contaminants hide in each case? How does washing the hand and leaf change how each looks?
3. Practice aseptic handling of tools. How do you pass scissors at home? How would you do this aseptically? Why might a scalpel be better for cloning work than a single-edged razor when both cut just fine?
4. Conduct a dry run of the cloning procedure using a spinach leaf. A spinach leaf bruises as easily as does a violet leaf, so it readily shows how gently it has been handled.

A Hood for Cloning

Materials:

60- by 80-in. sheet of 2-, 3-, or 4-mil clear plastic (painter's tarp)
bulldog clips (2- to 3-in. length)
support frame for hanging file folders

Assembly:

1. Place a file folder frame on the table with its arms facing you.
2. Fold the plastic sheet so it is two layers thick.
3. Drape the folded sheet with the foldline in front, so it overhangs the arms by about 2 inches.
4. Clamp the sheeting to the top of the arms to form a flat roof.
5. Straighten the sheeting to minimize creases.
6. Spray the inside of the hood and the work surface with 70 percent ethanol. Dry only the work surface, not the plastic.

The hood will look like a lean-to with a short curtain valance in front. This fits well on a student desk. The overhanging plastic on the sides and back create a larger work area than just the frame.

When working, stand over the hood. Do not breathe onto the cloning materials or work area.

BIO•TECH CONNECTION

The Search for Perfect Plants

The plants we use today for food, clothing, fiber, and ornamental use are quite different from those found in the wild. Domestic plants or plants grown for a specific use have generally been selected or bred to survive better, grow faster, look different, or in some way perform differently from their ancestors in the wild. However, it is becoming increasingly apparent to plant breeders that we must have wild plants that are not closely related to our favorite domestic species to inject new characteristics into our favored domestic plants.

Pest resistance is an area that requires a continuous reserve of foreign genetic sources. This is to be expected because the very pests that we breed plants to resist are constantly adapting to our plants through survival of the fittest among their kind. Insects and disease-causing pathogens have an amazing capacity to adapt to and eventually break crop resistance. Resistant varieties usually become obsolete in 3 to 10 years. It generally takes 8 to 11 years to breed a new variety to resist the changing individuals of a given pest. Therefore, plant breeding programs must function on a continuous basis to maintain our current capability to feed, clothe, and otherwise supply our population.

American farmers grow more than two hundred varieties of wheat, eighty-five varieties of cotton, two hundred varieties of soybeans, and many varieties of fruit, vegetables, and ornamental crops. Disease and insect resistance must be bred into each of these varieties. Many of them often thrive only in specific, limited growing areas, such as one part of one state. Keeping up with known and persistent pests is a rela-

Plant geneticist Keith Schertz examines grain sorghum bred for tropical climates. Bags prevent the sorghum flowers from cross-pollinating, so the plant breeder can control which plants provide the male pollen to fertilize the female part of any given plant. *(Courtesy USDA/ARS K-2492-13)*

tively manageable process, so long as our state and federal experiment stations are reasonably well funded. However, the sudden appearance or introduction of a pest without resistant varieties or natural biological enemies can be devastating.

For instance, the Russian wheat aphid first appeared in the United States in 1986 and has cost wheat growers hundreds of millions of dollars since its arrival. American wheat varieties have very little resistance to the Russian wheat aphid. Therefore, chemical insecticides have to be used until either resistance can be bred into our domestic wheats or biocontrol agents can be found or developed. Entomologist Robert Burton of the Plant Science Laboratory in Stillwater, Oklahoma believes the answer will be found by introducing selected genes from wheat varieties from Southwest Asia. It is estimated that it will take 5 to 10 years to breed in strong Russian aphid resistance.

Materials Preparation

To sterilize materials, autoclave for 15 minutes at 15 to 21 lb of pressure. Open only in a hood that has been surface-sterilized with ethanol.

1. Mix and sterilize plant nutrient agar. One type is Murashige African violet/gloxinia multiplication medium, available from Carolina Bilogical Supply Co. One pack makes thirty to forty plates. Stir 1 pack into 1 liter of distilled water. Add 30 g sucrose and 15 g agar. The agar will not dissolve until it is heated. Loosely cap the flask with aluminum foil and autoclave or otherwise heat for 15 minutes. Cool in hot tap water of about 122°F (50°C). Pour 25 to 30 ml of agar solution into each plastic petri dish (20 × 100 mm size). Gel at room temperature, then store in a refrigerator.

2. Sterilize a 100-ml beaker to hold disinfected leaves. Cap with a 4-in. square of heavy-duty aluminum foil.

3. Sterilize 150 ml of tap water in a foil-capped 250-ml Erlenmeyer flask.

4. Wrap the glass petri plate (cover and bottom assembled) in a double layer of white T-shirt rag, then autoclave. The inside of the cover will be used as the cutting surface and the rim of the bottom to support the tools as they drain.

5. Assemble, but do not sterilize:

   - 500-ml beaker for waste liquid
   - 20-by-150-mm test tube filled to the brim with 70 percent ethanol (support in a 250-ml Erlenmeyer flask)
   - curved 8-in. forceps and a 6-in. scalpel handle with #11 blade
   - 200-ml beaker containing 100 ml of 70 percent ethanol to dip the leaf
   - single-edged razor to cut the leaf from the plant
   - 100-ml beaker to hold the leaf during disinfection

6. Prepare the disinfectant. Mix a 20 percent solution of liquid chlorine bleach and add 1 drop of Joy detergent per 500 ml. Swirl gently to mix; too many bubbles will inhibit wetting of the leaf surface. (Remember those little hairs.)

7. Soak the forceps and scalpel in the ethanol tube for at least 5 minutes before use. Do not store in the ethanol because the blade will rust.

8. Place in the hood:

   - plant nutrient agar plate
   - sterile H_2O flask
   - sterile beaker
   - sterile glass petri dish
   - 500-ml waste beaker
   - test tube of ethanol
   - forceps and scalpel
   - 200-ml beaker with ethanol

1.

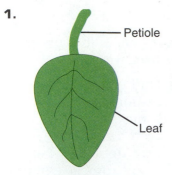

Petiole

Leaf

2.

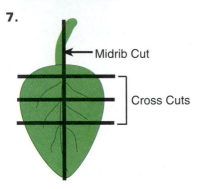

Beaker

Disinfectant

7.

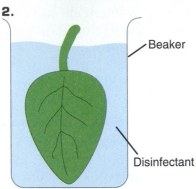

Midrib Cut

Cross Cuts

8.

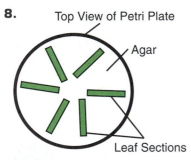

Top View of Petri Plate

Agar

Leaf Sections

9.

Petri Plate with Cover (side view)

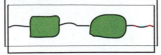

Cloning Procedure

1. Cut a young leaf so the petiole (stem) remains attached.
2. Put the leaf and petiole in the nonsterile beaker with disinfectant to remove dirt, mites, or other vermin. Leave it there for 10 minutes, but swirl it occasionally.
3. Working under the plastic cover, use the alcohol-soaked forceps to transfer the leaf, by the petiole, to the small sterile beaker. Pour the disinfectant into the waste beaker and remove the non-sterile beaker.
4. Rinse the leaf with 50 ml of sterile water. Swirl and pour the water into the waste beaker. Gently hold the leaf in the small beaker with the forceps.
5. Wash again.
6. Disinfect the leaf by dipping the leaf in the ethanol leaf soak, count to 10, and remove the leaf. Meanwhile, open the glass petri plate. Use the bottom plate to drain the forceps and blade. (Resoak the forceps before cutting.)
7. Cut the leaf into fragments. Use the lid of the glass petri plate as the operating table.
    a. Cut off the petiole. Do not plant it.
    b. Cut down the midrib firmly.
    c. Cut across each half into 4 pieces, being sure to cut through a branching vein.
8. Plant each fragment in the plant nutrient agar so each piece is in, not just on, the agar. Plant in a spoke arrangement to minimize spread of contaminants.
9. Put the lid back on and cover the plastic petri plate with cling wrap to keep moisture in.
10. Store in the dark for about one week, then place where the fragments have a daily light cycle and 68° to 77°F (20° to 25°C) temperatures.
11. Observe weekly. If part of a plate becomes contaminated, transfer healthy fragments to a fresh plate. Even with a commercial hood with sterile air, only about 50 percent of the plates remain completely free of contaminants.
12. Plantlets will appear on some fragments after about eight weeks. When plantlet leaves grow to about 0.5 cm, aseptically remove the fragment. Separate its plantlets and return the plate to incubate. Gently cut the plantlets, being sure to have some root and some shoot. Place the individual plantlets in growing medium to develop under aseptic conditions.
13. After about a month, the plants should be large enough to transplant into a pot in sterile potting medium.

Plants should be planted with at least two per pot and fertilized with African violet food.

A mini-greenhouse for maintaining high humidity around fragile plants can be made by setting the plants in a plastic sandwich bag and adding enough water to maintain some moisture on the inside of the bag.

Adapted from: *Science of Food and Agriculture,* January 1989, Council for Agricultural Science and Technology (CAST).

STUDENT ACTIVITIES

1. Write the Terms to Know and their meanings in your notebook.

2. List the crops that are grown commercially in your area. Visit a site or sites where plants are propagated and ask the grower to discuss the propagation methods used.

3. List the prevalent crops that are grown from seed in your area and the popular varieties of each. Study a seed catalog and determine the requirements for germination of each variety and why the varieties are popular in your community.

4. Plant a seed in a jar filled with medium. Place the seed near the edge of the jar so that you can see what is happening. Keep a journal of daily observations as the seed or plant changes. You may want to do this with several seeds in different jars and vary the amount of water, light, air, or temperature in each jar. Note your observations and the conditions regarding each seed. Write your conclusions about what is best for maximum germination results.

5. Experiment with different kinds of plants and various kinds of cuttings. Keep a journal to determine the best type of cutting for specific plants. Keep notes on different media, temperatures, light, and rooting hormones.

6. Practice making each type of graft discussed in this unit under the supervision of your instructor.

7. Research additional grafting methods.

8. Conduct an experiment with the tissue-culture method of propagation. Keep notes on the different kinds of plants used, the time required for root formation, and observations regarding the benefits of propagation by tissue culture.

9. Prepare a statement about the propagation methods best suited to your purpose.

SELF EVALUATION

A. Multiple Choice

1. Propagation is defined as
   a.   the union of an egg and sperm.
   b.   the process of increasing the numbers of a species.
   c.   a cheaper method of propagation than with seeds.
   d.   the only way to propagate some species and cultivars.

2. A seed consists of
   a.   a root, stem, and flower.
   b.   a root, seed coat, and endosperm.
   c.   a seed coat, endosperm, and embryo.
   d.   an embryo, cotyledons, and new plant.

3. A type of stem cutting used where stock material is limited and has alternate leaves is a
   a.   stem tip cutting.
   b.   cane cutting.
   c.   simple layering.
   d.   single-eye cutting.

4. A cutting that is usually made from a large-leaf plant with the veins split is
    a.   split-vein cutting.
    b.   leaf petiole cutting.
    c.   terminal tip cutting.
    d.   tissue propagation.

5. Grafting is
    a.   a type of sexual propagation.
    b.   a type of hybridization.
    c.   a method by which two plants are propagated.
    d.   a method of joining two parts of two different plants.

6. The most common type of grafting is
    a.   T-budding.
    b.   simple layering.
    c.   stem-tip propagation.
    d.   scion cut out of the stock plant.

7. Tissue culture may be used for
    a.   cloning.
    b.   disinfecting.
    c.   sexual reproduction.
    d.   sterilization.

8. With tissue culture, contamination may be a problem from
    a.   air.
    b.   hands.
    c.   tools.
    d.   all of the above.

9. The first stage of cloning African violets takes approximately
    a.   1 to 3 weeks.
    b.   4 to 5 weeks.
    c.   6 to 8 weeks.
    d.   10 to 12 weeks.

10. To make a very inexpensive hood for tissue culturing, the recommended covering is
    a.   aluminum.
    b.   plastic.
    c.   sheet steel.
    d.   wood.

B. Matching

_____ **1.** Outer seed coat   a.   A cutting from an end of a branch containing a terminal bud

_____ **2.** Germinate   b.   Best taken when the plant is dormant

_____ **3.** Imbibition   c.   The absorption of water into the young seed

_____ **4.** Tip cutting   d.   Functions as a protector for the seed

_____ **5.** Root cutting   e.   When plants are separated and then replanted

_____ **6.** Division   f.   When a seed starts to sprout

C. Completion

1. _____ propagation utilizes a part or parts of one parent plant.

2. The _____ will supply food to the young seedling until it is able to make its own.

3. A _____ might contain fungicide and is used to help plants produce roots more quickly.

4. _____ cuttings are normally made of a section containing one or two nodes.

5. Tissue culture is also known as _____.

6. A major aspect of tissue culture is that the area to be worked in must be _____ and _____.

D. True or False

_____ 1. A clone is almost like the parents.

_____ 2. The embryo is actually the young plant.

_____ 3. All seeds need light to germinate.

_____ 4. An advantage of tissue culture is that only one plant can be made from a disease-free plant.

_____ 5. Double-eye cuttings are often used when plants have opposite leaves.

_____ 6. Herbaceous plants are propagated by cuttings.

Crop Science

Crops in Space!

As humans push back the frontiers of outer space and consider the likelihood of space travel taking years instead of days, we face the age-old dilemma—how do we feed the crew? Early manned space voyages were accommodated by freeze-dried food and food in tubes. Now scientists are gearing up for food production in space. Not only is there a need to produce food, but there is a need to generate oxygen, get rid of carbon dioxide, and dispose of organic waste. The solution is a controlled ecological life support system (CELSS). A controlled ecological life support system is a system that will provide the basic components to sustain life without requiring inputs from external sources.

At Kennedy Space Center in Florida, scientists use a pressure chamber from a previous space craft to develop a controlled ecological life support system (CELSS), where plants recycle air, water, and waste to produce food. *(Courtesy of NASA #89-HC-130)*

Consider a crew in space. For basic survival they need food, water, and oxygen. The food could come from fruits, vegetables, grains, meat, milk, or eggs. Without a source of oxygen to replenish what is breathed in, the crew would soon die. In addition, if carbon dioxide were permitted to build up in the air, the crew would soon be poisoned. Similarly, if the human waste could not be decomposed, its accumulation would become unbearable.

Enter an effective controlled ecological life support system. Through photosynthesis, plants can take up nutrients from soil or water and utilize carbon dioxide from the air to manufacture food for themselves, as well as for animals. They give off oxygen as a by-product of this process. Fish and poultry are excellent sources of protein for humans and are exceptionally efficient converters of plant material into essential proteins. They are small, grow fast, and would be excellent candidates for a space farm. Like humans, they give off carbon dioxide for plants to utilize. Also, like humans, they produce organic wastes that can be decomposed to replace the nutrients in the soil and water, where plants are growing. Micro-organisms, worms, insects, and other forms of life would round out the system. Heat and combustible gases would be energy sources generated in the process.

Could it work as a closed system? Scientists believe it will. However, many questions must be answered and problems must be overcome first. Steven Britz, a plant physiologist in the Plant Photobiology Laboratory in Beltsville, Maryland, has been trying to ask the right questions for the scientific community to answer in order to create a CELSS in space. Some questions raised so far are as follows:

1. What is critical to producing a harvest when you must supply every need in a closed environment?

2. What kind of and how much light is needed, and how can it be supplied?

3. How much room is needed for plant roots, and what are the effects of root area restriction?

4. What type of medium is needed—soil, water, other?

5. What compounds will plants release in a closed environment, and are these useful or toxic?

6. What trace metals are picked up and transported in recirculated water?

7. What are the effects of little or no gravity on plant and animal growth and reproduction?

8. What light/dark cycles are best to optimize plant performance?

There are many more questions that need answering before we grow plants and animals in space. Can you suggest some? Answers to the questions about functioning in a space environment are helping us raise better plants today. A delegation to the United States from the Chinese Agricultural Ministry was very interested in the details of raising maximum yields in very confined spaces. In China, there is a growing industry raising hydroponic vegetables for the hotel trade. Using controlled environments lets us discover the optimum conditions for the functioning of a particular plant, which may not be its natural or traditional environment.

Controlled environment facilities are gaining acceptance as a viable means of commercially growing high-value crops under scheduled production management in an environment free from harmful insects and pollution.

Home Gardening

To plan, plant, and manage a home garden.

✓ grid-type paper

✓ pencil and eraser

✓ seed catalogs

After studying this unit, you should be able to:

- analyze family needs for homegrown fruits, vegetables, and flowers.
- determine the best location for a garden.
- plan a garden to meet family needs.
- establish perennial garden crops.
- prepare soil and plant annual garden crops.
- list recommended cultural practices for selected garden crops.
- protect the garden from excessive damage from drought and pests.
- harvest and store garden produce.
- describe the use of cold frames, hotbeds, and greenhouses for home production.

FIGURE 18-1 Some people seem to have a natural talent for gardening; however, the growth of plants and procedures for obtaining best results are based on scientific principle. *(Courtesy of USDA/ARS #K-5134-04)*

Gardening is an activity that can be enjoyed by all members of the family. It provides fresh fruits, vegetables, and flowers for immediate use or to sell for profit. Gardening is both an art and a science. It is demanding of the gardener, both in skill and creativity. A garden is alive and changing every day, presenting new challenges to the gardener (Figure 18-1).

ANALYZING A FAMILY'S GARDENING NEEDS

First, one must decide what vegetables and flowers the family likes. It would not make sense to plant sweet potatoes, green beans, and marigolds if no one in the family cared for these vegetables and flowers. The home gardener should provide a seasonal and continuous variety of vegetables and flowers. **Seasonal** means pertaining to a certain season of the year. Fresh vegetables are important to everyone's diet. Plan to have plenty available during the growing season and to store some for future use. Choose mostly those vegetables that will produce high yields (Figure 18-2).

FIGURE 18-2 The garden plan should provide for high-yielding fruits, vegetables, and flowers that meet the needs and preferences of the family. *(Courtesy of USDA/ARS #K-2282-1)*

When it has been decided what vegetables and flowers the family likes, it is time to figure our how much ground will be needed. It is better to have a smaller garden that is well cared for than to have a garden that goes to waste because it becomes too much to handle. A good rule of thumb for four grown people is to start with a plot 10 ft wide and 26 ft long, or 260 ft². A **square foot** is an area equal to 12 in. × 12 in.

THE GARDEN PLAN

A prospective gardener should make a sketch on paper detailing the amount and placement of the various crops. At this stage, it is important to consider **successive** plantings. These crops follow each other in the season so that the ground is occupied throughout the growing season. Fall crops can follow spring and summer crops in many areas. This can be easily done by planting perennial crops and different varieties of specific annual crops. **Variety** is a category within a species of plant. Allow adequate space between the rows for cultivation.

LOCATING THE GARDEN

Depending on where you live—in the city, suburbs, or a rural area—the location of the garden is an important consideration. The garden should be convenient to the house. It should also be accessible to a water supply; on loamy, well-drained soil; in a sunny area; and visible from the home, if possible. **Loamy** means a granular soil with a balance of sand, silt, and clay particles.

One must visualize where to locate the garden. What happens to the proposed spot when there is a heavy rain? Or, what happens when it is very dry? From the chosen spot, look up to determine whether trees or branches will cause problems by excessively shading the garden. When planting flower beds around the house, remember that along the south and west sides, the heat will be reflected onto these beds, and they may require extra water. Select the best site you can for both vegetable and flower gardens.

PREPARING THE SOIL
Conditioning the Soil

Garden soil should be loose and well drained. The ideal soil type should be granular—like coffee grounds—so that water will soak in rapidly. Few soils are originally found this way. Soil-building practices can improve the soil over time, and the gardener must prepare the soil to achieve the best results.

If the proposed site has not been previously planted, it is a good idea to add organic matter and plow, spade, or rototill it into the soil. The decayed organic matter and the freezing and thawing of the soil during the winter months will help improve the soil's physical condition. The application of fine organic matter in the spring can also improve soil condition and fertility.

Materials such as composted leaves and grasses, peat moss, composted sawdust, and sterilized weed-free manure are good soil condi-

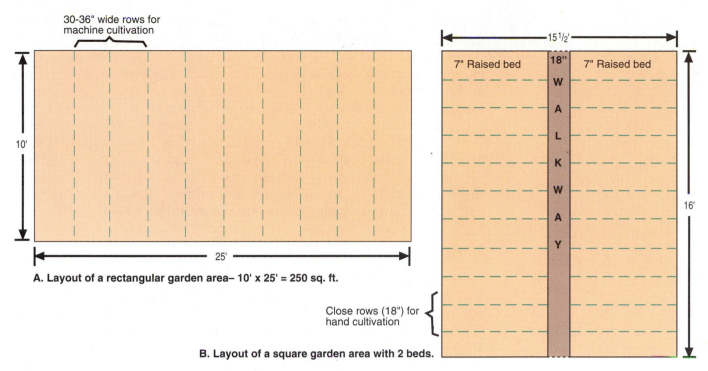

30-36" wide rows for machine cultivation

10'

25'

A. Layout of a rectangular garden area– 10' x 25' = 250 sq. ft.

15½'

7" Raised bed

18"

WALKWAY

7" Raised bed

16'

Close rows (18") for hand cultivation

B. Layout of a square garden area with 2 beds.

FIGURE 18-3 Layouts for rectangular and square gardens.

tioners. **Peat moss** is a soil conditioner made from spagnum moss. A good rate of application for organic matter is one pound of dry material per square foot of surface area.

Preparing to Plant

If the site selected has never been planted, it is best to remove the sod before tilling the soil. Then spread the organic matter over the soil. When moisture conditions are favorable, turn the soil with a shovel, spade, plow, or rototiller. When turning the soil, it is important to break up all clods. A **clod** is a lump or mass of soil.

After the spading (turning of the soil) is completed, make the planting beds. A good procedure is to heap the soil to make raised rows with grooves, or **furrows,** between them. Another method is to prepare raised beds that are 4 to 8 ft wide with sunken walks between them (Figure 18-3). The furrows or sunken walkways can drain off excess rain.

Next, level the raised beds with a rake, but do not push the soil back into the furrows. The next step is to walk over the beds or tamp them with the head of a garden rake to firm them and to eliminate air pockets. Finally, use the hoe and back of the garden rake to push and pull the soil to level it and to break any remaining clods into fine particles (Figure 18-4). The soil is now ready for planting.

COMMON GARDEN CROPS AND VARIETIES

Consider the Climate

When choosing varieties of vegetables and flowers to plant in the home garden, consider the climate. **Climate** refers to the weather conditions for a specific region. Do not plant crops outdoors until all danger of frost is gone. Except for a few mountainous areas and the

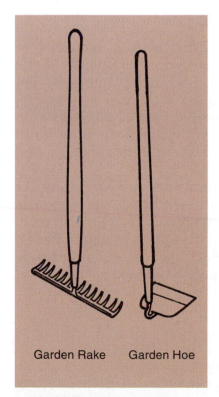

Garden Rake   Garden Hoe

FIGURE 18-4 The garden rake and hoe are used to break up clods and loosen, smooth, and level the soil.

northern tier of states, almost all areas of the United States are frost-free from June through August (Figure 18-5).

Plants grow rapidly in frost-free weather. In some areas there is not enough time for long-season vegetables, such as eggplant, cantaloupe, and watermelon, to mature reliably. Much of the country has an intermediate growing season, with 5 to 6 months of frost-free weather. Long-season crops can be grown in these areas. Growing seasons of 7 to 10 months are common across the mid-South, South, and low-elevation regions of the Southwest and West. In these regions, gardeners can grow both spring and fall crops in many varieties. However, these same regions are susceptible to summers that are so hot or dry that only a few crop varieties can survive the intense summer heat.

Consider the Variety

There are both warm-season and cool-season crops. Certain flowers and vegetables must have continuous cool weather to do well. Heat will quickly make the plants dry up or go to seed. Where the growing season is 5 months or longer, outdoor seeding in the late summer usually results in an excellent harvest during the cool fall season. Most of the cool-season plants can withstand light frosts with little damage.

There are two types of warm-season flowers and vegetables: those that mature quickly and those that require 4 months or more from planting to harvest. The quick-maturing types are almost always started in the garden when the soil is warm. The later-maturing types are usually started indoors and are transplanted to the garden after all danger of frost is past.

Annuals, Biennials, and Perennials

Flowers and vegetables are classified as either annuals, biennials, or perennials. An annual is a plant whose life cycle is completed in one growing season. Growth is very rapid. Practically all vegetables, except asparagus, rhubarb, and parsley, are annuals.

A biennial is a plant that takes two growing seasons or 2 years from seed to complete its life cycle. Some biennials bloom very little or not at all the first year, but they come into full bloom the second year and then go to seed.

A perennial is a plant that lives on from year to year. A gardener commonly treats some plants that are true perennials as either annuals or biennials. This is done because, when planted each year from seed, perennials produce higher quality blooms than do older plants that remain from year to year. Perennial crops, such as asparagus, artichokes, rhubarb, and some herbs and flowers, should be planted in one section of the garden, separate from the annuals. By such separation, the perennials do not interfere with the cultivation of the annuals.

Vegetables or Flowers

The kinds of vegetables and flowers to plant depend upon the individual tastes of the members of your family. There are certain top-ranking vegetables that are popular everywhere. Certain ones are popular only within certain regions of the country. The most popular vegetables are tomatoes, snap beans, onions, cucumbers, peppers, radishes, lettuce, carrots, corn, beets, cabbage, squash, and peas. Favorites of specific

Vegetable Planting Guide

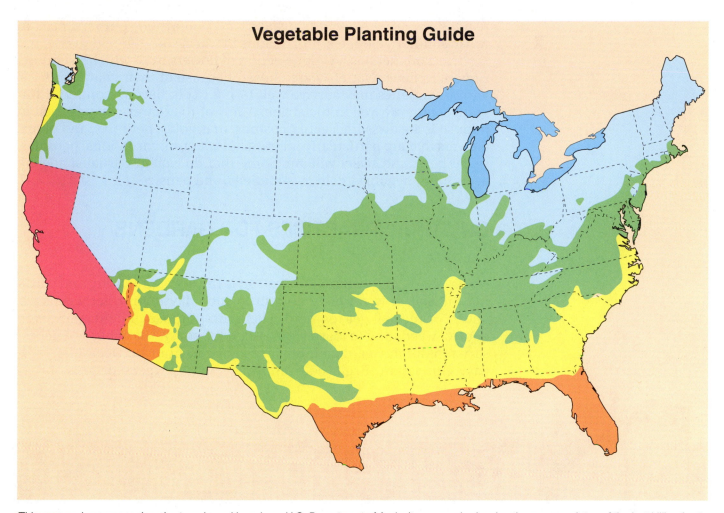

This map and accompanying chart are based largely on U.S. Department of Agriculture records showing the average dates of the last killing frosts in various parts of the country. Use this information to plan the planting and harvesting periods for your garden, but remember that these are averages, and individual conditions may vary.

	SUBTROPIC	WARM	MILD	COOL	CALIF.	PLANTING DEPTH	ROW SPACING	FIRST HARVEST	HARVEST LASTS
Beans	Apr.-Aug.	Apr.-June	May-June	May-June	Mar.-Aug.	1"	2-3 ft.	8 weeks	Until frost
Beets	Jan.-Dec.	Feb.-Oct.	Mar.-July	Apr.-July	Jan.-Dec.	1/2"	15-18"	50-78 days	6 weeks†
Broccoli	July-Oct.	Feb.-Mar.	Mar.-Apr.	Mar.-Apr.*	Sep.-Feb.	1/4"	3 ft.	65-70 days	To frost
Brussels sprouts	Feb.-May	Feb.-Apr.	Mar.-Apr.*	Mar.-Apr.*	Sep.-Feb.	1/2"	3 ft.	14-20 wks.	Past frost
Carrots	Jan.-Dec.	Jan.-Mar.	Mar.-June	Apr.-June	Sep.-May	1/4"	1 1/2-2 ft.	8 weeks	8 weeks†
Chard	Jan.-Dec.	Feb.-Sep.	Mar.-Aug.	Apr.-July	Jan.-Dec.	3/4"	1 1/2-2 ft.	8 weeks	To frost
Chives	Feb.-May	Mar.-May	Mar.-May	Apr.-June	Mar.-May	1/4"	1 1/2 ft.	8 weeks	To frost
Corn	Apr.-June	Mar.-June	May-July	May-July	Mar.-Aug.	1 1/2"	3 ft.	9-12 weeks	10 days
Cucumbers	Apr.-June	Apr.-June	Apr.-June	May-June	Mar.-Aug.	1/2"	6 ft.	7 weeks	5 weeks
Eggplant	Feb.-Mar.	Feb.-Apr.	Mar.-May*	Apr.-May*	Mar.-May	1/4"	3-4 ft.	10-14 wks.	To frost
Endive	July-Sep.	Aug.-Sep.	Mar.-May	Apr.-June	Jan.-Dec.	1/4"	2-3 ft.	10-12 weeks	7 weeks
Lettuce	Jan.-Dec.	Aug.-May	Mar.-June	Apr.-June	Sep.-May	1/4"	15"	6 weeks	6 weeks†
Mustard	Feb.-May	Feb.-May	Mar.-June	May-July	Mar.-Aug.	1/4"	1 1/2-2 ft.	40-50 days	2 weeks
Okra	Apr.-June	Apr.-June	Apr.-June	May-June*	Mar.-May	1/2-3/4"	3 ft.	55-60 days	To frost
Onions	Dec.-Mar.	Dec.-Apr.	Feb.-May	Mar.-June	Sep.-May	1/2"	18"	Variable	—
Parsley	Jan.-Dec.	Jan.-Dec.	Feb.-June	Mar.-June	Jan.-Dec.	1/4"	1 1/2-2 ft.	10 weeks	To frost
Parsnips	Mar.-June	Feb.-June	Apr.-June	May-June	Dec.-May	1/2"	1 1/2-2 ft.	14-17 wks.	Past frost
Peas	Jan.-May	Jan.-Apr.	Feb.-May	Mar.-June	Sep.-May	1 1/2"	2-3 ft.	9 weeks	2 weeks†
Peppers	Feb.-Mar.	Feb.-Apr.	Mar.-May*	Mar.-May*	Mar.-May*	1/4"	3 ft.	9 weeks	To frost
Potatoes	Jan.-Dec.	Feb.-Oct.	Mar.-Apr.	Apr.-May	Jan.-Dec.	5"	2 1/2-3 ft.	100 days	To frost
Pumpkin	Apr.-June	Apr.-June	Apr.-June	May-June	Mar.-May	1"	8-10 ft.	14-17 wks.	To frost
Radishes	Jan.-Dec.	Feb.-Oct.	Mar.-Aug.	Apr.-July	Sep.-May	1/4"	1-2 ft.	3-6 weeks	1-2 weeks†
Squash, Summer	Apr.-June	Apr.-June	Apr.-June	May-June	Mar.-Aug.	1"	4-6 ft.	60 days	To frost
Squash, Winter	"	"	"	"	"	1"	6-8 ft.	100 days	"
Tomatoes	Jan.-Mar.	Feb.-Mar.	Mar.-May*	Mar.-May*	Mar.-May*	1/4"	3-4 ft.	9-12 weeks	To frost
Turnips	Feb.-Mar.	Jan.-Mar.	Feb.-Apr.	Mar.-May	Jan.-Dec.	1/2"	1-2 ft.	6-10 weeks	3 weeks†

*Transplants recommended.
†Following harvest, space may be used for late planting of carrots, beets, or bush beans.

FIGURE 18-5 Planting times, planting depths, row spacing, and harvest times for common garden vegetables.

regions are artichokes in Louisiana and California, southern peas in warm southern climates, and melons in warm summer regions.

The gardener has a wide selection of flowers from which to choose. Generally, flower varieties are chosen because of their characteristics. Various varieties may survive well in poor soil, have a wonderful fragrance, or grow well in the shade. A few of the flowers that do well in poor soils are alyssum, cactus, cosmos, marigolds, petunias, and phlox. Flowers that are particularly fragrant are alyssum, carnations, petunias, sweet peas, and sweet william. Some that do well in partial shade are ageratums, begonias, coleus, impatiens, and pansies.

CULTURAL PRACTICES FOR GARDENS

Cultivating

Never put off cultivation. **Cultivation** is the act of preparing and working the soil. It is easier to do a little each day than to let weeds get ahead or the soil get too hard. The main purpose of cultivation is to control weeds and to loosen and aerate the soil. A sharp hoe is a necessary tool for any gardener. Buy one that can fit in-between plants for easy cultivation. Begin cultivation as soon as weeds or grass break

AGRI·PROFILE

CAREER AREAS: Gardener/Caretaker/ Homeowner/Market Gardener

Fruit and vegetable gardening has long been an important enterprise for providing fresh and wholesome food for rural and suburban families. Similarly, flower gardening provides a rewarding pastime and greatly increases the beauty of homes in urban as well as rural and suburban settings. Career opportunities exist for gardeners on estates, institutions, colonial farms, truck farms, and residential neighborhoods.

Nationally, there is a resurgence of small farms and truck gardens where roadside stands and farmers' markets have created new interest in farm-fresh fruits and vegetables. Pick-your-own operations provide opportunities for people to harvest their own field-fresh produce and save the producer labor costs.

People have a new awareness of the benefits of fresh food and pay premium prices where freshness is ensured. This has created new markets for garden produce and opened new career opportunities in production, processing, and marketing.

Home gardens, pick-your-own farms, road-side markets, and farmers' markets are all developed from basic gardening skills. *(Courtesy of USDA/ARS #K-4173-2)*

FIGURE 18-6 Plant materials, newspaper, and plastic films all make excellent mulches that help control weeds and conserve moisture. *(Courtesy of USDA/ARS #K-4175-13)*

through the soil. Do not wait until the weeds and grasses are tightly rooted and threaten to take over the garden. When cultivating, take care not to damage the roots of the vegetables and flowers. Use short shallow scraping motions instead of chopping deeply into the soil. Deep hoeing or cultivation will destroy desirable plant roots.

Weeding

Begin weeding as soon as the weeds appear. When weeds grow near the plants, their roots can become intertwined with the crop. Therefore take care to avoid pulling out both the weed and the plant. When weeding, select a time of day when the soil and plants are fairly dry, so the weeds wither when pulled. Be careful to shake the soil from weed roots so the weeds will die quickly. Rake persistent weeds such as purslane, crabgrasses, and Bermuda grasses out of the garden to prevent them from taking root again. Mulches of various types are excellent for controlling weeds (Figure 18-6.)

Weed-control chemicals called **herbicides** can be used in the home garden, with care. However, the use of herbicides is best suited to larger gardens. Chemicals are very specific in their actions in certain types of plants, so they must be used with caution to prevent injury to the vegetables and flowers being grown. Before using any herbicides, obtain detailed information from an extension agent or an informed garden center salesperson. Always read the label on the container and use the chemical strictly according to the label instructions.

Watering

During a growing season, there will probably be some days or weeks of dry weather when the garden must be watered. Most soils require at least 1 in. of water per week, either through rain or irrigation. If water is needed, make it a practice to soak the soil to a depth of 6 in. Frequent, light waterings tend to promote shallow root development and should be avoided.

When watering, run the water on the soil near the plants. A good sprinkler head, soaker hose, or water breaker is needed to ensure even distribution of water. Let the garden hose run for 15 to 30 minutes to water 100 ft^2 at a depth of several inches. You should water whenever the soil lacks moisture 1 to 2 in. below the surface.

Protection from Pests

Gardens can be damaged from a variety of insects and diseases. To help prevent severe damage, the following practices should be helpful:

1. Rotate crops so that the same or a related crop does not occupy the same area every year. This helps control soilborne diseases.
2. Watch closely for insects. Pick them off by hand or knock them off with a hard stream of water. Introduce predatory insects (Figure 18-7).
3. When using sprinklers, water early in the day, so the foliage can dry before nightfall. Diseases of leaves and fruit prosper in damp conditions.
4. Keep weeds out of the garden. Weeds may harbor diseases or insects and will interfere with spraying and dusting of crops.
5. Use enough fertilizer and lime to promote vigorous growth.

FIGURE 18-7 Lady beetles feed upon pea aphids and help keep the damaging insect population under control. *(Courtesy of USDA/ARS #K-4179-19)*

HARVEST AND STORAGE OF GARDEN PRODUCE

Harvesting

Harvest the vegetables and flowers at the peak of quality or at the stage of maturity preferred by the users (Figure 18-8). Some vegetables will lose their quality within a day or two, while others hold it for a week or more.

Vegetable	Maturity Level	Vegetable	Maturity Level
Asparagus	From seeds—3rd year. From roots—after 1st year. Cut when spears are 6" to 10" high, in spring.	Lettuce	Leaf—When tender and desired size. Head—When round and firm.
Beans, Pole	When pods are nearly full size.	Muskmelons	When stem separates easily from fruit.
Beans, Bush	When pods are nearly full size.	Mustard Greens	When large leaves are still tender.
Beans, Bush Lima	When tender, pods are nearly full size.	Okra	Cut pods when about 2 or 3 inches long.
Beets	When bulbs are 1¼" to 2" in diameter.	Onions	For Fresh use—When ¼" to 1½" in diameter.
Swiss Chard	Outer leaves can be harvested anytime.		For Storage—When tops shrivel at the bulb and fall over.
Broccoli	Before green clusters begin to open.	Parsley	Any time the outer leaves are desired size.
Brussels Sprouts	When sprouts are firm, pick from bottom up on stalks.	Parsnips	After hard frost. Can leave in ground all winter for spring use.
Cabbage	When heads are solid, before splitting.	Peas	When pods are well-filled, before seeds are largest.
Cauliflower	After blanching, when firm curds are 2" to 3" in diameter.	Peppers	When solid and nearly full size.
Carrots	When top of root is 1" to 1½" in diameter.	Pumpkins	When skin is hard and not easily punctured. Cut with some stem on.
Celery	When about ⅔ mature, harvest as needed.	Radishes	When desired size.
Collards	When leaves are large but tender. Can harvest up to winter.	Rhubarb	When stems are of desired size, twist off near base of plant and discard leaves. Do not pick more than ⅓ of plant during a season.
Corn	When kernels are filled out and milky. The silk at the tip of the ear should be dry and brown.		
		Spinach	When outer leaves are large enough.
Cucumbers	When slender and dark green.	Squash	Summer—When skin is soft. Winter—When skin is hard, cut with part of stem on, before frost.
Eggplant	When half grown, glossy, and bright.		
Endive	When leaves are tender and desired size. About 15" in diameter.	Tomatoes	When uniformly red and firm.
		Turnips	When 2" to 3" in diameter.
Kale	When young and tender.	Watermelons	When underside is yellow and thumping produces a muffled sound.
Kohlrabi	When 2" to 3" in diameter.		

FIGURE 18-8 Levels of maturity for harvesting vegetables for top quality.

Crop	Temper- ature °F	Relative Humidity Percent
Asparagus	32	85-90
Beans, snap	45-50	85-90
Beans, lima	32	85-90
Beets	32	90-95
Broccoli	32	90-95
Brussels sprouts	32	90-95
Cabbage	32	90-95
Carrots	32	90-95
Cauliflower	32	85-90
Corn	31-32	85-90
Cucumbers	45-50	85-95
Eggplants	45-50	85-90
Lettuce	32	90-95
Cantaloupes	40-45	85-90
Onions	32	70-75
Parsnips	32	90-95
Peas, green	32	85-90
Peppers, sweet	45-50	85-90
Potatoes	38-40	85-90
Pumpkins	50-55	70-75
Rhubarb	32	90-95
Rutabagas	32	90-95
Spinach	32	90-95
Squash, summer	32-40	85-95
Squash, winter	50-55	70-75
Sweet potatoes	55-60	85-90
Tomatoes, ripe	50	85-90
Tomatoes, mature green	55-70	85-90
Turnips	32	90-95

FIGURE 18-9 Recommended temperatures and relative humidity for storage of fresh vegetables. *(Courtesy University of Maryland Cooperative Extension Service)*

For the best results for flowers, pick them in the early morning or late afternoon. Immediately place the cut flowers in lukewarm water for a short time. Display the flowers in a cool place.

Storage

Many vegetables that are not canned or frozen for later use can be stored successfully if proper temperature and moisture conditions are met (Figure 18-9). Only vegetables that are of good quality and at the proper stage of maturity should be stored.

Warm Storage   Vegetables that tolerate higher storage temperatures than others are squash, pumpkins, and sweet potatoes. These vegetables may be stored on shelves in a furnace room or in an upstairs storage area. Damp basement areas are not recommended. Squash and pumpkins should be kept in a heated room, with a temperature between 75° and 85°F (23.9° and 29.4°C) for two weeks to harden the shells.

Cool Storage   Most vegetables require cool temperatures and relatively high humidity for successful storage. The storage area should be cool, dark, and ventilated. The room should be protected from frost, heat from a furnace, or high outdoor temperatures.

The suggested temperature and relative humidity for storage of crops is shown in Figure 18-9. This list will provide you with a general idea of the type of storage area that each vegetable requires.

COLD FRAMES, HOTBEDS, AND GREENHOUSES FOR HOME PRODUCTION

Cold Frames

Cold frames are an extremely convenient aid to the gardener. A **cold frame** is a bottomless wooden box with a sloping glass roof (Figure 18-10). It can be constructed by using plywood or common lumber and a window sash. The size of the sash should determine the dimensions of the box. The length and width should be 2 in. smaller than the overall dimensions of the cover sash.

To construct a cold frame, the following procedures should be helpful.

1. Make the front of the box approximately 8 in. high and the back of the box about 12 in. high.

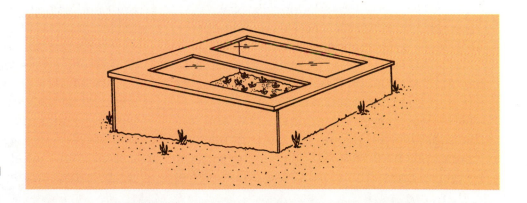

FIGURE 18-10 Cold frames and structures with glass or plastic covers used for starting garden plants.

2. Cut the two sides on an angle, from 12 in. at the back to 8 in. in the front.

3. Nail the four sides together.

4. Try the sash or top for size.

5. Hinge the sash to the back of the frame for easy use. The hinge allows you to prop the cover open slightly in warm weather for ventilation.

6. Place the cold frame in a southern exposure, with good protection from the wind and proximity to a water supply.

When the cold frame is built, it can serve three purposes:

1. It can be used as a protective home for seedlings that have been started indoors. The seedlings can continue to grow inside the cold frame and become "hardened off" before they are transplanted to the garden.

2. You can begin plants in the cold frames. Seeds can be planted directly into the soil in the cold frame. After they are 4 to 6 in. tall, they can be transplanted straight into the garden.

3. Vegetables, such as lettuce and endive, grow well into the fall. Sow seeds in early autumn. Cover the cold frame with a blanket or tarp during extremely cold nights.

BIO·TECH CONNECTION

Finding and Promoting the Good Ones

Across the road from the giant Chrysler automobile assembly plant in Newark, Delaware, is a different facility specializing in foreign imports—insects! The imports are flies, wasps, beetles, and other insects that parasitize, or feed upon, the insects that devour our crops, devastate our gardens, cause our lawns to turn brown, and infest our house plants. These insects have been found doing their "good deeds" in countries all over the world. Most of our recent imports have come from Argentina, Brazil, Canada, Chile, China, France, Germany, India, Indonesia, New Zealand, South Korea, and Russia and other previously Soviet nations.

The facility is called the USDA Beneficial Insects Introduction Research Laboratory. Its job is to see that insects brought into the United States are, indeed, beneficial and that the eggs, larvae, pupae, or adults of harmful insect species do not hitchhike along with the beneficial ones. The work of the entomologist at the Newark Laboratory includes picking up incoming shipments of beneficial insects at the Philadelphia airport; transporting them to the nearby Newark laboratory; taking them through a special receiving process; disinfecting all shipping materials that arrive with the insects; isolating the insects in special refrigeration units; exercising them daily; examining each insect under magnification to confirm identity; separating eggs, larvae, pupae, and adults for rearing purposes; increasing their numbers through reproduction; and shipping them under carefully controlled conditions to USDA and state research facilities throughout the United States and cooperating foreign nations.

Three sets of heavy steel doors lead into the quarantine's main work area. There, a large sink and autoclave are used to clean and sterilize materials used in the lab. Sterilization is essential before discarding materials to avoid accidental release of both

Hotbeds

A **hotbed** is simply a cold frame with a heat source. In many areas, a cold frame is usually adequate. However, in colder areas, some type of artificial heat is necessary. The hotbed should be located on well-drained land.

Electricity is a convenient means of heating hotbeds. Use either lead- or plastic-coated electric heating cables or 25-W frosted light bulbs. Temperature can be controlled with a thermostat. When using light bulbs, attach the electrical fixtures to strips of lumber and suspend the bulbs 10 to 12 in. above the soil surface. Allow one 25-W bulb per 2 ft^2 of space. When using a heating cable, a standard 60-ft length will heat a 6 × 6 ft or a 6 × 8 ft bed. Lay cable loops about 8 in. apart for uniform heating. Lay the cable directly on the floor of the bed, unless drainage material is needed.

Greenhouses

Growing flowers and vegetables in a greenhouse can be enjoyable as well as profitable. The conventional greenhouse is designed primarily to capture light and control temperature. It can be free standing, but most often it is attached to a building to provide convenient access,

beneficial insects and "hitchhiking intruders." It is interesting to note that insects must receive regular exercise to remain healthy, just like humans! Once a day, even on weekends and holidays, batches of insects are removed from cool storage and released in restricted, warm and lighted areas to warm up, stretch, soak up light, and move around in the greater space.

Many innovations in packaging and handling have evolved and have become standardized over the years. One of the more unusual ones is the parasite pill, or Trichocap. The parasite pill is a shipping capsule made of soft, gray cardboard and is about the size of an oyster cracker. It is used to ship insect eggs. One parasite pill holds about five hundred eggs of the Mediterranean flour moth. Inside each egg, there is one future parasitic wasp used as a biocontrol against the devastating European corn borer. The discovery and introduction of parasitic wasps have helped control some of our most damaging insects while reducing our reliance on chemical sprays.

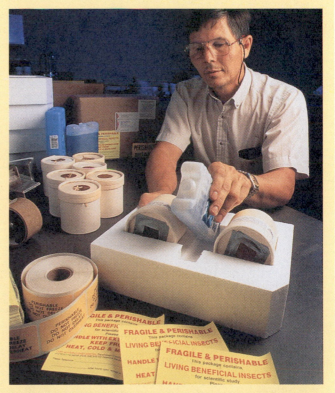

Entomologist and quarantine officer Larry Ertle at the Beneficial Insects Introduction Research Laboratory in Newark, Delaware, gently places beneficial insects in custom-designed packaging for shipment. *(Courtesy of USDA/ARS #K-4183-12)*

FIGURE 18-11 A home greenhouse can be attached to the house for convenience and efficiency.

simplified construction, and a potential source of supplemental heat if needed (Figure 18-11). The greenhouse can provide an environment for starting plants, hardening them off, or completely growing the plants.

In summary, gardening is both an art and a science. You can keep a garden for fun, reduce the cost of food for the family, generate income, or improve the beauty of the surroundings. If gardening is to be profitable, the operator must read extensively, get help occasionally with disease and insect problems, and perform garden chores in a timely fashion. For many people, the garden is a real source of satisfaction and creates a feeling of self-sufficiency and achievement.

STUDENT ACTIVITIES

1. Write the Terms to Know and their meanings in your notebook.

2. Discuss with family members the types of vegetables and flowers that they want in the home garden.

3. Select vegetable and flower varieties for the home garden from seed catalogs.

4. Sketch a garden plot containing both flowers and vegetables that are to be grown.

5. Make a calendar indicating the dates to plant and harvest the various crops in the garden.

6. Start and manage your own garden area. Use the flowers and vegetables at home or sell them for profit.

7. Learn to calculate area in square feet. A square foot is an area that is 1 ft long and 1 ft wide. An area that is 1 ft long and 2 ft wide contains 2 ft². Area (A) in square feet is calculated by multiplying length (L) in feet times width (W) in feet. Therefore, $A = L \times W$, or $A = LW$.

    a.   How many square feet are in a garden that is 10 ft long × 5 ft wide?

          A = _____ sq ft

    b.   How many square feet are in a lawn that is 40 × 70 ft?

          A = _____ sq ft

    c.   What is the area of a building lot that is 109 × 150 ft?

          A = _____ sq ft

    d.   Suppose the lot in item c, above, is covered with lawn except for the house, which is 28 × 42 ft. What is the area of the house?

          A = _____ sq ft

    e.   In item c, above, what is the area of the lawn?

          A = _____ sq ft

SELF EVALUATION

A. Multiple Choice

1. Gardening is both a
    a.   science and art.
    b.   chore and hard work.
    c.   science and hobby.
    d.   hobby and job.

2. A good rule of thumb for planning a garden for four people is to start with a plot that is
    a.   40 × 60 ft.
    b.   15 × 25 ft.
    c.   10 × 26 ft.
    d.   3 × 7 ft.

3. The south and west sides of a house may not be the best locations for a flower garden because they
    a.   are not warm enough.
    b.   reflect heat.
    c.   have poor exposure to the sun.
    d.   collect rainwater.

4. The furrow in a garden is a
   a.   sunken walkway.
   b.   hole.
   c.   place to plant seeds.
   d.   pile of weeds.

5. An annual is a plant whose life cycle is completed in
   a.   two growing seasons.
   b.   one growing season.
   c.   four growing seasons.
   d.   three growing seasons.

6. When cultivating, a gardener should use a _____ for best results.
   a.   rake
   b.   rototiller
   c.   shovel
   d.   hoe

7. The technical name for weed control chemicals is
   a.   pesticides.
   b.   fungicides.
   c.   herbicides.
   d.   weed killers.

8. After cutting flowers from the garden, it is best to immediately put them in
   a.   hot water.
   b.   cold water.
   c.   salty water.
   d.   warm water.

9. To heat a hotbed, frosted light bulbs or _____ are commonly used for heat sources.
   a.   a furnace
   b.   lead- or plastic-coated heating cables
   c.   a fire
   d.   a small stove

10. The conventional greenhouse is designed to
   a.   add more living space to the home.
   b.   keep plants warm.
   c.   capture light and control temperature.
   d.   protect plants from pests and diseases.

B. Matching

_____	**1.** Perennial	a.	12 × 12 in.
_____	**2.** Peat moss	b.	Fragrant flower
_____	**3.** Square foot	c.	More than two growing seasons
_____	**4.** Loamy	d.	Long-season vegetable
_____	**5.** Watermelon	e.	Persistent weed
_____	**6.** Sweet pea	f.	Soil type
_____	**7.** Bermuda grass	g.	Form of organic matter

C. Completion

1. A cold frame is a bottomless wooden box with a sloping _____ _____.

2. You should store only the vegetables that are of good quality and at the proper stage of _____.

3. You should plant only the vegetables and flowers that are liked by _____ _____.

4. Fertilizer and lime should be used to promote _____ _____.

5. Watering should be done to a depth of _____.

Vegetable Production

OBJECTIVE

To determine the opportunities in and identify the basic principles of vegetable production.

COMPETENCIES TO BE DEVELOPED

After studying this unit, you should be able to:

- determine the benefits of vegetable production as a personal enterprise or career opportunity.
- identify vegetable crops.
- plan a vegetable production enterprise and prepare a site for planting.
- describe how to plant vegetable crops and utilize appropriate cultural practices.
- list appropriate procedures for harvesting and storing at least one commercial vegetable crop.

MATERIALS LIST

- ✓ seed catalogs
- ✓ scissors, index cards, and glue
- ✓ pen, paper, eraser, and ruler
- ✓ lima bean seeds, containers, and soil mix
- ✓ hand trowels
- ✓ tomato seeds, soil mix, and flats

The study of vegetable production is **olericulture. A vegetable** is the edible portion of an herbaceous plant. **Herbaceous** means vegetables that both grow and die in a single growing season. The production of vegetables can be classified into three categories: home garden, market garden, or truck crop. **Home gardening** usually refers to the vegetable production for one family and does not involve any major selling of the crops. **Market gardening** refers to growing a wide variety of vegetables for local or roadside markets. **Truck cropping** refers to large-scale production of a few selected vegetable crops for wholesale markets and shipping.

VEGETABLE PRODUCTION FOR HOME PROFIT OR AS A CAREER

Home Enterprise

Growing vegetables in a home garden is enjoyed by millions of Americans. It has become a part of the lifestyle of most families with access to a little bit of ground. The vegetables produced can be used for fresh table consumption or stored for later use. Vegetable gardening not only produces nutritious food but also provides outdoor exercise from spring until fall.

The gardener who enjoys this type of activity can plant enough to provide for the family and harvest vegetables for sale on a small scale for extra income. Fresh, homegrown vegetables are usually superior in quality to those found in the supermarkets (Figure 19-1). In addition, the gardener can grow those vegetables that may be too expensive to buy.

FIGURE 19-1 Garden vegetables are fresh, nutritious, and good! *(Courtesy of USDA/ARS #K-5384-7)*

Career Opportunities

The vegetable industry is a very large and complex portion of the horticultural industry today. Even though there are millions of homeowners who garden, the majority of vegetables consumed by the public are grown commercially. The commercial vegetable industry is fast moving, intensive, and competitive. It is a business that is continually changing as the demands for certain vegetables fluctuate with the tastes of the American public.

There are numerous and varied career opportunities in the vegetable industry. These include being the owner of a small market gardening business, a member of a larger truck-crop business, a vegetable wholesaler or retailer, or a worker in a vegetable processing plant. With adequate education and training, a person can become an **olericulturist,** who develops pest-resistant strains and new varieties of vegetables, and does other specialized work. The opportunities are endless in the vegetable industry.

IDENTIFYING VEGETABLE CROPS

Vegetables can be identified in various ways—by their botanical classifications, according to their edible parts, or by the growing seasons required.

Botanical Classification

All vegetables belong to the division of plants known as **angiosperms.** These are plants with ovules and an ovary. From this division, the vegetables can be grouped into either Class I, **monocotyledons** (having only one seed leaf), or Class II, **dicotyledons** (having two seed leaves) (Figure 19-2). A vegetable can be further grouped into a family, a genus, a species, and sometimes a variety. One of the more popular vegetable families is Cruciferae—the mustard family, which contains brussels sprouts, cabbage, cauliflower, collards, cress, kale, turnips, mustard, watercress, and radish. Another is Leguminosae—the pea family, which contains bush beans, lima beans, cow peas, kidney beans, peas, peanuts, soybeans, and scarlet runner beans. Cucurbitaceae—the gourd or melon family—contains pumpkins, cucumbers, cantaloupe, casaba melons, and watermelons. Solanaceae—the nightshade family—contains eggplants, ground cherries, peppers, and tomatoes.

Edible Parts

Vegetables can also be classified by the part of the vegetable that is eaten. There are three groupings of vegetables according to their uses: (1) vegetables of which leaves, flower parts, or stems are used; (2) vegetables of which the underground parts are used; and (3) vegetables of which the fruits or seeds are used. See Figure 19-3 for a list of these categories and the vegetables in each one.

Growing Seasons

There are basically two growing seasons—warm and cool. Cool-season vegetable crops grow best in cool air and can withstand a frost or two. Some of these crops, such as asparagus and rhubarb, can even endure

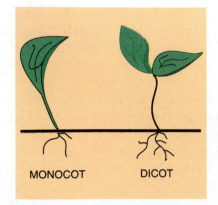

MONOCOT   DICOT

FIGURE 19-2 Monocots have one seed leaf, while dicots have two.

Plants of which the Fruits or Seeds are Eaten		Plants of which the Leaves, Flower Parts, or Stems are Eaten		Plants of which the Underground Parts are Eaten	
Family	**Vegetable**	**Family**	**Vegetable**	**Family**	**Vegetable**
Grass, *Gramineae*	Sweet corn, *Zea mays*	Lily, *Liliaceae*	Asparagus, *Asparagus officinalis* var. *altilis* Chives, *Allium schoenoprasum*	Lily, *Liliaceae*	Garlic, *Allium satvium* Leek, *Allium porrum* Onion, *Allium cepa* Shallot, *Allium ascalonicum* Welsh onion, *Allium fistulosum*
Mallow, *Malvaceae*	Okra (gumbo), *Hibiscus esculentus*	Goosefoot, *Chenopo-diaceae*	Beet, *Beta vulgaris* Chard, *Beta vulgaris* var. *cicla*		
Pea, *Leguminosae*	Asparagus or Yardling bean, *Vigna sequipedalis* Broad bean, *Vicia faba* Bush bean, *Phaseolus vulgaris* Bush Lima bean, *Phaseolus limensis* Cowpea, *Vigna sinensis* Edible podded pea, *Pisum sativum* var. *macrocarpon* Kidney bean, *Phaseolus vulgaris* Lima bean, *Phaseolus limensis* Pea (English pea), *Pisum sativum* Peanut (underground fruits), *Arachis hypogaea* Scarlet runner bean, *Phaseolus coccineus* Sieva bean, *Phaseolus lunatus* Soybean, *Glycine max* White Dutch runner bean, *Phaseolus coccineus*	Orach, *Atriplex hortensis* Parsley, *Umbelliferae* Sunflower, *Compositae* Mustard, *Cruciferae*	Spinach, *Spinacia oleracea* Celery, *Apium graveolens* Chervil, *Anthriscus cerefolium* Fennel, *Foeniculum vulgare* Parsley, *Petroselinum crispum* Artichoke, *Cynara scolymus* Cardoon, *Cynara cardunculus* Chicory, witloof, *Chichorium intybus* Dandelion, *Taraxacum officinale* Endive, *Chichorium endivia* Lettuce, *Lactuca sativa* Brussels sprouts, *Brassica oleracea* var. *gemmifera* Cabbage, *Brassica oleracea* var. *capitata* Cauliflower, *Brassica oleracea* var. *botrytis* Collard, *Brassica oleracea* var. *viridis* Cress, *Lepidium sativum* Kale, Borecole, *Brassica oleracea* var. *viridis* Kholrabi, *Brassica oleraceae* var. *gongylodes* Mustard leaf, *Brassic juncea* Mustard, Southern Curled, *Brassica juncea* Pak-Choe, Chinese Cabbage, *Brassica chinensis* var. *crispifolia* Pe-tsai, Chinese cabbage, *Brassica pekinensis* Seakale, *Crambe maritima* Sprouting Broccoli, *Brasica oleracea* var. *italica* Turnip, Seven Top, *Brassica rapa* Upland cress, *Barbarea verna* Watercress, *Rorippa nasturtium-aquaticum*	Yam, *Dioscoreaceae* Goosefoot, *Chenopo-diaceae* Mustard, *Cruciferae* Morning Glory, *Convolvulaceae* Parsley, *Umbelliferae* Nightshade, *Solanaceae* Sunflower, *Compositae*	Yam (true), *Dioscorea batatas* Beet, *Beta vulguris* Horseradish, *Armoracia rusticana* Radish, *Raphanus sativus* Rutabaga, *Brassica campestris* var. *napobrassica* Turnip, *Brassica rapa* Sweet Potato, *Ipomoea batatas* Carrot, *caucus carota* var. *sativa* Celeriac, *Apium graveolens* var. *rapaceum* Hamburg parsley, *Petro-selinum crispum* var. *radicosum* Parsnip, *Pastinaca sativa* Potato, *Solanum tuberosum* Black salsify, *Scorzonera hispanica* Chicory, *Chicorium intybus* Jerusalem artichoke, *Helianthus tuberosus* Salsify, *Tragopogon porrifolius* Spanish salsify, *Scolymus hispanicus*
Parsley, *Umbelliferae*	Caraway, *Carum carvi* Dill, *Anethum graveolens*				
Martynia, *Martyniaceae*	Martynia, *Proboscidea louisiana*				
Nightshade, *Solanaceae*	Eggplant, *Solanum melongena* Groundcherry (husk tomato), *Physalis pubescens* Pepper (bell or sweet), *Capsicum frutescens* var. *grossum* Tomato, *Lycopersicon esculentum*				
Gourd or Melon, *Cucurbitaceae*	Chayote, *Sechium edule* Cucumber, *Cucumis sativus* Cushaw, *Cucurbita moschata* Gherkin, *Cucumis anguria* Cantalope (Muskmelon), *Cucumis melo* Pumpkin, *Cucurbita pepo* Summer squash (bush pumpkin), *Cucurbita pepo* Squash, *Cucurbita maxima* Watermelon, *Citrullus lunatus* Winter melon, *Cucumis melo* var. *inodorus*				

FIGURE 19-3 Classification of vegetable crops by botanical family and crop use.

Cool-Season Crops		
• Asparagus	• Chinese cabbage	• Mustard
• Beet	• Chive	• Onion
• Broad bean	• Collard	• Parsley
• Broccoli	• Endive	• Parsnip
• Brussels sprouts	• Garlic	• Pea
• Cabbage	• Globe artichoke	• Potato
• Carrot	• Horseradish	• Radish
• Cauliflower	• Kale	• Rhubarb
• Celery	• Kohlrabi	• Salsify
• Chard	• Leek	• Spinach
• Chicory	• Lettuce	• Turnip

FIGURE 19-4 Some cool-season crops.

Warm-Season Crops	
• Cowpea	• Pumpkin
• Cucumber	• Snap bean
• Eggplant	• Soybean
• Lima bean	• Squash
• Muskmelon	• Sweet corn
• New Zealand spinach	• Sweet potato
• Okra	• Tomato
• Pepper, hot	• Watermelon
• Pepper, sweet	

FIGURE 19-5 Some warm-season crops.

winter freezing. This group of crops is planted early in the spring and late in the season for fall and winter harvest. This category contains mostly leaf and root crops. See Figure 19-4 for a list of cool-season crops.

The warm-season, or warm-weather, crops are those that cannot withstand cold temperatures, especially frosts. These vegetables and fruits require soil warmth to germinate and long days to grow to maturity. They must have very warm temperatures to produce their edible parts. The edible portions of these crops are basically what can be picked off the standing plant or the fruit. A list of warm-season crops can be found in Figure 19-5.

PLANNING A VEGETABLE-PRODUCTION ENTERPRISE

Prior to planting a vegetable garden or truck crop, the operator needs to plan. The gardener must decide where to plant as well as what to plant. Without some advance planning, the vegetable garden or truck farm is not likely to be successful.

Selecting the Site

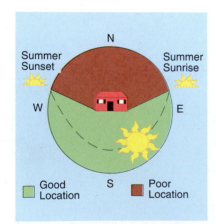

FIGURE 19-6 Avoid situating the garden in the shadow of a building. The location of a building's shadow changes throughout the day and throughout the year.

Choose a site that is convenient to a water supply. The site should also be exposed to the sun a minimum of 50 percent during the day. A minimum of 8 to 10 hours of direct sun is needed. Also consider the type of trees that are around the proposed site. Trees can provide excessive shade. They can also take away nutrients from the soil that are needed by the vegetable plants. Some kinds of trees can also produce toxins, which are harmful to certain vegetables. For example, the walnut tree is toxic to the tomato. Buildings and structures can also cast shade that can slow down or prevent growth of a vegetable crop. Avoid trying to grow vegetables closer than 6 to 8 ft from the northern side of a one-story structure—farther for higher structures. Both the south and west sides of a building have good light and often radiate heat late in the day (Figure 19-6).

Amounts of Organic Matter Needed to Cover 100 Square Feet at Various Depths	
Depth (inches)	To Cover 100 sq. ft.
6	2 cubic yards
4	35 cubic feet
3	1 cubic yard
2	18 cubic feet
1	9 cubic feet
1 cubic yard = 27 cubic feet	

FIGURE 19-7 Amounts of organic matter to apply per 100 sq ft of garden soil.

The type of soil in the area selected is important as well. The majority of vegetables grow best in a well-drained, loamy soil. Avoid ground that develops puddles after it rains. This is a sign of poor drainage. To test the drainage of a site, dig a trench 12 in. wide and 18 in. deep. Fill it with water and observe how long it takes for the water to drain away. If it takes one hour or less, the soil can be considered well drained. If the area selected has supported vegetation before, even if it has only been weeds, it will probably support a vegetable crop. If the site selected needs some alterations to its structure, adding organic matter can help. The organic matter can improve the drainage and allow air to move readily through the pores of the soil. If possible, the garden soil should be about 25 percent organic matter. To accomplish this, put a layer of organic matter 2 in. thick over the soil and work it in to a depth of at least 4 in. If necessary, repeat this procedure until the final mix contains approximately 25 percent organic matter (Figure 19-7).

Deciding on the Size

The size of the vegetable growing area will depend largely on the amount of ground that is available, the number of people served by

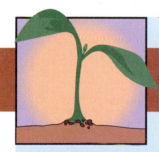

AGRI·PROFILE

CAREER AREAS: Vegetable Producer/Processor/Distributor/ Produce Manager

Career opportunities in vegetable production are similar to those in pomology. Vegetables and fruits are frequently grown, processed, and marketed by the same people. In addition to the careers mentioned in the Fruits and Nuts unit, many people find rewarding careers that pay well in the area of distributing and marketing produce. Here, truckers, wholesalers, and retailers move the product from producer to consumer.

Excellent jobs are available in supermarkets for produce stockers and foremen. Products include fruits, vegetables, and, possibly, ornamental plants. Tasks may include inventorying, ordering, handling, stocking, and displaying produce to keep it fresh and attractive. In large cities, as well as small towns, street vendors are frequently seen selling premium fruit and, occasionally, vegetables. Roadside stands provide opportunities for younger members of families to develop business skills while earning money for present and future needs.

Good fries require the cooperative work of plant breeders, potato growers, food specialists, and vegetable processors, among others. (Courtesy of USDA/ARS #K-4018-12)

the garden, and the use of the produce. A large garden 50 × 100 ft will produce enough vegetables for the annual needs of a large family. Figure 19-8 shows a suggested planting plan for a home garden 50 × 100 ft.

If this much land is not available, the fresh-vegetable needs for a medium-sized family may be met with a plot 40 × 50 ft. A planting

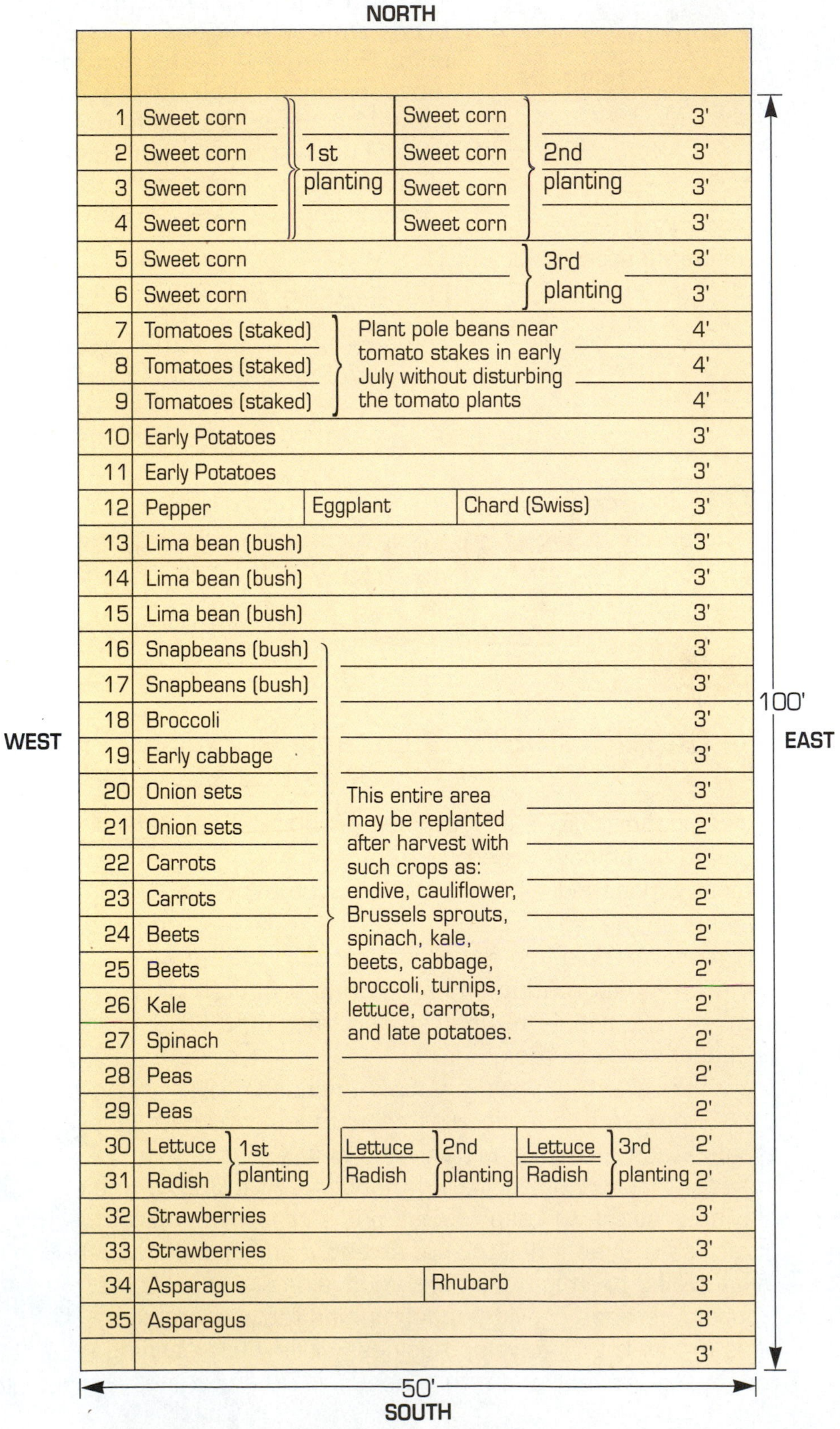

FIGURE 19-8 Example of a plan for a large home garden 50 × 100 ft.

Feet between Rows

Rows	Crop					1½'
1	Early Cabbage		(Turnips)			2'
2	Early Potatoes		(Late Cabbage)			2'
3	Peas	(Beans)		Peas	(Kale)	2'
4	Kale	(Carrots)		Turnips	(Beans)	2'
5	Parsley			Parsnips		2'
6	Onions and Radishes (same row)			Carrots	(Spinach)	2'
7	Swiss Chard	(Spinach)		Spinach	(Beets)	2'
8	Lettuce	(Beans)		Beets	(Kale)	2'
9	Beans	(Spinach)		Beans	(Lettuce)	2'
10	Peppers			Eggplant		2'
11	Lima Beans, Bush					2'
12	Tomatoes, Late					2'
13	Tomatoes, Early					2½'
14	Squash			Cucumbers		2½'
15						
16	Corn, Sweet interplanted with Pole Lima Beans					2½'
17	Corn, Sweet interplanted with Winter Squash					2½'
18						2'

40'

|← 50' →|

NOTE: Items in parentheses are succession crops planted after the first crop is harvested or planted between the rows of mature plants to permit germination and early growth before the first crop is removed.

FIGURE 19-9 Example of a plan for a medium size garden 40 × 50 ft.

guide for such a garden can be found in Figure 19-9. When deciding on the size, it would be wise to remember that the larger the vegetable production, the more care it will take. Vegetable production will be more efficient with better crops if the area is a smaller and better managed garden than if you take on a bigger area that is neglected and full of weeds.

Deciding What to Plant

There are several items that a potential vegetable gardener or truck cropper must consider when deciding which vegetables to plant. If for family use only, the first thing to determine is which vegetables the family likes and how often each vegetable will be served. Also, consider which vegetables will be stored. Select vegetable varieties that the family enjoys and concentrate on the kinds that are definitely better when they are freshly harvested. Examples are tomatoes, carrots, corn, snap beans, and cantaloupe. Second, consider the maintenance that some vegetables require. The easier they are to grow and maintain, the better.

A third factor to consider is the length of the harvest time for each vegetable. If the object is fresh consumption, then vegetables that can be harvested over a long period of time are best. For instance, carrots can be harvested for months, whereas cabbage should be picked at maturity. The last item to consider is the use for which the vegetables are planted. If the vegetables are going to be for fresh use, plant only for that purpose. Surplus vegetables can be stored in a root cellar or freezer, but plantings should match the estimated needs.

The truck cropper has a different set of questions to consider. First, which vegetables are in high demand in the area? Which ones will produce the highest return for the cost and labor involved? The truck-cropping enterprise is a business and must make a reasonable profit to be considered a good business. Second, what vegetables can be raised in the soil and climate? Truck cropping should be a high-income business, so high quality and quantity of produce is very important. Third, what combination of vegetables can be grown to keep the land occupied and provide cash flow from sales throughout the season?

Other questions to be explored are: (1) can the diseases and insect problems be managed? (2) will continuous use of the land for the same crops create pest buildup? (3) will the soil stand continuous cropping? and (4) what perennials may be included to reduce the labor involved?

BIO•TECH CONNECTION

Encouraging the Master Tillers

Earthworms are nature's master tillers. Always a favorite for fishing bait, the lowly earthworm is now regarded as an ally of homeowners, gardeners, farmers, landscapers, and conservationists. They enhance soil tilth and crop growth by consuming and digesting organic matter and mixing it with the mineral content of the soil. The new mixtures left behind by feeding worms are called *worm casts* and are rich in nutrients for feeding plants. Worm holes in the soil start at the surface and go as deep as the worm needs to go to escape winter cold and summer heat, perhaps three feet or more in the central area of the United States. This network of holes and tunnels collects rainwater as it falls, and the enriched soil absorbs and holds water for future plant use.

The North Appalachian Experimental Watershed at Coshocton, Ohio, has been the scene of no-till experiments for over 30 years. In a major experiment there, night crawlers (Lumbricus terrestris), usually 4 to 8 in. long and 3/8 in. in diameter, and other species were studied. It was found that earthworms may be encouraged by leaving crop residues on the surface or by incorporating them into the soil. However, if crop residues are removed from the soil, the earthworms do not have an adequate food source, and their populations do not expand as much.

Entomologist Edwin Berry of the National Soil Tilth Laboratory at Ames, Iowa, identified seven other species of earthworms. These all range in size from 2 to 6 in. in length and about half the diameter of night crawlers. Most of their activity occurs in the first foot of topsoil. They make temporary burrows as they pass through the soil, consuming, digesting, and mixing soil particles and organic matter, and excreting their rich casts.

PREPARING A SITE FOR PLANTING

Preparing the Soil

Before planting vegetables, the soil must be properly prepared. This will generally mean adding organic matter, lime, and fertilizer. The land should be plowed if it is in sod, and left for four to six weeks for the sod to decay.

Plowing   Land should be plowed or spaded to a depth of 6 to 8 in. The soil can be plowed either in the spring or the fall. Fall plowing has advantages First, the freezing and thawing of the winter months can improve the physical condition of the soil. Second, the exposure of the soil to the weather can result in a reduced population of insects. Spring plowing should be done only a few weeks prior to planting. Take care not to plow the soil when it is too wet. Doing so can destroy the physical structure of the soil, which results in hard clumps. The moisture content of the soil may be judged by pressing a handful of soil into a ball. If the ball crumbles easily, the soil is ready to plow. If the soil sticks together, it is too wet to plow.

Worm farming has become an important industry, and earthworms may be purchased for release for soil improvement. Worms grow and multiply the fastest where plenty of organic matter and moisture is available. Unfortunately, worms do not function as well in coarse, sandy soils as they do in soils where medium and fine particles are available. Gardeners, horiticulturists, and farmers should consider practices that will encourage the work of the master tillers.

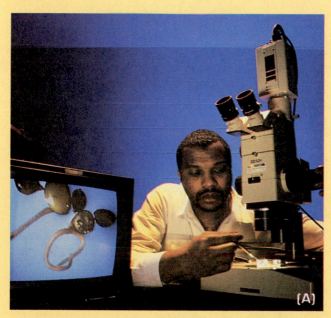

(A) Ken Ford at the Soil Tilth Research Lab in Ames, Iowa, uses a video camera adapted for microscopic work to study earthworm cocoons (top of monitor) and newly hatched earthworm larvae; and (B) adult earthworms—nature's master tillers. *(Courtesy of USDA/ARS # (A) K-4024-1 and (B) K-3237-1)*

Maintaining Organic Matter   Organic matter increases the water-holding and absorption capacity of the soil. Furthermore, it helps prevent erosion and promotes aeration in the soil. **Aeration** refers to the movement of air through soil. If it is available, and can be secured cheaply, animal manure is the best material for maintaining the organic content of the soil. It is also a good source of nutrients. Manure from animals other than sheep and poultry can be turned under at the rate of 15 to 20 tons per acre, or 20 to 30 bushels per 1,000 ft². Sheep and poultry manures can be used, but at half this rate. One note of caution—animal manures are low in phosphorus. Therefore, it may be necessary to use high-phosphorus fertilizer to balance the nutrients. For best results, apply fresh manure in the fall and well-rotted manure in the spring.

Another type of organic matter to use is green manure in the form of a cover crop, especially if it is a legume. **Green manure** is an active, growing crop that is then turned under to help build the soil. Green vegetation incorporated into the soil rots more quickly than does dry material. A cover crop should be planted at the end of the growing season. A **cover crop** is a close-growing crop planted to prevent erosion.

Liming   The need for lime and fertilizer should be determined through the use of soil tests. A pH greater than 7.0 indicates alkalinity, a pH of 7.0 is neutral, and a pH below 7.0 indicates acidity.

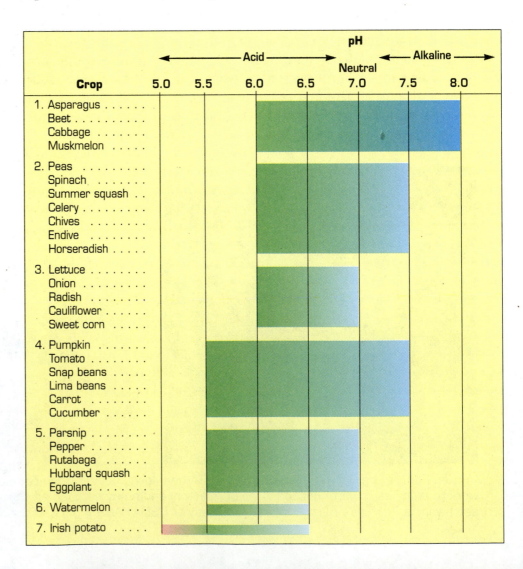

FIGURE 19-10 Optimum pH ranges for vegetable crops.

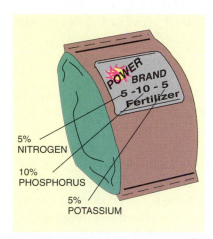

FIGURE 19-11 The analysis or ratio on a fertilizer bag indicates the percentages of nitrogen, phosphoric acid, and potassium in the fertilizer.

If the test indicates a pH of 6.0 or less, lime should be applied. The lime should be mixed into the top 3 to 4 in. of the soil. The amount of lime to add depends on the type of soil, the desired pH, and the form of lime used. Some plants, such as blueberries, require acidic soils. For such plants, special materials are available to make the soil acidic if the pH is too high.

Vegetable crops grow best at specific pH levels. Figure 19-10 lists vegetables and their optimum pH ranges for maximum growth.

Fertilizing   The amount and type of fertilizer to add can best be determined by a soil test. The test will help prevent applying too much or too litter fertilizer and determine the best timing. The application of commercial fertilizer is done to increase the amounts of nutrients available to the plants. Soils without additives are not likely to provide the best combination of conditions for all vegetables.

Commercial fertilizers usually contain nitrogen, phosphoric acid, and potassium. Each vegetable crop needs varying amounts of each of these three elements. In general, fertilizer ratios used for home gardens are 5–10–10, 10–10–10, and 5–10–5 (Figure 19-11). The first number in a fertilizer ratio or grade represents the percentage of nitrogen, the second shows the percentage of phosphorous, and the third indicates the percentage of potassium in the fertilizer. A rough guide for home gardeners is that most vegetables will need about 3 to 4 lb of fertilizer per 100 ft^2. Consult a county extension agent for exact fertilizer recommendations.

PLANTING VEGETABLE CROPS

Vegetable crops are initially grown from seeds. Some seeds can be sown where the plants will grow. Others will need to be grown into plants and the seedlings then transplanted to where the crop will grow. Transplants are plants grown from seeds in a special environment, such as a cold frame, hotbed, or greenhouse. The method used depends upon the vegetable's climatic requirements and the germinating characteristics of the seed.

Planting Seed

Depending on the size and scope of the garden and the quantity of seed to be planted, vegetable seeds are planted (1) by hand in hills or rows, (2) through broadcasting by hand or machine, (3) with one-row hand seeders, and (4) with single- or multiple-row, tractor-drawn seeders.

Broadcasting means scattering around. Regardless of the type of planting method used, the seeds should be planted at the proper depth. The soil should also be left smooth and firm over the seed. For most vegetables, seed must be fresh. Seed more than one year old will probably not germinate at the rate necessary to produce good yields.

Most vegetable seeds should be planted at a moderately shallow depth. It is best to plant in loamy-textured soil, with adequate moisture and no danger of frost. The seed will germinate best when it is planted at a depth no more than four times the diameter of the seed. Germinate means to produce a plant from a seed. Regardless of how deep the seeds are planted, the soil surface should be level and firmly pressed after the seed is planted. This will help prevent the seed from washing away or water from puddling above the seed after a rain.

Transplanting Vegetable Seedlings

There are three methods used in transplanting, depending on the number of plants that are being transplanted. The methods are: (1) hand setting, (2) hand-machine setting, and (3) riding-machine setting. In home gardening situations, hand setting is most appropriate.

Hand setting involves six steps: (1) dig a hole slightly bigger and deeper than the root ball of the plant being transplanted; (2) add some fresh soil to the hole; (3) place the plant in the soil a little deeper than it grew in its original container, trying to keep as much of the original soil as possible around the roots; (4) pull soil in and around the plant and firm it slightly; (5) add about one-half pint of water and let it soak into the soil; and (6) pull in some dry soil to level off and cover the wet area (which helps to prevent loss of moisture and baking of the soil).

The vegetable transplants should be set out on a cloudy day or late in the afternoon. Further, it is best to set plants just before or just after a rain.

CULTURAL PRACTICES

Cultivating

Cultivation, or intertillage of crops, is an old agricultural practice. The benefits of cultivation are: (1) weed control, (2) conservation of moisture, and (3) increased aeration. Cultivation increases the yield of most vegetable crops, mainly because the weeds are controlled.

All types of equipment are used to cultivate crops, from small hand tools to large tractors. The cultivation should be shallow and done only when there are weeds to be killed. Excessive cultivation is not beneficial and may even be damaging.

Controlling Weeds

Weeds are easier to control when they are small, using shallow cultivation. Weeds that are older and more established are more difficult to control. For them, deeper cultivation is required to rid the area of weeds. This can result in injury to the roots of vegetable crops (Figure 19-12).

Using herbicides is recommended for large home gardens and commercial vegetable production plots. The use of any herbicide depends on the registration of the herbicide by federal and state environmental protection agencies (EPAs). Do not use any herbicide unless it is clearly stated on the label that it is intended for a particular crop. Always seek the latest information for use of herbicides on vegetables.

Irrigation

Nearly all commercial vegetable operations irrigate their crops. Irrigation is especially important in California and other **arid** and **semi-arid,** or dry, regions of the West. Irrigation is important because nature's rainfall is rarely uniform and adequate for high yields.

FIGURE 19-12 Evaluating the weed-control effectiveness in plants passed over by the spyder (front), torsion weeder, side knife, and spring hoe attachments. *(Courtesy of USDA/ARS #K-5226-18)*

Water for irrigation can be obtained from a stream, lake, well, spring, or stored storm water. Definite laws and regulations guide the practice of irrigation in each state. There are several types of irrigation systems. These include sprinkler; drip surface, or furrow; and subirrigation.

Sprinkler systems are very versatile. They can be used in almost any situation. They can deliver shallow and light irrigation to promote seed germination and can be used for the application of fertilizers.

Drip irrigation uses a system of pipes and tubes to deliver the water to individual plants rather than wetting all the soil. It is gaining in popularity because of its water-conserving benefits.

Surface irrigation has the advantage of requiring a relatively low investment. The topography of the land is important for surface irrigation. The land must be gently sloping and uniform. This method is well suited for irrigating vast areas of land, but generally requires land leveling to work well.

Subirrigation requires large amounts of water. With this system, the water is added to the soil so that it permeates the soil from below. This method is expensive and suited only to commercial enterprises.

Trickle irrigation is a form of subirrigation where water is delivered under low pressure near the plants. The water seeps through porous hose under the rows. This system is expensive, but gives excellent results, with increased yields and with less water consumed than with other methods.

Mulching

A mulch is created whenever the surface of the soil is modified artificially by using straw, leaves, paper, polyethylene film, or a special layer of soil. Mulching helps control weeds, regulates soil temperatures, conserves soil moisture, and provides clean vegetables. The mulching material should be spread around the plants and between the rows.

Plastic film is frequently used with cantaloupes, cucumbers, eggplants, summer squash, and tomatoes. It is important that the plastic be laid over moist soil that has been freshly prepared. It is advisable to lay plastic mulch one to two weeks before planting the vegetables. This allows the soil to warm up underneath the mulch.

Pest Control

Good pest-control practices are essential for quality vegetables. A combination of cultural, mechanical, biological, and chemical methods can be used. Insecticides and fungicides can be effectively used by both home and commercial vegetable growers.

Insect and disease infestations bring about losses that can be devastating to vegetable crops (Figure 19-13). Types of losses include: (1) reduced yields, (2) lowered quality of produce, (3) increased costs of production and harvest, and (4) increased expenditures for materials and equipment. The use of chemicals is limited because of toxic and environmental considerations, so good biological controls are in great demand (Figure 19-14). There are many measures that can help the vegetable grower control insects and diseases. They are:

1. Dispose of old crop residues that serve to shelter pests, and thus enable them to overwinter or survive from one crop to another.

FIGURE 19-13 Insects such as the Colorado potato beetle and the Mexican bean beetle (shown above) can strip the leaves from vegetable plants in a matter of hours. *(Courtesy of USDA/ARS #K-1291-2)*

FIGURE 19-14 (A) New disease-resistant varieties of sweet potatoes, (B) spraying with non-toxic Bt to spread insect-fighting bacteria, and (C) a spined soldier beetle making a meal of a leaf-devouring Mexican bean beetle larva. *(Courtesy of USDA/ARS #(A) K-3461-14, (B) K-2785-2, and (C) K-1461-3)*

2. Rotate the location of individual crops.
3. Choose insect- and disease-resistant vegetable varieties.
4. Treat seeds with fungicides and insecticides.
5. Control weeds.
6. Use adequate amounts of fertilizer.
7. Follow recommended planting dates.

HARVESTING AND STORING VEGETABLES

Harvesting

For the best possible quality, vegetables should be harvested at the peak of maturity. This is possible when the vegetables will be used for home consumption, sold at a local market, or processed. The best harvesting time varies with each vegetable crop. Some will hold their quality for only a few days, while others can maintain quality over a period of several weeks. Frequent and timely harvest of crops is needed if the vegetable grower wishes to supply the market or the kitchen table with high-quality produce over a period of time (Figure 19-15). Specifications for harvesting each kind of vegetable crop should be consulted for the best results.

Harvesting is done by hand and through the use of machines. Mechanical harvesting is especially useful for the commercial grower. When harvesting a crop, care must be taken to prevent injury to the crop by machinery and handling.

Storing

When storing vegetables, proper stage of maturity is essential. There are specific temperatures and humidity levels for storage of vegetable crops. Most crops need 90 to 95 percent humidity. Most homeowners find it difficult to reproduce the exact temperatures and humidity levels, but commercial growers can accomplish storage through specialized facilities (Figures 19-16 and 19-17).

FIGURE 19-15 Lettuce harvested at the peak of quality—regular size on the right and the new individual serving, or mini-size, on the left. *(Courtesy of USDA/ARS #K-5384-02)*

Crop	Temperature °F	Relative Humidity Percent
Asparagus	32	85-90
Beans, snap	45-50	85-90
Beans, lima	32	85-90
Beets	32	90-95
Broccoli	32	90-95
Brussels sprouts	32	90-95
Cabbage	32	90-95
Carrots	32	90-95
Cauliflower	32	85-90
Corn	31-32	85-90
Cucumbers	45-50	85-95
Eggplants	45-50	85-90
Lettuce	32	90-95
Cantaloupes	40-45	85-90
Onions	32	70-75
Parsnips	32	90-95
Peas, green	32	85-90
Potatoes	38-40	85-90

FIGURE 19-16 Temperatures and humidities for storing vegetables.

Storage Conditions							
Vegetable	Temperature (°F)	Relative Humidity (%)	Storage Life	Vegetable	Temperature (°F)	Relative Humidity (%)	Storage Life
Pea, English	32	95-98	1-2 weeks	Squash, winter	50	50-70	___[4]
Pea, southern	40-41	95	6-8 days	Strawberry	32	90-95	5-7 days
Pepper, chili (dry)	32-50	60-70	6 months	Sweet corn	32	95-98	5-8 days
Pepper, sweet	45-55	90-95	2-3 weeks	Sweet potato	55-60[3]	85-90	4-7 months
Potato, early	___[1]	90-95	___[1]	Tamarillo	37-40	85-95	10 weeks
Potato, late	___[2]	90-95	5-10 months	Taro	45-50	85-90	4-5 months
Pumpkin	50-55	50-70	2-3 months	Tomato, mature green	55-70	90-95	1-3 weeks
Radish, spring	32	95-100	3-4 weeks	Tomato, firm ripe	46-50	90-95	4-7 days
Radish, winter	32	95-100	2-4 months	Turnip	32	95	4-5 months
Rhubarb	32	95-100	2-4 weeks	Turnip greens	32	95-100	10-14 days
Rutabaga	32	98-100	4-6 months	Water chestnut	32-36	98-100	1-2 months
Salsify	32	95-98	2-4 months	Watercress	32	95-100	2-3 weeks
Spinach	32	95-100	10-14 days	Yam	61	70-80	6-7 months
Squash, summer	41-50	95	1-2 weeks				

Adapted from R. E. Hardenburg, A. E. Watada, and C. Y. Wang, *The Commercial Storage of Fruits, Vegetables, and Florist and Nursery Stocks*, USDA Agriculture Handbook 66 (1986).

[1]Spring- or summer-harvested potatoes are usually not stored. However, they can be held 4-5 months at 40°F if cured 4 or more days at 60-70°F before storage. Potatoes for chips should be held at 70°F or conditioned for best chip quality.

[2]Fall-harvested potatoes should be cured at 50-60°F and high relative humidity for 10-14 days. Storage temperatures for table stock or seed should be lowered gradually to 38-40°F. Potatoes intended for processing should be stored at 50-55°F; those stored at lower temperatures or with a high reducing sugar content should be conditioned at 70°F for 1-4 weeks, or until cooking tests are satisfactory.

[3]Sweet potatoes should be cured immediately after harvest by holding at 85°F and 90-95% relative humidity for 4-7 days.

[4]Winter squash varieties differ in storage life.

FIGURE 19-17 Temperatures and humidities for storage and storage life of vegetables.

Vegetables differ in the amount of time that they can be stored. Some may be stored for several months, others for only a few days. For the vegetables that do store well, storage is essential to prolong the marketing period.

Fresh vegetables utilized for storage should be free of skin breaks, bruises, decay, and disease. Such damage or disease will decrease the life of the vegetable in storage.

Refrigeration storage is recommended. This type of storage reduces respiration and other metabolic activities; ripening; moisture loss and wilting; spoilage from bacteria, fungus, or yeast; and undesirable growths, such as potato sprouts.

One activity that can increase the successful storage of vegetables is precooling. **Precooling** is the process of rapid removal of the heat from the crop before storage or shipment. One method of precooling is termed **hydrocooling.** This is a method wherein vegetables are immersed in cold water and stay under water long enough to lower the temperature to the desired level.

STUDENT ACTIVITIES

1. Write the Terms to Know and their meanings in your notebook.

2. Using outdated seed catalogs, prepare flash cards of warm- and cool-season vegetables.

3. Bring in an unusual vegetable to show to the class. Be prepared to tell the class which part is edible.

4. Select a site for planting a vegetable crop and prepare a planting plan to scale.

5. Take a soil sample from the site selected in the previous student activity. Submit the sample to the local county extension service for analysis and lime and fertilizer recommendations for the vegetables planned.

6. Plant lima bean seeds in containers to observe the growth of the stem and leaves. Determine whether the plant is a monocotyledon or a dicotyledon.

7. Plant tomato seeds in a greenhouse for later transplanting.

8. Develop a chart with pictures illustrating the various insects that attack vegetable plants. Find pictures of insects in vegetable catalogs or extension service bulletins.

9. Store vegetables under different conditions (light/dark, dry/humid, warm/cool) in the classroom or at home. Record what happened under each set of conditions. Note which conditions are best for each vegetable.

SELF EVALUATION

A. Multiple Choice

1. The part of horticulture that has to do with the production of vegetables is called
   a. viticulture.
   b. olericulture.
   c. floriculture.
   d. pomology.

2. Vegetable gardening produces fresh, nutritious food and provides the gardener with
   a. exercise.
   b. weeds.
   c. pests.
   d. diseases.

3. The majority of vegetables grow best in
   a. well-drained, clay soil.
   b. well-drained, sandy soil.
   c. well-drained, silt soil
   d. well-drained, loam soil.

4. For vegetable gardening, the soil should be plowed to a depth of
   a. 1 to 2 in.
   b. 3 to 5 in.
   c. 6 to 8 in.
   d. 10 to 12 in.

5. The first number in the analysis of a fertilizer represents the percentage of _____ in the bag.
   a. nitrogen
   b. calcium
   c. phosphoric acid
   d. potash

6. The best time to transplant vegetable plants is on
   a. a bright, sunny day.
   b. a rainy day.
   c. a cloudy day.
   d. a snowy day.

B. Matching

_____ **1.** Cultivation
_____ **2.** Herbicides
_____ **3.** Polyethylene film
_____ **4.** Olericulturalist
_____ **5.** Monocotyledon
_____ **6.** Precooling

a. Weed-control chemicals
b. Rapid removal of heat from vegetables
c. A person involved in the study of vegetable production
d. Intertillage of crops
e. Having only one seed leaf
f. A type of mulching material

C. Completion

1. The scientific name for the mustard family of vegetables is _____.

2. A site for the vegetable garden should be one that is exposed to the sun a minimum of _____ percent of the day.

3. To correct the acidity level in the soil, _____ is commonly added.

4. The type of irrigation that allows water to permeate the soil from below is called _____ irrigation.

5. Whenever the surface of the soil is artificially modified through the use of straw, leaves, plastic film, or paper, a _____ is created.

6. Tomatoes for use in the local market are best harvested in the _____ stage of maturity.

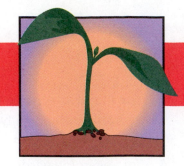

Fruit and Nut Production

OBJECTIVE

To determine the opportunities in and identify the basic principles of fruit and nut production.

COMPETENCIES TO BE DEVELOPED

After studying this unit, you should be able to:

- determine the benefits of fruit and/or nut production as a personal enterprise or career opportunity.
- identify fruit and nut crops.
- plan a fruit or nut production enterprise and prepare a site for planting.
- describe how to plant fruit and nut trees and utilize appropriate cultural practices in fruit and nut production.
- list appropriate procedures for harvesting and storing at least one commercial fruit or nut crop.

MATERIALS LIST

✓ paper, pencil, and eraser

✓ old seed catalogs

✓ pruning shears

✓ scissors, glue, and index cards

H ome fruit operations can provide high-quality and bountiful varieties of fruit and nuts. The fruits grown at home can be enjoyed at the peak of ripeness. This quality is difficult to obtain from supermarket fruits. Home fruit and nut operations can be enjoyed as hobbies or turned into profitable commercial enterprises.

CAREER OPPORTUNITIES IN FRUIT AND/OR NUT PRODUCTION

Like other agricultural enterprises today, the production of fruit and nut crops is more a business than a way of life. The orchardist or small-fruit grower has to be a good financial manager, as well as a scientist and farmer. Producing food, however, can be a very enjoyable and rewarding profession.

Career Descriptions

The production, harvest, and marketing of fruits is a large industry. The United States accounts for a sizable part of the combined world crops of apples, pears, peaches, plums, prunes, oranges, grapefruit, limes, lemons, and other citrus fruits. There is a shortage of trained pomologists today in the fruit industry and profession. A **pomologist** is a fruit grower or fruit scientist.

Career opportunities exist at all levels of the industry. One can be an owner/grower, orchardist, plant manager, foreman, or technician. People who work in this field must be able to propagate fruit and nut trees and vines, as well as plant and transplant, prune, thin, train, and fertilize them. They must also be able to provide protection and control of diseases and insects.

FIGURE 20-1 (A) Seven-year-old semi-dwarf and (B) twelve-year-old dwarf apple trees. *(Courtesy of USDA/ARS # (A) K-4048-17 and (B) K-4048-14)*

IDENTIFICATION OF FRUITS AND NUTS

Types of Fruits

There are three broad categories of fruit—tree; small bush and cane; and vine. These categories describe the growth habit of the fruit plant. Each category includes numerous kinds of fruit.

Tree Fruits Tree fruits are popular with home gardeners as well as large producers. One drawback, however, is that some types of trees can take several years before producing the first harvest. There are three types of fruit trees based on growth habit: standard, semi-dwarf, and dwarf. These three terms refer to the size of the tree and the length of time it takes from planting to producing the first crop (Figure 20-1).

A standard tree is one that has its original rootstock and reaches normal size. A standard apple tree can grow to be 30 ft high. However, **semi-dwarf** and **dwarf** trees are smaller. They are standard varieties of trees, but they are grafted onto dwarfing rootstocks. Such rootstocks cause the tree to produce less annual growth. Thus, it remains smaller throughout its life. The semi-dwarf tree averages 10 to 15 ft in height, while dwarf trees average 4 to 10 ft. Dwarf trees are better suited to the home garden because they bear earlier and take up less space than semi-dwarf or standard trees (Figure 20-2).

Common fruits include apples, cherries, grapefruit, oranges, peaches, pears, plums, and quince. There are two types of tree fruits: pome and drupe. The common pome fruits are apples and pears. A **pome** is a fruit with a core and embedded seeds. A **drupe** is a fruit with a large, hard seed called a *stone*. Stone fruits include peaches, plums, quince, and cherries.

When growing tree fruits, particular attention must be paid to the climate of the area. Peaches, plums, and cherries are especially sensitive to climatic conditions (Figure 20-3).

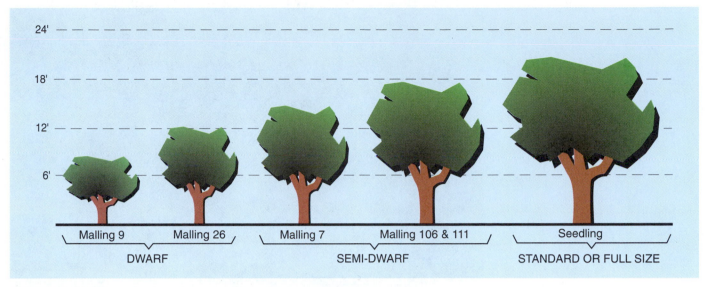

	DWARF		SEMI-DWARF		STANDARD OR FULL SIZE
	Malling 9	Malling 26	Malling 7	Malling 106 & 111	Seedling

FIGURE 20-2 Rootstocks control sizes of fruit trees.

Small Bush Fruits   These fruits are excellent selections for home gardens and small farms. The best choices are strawberries, blueberries, raspberries, and thornless blackberries. Small bush fruits grow either low to the ground or only 3 to 4 ft high. They require less maintenance than do tree fruits or vine fruits. They tend to bear quickly—usually between nine months and one year after planting.

Small bush and cane fruits that are popular with gardeners and commercial operations are strawberries, blueberries, red raspberries, currants, gooseberries, and thornless blackberries (Figure 20-4). A factor to consider in growing bush fruits is to buy plants from a reliable nursery. Some of these fruits will bear a harvest in the summer and again in the fall.

Vine Fruits   The best known vine fruit is the grape. Grapes will occupy the land for many years, so site selection is important. The vines need to be trained and pruned for adequate production and yield. Grapes require a growing season that has at least 140 frost-free days.

FIGURE 20-3 Peach-growing research at the Appalachian Fruit Research Station in Kearneysville, West Virginia. (Courtesy of USDA/ARS #K-3275-8)

FIGURE 20-4 Trying to mechanize the harvest, agricultural engineer David Peterson watches his spiked-drum shaker as it traverses a row of blackberry bushes. (Courtesy of USDA/ARS #K-2949-6)

FIGURE 20-6 Nut trees work well in landscaping. They provide shade, food, and wood. *(From Bridwell,* Landscape Plants: Their Identification, Culture, and Use. *Copyright 1994 by Delmar Publishers)*

FIGURE 20-5 Grapes are popular as fresh fruit, dried fruit (raisins), juice, and wine. *(Courtesy of USDA/ARS #K-3681-7)*

Grapes are also the most prevalent vine fruit. They are easily grown and have a wide range of flavors. The disadvantage is the three- to four-year wait for the vines to reach maturity before they produce fruit (Figure 20-5).

Types of Nuts

Nut trees are planted for numerous purposes. They can be useful for shade, landscaping, food, and wood. Wherever trees grow, a nut tree can also be grown. Popular nut trees grown in the United States are pecan, black walnut, filbert, hickory, almond, English walnut, chestnut, and macadamia (Figure 20-6).

PLANNING FRUIT AND NUT ENTERPRISES
Selecting Fruit or Nuts to Plant

Climatic Considerations   Certain varieties and **cultivars** of fruits or nuts are more suited to one type of climate than another. For example, citrus fruits can be grown only in southern-type climates such as Florida, Texas, and California. This information is a key factor when deciding which fruit to grow. Even though you may absolutely love nectarines, they may not grow successfully in your area. Recommendations for each fruit crop for a given area can be obtained from your local county agricultural extension agent.

Rootstock Selection   Because many fruit crops are propagated by grafting, selection of proper **rootstock** is important. Select one that is true to its variety and is hardy enough to support the tree. Most nurseries will readily give this information about the rootstock. The rootstock will determine whether the tree is a standard, semi-dwarf, or dwarf tree.

Frost Susceptibility   Locate crops in areas as protected from frost as possible. Also, select varieties that are late bloomers if your area is **susceptible** to late frosts. Planting areas on hillsides, near ponds, or near cities will remain warmer on clear, frosty spring nights. Such locations decrease the amount of damage caused by frosts. Apricots and sweet cherries are especially susceptible to late frost damage. The home gardener can protect some fruit crops, such as strawberries, from frost by remulching or irrigating and by placing cloth or plastic sheets over trees. Commercial growers use irrigation sprinklers, heaters, and fans to help reduce frost damage.

Fertility   The production of high-quality fruit is dependent upon the fertility of the soil. On the average, soil should have a pH of 5.5 to 6.0. However, for growing blueberries, the soil should be more acidic—a pH of 5.2 or less. Fruit crops should be fertilized annually after they are planted. However, too much fertilizer can damage roots and create other problems. One pound of 10–10–10 fertilizer per 100 ft^2 of surface soil should give good results.

Planning for Pollination   Most fruit crops require pollination. **Pollination** is defined as the transfer of pollen from the male stamen to the stigma, or female part of the plant. Pollination is necessary for the fruit to set, or form. The gardener should choose crops that are known for having high-quality and abundant pollen. The pollen is distributed by wind or insects. Sometimes it is necessary to have several varieties that bloom at the same time for adequate cross-pollination. Similarly,

AGRI·PROFILE

CAREER AREAS: Pomology/Fruit Grower/ Nut Grower/Packer

Pomology is the scientific study and cultivation of fruit, including nuts. Careers in pomology include work on farms, and in nurseries, orchards, and groves. Fruit and nut specialists may be plant breeders, propagators, educators, food scientists, or consultants. Technical-level jobs are available for graders, packers, supervisors, and managers.

In the past, fruit production meant hard labor because of hand picking. However, the need for mechanical harvesting has opened new opportunities for agricultural engineers, electronics technicians, plant breeders, systems managers, food scientists, and inventors. Mechanical harvesters range from hydraulically controlled buckets, to tree shakers, to sensor-directed robots.

Development of pecan crosses are monitored continuously during the growing season. *(Courtesy of USDA/ARS #K-3318-8)*

FIGURE 20-7 Honeybees are essential for good pollination of some fruit trees, and the sale of honey can provide another source of income for the fruit grower. *(Courtesy of USDA/ARS #K-3438-1)*

it may be necessary to plant a tree just for its pollen quality and availability. Many fruit growers keep honeybees to help ensure good pollination (Figure 20-7).

When favorable conditions exist, just two varieties of a particular fruit crop are necessary for cross-pollination. Choose varieties that bloom at the same time or, at least, overlap for a period of time. Varieties of pears, plums, and cherries all tend to bloom at approximately the same time. However, apple varieties differ greatly in their times of bloom. Careful selection of apple varieties is very important to ensure good cross-pollination.

Growing and Fruiting Habits   The various types of fruit and nut crops have different kinds of growth patterns and fruiting dates. This fact is especially true of fruit trees. Some varieties of apple tree require as long as eight years before they bear fruit. Newly planted peach and sour cherry trees can bear fruit in as little as three years. It pays to understand what to expect from the variety you have chosen. Most fruit trees go through two stages of growth—a vegetative period and a productive period. The **vegetative** period consists of rapid, vigorous growth of the tree. In younger trees, 4 to 5 ft of growth may occur in a single season. When the tree enters the productive period of fruit bud formation, a change is noticeable. Fruit buds are larger and more plump. On apple, pear, cherry, and plum trees, short, spurlike growths occur (Figure 20-8). On the peach tree, the peach buds become larger and occur in clusters, with two large ones surrounding a center one. The vigorous **terminal** growth slows down to only 18 to 20 in. in a season.

SOIL AND SITE PREPARATION
Selecting the Site

Because many fruits require full sunlight for maximum production and yield, choose a site that allows for maximum exposure to the sun.

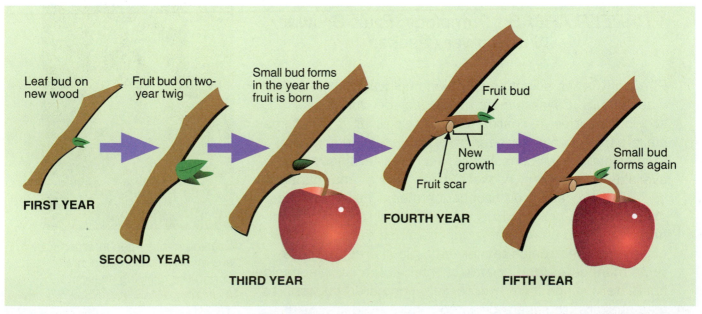

Leaf bud on new wood

Fruit bud on two-year twig

Small bud forms in the year the fruit is born

Fruit bud

New growth

Fruit scar

Small bud forms again

FIRST YEAR

SECOND YEAR

THIRD YEAR

FOURTH YEAR

FIFTH YEAR

FIGURE 20-8 Growth of an apple fruiting spur.

A southern exposure is best. To help prevent frost damage, select land with a slight slope and good air circulation. The soil for fruit crops should be well drained and of medium texture.

Soil Preparation

The soil for fruit and nut trees should be limed, fertilized, and deeply plowed. These root systems grow deep into the soil. Deep plowing can be accomplished the year before the trees are planted. This permits the freezing and thawing action of the winter months to improve the texture of the soil. A soil sample should be taken a year in advance. Because many fruit and nut crops have specific pH and fertilizer requirements, a soil sample tested by a reputable testing laboratory will provide important information.

Any additional amendments, such as organic matter, lime, or fertilizer, should be added and disced into the soil prior to planting. Care should be taken when planting small bush and cane fruits in soils that have been planted with vegetable crops, such as peppers, tomatoes, and potatoes. These vegetables may leave residues in the soil that cause diseases.

PLANTING ORCHARDS OR SMALL FRUIT AND NUT GARDENS

When to Plant

Tree stock from reputable nurseries can be planted either in the spring or fall. There are advantages to planting in both seasons. Fall planting is advisable if the soil will be less than desirable or wet in the spring. Planting in the fall also allows the crop to get its root system established before rapid top growth occurs in the spring. Finally, weather conditions may be more stable in the fall than in the spring.

The advantages of planting in the spring are:

1. The crop can make considerable top and root growth before the stress of the winter season.
2. The threat of damage by rodents is reduced.
3. The threat of winter kill is reduced.

Laying Out the Orchard

Each fruit or nut species and type will have its own planting distance. It is important to follow the specific planting distance for each variety, too. When planting small trees, the distance seems enormous. However, the trees will grow and fill the spaces in time. It is important to plant the trees in a straight line. A crooked row of trees will be obvious and reflects badly on the owner or manager.

Planting the Fruit or Nut Crop

Each fruit or nut variety has its own specifications for planting. With trees, make the hole just wide enough to accommodate the roots without crowding them. The hole should be deep enough to allow the tree to be set 2 to 3 in. deeper than it was in the nursery. The two sides of the hole should be parallel, and the bottom should be as wide as the top. By making the hole to these specifications, the soil can be packed

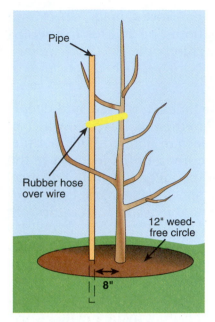

FIGURE 20-9 Providing support for the newly planted tree.

Labels: Pipe, Rubber hose over wire, 12" weed-free circle, 8"

uniformly around the roots to avoid leaving air spaces. Air spaces permit the roots to dry out.

When planting trees, add some prepared soil to the hole. Then spread the roots, adding more soil until all roots are covered. Move the tree up and down several times, while firming the soil around the roots. This procedure allows all roots to be in contact with the soil and avoids air pockets.

Before the hole is completely filled, add about two gallons of water. After the water soaks in, add the final soil. Most trees will need to be staked for good support. Metal pipe may be driven 8 in. from the tree and the tree tied to the pipe by using wire strung through a piece of rubber hose. (Figure 20-9). After planting, keep a 12 in. circular weed-free and grass-free area around the tree.

CULTURAL PRACTICES IN FRUIT AND NUT PRODUCTION

Fertilization

A complete fertilizer, such as a 5–10–10 or a 10–10–10, should be used annually on fruit and nut trees. After filling the hole of a newly planted

BIO·TECH CONNECTION

The Ripening Gene—Can it be Controlled?

Inside the fruits and vegetables growing in gardens, vineyards, orchards, groves, farms, and in the wild, microscopic cells manufacture a gas known as ethylene. This colorless and odorless gas is known to be a major manager of the ripening process. The more ethylene the plant produces, the faster the ripening process will be. The work of this compound is essential and good, and transforms inedible, green produce to ripe and delicious food. However, there is a down side—that of changing delicious fruit into soft and mushy trash!

Molecular biologist Athanasios Theologis at Agricultural Research Service/University of California Plant Gene Expression Center in Albany, California, believes that much produce spoilage in and out of the field is the work of ethylene gas. He believes that it does not have to be that way. But rather, perhaps we can find the way to manage the production of ethylene in plants. If so, we can control when the fruit of a given crop will ripen. The dream is to modify plants through biotechnology to ripen at our convenience.

Theologis and his colleague, Takahide Sato at Chiba University in Japan, brought molecular biologists a step closer to the dream when they cloned a gene that plants apparently must have to manufacture ethylene gas. It is hoped that scientists will eventually find ways to turn down or turn off the effect of the gene and thereby gain more control of the ripening process. Such a feat would enable produce to arrive at the supermarkets in fresher, better-tasting condition. Such produce would be better suited for export as well.

One possible new destination for American fresh produce could be Japan's Mom-and-Pop-style produce stands. Here produce

tree, apply one pound of the 5–10–10 or one-half pound of 10–10–10 fertilizer around each tree. Do not place the fertilizer closer than 8 to 10 in. from the trunk. Excessive fertilization can result in damage or death to young trees. After the first year, apply fertilizer at the rate of one pound times the age of the tree. After the tree is six to eight years old, do not increase the rate of application. As the tree becomes larger, fertilize all of the ground area under the limb spread.

The above rates are useful for all tree fruits except pears. Pears should not be fertilized, because increased susceptibility to fire blight more than offsets the benefit from fertilizing.

Fertilization requirements for the various bush and cane fruits vary considerably. Therefore, it is advisable to consult a grower's guide for each individual variety.

Pruning

The purpose of pruning and training young fruit trees is to establish a strong framework of branches that will support the fruit. This framework will resemble the shape best suited for the tree and for the fruit it will bear. Each fruit tree has a specific way that it should be pruned. For example, peach trees are pruned for an open center, V-shape

must be fresh, ripe, and edible on the spot. Without better control of the ripening process, we do not have the technology to move large volumes of produce from field to consumer fast enough to serve this market. Ethylene gas also causes plant parts to

wither. The control of ethylene in the plant would slow down the wilting and withering processes in leafy vegetables and cut flowers. Imagine, perhaps some day, these commodities may be stored or shipped with little or no refrigeration!

 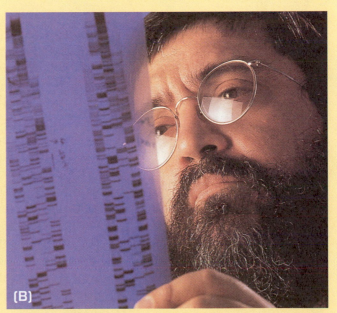

(A) Jean Murphy at the ARS/University of California Plant Gene Expression Center examines the different stages of ripening of tomatoes, while (B) molecular biologist Athanasios Theologis uses electrophoresis to separate and match nucleotide sequences of DNA fragments. *(Courtesy of USDA/ARS # (A) K-5335-01 and (B) K-3489-6)*

FIGURE 20-10 A peach tree pruned to the open center of V-shape pattern.

FIGURE 20-11 An apple tree showing the cuts to be made for proper pruning.

1

Sketch of a young apple tree after one year's growth. All limbs with broken lines should be removed. The central leader (CL) should be tipped if it is more than two feet long.

2

The same tree as in 1, after pruning.

3

Sketch of how the tree in 2 may look after the second year. Branches with broken lines should be removed. The central leader (CL) should be tipped if it grew more than two feet.

4

The same tree as in 3, after pruning and spreading of branches. All limbs in the first tier of braches (A, B, C, & D) have been spread with wooden spreaders with a sharp-pointed nail in each end, to illustrate the beginning of the Christmas-tree shape. Limbs E and F only, in the second tier of branches, have been spread with wire with sharpened ends.

(Figure 20-10). However, apple trees should have a strong central leader and be pruned into a Christmas-tree scaffold (Figure 20-11).

Pruning a grapevine depends on its fruiting habit. Next year's grape clusters are on shoots produced during the current growing season. Therefore, this new growth should be carefully pruned. When the shoots have fruited and the leaves have fallen, the shoots are called **canes.** The canes will produce the new shoots and next year's crop.

A common method of training grapevines is the Four-Arm Kniffin System. In this system, the vine is pruned to form a double T on a two-wire trellis system. Two arms originating at the two wires are trained to grow in opposite directions from the trunk. The vines are pruned to remove all canes except those that are saved for fruiting and renewal growth (Figure 20-12).

Pruning and thinning small bush and cane fruits should be accomplished according to the specific standards for each fruit. Pruning, thinning, and training these smaller fruits is an essential practice for high-quality yields. These practices also help prevent pests and diseases.

Disease and Pest Control

A great deal of the time and expense in fruit and nut production is spent controlling diseases, insects, and other pests. Control of diseases

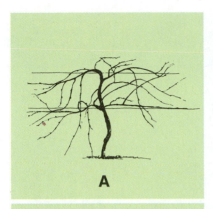

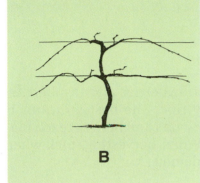

FIGURE 20-12 A grape vine trained to the four-arm Kniffin system (A) before pruning and (B) after pruning. *(Courtesy Maryland Cooperative Extension Service)*

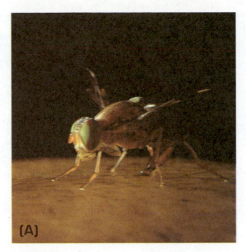

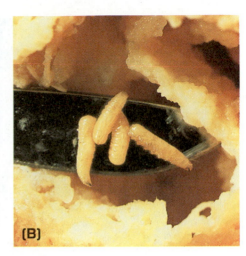

FIGURE 20-13 (A) A Caribbean fruit fly laying eggs in the peel of a grapefruit. (B) Fruit fly larvae then hatch and devour the pulp. *(Courtesy of USDA/ARS # (A) K-898-8 and (B) K-1674-7)*

and insects is a major component of fruit and nut production (Figure 20-13). The type of pesticide needed and how often it is applied varies tremendously among all the species of fruits and nuts.

Each tree fruit, nut, and small fruit variety must be treated individually. What may work on one variety may not be appropriate for another. Most orchardists will have a spray schedule worked out for each of their crops. Tree fruits require more pesticide than do any other fruits. Before applying any pesticide, however, read the label thoroughly.

HARVEST AND STORAGE

Harvesting

Harvesting fruits can be a daily practice. The advantage of homegrown fruit is that it can be picked at its peak of ripeness. However, fruits for commercial use must be picked several days ahead so they will withstand shipping and handling.

For the best quality of fruit, harvest apples, pears, and quince when they begin to drop, soften, and become fully colored. Some varieties will ripen over a two-week period, requiring picking every day. Other varieties will ripen all at once.

Peaches, plums, apricots, and cherries should be harvested when the green disappears from the surface skin of the fruit. A yellow undercolor should be developed by this time. The fruit should be soft when pressed lightly in a cupped hand.

The nut varieties ripen from August to November. Harvest nuts immediately after they fall from the tree. Most of the nuts that do not fall can be knocked off the tree with poles or mechanical harvesters.

The nut varieties of pecan, hickory, chestnut, and Persian walnut will lose their husks when ripe. However, the husk of black walnuts and other types of walnuts will need to be removed by hand.

Grapes should be harvested only when fully ripe. The best way to judge when the grape is at full maturity is to sample an occasional grape.

| Commodity | Freezing Point | Place to Store | Storage Conditions | | Length of Storage Period |
			Temperature	Humidity	
	°F		°F		
Fruits:					
Apples	29.0	Fruit storage cellar	Near 32° as possible	Moderately moist	Through fall and winter
Grapefruit	29.8	........do	........do	........do	4 to 6 weeks
Grapes	28.1	........do	........do	........do	1 to 2 months
Oranges	30.5	........do	........do	........do	4 to 6 weeks
Pears	29.2	........do	........do	........do	4 to 6 weeks

FIGURE 20-14 Storage recommendations for fruits. *(Courtesy USDA)*

Small bush and cane fruits, such as strawberries and raspberries, should be harvested when they are fully ripe. The best indicator is when the fruit is fully colored. When these fruits and berries reach maturity, they should be picked every day. When harvesting, pick when the fruit is dry to avoid mildew and mold damage.

Storage

Fruit storing is limited to those varieties that mature late in the fall or those that can be purchased at the market during the winter months. The length of time fruits can be stored depends on the variety, stage of maturity, and soundness of the fruit at harvest (Figure 20-14).

For long-term storage of apples, the temperature should be as close to 32°F as possible (Figure 20-15). Apples can be stored in many ways, but they should be protected from freezing. A cellar or other area below ground level that is cooled by night air, is a good place for apple storage. There should be moderate humidity in the storage area. Pears have the same storage requirements as apples.

Nuts should be air dried before they are stored. Nuts, especially pecans, keep longer if they are left in the shell and refrigerated at 35°F. They can also be frozen if they are kept in their shells. Chestnuts have special requirements. They should be stored at 35° to 40°F and high humidity, shortly after harvest.

Small bush and cane fruits, including grapes, are not suitable for storage. These fruits are too perishable to keep very long and are best utilized as soon as they are harvested.

FIGURE 20-15 Red Delicious apples ready for harvest in an orchard near Wenatchee, Washington. *(Courtesy of USDA/ARS #K-3853-5)*

STUDENT ACTIVITIES

1. Write the Terms to Know and their meanings in your notebook.

2. Develop a bulletin board display explaining opportunities in fruit- and nut-related occupations. Include those that can be found on the local, state, and national levels.

3. Bring a specimen of some type of unusual fruit or nut to class. Discuss its origin and the techniques for growing the fruit or nut.

4. Prepare fruit and nut flash cards using pictures from seed catalogs.

5. Prepare a planting plan for fruits and nuts, drawn to scale. Indicate the types, quantity, and location of plants to be grown.

SELF EVALUATION

A. Multiple Choice

1. Most fruits and nuts require _____ to grow properly.
   - a. full sunlight
   - b. partial sunlight
   - c. partial shade
   - d. no sunlight

2. The stock for a fruit or nut crop can be planted in the
   - a. fall or spring.
   - b. spring only.
   - c. summer only.
   - d. fall only.

3. When planting a fruit tree, you should make the hole deep enough to allow the tree to be set _____ deeper than in the nursery.
   - a. 3 to 5 in.
   - b. 2 to 3 in.
   - c. 1 to 2 in.
   - d. 6 to 8 in.

4. Nuts can be stored in the refrigerator at the optimum temperature of
   - a. 40°F.
   - b. 32°F.
   - c. 38°F.
   - d. 35°F.

5. The pear is an example of a _____ fruit.
   - a. drupe
   - b. pome
   - c. bramble
   - d. aggregate

B. True or False

1. A standard tree is one that has its original rootstock and grows to normal size.

2. A semi-dwarf tree averages a mature height of 30 to 40 ft.

3. Small bush and cane fruits tend to bear quickly, usually between nine months and one year after planting.

4. Rootstock selection is important to fruit and nut crops, because most of these crops are not propagated by seed.

5. Frost susceptibility is not a factor to fruit crops.

6. Too much fertilizer can be a hazard to fruit trees.

7. Control of diseases and insects is a major component of fruit and nut production.

8. An apple is an example of a drupe fruit.

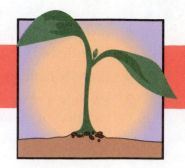

Grain, Oil, and Specialty Field-Crop Production

OBJECTIVE

To determine the nature of and approved practices recommended for grain, oil, and specialty field-crop production.

MATERIALS LIST

✓ sample of field crops and products made from field crops

✓ bulletin board materials

✓ crop magazines

✓ reference materials on crops

COMPETENCIES TO BE DEVELOPED

After studying this unit, you should be able to:

- define important terms used in crop production.
- identify major crops grown for grain, oil, and special purposes.
- classify field crops according to use and thermo requirements.
- describe how to select field crops, varieties, and seed.
- prepare proper seedbeds for grain, oil, and specialty crops.
- plant field crops.
- describe current irrigation practices for field crops to meet their water needs.
- control pests in field crops.
- harvest and store field crops.

The cultivation of the land and the growing of crops began about 10,000 years ago. The need to produce food for the animals that humans had captured and begun to domesticate caused early humans to change from hunters to farmers. There were no guidelines to follow in selecting plants. Early agriculturists had to rely on observing what the animals were eating to decide which plants to grow. Trial and error and thousands of years of selection have led to the crops being grown today. New types, varieties, and uses of plants continue to be developed in response to current needs and in anticipation of future demands for food and plant fiber by an ever-increasing world population.

In the United States, the production of grain, oil, and specialty crops occupies more than 450 million acres. These crops are normally called **field crops.** This acreage represents nearly 20 percent of the land mass in the United States. American agriculturists are among the most efficient in the world, producing enough food for themselves and many other people. As a result, about 2 percent of American workers are engaged in the production of food and fiber for the rest. The efficiency of the American agriculturist allows the U.S. population to spend only about 11 percent of their income on food. It also allows for sizable exports of food crops all over the world. This has helped maintain a favorable balance of trade for the United States in recent years. Today we have so many products from agriculture that we no longer are aware of the plant and animal origins of common nonfood items (Figure 21-1).

FIGURE 21-1 New products from plants and animals through agriscience: Top row—plastics from vegetable oils and industrial uses for corn starch; Second row—cocoa butter substitute, improved cotton processing; Third row—tallow-based soap, detergents with improved surfactants, more uses for corn starch; and Fourth row—starch-based rubber products, plastics from vegetable oil, starch-based rubber. *(Courtesy of USDA/ARS #K-4796-20)*

MAJOR FIELD CROPS IN THE UNITED STATES
Grain Crops

There are seven major grain crops in the United States. **Grain crops** are grasses that are grown for their edible seeds. These crops are corn, wheat, barley, oats, rye, rice, and grain sorghum.

Corn Corn is the most important of the field crops grown in the United States. Corn is well adapted, high yielding, and is an important crop in most states. Thirty-five to forty percent of the corn produced annually in this country is grown in the midwestern states, commonly called the Corn Belt. These states include Iowa, Illinois, Nebraska, Minnesota, Indiana, and Ohio, in descending order of production. The United States accounts for nearly 50 percent of the corn produced in the world (Figure 21-2).

Originating in Central America, corn served as a major part of the diet of the Indians of the area for 3,000 years before the settlement of the New World by Europeans. Corn was unknown to the rest of the world until explorers found Native Americans growing corn.

Less than 10 percent of the corn grown in the United States is for human consumption. The rest is used for livestock feed, alcohol production, and hundreds of other products.

The major types or classifications of corn are dent corn; flint corn; popcorn; sweet corn; flour, or soft corn; and pod corn. Most of the corn grown in the United States is dent corn.

A general discussion of cultural practices for crops appears later in this unit. However, the following list of approved practices are useful for growing corn.

1. Select a hybrid that will mature during the growing season.
2. Obtain a current soil test.
3. Lime and fertilize according to soil test and desired production.
4. Select a field with deep, rich, well-drained soil.
5. Select the proper tillage method.

FIGURE 21-2 Corn is high-yielding in grain for humans and animals, and silage for livestock. *(Courtesy of Michael Dzaman)*

6. Prepare a firm seedbed or use proper no-till planter.
7. Plant corn when soil temperature is 50°F or above.
8. Calibrate planter for proper plant population.
9. Match plant population to soil-yield potential.
10. Adjust planter for proper depth of seed placement.
11. Calibrate sprayer for proper application of materials.
12. Use selected herbicides for controlling problem weeds.
13. Cultivate to help control weeds.
14. Use insecticides necessary for proper control.
15. Apply fungicides when needed.
16. Make a yield check.
17. Check moisture content when harvested.
18. Harvest crop as soon as moisture permits, to reduce losses.
19. Take a forage analysis on ensiled corn.
20. Store ensiled or high-moisture corn in sound structures.
21. Track market trends and habits.
22. Market crop at optimum time.
23. Keep accurate enterprise records.
24. Summarize and analyze records.

FIGURE 21-3 Worldwide, wheat is the most important grain crop. *(Courtesy of USDA/ARS #K-3597-18)*

Wheat   The most important grain crop in the world is wheat (Figure 21-3). In the United States, wheat ranks second to corn in bushels produced. Leading states in the production of wheat are Kansas, Oklahoma, Washington, Texas, and Montana.

Wheat is used primarily for human consumption. The wheat is ground into flour, which is then made into products such as bread, cakes, cereal, macaroni, and noodles. Other uses of wheat include the manufacture of alcohol and livestock feed.

Types of wheat grown in this country include common, durum, club, Poulard, Polish, Emmer, and spelt. Most of this wheat is of the common type. Classes of common wheat include hard red spring, hard red winter, soft red winter, and white.

Barley   The leading states in the production of barley in the United States include North Dakota, Montana, Minnesota, Idaho, South Dakota, and Washington (Figure 21-4). Barley ranks fifth among the grain crops produced in the United States.

Most barley is used for livestock feed, and it has about the same food value as corn. The production of barley for malting is also important. **Malting** is the process of preparing grain for the production of beer and other alcoholic beverages.

The cultural practices for growing wheat, barley, oats, and rye are very similar. The following list of approved practices for growing small grains should be useful for growing any of these crops.

1. Select a variety adapted to your conditions.
2. Buy certified seed.
3. Conduct a germination test if planting home-grown seed.
4. Obtain a current soil test.
5. Lime and fertilize according to soil test and desired production.
6. Select a field suitable for your grain choice.

FIGURE 21-4 Most of the barley grown in the United States is used for livestock feed.

7. Select the proper tillage method.
8. Prepare a firm and smooth seedbed.
9. Calibrate drill for proper plant population.
10. Match plant population to soil-yield potential.
11. Adjust drill for proper depth of seed placement.
12. Calibrate sprayer for proper application of materials.
13. Use selected herbicides for controlling problem weeds.
14. Use insecticides necessary for adequate control of insects.
15. Apply fungicides when needed.
16. Make a yield check.
17. Adjust combine for proper harvesting.
18. Check moisture content when harvested.
19. Harvest crop as soon as moisture permits, to reduce losses.
20. Store grains in disinfected grain storage tanks.
21. Bale straw as soon as moisture levels allow.
22. Track market trends and habits.
23. Market crop at optimum time.
24. Keep accurate enterprise records.
25. Summarize and analyze records.

Oats   Oats as a grain crop are fourth in acres produced in the United States (Figure 21-5). Major oat-producing states are South Dakota, Minnesota, Wisconsin, Iowa, and North Dakota.

The value of oats in adding bulk and protein to the diets of livestock is well documented. However, about 5 percent of the oats produced in the United States are made into oatmeal and cookies. Oats are also used in the production of plastics, pesticides, and preservatives. In addition, they are important in the paper and brewing industries.

Rye   Although rye is grown in nearly every state in the United States, it is the least economically important grain crop. It does, however, have many uses (Figure 21-6). Most rye is grown in South Dakota, Georgia, Minnesota, North Dakota, and Nebraska.

FIGURE 21-5 Oats provide bulk and protein to the diets of animals as well as food for humans.

FIGURE 21-6 Rye provides grain for livestock, feed, and flour; however, it is the least economically important grain crop.

FIGURE 21-7 Rice is the only commercially grown grain crop that will grow in standing water. *(Courtesy of USDA/ARS #K-3444-1)*

FIGURE 21-8 Sorghum with the grain head forming. *(Courtesy of National FFA)*

About 25 to 35 percent of rye acreage is used for grain. The rest is used for forage, as a cover crop, or as a green manure crop. A **forage** is hay or grass grown for animals. **Cover crops** are planted to protect the soil from erosion. **Green manure crops** are grown to be plowed under to add nutrients and organic matter to the soil. The rye grown for grain is used for livestock feed, flour, whiskey, and alcohol production.

Rice   Rice is the major grain crop grown for food for almost half of the people in the world. Most of the rice in the United States is grown in Arkansas, California, Louisiana, Texas, and Mississippi. Rice is the only commercially grown grain crop that can grow and thrive in standing water (Figure 21-7). The types of rice grown in the United States are short grain, medium grain, and long grain.

The majority of the rice grown in this country is used for human consumption. The excess that is produced is exported to other countries of the world.

Sorghum   Sorghum grown in the United States is used primarily for livestock feed. It is about equal to corn in food value. Other uses of sorghum include forage, the manufacture of syrup or sugar, and the making of brooms. Leading states in the production of sorghum include Kansas, Texas, Nebraska, South Dakota, and Oklahoma. Based on acres harvested, sorghum is the third most important grain crop in the United States.

The five types of sorghum are grouped according to use. They are grain, forage, syrup, grass, and broomcorn (Figure 21-8).

Oilseed Crops

Crops that are grown for the production of oil from their seeds are called **oilseed crops.** These crops are growing in importance each year, as Americans rely more and more heavily on vegetable oils and less on animal fats in their diets. Important crops grown for the oil contained in their seeds are soybeans, peanuts, corn, cottonseed, safflower, flax, and sunflowers. Corn and cottonseed are discussed in other sections of this unit.

Soybeans There are approximately 60 million acres of soybeans grown in the United States each year. With an average yield of about thirty-four bushels per acre, the production of soybeans grosses more than $11 billion each year (Figures 21-9 and 21-10).

Oil and grain products are the major uses of soybeans. The meal resulting from the extraction of oil from soybeans is an important source of protein in livestock feeds. Soybeans are also used for hay, pasture, and other forage crops. Research has led to the development of hundreds of other uses for soybeans.

Large centers of production include the midwestern states of the Corn Belt. Major soybean-producing states include Illinois, Iowa, Minnesota, Indiana, and Missouri. The approved practices for raising soybeans are similar to those for raising corn, except that soybeans and other legume seeds should be treated with the proper inocculant bacteria to ensure good nitrogen fixation.

Peanuts The peanut is actually a pea rather than a nut, in spite of its nutlike taste and shell. It is grown primarily in the South, where warm temperatures and a long growing season are keys to success. Leading peanut-producing states are Georgia, Texas, Alabama, North Carolina, Oklahoma, and Florida.

One ton of peanuts in the shell will yield about 500 lb of peanut oil and 800 lb of peanut-oil meal. The remaining 700 lb is mostly shells. The oil meal is used for livestock feed and as a good protein source in

FIGURE 21-9 Sometimes called the Soybean Doctor, agronomist Edgar E. Hartwig has devoted half a century to soybean research. His focus has been on developing productive plants with built-in resistance to insects, nematodes, and diseases. *(Courtesy of USDA/ARS #K-5272-1)*

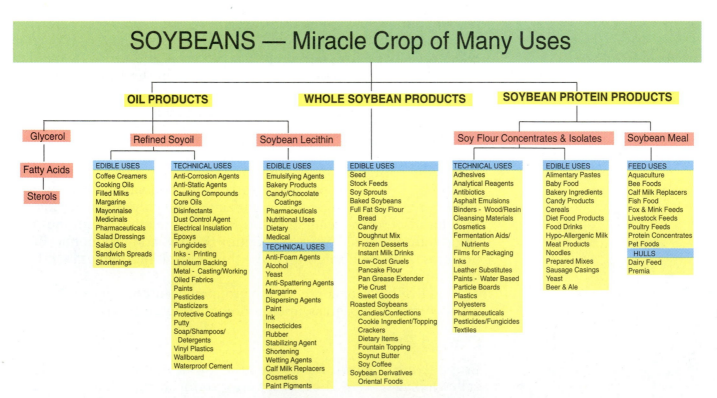

FIGURE 21-10 The soybean has become famous for its many products and is growing in importance in world markets. *(Courtesy of American Soybean Association)*

FIGURE 21-11 George Washington Carver, the famous agricultural research pioneer, developed over three hundred products from peanuts. *(Courtesy of USDA/ARS #K-4297-18)*

human diets. Other food products produced from peanuts include peanut butter and dry roasted peanuts (Figure 21-11).

Safflower The production of safflower for oil occurs mainly in California. Safflower plants grow 2 to 5 ft in height and have flower heads that resemble Canadian thistles. The oil comes from the wedge-shaped seeds that the plant produces. The seeds may contain from 20 to 35 percent oil.

Safflower oil is used in the production of paint and other industrial products. It is also used for cooking oil and for low-cholesterol diets.

Flax Originally, the production of flax was mostly for fiber. Flax fibers contained in the stems of plants are used to produce **linen** (a fabric made from flax fibers).

The oil produced from the seed of the flax plant is called **linseed oil.** It is an important part of many types of paint and has hundreds of uses in industry. The linseed-oil meal that is left after the extraction of the oil is an excellent source of protein for animal feeds.

Most flax is grown in North Dakota, South Dakota, Minnesota, and Wisconsin.

Sunflowers The production of oil-type sunflower seed has been important in the United States in recent years. Most of the sunflower production is located in North Dakota, South Dakota, Kansas, Minnesota, and Texas.

There are two types of sunflowers grown commercially in the United States—oil-type and non-oil-type. About 90 percent of sunflower production is of the oil-type. Oil-type sunflower seeds contain 49 to 53 percent oil. The meal left after the oil has been removed is high in bulk and contains 14 to 19 percent protein. It is used for livestock feed. The oil is used for margarine and cooking oil. Sunflower oil can also be used as a substitute for diesel fuel in tractors (Figure 21-12).

Specialty Crops

Fiber crops, sugar crops, and stimulant crops are grouped into the category called specialty crops. Some specialty crops grown in the United States are cotton, sugar beets, sugarcane, and tobacco.

FIGURE 21-12 A field of sunflowers in bloom is a sight to behold before the development of its oil-rich seed. *(Courtesy of USDA)*

FIGURE 21-13 Cotton has been an important crop since colonial times, but vigilance is necessary to protect the plants from insects and diseases.

Cotton   Cotton originated in Central and South America. It has been an important crop in the South since colonial days. The cotton plant requires warm temperatures and a long growing season in order to reach maturity. The top cotton-growing states include Texas, Mississippi, California, Louisiana, and Arkansas. The extensive irrigated acreages of cotton crops, which can produce up to three crops per year on the same land, also places Arizona among the high-producing states.

Over 15 million bales of cotton are harvested per year in the United States. Because only about 9 million are needed by the textile industry, the rest is exported to other countries of the world (Figure 21-13).

The seed must be removed from cotton after it is harvested. This process is called **ginning.** The cotton seed is then processed to remove its oil, which is a major contributor to the vegetable-oil needs of the United States. The seed, after the extraction of the oil, is ground into a high-protein animal feed.

Sugar Beets   The production of sugar beets for sugar accounts for about 35 percent of the refined sugar produced in the United States. This crop is grown for its thick, fleshy storage root in which sugars are accumulated. Centers of production of sugar beets are the western states and the upper Midwest.

Sugar Cane   Sugar cane production in the United States is concentrated in subtropical areas of Florida, the Gulf Coast states, and Hawaii. Sugar cane accounts for about 65 percent of the sugar refined in the United States.

This crop, which is a grass, is grown from sections of stalk called **seed pieces** rather than from seed. It takes about 2 years for sugar cane to reach the harvesting stage in Hawaii. Compared to production in Hawaii, in the southern states sugar cane is harvested about 7 months after planting, with a corresponding loss of yield. The same field of sugar cane can be harvested several times before needing to be replanted (Figure 21-14).

Tobacco   Tobacco is an original North American product that was used by the Native Americans in religious rites. It is produced in the southeastern states, predominantly in North Carolina, Kentucky, Vir-

FIGURE 21-14 Sugar cane fields in Hawaii. *(Courtesy of Elmer Cooper)*

ginia, South Carolina, Tennessee, and Georgia, as a cash crop. A **cash crop** is grown strictly to sell for cash. Production of tobacco dropped in the United States in the 1980s as cigarette smoking and other uses of tobacco declined, but it increased again in the 1990s as export markets expanded.

Tobacco production requires large amounts of labor and is therefore best adapted to small farming operations. Warm temperatures and plenty of rainfall are required for optimum production of high-quality tobacco.

Classification of Field Crops

There are a number of ways of classifying field crops. Three ways are by use, thermo requirements, and life span. **Thermo** refers to heat requirements or length and characteristics of the growing season required.

The classification of field crops according to use is as follows:

1. **Cereal crops** are grasses grown for their edible seeds. They include corn, wheat, barley, oats, rice, rye, and sorghum.

2. **Seed legume crops** are nitrogen-fixing crops that produce edible seeds. Included in this class are soybeans, peanuts, field peas, field beans, and cowpeas.

3. **Root crops** are grown for their thick, fleshy storage roots. Beets, turnips, sweet potatoes, and rutabagas are root crops.

4. Forage crops are grown for hay, silage, or pastures for livestock feed. Examples of forage crops include alfalfa, clover, timothy, orchard-grass, and many other crops used for their stems and leaves.

5. **Sugar crops** are grown for their ability to store sugars in their stems or roots. They include sugar cane, sugar beets, and sorghum. Corn is also used to produce sugar.

6. Oil crops are produced for the oil content of their seeds. Examples are soybeans, peanuts, cottonseed, flax, rapeseed, and castor beans.

7. **Tuber crops** are grown for their thickened, underground storage stems. Potatoes and Jerusalem artichokes are examples of tuber crops.

8. **Stimulant crops** are grown for their ability to stimulate the senses of the user. Examples of stimulant crops are tobacco, coffee, and tea.

Crops can also be classified according to their thermo requirements. The two major thermo groups are warm season and cool season.

Warm-season crops must have warm temperatures in order to live and grow. They are adapted best to areas where freezing or frost seldom occur. They also normally require longer growing seasons than do cool-season crops. Examples of warm-season crops are cotton, tobacco, and citrus.

Cool-season crops are normally grown in the northern half of the United States, where temperatures below freezing are normal. These crops often need a period of cool weather in order to attain maximum production. Most of the grains, tubers, and apples are cool-season crops.

Crops can also be classified according to their life spans. They may be annual, biennial, or perennial. An example of an annual crop is corn; red clover is a biennial crop; and alfalfa is a perennial crop.

Selection of Field Crops

There are a number of factors to consider when selecting which field crops to grow. Some of these factors are:

1. Crops that will grow and produce the desired yields under the type of climate available. Be sure to consider length of growing season, average yearly rainfall, average temperatures, humidity, and prevailing winds.
2. Crops that are adapted to the type of soil available. Consider soil pH, soil type, soil depth, and soil response to fertilizers.
3. Demand or markets available for the crop to be produced.
4. Labor requirements and availability of labor for the crop.
5. Machinery and equipment necessary to grow the crop.
6. Availability of enough land to justify production of the crop.

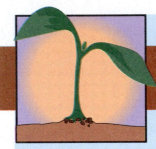

AGRI·PROFILE

CAREER AREAS: Broker/Elevator Manager/Grain Handler/Agronomist/Producer

Corn; wheat and other small grains; sugar cane; soybeans; sugar beets; and other specialty crops are grown on large acreages in the United States, Canada, and many nations of the world. In the United States, grain, oil, and specialty crops account for large amounts of exports and do much to help maintain our balance of payments in foreign trade. Grain brokers, futures brokers, market reporters, grain elevator operators, crop forecasters, farmers, and others owe their jobs to these crop enterprises.

Custom combine operators, haulers, maintenance crews, cooks, and other service workers follow the grain harvest from Mexico to Canada. At season's end, the crews return south and prepare for the next season, when the cycle is repeated. At the same time, smaller farm owners and operators grow, harvest, and market millions of acres of field crops as part, or all, of the farm operations.

Along with crop enterprises are jobs in building and storage construction; systems engineering; machinery sales and service; welding; irrigation; custom spraying; hardware sales; agricultural finance; chemical sales; and seed distribution.

Plant technicians, as well as research specialists, are necessary to plan research and grow and manage crops like this pearl millet to achieve crop improvement. *(Courtesy of USDA/ARS #K-3882-2)*

If The Soil Is:	The Benefits Will Be
Fine, pulverized	Seed in direct contact with soil particles
Firm	Prevents drying. Good root hair medium
Loose and mellow	Permits air circulation and seedling growth
Free of balky trash and weeds	Trash and weeds shredded and distributed
Fertile	Provides adequate plant nutrients
Moist	Absorbs and holds water

FIGURE 21-15 Characteristics of a good seedbed.

7. Pest-control problems.

8. Expected yields.

9. Anticipated production costs. Can a reasonable profit be expected?

Seedbed Preparation for Field Crops

The purpose of seedbed preparation for field crops is to provide conditions that are favorable for the germination and growth of the seed to be planted. Not only does the seedbed need to be prepared for seed germination, but the area under the seedbed must also be prepared for the root growth of the crop (Figure 21-15).

Eliminating competition from weeds and crop residues is a consideration when preparing a seedbed for planting. Proper seedbed preparation can also increase the availability of soil nutrients to plants.

Seedbeds should not be overworked. The texture of the soil should be porous and allow for free movement of air and water. Small seed requires a seedbed with a finer texture than larger seeds require. The seedbed should contain enough fertility to encourage germination and growth until additional fertilizer can be applied. The control and elimination of weeds, insects, and diseases is also an important consideration when preparing seedbeds for field crops.

There are several methods of properly preparing seedbeds for field crops. They can be divided into three general categories: conventional tillage; reduced, or minimum, tillage; and no-tillage.

In **conventional tillage,** the land is plowed with a moldboard plow turning under all of the residue from the previous crop. The soil is then disked or harrowed to smooth and further pulverize the soil for the seedbed (Figure 21-16).

Reduced or **minimum tillage** is a system of seedbed preparation that works the soil only enough so that the seed can make contact with the soil and germinate. A chisel plow is frequently used (Figure 21-17). Minimum-tillage systems usually combine several operations into one pass across the field. This method reduces the amount of soil

FIGURE 21-16 With conventional tillage, the soil is turned via a moldboard plow so all crop residues, livestock manure, lime, and fertilizer are mixed through the plow layer. [Courtesy of Elmer Cooper]

FIGURE 21-17 A chisel plow loosens the soil but leaves the crop residue on the surface. [Courtesy of Elmer Cooper]

FIGURE 21-18 This implement cuts trash and loosens surface soil via the discs, loosens the deep soil via chisel points, and leaves a fine seedbed on the surface. *(Courtesy of Elmer Cooper)*

FIGURE 21-19 No-till planting leaves crop residues on the surface for better erosion control. *(Courtesy Dr. Allen Hammer)*

compaction, conserves soil moisture, and usually provides less opportunity for soil erosion (Figure 21-18).

No-till preparation of seedbeds involves planting seeds directly into the residue of the previous crop, without exposing the soil. Seed is usually planted in a narrow track opened by the seed planter. It is extremely important that good management practices be employed when using the no-till method of seedbed preparation. Such practices will ensure control of insects and diseases and eliminate competition from previous crop residues (Figure 21-19).

Planting Field Crops

The invention of the seed planter was one of the most important events for American agriculture. There are three general types of planters used in planting field crops today. They are row crop planters, drill planters, and broadcast planters.

Row crop planters plant seeds in precise rows and with even spacing within the rows. Three types of row crop planters are drill planters, hill-drop planters, and checkrow planters. Drill planters drop seeds individually in a row at set distances apart. Hill-drop planters drop two or three seeds together in rows. Checkrow planters plant several seeds together in a checkered pattern in the field to permit cross-cultivation. Row crop planters are used to plant corn, soybeans, sorghum, and cotton.

Drill planters plant seeds in narrow rows at high population rates. Drills are available in many row spacings and planter widths. Seeding rates are less accurate than with row crop planters. Some field crops that are planted with drill planters are wheat, oats, barley, rye, and many grasses. Fertilizer and pesticides are usually applied at the same time the seed is planted with drill planters.

Broadcast planters scatter the seed in a random pattern on top of the seedbed. The accuracy of seeding using this type of planter is the poorest of the planting methods. Broadcast seeders cover wide areas and usually plant seeds much faster than other methods. They are sometimes employed when weather conditions make it difficult to get machinery into the fields being planted. Airplanes are often used as broadcast planters. Knapsack seeders and spinners are other types of broadcast seeders (Figure 21-20).

One disadvantage of using broadcast seeders is that some seed needs to be covered in order to germinate and to protect it from loss. This means that a second trip often must be made over the field to cover the seed. Small grains, grasses, and legumes are often planted by broadcasting.

There are other considerations in planting field crops. These include the date to plant, germination rate of seeds, uniformity of seed, weather conditions, and insect- and disease-control problems.

Meeting Water Needs of Crops

The soil that plants grow in acts as a storage vat for the water needed by the plant. Under ideal soil conditions, approximately one-half of the pore space is filled with water. About one-half of this water is available for use by plants. Unfortunately, this ideal condition seldom exists; often, much less water is available for plant use than is needed. Factors that affect the water available for crops include the type of soil, natural rainfall, water-table levels, and prevailing winds.

When conditions are such that sufficient water is not available for the crop being grown, irrigation may be the answer to obtaining profitable yields. **Irrigation** is the artificial supplying of water to crops.

The irrigation of crops has been practiced for more than 5,000 years. The Nile River was used by the Egyptians to irrigate crops grown in the fertile deserts of the area. The Chinese diverted the water from many rivers to irrigate their rice fields. Even the Native Americans of the American West used irrigation to allow for the production of corn in arid areas.

The major methods of supplying irrigation water to crops are sprinklers, surface irrigation, and subsurface irrigation.

Sprinklers spray water through the air, much like rainfall. This is the least efficient means of using supplemental water. Much of the water evaporates before it can be used by the crop being irrigated. Some of the water is also placed where it is not available to crop roots.

In **surface irrigation,** water gets to the crop by gravity, flowing over the surface of the soil or in ditches or furrows. It is an inexpensive means of providing water to crops. However, the cropland may need to be leveled before it can be irrigated. Containment borders may also interfere with other farming operations.

Subsurface irrigation supplies water to the roots of crops underground. Use of pipes under the ground to deliver the water to the crops makes subsurface irrigation expensive to set up. It has the advantage of low operating costs when in operation, however, and it permits the most efficient use of water (Figure 21-21).

Pest Control in Field Crops

The control of pests in field crops is often the factor that determines whether or not a profit is made. Pests of field crops may include diseases, weeds, insects, and animals. They may destroy the seed before it germinates; attack the growing crop; or render the harvested crop

FIGURE 21-20 Three types of broadcast seeders are knapsack; spinner, or cyclone; and airplane.

FIGURE 21-21 Tomatoes thrive well with subsurface drip irrigation in California's San Joaquin Valley. *(Courtesy of USDA/ARS #K-3231-17)*

FIGURE 21-22 A few lesser grain borers left unchecked soon become a devastating horde. *(Courtesy of USDA/ARS #K-3965-17)*

unsalable, unusable, or even consumed. Economic losses from plant pests total billions of dollars each year (Figure 21-22).

There are three main categories of losses from plant pests. They are reduced yields, reduced quality, and spoilage.

Reduced yields occur when weeds germinate and grow faster than the crop being grown. Weeds compete successfully with the crop for moisture and nutrients, often causing the crop plant to be unhealthy. Parasitic plants actually feed on the host crop and may cause it to break off and die.

Damage from insects also causes the crop to yield less than expected. The damage may occur during the insect's feeding or while it lays eggs in and on the crop. Reduction of yield may also occur as insects spread disease from plant to plant.

Diseases can reduce yields by interfering with the plant's ability to manufacture food. They can also cause other plant processes to be changed, affecting the health and welfare of the plant.

The reduction of quality may result from such things as weed seeds or rodent hairs and droppings intermixed with the crop. Foreign materials may cause flavors that are objectionable to users of the crop. Most foreign materials must be removed before the crop can be used.

Damage from insects and diseases can make the crop less desirable in appearance and can increase processing costs tremendously. Food crops may be deemed unsuitable for human consumption and a total loss if they are too severely infested with insects and diseases.

Spoilage of crops results when weeds hinder the drying process. Insects may also cause stored crops to overheat and mold.

Methods of controlling pests in field crops include mechanical control, cultural control, biological control, genetic control, and chemical control.

Mechanical pest control refers to anything that affects the environment of the pest or the pest itself. Cultivation is the normal mechanical control of weeds (Figure 21-23). Cultivation of the soil may also expose insects and soilborne diseases to the air. The effect of the suddenly raised or lowered temperatures as a result of exposure to the Sun and air often proves fatal to many pests. Other types of mechanical control of pests include the pulling or mowing of weeds, and the use of screens, barriers, traps, and electricity.

FIGURE 21-23 The use of this ridge tillage machine for cultivation reduces or eliminates preplant tillage. *(Courtesy of USDA/ARS #K-3239-1)*

BIO•TECH CONNECTION

New Crops for New Uses

As the medical profession scrambles to find new drugs and treatments for new and old medical problems, agriscience is becoming an increasingly important player. At the Southern Weed Science Laboratory in Stoneville, Mississippi, scientists are developing a weed-control system to ensure the supply of a medicine for malaria patients. Annual wormwood has been used for 2,000 years in China to treat malaria. Now a new, refined drug derived from the plant is in demand worldwide. The World Health Organization is seeking ways to increase the production of wormwood sufficiently to meet the need.

Annual wormwood grows throughout the United States as a weed. The woody-stemmed, cone-shaped plant reaches a height of 4 to 8 ft and could become a cash crop for farmers in the future. Wormwood is harvested with a sickle-type machine similar to that used to cut kenaf and sugar cane. Cut plants must be protected from prolonged exposure to the sun, because ultraviolet light breaks down artemisinin—the antimalarial substance.

Artemisinin is found in the leaves of growing tips, and it takes 2.2 lb of the leaves to process 1 gram of artemisinin. Authorities indicate that it would take thousands of acres of wormwood to meet the world demand for the drug. However, before commercial production of wormwood could occur, farming methods, including weed control, must be developed. The World Health Organization estimates that 110 million people throughout the world contract malaria each year.

Another newly emerging crop is lesquerella. Lesquerella has been developed

Checking lesquerella for seed set. New crops, such as lesquerella and wormwood, promise new advances in fabrics, plastics, lubricants, corrosion inhibitors, cosmetics, and medicines. *(Courtesy of USDA/ARS #K-4692-2)*

from *Lesquerella fendleri*, a wild plant native to Arizona, New Mexico, Texas, and Oklahoma. The University of Arizona and two commercial firms joined with the U.S. Department of Agriculture to develop the crop. Machinery used for other grain and oil seed crops is suitable for lesquerella.

Cosmetic manufacturers and other companies are looking for the oils extracted from lesquerella seed. The oils can be used in resins, waxes, nylons, plastics, high-performance lubricants, corrosion inhibitors, coatings, and cosmetics such as hand soap and lipstick. The oil meal left after the oils are extracted is suitable as a high-protein livestock feed.

FIGURE 21-24 Parasitic wasps lay eggs in the insect larvae of specific insect pests, which reduces the damage from and terminates the life cycle of those pests. *(Courtesy of USDA/ARS #K-5035-20)*

FIGURE 21-25 Environmentally friendly, ultra-low-volume herbicide application methods can significantly reduce the amount of agricultural chemicals used. *(Courtesy of USDA/ARS #K-5287-4)*

Cultural control refers to adapting farming practices to control pests. Some cultural controls include timing farming operations to eliminate pests, rotating crops, planting resistant varieties, and planting trap crops that are more attractive to insects than is the primary crop.

Biological control of plant pests involves the use of predators or diseases as the control mechanisms. The release of sterile male insects and the use of baits and repellents are also examples of this type of pest control. When using insects or diseases to control crop pests, it is important that the control be specific to the intended pest (Figure 21-24).

The development of varieties of crops that are resistant to pests is called **genetic control.** This may involve making the crop less attractive to the pest because of taste, shape, or blooming time. Developing more rapidly-growing crops that crowd out weeds is also an example of genetic pest control. Crops with resistance to diseases also fall into this category.

Chemical control of plant pests involves the use of pesticides to control pests of field crops. Excellent management practices must be exercised when using chemicals to control pests. Care should be taken to correctly identify the pest to be controlled and the chemical to be used. Dosage, runoff, and pesticide residues need to be carefully monitored (Figure 21-25).

Harvesting and Storing Field Crops

Harvesting field crops at the proper stage of maturity is a key to maximizing profits. The harvest of the crop is the culmination of a growing season of work and anticipation of the rewards of a job well done.

The development of mechanical harvesting equipment allowed field-crop producers to harvest thousands of bushels of grain daily and with less labor than previously required. This allowed for tremendous increases in the amount of food available for people and animals and a greatly improved standard of living.

FIGURE 21-26 Late corn harvest using a combine near College Park, Maryland. *(Courtesy of Elmer Cooper)*

FIGURE 21-27 Weather- and rodent-proof grain storage bins near Savannah, Georgia, are being used to find optimum storage conditions to minimize insect and mold damage to stored grain. *(Courtesy of USDA/ARS #K-4289-1)*

The primary harvesting machine for field crops is the combine (Figure 21-26). It performs the tasks of cutting the crop, threshing it, separating it from debris, and cleaning it. **Threshing** refers to the separation of grain from the rest of the plant materials. There are many types of combines adapted to the harvest of specific crops.

The proper storage of crops after harvesting is also important. The threats to the quality of stored crops include heat, moisture, fungi, insects, and rodents (Figure 21-27).

Drying grain to reduce moisture and heat is important for successful long-term storage. Much grain is harvested with a moisture content that is much too high. If stored without drying, the grain may heat up and encourage the growth of fungi, causing the grain to spoil. Foreign materials, such as weed seeds, may also cause stored crops to spoil.

Stored crops must also be protected from insects and rodents if quality is to be maintained. Rodent droppings, hair, and urine, as well as insect parts, may render crops unfit for human consumption. Reduction of food value and spoilage are other hazards of stored crops when insects and rodents are not controlled.

The production of field crops generates more income for American agriculturists than does any other production enterprise. With more than 20 percent of the land in the United States currently being used for growing crops, and with the excess production contributing to a more favorable balance of trade, crop production is likely to remain very important in the future.

STUDENT ACTIVITIES

1. Write the Terms to Know and their meanings in your notebook.
2. Compile a list of the field crops grown in your area.
3. Write a report on a field crop of interest to you.
4. Select a field crop and determine as many uses for it as possible.
5. Visit a local crop farm and talk to the operator about the advantages and disadvantages of growing a particular crop.
6. Prepare an advertisement to promote a field crop.
7. Construct a bulletin board about field crops and products made from them.
8. Visit a farm-machinery dealer. Make a list of all the equipment sold there that is used in the production and harvesting of field crops.
9. Make a collection of as many different field crops as you can.
10. Do a germination test on the seeds of several types of field crops. Observe the number of days required for germination to take place and the percentage of the seeds that germinate. Also conduct germination tests under a variety of environmental conditions, such as warm and cool, wet and dry, or with and without light. Compare the results to determine optimum conditions for the germination of crop seeds.

SELF EVALUATION

A. Multiple Choice

1. Flax is an example of a/an _____ crop.
   - a.   oilseed
   - b.   grain
   - c.   sugar
   - d.   fiber

2. The most important grain crop in the world is
   - a.   corn.
   - b.   wheat.
   - c.   rice.
   - d.   barley.

3. The moldboard plow is the primary tillage machine for the _____ tillage system.
   - a.   no-till
   - b.   minimum
   - c.   conventional
   - d.   none of the above

4. About 65 percent of the refined sugar produced in the United States comes from
   - a.   sugar cane.
   - b.   sugar beets.
   - c.   corn.
   - d.   sorghum.

5. The use of airplanes is an example of _____ seeding.
   - a.   row crop
   - b.   drill
   - c.   aerial
   - d.   broadcast

6. Grasses grown for their edible seeds are _____ crops.
   - a.   grain
   - b.   legume
   - c.   oil
   - d.   sugar

7. Ginning is the process used to
   a.   prepare a seedbed for planting.
   b.   prepare crops for storage.
   c.   remove seeds from cotton.
   d.   prepare grain for use as alcohol.

8. An example of genetic control of pests is
   a.   planting a crop when insects are not present.
   b.   releasing sterile male insects.
   c.   releasing an insect that feeds only on a certain weed.
   d.   planting a variety of a crop that grows more rapidly than do weeds.

9. Tobacco falls into the category of crops called
   a.   fiber.   c.   thermo.
   b.   biennial.   d.   stimulant.

10. Soybeans are grown for
    a.   oil.   c.   rain.
    b.   hay.   d.   all of the above.

B. Matching

_____ **1.** Most important U.S. crop   a.   Malting
_____ **2.** Only crop to grow in standing water b.   Linen
_____ **3.** Irrigation   c.   Combine
_____ **4.** Preparing barley for alcohol production d.   Sugar cane
_____ **5.** Primary harvesting machine   e.   Corn
_____ **6.** Most important world grain crop   f.   Sprinkler
_____ **7.** Oil crop   g.   Peanut
_____ **8.** Cloth made from flax   h.   Rice
_____ **9.** Warm-season crop   i.   Threshing
_____ **10.** Separation of grain from plant   j.   Wheat

C. Completion

1. Heat, moisture, fungi, rodents, and insects are all problems to be dealt with in the _____ of crops.

2. _____ crops protect the soil from erosion.

3. The primary use for wheat is for _____ _____ .

4. Cotton is grown for fiber and _____.

5. Mowing is a means of _____ control for weeds.

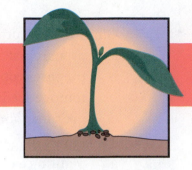

Forage and Pasture Management

COMPETENCIES TO BE DEVELOPED

After studying this unit, you should be able to:

- define important terms used in forage and pasture production and management.
- identify major crops grown for forage and pasture.
- select varieties for forage and pasture.
- prepare proper seedbeds for forage and pasture crops.
- plant forage crops and renovate pastures.
- control pests in forage crops and pastures.
- harvest and store forage crops.

MATERIALS LIST

✓ several specimens of forage and pasture crops

✓ quart jars for Student Activity

✓ bulletin board materials

✓ samples of lactic and butytric acids

✓ agriscience magazines

Forages

Hay

Silage

Pasture

Rhizomes

Noxious weed

Nurse crop

Overseeding

Carrying capacity

Haylage

Silo

Forage and pasture crops are the most important category of crops grown in the United States, if acres in production is the criterion used in rating. There are more than 475 million acres of pasture and range land. Another 61 million acres are used to produce hay. The importance of forages is further emphasized when you consider that half of the pasture and range land in the United States is not suited for the production of cultivated field crops. **Forages** are crop plants that are produced for their vegetative growth.

Forage production can be divided into three general categories: hay, silage, and pasture. **Hay** is forage that has been cut and dried so it contains a low level of moisture. **Silage** is green, chopped forage that has been allowed to ferment in the absence of air. **Pasture** is forage that is harvested by livestock itself. Forages are generally planted or maintained with one of these uses specifically in mind.

FORAGE AND PASTURE CROPS

Legumes

Alfalfa   The most important forage crop in the United States is alfalfa (Figure 22-1). It is often called the "queen of the forages." Alfalfa is a

FIGURE 22-1 Cutting and conditioning alfalfa with a hay harvester.

FIGURE 22-2 Alfalfa and other plants develop and then lose microscopic roots as they grow. *(Courtesy of USDA/ARS #K-93-07)*

legume that adds nitrogen to the soil. It is high in protein and other nutrients and very productive in fertile soils (Figure 22-2). It is also one of the oldest cultivated forage crops, and was mentioned in the Bible and in other early writings.

When properly harvested and stored, alfalfa is the most economical source of nutrients for ruminant animals. With production of alfalfa and alfalfa mixtures accounting for nearly 40 percent of the hay produced in the United States, alfalfa production ranks third to corn and soybeans in dollar value of product produced.

The North Central states account for about two-thirds of the yearly alfalfa production in the United States. However, alfalfa can be grown in nearly every state. Wisconsin ranks number one in alfalfa production, followed by South Dakota, Iowa, Minnesota, Nebraska, North Dakota, and Montana.

True Clovers True clovers include about three hundred species. However, only about twenty-five have agricultural importance. Clovers of economic importance in the United States include red clover, white clover, crimson clover, ladino clover, and alsike clover (Figure 22-3).

Clover is the second largest group of the hays grown in this country. It is a legume with the ability to add nitrogen to the soil. Clover has the disadvantage of lower yields than alfalfa under similar growing conditions. It is popular as hay, pasture, and silage and grows well in combination with many forage grasses. The Northeast and North Central states produce most of the clover grown in the United States.

Sweet Clover Sweet clover is used most often in areas that are hot and/or drought stricken, where its ability to survive and produce a crop is unsurpassed. It is used as hay, pasture, and sometimes as a green manure crop. Sweet clover is also used in Texas in rotation with cotton to help control a cotton-root disease. It is also an excellent source of nectar, which honeybees make into honey.

There are three species of sweet clover grown in the United States—biennial yellow, biennial white, and annual. Biennial white sweet clover yields more than the other species, although biennial yellow sweet clover is usually of higher quality. Sweet clover will grow in all areas of the United States. Areas of large production include the Dakotas, Minnesota, Wisconsin, Michigan, and the central states to Texas.

FIGURE 22-3 Clover is an important hay and pasture crop in the United States. *(Courtesy of Sharon Rounds)*

FIGURE 22-4 Bird's-foot trefoil is a legume containing tannin, which reduces the threat of life-threatening bloat to grazing cattle. *(Courtesy of USDA/ARS #K-2610-10)*

Bird's-foot Trefoil   This is a comparatively new crop in the United States. It originated in Europe, where it has been a forage crop for 300 years. There are approximately 2 million acres in production in the United States. Bird's-foot trefoil is used as pasture in most cases. Some is also grown for hay (Figure 22-4).

The food value of bird's-foot trefoil is about equal to that of alfalfa. Because of smaller yields, bird's-foot trefoil is unlikely to seriously challenge alfalfa in importance. It does have the advantage over true clovers in being much longer lived. States that report large acreages of bird's-foot trefoil include California, Ohio, Iowa, New York, and Pennsylvania.

Lespedeza   This legume is grown primarily in the South, where 1 to 2 million acres are harvested as pasture and hay each year. It can grow and thrive in soils that are low in fertility. Because lespedeza has a lower nutritive value than true clovers and alfalfa, it is recommended for feed for beef cattle but not for dairy cattle.

Types of lespedeza include annual and perennial. Most of the lespedeza grown in the United States is of the annual type (Figure 22-5).

Grasses

Bromegrass   Bromegrass is an important forage grass throughout the northern half of the United States. It is extremely hardy and grows to a height of 2 to 3 ft (Figure 22-6). Because bromegrass produces many rhizomes, thin stands rapidly thicken with age. **Rhizomes** are horizontal

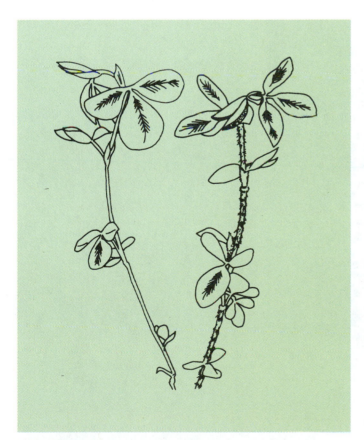

FIGURE 22-5 Annual lespedeza may be grown as forage, while Korean lespedeza is used as a soil-stabilizing plant.

FIGURE 22-6 Bromegrass is grown throughout the northern half of the United States.

FIGURE 22-7 Orchard grass is known for its rapid early spring growth.

underground stems from which new plants arise. Fertile soil is required for the production of bromegrass.

Orchard Grass   This grass is known for its rapid germination and early spring growth (Figure 22-7). It also recovers quickly after being harvested. Timing of the harvest of orchard grass is very important because quality drops rapidly after the plant reaches maturity. Orchard grass makes excellent pasture and is often used for hay in combination with legumes.

Timothy   Timothy is a cool-season grass that grows best when temperatures are between 65° and 72°F (Figure 22-8). The use of timothy as a forage grass had declined in recent years because bromegrass and orchard grass outyield it where all three species are adapted. Most timothy is grown in the northeastern part of the United States.

Reed Canarygrass   In areas that are wet or poorly drained, reed canary grass is often the only answer to producing a forage crop (Figure 22-9). It can produce more than four tons of forage per acre in the cool, damp areas where it thrives. This grass grows as tall as 7 ft and can produce high-quality feed if harvested before reaching maturity. Most reed canary grass is grown in Washington, Oregon, California, Iowa, and Minnesota.

Kentucky Bluegrass   Kentucky bluegrass is grown over much of the United States, even though it is best adapted to the Northeast. It is the major grass of many pastures, lawns, and even golf courses in areas

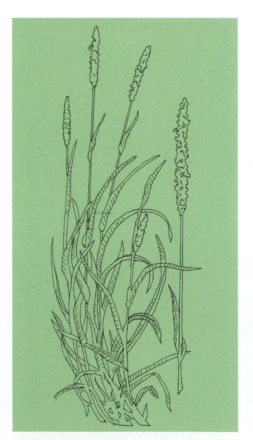

FIGURE 22-8 Timothy grows best when temperatures are between 65° and 72°F.

FIGURE 22-9 Reed canarygrass grows in areas too wet for the production of other grasses.

where summers remain fairly cool. Kentucky bluegrass is a good native, permanent pasture grass, but is outyielded by other grasses. It is not practical to use for hay because of low production. It also goes dormant during hot weather, making it necessary to use bluegrass in combination with other grasses to have forages available all summer (Figure 22-10).

Fescue   Tall and meadow fescues are perennial grasses that are used for pasture and hay, usually in combination with other grasses and legumes. Fescue is best adapted to the Southeast, where 10 to 20 million acres are grown each year. The quality and palatability of the fescues are lower than that of most other forage grasses.

Bermuda Grass   This is a warm-season grass that is dormant during cool weather. It is adapted to pastures and lawns because it grows only 6 to 12 in. tall. Common Bermuda grass is considered to be a weed in some areas, because it tends to crowd out other grasses. However, improved varieties of Bermuda grass make good pastures in the Southeast (Figure 22-11).

Dallis Grass   Dallis grass grows 2 to 4 ft tall in the warm areas of the South. It cannot stand continuous use as pasture because it needs to be able to recover from close grazing. However, it is productive earlier in the spring than other warm-season grasses.

Johnson Grass   This grass is so aggressive that it has been declared a noxious weed in many states of the country. A **noxious weed** is a plant that is prohibited by state law. It is very coarse and grows to a height of 6 ft or more. It spreads quickly by rhizomes and often crowds out most other plant species.

Selection of Forages

There are many considerations to be made when selecting forages for hay, silage, and pasture. Some of these are:

1. the intended use of the forage.
2. the expected or desired yield.
3. the nutrient value of the crop.
4. the climatic conditions under which the forage will be grown—warm season versus cool season, summer and winter temperatures, humidity, soil type, anticipated rainfall, nutrient level, length of growing season.
5. the pest-control measures required.
6. the methods of establishment required.
7. the compatibility with other forages when grown in mixtures.
8. the expected and desired life of the crop.
9. the care and maintenance required.
10. the equipment and labor necessary for growing, harvesting, and storing the crop.

Forage crops that are adapted to the production of hay include alfalfa, clover, bromegrass, orchard grass, timothy, and fescue. Forages used for pastures are clover, lespedeza, Kentucky bluegrass, and Bermuda grass. Almost any legume or grass crop can be used for silage. In addition, corn and most small-grain crops make excellent silage when harvested at an immature stage.

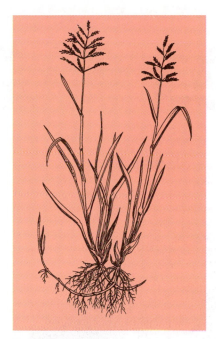

FIGURE 22-10 Kentucky bluegrass is a grass found in the pastures and lawns of the Northeast.

FIGURE 22-11 Improved varieties of Bermuda grass make good pastures in the Southeast.

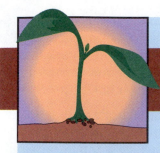

AGRI·PROFILE

CAREER AREAS: Agronomist/Plant Physiologist/ Forage Manager/Range Manager

Forage and pasture provide a variety of career options, ranging from farm and ranch to plant breeding and physiology. Grains, such as corn, cut for silage provide enormous volumes of feed for dairy and beef cattle. Crops cut for silage, as well as those cut for hay, are numbered among the forages. Similarly, pasture and range plants are forages. Those specializing in the science of forage growth and utilization are called *agronomists*.

Forage specialists are hired by universities and agricultural research centers as well as by large farms and businesses. In certain parts of the United States, hay businesses are on the increase, as more people do part-time farming, especially those with pleasure horses. Additionally, race tracks need hay and straw, which creates a strong market for these forages in many areas.

Hay business managers, truckers, dealers, and the like also conduct thriving businesses hauling straw and manure from horse barns at race tracks to mushroom farms, where the materials are used as the media for growing mushrooms. Recent droughts in sections of the United States created markets for cross-country shipping of hay, a practice once regarded as too expensive to be worthwhile.

Summer employee Aimee Crago stains an alfalfa cotyledon for microscopic examination. [Courtesy of USDA/ARS #K-4264-8]

Seedbed Preparation

In general, preparing a seedbed for forages is much the same as it is for the production of field crops. Residues from previous crops must be dealt with. The soil must be prepared for the planting of seed or the vegetative pieces used in the propagation of some grasses. Necessary seedbed pest-control measures must be practiced. The soil must be amended so that its pH and fertility are suitable for the germination and growth of the intended forage crop. And with some crops, the moisture level of the seedbed may need to be regulated to ensure germination.

One difference between preparing a seedbed for most forages and one for field crops is that the texture of the soil in the seedbed must be finer and the seedbed somewhat firmer than is necessary for most field crops. This is because most forages have very small seeds that have difficulty making firm contact with the soil in seedbeds that are not finely textured.

FIGURE 22-12 The no-till grassland drill places seed, fertilizer, and pesticides in one pass across the field. *(Courtesy of Elmer Cooper)*

Conventional tillage has traditionally been used when preparing seedbeds for starting grass and many legume crops for forage. The residues from previous crops are plowed down or shredded. The soil is then pulverized and leveled by disking and/or harrowing. For very small seeds, further preparation may be necessary. Final preparation of the seedbed should occur immediately before planting the seed. New seeding equipment has permitted grass seeding into established sod. This greatly reduces soil losses and provides protective vegetation while the seedlings become established (Figure 22-12).

Planting Forages and Renovating Pastures

Forage crops are usually planted by drilling or by broadcasting. The same grain drills that are used for planting small grains may be used to plant forages. Usually there is a separate seedbox on the drill. The forage seed is dropped on the seedbed at the same time that the grain is planted. Many grasses are planted in this way. The grain crop germinates faster than the forage and acts as a nurse crop until the forage crop can become established. A **nurse crop** is used to protect another crop until it can get established.

Many forage crops are also planted alone, using either drills or broadcast planters. With many forages, the seed must be covered after it is broadcast onto the soil by broadcast planters. This is done to protect the seed from pests and the drying effects of sun and wind. Covering the seed may be accomplished by a light disking or by the use of a cultipacker, which has corrugated wheels to press the seed into the seedbed (Figure 22-13). Some types of forage seeds are broadcast into growing crops. They must be able to germinate fairly easily, because contact with the seedbed is often minimal. Red clover is often overseeded into stands of forage grasses to make mixed hay. **Overseeding** is the practice of seeding a second crop into one that is already growing. This is usually done during late winter and early spring, when freezing and thawing of the soil helps provide contact between seedbed and seed.

Some forages are being planted with no-till planters in live or killed crop residues. The no-till planter opens a narrow furrow in which the

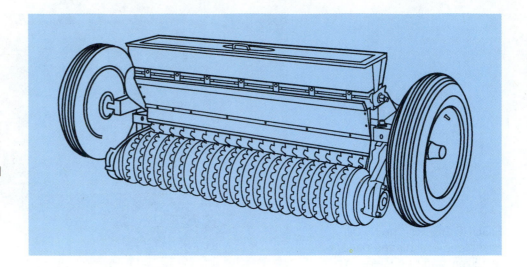

FIGURE 22-13 Fine grass and legume seeds must be slightly covered and left in a compact seedbed. The cultipacker presses the seeds into the seedbed.

seed is planted. Advantages include fewer trips across the field with heavy tractors and tillage equipment. This means that there is less compaction of the soil. There is also little exposure of the seedbed to erosion. Care must be taken to control pests in no-till planting. Another concern is to keep competition for water and nutrients from other plants growing in the seedbed to a minimum.

There are several general methods of renovating pastures in the United States. The existing pasture can be killed with a herbicide, the soil disked heavily to prepare a fine seedbed, and the pasture then reseeded with the desired types of grasses and/or legumes. The pasture is usually treated for insects and diseases at this time if necessary. It is also an ideal time to fertilize and lime the pasture.

Another method of renovating a pasture involves using selective herbicides to kill unwanted species of plants. The pasture is then disked to break up the existing sod. This allows easier entry of moisture and nutrients into the soil. The pasture may or may not be overseeded with desirable species to improve yield. It is also fertilized and limed at this time, according to soil test results.

No-till grassland seeders may be used to place new seed in existing sod, which permits the thickening of existing seedings. These seeders can also introduce new, improved, and aggressive species or varieties of forage crops.

Pest Control in Forage Crops

The control of weeds, insects, diseases, and rodents in forage crops helps result in optimum yields. Because many forage crops grow for more than one growing season, pest control is usually an ongoing part of forage crop management (Figure 22-14).

The proper identification of the pest or pests affecting the crop is very important. To that end, personal experience and the use of trained professionals are often necessary.

The actual methods of pest control are many, and they were discussed in previous units. Chemicals, cultural practices, biological control, genetic control, and the timing of crops are all tools to be used in controlling the pests of forage crops. Not to be overlooked in pest-control management is the control of pests in stored forages. When chemicals are used to control pests, care must be taken to ensure that the pesticide is properly applied in the recommended dosages. Timing

FIGURE 22-14 One of more than 600 species of grasshoppers found in the United States. *(Courtesy of USDA/ARS #K-1066-2)*

the application of the pesticide for maximum effect is also essential. Care must also be taken to ensure that overspray and run-off of pesticide materials do not adversely affect other than the intended pest. Concern that the residue of pesticides does not end up in food sources is of utmost importance.

Harvesting and Storage of Forage Crops

The proper harvesting and storage of forage crops is extremely important in the management of the forage enterprise for maximum profits. Most of the forages grown on American farms and ranches are fed to livestock grown on the same farm. Forages are usually the least expensive sources of nutrients available for cattle and sheep.

Pastures   The harvesting of pastures involves several factors. One of the most important is that the carrying capacity of the pasture must be determined in order to figure out how many animals can be fed by the pasture available. The **carrying capacity** of a pasture refers to the number of animals for which it will provide feed.

Pastures also need to have time to recover from the ravages of the animals that crop them. The rotation of pastures, or actually the rotation of the animals using the pastures, is important if the pastures are to yield their potentials. Pasture rotation also helps to control parasites of the animals grazing on them by breaking the pests' life cycles.

Hay   Harvesting hay involves several operations that must be timed fairly accurately if a quality product is to be harvested. The hay must be cut at the optimum time, with an eye on the weather forecast for the next several days. Hay is normally dried in the field by the sun. Nothing ruins hay faster than rain on the crop after it starts to dry and cure. The maturity of the forage being harvested for hay is also critical (Figure 22-15). The more mature the forage becomes, the lower the quality and food value of the hay produced.

There are several types of machines that are designed especially for cutting forages intended for hay. These include the sickle-bar-type mower, rotary mower, flail-type mower, and variations of these types.

The sickle-bar-type mower cuts with the same action as a pair of scissors. These types of mowers are best adapted for forages that are standing up straight.

The rotary mower cuts with blades that move in a circular motion parallel to the ground. This type of mower will cut any type of forage, although higher amounts of horsepower are required when the stands of forage are extremely thick and heavy.

The flail-type mower is most commonly used when harvesting forages for silage, although it will cut crops intended for hay. This type of mower features a cutter head that rotates in a circular motion vertically to the ground. It has the disadvantage of chopping up the hay more than is usually desirable.

Often included in the cutting operation is crushing or mashing the stems of the forages to hasten their drying. Windrowers are also a part of some mowing operations. They leave the cut forages in rows to make it convenient for gathering by the harvesting machines.

In areas of high humidity, chemicals to speed the drying of the forage are sometimes sprayed on the forage as it is cut.

Type of Hay	When to Harvest
Alfalfa	Pre-bud to 1/10 bloom stage
Clover	1/4 to 1/2 bloom stage
Birdsfoot trefoil	1/4 bloom stage
Sweetclover	Start of the bloom stage
Bromegrass	Medium head stage
Timothy	Boot stage to early bloom stage
Lespedeza	Early bloom stage
Orchardgrass	Full head but before blooming
Reed canary-grass	When the first head appears

FIGURE 22-15 Maturity levels of harvesting selected forages for hay.

FIGURE 22-16 Hay is typically cut, conditioned, air dried, and then baled. *(Courtesy of USDA)*

FIGURE 22-17 Many machines are available to help mechanize hay making. *(Courtesy of USDA)*

The hay is usually raked at least once before it is harvested, to allow it to dry more evenly and more rapidly. Legume-type hay should be raked during the part of the day when humidity is highest, to keep the loss of leaves to a minimum. Raking hay also puts it into windrows so that it can be gathered more easily.

When the hay has dried to the desired level of moisture, it is harvested by one of three basic methods. It may be baled, harvested loose with stackers, or cubed.

Hay may be harvested into square or rectangular bales or into round bales of various sizes (Figure 22-16). The baler gathers the hay from the ground, compresses it into a bale, and ties twine or wire around the bale to hold it together and allow for handling. The bale is then expelled from the baler. It may be allowed to drop on the ground for later removal to storage or, in the case of square bales, ejected directly into a wagon.

There are a number of machines that are designed to move baled hay from the field to the storage site. These include wagons, bale loaders, bale handlers, and bale stackers (Figure 22-17).

Hay may also be harvested with machines called stack wagons. These machines gather forages from the field and form dense stacks or loaves of weather-resistant hay. Less labor is required for the production of hay stacks than for baled hay. With special handling equipment, stacks can be moved with few problems.

Hay cubes were developed to allow for the full mechanization of the feeding of hay. The cuber gathers the dry forage and compresses it into cubes about 1¼ in. × 1¼ in. × 2 to 3 in. Normally, only legumes are cubed because of difficulty in getting grass cubes to bind together. Artificial binders are sometimes used.

Hay may be stored in buildings or in the open field. Regardless of the method of storage, care must be taken to ensure that quality is maintained. Hay that is stored in buildings must be kept free from pests, particularly rodents. Care must also be taken to ensure that the moisture level of stored hay is low enough that spoilage does not occur.

When hay is stored outdoors, it should be placed on land that is well drained and sloped to drain water away from the hay. The stacks or bales of hay stored outside must be dense enough so that little moisture can penetrate them and cause spoilage. Hay stored outside is often covered with plastic or other materials to protect it from adverse weather conditions.

Silage   Harvesting forages for silage is far more simple than making hay. Less equipment and labor is required, and the entire harvesting operation is usually completed in one pass over the field. A forage harvester cuts the green forage, chops it into small pieces, and deposits it

BIO•TECH CONNECTION

Fiber Giant and Super Forage?

The object of extensive research for several decades, the bamboo-like kenaf plant originated in Africa, but promises to be an important cash crop in the United States. By the early 1990s, the Kenaf Paper Company of Texas was planning a kenaf-based newsprint mill in south Texas to process kenaf grown by area farmers. The $35-million plant was designed to process kenaf into fibers for a variety of purposes. These could include fibers for making newsprint; bast for carpet backing or mats for seeding yards and slopes; and core fibers for cleaning up oil spills, with oil and fibers useable as fuel. Another promising product is furniture made from a lightweight fiber board. However, that's not all kenaf can do!

Growing 12 to 15 ft tall, kenaf promises to be the super crop of the future for fiber and possibly forage. *(Courtesy of USDA/ARS #K-2975-9)*

The same tropical plant that may provide backing for your dining room rug and newsprint for your morning paper could also help grow your dinner steak, scientists say. Kenaf, or Hibiscus cannabinus, is considered an important contender against Southeast Asian jute for fiber and rope. It is also being researched for its usefulness as a roughage and protein source for sheep and cattle. Research at the Agricultural Research Service Forage and Livestock unit at El Reno, Oklahoma indicates that animals will eat the leaves and stems, and the digestibility of certain parts of the plant is about the same as that of alfalfa. Further, kenaf leaves contain 30 to 35 percent crude protein. This combination of high fiber and high protein has attracted interest in the plants as cubes or pellets for cattle. The leaves are of growing interest as a small animal feed.

Kenaf grows very rapidly and could produce two crops in a season. When cut, two or more shoots will spring from each stalk to produce new plants. The potential is to use the stalks for fiber and the tops for forage, or simple to grow two forage crops in one season. Being an annual, the crop is easy to establish and has few or no insect and disease pests. Production in the United States had reached over 40,000 acres by 1994.

FIGURE 22-18 Forage harvesters cut the green crop, chop it up, and deposit it into a wagon or truck. *(Courtesy of Michael Dzaman)*

into a wagon or truck to be hauled to the storage facility. Cutter heads on the forage harvester may be of either the cutter type or flail type (Figure 22-18).

Flail-type forage harvesters are used to harvest low-moisture silage called **haylage.** Forages used for haylage are cut and allowed to dry to a moisture content of 40 to 55 percent before they are placed in a silo. A **silo** is an airtight storage facility for silage or haylage.

Silage may be stored in silos or in trenches or bunkers. It is also sometimes stored in piles on top of the ground and sealed with plastic or other materials to make it airtight. The production of silage is dependent on storage in the absence of air or oxygen. The fermentation that occurs in the absence of air preserves the silage. Spoilage or rot occurs when green forages are stored in the presence of air (Figure 22-19).

The Fermentation Process Silage-making is a fermentation process wherein bacteria break down sugars and starches to produce acids, which preserve the forage. Silage-making and silage-keeping require that several conditions be met and maintained. First, the crop must be chopped into short pieces and have sufficient moisture to pack tightly enough to exclude most of the air. Secondly, the crop must contain a sufficient amount of sugars and starches for the fermentation process to occur. Thirdly, the silo or silage container must be airtight, so air cannot re-enter the silage after it has fermented.

What happens in the silo to change the taste, texture, and chemical content of the forage? If the forage being ensiled contains grain, such as corn, sorghum, wheat, barley, oats, rye, or other seeds, it will ferment when packed in an airtight container or silo with about 60 percent or more moisture. If the mixture does not have sufficient sugars and starches, or if the moisture content is not within the acceptable range, then grains or preservatives must be added to the forage for good silage. If the moisture content is less than 60 percent and the silo sufficiently tight, a low-moisture silage known as haylage may be made with the addition of certain preservatives.

Immediately after suitable forage has been chopped and packed in an airtight space, bacteria that live in the presence of oxygen in the air that fills the pores of the forage start to multiply. They feed on the sug-

FIGURE 22-19 Silos are airtight structures used to preserve and store silage. *(Courtesy of USDA)*

ars and starches in the forage and consume the oxygen in the process. They feed and multiply until the oxygen is consumed, so long as there are sufficient sugars and starches in the forage to keep them feeding. The bacteria make lactic and acetic acids, which smell good and give the forage a pleasant taste for livestock.

The frantic burst of bacterial activity in freshly packed forage causes the temperature of the forage to rise within hours after the process starts. This continues until the oxygen is consumed and the bacteria die. The resulting level of acid in the forage, now called silage, prevents any other bacterial action so long as oxygen-laden air does not re-enter the silage. Without bacterial action, the silage cools down and is preserved until air enters the silage again. In practice, air may get into silage and cause spoilage around poorly fitted silo doors, holes left in the plastic film used to seal silos, or movement through poorly packed or excessively dry silage. Spoiled silage can be detected by the presence of mold or the foul odor of butyric acid, which forms when undesirable fermentation takes place.

Approved Practices for Forage Crops

Many crops are useable as forages. Forages may be utilized in the form of grazing of young plants by livestock; cut and dried as hay; or harvested and stored in silos as medium- or high-moisture haylage or silage. Some approved practices for the production of forages are as follows.

1. Select a perennial species for long-term stands.
2. Select a perennial or annual species for short-term stands.
3. Buy certified seed of the selected variety.
4. Obtain a current soil test.
5. Lime and fertilize according to soil test and desired production.
6. Select a field with soil suitable for your forage choice.
7. Select the proper tillage method to prepare a firm and smooth seedbed.
8. Calibrate drill for proper plant population.
9. Match plant population to soil-yield potential.
10. Adjust drill for proper depth of seed placement.
11. Decide if forage will best be grown with a companion crop.
12. Calibrate sprayer for proper application of materials.
13. Use selected herbicides for controlling problem weeds.
14. Use insecticides if necessary.
15. Decide the stage of maturity to harvest the crop.
16. Make a yield check.
17. Keep harvesting equipment in good repair to save down-time.
18. Check moisture content when harvested.
19. Harvest crop as soon as moisture permits, to reduce losses.
20. Store hay in a clean, dry structure.
21. Store silage in a structure best adapted to your situation.
22. Obtain a forage analysis.
23. Track market trends and habits.
24. Market crop at the optimum time.

25. Keep accurate enterprise records.

26. Summarize and analyze records.

The production of forages in the United States accounts for about half of all the land in agricultural use. Much of this land is unsuitable for production of other crops. With the production of hay ranking third only to corn and soybeans, forages are extremely important to American agriculture. Because forages are the least expensive sources of nutrients for cattle, sheep, and horses, they are likely to remain essential for years to come.

STUDENT ACTIVITIES

1. Write the Terms to Know and their meanings in your notebook.

2. Put fresh-cut forage in each of two quart jars. Pack the forage tightly into the jars. Seal one jar and leave the other open to the air. Compare the contents of the two jars after ten days.

3. Make a collection of as many different forage seeds as you can.

4. Write a report on a forage of economic importance in your area.

5. Make a collection of local forages.

6. Have a forage producer speak to the class about production of forage crops.

7. Visit a machinery dealership to learn about the types of seedbed preparation, planting, and harvesting equipment for forages.

8. Make a bulletin board about forages.

9. Obtain samples of hay and silage and compare them for firmness of leaves and stems, odor, and color.

10. Ask your instructor to obtain samples of lactic acid (found in good silage) and butyric acid (found in spoiled silage) from the science laboratory, and compare the odors of these acids.

SELF EVALUATION

A. Multiple Choice

1. The type of mower that cuts with a scissors-type action is a _____ type mower.
   a. rotary
   b. flail
   c. sickle-bar
   d. vibrating

2. A silo is used to store
   a. silage.
   b. hay.
   c. pasture.
   d. cubes.

3. The most important legume hay in the United States is
   a. clover.
   b. lespedeza.
   c. bird's-foot trefoil.
   d. alfalfa.

4. A forage that grows in wet or poorly drained soil is
   a. reed canarygrass.
   b. timothy.
   c. fescue.
   d. bird's-foot trefoil.

5. _____ is suitable for only pasture and lawns because it grows only 6 to 12 in. high.
   a.   Dallis grass
   b.   Sweet clover
   c.   Kentucky bluegrass
   d.   Orchard grass

6. Forage that is cut and allowed to dry to 40 to 55 percent moisture before storing is called
   a.   silage.
   b.   haylage.
   c.   hay.
   d.   cubes.

7. The harvesting of _____ requires the most labor of any of the forages.
   a.   pasture
   b.   hay
   c.   silage
   d.   haylage

8. The production of forages in the United States equals _____ percent of the land used for crops.
   a.   25
   b.   35
   c.   50
   d.   65

9. A warm-season grass is
   a.   sweet clover.
   b.   bromegrass.
   c.   reed canarygrass.
   d.   Bermuda grass.

10. Hay is raked to
   a.   allow even drying.
   b.   allow the hay to ferment.
   c.   reduce its maturity.
   d.   remove leaves and weeds.

B. Matching

_____ **1.** Timothy
_____ **2.** Alfalfa
_____ **3.** Bromegrass
_____ **4.** Sweet clover
_____ **5.** Orchard grass
_____ **6.** Lactic acid
_____ **7.** Butyric acid

a.   Number one grass hay crop
b.   Excellent source of nectar
c.   Grows very early in the spring
d.   Number one forage crop
e.   Popular in mixes with clover
f.   Indicates silage is spoiled
g.   Gives a good smell to silage

C. Completion

1. A disadvantage of _____ type mowers is that they tend to chop up the forage.

2. Sweet clover types include annual and _____.

3. Seedbeds for forages should be _____ than those for most other crops.

4. _____ is forage harvested by livestock.

5. Horizontal underground stems are called _____.

6. A _____ weed is prohibited or banned by state law.

7. Crops that are harvested for their vegetative growth are called _____.

8. Two methods of planting forages are _____ and broadcasting.

9. A _____ gathers dry forage, compresses it, and ties it with twine and wire.

10. The more mature the forage, the _____ its quality.

11. _____ converts sugars and starches to acids in the fermentation process.

12. Silage will spoil if _____ is permitted to re-enter.

Ornamental Use of Plants

Ornamental Plants and the American Dream

(Courtesy of Michael Dzaman)

When one considers the fastest growing segment of American agriculture, the image of great fields of golden grain, corn "as high as an elephant's eye," or feedlots of cattle as far as the eye can see come to mind. Many would not envision greenhouses of flowers, golf courses with gorgeous turf, or suburban communities of beautifully landscaped homes. But flowers, potted plants, and landscape trees and shrubs are some of the hottest agricultural products today. They not only beautify our surroundings, but they also help purify the air.

A conservative estimate puts the worth of the floral and landscape industries in the United States at about $8 billion at the grower level, according to the USDA's Economic Research Service. During the 1980s, floral crops and nursery production grew at about 10 percent per year and continued growing even during the recession of the early 1990s. In 1992, flowers and house and nursery plants advanced into the sixth largest commodity group in the United States. In twenty-one states, this segment placed among the top five agricultural commodities.

People tend to take our decorative plants for granted and assume they are simply gifts from nature. This is far from the way it is, however. Years of painstaking research and development have gone into most ornamental plants, and research must continue on an ongoing basis to stay ahead of pests of ornamental plants and the public appetite for new decorations. For the most part, we seldom see plants from the wild—nearly all that are in common use are products of plant improvement efforts.

Consider the poinsettia, for example. Research and genetic improvements contributed greatly to this plant's phenomenal growth in commercial value during the last two decades. The poinsettia industry grew from a wholesale value of $38 million in 1976 to $170 million by 1991. At that time, the poinsettia became America's number-one potted plant, based on numbers sold, even though its annual sales period is only six weeks.

Most of the USDA's work on the poinsettia and many other ornamentals is centered at the National Arboretum in Washington, D.C., and the Agricultural Research Service's Florist and Nursery Crops Laboratory in nearby Beltsville, Maryland. Scientists there have developed many new ornamentals and are constantly improving others for commercial use. An ongoing project investigates the biochemical and physical bases for color in plants and our perception of it. Such information may help make development of new kinds of flowers easier and may provide a scientific basis for variety identification.

Applied research answers immediate needs of growers, such as solving pressing problems occurring in today's commercial ornamental enterprises. However, the long-range research goals are to stay well ahead of the current technology. From all indications, the ornamental use of plants will play an increasing role in the health and well-being of our citizens as we all pursue the great American Dream.

Indoor Plants

COMPETENCIES TO BE DEVELOPED

After studying this unit, you should be able to:

- identify plants that grow well indoors.
- select plants for various indoor uses.
- grow indoor foliage plants.
- grow indoor flowering plants.
- describe elements of design for indoor plantscapes.
- describe career opportunities in indoor plantscaping.

MATERIALS LIST

✓ indoor plant reference books

✓ seed catalogs

✓ pencil, grid paper, and eraser

✓ indoor plants

✓ pots of various sizes

Floriculture
Succulent
Foliage
Variegated
Herb
Foot-candle
Relative humidity
Dormant

Rootbound
Plantscaping
Basic color
Accent color
Texture
Accent
Sequence
Balance

Formal
Symmetrical
Informal
Asymmetrical
Scale
Circulation
Habitat
Development limitations

The world of indoor plants is a fascinating one. There are plants that have magnificent blooms, unusual shapes, fancy foliage, and fragrant smells. If selected and cared for properly, they can last for years and may even be passed from generation to generation. The industry of indoor plants is the floriculture part of the ornamental horticulture field. **Floriculture** involves the production and distribution of cut flowers, potted plants, greenery, and flowering herbaceous plants. Indoor plants present a challenge to the homeowner who wishes to invest the rooms of the home with living color. They also challenge those who decorate public areas (Figure 23-1).

FIGURE 23-1 Indoor plants add beauty and atmosphere to homes and public areas. *(Courtesy of Michael Dzaman)*

FIGURE 23-2 African violet *(Saintpaulia ionantha)*. *(Courtesy of Michael Dzaman)*

FIGURE 23-3 Fuchsia *(Fuchsia triphylla)*. *(Courtesy of Michael Dzaman)*

PLANTS THAT GROW INDOORS

Almost all plants can be grown indoors. There are plants, however, the favor indoor conditions. Many of the trees and shrubs do better when grown outdoors. Smaller and more succulent plants are best for indoor use. **Succulent** means having thick, fleshy leaves or stems that store moisture. There are a wide variety of shapes and sizes of indoor plants. These plants can be divided into two major groups—those that flower and those that are grown only for their foliage. **Foliage** means stems and leaves.

Popular and Common Indoor Flowering Plants

African Violets   One of the most popular and common indoor flowering plants is the African violet *(Saintpaulia ionantha)*. This plant can be recognized by its small size and hairy leaves. The leaves are oval-shaped, dark green, and covered with soft, short hairs. The flowers contain four to five petals arranged in a clover pattern. They vary in color from deep purple to brilliant white (Figure 23-2).

Fuchsias   Fuchsias *(Fuchsia triphylla)* are plants with very colorful flowers that cascade from the plant. Most have flowers that are two-tone pinks and reds. The foliage is dark green and the leaves tend to be long and oval in shape, with a bronzy hint of color (Figure 23-3).

Gardenias   Particularly fragrant, flowering indoor plants, gardenias *(Gardenia jasminoides)* have deep-green shiny foliage and pure white flowers. The leaves are in clusters of three and are pointed (Figure 23-4).

Geraniums   One of the most versatile flowering indoor plants is the geranium *(Pelargonium zonale)*. These indoor plants are one of the oldest. The leaves are a rounded, yellowish green with scalloped edges. The flower is borne on a stem and consists of many petals in a cluster shaped like a ball. The flower color ranges from the most popular red, to white and pink (Figure 23-5).

Impatiens   If an indoor plant with many blooms is desired, the impatiens *(Impatiens wallerana sultanii)* is a good choice. The flowers are small and rounded, with five petals. One petal is shaped like a tube

FIGURE 23-4 Gardenia *(Gardenia jasminoides)*. *(Courtesy of H. Edward Reiley)*

FIGURE 23-5 Geranium *(Pelargonium zonale)*. *(Courtesy of H. Edward Reiley)*

FIGURE 23-6 Impatiens *(Impatiens wallerana)*. *(Courtesy of H. Edward Reiley)*

that protrudes from the underside of the flower. Flower colors include white, pink, salmon, coral, lavendar, purple, and red. The leaves are lance-shaped and have succulent stems (Figure 23-6).

Others   There are many other varieties of plants that can be grown indoors for their flowers. Some flowering plants are both indoor and outdoor, such as wax begonias, ageratums, verbenas, and petunias.

Popular and Common Indoor Foliage Plants

Indoor foliage plants can be divided into five groups for easy identification. The groups are ferns; indoor trees; vines; cacti and succulents; and others. A foliage plant is grown for the appearance of the leaves and stems (Figure 23-7).

Ferns   Ferns come in a variety of types. Some of the most popular indoor-type ferns are Boston fern *(Nephrolepis exaltata)* (Figure 23-8), asparagus fern *(Asparagus sprengeri)* (Figure 23-9), maidenhair fern *(Adiantum capillusveneris)*, sword fern *(Nephrolepis cordifolia)*,

FIGURE 23-7 Businesses and public areas are generally landscaped with foliage plants. *(Courtesy of Michael Dzaman)*

FIGURE 23-8 Boston fern *(Nephrolepis exaltata)*. *(Courtesy of H. Edward Reiley)*

FIGURE 23-9 Asparagus fern *(Asparagus sprengeri).* *(Courtesy of H. Edward Reiley)*

rabbit's-foot fern (*Davallia canariensis*), and staghorn fern (*Platycerium biforcatum*). Ferns are categorized by their long and often multicut leaves. Most ferns are feathery in appearance. Some ferns are used extensively as greens in floral arrangements.

Indoor Trees An indoor tree can be an excellent accent to a room or an attractive addition to a patio or hallway. Trees can grow to be 6 to 7 ft tall. Some of the more popular indoor trees are Norfolk Island pine (*Araucaria excelsa*) (Figure 23-10), fiddleleaf fig (*Ficus lyrata*), umbrella plant (*Schefflera actinophylla*) (Figure 23-11), rubber plant (*Ficus elastica*) (Figure 23-12), fragrant dracaena (*Dracaena fragrans*), weeping fig (*Ficus benjamina*) (Figure 23-13), and croton (*Codiaeum variega-*

FIGURE 23-10 Norfolk Island pine *(Araucaria excelsa). (Courtesy of H. Edward Reiley)*

FIGURE 23-11 Umbrella plant *(Schefflora actinophylla). (Courtesy of Michael Dzaman)*

FIGURE 23-12 Rubber plant *(Ficus elastica). (Courtesy of Michael Dzaman)*

FIGURE 23-13 Weeping fig *(Ficus benjamina)*. *(Courtesy of H. Edward Reiley)*

FIGURE 23-14 Philodendron *(Philodendron species)*. *(Courtesy of H. Edward Reiley)*

FIGURE 23-15 Wandering Jew *(Tradescantia flumenencis)*. *(Courtesy of Michael Dzaman)*

tum). Indoor trees vary in the type and size of their foliage, but all have woody-type stems.

Vines The vines are characterized by their habit of climbing or draping from the sides of the pot. Some of the more widely recognized vine-type indoor plants are philodendrons *(Philodendron species)* (Figure 23-14), Wandering Jew *(Tradescantia fluminensis "Variegata")* (Figure 23-15), grape ivy *(Rhoicissus rhomboidea)*, and English ivies *(Hedera species)*. These plants can be trained to climb up a piece of wood, around a planter box, or up a wall.

Cacti and Succulents This unique group of indoor plants originated in desert areas. They are among the easiest plants to grow indoors. This group of plants tend to hold water within their stems and leaves. Cacti and succulents survive dry heat, low humidity, and varying temperatures. Cacti usually have some type of prickly needles, whereas succulents do not. One of the most popular succulents is the jade plant *(Crassula arborescens)* (Figure 23-16).

FIGURE 23-16 Jade plant *(Crassula arborescens)*. *(Courtesy of H. Edward Reiley)*

FIGURE 23-17 Peperomia *(Peperomia caperata)*. *(Courtesy of Michael Dzaman)*

FIGURE 23-18 Passion plant *(Gynura aurantica)*. *(Courtesy of H. Edward Reiley)*

FIGURE 23-19 Snake plant *(Sansevieria trifasciata)*. *(Courtesy of H. Edward Reiley)*

Others, or Specimens   There are numerous other types of indoor plants that do not fit into the categories of ferns, indoor trees, vines, and cacti and succulents. These indoor plants tend to have special characteristics or features that people like to display—thus they are called specimen plants. A few examples of more popular specimens are the spider plant *(Chlorophytum elatum vittatum)*, which shoots off baby plants or "spiders"; peperomia *(Peperomia caperata)*, which has pale pink to red stems with deeply grooved heart-shaped leaves (Figure 23-17); purple passion plant *(Gynura aurantica)* (Figure 23-18), which has rich royal-purple leaves that are covered with velvet-type hair; bromeliads, one variety of which has a deep pink color at the center; and snake plant *(Sansevieria trifasciata)*, which has long, spikelike, thick leaves that are variegated with gold (Figure 23-19). **Variegated** means having streaks, marks, or patches of color.

SELECTING PLANTS FOR INDOOR USE

There are two rules of thumb when selecting plants for use indoors. The first is to be selective when you purchase the plant. The second is to choose the right plant for the growing conditions available in the location where you want the plant to live. Careful attention to these two rules will greatly increase the chance for success with indoor plants.

Purchasing an Indoor Plant

When buying an indoor plant, look for one that appears to be healthy. Look closely at the plant for insects, being careful to check the undersides of the leaves where many insects hide or lay their eggs. Select plants with even green color, no yellowing leaves, and no spots or blotches on the leaves. Avoid plants that have spindly growth or appear wilted. If possible, purchase the plant during its growing season. Finally, look for new growth, such as leaf or flower buds.

AGRI·PROFILE

CAREER AREAS: Plantscape Designer/Plantscape Contractor/ Greenhouse and Plant Technician/Florist

Foliage plants including small trees have become popular decor for interior public areas such as shopping malls, office buildings, institutions, and housing complexes. Flowering and foliage plants have long been part of the interiors of attractive homes and have provided excellent hobbies for many. However, their extensive use in commercial settings is a relatively recent development that has stimulated many new job opportunities in ornamental horticulture.

Youth in suburban settings may obtain jobs raising and caring for plants, installing interior plantscapes, rotating plants between growing areas and display areas, and contracting to maintain interior plantscapes. Early experiences frequently lead to life-long careers.

With more experience and education, one may become a plantscape designer, business owner-operator, grower, wholesaler, plant doctor, or extension specialist. Ornamental horticulture specialists are hired by universities, research institutes, and business firms to develop improved plants and horticultural practices. The florist industry provides cut flowers and floral arrangements for all occasions and provides jobs in design, arranging, care, and delivery.

The florist industry provides cut flowers and floral arrangements for all occasions. Interior plants are grown in greenhouses, nurseries, and fields and become part of floral and interior arrangements and plantscapes. *[Courtesy of Michael Dzaman]*

Choosing the Right Plant

Before selecting an indoor plant, decide where the plant will be placed. Note particularly the light intensity and duration, as well as the temperature of the area where the plant will be placed. Each species of indoor plants has specific conditions for optimum growth. The plant should match the location for best results.

Light The amount of light available for a plant is a factor that is difficult to control. A shadow test can help determine this. To conduct this test, hold a piece of paper up to the light (window or lamp) and note the shadow it makes. A sharp shadow means you have bright or good light. However, if there is barely a shadow visible, the light is dim or poor. It is important to know how much light the plant needs. If a plant needs direct or full sun, then exposure of the sun is needed for at least half of the daylight hours. When a plant needs indirect or partial

FIGURE 23-20 Chinese evergreen *(Aglaonema modestum)*. *(Courtesy of H. Edward Reiley)*

FIGURE 23-21 Dracaena *(Dracaena fragrans). (Courtesy of H. Edward Reiley)*

sun, the light should be filtered through a curtain or slats. Plants that fit into this category are Chinese evergreen *(Aglaonema modestum)* (Figure 23-20) and dracaena *(Dracaena fragrans)* (Figure 23-21). Even for plants preferring no direct sunlight, the room should be bright and well lighted. Plants that need shade should be kept in a well-shaded part of the room. In summary, too much or too little light can greatly affect the health of the plant.

Temperature　There are three temperature categories in which indoor plants are grouped. These are cool, moderate, and warm. The cool temperature range is 50° to 60°F, with temperatures not falling below 45°F. The moderate temperature range is from 60° to 70°F, with a minimum of 50°F. The warm temperature range is from 70° to 80°F, with a bottom limit of 60°F.

USES OF INDOOR PLANTS

The uses of indoor plants are varied. Plants can be used as room dividers or to brighten up a dull spot in the kitchen or bathroom. Plants can be used to divide a room into specific living areas. They may be placed in containers with special watering devices, or several plants may be grouped together to form a natural barrier. Containers are available in a number of sizes and shapes, some with castors on the bottom so they can be moved easily. Indoor trees and climbing plants are best used as living-area dividers (Figure 23-22).

Specimen or showy plants, such as the Boston fern or date palm *(Phoenix dactylifera)*, are good for brightening up dull areas (Figure 23-23). Plants may be placed on ornate plant stands or in decorative pots. A series of wall shelves may also be used to display interior plants. Plants on shelves should be compact, and perhaps trailing, for a more dramatic effect.

Hanging baskets filled with indoor plants are useful when space is at a premium. Because baskets can be easily suspended from the ceiling, they can utilize the unused space above head level. Care should be used when selecting plant containers so that water does not drip on

FIGURE 23-22 Decorative plants make excellent area dividers. *(Courtesy of Michael Dzaman)*

FIGURE 23-23 Date palm *(Phoenix dactylifera). (Courtesy of H. Edward Reiley)*

FIGURE 23-24 Hanging baskets are versatile and attractive. *(Courtesy of Michael Dzaman)*

FIGURE 23-25 The high humidity of bathrooms can provide an excellent environment for plants if light is adequate. *(Courtesy of Elmer Cooper)*

the furniture or the floor. Select baskets that are designed with drip trays attached to the bases.

Some interesting areas for hanging baskets are stairwells, offices, and foyers. One other element that is unique to hanging baskets is the hook securing the basket to the ceiling. It should be secured firmly enough to hold the basket when the soil is wet and when the plant is being cared for (Figure 23-24).

Bathrooms are excellent places for indoor plants. These rooms are humid and warm, which provides a good atmosphere for plants. They can be placed on windowsills or around the tub. Shelves can also be installed in a bathroom to display indoor plants. Plants will add a touch of color to a bathroom, especially today as the bathroom is expanding to hold a hot tub or jacuzzi. Philodendrons are excellent in bathrooms (Figure 23-25).

The kitchen is another room of the home where plants can add a beautiful touch. Refrigerators, other appliances, and kitchen windowsills all make good places for plants that prefer cooler temperatures. Most of the plants in the kitchen should be relatively small, as space is usually at a premium. Several plants suitable for the kitchen are aloes *(Aloe variegata)* (Figure 23-26), maidenhair ferns, peperomias, and even a variety of herbs. An **herb** is a plant kept for aroma, medicine, or seasoning.

GROWING INDOOR PLANTS

Growing indoor plants is much like growing outdoor plants. Consideration needs to be given to the environment in both cases. Tomatoes will not grow and produce in the desert, nor will sweet corn grow in the shade of a tree. Both need full sun, but cannot survive in extreme tem-

FIGURE 23-26 Aloe *(Aloe variegata). (Courtesy of H. Edward Reiley)*

peratures. Aspects of the environment that must be considered are light, temperature, water, drainage, and feeding conditions. While indoor plants are not harvested like tomatoes or sweet corn, they need to be similarly maintained. Some aspects of maintenance that need to be addressed are grooming, repotting, and propagation.

Light

Light is one of the most crucial factors to consider when growing indoor plants. Light is measured in foot-candles. A **foot-candle** is the amount of light found one foot from a burning standard candle, known as a candela. Two aspects of light that need to be determined are intensity and duration. The intensity of light varies at different times of the day and through different windows. A plant near a window with a southern exposure receives more intense sun, and for a longer period of time, than does a plant in a window with a northern exposure. Pulling shades across a window can reduce the intensity of light, and a tree outside a window can reduce both the intensity and duration of light that enters the window. The season of the year also affects the light's intensity and duration. Summer sun is much more intense than winter sun, because the sun is closer to the Earth during the summer.

Too little light will cause some indoor plants to grow tall and leggy and lose leaves because the long, thin, weak stems can no longer support the plants due to limited photosynthesis. On the other hand, too much light causes indoor plants to wilt and lose their vibrant, green colors. The youngest leaves on plants are affected first by unfavorable conditions. Another item concerning light is the tendency of plants to grow toward the strongest source of light. If you notice that a plant is leaning toward the light, you may need to rotate the plant periodically so that it maintains a balanced shape (Figure 23-27).

There are five basic light categories for indoor plants. These are full sun, some direct sun, bright indirect light, partial shade, and shade. Full sun is a location that receives at least five hours of direct sun a day. This amount of sun can be found in areas that have a southern exposure. Some direct sun occurs in areas that are brightly lit but receive less than five hours of direct sun a day. These areas are windows facing east and west.

Bright indirect light describes areas that receive a considerable amount of light but no direct sun. An example is an area 5 ft away from a window that receives full sunlight. Partial shade refers to areas that receive indirect light of various intensities and durations. Areas 5 to 8 ft away from windows that receive direct sun are partial-shade areas. Shade refers to poorly lit sections away from windows that receive direct sun. Very few plants can survive this low-intensity light.

Temperature

Plants survive best at constant temperatures. Temperatures that go up and down are not ideal for plant growth. Temperature interacts with light, humidity, and air circulation. It is best to maintain temperatures in a range from 60° to 68°F for optimum indoor plant growth. As temperatures increase beyond this range, the air gets hotter, and the available moisture in the air decreases. Although the thermostat in a house reads one temperature, each room usually varies as much as five de-

FIGURE 23-27 Plants generally must be moved or rotated periodically to provide correct lighting to all parts of the plant. (Courtesy of Michael Dzaman)

FIGURE 23-28 Plants need to receive water on a regular basis, but overwatering is a frequent problem of the inexperienced. *(Courtesy of Michael Dzaman)*

grees in one direction or the other. Because plants have various temperature preferences, it is advisable to place plants in rooms that match their specific temperature requirements.

Water

Water is an essential ingredient for growth of any living organism. The amount of water needed by indoor plants is usually not as much as one would think. The most likely problem with unhealthy indoor plants is too much water. More plants die from overwatering than from any other cause. As with other environmental factors such as light and temperature, each plant varies in its requirements for water. Unless you know the particular needs of the plant, it is best to water when the soil around the plant is a little on the dry side. Most plants do best if allowed to dry out between waterings. When indoor plants are watered properly, the roots remain more active than if the soil becomes waterlogged or excessively wet.

There are numerous ways to water plants. The basic rule about watering is to use water that is not hot, nor cold. Water should be tepid, or at moderate temperature. One way to water is to soak the pot in a bucket of warm water for half an hour, remove the pot, and drain it. A second way is to pour the water on top of the soil slowly, filling the pot to the top with water. Allow it to absorb the water until the excess drains from the hole in the bottom of the pot. Do not rewater until the soil first becomes dry to the touch (Figure 23-28).

Besides water in the media around their roots, plants also need moisture in the air. The moisture content in the air is referred to as humidity. It is expressed as **relative humidity** (a relationship between the amount of water vapor in the air compared to the maximum moisture the air will hold at a given temperature). Almost all indoor plants prefer 50 percent relative humidity. The simplest method for humidifying the air around plants is to set pots in trays filled with gravel and add water to just cover the gravel. Misting is a good shortcut way to add humidity, but it should be done several times a day to be effective. The humidity around plants can also be increased by grouping them together. However, plants should have enough space around them to allow for adequate air circulation.

Drainage

Good drainage is achieved by using pots with porous materials in the bottom and drainage holes in the pot. Good drainage is essential for indoor plants. Adding coarse material such as sand or perlite to the soil will improve soil drainage. The addition of gravel or bits of broken clay-pot material to the bottom of a pot before adding the soil is a substantial aid to good drainage.

Fertilizing

Plants need nutrients on a regular basis for good health. A balanced fertilizer, such as a 5–5–5, should be applied at regular intervals. Fertilizers with a higher proportion of nitrogen then of phosphorus and potash are often used to keep foliage plants green and healthy looking. Plants should be fertilized at two- to six-week intervals, depending on the type of fertilizing material. Slow-release-type fertilizers dissolve

FIGURE 23-29 Plants need to be fertilized with carefully selected materials. *(Courtesy of Michael Dzaman).*

slowly and release nutrients evenly over weeks or months. On the other hand, liquid fertilizers suitable for foliar application are used by the plant within a few days. They may also be applied at weekly or bi-weekly intervals (Figure 23-29).

Plants should be fertilized while actively growing. Fertilization should be discontinued when the plant is **dormant,** or in a resting stage. Flowering plants need more fertilizer. Addition of fertilizer once every two weeks is recommended from the time flower buds first appear until the plants stop blooming. Avoid fertilizing plants when the soil is excessively dry. Under such conditions, the fertilizer solution is likely to be too highly concentrated and may cause burned leaves or roots.

Grooming

Grooming plants is important even with the best combination of light, temperature, humidity, water, and drainage. Grooming should be done weekly. This task includes removing wilted or withering leaves, flowers, and stems using sharp scissors or shears (Figure 23-30). The plant should be observed to determine if it is getting leggy or thin. If so, pinch out new growth to force the plant to branch out. To avoid crooked stems, stake plants when they are young. Climbing or trailing plants need a stake made of bark to enable them to climb and cling to the stake.

Once a week, the plant should be dusted. Dust accumulates on the leaves and blocks the stoma so that the plant cannot breathe or transpire as well as it should. The soil around the base of the plant should be loosened with a fork or small spade to allow air to enter the soil and water to percolate through. To help control insects, mist infested plants with a diluted solution of mild dishwashing soap and water.

Repotting

FIGURE 23-30 Plants need to be examined and groomed weekly. *(Courtesy of Michael Dzaman)*

In time, plants develop root systems that are restricted by the pot or container. When this happens the plant is said to be **rootbound.** There is no place for the roots to continue growth, so repotting is necessary for plant health. In general, repotting is done in the spring or the fall. A

FIGURE 23-31 Repotting at appropriate times prevents the plant from becoming root-bound and promotes good plant health. *(Courtesy of Michael Dzaman)*

good rule of thumb for determining pot size is to use one with a diameter at the top of the rim equal to one-third to one-half the height of the plant. This rule does not apply to plants whose growth habit is tall and slender as opposed to a balanced top growth. When repotting, it is not desirable to move an established plant to a new pot that is more than 2 in. wider than the original pot. Excessively large pots result in wasted soil, water, and nutrients (Figure 23-31).

Flowering plants are best repotted after the flowers have faded. During repotting, check the roots for insects and root damage. Remove any roots that look or feel unhealthy.

Propagation

Propagating most indoor plants is relatively easy. The method chosen depends upon the type of plant (whether it is herbaceous or woody, flowering or foliar). Both sexual and asexual methods of propagation are utilized.

In sexual propagation seeds may be started in containers with a good potting soil, plenty of moisture, and adequate air circulation. Keeping pots in warmer areas of the house increases the speed of seed germination. Covering the pots with glass or plastic held up by stakes will also aid in the germination process.

Some popular methods of asexual propagation of indoor plants are leaf and stem cuttings, removal of plantlets from parent plants, and air layering. Each of these procedures results in the multiplication of the plant.

Flowering Plants

Indoor flowering plants may take some extra, special care to ensure blossoming. These plants are more sensitive to the availability of light, so some artificial lighting may be needed. They are also more sensitive to temperature changes. Flowering plants have particular seasons in which they flower. It is important to know when to expect the plants to bloom. Some examples of plants with specific bloom times are the Christmas cactus (*Schlumbergera x buckleyi*), which is expected to bloom between October and late January, and the Easter cactus (*Rhipsalidopsis gaertneri*), which blooms in April or May. These two plants are very similar in their leaf types and flowers, but they bloom in opposite seasons of the year. Indoor plants may flower in the winter, spring, summer, or year-round.

Foliage Plants

There is such a wide variety of foliage plants grown in homes and offices that specific instructions for their care compose entire books. However, the correct management of light is probably the greatest single factor for growing foliage plants. Light is the source of energy for the process of photosynthesis, wherein the leaves produce sugars and starches to feed all parts of the plant. Different foliage plants have different light requirements. Therefore, it is recommended that a good reference book on indoor plants be used to determine the particular requirements for any given plant. Regardless of the amount and duration of light, it is desirable to rotate plants so that each side receives the same amount of light over time.

INTERIOR LANDSCAPING OR PLANTSCAPING

Plantscaping is the design and arrangement of plants and structures in indoor areas. This design and arrangement is an art. Interior plantscaping is an activity that is fun. However, it should be approached seriously,

BIO•TECH CONNECTION

Like the Season's First Snowfall

With adults not much larger than the head of a pin, whiteflies suck the life from more than 600 plants including fruits, vegetables, flowers, shrubs, and field crops. Control of and damage by the pest run an annual tab in the millions of dollars. Greenhouse growers, nurseries, and retail outlets wage a continuous battle against whiteflies in an effort to save crops from damage and infestations that would prevent the sale of these products. Similarly, homeowners and interior plantscapers must be on a constant vigil, lest unobserved eggs hatch into an infestation of an unknown environment. Like the

A serious threat to ornamentals and nonornamentals alike, whiteflies settle on plants, like the season's first snow. *(Courtesy of USDA/ARS #K-4853-3)*

just as the arrangement of furniture or other interior decorations must be done carefully. The indoor plants should be used to complement people-oriented spaces. The more creative you are in designing, the more distinctive indoor plantscapes will be. There is no right or wrong way to design with indoor plants, but there are elements that enhance the interior plantscaping process.

Design

Before beginning a plantscape design, the designer must know the purpose or intent of the plants. Are the plants to be used as a space divider? Are they being used to accent existing furnishings? Will they fill empty space? Answers to these questions will result in an organized design, rather than a happenstance.

Another question to consider is the function or functions of the plants. Are they to create a specific shape, emphasize a specific area, or support a specific architectural feature of the room? If the function of the plantscape is not considered, the end result is not likely to be successful.

When selecting plants for a plantscape, consider the physical characteristics of color, form, and texture. These are determined in concert with the perceptual characteristics of accent, sequence, balance, and scale. These characteristics are the basic tools a designer uses to create the interior plantscape (Figure 23-32).

Color   This is the most important physical characteristic. It can influence emotions, create specific feelings, and add beauty and harmony to the environment. There are two types of color that must be considered by the designer in creating an interior plantscape. First is the

season's first snowfall, the eggs can suddenly appear as white specs all over the plant. When the plant is moved or the insects disturbed, the masses of adults can take flight and create the appearance of a snowstorm in the air!

First observed in Florida in 1986, an especially viral strain of the whitefly spread rapidly from Florida to California. It is challenging scientists and growers alike, as it spreads nearly unchecked by either pesticides or natural enemies. Around the country, researchers are showing the textile industry how to wash off the sticky goo the insects leave on cotton fiber. They are also uncovering ways to deal with plant viruses left in the wake of the troublesome, sucking insects. They are testing legions of environmentally friendly controls such as insect parasites, predators, fungi, plant extracts, and dish detergent and vegetable oil mixtures.

Scientists have identified over thirty predators and twenty-five parasites. So far, the big-eyed beetle is the most promising. It secretes a sticky substance from its mouth to stick the whitefly to the leaf, then leisurely eats the fly. Scientists have released some big-eyed beetles on a pilot basis. Another tiny but promising insect that has been identified devours an estimated ten thousand whitefly eggs or seven hundred nymphs in its six- to nine-week lifetime. However, not enough is yet known about this insect to release it. Similarly, a number of parasitic wasps are under examination. One of the more promising spray materials is a biosoap, which provides some control over whiteflies as well as other damaging insects.

FIGURE 23-32 Interior plantscaping is an important element in shopping malls and other public areas. *(Courtesy of Michael Dzaman)*

basic, or background, color. **Basic color** is the color of the walls, ceiling, and floor. These colors should influence the selection of the flowering characteristic and foliage of plant material. The second type of color is accent. **Accent color** is the color of the plants or other attention-getting objects.

Form Form refers to shape. There are different forms that plants possess naturally. The most common shapes are round; oval; weeping, or drooping; upright; spiky; and spreading, or horizontal. These shapes can be used individually or grouped to form an aritificially sculptured shape and form. Each plant shape or cluster shape adds its own particular feature to the design.

Texture Plant **texture** refers to the visual or surface quality of the plant or plants. Texture is influenced by the arrangement and size of leaves, stems, and branches. It is described in terms of coarseness or fineness, roughness or smoothness, heaviness or lightness, and thickness or thinness. Coarse-textured plants, such as fiddleleaf fig or prickly pear cactus, should be used in large spaces, whereas a maidenhair fern should be placed in small spaces, such as on shelves.

Accent and Sequence **Accent** means a distinctive feature or quality. An accent captures the attention of the viewer. It has a dramatic effect on the visual appearance of the room or part of the room. You can create an accent with the use of color, form, or texture. It can also be created through the use of a sequence. **Sequence** refers to a related or continuous series. Sequence is created with plant material by repeating the same color, texture, or form. The overuse of accents, however, will detract from their function to capture the attention of the viewer.

Balance and Scale **Balance** is the state of equality and calm between items in a design. In a design, two types of balance are utilized—formal, or symmetrical, and informal, or asymmetrical. **Formal,** or **symmetrical,** balance occurs when the items are equal in number, size, or texture on both sides of the center of the design. **Informal,** or **asymmetrical,** balance occurs when the items are not equal in number, size, or texture on the two sides of the center. **Scale** refers to the size of items. A large patio with African violets as accents would be out of scale. Plant materials need to complement the size and, therefore, the scale of the room. A better selection for a patio would be Norfolk Island pines and dracaenas.

A Design Process

There are various types of design processes. One that is good to use has three phases. The three-phase process emphasizes the how and why of using plants for the particular space.

Phase One   The first phase of the process is preplanning, which includes three steps. The first step is to develop the design objectives by determining the purpose of the plantscape. The second step is to determine the space capacities, which include the habitat and the circulation of people. It is best to know how people will move in the area, referred to as **circulation,** and the amounts of light, temperature, and humidity in the room, referred to as **habitat.** The third step is to determine the development limitations. **Development limitations** include the amount of money available; room and space characteristics; and habitat limitations created by light, temperature, people, pets, and others.

Phase Two   The second phase is developing the plan. The plan will include the basic design and arrangement of plants and materials. The physical and perceptual characteristics of the plants are very important in this phase. Tentative selection of plants with visual placement in the room is part of this step. The planning is done on paper, and many alternatives are examined. The final plan is drawn up with all the plants identified.

Phase Three   Implementation is the third phase of the interior plantscaping process, where the design comes to life. There are three steps to this phase. The first step is the preparation of the documents. If there is any construction involved, such as platforms, decks, or planter boxes, drawings for these objects must be done. Drawings and specifications for the installation and maintenance of water lines and fixtures, electrical devices, and the plants should also be listed at this point. This step is very important if you are designing for someone other than yourself. The second step in this phase is the installation of all physical modifications. Selection of plant containers and the planting of the indoor plants happens in this phase. The plants are actually put in place. The final step is evaluation of the project. A thorough look at the plants, their containers, and their position in the room or area must be done. If all looks balanced, no changes need to be made. If not, minor adjustments may be made. The interior plantscape is now complete.

Plantscape Maintenance

Maintenance of the plantscape with attention to light, moisture, humidity, and grooming is important. These factors are discussed in preceding parts of this unit. A major segment of the plantscaping picture is the nursery and greenhouse industry that provides the plants (Figure 23-33).

FIGURE 23-33 Greenhouses, nurseries, florists, and suppliers are needed to provide a constant supply of plant materials. *(Courtesy of Michael Dzaman)*

CAREERS IN INDOOR PLANTSCAPING

Interior plantscaping is a career field within the large industry of horticulture. Horticulture is the study of plants, especially garden crops, both indoor and outdoor, for human consumption, for aesthetic purposes, or for medicinal purposes. Interior plantscaping is a part of floriculture, which is a division of horticulture. Plantscaping is both a science and an art. It is a career area that allows for creativity and use of scientific knowledge and technology.

The opportunities for a career in indoor plantscaping are numerous. An individual may be involved in growing indoor plants, designing interior plantscapes, installing plantscapes, maintaining plantscapes, or selling or servicing plantscape materials. The plantscapes can be small-scale within a residence or home, or large-scale in public areas such as shopping malls or office buildings. This area of horticulture offers many individuals the chance to be entrepreneurs with relatively low investments. The education required for a career in indoor plantscaping varies. Persons with high school diplomas, two-year college degrees, or four-year college degrees can be successful in this area.

STUDENT ACTIVITIES

1. Write the Terms to Know and their meanings in your notebook.
2. Take an inventory of species, type, and number of indoor plants in your home.
3. Select indoor plants from reference books and seed or plant catalogs that are suitable for your home.
4. Develop a plan for an interior plantscape of a room in your home, the school office, or another location.
5. Repot an indoor plant.
6. Propagate indoor plants using one or more methods.
7. Perform a shadow test at several windows in your home or classroom to determine the intensity of the light available.
8. Prepare hanging baskets for use in the home or office.
9. Take a trip to a local nursery, florist shop, or grocery store and identify the indoor plants.

SELF EVALUATION

A. Multiple Choice

1. One of the most popular flowering indoor plants is the
   a. spider plant.
   b. petunia.
   c. African violet.
   d. Norfolk Island pine.

2. An example of an indoor tree is a
   a. philodendron.
   b. cactus.
   c. fiddleleaf fig.
   d. zebra plant.

3. When purchasing an indoor plant, check the underside of the leaves for
   a. powdery mildew.
   b. insects.
   c. price tags.
   d. brown spots.

4. The moderate temperature range for indoor plants is
   a. 60° to 70°F.
   b. 45° to 55°F.
   c. 90° to 100°F.
   d. 35° to 45°F.

5. If an indoor plant needs direct or full sun, then the plant will need sun for at least
   a. the entire day.
   b. one hour.
   c. one-fourth of the daylight hours.
   d. one-half of the daylight hours.

6. The place in a home not usually considered an ideal location for an indoor plant is the
   a. bathroom.
   b. basement.
   c. kitchen.
   d. living room.

7. Which of the following is not considered an environmental factor when growing indoor plants?
   a. weather
   b. light
   c. temperature
   d. humidity

8. When watering plants, the ideal temperature for the water is
   a. ice cold.
   b. hot.
   c. tepid.
   d. cold.

9. Which of the following is not a physical characteristic of an indoor plant?
   a. texture
   b. form
   c. color
   d. balance

10. Interior plantscaping is a career area in which area of horticulture?
    a. agronomy
    b. floriculture
    c. forestry
    d. arboriculture

B. Matching

_____ **1.** *Ficus lyrata*   a. A fern-type foliage plant
_____ **2.** Gardenia   b. Has unusually colorful leaves
_____ **3.** Croton   c. A succulent-type foliage plant
_____ **4.** Wandering Jew   d. A fragrant, flowering indoor plant
_____ **5.** jade plant   e. A tree-type indoor plant
_____ **6.** *Aspargus sprengeri* f. A vine-type foliage plant

C. Completion

1. The two environmental factors that are most important to the survival of indoor plants are _____ and _____.

2. An indoor plant that is a tree can be identified by its usually _____ stem.

3. The bathroom is a great place for indoor plants because of the _____ usually found in a bathroom.

4. Indoor plants survive best at a _____ temperature.

5. More indoor plants die from _____ than any other cause.

6. Adding a coarse material such as sand to the soil will improve an indoor plant's _____.

7. A 5–5–5 fertilizer is an example of a _____ fertilizer.

8. When repotting an indoor plant, do not move a plant to a pot that is more than _____ wider than its original pot.

UNIT: 24

Turfgrass Use and Maintenance

MATERIALS LIST

✓ writing materials

✓ samples of grass seed

✓ various types of turf samples

✓ brochures from lawn-care services

✓ lawn equipment catalogs

COMPETENCIES TO BE DEVELOPED

After studying this unit, you should be able to:

■ identify and describe careers available in the turfgrass industry.

■ identify turfgrass plant parts.

■ select turfgrass species for various purposes and locations.

■ state the basic cultural practices for turfgrass production and maintenance.

■ list the basic steps for turfgrass establishment.

FIGURE 24-1 Sports turf requires maintenance practices that provide an acceptable level of playability. *(Courtesy of Albany Country Club. Photo by Michael Dzaman)*

THE TURFGRASS INDUSTRY

It has been estimated that the value of the turfgrass industry in the United States is $25 billion. This industry is large and diverse. Career opportunities in this field expanded rapidly in the last decade and are projected to continue growing in the years ahead. **Turfgrasses** are grasses that are mowed frequently to maintain a short and even appearance.

Golf course superintendents and other athletic-field managers maintain turfgrass at a certain level of playability. **Playability level** means suitability for the intended use. It is determined by the type of sporting or recreational event. For example, **putting greens** are areas used for playing golf, and the turfgrass is very short. These areas are maintained to provide a surface for consistent, yet adequate putting speeds (Figure 24-1). A football field may be managed to offer secure footing and sufficient turfgrass resiliency for player safety.

Lawn-care services offer many job positions and are one of the largest employers within the industry. Job opportunities include lawn-care specialist, branch manager, owner-operator, and others (Figure 24-2).

The maintenance of turfgrasses at large government, apartment, university, commercial, and private complexes requires personnel trained in turfgrass management.

FIGURE 24-2 Lawn-care services offer many jobs and are one of the largest employers within the turfgrass industry. *(Courtesy of Elmer Cooper)*

FIGURE 24-3 Sod production is an important part of the turfgrass industry. *(Courtesy of USDA)*

FIGURE 24-4 Equipment demonstration by a sales representative. *(From Emmons, Turfgrass Science and Management, 1995, Delmar Publishers)*

Sod-production farms and landscaping businesses are involved in the establishment and installation of turfgrass for lawns. Turfgrass specialists are also needed for these segments of the industry (Figure 24-3).

The turfgrass industry supports a substantial sales force. Companies that produce or distribute seed, fertilizer, pesticides, and turfgrass equipment require an extensive support and sales staff (Figure 24-4).

Federal and state governments and private companies hire turf specialists and scientists with advanced degrees. Career opportunities in these areas offer challenging positions in research, teaching, and extension.

TURFGRASS GROWTH AND DEVELOPMENT

Turfgrasses are plants grouped into the *Poaceae* family. These grasses differ from other grass plants because they can withstand mowing at low heights. They can also tolerate vehicle and foot traffic. Turfgrasses are frequently used as soil-holding as well as ornamental plants. These traits have made turfgrasses the most widely used ornamental crop in the United States. To properly maintain turfgrasses, an understanding of their growth and development is required.

THE TURFGRASS PLANT

The grass plant can be divided into two broad areas known as the root and shoot systems. The root system consists of adventitious and seminal roots. The shoot system includes the stem and leaves of the plant.

Root System

The **seminal roots** develop from the seed during seed germination. They initially anchor the seed into the soil. The seminal root system will be active for six to eight weeks. The **adventitious roots** develop from the nodes of stem tissue. They usually compose the entire root system of a mature turfgrass stand.

Turfgrass roots are multibranching and fibrous. They are responsible for nutrient and water absorption. They also prevent soil erosion, by effectively stabilizing and anchoring the plant into the soil.

Seasonal change in root growth is dependent upon soil temperature and moisture. Active root growth for warm-season turfgrasses occurs in the summer. A **warm-season turfgrass** is one of a group of grasses adapted to the southern region of the United States. A **cool-season turfgrass** is a plant adapted to the northern region of the United States. These turfgrasses have active root growth in the fall and early spring. Optimum growth occurs at temperatures from 60° to 75°F.

Rooting depth is affected by plant species, soil factors, and cultural or maintenance practices. Average rooting depth for turfgrasses is 6 to 12 in. The warm-season grasses have deeper root systems than do the cool-season grasses. Well-drained, sandy-loam soils with neutral soil pH constitute an optimum growing medium for turfgrass roots.

Cultural practices that influence rooting depth and growth are mowing, fertilization, and irrigation. Frequent mowing and low mowing heights reduce rooting depth. In addition, fertilization programs

AGRI·PROFILE

CAREER AREAS: Golf Course Superintendent/Turfgrass Grower/Groundskeeper/Landscape Maintenance Technician

In the last three decades, turfgrass production and management have become big business in America. In one eastern state, turfgrass recently became the number one crop based on acres covered. Starting in the 1950s, growth and development of golf courses stimulated the turfgrass industry with high-paying salaries for golf course superintendents and other turfgrass specialists.

Today, career opportunities in turfgrass production, management, service, supervision, research, and consultation are extensive. Turfgrass technicians and specialists generally work in attractive and appealing surroundings. Many work outdoors in sunny weather and indoors when the weather is bad. In many localities, salaries have become quite attractive even for laborers.

Educational programs in turfgrass are available in high schools, technical schools, colleges, and universities. With the movement toward urbanization, interest in open spaces, concern for the environment, and increasing population, the outlook for careers in turfgrass production and management is excellent.

Turfgrass production, establishment, and maintenance provide many exciting jobs and create attractive surroundings for home, business, and recreation. (Courtesy of Michael Dzaman)

that emphasize only shoot growth impair root growth. Light and frequent irrigation causes shallow-root grasses. Heavy but infrequent irrigation is more conducive to deep root growth.

Turfgrass maintenance practices should attempt to optimize the rooting potential of turfgrass plants. An extensive root system allows a plant to recover from drought and other stress conditions more rapidly.

Shoot System

The shoot system consists of stems, leaves, and seed head, or inflorescence.

Stems   Turfgrass stems include the crown, tillers, rhizomes, stolons, and seed culms. The crown is the major growth, or meristematic, tissue of the grass plant. The **crown** is a stem with the nodes stacked on top of each other (Figure 24-5). All root, leaf, and other shoot growth originates from this area. The crown is located at the base of the grass plant in the soil surface area.

Rhizomes and stolons are horizontal stems. A **rhizome** is a creeping underground stem, whereas a **stolon** is an aboveground stem (Figure 24-6). They both originate from an axillary bud on the crown and will penetrate through the lower leaf sheath. This type of growth is referred to as **extravaginal growth.** Rhizome and stolon growth allows for vegetative spreading of turfgrasses. **Vegetative spreading** means reproduction by plant parts other than seed.

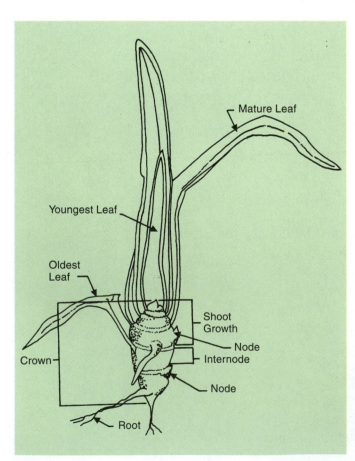

FIGURE 24-5 The crown of a turfgrass plant.

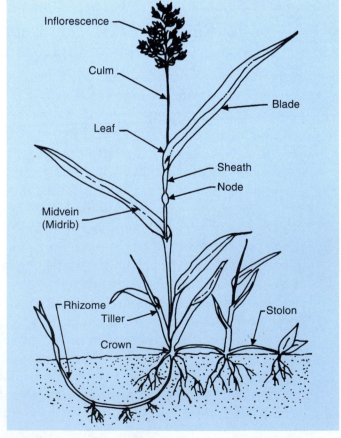

FIGURE 24-6 The major parts of a grass plant.

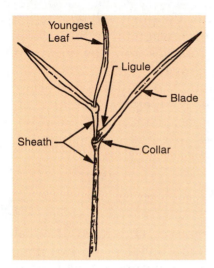

FIGURE 24-7 The leaf blade and leaf sheath of a grass plant.

Tiller **Tillers** are new shoots of a grass plant that develop at the axillary bud of the crown. They form within the lower leaf sheath of the plant. This type of growth is referred to as **intravaginal growth.** Increased tillering will enhance turfgrass density. All turfgrasses produce tillers. Optimum tillering for the cool-season grasses occurs in the spring and fall months. Warm-season grasses have optimum tillering during the summer. Under low-moisture and high-temperature stress conditions, tiller, rhizome, and stolon development is reduced.

Seed Culm and Inflorescence The **seed culm,** or seed stem, supports the inflorescence of the plant. The seed culm originates at the top of the crown. **Inflorescence** is the arrangement of and flowering parts of a grass plant. Cool-season grasses produce their inflorescence in the spring. Warm-season turfgrasses produce their inflorescence in the late summer.

Flower **induction** or initiation is caused by several environmental conditions. Temperature and photoperiod are the two induction processes for grasses. When seed heads form, they will cause a decrease in playability and appearance of the turfgrass stand. Mowing will remove the seed head and thus improve turfgrass quality.

Leaf The turfgrass leaf consists of the sheath and blade (Figure 24-7). The **sheath** is the lower portion of the leaf and may be rolled or folded over the shoot system. The **blade** is the upper portion of the leaf. At the junction of the blade and sheath is the collar and ligule. The **ligule** is located on the inside of the leaf and is a membranous or hairy structure. The **collar** can be found on the outside of the leaf and is a light-green or white-banded area (Figure 24-8). These two features are important vegetative traits for turfgrass identification.

Turfgrass growth is dependent upon the production and utilization of carbohydrates. The turfgrass leaf is responsible for photosynthesis and, ultimately, carbohydrate production. Reserve carbohydrates will be stored in crown, rhizome, and stolon tissue.

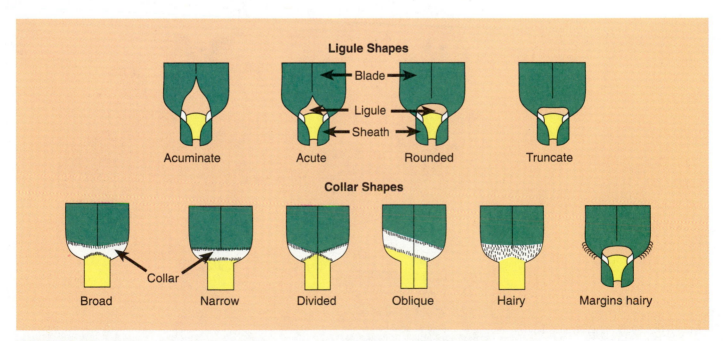

FIGURE 24-8 Ligules and collars of a grass.

During unfavorable growth conditions, the plant will go into dormancy. All leaf tissue will die back. However, when favorable environmental conditions occur, carbohydrate reserves from the crown and other stem tissue will be used for plant regrowth.

Turfgrass maintenance programs attempt to optimize root growth and carbohydrate accumulation. Greater recuperative potential and plant persistency occur when these two basic concepts of growth are understood and managed. **Recuperative potential** is the ability of a plant to recover after being damaged.

TURFGRASS SPECIES

There are some 7,500 plants classified as grasses. Only a few dozen are considered for turfgrass use. Turfgrasses are divided into two major groups based on climatic adaptation. The two groups are cool-season and warm-season grasses. The United States may be divided into a number of zones based on temperature and moisture conditions. It is helpful to be aware of these zones when selecting turfgrass (Figure 24-9).

Cool-Season Turfgrasses

These turfgrasses originated in Europe and Asia and have optimum growth at temperatures from 60° to 75°F. They predominate in the northern and central regions of the United States. Species adaptation within the cool-season group is determined by rainfall, soil fertility, and turf use. Major cool-season turfgrasses in the United States include Kentucky bluegrass, tall fescue, red fescue, perennial rye grass, creeping bentgrass, and crested wheatgrass (Figure 24-10).

Kentucky Bluegrass (*Poa pratensis* L) Kentucky bluegrass is used extensively in residential and commercial lawns, in recreational facilities, and along highway rights-of-way. It performs well with moderate

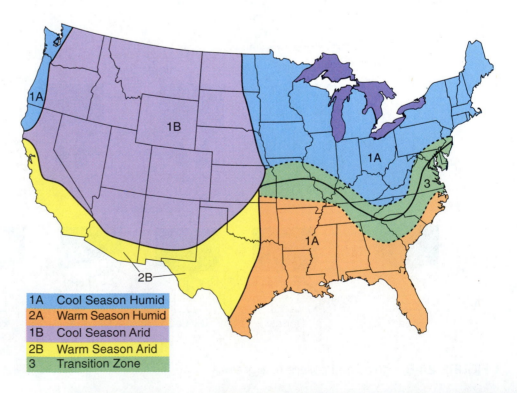

1A	Cool Season Humid
2A	Warm Season Humid
1B	Cool Season Arid
2B	Warm Season Arid
3	Transition Zone

FIGURE 24-9 The major turfgrass adaptation zones.

Cool-Season	Warm-Season
Colonial Bentgrass (*Agrostis tenuis*)	Bermudagrass (*Cynodon dactylon* L)
Creeping Bentgrass (*Agrostis palustris*)	Zoysiagrass (*Zoysia japonica* Steud)
Kentucky Bluegrass (*Poa pratensis*)	Buffalograss (*Buchloe dactyloides*)
Tall Fescue (*Festuca arundinacea*)	St. Augustinegrass (*Stenotaphrum secundatum*)
Fine Fescue (various *Festuca* species)	
Red Fescue (*Festuca rubra* L)	
Perennial Ryegrass (*Lolium perenne*)	
Crested Wheatgrass (*Agropyron cristatum*)	

FIGURE 24-10 Some major warm- and cool-season turfgrasses in the United States.

levels of maintenance. The plant has a medium leaf texture and an extensive rhizome system. **Texture** refers to leaf width. **Fine texture** turf contains grasses with narrow blades, whereas **coarse texture** turf consists of wide-blade grasses.

Kentucky bluegrass grows best with full sun, moist and fertile soil, and a mowing height of 1½ to 2½ in. More than one hundred cultivars of Kentucky bluegrass have been developed for certain geographic areas and specific maintenance conditions. A **cultivar** is a plant of the same species that has been discovered and propagated because of its unique characteristics. Cultivar differences exist with respect to disease tolerance, leaf width, color, and other traits.

Tall Fescue (*Festuca arundinacea* Screb) Tall fescue is a coarse-textured, bunch-type grass used in home lawns or as a utility-type turf. A **bunch-type grass** grows in clumps rather than spreading evenly over the ground. However, the bunching tendency of tall fescue can be overcome by heavy seeding, which produces thick stands of grass. A **utility-type grass** refers to a turfgrass adapted to lower maintenance levels. Tall fescues have extensive root systems and are among the most drought-tolerant, cool-season species. However, they are prone to winter injury in the northern range of the cool-season zone.

Recent breeding work has introduced many new and improved cultivars of tall fescue. These new cultivars have medium leaf texture with more aggressive rhizome development. This turfgrass will not tolerate low mowing heights and should be mowed at 2½ to 3 in. Insect resistance and plant persistency is excellent for tall fescues infected with fungal endophytes. **Fungal endophytes** are microscopic plants growing within a plant.

Red Fescue (*Festuca rubra* L) Red fescue is a fine-textured turfgrass well adapted to shady, dry locations. It has excellent drought tolerance and can persist on rather infertile soils and under acidic soil conditions (pH 5.5 to 6.0).

Red fescue is often seeded in mixtures with Kentucky bluegrass for lawn turf. It is not used on athletic fields as a permanent turf because it has poor recuperative potential. However, in the South it may be overseeded into dormant Bermudagrass greens to provide winter play endurance and color.

Red fescue functions satisfactorily if mowed at 1½ to 2½ in. and provided with minimal levels of nitrogen fertilizer and water. The major pest problem of red fescue is leaf spot disease.

Perennial Ryegrass (*Lolium perenne* L)   This grass has a medium leaf texture and is most often used in seed mixes with other turfgrasses. It has a rapid germination and excellent seedling vigor. It is often used in seed mixes to provide soil stabilization during the establishment period.

The perennial ryegrasses are used extensively for recreational turf. They have good wear tolerance and can be rapidly established if the turfgrass stand is damaged. They are also utilized in winter overseeding programs for warm-season turfgrasses.

Perennial ryegrasses require a moderate level of maintenance to form an attractive turf. As a group, they have poor disease resistance. Improved plant persistence and insect resistance occur if they are infected with fungal endophyte.

Creeping Bentgrass (*Agrostis palustris* Huds)   Used in close-cut, high-maintenance areas, this is an extremely attractive turfgrass and

BIO•TECH CONNECTION

One Plus One Equals One?

Each summer millions of people visit the museums, monuments, and Capitol Mall in Washington, DC. The daily foot traffic of Washington's employed, as well as the continuous flow of tourists, mean many thousands of feet pounding the turfgrass sections throughout the summer months. Further, the seasonal use of the Mall for national rallies, concerts, inaugurations, and demonstrations pound and punish the turfgrass and soil like no other place. The end of summer generally leaves the turfgrass thin and vulnerable at best, and bare, muddy, and void at worst. The best of traditional turfgrasses have not been up to the task, and the most heavily traveled areas have been converted to gravel and concrete.

Turfgrass is still the preferred surface for large areas from the standpoint of both beauty and function. The unique needs of the Mall have stimulated the development of a turfgrass mixture that fills the need for toughness, sustainability, and year-round

Turf created by a mixture of zoysiagrass and tall fescue. *(Courtesy of the D.C. Committee to Promote Washington)*

fits well when used on putting greens or bowling greens mowed at ⅛ in. It can be maintained at higher mowing heights (½ to ¾ in.) for use on football fields; golf course fairways and tees; lawns; and formal gardens.

The plant has extensive stolon growth and is a fine-textured turfgrass. It is best adapted to slightly acidic soil with a pH of 5.5 to 6.0 that has good internal drainage and is not prone to compaction. Creeping bentgrass has excellent cold tolerance, but poor heat tolerance.

Creeping bentgrass requires a high level of maintenance to produce a quality turf. It requires proper irrigation, disease control, mowing, and cultivation practices when grown as a sports turf. Because of these high-maintenance requirements, creeping bentgrass is not recommended as a lawn turf.

Crested Wheatgrass (Agropyron cristatum)   Crested wheatgrass is also known as fairway wheatgrass and is used on nonirrigated lawns and fairways in drier, colder regions. It is a coarse-textured, noncreeping, bunch grass. Crested wheatgrass is not considered a high-quality turf, but it is durable in semiarid, northern areas of the Great Plains.

green color. This is a large order in the Washington, DC, climate, which may swing from below 0°F in the winter to above 100°F in summer.

Enter USDA agronomists with an idea for a solution. Why not use one high-quality turfgrass with exceptional durability in hot weather in a mixture with one having exceptional performance in cold weather for a high-quality, tough, year-round turf?

For example, zoysiagrass forms a dense, tough, fine-textured, attractive, and highly competitive turf from May to October. Unfortunately, it turns brown with October's first frost and stays brown until late April or early May. It is usually propagated with plugs.

On the other hand, the new, improved tall fescue turfgrasses shine in the cooler weather of fall and spring, and generally stay green through Washington's winter months. They are propagated with seed. In fact, the tall fescue seed germinates with exceptional ease; and the seedlings become established and grow in very competitive environments.

However, both grasses are tough competitors and tend to crowd out other plants. Could new strains, varieties, or cultivars of each species be found or bred that would survive together and combat the common enemies—excessive heat, cold, wet, dry, and the constant pounding of foot traffic in all kinds of weather?

The agronomists screened hundreds of varieties of grass in search of a fine-bladed fescue that could hold its own against zoysiagrass. Eventually they were able to bring together a mixture of fescue and zoysiagrass that was worth field trials. Some 200,000 square feet, or the area of four football fields, were eventually sodded with zoysia and overseeded with tall fescue on the Washington Mall. Only time will tell if the mixture will be the answer there.

Meanwhile, a zoysiagrass that will produce seed has been developed, and the turfgrass industry has picked up on the new products. Similarly, agronomists of the National Turfgrass Evaluation Program have established a testing program for zoysiagrass and mixtures at over twenty-five locations throughout the country. The greatest promise for the zoysia-fescue mixture is in the transition zone that stretches from New Jersey to Georgia and as far west as Kansas. Perhaps one plus one can equal one—one warm-season turfgrass plus one cool-season turfgrass equals one exceptional year-round turfgrass!

Its extensive, deep root system enables the plant to survive lengthy drought conditions without irrigation.

Recommended mowing height for crested wheatgrass is 1½ to 2½ in., and low-to-moderate fertility is required. Heavy watering will stress the plant, so irrigation should be minimal.

Warm-Season Turfgrasses

These turfgrasses originated in Africa, North and South America, and southeastern Asia, but they are adapted to the southern United States. They make optimum growth at temperatures of 80° to 95°F. These grasses go into winter dormancy as temperatures drop below 50°F. There are some fourteen species of warm-season turfgrasses found throughout the world.

Some major warm-season turfgrasses of the United States include Bermudagrass, zoysigrass, St. Augustinegrass, and buffalograss.

Bermudagrass (*Cynodon dactylon* L)   This grass is considered the most important and widely used warm-season turfgrass in the United States. It is principally used as a lawn turf and sports turf. Improved breeding lines have provided fine-textured Bermudagrasses capable of being used on putting greens and fairways. The common type of Bermudagrass is a medium-textured turfgrass used for airport runways, rights-of-way, and other low-maintenance areas.

The plant spreads by both stolon and rhizome growth. It has excellent wear resistance, recuperative potential, and drought tolerance. Bermudagrass can persist on a wide range of soil types and soil pH. It has poor shade tolerance and low winter hardiness.

The improved types of Bermudagrass require a high level of maintenance and must be vegetatively established. Recommended mowing heights for these grasses range from ¼ to 1 in. The common type of Bermudagrass is established by seed, requires a higher mowing height than the improved cultivars, and is often considered a weed in certain situations.

Zoysiagrass (*Zoysia japonica* Steud.)   Zoysiagrass has excellent low-temperature hardiness and is found in home lawns as far north as New Jersey. It can also be found on golf course fairways and tees within the transition zone. The **transition zone** is a geographic area of the United States where the warm-season and cool-season adaptation zones overlap.

Zoysiagrass has good drought and shade tolerance. It can survive in a wide range of soil types. However, this plant will not perform well in poorly drained soils that remain waterlogged. Though zoysiagrass has excellent wear resistance, it has a low recuperative potential. It is not as aggressive as Bermudagrass.

Zoysiagrass requires a low-to-moderate level of maintenance. It is vegetatively established, because seed germination is poor. New techniques to increase seed germination are presently being evaluated. This grass should be mowed at a height of between 1 and 2 in. Two major pest problems of zoysia are nematodes and billbugs.

Buffalograss (*Buchloe dactyloides*)   Buffalograss is adapted to the dry, semiarid Great Plains region of the United States. It is native to

Oklahoma, Texas, Arizona, Kansas, Colorado, North and South Dakota, and Montana. Regarded as the most drought-tolerant turfgrass in the United States, it survives both high and low temperature extremes and persists through droughts without irrigation. However, its shade tolerance is very poor.

Buffalograss makes a gray-green turf with fine texture. Its vertical growth is slow, but it spreads by stolons, creating a continuous turf. It tolerates alkaline soil and prefers soils of the heavier, fine-textured type. It is used on nonirrigated lawns and golf fairways with cutting heights ranging from ½ to 2 in. Buffalograss is also used along roads and other low-maintenance areas for erosion control. Little or no fertilizer is necessary for survival, but turf quality may be improved by light applications of fertilizer. The grass is propagated by seed or plugs. It is virtually pest-free if not overfertilized or overirrigated.

St. Augustinegrass (*Stenotaphrum secundatum*)   St. Augustinegrass is a major lawn grass in the deep south and is used in the warmest area of the subtropical zone. It is adaptable to many soil conditions, but does best on moist, well-drained sandy soils. Drought resistance is only fair, so irrigation is required in dry weather. The grass has thick stolons and produces a coarse turf of good color, medium density, and excellent shade tolerance.

St. Augustinegrass is propagated with sprigs or sod and has a good establishment rate. The grass has a vigorous growth rate with moderate maintenance requirements. Medium fertility is required and acceptable cutting heights range from ½ to 3 in. Thatch buildup, chinch bugs, and the St. Augustine decline virus are some of the problems encountered with this grass.

TURFGRASS CULTURAL PRACTICES

Mowing, irrigation, and fertilization are the most common and most important cultural practices performed to maintain turfgrass stands. These practices have tremendous effects on turfgrass quality and persistency. Improper mowing, fertilization, and irrigation are major causes of poor lawns (Figure 24-11).

Mowing

Mowing will influence the functional use, persistency, and aesthetic value of a turfgrass. Grasses used for recreational purposes must be playable. The playability of an athletic-field turf is principally determined by its mowing height. A uniform turf surface, fine leaf texture, and freedom from weed encroachment require proper mowing height and frequency.

Turfgrasses are capable of being mowed because their growing point, the crown, is located just below or at the soil surface. However, mowing does have several adverse effects on the grass plant. Reduced rooting depth and decreased carbohydrate reserves occur in mowed turfs. Improper mowing practices only accentuate these and many other adverse effects, causing a decline in turfgrass quality.

Causes of Poor Quality Lawns

- Using the wrong turfgrass species or cultivars.
- Using poor quality seed.
- Mowing the lawn too closely.
- Permitting excessive growth between mowings.
- Using too little or too much lime or fertilizer.
- Improper watering.
- Too much shade.
- Droughty or poorly drained soils.
- Too much traffic.
- Damage by insects or disease.
- Improper use of chemicals.

FIGURE 24-11 Major causes of poor quality lawns.

Recommended Mowing Heights	
Species	Mowing height range in inches
Bahiagrass	2-4
Bermudagrass	
Common	0.5-1.5
Hybrids	0.25-1
Carpetgrass	1-2
Centipedegrass	1-2
St. Augustinegrass	1.5-3.0
Zoysiagrass	0.5-2.0
Creeping bentgrass	0.2-0.5
Colonial bentgrass	0.5-1.0
Fine fescue	1.5-2.5
Kentucky bluegrass	1.5-2.5
Perennial ryegrass	1.5-2.5
Tall fescue	1.5-3.0
Crested wheatgrass	1.5-2.5
Buffalograss	0.7-2.0
Blue grama	2.0-2.5

FIGURE 24-12 Recommended mowing heights for different turfgrasses.

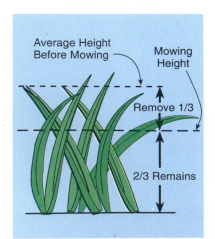

FIGURE 24-13 No more than one-third of the top growth should be removed per mowing.

Mowing Height The recommended ranges for mowing height are listed in Figure 24-12. If one mows below or above these ranges, various problems will occur. Mowing below the desired range will reduce photosynthesis, thus preventing carbohydrate production within the plant. This will cause a reduction in rooting and decrease the recuperative potential of the plant.

Mowing above the recommended height will increase thatch buildup and leaf texture. It will also decrease turf density and appearance. **Thatch** is the buildup of organic matter on the soil around the turfgrass plants. Excessive thatch will cause poor water infiltration, increased disease activity, and decreased rooting depth.

Mowing Frequency Frequency of mowing is determined by the mowing height and the growth rate of the plant. No more than one-third of the top growth should be removed per mowing (Figure 24-13). This will allow sufficient leaf area for photosynthesis after mowing. The lower the mowing height, the more frequent the mowing if the one-third rule is observed. For example, creeping bentgrass mowed at ¼ in. will be mowed five to six times per week. However, tall fescue cut at 3 in. will require only one mowing per week. Under ideal growing conditions, grasses will require more frequent mowing than they will under poor growing conditions.

Fertilization

Fertility programs attempt to supply adequate levels of plant nutrients to allow for favorable plant growth. Proper fertilization should increase turfgrass density and color. Lush or succulent growth caused by excessive fertilization should be avoided.

A complete fertilizer is often recommended, with the rate of application based on the amount of nitrogen required per 1,000 ft^2 (Figure 24-14). A **complete fertilizer** consists of nitrogen, phosphorus, and potassium. The **fertilizer analysis** states the percentage of nutrients by weight in the fertilizer. A 10–6–4 fertilizer consists of 10 percent nitrogen, 6 percent phosphorus, and 4 percent potassium. A formula can be used to determine fertilizer amounts based on nitrogen-rate recommendations (Figure 24-15). Nitrogen is the most important element in turfgrass fertilization.

Proper timing and application rate are the keys to a successful fertilization program. Timing of fertilization will vary depending upon turfgrass species and geographic location. For cool-season turfgrasses, the fall is the desired time period. Optimum root growth occurs during this time. On the other hand, fertilization of warm-season turfgrasses should be done during late spring or the summer months (Figure 24-16).

Improper fertilization will reduce turfgrass quality. Turfgrasses receiving insufficient amounts of fertilizer will lack color, density, and recuperative potential. Excessive fertilization will reduce heat and drought tolerance, increase disease and insect damage, and cause excessive top growth. Appropriate nitrogen rates range from ¼ to 1 lb of nitrogen per 1,000 ft^2. The rate is dependent on growth and environmental conditions. Phosphorus and potassium rates should be determined by soil testing.

Kentucky Bluegrass		
Time to Apply	Nitrogen Applied Per 1,000 Square Feet	Acceptable Fertilizer Per 1,000 Square Feet of Area
September	1 lb	8 to 10 lb 12-4-8 or 10 lb 10-6-4 or 5 lb 20-10-10
October	1 to 1½ lb	8 to 12 lb 12-4-8 or 10 to 15 lb 10-6-4 or 5 to 8 lb 20-10-10
November/December or February/March	1 to 1½ lb	8 to 12 lb 12-4-8 or 10 to 15 lb 10-6-4 or 5 to 8 lb 20-10-10
May-June 20	0 to ½ lb	0 to 4 lb 12-4-8 or 0 to 5 lb 10-6-4 or 0 to 3 lb 20-10-10

FIGURE 24-14 Fertility requirements and types of fertilizers for Kentucky bluegrass.

Percentage of nutrient(s) in the fertilizer times weight of the fertilizer equals the weight of the nutrient(s) contained.

Question #1: How much nitrogen is there in a 50 lb bag of 10-6-4 fertilizer?

Solution: % nutrient(s) x Weight of fertilizer = Weight of nutrient(s)

or

.10 x 50 lb = 5 lb of nitrogen

Question #2: How many pounds of 10-6-4 fertilizer must be used to apply 1 pound of nitrogen per 1,000 ft²?

Solution: Pounds of nutrient needed divided by % nutrient equals pounds of fertilizer needed.

or

5 lb N divided by .10 = 50 lb of 10-6-4

FIGURE 24-15 Determining nutrient content of fertilizer and calculating fertilizer application rate to meet nutrient needs.

FIGURE 24-16 Fertility requirements and some recommended fertilizers for Bermudagrass and zoysiagrass.

Bermudagrass and Zoysiagrass		
Time to Apply	Nitrogen Applied Per 1,000 Square Feet	Acceptable Fertilizer Per 1,000 Square Feet of Area
April or May	1 lb	8 lb 12-4-8 or 10 lb 10-6-4 or 5 lb 20-10-10
June	1 lb	8 lb 12-4-8 or 10 lb 10-6-4 or 5 lb 20-10-10
July	1 lb	8 lb 12-4-8 or 10 lb 10-6-4 or 5 lb 20-10-10
August	1 lb	8 lb 12-4-8 or 10 lb 10-6-4 or 5 lb 20-10-10

Irrigation

The application of water to turf may accomplish several different objectives. First, sufficient soil moisture will allow for optimum plant growth. Irrigation is also used to establish turfgrass, reduce plant surface temperatures, and rinse in fertilizer and pesticide applications. **Syringing** is a light application of water to a turfgrass. It may be used to reduce plant temperatures or to remove dew and frost from the turfgrass leaf.

The amount of irrigation needed to maintain optimum plant growth is dependent upon many factors. Some factors are turfgrass species, geographic location, soil type, weather conditions, and turfgrass use. General recommendations are to apply 1 in. of water per week during the summer months. It takes approximately 620 gallons of water per 1,000 ft² to provide 1 in. of water. This amount of water will produce a green and actively growing turfgrass stand.

Heavy but infrequent irrigation will force root growth deep into the soil. Light and frequent irrigation will keep the surface soil moist, thus encouraging shallow rooting.

The best time to irrigate is at night, when evaporation and wind are low. However, an increase in disease activity will occur at this time. Therefore, early morning watering is often selected because evaporation, wind, and disease activity are at a minimum.

TURFGRASS ESTABLISHMENT

Turfgrasses may be established by seeding or vegetative propagation. Regardless of the method used, proper establishment practices, including site preparation, should be performed. Correct establishment practices will ensure adequate turfgrass quality and persistency.

Turfgrass Selection

Selection of the correct turfgrass species is one of the most important decisions in the establishment process. Selecting a turfgrass seed blend or seed mixture should be based on the intended use and performance data of the turfgrasses. A **seed blend** is a combination of different cultivars of the same species. A **seed mixture** is a combination of two or more species. State universities evaluate different turfgrass species and new cultivars and provide information on their performances (Figure 24-17).

Kentucky Bluegrass Cultivar	Quality Rating*					
	April 1	May 4	June 8	July 15	Sept. 2	Oct. 17
Adelphi	4.0*	5.6	5.4	4.5	5.9	5.7
Baron	3.8	5.1	4.8	4.4	5.4	5.5
Bensun (A-34)	3.6	5.3	6.0	5.9	6.1	5.4
Cheri	3.2	5.2	4.7	4.9	5.8	6.0
Emmundi	4.0	4.9	4.6	4.5	5.4	5.7
Glade	4.2	5.1	5.4	5.5	6.3	6.3
Majestic	3.9	6.1	6.0	5.8	6.0	5.8
Newport	3.7	3.6	3.8	4.0	4.2	4.9
Parade	4.5	6.8	5.9	4.4	5.7	5.8
Ram 1	4.3	6.9	5.9	6.0	6.2	6.4
Sydsport	3.8	6.2	6.0	5.9	6.4	6.1
Touchdown	4.1	6.3	5.9	5.7	6.2	6.5

FIGURE 24-17 Quality ratings for Kentucky bluegrass cultivars.

*It is important to note that quality ratings can vary significantly from one region or location to another. A rating of 1 = no live turf, 9 = ideal turf, >5 = acceptable quality.

Site Preparation

Proper site preparation may include any or all of the following activities:

1. Debris removal
2. Nonselective weed control
3. Installation of a subsurface drainage and/or irrigation system
4. Soil modification, tillage, and grading

The removal of woody vegetation, leftover construction debris, and any large stones will reduce future maintenance problems. If woody debris is buried on the site rather than removed, the soil will settle and diseases such as fairy ring may result. Excessive amounts of stone and rock present in the seedbed will cause localized dry spots and interfere with future cultivation practices.

Nonselective weed control may be necessary if perennial grassy weeds or difficult-to-control weeds are present. A nonselective herbicide or a soil fumigant should be applied prior to seeding. The herbicide glyphosate (Round Up) is often used to provide nonselective weed control. A soil fumigant, such as metham, may also be used.

If the site is to be used as a sports turf, the installation of a drainage and an irrigation system may be required. This should be done prior to final grading.

Soil modifications, followed by tillage and grading, will be the last steps in site preparation. Soil amendments, such as organic matter and sand, may be incorporated to improve nutrient retention and drainage. Topsoil may also be used to provide a favorable growth medium for turfgrasses. The soil should be worked to a depth of 6 in. and then lightly tilled to provide a uniform seedbed and adequate surface drainage (Figure 24-18). Incorporation of lime and fertilizer should be done during this time. Excessive tillage will destroy soil tilth. A seedbed with soil aggregates of ¼ to 1 in. in diameter is ideal.

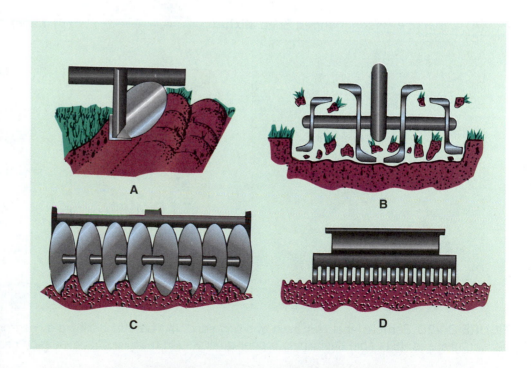

FIGURE 24-18 Tillage equipment frequently used for turfgrass establishment: (A) plow, (B) rototiller, (C) disk cultivator, and (D) harrow.

Seeding Rates

Species	Pounds of Seed per 1,000 ft²
Bahiagrass	3-8
Bentgrass	
Colonial	0.5-1.5
Creeping	0.5-1.5
Bermudagrass	
(hulled)	1-2
Bluegrass	
Kentucky	1-2
Buffalograss	3-7*
Carpetgrass	1.5-5
Centipedegrass	0.25-2*
Fescue	
Fine	3-5
Tall	5-9
Grama, blue	1.5-2.5
Ryegrass	
Annual	5-9
Perennial	5-9
Wheatgrass, crested	3-6
Zoysiagrass	1-3

*The higher rates are best, but lower rates are commonly used because the seed is expensive.

FIGURE 24-19 Seeding rates for the major turfgrass species.

Planting

Turfgrasses can be established by seeding, sodding, stolonization, sprigging, and plugging. The last four methods are vegetative means of planting. Vegetative establishment may be done if the grasses produce infertile seed or have low seed yields. Also, if quick establishment is required, sod may be installed.

Seeding Seeding is the principal method for turfgrass establishment, because it is the least expensive. Figure 24-19 shows the seeding rates for the different grasses. These rates will vary depending on seed size and percentage of seed germination.

The seed label informs the buyer or consumer of the quality and ingredients of a seed blend or mixture. Information on purity percentage, germination, weed seed content, inert matter, and test date is present on the label (Figure 24-20). Certification programs are available for turfgrass seed. **Seed certification programs** are administered by state governments, and they ensure that seed is true to type. They also require that the seed meet other minimum quality standards concerning percentage of germination and weed seed content.

The best time to seed turfgrass will vary, depending on the type of turfgrass. The optimum time for establishing cool-season grasses is in the fall. Ideal weather conditions and less annual weed competition are the principal reasons for fall establishment. The plant will also have sufficient time to develop an adequate root system prior to summer stress conditions. Spring seeding is less desirable because there is insufficient time to develop a mature stand capable of competing with undesirable grasses.

Warm-season turfgrasses are established during the spring and summer months. The best time is in late spring. The new grass then has the longest time period for optimum growth and development after establishment.

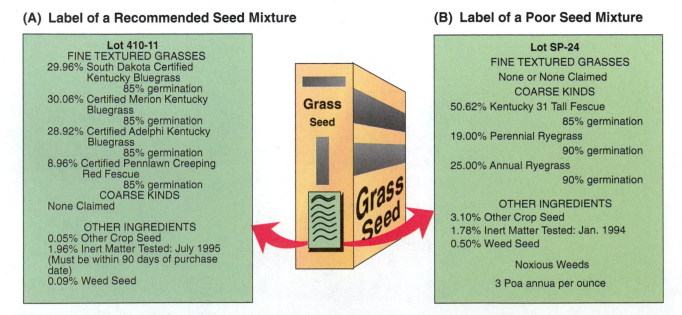

(A) Label of a Recommended Seed Mixture

Lot 410-11
FINE TEXTURED GRASSES
29.96% South Dakota Certified Kentucky Bluegrass
85% germination
30.06% Certified Merion Kentucky Bluegrass
85% germination
28.92% Certified Adelphi Kentucky Bluegrass
85% germination
8.96% Certified Pennlawn Creeping Red Fescue
85% germination
COARSE KINDS
None Claimed

OTHER INGREDIENTS
0.05% Other Crop Seed
1.96% Inert Matter Tested: July 1995
(Must be within 90 days of purchase date)
0.09% Weed Seed

Grass Seed

(B) Label of a Poor Seed Mixture

Lot SP-24
FINE TEXTURED GRASSES
None or None Claimed
COARSE KINDS
50.62% Kentucky 31 Tall Fescue
85% germination
19.00% Perennial Ryegrass
90% germination
25.00% Annual Ryegrass
90% germination

OTHER INGREDIENTS
3.10% Other Crop Seed
1.78% Inert Matter Tested: Jan. 1994
0.50% Weed Seed

Noxious Weeds

3 Poa annua per ounce

FIGURE 24-20 Sample labels for lawn and turf seed. (A) Label describing a recommended seed mixture of cool-season grasses. (B) Label describing a poor-quality seed mixture of cool-season grasses.

FIGURE 24-21 Use of a cultipacker or hydroseeder leaves the seed in touch with soil particles, which is critical for moisture absorption and germination. *(Courtesy of Michael Dzaman)*

FIGURE 24-22 Cutting sod for placement in a new location as turfgrass. *(Courtesy of H. Edward Reiley)*

Different types of equipment may be used to apply the seed. Fertilizer drop spreaders, overseeders, hydroseeders, and cultipacker seeders are some types of equipment used for seeding. Shallow seed placement (¼ in.) is important for proper germination. This can be done by hand raking for small areas or the use of a drag mat for larger areas. Specialized seeding equipment, such as a hydroseeder or cultipacker, will place the seed at the right soil depth (Figure 24-21).

After seeding, the use of a mulch will provide a more favorable environment for seed germination. Mulch can conserve soil moisture and prevent soil erosion. This is extremely important, because adequate surface moisture must be present for seed germination. A straw mulch is preferred and is usually applied at the rate of two bales per 1,000 ft². Other mulches are wood and paper byproducts; net or fabric; and peat moss.

Sodding   **Sodding** refers to removing a rectangular piece of grass and a slice of the soil beneath it and moving it to another location. The grass and the soil immediately beneath are referred to as **turf.** Sodding offers the ability to establish turf at any time of the year. It also provides an instant cover. However, the cost is extremely high, and the sod must be installed shortly after harvesting. Specialized equipment for sod harvesting has been developed (Figure 24-22). Sod is cut into 12 to 24 in. widths and lengths of up to 3 ft, or in carpetlike pieces. The depth of cut will vary from 0.3 to 0.5 in. The cut sod is then rolled or folded for transport and later placed on a carefully prepared seed bed. The soil is then rolled to ensure good contact with the soil, and watered thoroughly.

Sprigging   **Sprigging** is the planting of a section of a rhizome or stolon, referred to as a sprig. A sprig may be up to 6 to 8 in. long. For successful establishment, a section of the sprig with several nodes must be inserted into the soil (Figure 24-23). After sprigging, irrigation must be applied to prevent desiccation, or drying out. Equipment for

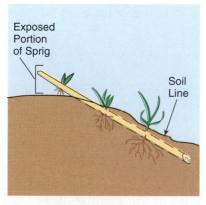

FIGURE 24-23 When sprigging, two or more nodes with shoots should be placed in the soil and watered well.

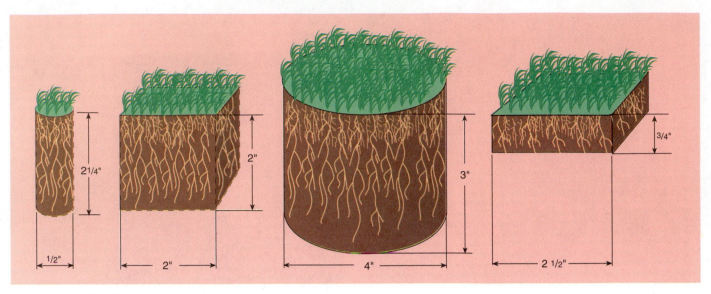

FIGURE 24-24 Turfgrass plugs used for vegetative establishment.

sprigging has been developed and will plant sprigs in rows spaced 6 to 18 in. apart.

Stolonizing   Stolonizing is similar to sprigging, in that 6 to 8 in. sections of rhizomes or stolons are used. However, in stolonizing, sprigs are broadcast onto the soil surface. These sprigs may be lightly topdressed with soil and rolled prior to irrigation. Survival rates are lower than with sprigging. Therefore, stolonizing requires greater quantities of sprigs.

Plugging   Plugging is the establishment of a turfgrass stand by using plugs, or small pieces of existing turf. Plug sizes will vary, as noted in Figure 24-24. Spacing of plugs can be on 6 to 18 in. centers, depending on how quickly you need the turfgrass established. Plugging can be done by hand or with specialized equipment. It will require a longer time to cover the soil than will the other establishment techniques.

In summary, the turfgrass industry is expanding. As more space is used for nonfarm purposes, the use of turfgrass for conservation, recreation, and beauty will continue to grow. Excellent career opportunities exist, and suitable educational programs are available to prepare new workers for this increasingly popular area of agriscience.

STUDENT ACTIVITIES

1. Write the Terms to Know and their meanings in your notebook.

2. Determine if your community is located in a warm-season, transitional, or cool-season zone.

3. Visit or call your county cooperative extension office and obtain copies of the various publications available on turfgrass or lawn production and maintenance.

4. Determine the recommended cultural practices for establishing and maintaining a major turfgrass species or mixture for your locality. Report your findings to the class.

5. Arrange to visit a golf course and discuss with the superintendent the duties of turfgrass workers on the golf course.

6. Obtain a map of your county and attach it to the bulletin board. Place pins at the locations of all businesses, schools, government buildings, and institutions that have extensive lawn or turfgrass areas around them. Ask your classmates to help identify them. Note: Use different colored pins to identify different types of institutions.

7. Obtain a grass identification key from your teacher or the cooperative extension service office. Collect ten specimens of different turfgrasses and identify them by using the identification key.

8. Work with your teacher or cooperative extension service to set up turfgrass demonstration plots. Show the effect of various fertilizer applications and mowing heights on turfgrass health and vigor.

SELF EVALUATION

A. Multiple Choice

1. The major meristematic tissue of the turfgrass plant is the
  a. inflorescence.
  b. leaf blade.
  c. rhizome.
  d. crown.

2. The root system of a cool-season turfgrass is actively growing during the
  a. summer.
  b. fall and winter.
  c. fall and spring.
  d. late spring.

3. The estimated value of the turfgrass industry in the United States is
  a. $25 billion.
  b. $50 billion.
  c. $100 billion.
  d. $200 billion.

4. Which irrigation practice will develop a deeper, more extensive root system?
  a. frequent and light applications
  b. infrequent and heavy applications
  c. frequent and heavy applications
  d. infrequent and light applications

5. Turfgrasses are in the _____ family.
  a. *Compositae*
  b. *Cyperacese*
  c. *Cruciferue*
  d. *Poacea*

6. _____ is a warm-season turfgrass.
   a. Kentucky bluegrass
   b. Zoysiagrass
   c. Red fescue
   d. Perennial ryegrass

7. Which cultural practice will have an adverse effect on rooting depth?
   a. low mowing
   b. heavy, infrequent irrigation
   c. moderate fertilizer applications during the time of root growth
   d. high mowing

8. A tiller is an example of
   a. rhizome growth.
   b. extravaginal growth.
   c. stolon growth.
   d. intravaginal growth.

9. Which of the following is not a turfgrass stem?
   a. rhizome
   b. stolon
   c. leaf sheath
   d. crown

10. Which cool-season turfgrass is adapted to shady, dry locations?
    a. red fescue
    b. Kentucky bluegrass
    c. tall fescue
    d. creeping bentgrass

B. Matching

_____ **1.** Rhizome
_____ **2.** Stolon
_____ **3.** Creeping bentgrass
_____ **4.** Sprigging
_____ **5.** Photoperiod
_____ **6.** Tall fescue
_____ **7.** Coarse texture
_____ **8.** Syringing
_____ **9.** 10-10-10
_____ **10.** Thatch

a. An induction process for inflorescence development
b. A turfgrass stem that grows horizontally above ground
c. A cool-season turfgrass that is very drought tolerant
d. A wide-leaf blade
e. A cool-season turfgrass used on putting greens
f. A turfgrass stem that grows horizontally below ground
g. A buildup of organic matter on the soil around turfgrass plants
h. A complete fertilizer
i. Light application of water to a turfgrass
j. Vegetative establishment

C. Completion

1. _____ is a warm-season turfgrass having good winter hardiness.

2. _____ roots are present at the time of seed germination.

3. _____ is an example of intravaginal growth.

4. A seed _____ consists of different cultivars of the same species.

5. _____ is a turfgrass species with rapid seed germination.

6. As mowing heights decrease, mowing frequency will _____.

7. _____ is considered the most important element in turfgrass fertilization.

8. A planting depth of _____ is recommended for turfgrasses.

9. _____ is an establishment practice of applying the sprig to the soil surface.

10. A _____ mulch is preferred over other types of mulch for seeding turfgrasses.

UNIT: 25

Trees and Shrubs

OBJECTIVE

To use trees and shrubs for beautification, improved air quality, and pollution control.

MATERIALS LIST

✓ nursery catalog

✓ pictures of plants commonly grown in an area

✓ list of insects and diseases of trees and shrubs that are normally problems in the area

COMPETENCIES TO BE DEVELOPED

After studying this unit, you should be able to:

■ identify ornamental trees and shrubs.

■ select trees and shrubs for appropriate landscape use.

■ classify trees and shrubs according to growth habit, growth habitat needs, and other requirements.

■ identify trees and shrubs by using proper nomenclature.

■ purchase plant material for installation in a landscape.

■ plant and maintain plant material.

Trees and shrubs are major components of the environment. They are important in providing natural beauty and in providing oxygen to humans and animals. A well-planned landscape will improve the economic value of an area. The need for trees and shrubs has increased in recent years. This need has increased the requirement for horticulture technicians who understand the production and care of landscape plants.

VALUE OF TREES AND SHRUBS

A well-designed landscape increases the value of a property (Figure 25-1). Residential properties are much more attractive when trees and shrubs are used to enhance their beauty and balance.

FIGURE 25-1 A well-designed landscape adds economic value to a property. *(From Emmons, Turfgrass Science and Management, 1995, Delmar Publishers. Photo by Michael Dzaman.)*

FIGURE 25-2 Trees and shrubs add beauty, provide sound insulation, purify the air, and provide privacy. *(Courtesy of Michael Dzaman)*

Trees and shrubs provide more than natural beauty. An acre of good, healthy trees and/or shrubs will produce enough oxygen to keep from sixteen to twenty people alive each year. These plants are also valuable in helping to keep our air clean by using the carbon dioxide that is produced by people, automobiles, and factories.

Trees and shrubs cut noise pollution by acting as barriers to sound. They can deflect sound as well as absorb it. When used properly in the landscape, trees and shrubs can provide shade and act as insulators, to keep the house cooler in the summer and warmer in the winter.

Many urban areas depend on trees and shrubs to soften the concrete, black top, and steel environment. In fact, many of these areas now employ urban foresters to help design, install, and maintain trees and shrubs in the large cities (Figure 25-2).

PLANT SELECTION

Trees and shrubs are separate groups of plant material. **Trees** are woody plants that produce a main trunk and a more or less distinct and elevated head (a height of 15 ft or more). **Shrubs** are woody plants that normally grow low, produce many stems or shoots from the base, and do not reach more than 15 ft high.

USE

Ornamental trees and shrubs in the landscape are both beautiful and functional. They are used as specimen plants, border plantings, or groupings. A **specimen plant** is used as a single plant to highlight or provide some other special feature to the landscape. A **border planting** is used to separate some part of the landscape from another, or to serve as a fence or a windbreak. A **group planting** is when a number of trees or shrubs are planted together so they point out some special feature, provide privacy, or create a small garden area.

The location of plant material in the landscape will play an important part in the selection process. Color of the leaves, texture of the plant, and color of the flowers are just some of the factors that must be considered before purchasing or planting trees or shrubs.

Geographic Location

The location in a country is an important factor in the selection of a tree or shrub. Some plants are not **hardy,** that is, they may not be able to survive or even grow properly in a given environment. The hardiness of a plant is affected by the intensity and duration of sunlight, the length of the growing season, minimum winter temperatures, the amount of rainfall, summer droughts, and humidity.

The U.S. Department of Agriculture has issued a **plant hardiness zone map** (Figure 25-3). There are eleven zones in the United States, counting Alaska and Hawaii. Each represents an area of winter hardiness that is based on the average annual minimum winter temperatures. It is possible that local climates may vary from the general zone map. They may be colder or warmer than is indicated on the map. Local nursery personnel are generally willing to help in the selection of the best plant material. Most nursery catalogs list the plant for the coldest zone in which it will grow normally. Two examples of such catalog listings are:

Example 1: *Cornus florida* (Flowering Dogwood)—Zone 5
A low-branched, flat-topped tree. It has a horizontal branching habit. Will grow 20 to 30 ft high and 25 to 35 ft wide. Growth rate is slow to medium, and the texture is fine to medium. White flowers appear in April to May. Red berries appear in the fall.

Example 2: *Pyrus calleryana* "Bradford" (Bradford Pear)—Zone 4
A dense, pyramidal tree that becomes brittle with age. Will grow 30 to 50 ft high and 30 to 35 ft wide. The growth rate is medium, and the texture is medium-fine. Makes an excellent all-purpose tree, useful as

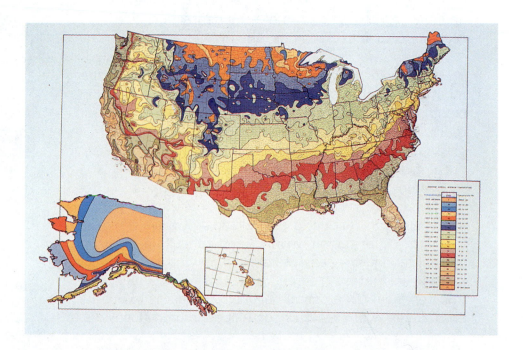

FIGURE 25-3 The USDA plant hardiness zone map.

a shade or street tree. Profuse white flowers ⅓ in. in diameter appear in late April or early May in clusters 3 in. in diameter.

A plant can also be expected to live in a warmer zone than indicated if rainfall, soil, and summer conditions are comparable. In some cases it may be necessary to adjust these conditions via irrigation, correction of soil condition, wind protection, and alteration of shade or sun exposure. It is also possible to grow some plants in areas north of the indicated zone. Such plants may need special attention to protect them from wind or cold. Without such protection, they may not perform normally and are likely to suffer winter injury.

Site Location

When planning the location for planting trees and shrubs, many factors must be given consideration. Some are:

■ Flower color. Is it compatible with the house, fence, patio, and any other plants in the area?

AGRI·PROFILE

CAREER AREAS: Landscape Architect/Landscape Technician/ Landscape Contractor/Plant Specialist

Ornamental trees and shrubs vary greatly in size, shape, temperature preference, light preference, fertility needs, and pest tolerance. This variety stimulates diversity in jobs and specialties in the area of ornamental horitculture. Career opportunities span both the arts and the sciences.

Landscape architects practice the art of design in that they plant landscapes pleasing to their clients. They must know plant species and plant materials to develop plans that are attractive yet functional in the given environment. Landscape architects generally have bachelor's degrees in landscape design and considerable experience in ornamental horticulture.

Horticulture specialists and technicians have careers centering on nursery or greenhouse management, landscape contracting, groundskeeping, wholesaling, retailing, public information, writing, and consulting. They may specialize in just trees, just shrubs, or both. Some specialize in pruning, tree planting, pest control, or landscape maintenance. Ornamental horticulture seems to provide an endless array of career opportunities in many localities.

Plant care and management require many skills including the diagnosis of disease and insect problems. Here, a plant specialist evaluates the extent of damage by scale insects. *(Courtesy of Michael Dzaman)*

■ Fruit size and type. Many trees, such as some crabapples or cherries, drop messy fruit. Therefore, they should not be placed near an area that will be walked on or used by people. Such areas include driveways, walks, patios, and swimming pools.

■ Other structures. Do not plant trees directly in front of doors; near wells, cesspools, or field drains; or under utility lines, where interference is likely when the plant matures.

■ The plant's ornamental characteristics. Flowering time, shape, foliage texture, fall color, pest resistance, landscape suitability, and mature size are all part of any consideration when selecting a plant.

Type and Growth Habit

Plants have many different types of growth habits. The type of growth habit is an important consideration when selecting plants (Figure 25-4). A good landscape planner or designer will be aware of the type of growth habit for a given plant used in a given area.

Plant size in relation to the structure that is being landscaped is very important. Maintenance and pruning of plants is often expensive. Pruning of shrubs is frequently done without climbing, while most trees need to be climbed when pruned. The use of a tree that is not in proportion to nearby structures can result in an expensive, frequent pruning program. Figure 25-5 illustrates some common trees and their mature sizes. A tree that is too tall will require extra work to keep it in proper relationship to adjacent structures. Such a design error would cancel out the goal of property enhancement.

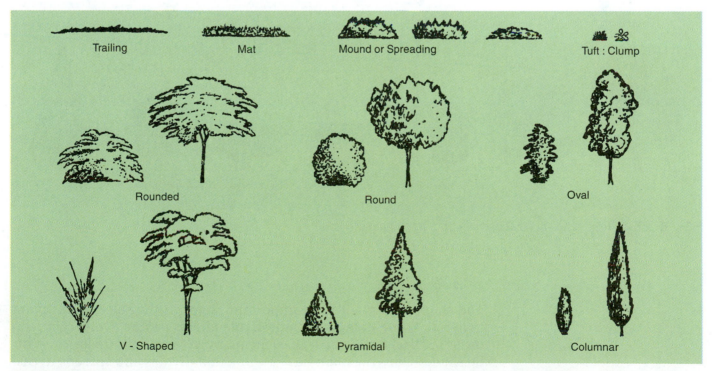

Trailing   Mat   Mound or Spreading   Tuft : Clump

Rounded   Round   Oval

V - Shaped   Pyramidal   Columnar

FIGURE 25-4 Types of growth habit.

FIGURE 25-5 Tree sizes in relation to a two-story house.

Dogwood   Ginkgo   White spruce   Catalpa   Black birch   Sugar maple

Red maple   Paper birch   Yellowwood   Red mulberry   White oak

Willow oak   Horsechestnut   Chestnut oak   Sweetgum

American birch   American linden   Black oak   Shellback hickory

Tuliptree   Scarlet oak   Red cedar   Mimosa   Hackberry   Redwood

Shape of the Plant

The shape of a plant is an important factor in the selection of plant material. Some important undesirable effects can be avoided by asking some basic questions. Does the plant grow straight up and give little shade? Does the plant have a trunk that divides and spreads to cast unwanted shade? Is the plant a low-growing tree that gives very little

FIGURE 25-6 Typical shapes of landscape trees.

shade? Consider which shapes of trees and shrubs are needed in the landscape. Then select the appropriate plant to achieve the objective (Figure 25-6).

Before plants are selected, research is necessary to better understand the types of plants used in a given area. To do this, study the types of plants sold and used in the locality and note the various characteristics of each plant. Catalogs are available from nurseries that sell plants in the locality. Also, a visit to local garden centers and nurseries will pay dividends in gathering information on plant selection.

PLANT NAMES

Trees and shrubs must be ordered by their proper names. Common names such as flowering dogwood and upright juniper are not governed by any formal code of nomenclature. **Nomenclature** is a systematic method of naming plants or animals. The botanical or scientific name is recognized internationally. Scientific designations are always written in Latin and consist of two names. The first name is the genus, and the second is the species. The genus name always begins with a capital letter and is a noun. The species name is usually written in all lowercase letters and is an adjective.

The **genus** (plural is genera) is defined as a group of closely related and definable plants composing one or more species. The common, definable characteristics are fruit, flower, and leaf type and arrangement. The **species** (plural is also species) is the basic unit in the classification system whose members have similar structures and common ancestors, and maintain their characteristics.

Variety (var.) is a subdivision of species that has various inheritable characteristics of form and structure that are continued through both sexual and asexual propagation. The varietal term is written in lowercase letters and <u>underlined</u> or *italicized*. Often in catalogs and on plant labels the abbreviation **var.** is used. An example of how this is used might be *Cornus florida rubra* or *Cornus florida var. rubra.*

A **cultivar** (cv.) is a group of plants within a particular species that has been cultivated and is distinguished by one or more characteristics and that, through sexual or asexual propagation, will keep these characteristics. The term is written inside single quotation marks. An example is *Pyrus calleryana* 'Bradford' or *Pyrus calleryana* cv. 'Bradford.'

It is important to become familiar with this plant-naming system because common names will vary from area to area. There are no standards for creating common names. Professional horticulturists order plants by the method described above. Landscape architects place scientific names on landscape drawings. Finally, nurseries use the scientific names in their catalogs to avoid confusion by anyone who orders plant material.

OBTAINING TREES AND SHRUBS

After determining the specific plants needed in a landscape, the plants must be purchased. Plants are normally dug and shipped as bar rooted, balled and burlapped (B & B), or container grown (Figure 25-7).

Bare-Rooted Plants

Bare-rooted plants are dug and the soil is shaken or washed from the roots. Normally only deciduous trees, shrubs, and trees with taproots are shipped this way. They are dug while the trees are dormant. Plants that are ordered from nursery catalogs and can survive bare rooted are shipped this way because of lower transportation costs and easier handling. It is not economical to ship soil with the roots because it is so heavy.

Bare-rooted plants are often planted while they are dormant; however, the roots must be protected to keep them from dying out. If the plant cannot be planted when received, the roots must be protected by putting them in a container of water, wrapping them in plastic after a good watering, or placing wet newspaper around them. If plants cannot be planted for many days, the plant may be healed in. **Healing in** is accomplished by simply digging a trench in the soil deep enough to hold the roots of the plants, placing the plants in the trench, and covering them with soil. It is important to wet the soil well after healing in. Frequently, plants that are bare rooted will need additional pruning at planting time (Figure 25-8).

Balled and Burlapped Plants

Balled and burlapped (B & B) plants are dug with a ball of soil remaining with the roots. This is wrapped with burlap and laced with twine. Normally, these plants have been root pruned. **Root pruning** is a process whereby roots are cut close to the trunk so that a good root system develops close to the trunk before the plant is dug. The result is that when the ball is dug, the tree or shrub will have a compact root system in the ball. This gives the plant a good chance to reestablish itself in a new environment.

Bare Rooted

Twine Ties　Burlap Bag

Balled and Burlapped

Containerized

FIGURE 25-7 Methods of preparing plants for shipping and handling.

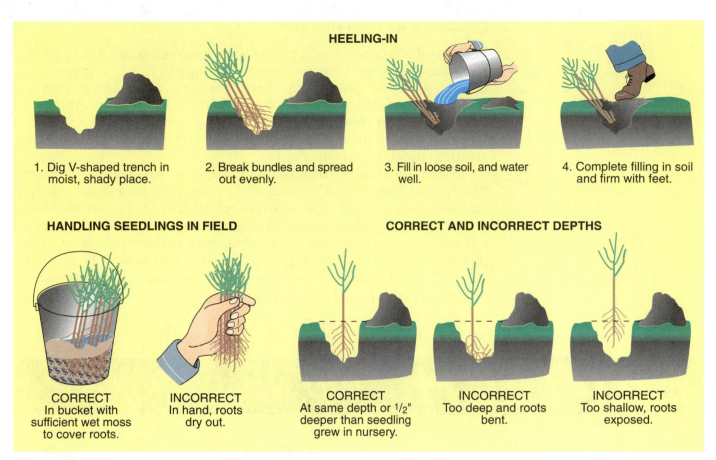

HEELING-IN

1. Dig V-shaped trench in moist, shady place.

2. Break bundles and spread out evenly.

3. Fill in loose soil, and water well.

4. Complete filling in soil and firm with feet.

HANDLING SEEDLINGS IN FIELD

CORRECT
In bucket with sufficient wet moss to cover roots.

INCORRECT
In hand, roots dry out.

CORRECT AND INCORRECT DEPTHS

CORRECT
At same depth or 1/2" deeper than seedling grew in nursery.

INCORRECT
Too deep and roots bent.

INCORRECT
Too shallow, roots exposed.

FIGURE 25-8 Handling and planting seedling trees.

It is important that transplanted trees and shrubs have a good chance to grow after transplanting. Plants typically balled and burlapped for transplanting are deciduous trees with branching root systems, conifers, azaleas, rhododendrons, and other plants that have fibrous root systems.

As is true with bare-root plants, B & B plants cannot have their root systems dry out. However, because there is soil around the roots, the drying-out process is more gradual. When purchasing B & B plants, avoid those that do not have a sound ball. Handling tends to break the ball up and strip the hair roots from the plants. Therefore, a good root ball will help substantially to avoid damage when handling. Plant material that is shipped as B & B may be planted anytime, as long as the soil can be worked.

The mechanical tree spade is becoming more popular in the industry today. A **tree spade** is an expensive piece of equipment that will dig a tree in a matter of a few minutes with a very specific-sized ball. Additionally, it is used to dig holes for trees and shrubs quickly and efficiently. The tree spade saves many hours in digging and planting trees. Frequently, B & B trees and shrubs are shipped in burlap with wire cages, specially prepared baskets, or other containers. The tree spade has helped make this an efficient way to ball and burlap.

These types of plants require very little pruning at planting time. This makes B & B a popular way of handling plants by the mechanized professional horticulturist.

Container-Grown Plants

The use of container-grown plants is increasing in the nursery industry. **Container grown** means grown and shipped in a pot or can. Normally, the smaller types of plants are handled in this manner. They are grown for a reasonable period of time in the containers in which they are shipped and purchased. They are easy to handle in the field, require little or no pruning when they are planted, and can be planted any time the soil can be worked.

The disadvantage of using container-grown plants is that they need to be planted carefully. The roots of the plant have developed in a very limited space and are generally rootbound. The roots may also have grown back around the trunk. To prevent the plant from strangling itself and also to encourage the root to leave the confined area, these plants must have the "container ball" broken. This is done by placing a sharp shovel through the root mass or by breaking and spreading the root mass as it is planted in the new hole.

BIO•TECH CONNECTION

Biocontrol—Tamer of the Euonymus Scale?

"Scale insects that attack euonymus plants rank among the most insidious enemies of trees and bushes in the United States," so says entomologist John J. Drea of the USDA Insect Biocontrol Laboratory in Beltsville, Maryland. A hitchhiker and intruder from Asia, *Unaspis euonymi* is commonly known as the euonymus scale. It has set up housekeeping and multiplied in the United States without the normal hassle of natural enemies.

Euonymus plants are said to rank twelfth among the top twenty plants used by the multibillion-dollar landscaping industry. They are used as ground covers, hedges, vines, shrubs, and trees. Yet scale insects have become such a problem that many nurseries have stopped selling euonymus plants. Further, the euonymus scale attacks many other trees and shrubs, such as pachysandra, hibiscus, and camellia. In one survey of suburban properties, nearly 70 percent of the euonymus plants had problems, due mostly to the scale.

The female euonymus scale is brownish in color, about $1/16$ in. long, and shaped like an oyster shell. It is possible for the scale to have three generations in one year, so the buildup can occur at a rapid rate. The female forms an armor-like covering of wax, inserts her mouth parts, and settles into uninterrupted feeding and egg laying until she dies. Pesticide sprays cannot get to her. After the eggs hatch, the flying males live about one day—sufficient time to mate. The fertile females then form their waxy shelters, feed, lay eggs, and start another generation. Controls based on sprays and dusts are not very useful, in view of the very brief window of insect exposure during the life cycle. Also, the use of pesticides on plants infested with scale insects tends to kill any natural enemies of the scale that might exist. Therefore, biocontrol such as the use of natural enemies is about the only likely solution to the problem.

The search for the scale's natural enemies in Asia started in the early 1980s. Finding ones that could be introduced to the

PLANTING TREES AND SHRUBS

This may be done in spring, summer, or fall, as long as soil can be worked. A good practice for preparing the hole is to dig it about one-third wider than the ball or container. If the soil is hard, compacted, or of poor quality, it is advisable to dig the hole 4 to 5 in. deeper than the ball. This will allow peat moss or another soil conditioner to be added to the soil and placed under and around the root ball. Fill the hole with enough good soil to allow the ball to be the proper depth when it is put in the hole. The ball or container should be about 2 to 5 in. above the top of the hole. After cutting the twine that holds the ball together, peel back and remove the burlap. Backfill the hole around the plant, tamping the soil lightly to ensure removal of all air pockets. Continue backfilling the hole until it is level with the surrounding soil. Place a ring of soil about 4 in. high around the backfill to retain water.

In the case of a container plant, remove the plant from the container and slice the root zone with a knife or shovel. Then place the plant in the hole and backfill carefully.

United States without becoming pests themselves, and testing such organisms for effectiveness and safety here took over ten years. Further, producing and distributing most natural or biocontrol agents adds years to such a project.

Two species of beetles were found and tested, and are in the process of being multiplied, distributed, and observed. They are the red-spotted, black Asian lady beetle, *Chilocorus kuwanae,* and the $\frac{1}{25}$-in. nitidulid, *Cybocephalus* prob. *nipponicus.* Both lay their eggs under the body of the euonymus scale and in cracks of the bark and other protected places on the host plant. When the larvae hatch, they feed voraciously on the scale insects and their offspring. Further, the powerful jaws of Asian lady beetles enable them to chew through the scale's armor, burrow under the protective cover, and devour the scale insects.

The USDA and nursery and landscaping industry are hopeful that these and other predators will eventually bring the euonymus scale under control. Predatory wasps and other natural enemies are also under study. Further, it is hoped that these and other biological controls will help solve the problems with white peach scale and San Jose scale—both serious pests of the peach industry.

(A) A red-spotted ladybug devours euonymus scale insects (B) and deposits an egg to start a new generation of predators. *(Courtesy of USDA/ARS (A) #K-2848-4, and (B) K-2848-12)*

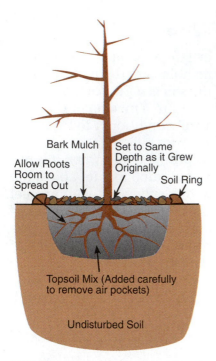

FIGURE 25-9 Tree planted and ready for guy wires and water.

After backfilling, fill the ring with several inches of water to saturate the soil to the bottom of the backfill. This will settle the soil and provide moisture for the plant (Figure 25-9).

MULCHING

Trees and shrubs will benefit from the addition of mulch around the planted area. Mulch in the form of shredded hardwood or pine straw will reduce evaporation and help hold in the moisture. A newly planted tree will take about fifteen to twenty gallons of water about twice a week in a hot, dry environment. The conservation of water is important. Other advantages of mulch is that it will help provide a more even soil temperature, help control weeds, prevent erosion, and help prevent soil compaction. Some species of plants have shallow root systems and will compete with any ground cover for nutrients. The addition of mulch will reduce this competition.

The use of mulch will prevent damage from lawn mowers or string weeders by keeping grass and ground covers away from the trunk. Mulch should be applied 3 to 4 in. deep. For estimating the mulch needs for a bed of plants, use the rule of thumb that 1 yd^3 of mulch can be spread over 100 ft^2 of area. Mulch will last for one to three years, depending on the type applied. Pine bark will need replacing annually, while shredded hardwood mulch will not need additional mulch for about three years.

As an additional bonus, consider the use of the new landscape fabrics to help control weeds. Such materials are placed under the mulch. The fabric is made of fiberglass and will last many years. It is better than sheet plastic because it will let water and fertilizer move through, whereas the plastic sheet will not.

STAKING AND GUYING

Newly planted trees and shrubs need to be staked or guyed. This is to avoid loosening and disfiguration of the plant by the wind. **Staking** is driving a wooden or metal pole into the ground near the plant and tying the upper part of the plant to the stake. **Guying** is tying a tree to stakes with wire or rope (Figure 25-10). Stakes and guys should be left in place for at least one growing season. When using guy wires, make sure the wires do not come in contact with the trunk. Use of old water hose with the wire running through is a good practice to protect the trunk. Another good practice is to tighten the wires at the stake and not at the trunk. The best method is to use a double strand of wire twisted together for tightening. Guy wires should be checked for tightness several times during the growing season.

FERTILIZING

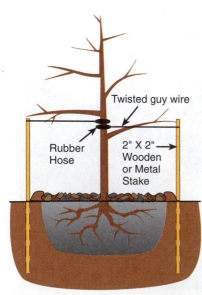

FIGURE 25-10 Newly planted tree with stakes and guy wires in place.

Plants need nutrients to maintain their vigor and to make healthy new growth. If the soil is fertile, it is not necessary to add fertilizer at planting time. Trees and shrubs should not be fertilized during the first year of growth. This practice is followed to prevent the plant from developing too much top growth in relation to the root growth. Plants should be fertilized every three to five years, starting with the growing season after the first year. Generally speaking, the best time is in the early spring. It is not a good practice to fertilize a plant after middle to late July. This practice would force new growth that will not mature enough to escape damage by the winter cold.

Unless the plants are in a bed, it is not a good practice to place fertilizer on the surface of the soil. It will wash away in the rains or will not penetrate into the root-zone area. To fertilize a tree properly, it is necessary to place the fertilizer into a hole 1 in. in diameter and 18 in. deep. Care should be taken to avoid underground utility wires and pipes, such as electric lines, telephone wires, gas pipes, and water pipes.

Holes for fertilizer are made using a 1-in. steel stake placed in concentric circles starting about halfway out from the trunk and the drip line. The **drip line** is the outer edge of the tree where the branches stop. Holes should be about 24 in. apart, and the circles should be about 24 in. apart (Figure 25-11). Each hole receives the appropriate amount of fertilizer and is sealed with soil or by closing the hole with a heel of the foot.

The amount of fertilizer needed by a tree is determined by measuring the trunk about 4½ ft above the ground. If the trunk of the tree is under 8 in., multiply the number of inches of diameter by 3. If the trunk is over 8 in., multiply by 6. This will give the number of pounds of fertilizer the tree needs. Divide the number of pounds of fertilizer by the number of holes to determine the amount of fertilizer to put in each hole.

For example, if the tree measures 6 in. at 4½ ft, multiply 6 × 3. The answer is 18 lb of fertilizer. If the tree measures 12 in. at 4½ ft, then 12 × 6 = 72 lb of fertilizer. If the tree has more than one trunk,

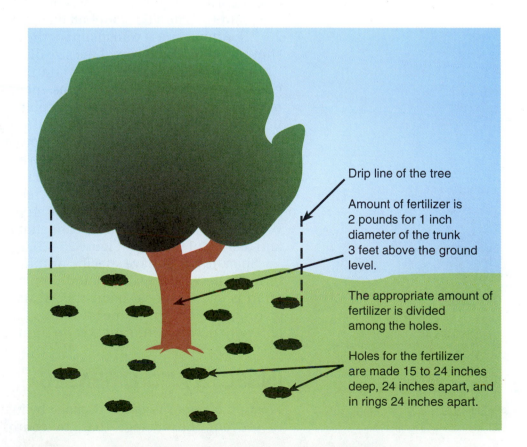

Drip line of the tree

Amount of fertilizer is 2 pounds for 1 inch diameter of the trunk 3 feet above the ground level.

The appropriate amount of fertilizer is divided among the holes.

Holes for the fertilizer are made 15 to 24 inches deep, 24 inches apart, and in rings 24 inches apart.

FIGURE 25-11 Pattern of holes for fertilizing a tree.

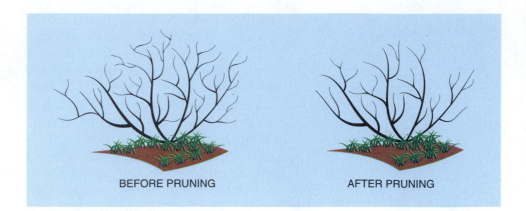

BEFORE PRUNING   AFTER PRUNING

FIGURE 25-12 Careful selection of branches for removal is important when pruning deciduous shrubs.

combine the diameters of all the trunks and multiply by the appropriate factor.

A fertilizer with an analysis of 10–6–4 or 10–10–10 is generally acceptable. Never apply more than 100 lb of fertilizer to any given tree in a year. If the rate to be applied is over 100 lb, make the application over a two-year period. Trees should be fertilized every three to five years.

PRUNING

Pruning is the removal of dead or undesirable limbs from a tree or shrub. Removing dead, broken, diseased, and insect-infested wood helps to protect the plant from additional damage. Trees may have waterspouts, bad-angle crotches, branches that cross over each other, or branches that form an asymmetrical habit. These need to be taken out or corrected. Sometimes it is necessary to reduce the top growth of the plant to match the root ball on a newly transplanted plant (Figure 25-12). After a tree is transplanted, the size of the **canopy,** or branches, should be reduced to equal the size of the root ball.

Proper pruning techniques and timing are important in growing good ornamental trees and shrubs. Incorrect pruning can leave a plant in worse condition than before. A plant in weak condition is susceptible to insects and diseases. Poor pruning can cause the loss of a season of flowers or fruit. In general, flowering plants should be pruned just after they **bloom,** or produce flowers. This is done so that the next season's flowering wood is not cut off.

It is easier to prune deciduous trees and shrubs in the late fall or winter, after the leaves drop. The framework is bare and easier to see. When removing large limbs, care should be taken to avoid splitting and tearing of the limb (Figure 25-13).

Specific plants have specific pruning requirements. It is best to determine how each plant should be pruned before the pruning operation is started. Local plant specialists are usually willing to help or give suggestions for proper pruning. Good reference books are available with accurate information on pruning particular plants.

First cut part way through the branch at A, then cut it off at B. Make the final cut, just outside the branch collar, at C.

FIGURE 25-13 Order of cuts when removing limbs.

INSECTS AND DISEASES

Most plants are subject to damage by insects and diseases. It is easier to prevent the damage than to control it. Pest-resistant varieties should be used, and they should be kept healthy and vigorous. Weakened trees and shrubs are easily attacked and damaged by insects and diseases (Figure 25-14).

It is important that plants be selected that are suited to the environment in which they are to be planted. Plants cannot tolerate stressful conditions continuously. Plants will adapt to various conditions, but some plants adapt better than others. Some plants can withstand air pollution, drought, heat, or even wet soils. Other types of plants can tolerate infertile conditions, but not hot, dry weather. Still others may not survive in heavy shade and/or in full sun.

In its lifetime, any plant can be expected to experience attack by some insects and diseases. It is wise to become aware of the plants in your area and to consult the cooperative extension service for more specifics on the types of insects and diseases that can be expected to infect various plants in a given locality.

FIGURE 25-14 Insect infestations can build up fast on susceptible plants. *(Courtesy of USDA/ARS #K-2849-11)*

STUDENT ACTIVITIES

1. Write the Terms to Know and their meanings in your notebook.

2. Contact a local nursery and obtain a nursery catalog.

3. Visit a garden center or nursery and compile a list of trees and shrubs that are available and recommended for your community.

4. Prepare a chart of popular ornamental trees and shrubs for your community. For each plant list the scientific name, common name, mature size of plant, time of flowering, color of flowers, spring leaf color, and fall leaf color.

5. Survey your home or other assigned area and make a list of trees and shrubs present. Use books, nursery catalogs, and other resources to help identify the plants.

6. For a given area, determine the diameter of each tree 4½ ft above the ground. Work up a recommended fertilizer program for the trees, including amount of fertilizer per tree, local fertilizer prices, cost of fertilizer for each tree, time of year to fertilize each, and the total cost of fertilizer for the area.

7. Prepare a pruning schedule for plants around your home or other area based on local recommendations. Include in the schedule the common and scientific names of each plant, time of pruning, and special requirements for pruning.

8. Prepare a sketch or a scale drawing of your home property or other assigned area. Locate the existing trees and shrubs on the sketch, using a circle with the initials of the scientific name to represent each plant. Add new plants that you believe will enhance the property you have surveyed.

SELF EVALUATION

A. Multiple Choice

1. An acre of plant material can produce enough oxygen to keep
   a.   5 to 10 people alive each year.
   b.   10 to 16 people alive each year.
   c.   16 to 20 people alive each year.
   d.   nobody alive. It does not produce enough oxygen to be of any value.

2. A border planting
   a.   is used as a single plant to highlight a fence or some other special feature of the landscape.
   b.   is used to separate some part of the landscape from another.
   c.   is a number of trees or shrubs planted together as a point of interest.
   d.   is a collection of plants that are placed in the landscape as needed.

3. When planning the location for planting trees and shrubs, which is not a major consideration?
   a.   fruit size and type   c.   other structures
   b.   flower color   d.   bare-rooted plants

4. The term *Cornus florida rubra* is a
   a.   common name.
   b.   scientific name.
   c.   name that was developed in Florida.
   d.   type of annual deciduous plant.

5. Planting of trees and shrubs may be done
   a.   in spring, summer, or fall.   c.   in fall only.
   b.   in spring only.   d.   in spring and summer.

6. Mulch should be applied
   a.   1 to 2 in. deep.   c.   3 to 4 in. deep.
   b.   2 to 2½ in. deep.   d.   6 in. deep to keep out the weeds.

B. Matching

_____ **1.** Urban foresters   a.   The removal of dead, broken, unwanted, diseased, and insect-infested wood

_____ **2.** Shrubs   b.   Used as a single plant to highlight it or some other special feature of the landscape

_____ **3.** Specimen plant   c.   A systematic method of naming plants

_____ **4.** Nomenclature   d.   Woody plants that normally grow low and produce many stems or shoots from the base

_____ **5.** Pruning   e.   The top of the plant; has the framework and leaves

_____ **6.** Canopy   f.   Help install and maintain trees in large cities

C. Completion

1. Plant material can cut noise pollution by _____.

2. A rule of thumb is that 1 yd³ of mulch can be spread over a _____ ft² area.

3. Newly planted trees may have to be _____ to prevent wind damage.

4. Plants should be fertilized every _____ to _____ years, starting with the growing season after the first year.

5. Proper pruning techniques and timing are important in growing good ornamental trees and _____.

6. _____ _____ is the systematic cutting of the roots by hand or machine to encourage the roots to develop close to the trunk.

D. True or False

1. Trees and plants are valuable only for beauty.

2. Trees can act as insulators, keeping a house cool in the summer and warm in the winter.

3. A well-designed landscape does not increase the value of a property.

4. All plants have the same type of growth habit.

5. It is not a good practice to fertilize trees after middle to late July.

6. Most plants are subject to damage from insects and/or diseases at some time.

Animal Sciences

Animals of the Future

Making Better Sheep

Fifty years ago, desert range sheep averaged a 70 percent lamb crop—that is, each one hundred females averaged a total of seventy live lambs per year. Today that figure is 85 percent, and Hudson Glimp, Director of the U.S. Sheep Experiment Station at Dubois, Idaho, believes the figure will be 150 percent in fifty years. Further, he projects that under intensive confinement production, it is reasonable to expect a national average of 400 percent. How will we do this? The Booroola Merino sheep, which originated in Australia, has some individual females that have five or six lambs instead of the typical one or two. There is hope that the gene responsible for the multiple lambings can be isolated and bred into the ge-

netic makeup of our most productive breeds. That would solve the numbers game, but how will the sheep and their managers keep so many newborn lambs alive and growing?

More Milk from Fewer Cows

Fifty years ago, the United States had 25 million dairy cows and the national average milk production per cow was 4,600 lb. In 1995, the figures were 10.1 million cows averaging over 14,000 lb of milk each. Some believe that in fifty years we will have herds averaging 40,000 lb per cow per year.

Will great increases in the future necessitate bigger cows? Larry Satter, dairy scientist at the USDA Dairy For-

Improvements in animal production are due more and more to genetic engineering and other laboratory procedures. *(Courtesy USDA/ARS #K-4134-6)*

age Research Center in Madison, Wisconsin, does not believe so. Better feed could make conversion of feed to milk a more efficient process. Currently, only 30 to 40 percent of the material in plant cell walls is available to the animal. The rest simply passes through. If plant cell walls could be modified to make them more digestible without diminishing plant health and performance, then the animal could extract more nutrients from a given amount of feed.

The discovery of the hormone bovine somatotropin (BST) is already affecting the cow's partitioning of calories between milk and other body functions. The use of BST causes the cow to shift more of her calories into milk, thus increasing her conversion rate. The widespread use of BST could cause an immediate increase in milk production by 15 percent or more. The future will bring more progress in this and other areas.

Preventing Poultry Diseases

Devastating diseases, such as avian influenza and Newcastle disease, still threaten to eliminate whole flocks in the poultry industry. Both viruses have strains with disease-producing capability that range from ones not killing any birds to ones that will kill 100 percent of the birds. So, upon the diagnosis of a few birds with the disease, the entire flock is destroyed. The movement of people and poultry in the locality, and even state or region, may be restricted. The economic loss is staggering. Scientists struggle to keep ahead of such diseases by developing vaccines and practicing strict quarantines.

Charles Beard, veterinary virologist at the USDA Southeast Poultry Research Laboratory at Athens, Georgia, hopes to find a better way to manage these poultry pathogens. The hope is to determine the fundamental genetic traits of the viruses that will enable prediction of their behavior in poultry. With genetic markers, the scientists could determine if a given virus in a bird or sample of birds is a mild one or a lethal one. The decision about destroying birds can then be made more intelligently. DNA probes for examining bird tissue promise to be a useful diagnostic tool to aid in the process.

Occupying the Parking Spot

Eradicating the African swine fever is a dream of the swine industry. So far, it has been an elusive dream. However, a genetic engineering technique using what is referred to as "the nonsense gene" has some promise.

Certain viruses, such as that of the African swine fever, have the ability to incorporate themselves into the genetic material of an animal and be carried into the animal's offspring. However, the virus has to establish itself in a particular spot on the genetic material. If a modified version of the virus that cannot cause the disease can be engineered, then animals can be infected with the safe virus. The safe virus would become "nonsense genes," occupying the "parking spot" on the genetic material and preventing the disease-causing virus from establishing itself.

In summary, the animals of the future will continue to improve in productivity and function. Much research on farm animals focuses on increasing productivity to help feed the world's expanding population. Additionally, research focuses on the feeding and management of pets and other animals for recreational uses. Research deals increasingly with learning the needs and behavior of animals in the wild and providing improved habitat according to the unique needs of wildlife.

Animal Anatomy, Physiology, and Nutrition

OBJECTIVE

To determine the nutritional requirements of animals and how to satisfy those requirements.

MATERIALS LIST

✓ commercial feed tags from various feeds

✓ charts of various animal digestive systems

COMPETENCIES TO BE DEVELOPED

After studying this unit, you should be able to:

- compare animal digestive systems.
- understand the basics of animal physiology.
- understand how nutrients are used by animals.
- identify classes and sources of nutrients.
- identify symptoms of nutrient deficiencies.
- explain the role of feed additives in livestock nutrition.
- compare the composition of various feedstuffs.

Nutrition	Lymph glands	Maltose
Scurvy	Carbohydrates	Lactose
Anorexia	Respiratory system	Starch
Obesity	Central nervous system	Cellulose
Ration	Peripheral nervous system	Fat
Balanced ration	Urinary system	Supplement
Deficiency diseases	Endocrine or hormone system	Synthetic nutrients
Vitamins	Hormones	Oil meal
Minerals	Digestive system	Urea
Physiology	Ruminants	Feed additive
Anatomy	Rumen	Antibiotics
Skeletal system	Roughage	Dry matter
Bone	Monogastric	TDN
Bone marrow	Concentrates	By-product
Muscular system	Lactation	Hay
Voluntary muscles	Glucose	Legume
Involuntary muscles	Fructose	Green roughages
Proteins	Galactose	Silage
Circulatory system	Sucrose	

FIGURE 26-1 Animal health, growth, and reproduction all are directly related to nutrition. (*Courtesy of USDA/ARS #K-4597-7*)

eed is animal food. It represents the largest single cost item in the production of livestock. Therefore, it is important to understand the complex nature of animal nutrition and how animals use the feed that they eat. **Nutrition** is the process by which animals eat food and use it to live, grow, and reproduce (Figure 26-1).

"You are what you eat" is an axiom that is true to a considerable extent, especially as it relates to good health of both humans and animals. This unit explores the relationship between good nutrition and good health.

NUTRITION IN HUMAN AND ANIMAL HEALTH

The relationship between proper nutrition and health has long been recognized. Early sailors stocked their sailing vessels with limes when going to sea for long periods. This was to prevent the dreaded disease, scurvy. **Scurvy** is a disease of the gums and skin caused by a deficiency of vitamin C in the diet. Even today, the effects of poor nutrition are seen in the human problems of anorexia and obesity. In simple terms,

anorexia is a result of too little nutrition, and **obesity** the result of too much or improper types of food being eaten.

In animals, proper nutrition is just as important as it is in humans. Feed efficiency, rate of gain, and days to market weight are all uppermost in the minds of those people who raise livestock for meat. Proper nutrition is just as important for animals being grown for milk, wool, or fur production. Slow growth, poor reproduction, lowered production, and poor health are generally the result of less-than-adequate animal rations. The amount and content of food eaten by an animal in one day is referred to as the animal's **ration.** When the amount of feed consumed by an animal in twenty-four hours contains all of the needed nutrients in the proper proportions and amounts, the ration is referred to as a **balanced ration.**

Numerous diseases may result from improper amounts or balances of vitamins and minerals. Such diseases are called **deficiency diseases. Vitamins** are complex chemicals and **minerals** are elements essential for normal body functioning of humans and animals alike. Not all types of animals require the same vitamins and minerals to maintain good health.

AGRI·PROFILE

CAREER AREAS: Animal Nutritionist/Feed Formulator/ Animal Manager/Physiologist/Veterinarian

Careers in animal nutrition are varied and interesting. One may work essentially as an organic chemist or technician in a laboratory where complex equipment is used for research, analysis, and discovery of animals' needs and the nutritional values of feedstuffs. One also may be a business person selling feeds, may produce feedstuffs, or may raise animals and manage nutrition, along with performing other management practices.

Certain elements of nutrition are basic, such as the composition of feed grains, animal by-products, and the basic nutrients needed by animals. However, the nutritive content of forages and other feedstuffs vary considerably according to stage of growth, condition, and quality; and the digestive capabilities of animals vary considerably from species to species. Additionally, nutritional needs of animals vary with age, stage of development, production, and pregnancy.

Careers in nutrition may be predominantly in the basic sciences or may be applied. They may focus on fish, small animals, pets, equines, poultry, livestock, dairy, or wild animals.

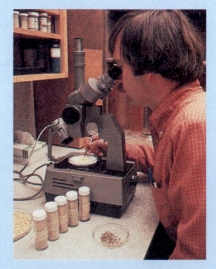

Nutrition is one of the most critical factors in animal production. Nutritional research, feed analysis, and animal evaluation are ongoing processes with production animals and pets alike. *(Courtesy of Jim Strawser)*

ANIMAL ANATOMY AND PHYSIOLOGY

The internal functions and vital processes of animals and their organs is referred to as animal **physiology.** The various body systems, such as the skeletal, muscular, circulatory, respiratory, nervous, urinary, endocrine, digestive, and reproductive systems, must all be properly fed and working together in order for the animal to be healthy and productive. To this end, proper nutrition is a must. The various parts of the body are collectively known as **anatomy.**

Skeletal System

The **skeletal system** (Figure 26-2) is made up of bones joined together by cartilage and ligaments. The purpose of the skeletal system is to

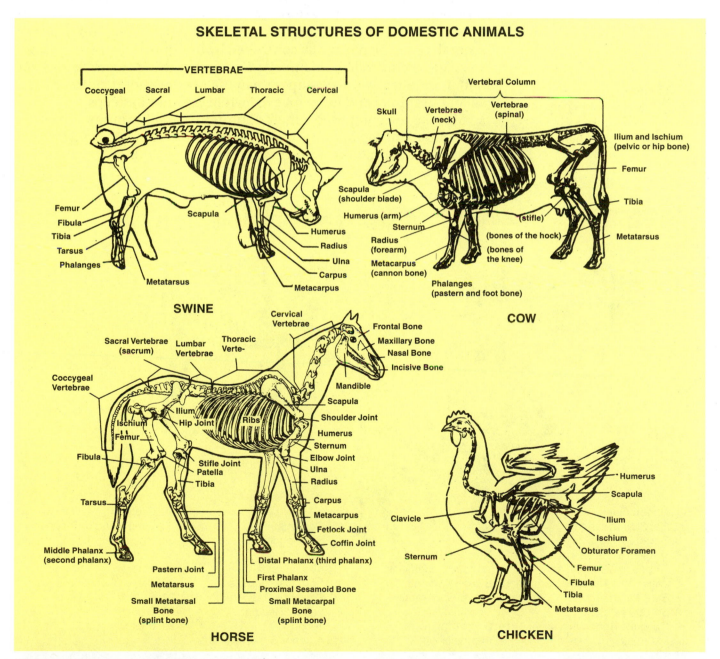

FIGURE 26-2 The skeletal system provides support for the body and protection for the soft organs. *(Courtesy of IMS, Texas A & M University)*

provide support for the body, and protection for the brain and other soft organs of the body.

Bone is the main component of the skeletal system. It is composed of about 26 percent minerals. This mineral material is mostly calcium phosphate and calcium carbonate. Another 50 percent of bone is water, 20 percent protein, and 4 percent fat.

The material inside bones is called **bone marrow,** and it produces blood cells. The growth and strength of bones are greatly affected by minerals and vitamins in animal rations.

Muscular System

The **muscular system** (Figure 26-3) is the lean meat of the animal and the part of the body that is used for human consumption (food). The purposes of muscles are to provide for movement in cooperation with the skeletal system and to support life.

Muscles may be voluntary or involuntary, depending on whether or not they can be physically controlled by the animal. **Voluntary muscles** can be controlled by animals to do such things as walk and eat food. **Involuntary muscles** operate in the body without control by the will of the animal and function even while the animal sleeps.

Muscles are composed largely of protein. Large amounts of protein are required for the maintenance of the animal and for growth and reproduction. **Proteins** are nutrients made up of amino acids and are building blocks of muscles.

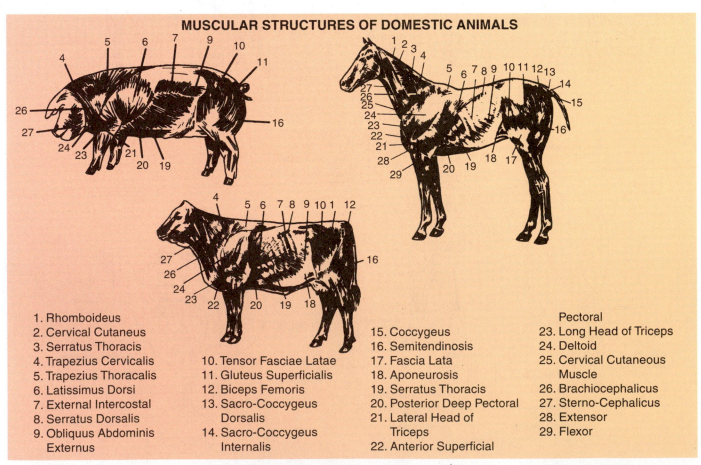

MUSCULAR STRUCTURES OF DOMESTIC ANIMALS

1. Rhomboideus
2. Cervical Cutaneus
3. Serratus Thoracis
4. Trapezius Cervicalis
5. Trapezius Thoracalis
6. Latissimus Dorsi
7. External Intercostal
8. Serratus Dorsalis
9. Obliquus Abdominis Externus
10. Tensor Fasciae Latae
11. Gluteus Superficialis
12. Biceps Femoris
13. Sacro-Coccygeus Dorsalis
14. Sacro-Coccygeus Internalis
15. Coccygeus
16. Semitendinosis
17. Fascia Lata
18. Aponeurosis
19. Serratus Thoracis
20. Posterior Deep Pectoral
21. Lateral Head of Triceps
22. Anterior Superficial Pectoral
23. Long Head of Triceps
24. Deltoid
25. Cervical Cutaneous Muscle
26. Brachiocephalicus
27. Sterno-Cephalicus
28. Extensor
29. Flexor

FIGURE 26-3 The muscular system provides for movement and supports life. *(Courtesy of IMS, Texas A & M University)*

CIRCULATORY SYSTEMS OF DOMESTIC ANIMALS

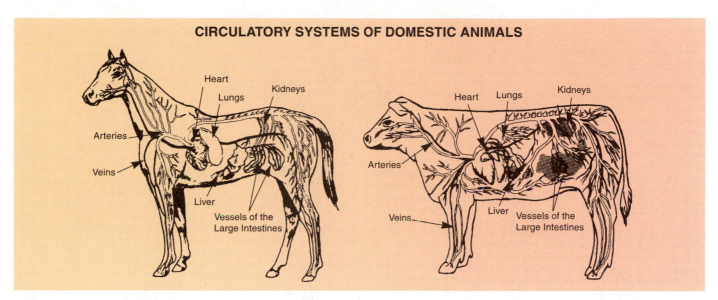

FIGURE 26-4 The circulatory system carries food and oxygen to the body. *(Courtesy of IMS, Texas A & M University)*

Circulatory System

The heart, veins, arteries, and lymph glands compose the **circulatory system** (Figure 26-4). This system provides food and oxygen to the cells of the body and filters waste materials from the body. **Lymph glands** secrete disease-fighting materials for the body.

Vitamins, minerals, proteins, and carbohydrates are all essential for the smooth running of the circulatory system. **Carbohydrates** are sugars and starches that supply energy to the animal.

Respiratory System

The **respiratory system** provides oxygen to the blood of the animal. This system is composed of the nostrils, nasal cavity, pharynx, larynx, trachea, and lungs (Figure 26-5). This system allows for breathing and makes use of the muscular and skeletal systems to draw air in and out of the lungs. Oxygen is then taken from the lungs and distributed to the cells of the body by the circulatory system.

RESPIRATORY SYSTEMS OF DOMESTIC ANIMALS

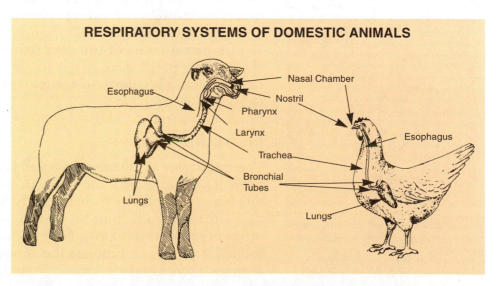

FIGURE 26-5 The respiratory system provides oxygen to the blood. *(Courtesy of IMS, Texas A & M University)*

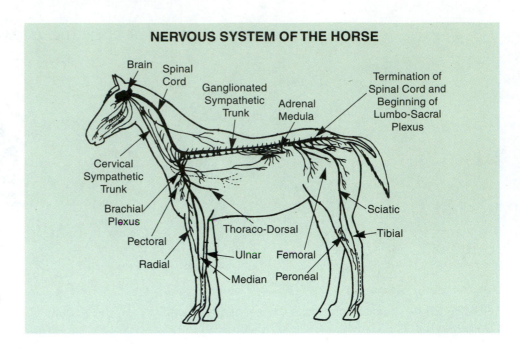

NERVOUS SYSTEM OF THE HORSE

FIGURE 26-6 The nervous system coordinates the senses of animals. *(Courtesy of IMS, Texas A & M University)*

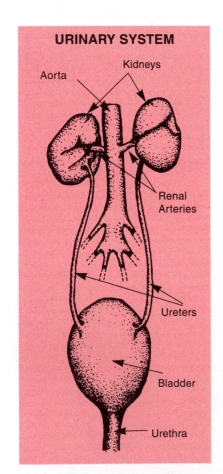

URINARY SYSTEM

FIGURE 26-7 The urinary system removes waste materials from the body. *(Courtesy of IMS, Texas A & M University)*

Nervous System

The nervous system of animals is composed of the central nervous system and the peripheral nervous system (Figure 26-6).

The **central nervous system** includes the brain and the spinal cord. It is responsible for coordinating the movements of animals and also responds to all of the senses. The senses are hearing, sight, smell, touch, and taste.

The **peripheral nervous system** controls the functions of the body tissues, including the organs. The nerves transmit messages to the brain from the outer parts of the body.

Because the nervous system is composed primarily of soft tissues, proteins are particularly important in maintaining its health.

Urinary System

The function of the **urinary system** is to remove waste materials from the blood. The primary parts are the kidneys, bladder, ureters, and urethra (Figure 26-7). The kidneys also help regulate the makeup of blood and maintain other internal systems.

Abnormal levels of proteins fed to animals have been known to cause stress to the urinary system, which rids the body of excess protein. Similarly, higher-than-recommended levels of minerals may also cause kidney problems.

Endocrine System

The **endocrine or hormone system** is a group of ductless glands that release hormones into the body. **Hormones** are chemicals that regulate many of the activities of the body. Some of these are growth, reproduction, milk production, and breathing rate. Hormones are needed in only very minute amounts. For example, only 1/100,000,000 g of oxytocin hormone will stimulate the almost immediate letdown of milk in females. Oxytocin is a hormone from the hypothalmus gland.

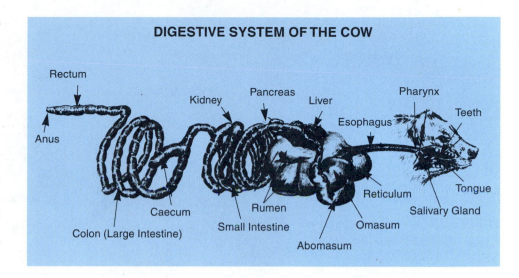

DIGESTIVE SYSTEM OF THE COW

FIGURE 26-8 The ruminant digestive system can utilize large amounts of roughages. *(Courtesy of IMS, Texas A & M University)*

Proper levels of all nutrients, especially minerals, are important for the proper functioning of the endocrine system.

Digestive System

The **digestive system** of animals provides food for the body and for all of its systems. This system stores food temporarily, prepares food for use by the body, and removes waste products from the body.

There are three basic types of digestive systems in animals of agriscience importance. They are polygastric, or ruminant; monogastric; and poultry.

Polygastric, or Ruminant, System   **Ruminants** are a class of animals that have stomachs with more than one compartment (Figure 26-8). Cattle and sheep are ruminants and have multicompartment stomachs. The very large compartment is called the **rumen.** The rumen can store large amounts of roughages. A **roughage** is grass, hay, silage, or other high-fiber feed. Ruminants have the ability to break down plant fibers and to use them for food far better than can animals that are not ruminants.

Manufactured in the digestive systems of ruminant animals are B-complex vitamins. Such vitamins need not be added to the diets of these animals, even though they are required by the body. It should be noted that calves do not develop true rumens until they are several months old; therefore, they need to be fed complete rations like nonruminant animals until their rumens develop.

Monogastric System   The digestive systems of swine, horses, and many other animals are called monogastric (Figure 26-9). **Monogastric** means having a stomach with one compartment. The stomachs of swine and horses are relatively small and can store only small amounts of food at any one time. Most of the digestion takes place in the small intestines. Monogastric animals are unable to break down large amounts of roughage. Therefore, their rations must be higher in concentrates. **Concentrates** are grains low in fiber and high in total digestible nutrients. Also included in the diets of monogastric animals must be B-complex vitamins, because they cannot make such vitamins in their digestive systems.

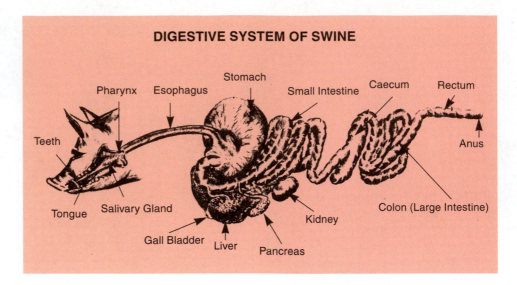

FIGURE 26-9 The monogastric digestive system has a simple stomach. *(Courtesy of IMS, Texas A & M University)*

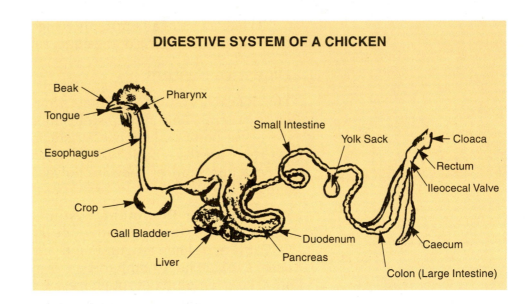

FIGURE 26-10 The poultry digestive system has no true stomach. *(Courtesy of IMS, Texas A & M University)*

Poultry Digestive System   Although poultry have monogastric digestive systems, their digestive systems are different enough to treat them separately (Figure 26-10). Chickens have no teeth and must swallow their food whole. The food is stored in a crop and passed on to the gizzard, where it is ground up. It then passes on to the small intestine for digestion. Poultry rations must be high in food value because poultry have no true stomachs and have very little room for storage of food that has been eaten.

MAJOR CLASSES OF NUTRIENTS

Water

Water is the largest component of nearly all living things. Growing plants are usually 70 to 80 percent water. Similarly, the muscles and internal organs of animals contain 75 percent or more water.

Water is the solution in which all nutrients for animals are dissolved or suspended for transport throughout the body. Water reacts with many chemical compounds in the body to help break down food into products usable by the body.

Water provides rigidity to the body, allowing it to maintain its shape. The liquid solution in each cell is responsible for this rigidity.

Water is also important in regulating the body temperature of animals, through perspiration and evaporation. Because of the ability of water to absorb and carry heat, body temperatures of animals rise and fall slower than would be possible otherwise.

Water is the least expensive nutrient for animals. However, most animals can live only a matter of days if they do not have access to it.

Protein

Protein is the major component of muscles and tissues. Proteins are very complex materials and are made of various nitrogen compounds called *amino acids.* Some amino acids are essential for animals, and some are not. Therefore, the quality of proteins fed to animals must be considered.

Monogastric animals need very specific amino acids, so it is important that they receive high-quality proteins containing the appropriate amino acids. However, in ruminant animals, quantity of protein is more important than quality. Ruminants can convert amino acids in their rumens to different amino acids to meet their needs.

Protein is used continuously by animals to maintain the body as body cells are continually dying and being replaced. In young animals, large amounts of protein are used for body growth. Protein is also important for healthful reproduction.

Carbohydrates

These are a class of nutrients composed of sugars and starches. They provide energy and heat to animals. Carbohydrates are composed primarily of the elements carbon, hydrogen, and oxygen.

The energy obtained from carbohydrates is used for growth, maintenance, work, reproduction, and **lactation** (milk production). Carbohydrates come in several forms, with the sugars being the simplest. Examples of simple sugars used in animal feeds are **glucose, fructose,** and **galactose.** Compound sugars include **sucrose, maltose,** and **lactose.** More complex forms of carbohydrates include **starch** and **cellulose.**

Carbohydrates make up about 75 percent of most animal rations, but there is very little carbohydrate in the body at any one time. Carbohydrates in the diet that are not used quickly are converted to fat and stored in the body. **Fat** has 2¼ times as much energy per gram as do carbohydrates.

Minerals

The functions of minerals in the animal are many. The skeleton is composed mostly of minerals. They are important parts of soft tissues and fluids in the body. The endocrine system is heavily dependent on various minerals, as are the circulatory, urinary, and nervous systems.

There are fifteen minerals that have been identified as being essential to the health of animals. These are calcium, phosphorus, sodium, chlorine, potassium, sulfur, iron, iodine, cobalt, copper, fluorine, manganese, molybdenum, selenium, and zinc. In the past, most of these minerals were provided naturally by feeds grown on fertile soils and by contact with the soil itself. Today, it is increasingly important to provide additional mineral matter to the diet of animals. Mineral supplements are especially important for animals that spend their lives in confinement. Minerals fed as a separate feed are called a **supplement.**

Vitamins

Vitamins are acquired by animals in several different ways. Some are available in roughages and concentrates. Some are available in feeds made from animal by-products. Finally, some are made by the body itself.

Vitamins are required in only minute quantities in animals. Vitamins act mostly as catalysts for other body processes. There are large variations in the necessity for vitamins in various species of animals important to agriscience.

Some of the specific ways that vitamins are used in animals include clotting of blood, forming bones, reproducing, keeping membranes healthy, producing milk, and preventing certain nervous-system disorders.

Fat

Only small amounts of fat are required in most animal diets. The addition of fat to the diets of animals improves the palatability, flavor, texture, and energy levels of feed. The addition of small amounts of fat to the diet has also been shown to increase milk production and to aid in the fattening of meat animals. Fats are also necessary in the body as carriers of fat-soluble vitamins.

SOURCES OF NUTRIENTS

The sources of nutrients for animals are many and extremely varied. Important animal feed components include roughages; concentrates; animal by-products; minerals from mineral deposits; and nutrients made chemically, called **synthetic nutrients.**

Proteins

The major sources of protein for animals include oil seeds such as soybeans, peanuts, cottonseed, and linseed. These seeds are processed by cooking and other procedures to remove the bulk of the oil from them. The remainder of the seed content is then dried and ground up for feed. Feed consisting of ground oil seeds with the oil removed is called **oil meal.**

Cereal grains provide lesser amounts of protein than does oil meal, but they are also important protein sources. Good-quality legume hay, such as alfalfa or clover, is also a good plant source of protein for ruminant animals.

Animal protein is generally of higher quality than plant protein. More specifically, animal protein usually contains more of the essential amino acids than does protein from plants. Sources of animal protein include tankage, fish meal, blood meal, skim milk, feather meal, and meat scrap.

Nonprotein nitrogen in the form of urea can be used as a substitute for protein for ruminant animals. **Urea** is a synthetic source of nitrogen made from air, water, and carbon. The feeding of urea should be limited to not more than 1 percent of the total dry matter fed. Young ruminants and all nonruminants are unable to digest urea.

Carbohydrates

Carbohydrates are found in all plant materials. The major sources of carbohydrates for animal feed are the cereal grains. Corn is the most important of these in the United States, followed by wheat, barley, oats, and rye.

BIO•TECH CONNECTION

Feeding for Two

The nutrient needs of dairy fetuses have become the research focus of animal physiologist William House of the USDA Plant, Soil, and Nutrition Laboratory of Ithaca, New York, and of animal science professor Alan Bell of Cornell University. Their immediate goal is to produce a complete profile of fetal mineral content for dairy cattle, including a look at how the proportion of various minerals changes as the fetus progresses through gestation. Such information will help producers feed the correct amounts of minerals to pregnant dairy cows to ensure the birth of healthy calves.

Healthy cows absorb about 45 percent of the calcium they eat. Therefore, they must eat nearly twice the combined amount needed for their own bodies and that of their fetuses. Before the team began their study, only calcium and phosphorus content of dairy fetuses had been measured. In their current study, they have determined the levels of potassium, sodium, magnesium, iron, zinc, copper, and manganese, as well as calcium and phosphorus taken up by the fetus, fetal membranes, fluids, and placement tissue.

House and Bell have confirmed that the amounts of calcium, phosphorus, and magnesium increase as the fetus grows. However, surprisingly, they discovered that the

USDA animal physiologist William House and animal science professor Alan Bell of Cornell University display a two-week-old calf born from a cow in their nutrient study. *(Courtesy of USDA/ARS #K-5177-1)*

amounts of potassium and sodium decrease near the end of the term. During the latter stages of gestation, it was concluded, a fetus each day accrues an average of 0.2 g magnesium, 1 g potassium, 1.4 g sodium, 18 mg iron, 11.7 mg zinc, 1.6 mg copper, and 0.3 mg manganese. This information, combined with knowledge about the mother's body and production requirements, will enable managers to better meet the nutritional needs of dairy cows during their gestation periods.

Other sources of carbohydrates include nonlegume hays such as orchard grass, timothy, other grasses, and molasses.

Normal animal rations generally contain adequate levels of carbohydrates.

Fats

Because fats are needed in fairly small amounts in the diets of animals, it is seldom necessary to identify specific sources of dietary fat. Most sources of protein are also sources of fat. This is especially true for the oil seeds and animal by-products.

Vitamins and Minerals

Vitamins and minerals are part of all the normal feeds of animals. Ruminants manufacture B-complex vitamins in their rumens. Exposure to sunlight allows the body to manufacture vitamin D. Contact with the soil, coupled with other feeds grown on fertile land, provides most of the mineral requirements of animals that have access to pasture and high-quality feeds.

However, it is sometimes necessary to supplement natural sources of vitamins and minerals. Commercial vitamin and mineral supplements are formulated for specific classes of animals and their special needs. Such supplements are available wherever animals exist in the developed countries of the world.

SYMPTOMS OF NUTRIENT DEFICIENCIES

Animals must be fed appropriate types and amounts of feed regularly to remain healthy and to produce milk, meat, wool, eggs, fur, work, or healthy young. Shortages, or deficiencies, of various nutrients will generally produce observable effects in animals. Some common symptoms of nutrient deficiencies are described in Figure 26-11.

FEED ADDITIVES

A **feed additive** is a nonnutritive substance that is added to feed to promote more rapid growth, to increase feed efficiency, or to maintain or improve health. Feed additives fall into two major groups—growth regulators (mostly hormones) and antibiotics. **Antibiotics** are substances used to help prevent or control diseases.

Some common growth regulators include stilbestrol, progesterone, and testosterone. They increase growth rates and feed efficiency by as much as 5 percent.

There are a wide range of antibiotics that are added in very low levels to the diets of animals such as swine and poultry. Antibiotics keep certain low-grade infections at bay. Antibiotics added to feed allow growing animals to gain weight at their highest potential.

In recent years, there has been a good deal of controversy over including antibiotics and growth hormones in the feed of animals. Of major concern is the possibility of these substances remaining in the meat of animals slaughtered for human consumption. To reduce this possibility, the substances generally must be removed from the feed well before the animals are marketed.

Disorders Caused by:

Vitamin Deficiency	Mineral Deficiency

Vitamin A
1. Night blindness
2. Loss of young
3. Poor growth
4. Nasal discharge
5. Diarrhea

Vitamin C
1. Scurvy
2. Gum inflammation
3. Hemorrhages
4. Slow healing of wounds

Vitamin D
1. Bone weakness and deformities
2. *Rickets* and *osteomalacia* are bone disease in young and old animals, respectively
3. Thin egg shells
4. Weak, deformed young
5. Lowered milk production

Vitamin E
1. Reproductive failures
2. Degeneration of certain muscles
3. *Stiff lamb disease*, or muscle degeneration in lambs
4. *White muscle disease*, or muscle degeneration in young calves
5. Poor egg hatchability

Vitamin K
1. Poor blood clotting
2. Internal hemorrhages

Thiamine
1. Poor appetite
2. Slow growth
3. Weakness
4. Nervousness

Riboflavin
1. Slow growth
2. *Dermatitis*, or skin disorder
3. Eye abnormalities
4. Diarrhea
5. Weak legs in pigs

Niacin
1. Dermatitis
2. Retarded growth
3. Digestive troubles

Pyridoxine
1. *Anemia*, or low red-blood-cell count
2. Poor growth
3. Convulsions in pigs

Pantothenic Acid
1. "Goose-stepping" in pigs
2. Unhealthy appearance
3. Digestive problems

Biotin
1. Dermatitis
2. Loss of hair
3. Retarded growth

Choline
1. Poor coordination
2. Poor health
3. Fatty liver
4. Poor reproduction in swine

Folic acid
1. Blood disorders
2. Poor growth

Vitamin B$_{12}$
1. Slow growth
2. Poor coordination
3. Poor reproduction

Calcium
1. Rickets
2. Poor growth
3. Deformed bones
4. Milk fever

Phosphorus
1. Lameness
2. Stiff joints
3. Rickets
4. Poor milk production

Sodium Chloride
1. Lack of appetite
2. Unhealthy appearance
3. Slow growth

Potassium
1. Slow growth
2. Joint stiffness
3. Poor feed efficiency

Sulfur
1. General unthriftiness (Lack of strong growth)
2. Poor growth

Iron
1. Anemia
2. Labored breathing
3. Edema (swelling) of the head and shoulders
4. Flabby, wrinkled skin

Iodine
1. *Goiter*, or enlarged thyroid gland in the neck
2. Weak or dead offspring at birth
3. Hairlessness
4. Infected navels, especially in foals

Cobalt
1. Delayed sexual development
2. Poor appetite
3. Slow growth
4. Decreased milk and wool production

Copper
1. Abnormal wool growth
2. Poor muscular coordination
3. Anemia
4. Weakness at birth

Fluorine
Poor teeth

Manganese
1. Poor fertility
2. Deformed young
3. Poor growth

Molybdenum
Poor growth rate

Selenium
1. Muscular degeneration
2. Heart failure
3. Paralysis
4. Poor growth

Zinc
1. Poor growth
2. Unhealthy wool or hair
3. Slow healing of wounds
4. *Parakeratosis*, or a skin disease similar to mange

FIGURE 26-11 Some disorders caused by vitamin and mineral deficiencies in animals.

COMPOSITION OF FEEDS

All feeds are composed of water and dry matter. The material left after all water has been removed from feed is **dry matter.** Water makes up 70 to 80 percent of most living things. However, dry feeds generally contain only 10 to 20 percent water.

Dry matter is made up of organic matter and ash or mineral. The organic-matter portion of animal feed consists of protein; carbohydrates, such as starch and sugar; fat; and some vitamins. The proportion of these materials varies widely among different feeds.

CLASSIFICATION OF FEED MATERIALS

In general, feed for animals is classified into two types—concentrates and roughages. Concentrates are low in fiber and high in total digestible nutrients, abbreviated as **TDN.** Total digestible nutrients include all of the digestible protein, digestible nitrogen-free extract, digestible crude fiber, and 2¼ times the digestible fat contained in the ration. On the other hand, roughages are high in fiber and low in TDN.

Concentrates

Included under the classification of concentrates are the feed, or cereal, grains. These include corn, wheat, oats, barley, rye, and milo, as well as many others. These grains make up the bulk of most concentrates.

Grain by-products, such as wheat bran, wheat middlings, brewer's grain, and distiller's grain, are concentrates. They are materials left over from the production processes used in making flour and alcohol. A **by-product** is a secondary product resulting from the production of a primary commodity.

The oil meals are by-products resulting from the processing of vegetable oil from oil seeds. Both oil meals and sugar in the form of cane molasses and beet molasses are considered concentrates used in animal feeds.

Finally, animal by-products are important concentrates. They include tankage, fish meal, meat scraps, blood meal, feather meal, and dried dairy products. These products are some of the by-products of the meat- and dairy-processing industries.

Roughages

Roughages can be divided into three categories—dry, green, and silage. The most important of the dry roughages is **hay.** Some types of legume hay are alfalfa, clover, lespedeza, soybean, and peanut. A **legume** is a plant in which certain bacteria can transform nitrogen in the air to nitrogen that plants can use. Grass hays include timothy, orchard grass, bromegrass, Bermudagrass, and others. The hulls of cottonseed, peanuts, and rice also fall into the category of dry roughages.

Green roughages are plant materials with high moisture content, such as grasses in pastures and root plants (including sugar beets, turnips, and rutabagas). Tubers such as potatoes are also a green roughage.

Silage is the feed that results from the storage and fermentation of green crops. The fermentation takes place in the absence of air. Corn silage is the most important member of this group. Other examples are grass, legume, and small-grain silages.

The successful production of animals for fun, profit, or sport requires proper animal nutrition. A knowledge of animal physiology, feed materials, and nutrition helps one keep animals healthy and productive. Further, a knowledge of nutrition-deficiency symptoms will permit the animal manager to take corrective steps when the animal suffers from improper nutrition.

STUDENT ACTIVITIES

1. Write the Terms to Know and their meanings in your notebook.

2. Obtain several different commercial feed tags and compare the percentages of protein, fat, TDN, and other nutrients listed on the tag. Also make note of any vitamins and mineral supplements that are part of the ingredients. Types of feed additives should also be noted. If the price is known, try to determine what causes variations in the prices of the various feeds.

3. Compare the digestive systems of ruminants, nonruminants, and poultry. Note the parts that are alike and those that are different.

4. Arrange to have the digestive tracts of ruminant and nonruminant animals dissected and compare the contents of each.

SELF EVALUATION

A. Multiple Choice

1. _____ are most important for the formation of bone.
   a. Vitamins
   b. Proteins
   c. Minerals
   d. Carbohydrates

2. Sugar, starch, and cellulose are all examples of
   a. carbohydrates.
   b. minerals.
   c. fats.
   d. proteins.

3. The purpose of additional fat in the diet is to
   a. improve palatability.
   b. increase energy levels.
   c. improve feed texture.
   d. all of the above.

4. Night blindness may be caused by a deficiency of
   a. zinc.
   b. vitamin A.
   c. protein.
   d. biotin.

5. Improved feed efficiency can be accomplished by adding _____ to animal feeds.

    a.   fats   c.   additives

    b.   minerals   d.   roughages

6. Grain, oil meal, molasses, and meat by-products are all forms of

    a.   roughage.   c.   carbohydrate.

    b.   concentrate.   d.   dry matter.

7. Dry matter is made up of organic matter and

    a.   mineral.   c.   fat.

    b.   water.   d.   concentrate.

B. Matching (Group I)

_____	**1.** Goiter	a.	Iron
_____	**2.** Parakeratosis	b.	Vitamin D
_____	**3.** Rickets	c.	Vitamin E
_____	**4.** White muscle disease	d.	Zinc
_____	**5.** Anemia	e.	Iodine
_____	**6.** Dermatitis	f.	Niacin

Matching (Group II)

_____	**1.** Ration	a.	Feed high in TDN, low in fiber
_____	**2.** Ruminant	b.	Disease-control substance
_____	**3.** Carbohydrate	c.	Essential element
_____	**4.** Mineral	d.	Amount of feed fed in one day
_____	**5.** Antibiotic	e.	Fermented green roughage
_____	**6.** Silage	f.	Sugars and starches
_____	**7.** Concentrate	g.	Multicompartment stomach

Animal Health

OBJECTIVE

To determine how to best maintain animal health.

COMPETENCIES TO BE DEVELOPED

After studying this unit, you should be able to:
- identify signs of good and poor animal health.
- identify symptoms of animal diseases and parasites.
- understand how to prevent animal health problems.
- explain various methods of treating animal health problems.

MATERIALS LIST

✓ various gauges of needles and syringes

✓ various containers or labels from containers of animal drugs

✓ discarded ears from a meat processing plant

Diseases
Parasites
Syringe
Disinfectant
Feedlots
Host animal
Contagious
Noncontagious
Abortion
Roundworms
Flukes

Protozoa
Secondary host
Mange
Vaccination
Balling gun
Drenching
Injection
Intravenous
Intramuscular
Subcutaneous
Intradermal

Intraruminal
Intraperitoneal
Infusion
Udder
Teats
Cannula
Dipping
Rectum
Immune
Veterinarian

Maintaining animal health is the key to a profitable and satisfying animal enterprise. There are several considerations that need to be made in dealing with the health of animals. These include being able to recognize signs of good and poor health; maintaining a healthy environment; being able to identify animal diseases and parasites; and knowing how to treat health problems when they occur. These items will be explored in this unit. **Diseases** are infective agents that result in lowered health in living things. **Parasites** are animals that live on other animals and derive their food from their hosts.

SIGNS OF GOOD AND POOR ANIMAL HEALTH

Having the ability to recognize the signs of good health or the symptoms of health problems is the single most important key to being efficient in maintaining good animal health. A keen sense of observation is important, as well as the innate ability to know when something is not right with an animal (Figure 27-1).

Signs of Good Health

One of the best signs of good health is simply a contented animal. Of course, a good deal of experience in dealing with animals is necessary to recognize contentment. Alertness and the chewing of the cud in

503

FIGURE 27-1 The good animal manager develops a keen sense of observation and learns to recognize the symptoms of an animal under stress. *(Courtesy of USDA)*

ruminant animals is a good sign. A shiny hair coat, bright eyes, and pink membranes are other signs that an animal is healthy. Normal body discharges of urine and feces are further evidences that animals are not suffering from serious health problems. On the technical side, a healthy animal should have a normal body temperature, pulse rate, and respiration, or breathing, rate (Figure 27-2).

Signs of Poor Health

Often it is easier to tell when an animal is sick than to tell when it is healthy. A rough hair coat and dull, glassy eyes are often the first signs that an animal is not well. Sick animals usually stay alone with their heads down. Such animals may be drawn up and may walk slowly when forced to walk. Abnormal feces, either too hard or too soft, as well as discolored urine, may also indicate that an animal is suffering from some health problem. Lowered production, especially in dairy cattle, is often the first sign that the animal is not well. High temperatures, labored breathing, and rapid pulse rates are other indications of poor health in animals.

HEALTHFUL ENVIRONMENTS FOR ANIMALS

Maintaining a healthy environment for animals is a key factor in a complete animal-health program. It is often much less expensive to maintain a healthy environment for animals than it is to treat animals that are unhealthy due to poor conditions that occur (Figure 27-3).

Sanitation

Good sanitation is important to good health. Factors related to good sanitation include keeping facilities for animals clean. Sanitation also requires the use of clean equipment when dealing with animals. This includes feed containers; milking equipment; artificial-breeding equipment; needles and syringes; and surgical equipment. A **syringe** is an

Class of Livestock or Poultry	Degree F Average	Degree F Range
Cattle	101.5	100.4-102.8
Sheep	102.3	100.9-103.8
Goats	103.8	101.7-105.3
Swine	102.6	102.0-103.6
Horses	100.5	99.9-100.8
Poultry	106.0	105.0-107.0

FIGURE 27-2 Normal body temperatures for animals.

FIGURE 27-3 A grim reminder that untreated disease or parasites; poisonous plants; or predators can quickly take the life of an expensive animal. *(Courtesy of USDA/ARS #K-4382-2)*

AGRI·PROFILE

CAREER AREAS: Veterinarian/Animal Health Technician

Animal pathologists, animal behaviorists, physiologists, biologists, zoologists, microbiologists, geneticists, nutritionists, and others must work together to understand the complexities of animals. Animals exist as pets, production animals, work animals, pleasure animals, fish, fowl, birds, wild animals, and specimen animals in zoos. The need for health services for animals varies with the species and use of the animal.

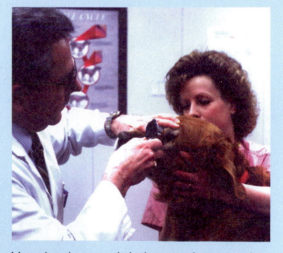

The desire to become veterinarians has been high among youth in recent years. This interest has permitted the supply of veterinarians to increase despite the rigor of the college curriculum and competition to get into veterinarian schools. Colleges of animal sciences offer curriculums in other career areas in animal sciences, such as nutrition, breeding, education, and production.

Veterinarians and their associates work constantly to treat animals having disorders and to improve animal health. *(From Warren, Small Animal Care & Management, 1995, Delmar Publishers)*

In urban and suburban areas, animal shelters, hospitals, kennels, and pet stores provide many career opportunities for those interested in animal health at the technician level. In rural areas, large-animal veterinarian, veterinary assistant, laboratory veterinarian, and laboratory technician are typical positions in the animal-health industry.

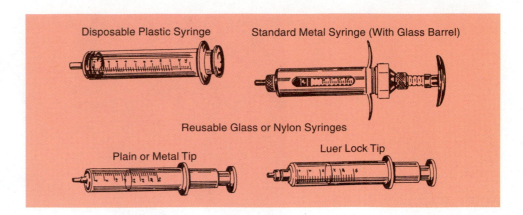

FIGURE 27-4 Types of syringes.

instrument used to give injections of medicine or to draw body fluids from animals (Figure 27-4). Simple on-farm surgical procedures should always be performed with the strictest sanitation possible. The liberal use of disinfectants in dealing with animals is also important. A **disinfectant** is a material that kills disease-causing organisms.

Housing

Maintenance of proper housing is an important consideration in maintaining good animal health. Housing should be clean and free from cold drafts. However, good air circulation throughout the housing is important to help lower high temperatures in the summer and reduce humidity in the cold of winter. Extremely dry and dusty conditions are to be avoided whenever possible. Proper maintenance of animal housing is important too. Loose boards, roofing materials, and nails often pose problems in poorly maintained facilities.

Handling Manure

Piles of manure, dirty pens, and dirty feedlots are often sources of serious health problems in animals. It is important that manure not be allowed to accumulate in areas frequented by animals (Figure 27-5). Ma-

FIGURE 27-5 An efficient manure handling system helps control disease and prevent parasites. *(Courtesy of Jim Strawser)*

nure piles often harbor diseases and parasites. They also attract flies, which may spread diseases. Cages and pens soiled continually with animal waste products may also lower the quality of the air breathed by the animals. Wet, poorly drained, manure-soiled feedlots usually reduce the rate of gain in beef cattle and swine. **Feedlots** are areas in which large numbers of animals are grown for food. Feet and leg problems can often be traced to poorly maintained feedlots.

Controlling Pests

The control of pests and parasites is also an important consideration in the maintenance of animal health and welfare. Regular use of disinfectants to control parasites such as lice and flies is necessary in a good disease-prevention program. Regular, close observation of animals may be necessary to determine when outbreaks of parasites occur. Prevention of such parasites is preferable to controlling outbreaks that occur. To that end, the development of a good prevention program is a wise decision (Figure 27-6).

The control of other pests, such as birds and wild animals, is also part of a good animal-health program. Many birds carry parasites on their bodies and in their droppings. When the parasites move from infected animals to healthy ones, they often carry diseases and parasites with them. Wild animals and pets may also cause serious health problems when allowed to roam freely around farm animals. Dogs and coyotes will often chase animals and cause them to injure themselves. Bites from these animals may also cause infection and other health problems. Just the presence of pets around farm animals may cause the farm animals to be nervous and may affect how they grow and produce.

Isolation

The isolation of animals new to a herd is an important part of any good health-prevention program. Such animals may be harboring disease or parasites that are not readily apparent. It is wise to keep them isolated from other animals for a period of time, usually a minimum of thirty

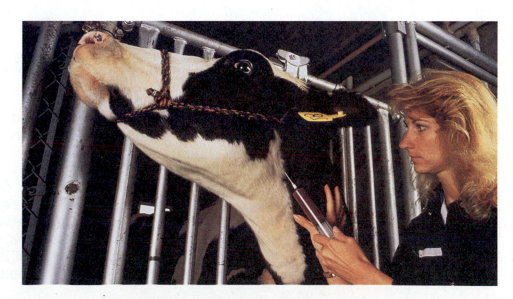

FIGURE 27-6 The drawing and analyzing of blood samples is a technique for diagnosing and treating many diseases and disorders in animals. *(Courtesy of USDA/ARS #K-4135-11)*

FIGURE 27-7 Isolation of animals when they are first brought on the premises can head off serious infection of large numbers of animals. *(Courtesy of USDA/ARS #K-4702-6)*

FIGURE 27-8 Pasture rotation is necessary to break the life cycles of diseases, insects, and internal parasites, and pays additional dividends in greater pasture production. *(Courtesy of USDA/ARS #K3714-2)*

days. This gives the new owner time to observe the isolated animals closely for health problems (Figure 27-7).

Similarly, isolation of diseased animals is important. Animals with contagious diseases that can be spread by contact should never be allowed to come into contact with healthy animals. It is difficult to treat unhealthy animals when they are living with large groups of animals. Healthy animals tend to pick on unhealthy ones, making it especially difficult for such animals to regain health.

Pasture Rotation

The rotation of pastures is a consideration in maintaining a healthy environment for animals and in preventing health problems. Many diseases of animals are harbored in the soil and are killed only by not being able to come into contact with host animals for extended periods of time. A **host animal** is an animal in or on which diseases or parasites can live. Moving animals to different pastures on a regular basis also allows for the breakdown of animal wastes and for pasture regrowth (Figure 27-8).

ANIMAL DISEASES AND PARASITES

Diseases

The diseases of animals can be divided into two major classes—contagious and noncontagious. **Contagious** diseases are those that can be passed on to other animals. **Noncontagious** diseases cannot be spread to other animals.

The handling of these two classes of diseases varies somewhat. It is important that animals suffering from contagious diseases be isolated from other animals in the herd as soon as the disease is identified. Because some contagious diseases of animals can be transmitted to humans, care must be taken when handling animals so infected. Similarly, humans handling animals should become familiar with the

BIO•TECH CONNECTION

The Unseen Harvesters

The control of internal parasites in production animals and pets is probably the most persistent animal health problem. Generally out of sight and frequently microscopic in size, parasites are small animals that live in the flesh or internal organs of larger animals and draw nutrients from the body of the hosts. Animals can become infected with the parasite and may be unaware of its presence or be helpless to do anything about it when symptoms start to appear. Fortunately, the body has some ways of helping to keep such intruders in check. If otherwise healthy and free of reinfection, the host can sometimes function fairly well with the freeloaders aboard and can sometimes even rid their systems of the parasites. Generally, however, the host needs outside help.

Farmers, ranchers, veterinarians, and scientists have long known that animals with parasites grow poorly and never reach their full potential. No matter how much they eat, they are undersized and underachievers. After all, who can do well when they have to split every meal with freeloaders? Further, research by zoologist Ron Fayer and endocrinologist Ted Elsasser of the Beltsville Agricultural Research Center, and of others, confirms that internal parasites do more than just consume nutrients intended for the host.

The animal's immune system responds to invading organisms by producing chemical signals that modify the host's metabolism. These immune-response signals are small proteins called *cytokines* that manipulate the hormones that regulate feed intake, nutrient use, and, ultimately, growth of the animals. Dairy and beef calves infected with a protozoan parasite called *Sarcocystics* run fevers, lose their appetites, and become emaciated. Even after an animal has been treated and the body ridden of the parasite, some calves still do not grow normally.

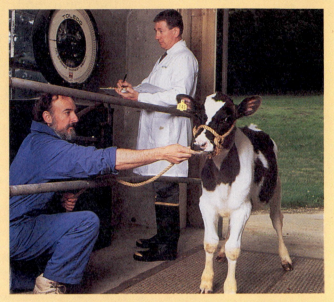

Animal scientist Ted Elsasser and zoologist Ron Fayer weigh a parasite-infected calf that has been treated. *(Courtesy of USDA/ARS #K-4736-3)*

Research was conducted on calves at Beltsville to find some explanations. Before, during, and after acute infection with the parasite, the concentration of growth-regulating hormones in the blood was measured. It was found that after acute infection, the concentration of a hormone essential for growth decreased and the concentration of another hormone that blocks growth-hormone secretion increased. Further, these hormone changes persisted in the infected calves even after the symptoms of infection were gone.

It is thought that the body implements a survival strategy in response to the invasion of the parasites. The strategy is one of growth restriction, so more energy can be available to fight off the intruders. It is therefore essential that every effort be made to keep animals in surroundings free from parasites and to use precautionary measures whenever applicable. Further, when symptoms appear, it is important to get a professional diagnosis and promptly administer proper treatment

proper techniques, vaccinations, and precautions to avoid human disease and parasitic infections from animals.

Noncontagious diseases pose no threat to humans or other animals, except to the animals having the diseases. Therefore, there is more leeway in dealing with animals suffering with noncontagious diseases. It is still a good idea to isolate these animals from the herd for their own good, however.

Causes Contagious diseases are caused mostly by bacteria and viruses. They can be spread by direct contact with infected animals, from shared housing, or from contaminated feed or water. In some cases, the spread of infectious diseases takes place through intermediary hosts, such as birds, rodents, or insects.

Noncontagious diseases may be caused by nutrient deficiencies or nutrient excesses. Poisonous plants and animals, injection of foreign material, and open wounds may cause or lead to noncontagious disease.

Symptoms General symptoms of disease are extremely varied and may include:

1. poor growth and/or reduced production.
2. reduced intake of feed.
3. rough, dry hair coat.
4. discharge from the nose or eyes.
5. coughing or gasping for breath.
6. trembling, shaking, or shivering.
7. unusual discharges, such as diarrhea or bloody feces or urine.
8. open sores or wounds.
9. unusual swelling of the body, including lumps and knots.
10. **abortion,** or the loss of a fetus before it is born.
11. peculiar gait, or walking pattern, or other odd movements.

Some diseases may have little or no external symptoms and may even progress so rapidly that death of the animal precedes the occurrence of noticeable symptoms.

Parasites

Parasites may also be grouped into two general classifications. These are internal—inside the animal—and external—living on the outside of the animal.

The most important group of internal parasites that infest animals are the **roundworms** (slender worms that are tapered on both ends). Other types of internal parasites include flukes and protozoa. **Flukes** are very small, flat worms, and **protozoa** are microscopic, one-celled animals. Most internal parasites spend at least some of their life cycles outside of the host animal (Figure 27-9). It is during this period that the parasite may most easily be spread to other animals. Contact with discharges from infested animals, contaminated feed, water, housing, or secondary hosts may result in the spread of internal parasites. A **secondary host** is a plant or animal that carries a disease or parasite during part of the life cycle of that disease or parasite. Some internal parasites are also spread by insects such as flies and mosquitoes.

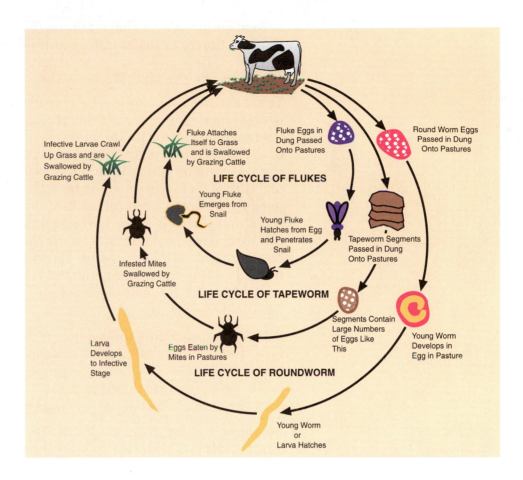

FIGURE 27-9 Life cycles of some common internal parasites.

External parasites include flies, ticks, mites, and fleas. They are spread in the same ways as are internal parasites.

Symptoms of parasite infestation may include:

1. poor growth.
2. weight loss.
3. constant coughing and gagging.
4. anemia.
5. lowered production and reproduction.
6. diarrhea or bloody feces.
7. worms in the feces.
8. swelling under the neck.
9. poor stamina.
10. loss of hair and **mange,** the presence of a crusty skin condition caused by mites.
11. visibility of the parasite itself.

PREVENTING AND TREATING ANIMAL HEALTH PROBLEMS

There are a number of activities and procedures that are used to prevent and treat health problems found in animals. Some of these include administering drugs, dipping, and restraining animals. The roles of feed additives and vaccination will also be explored. **Vaccination** is the injection of an agent into an animal to prevent disease.

Administering Drugs

There are several factors to be considered before administering drugs to an animal. These include determination of the amount to be administered, type of drug to use, purpose of the drug, site of administration of the drug, and type of animal to be treated. Most of this information can be found on the drug container. It is important that the drug manufacturer's recommendations be followed closely. Another factor that needs to be considered is the amount of time that the drug remains in the animal. This is important when determining how long milk from the animal will be contaminated by the drug. Contaminated milk must be discarded. Also, it must be determined how long to wait before a treated animal can be slaughtered for meat.

Drugs may be manufactured and sold as pills, powder, paste, or liquid (Figure 27-10).

Pills   The procedure for giving a pill to an animal is to restrain the animal and lift its head so that the mouth opens. Force the pill as far down the side of the mouth as possible, using either your hand or a **balling gun** (a device used to place a pill in an animal's throat). Massage the animal's throat until it swallows the pill.

Powders   These drugs are normally mixed in the feed or water of the animal. Often it is necessary to withhold feed or water for a period of time before administering the drug. Otherwise, the animal may refuse to eat or drink the drugged food.

Paste   Paste is normally used for treating horses for worms. The preparation is placed on the back of the horse's tongue with a caulking gun, and the horse is forced to swallow. Pastes are used for horses because it is often nearly impossible to treat them for worms by any other method.

Liquids   Liquid drugs administered orally (by mouth) are often placed directly in the animal's stomach by drenching. **Drenching** is the process of administering fairly large amounts of liquid to an animal. A syringe or

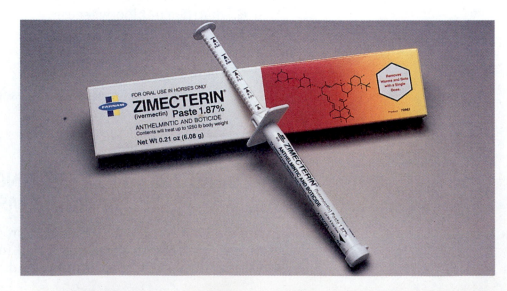

FIGURE 27-10 Pills, liquids, powders, and pastes are methods of getting medicine to animals. It is important that the drug manufacturer's recommendations be followed closely. *(Courtesy of Michael Dzaman)*

drenching gun is used. In the process, the animal is restrained, with the head held level. The upper lip of the animal is lifted, and the tube is inserted along the side of the tongue. The drug is released, and the animal is allowed to swallow. Care must be taken not to get the drug into the animal's lungs.

The injection of drugs into animals takes many forms, based on the location of the injection. **Injection** is the process of administering drugs by needle and syringe. Some of the injection sites include **intravenous** (in a vein), **intramuscular** (in a muscle), **subcutaneous** (under the skin), **intradermal** (between layers of skin), **intraruminal** (in the rumen), and **intraperitoneal** (in the abdominal cavity). One determining factor as to where injections are made is how fast the drug needs to work. A drug injected into the blood is available faster than one injected under the skin. Often it is desirable for drugs to be released slowly over a long period of time. Growth hormones are generally administered in this way.

The procedure for giving an injection is to:

1. restrain the animal.
2. select the location for the injection.
3. fill the syringe, making sure that all air is removed.
4. disinfect the area to be injected.
5. if the injection is to be made intradermally, clip the hair from the area to be injected.
6. insert the needle in the desired area without the syringe attached. (This prevents the loss of the drug if the animal jumps.)
7. attach the syringe to the needle and inject the liquid.

Infusion   **Infusion** is another method of getting drugs to the site of the infection. It is used most often to treat dairy animals with udder and teat problems. The **udder** is the milk-secreting gland of the animal. **Teats** are the appendages of an udder. A sterile **cannula** (blunt needle) is inserted into the opening of the teat, and the drug is forced into the teat canal (Figure 27-11).

Dipping   **Dipping** is a process for treating animals, mostly cattle and sheep, for external parasites. It involves filling a vat with medicated water and forcing the animal to walk or swim through it. This process

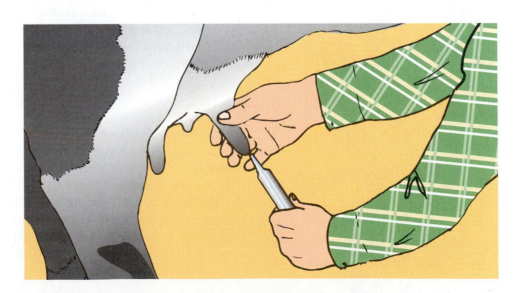

FIGURE 27-11 Infusion is a method of treating udder problems.

is also used to treat dogs for ticks and fleas. Dipping is popular where large numbers of animals must be completely covered with the medication.

Taking Temperatures

Taking an animal's temperature is basically the same as taking a human's temperature. It is usually taken in the rectum. The **rectum** is the last organ in the digestive tract. Animal thermometers are normally longer and heavier than those used in human medicine. An animal thermometer also has an eye at one end and should have a string attached to it to prevent loss of the thermometer in the body cavity.

To use an animal thermometer, first shake down the column of mercury. Coat the thermometer with sterile jelly to make insertion easier. Do not force the thermometer into the rectum. If there is resistance, injury may result. Correct the conditions that are causing the resistance and then reinsert the thermometer. After several minutes, remove the thermometer and read the temperature on the scale. Thermometers with digital readouts are also available.

Determining Pulse and Respiration Rates

The pulse rate for a large animal can be taken by holding your ear against the animal's chest and listening to the heartbeat. The number of heartbeats in one minute is the pulse rate.

The respiration rate of an animal can be determined by watching its rib cage move. Counting the number of breaths that the animal takes in one minute will indicate the rate of respiration.

Restraining Animals

There are a number of ways to restrain animals for observation and treatment of diseases and parasites. These include head gates, squeeze chutes, halters, twitches, nose leads, and casting harnesses. Head gates trap the heads of large animals, whereas squeeze chutes hold the whole animal. When halters are used, they are usually tied to a post or something else substantial to hold the animal. Twitches hold the tender lip of a horse (Figure 27-12). Nose leads hold cattle by the nose (Figure 27-13). Sometimes, large animals must be lying down in order to be examined. An easy way to accomplish this is by using a casting harness (Figure 27-14). Properly applied, a casting harness will cause an animal to fall down with just a gentle tug of the rope.

Vaccination

The prevention of diseases is nearly always less expensive than treating animals once they have diseases. A good disease-prevention program should include vaccination of animals. Vaccination is the injection of an agent into an animal to prevent disease. The agent causes the animal's body to become immune to the disease. **Immune** means not affected by something. Vaccination programs are usually part of the services of a **veterinarian** (an animal doctor). Vaccination programs vary widely with the type of animal and area of the country.

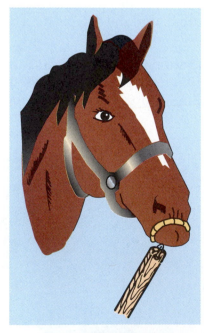

FIGURE 27-12 A twitch is used to obtain the attention of the horse so it will hold still for diagnosis or treatment.

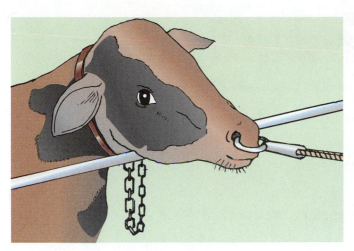

FIGURE 27-13 A nose lead used to restrain cattle.

FIGURE 27-14 A properly used casting harness makes putting a large animal down a simple task.

Feed Additives

These additives are used primarily to control the incidence of low-level infections in growing animals. The materials used are primarily antibiotics that help increase feed efficiency and rate of gain, as well as control disease. Feed additives are sometimes used to control internal parasites. Caution must be taken to always follow the manufacturer's recommendations concerning the use of these materials. Failure to do so may lead to contamination of animal products used for human consumption.

VETERINARY SERVICES

The veterinarian is an essential part of any good health program for animals. It is important to know when to call the veterinarian for help and when to deal with a problem yourself. There are no hard and fast rules in this regard, and it will vary greatly depending on the experience of the individual. In general, it pays to call the veterinarian any time that you are not absolutely sure of the problem and how to handle it.

A veterinarian should normally be consulted when you are planning and executing a disease-prevention program. Any time that an animal is having reproductive problems, a veterinarian should be consulted. Such problems include failure to conceive, abortion, or great difficulty in giving birth. When an animal dies suddenly and there is no apparent reason, a veterinarian should also be consulted to determine the cause of the death. A veterinarian should also be contacted when animals have symptoms of a contagious disease. This will help minimize spread of the disease.

STUDENT ACTIVITIES

1. Write the Terms to Know and their meanings in your notebook.

2. Compare the methods of administration, dosage rates, and times of withdrawal of several drugs for animals.

3. Practice giving injections of water using discarded animal ears from a meat processing plant.

4. Develop a complete disease-prevention program for a livestock operation.

5. Check the housing of production animals for animal safety and proper sanitation at home, on someone else's farm, or at someone's business.

6. Visit a farm supply store or a pet center and record the names of animal medicines, animal pest controls, and disinfectants that are for sale there. Also list the use(s) of each item.

7. Interview the manager of a small animal or livestock operation concerning the disease-prevention and health-maintenance practices used. Report your findings to the class.

8. Invite a veterinarian to talk to the class about animal health and precautions to avoid diseases and parasites that may be transmitted to humans from animals.

SELF EVALUATION

A. Multiple Choice

1. Subcutaneous injections are made
   a. in a vein.
   b. under the skin.
   c. between layers of skin.
   d. in the body cavity.

2. An animal that lives and feeds on other animals is a
   a. parasite.
   b. disease.
   c. vaccination.
   d. host.

3. When taking the temperature of an animal, use a/an
   a. catheter.
   b. oral thermometer.
   c. rectal thermometer.
   d. syringe.

4. Which of the following is *not* a means of administering drugs orally?
   a. drench
   b. balling gun
   c. pills
   d. infusion

5. The purpose of vaccination is to
   a. prevent parasites.
   b. prevent diseases.
   c. control parasites.
   d. treat diseases.

6. Which of the following is *not* a sign of good health in animals?
   a.   smooth hair coat   c.   alertness
   b.   elevated pulse   d.   contentment

7. The purpose of isolating sick animals is to
   a.   prevent spread of contagious diseases.
   b.   keep other animals from hurting the sick animal.
   c.   allow for easier treatment of the problem.
   d.   all of the above.

8. Which of the following is *not* an internal parasite?
   a.   fluke   c.   stomach worm
   b.   mite   d.   protozoa

9. Call a veterinarian when the animal
   a.   has bright eyes.   c.   aborts.
   b.   chews its cud.   d.   starts to lose hair.

10. A twitch is used to
   a.   restrain horses.   c.   give injections.
   b.   treat internal parasites. d.   isolate infected animals.

B. Completion

1. _____ diseases cannot be passed from one animal to another.
2. An _____ injection is one that is made in the vein of an animal.
3. Parasites that are very small, flat worms are called _____.
4. _____ refers to the number of heartbeats in one minute.
5. An agent that prevents the growth of a disease organism is called an _____.
6. Protozoa are a form of _____ parasites.
7. Oral administration of medicine refers to administering the drug through the animal's _____.
8. Vaccination is a form of disease _____.
9. An animal doctor is called a _____.
10. A nose lead is used to _____ animals.

28

Genetics, Breeding, and Reproduction

Artificial insemination

Eggs

Sperm cells

Embryo transfer

Zygotes

Geneticist

Genetics

Heredity

Breed

Gene

Chromosome

Gamete

Cell

Mitosis

Meiosis

DNA

Homozygous

Heterozygous

Dominant

Recessive

Incomplete dominance

Genotype

Phenotype

Sex-linked

Mutations

Polled

Lethal

Inheritability

Genetic engineering

Testes

Testosterone

Estrogen

Estrus

Progesterone

Mammary system

Ovaries

Ovulation

Embryo

Parturition

Fetus

Gestation

Freemartin

Sterility

Purebred

Registration papers

Crossbreeding

Hybrids

Hybrid vigor

Grade

Grading up

Progeny

Inbreeding

Sire

Dam

Closebreeding

Linebreeding

Outcrossing

Natural service

Pasture mating

Hand mating

Artificial insemination

Semen

Ejaculate

In vitro

Pedigree

Probably the fastest growing area of technology in agriscience is in genetics and reproduction. Artificial insemination has allowed for more improvement in milk production in the last thirty years than had occurred in the previous two hundred years. **Artificial insemination** is the placing of sperm cells in contact with female reproductive cells, called **eggs,** by a method other than natural mating. **Sperm cells** are male reproductive cells. Artificial insemination allows for use of a superior male to father many times more offspring than would be possible naturally. Embryo transfers have made it possible for superior females to produce far more offspring than would be possible otherwise. **Embryo transfer** is a process that removes fertilized eggs, or **zygotes,** from a female and places them in another female who carries them until birth. Embryo transfers are even being used to reproduce endangered species of animals faster than would normally be possible (Figure 28-1).

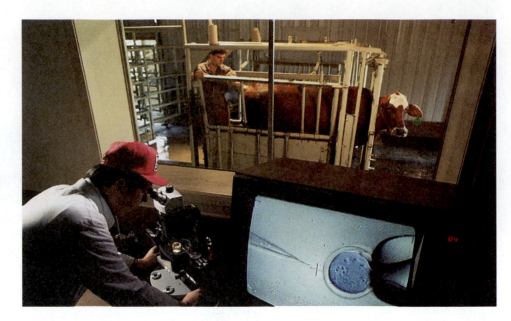

FIGURE 28-1 Successful embryo transfer requires the services of a highly trained veterinarian backed up by sophisticated laboratory equipment. (*Courtesy Gillespie,* Modern Livestock & Poultry Production, 5th ed., *Delmar Publishers,* copyright 1997)

AGRI·PROFILE

CAREER AREAS: Geneticist/Genetic Engineer/Animal Physiologist/ Veterinarian/Insemination Technician

The greatest improvements in animal production and performance have come about as a result of selection and breeding. Recent advances in the field of genetics and the ability to stimulate multiple ovulation, to transplant embryos to surrogate mothers, and to freeze embryos for future incubation have greatly increased the genetic output of superior individuals.

However, genetic engineering, where the genetic coding of cells can be manipulated, holds forth the greatest opportunities for changing animal characteristics and capabilities in the future. These changes are products of biotechnology.

A technician removes frozen semen from liquid nitrogen for artificial insemination of a farm animal.

In the foreseeable future, the greatest number of new positions in agriscience is expected to be for scientists, engineers, and related specialists. The animal industry will provide many of these positions.

Genetic engineering has made it possible to increase resistance to diseases, improve production, and improve efficiency of animals. Gene splicing, recombinant DNA, and biotechnology are common terms used by geneticists today. A **geneticist** studies **genetics,** or heredity. **Heredity** is the passing on of traits or characteristics from parents to offspring.

This unit explores some new technologies as well as the basics of animal breeding that have always been true. Also to be noted are current directions of research in animal breeding.

THE ROLE OF BREEDING AND SELECTION IN ANIMAL IMPROVEMENT

Robert Bakewell, from England, is generally credited with being the father of animal husbandry. His work in selection of Merino sheep for fine wool production and quality encouraged other farmers of his era to try to improve their livestock. Bakewell and others took great pains to always cross the most desirable females with the best males, with the expectation that the offspring would be as good as or superior to their parents. These practices have continued through the years and have resulted in advances in animal agriscience that were not imagined even thirty years ago (Figure 28-2).

By continually selecting animals for a specific type or characteristic, the resulting generations of animals tend to conform to the characteristics for which they were selected. For example, two hundred years ago cattle were not separated into dairy and beef types. Through careful selection of those animals with superior milk production and those types with excellent meat production, two distinct types of animals emerged from the same ancestors. The many breeds of animals have been developed in the same way. A **breed** is a group of animals having similar physical characteristics that are passed on to their offspring. It should also be noted that selection is an extremely important part of animal agriscience today. This is especially true as consumer demand for animal products changes, and the margin of profit continues to decline.

FIGURE 28-2 Crossing the most desirable females with the best males results in offspring as good or superior to their parents. *(Courtesy of USDA/ARS #K-2681-13)*

PRINCIPLES OF GENETICS

Gregor Mendel, an Austrian monk, is generally given credit for having discovered the basic principles of genetics. He did this through keen observation as he raised peas in his garden. These principles have become the foundation of modern genetics. They are summarized as follows:

1. In every living thing there is a pair of genes in every cell. These genes determine every trait in that individual. A **gene** is a unit of hereditary material located on a chromosome. A **chromosome** is the rodlike carrier for genes.

2. Individuals receive one gene for each trait from each parent.

3. Genes are transmitted from parent to offspring as unchanging units.

4. In the production of reproductive cells, gene pairs separate; only one gene for each trait is contained in each gamete. A **gamete** is a reproductive cell.

5. When an individual has different genes for a trait, one usually shows while the other does not.

Cells and Cell Division

Cells are the basis of all genetic activity. A **cell** is a unit of protoplasmic material with a nucleus and wall (Figure 28-3). It is the basic structure of all living things. Cells are microscopic in size. All plant and animal life begins as a single cell. The nucleus of the cell contains pairs of chromosomes on which rest genes at specific locations. The gene for a specific trait is always located in the same place on the same pair of chromosomes in a given plant or animal species.

Animal growth and reproduction takes place via cell division. In simple cell division for growth, called **mitosis,** each chromosome first divides in two. The wall of the nucleus disappears, and the chromo-

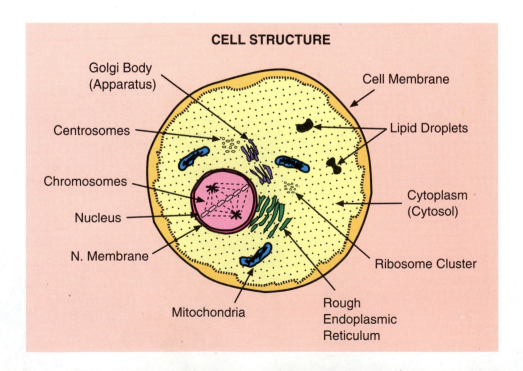

CELL STRUCTURE

FIGURE 28-3 An animal cell and some of its major parts.

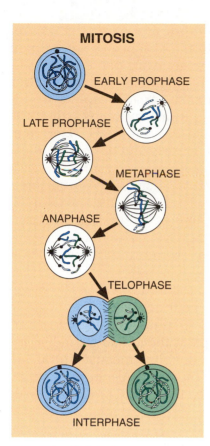

FIGURE 28-4 Mitosis is the process of division or duplication of a typical cell.

somes move to opposite sides of the cell. A new nucleus wall forms around each of the groups of chromosomes. The walls on the two sides of the cell then move toward each other until they divide the cell into two new cells complete with nuclei and pairs of chromosomes (Figure 28-4).

The cell division that results in the formation of gametes is called **meiosis** (Figure 28-5). It differs from mitosis primarily in that instead of the chromosomes dividing and moving in pairs to the opposite sides of the cell, they separate and move individually to the cell walls. When the new cells are formed, each cell contains only one of each chromosome rather than pairs of each. Meiosis occurs only in the reproductive organs of animals. When an egg is fertilized, the sperm contributes one of each chromosome pair; the egg contributes the other.

Genes

Genes are the units of genetic material that are responsible for all of the traits, or characteristics, of animals. Genes occur at specific locations on chromosomes. Chromosomes control certain enzyme and protein production that controls some traits in animals. The chromosomes themselves are composed of a protein covering surrounding two chains of **DNA,** or deoxyribonucleic acid. This substance serves as the coding mechanism for heredity.

In pairs of genes on matching chromosomes, the genes may be either alike or different. Pairs of genes that are alike are said to be **homozygous,** whereas those pairs that are different are called **heterozygous.** When the two genes in a pair are different, one gene usually expresses itself, and the expression of the other remains hidden. The

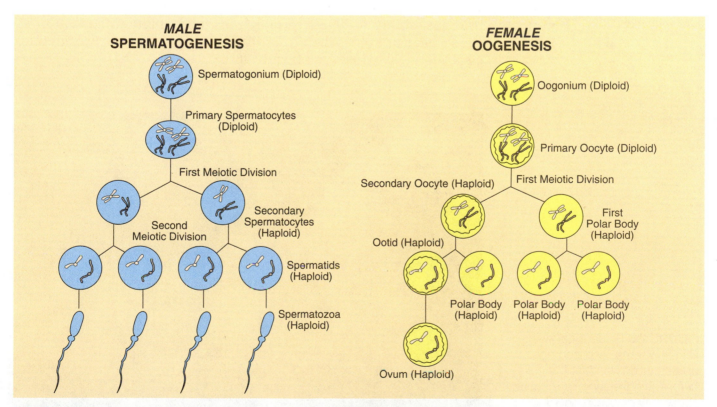

FIGURE 28-5 Meiosis is the division or duplication of egg and sperm cells.

gene that expresses itself is referred to as **dominant.** The gene that remains hidden and expresses itself only in the absence of a dominant gene is called **recessive.** Sometimes neither gene of a pair expresses itself to the exclusion of the other. When this happens, the gene pair is referred to as expressing partial dominance, or **incomplete dominance.** The actual configuration of genes in an animal is called the **genotype,** whereas the physical appearance of the animal is referred to as **phenotype.** All of this is important when exploring the basics of genetics and the use of genetics in animal breeding.

Some traits are controlled by genes that are located on the chromosomes that control the sex of the animal. These are called **sex-linked** traits. The chromosomes that control sex in most animals are not perfectly matched. The result is that not all of the genes on these chromosomes occur in pairs. When this happens, some traits show only in males and some only in females.

Genes normally duplicate themselves accurately. However, sometimes accidents or changes occur. These genetic accidents or changes in genes are called **mutations.** Sometimes these mutations result in desirable changes in animals. One such example is the polled characteristic in breeds of cattle that are normally horned. **Polled** means naturally or genetically hornless. In other cases, the mutation results in a **lethal** characteristic, which causes an animal to be born dead or to die shortly after birth.

Genetics in the Improvement of Animals

The improvement of animals through genetics can be either natural or planned. In natural selection, the "survival of the fittest" occurs. In other words, as changes in genes occur naturally, only those animals experiencing gene changes that make them better adapted to their environments will survive. Popular examples include protective colorations, ability to digest certain feeds, and ability to survive in extreme heat or cold. When a fertilized egg divides into two identical eggs, identical twins result (Figure 28-6).

FIGURE 28-6 Identical twins result when a fertilized egg divides into two identical cells before progressing to mitosis. *(Courtesy of USDA/ARS #K-4323-18)*

In planned or artificial selection, people decide which traits they want in animals. They then use the animals with the desirable traits in the breeding program. Over a period of time, the animals that result from such selection show more and more of the desired traits.

Unfortunately, most of the traits for which people are selecting animals are the result of a combination of many pairs of genes. Because of this, few traits are 100 percent inheritable from parents. For example, the extent of inheritability for loin-eye size in pigs is 50 percent. **Inheritability** means the capacity to be passed down from a parent to offspring. A boar with a 6-in. loin-eye is crossed with a sow that has a 5-in. loin-eye. The expected average loin-eye size for the resulting offspring would be 5½-in. if loin-eye size were 100 percent inheritable. However, because loin-eye size is only 50 percent inheritable, the offspring can only be expected to have 5¼-in. loin-eyes.

The percent inheritability rates of other traits can be found in Figure 28-7. These rates should be used as a guide only when attempting to improve animals through genetics.

Environmental factors often play a part in the expression of genetic traits, masking to some extent the true potential of the animal. For example, an animal that is improperly fed or cared for may never reach the size or weight that its genetic potential would permit.

Estimated Percent Heritability						
Trait	Cattle	Sheep	Swine	Poultry	Rabbits	Horses
Fertility	0-15	0-15	0-15	0-15	–	Low
Number of young weaned	10-15	10-15	10-15	–	3	–
Weight of young at weaning	15-25	15-20	15-20	–	35	–
Postweaning rate of gain	50-55	50-60	25-30	–	60	–
Postweaning gain efficiency	40-50	20-30	30-35	–	–	–
Fat thickness over loin	45-50	–	40-50	–	–	–
Loin-eye area	50-60	–	45-50	–	60	–
Percent lean cuts	40-50	–	30-40	–	60	–
Milk production (lb)	25-30	–	–	–	–	–
Milk fat (lb)	25-30	–	–	–	–	–
Milk solids, nonfat (lb)	30-35	–	–	–	–	–
Total milk solids (lb)	30-35	–	–	–	–	–
Body weight	–	–	–	35-45	40	–
Feed efficiency	–	–	–	20-25	–	–
Total egg production	–	–	–	20-30	–	–
Age at sexual maturity	–	–	–	30-40	–	–
Viability	–	–	–	5-10	–	–
Speed	–	–	–	–	–	25-50
Wither height	–	–	–	–	–	25-60
Body length	–	–	–	–	–	25
Heart girth circumference	–	–	–	–	–	34
Cannon bone circumference	–	–	–	–	–	19
Points for movement	–	–	–	–	–	40
Temperament	–	–	–	–	–	25

FIGURE 28-7 The rates of inheritability for certain traits of domestic animals.

Genetic Engineering

Genetic engineering has much potential for improving animals for the use of humans. **Genetic engineering** is the process of transferring genes from one individual to another individual or organism without mating male and female cells. Geneticists have been able to link specific genes to specific traits. They have also developed procedures for removing the genes from the cells of one animal and inserting them into the cells of another animal.

The potential for change in animals is tremendous using genetic engineering. For example, if a species of animal is genetically resistant to a certain disease, genes that make that animal resistant could be inserted into cells of an animal species that is not resistant. Because genes are passed on to offspring from parents, resulting generations of animals would be resistant to that disease.

Some of the areas being explored by geneticists working with genetic engineering include disease resistance; cancer research; vaccines; increased growth and production; and immunology.

REPRODUCTIVE SYSTEMS OF ANIMALS
Male

The male reproductive system functions to produce, store, and deposit sperm cells. Secondary functions include production of male sex hormones and elimination of urine from the body (Figure 28-8).

The actual structural makeup of the male reproductive system varies widely with different species of animals. The **testes** are the organs that produce sperm cells. They also produce **testosterone,** the male sex hormone. The testes are attached to the body by the spermatic cord and are protected by the scrotum. A coiled tube called the epididymis stores and transports the sperm cells. The vas deferens

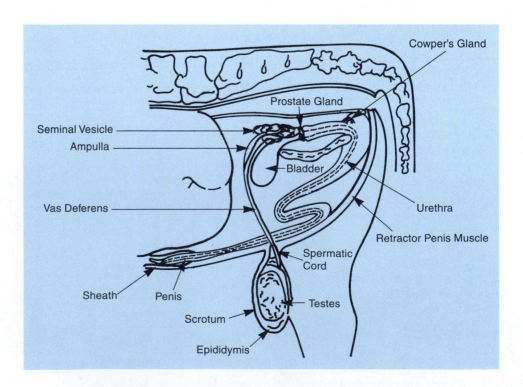

FIGURE 28-8 Reproductive system of a bull. *(Courtesy of Maryland State Instructional Guide, University of Maryland)*

carries the sperm cells to the urethra, which extends through the penis. Other glands of the male reproductive system include the seminal vesicles and the prostate and Cowper's glands. The seminal vesicles secrete seminal fluid. The prostate gland provides nutrition to the sperm cells, and the Cowper's gland prepares the urethra for the passage of the sperm cells. The penis serves to deposit the sperm cells into the female reproductive tract.

Female

The female reproductive system produces the egg and the female sex hormones estrogen and progesterone (Figure 28-9). **Estrogen** regulates the heat period, **estrus,** whereas **progesterone** prevents estrus during pregnancy and causes development of the mammary system. The **mammary system** produces milk.

The parts of the female reproductive system include two **ovaries,** which produce eggs. Funnel-shaped devices called the infundibulums catch the eggs during **ovulation** (the process of releasing a mature egg from the ovaries). The egg then passes to the fallopian tubes, also call oviducts. The fallopian tubes are where fertilization of the egg takes place. The fertilized egg, or **embryo,** travels to the uterine horn where it attaches to the wall of the uterus and remains until birth, or **parturition.** When the embryo attaches itself to the uterine wall, it becomes known as a **fetus.** The period of time between fertilization of the egg by a sperm cell and birth is called the **gestation** period. The vagina, which is separated from the uterine horn by the cervix, serves as the passageway for the sperm cells. The external opening, or vulva, protects the rest of the female reproductive system from infection from the outside.

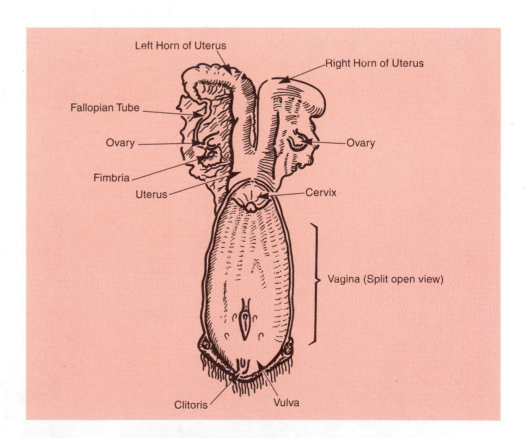

FIGURE 28-9 Reproductive system of a female cow. *(Courtesy of Maryland State Instructional Guide, University of Maryland)*

Reproductive Problems

There are a number of conditions that may result in reproductive problems or failures. Some of these problems are either physical or genetic. Examples include: (1) **Freemartin,** a sterile female born as a twin of a male in cattle, (2) scrotal hernia (a muscle tear), (3) undeveloped or missing ovaries, and (4) malformed penis.

Infections and diseases are also important causes of sterility in animals. **Sterility** is the inability of an animal to reproduce. Physical damage to the reproductive system and nutritional deficiencies may also contribute to reproductive failures.

SYSTEMS OF BREEDING

There are a number of breeding systems that are important to animal agriscience. Which system or systems to use depends on many factors. Some of the considerations include type of operation, markets available, resources available, climatic conditions, size of operation, goals of the breeder, and personal preference.

Commonly recognized systems of breeding include purebreeding, crossbreeding, grading up, inbreeding, and outcrossing (Figure 28-10).

	Relationship of Mates	Advantages	Disadvantages
Purebreeding	Unrelated	1. Concentration of selected traits 2. Breed assn to create demand	1. May result in less desirable traits 2. Loss of hybrid vigor
Crossbreeding	Unrelated	1. Increased growth 2. Increased prod. 3. Increased hybrid vigor 4. Higher fertility 5. Disease resistance	1. Less uniformity of offspring 2. Not eligible for registry
Grading up	Unrelated	1. Herd improvement w/o purchasing purebreds 2. Develop uniformity	Slow process of improvement
Closebreeding	Sire-mother Son-dam Brother-sister	Concentrates desirable traits	1. Concentrates undesirable traits 2. Expression of abnormal traits
Linebreeding	Not closer than ½ brother to ½ sister	Concentrates desirable traits of one individual	1. Concentrates desirable traits 2. May result in expression of abnormal traits
Outcrossing	Unrelated	Produces hybrid vigor within a breed	

FIGURE 28-10 Systems of breeding for animal improvement.

Purebreeding

Purebreeding occurs when a purebred animal is bred to another pure-bred animal. A **purebred** animal is one of a recognized breed and one having registration papers. **Registration papers** are records of ancestry.

Although there are no guarantees, purebred animals are usually considered to be superior to nonpurebreds. They are used as show animals and are important parts of the crossbreeding and grading-up systems of breeding.

Persons who elect to use the purebreeding system of breeding should have ample resources and a good knowledge of genetics. They should also be good salespersons capable of marketing the purebred animals at premium prices that reflect the superior qualities of the animals.

Crossbreeding

Crossbreeding is the breeding of one recognized breed of animals to another recognized breed. The resulting offspring are called **hybrids.**

There are a number of advantages to hybrid animals. They tend to be faster growing, stronger, and higher producing as a result of the combination of desirable traits from the two breeds. This is called **hybrid vigor.** They also tend to be more fertile and more disease resistant.

Crossbreeding is generally used by commercial producers who are more interested in offspring that are efficient producers than in maintaining a specific breed of animals.

Grading Up

Livestock producers who are raising animals that are not purebred often use grading up to improve their herds. When a nonpurebred female animal, called a **grade,** is mated with a purebred male, the process is called **grading up.** The idea is that the purebred male should be superior to the grade female and that the resulting offspring should be superior to their mother. Succeeding generations of females are also mated to purebred males of superior quality.

The purposes of grading up include the improvement of quality and production in the offspring. Offspring are also called **progeny.** The development of uniformity in the herd is also a reason for grading up animals.

Inbreeding

In the simplist terms, **inbreeding** is the crossing or mating of animals that are related. The purpose of inbreeding is to intensify the desirable characteristics of a particular animal or family of animals. Unfortunately, inbreeding also intensifies the undesirable and abnormal characteristics.

There are various degrees of inbreeding based on how closely related are the individuals being mated. When a father, or **sire,** is mated with his daughter; a son is mated with his mother, or **dam;** or a brother is mated to his sister; the term **closebreeding** is used.

Linebreeding is the mating of less closely related individuals that can be traced back to one common ancestor. Normally, the most closely related cross made in linebreeding is half brother to half sister.

Inbreeding must be used carefully, because inbred animals tend to exhibit more undesirable characteristics than animals produced by other systems. Unless a breeder is willing to carefully select the outstanding individuals resulting from inbreeding, this system should be avoided.

Outcrossing

Outcrossing is the mating of unrelated animal families in the same breed. This is probably the most popular system of breeding used in purebred herds of animals. It also has many of the advantages of crossbreeding, including increased production and improved type.

BIO•TECH CONNECTION

Embryo Co-culture

Sheep and cattle embryos can be kept alive and growing outside of the mother's womb for up to six days when tissue-cultured oviduct cells are used for nourishment. This technique was first developed for sheep and is known as *co-culture.* The technique permits embryos to be held for observation and checked for success of genetic engineering before being implanted into surrogate mothers.

Single-celled embryos that have had a gene inserted are placed in cultures of cells from the oviduct. Scientists believe that certain nutrients from the cultured cells keep the embryos alive and growing. Microscopic examination of the cultured embryos is used to determine if they are developing properly. An embryo cultured for three days should have more than eight cells. If there are eight or less cells, it signals that development is not progressing normally and implanting should not be performed.

Another technique that can be used is to place gene-implanted embryos into rabbits for five days and then remove them for implanting. Greater labor requirements and other disadvantages related to the use of rabbits makes the use of oviduct cells the more promising of the techniques, however.

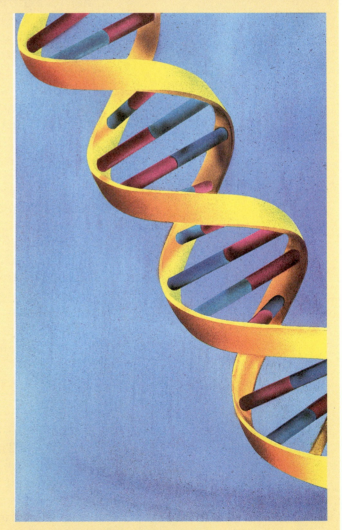

Various genetic-engineering techniques are being used for embryic development. *(Courtesy Herren, The Science of Animal Agriculture, Copyright 1994, Delmar Publishers)*

METHODS OF BREEDING

There are three general methods of breeding animals—natural service, artificial insemination, and in vitro. **Natural service** occurs when the male is allowed to mate directly with the female. There are several ways in which this may be accomplished, depending on the type of operation, amount of labor available, number of animals in the herd, requirements of breed associations, and personal preference.

Pasture mating is a system of natural service where the male animal is allowed to roam freely with the females in the herd. The male is responsible for detecting the heat period of each female in the herd and for mating with her at the appropriate time. There are some disadvantages to this system. One male can mate with only a limited number of females. Also, if more than one female is in heat at the same time, all may not get bred. Similarly, breeding records are more difficult to keep. Finally, the male may become sterile and the females not get bred at all.

Hand mating is the bringing of the male to the female for mating. More labor and better management is required, because someone must determine when the female is in heat and get the male to the female so mating can take place. Advantages of hand mating include being able to keep more accurate breeding records and being able to mate more females to a single male.

There are numerous advantages of mating animals via artificial insemination. There are a few disadvantages as well. **Artificial insemination** involves collecting semen from a male animal and placing it in the reproductive tract of the female. **Semen** is the sperm cells and accompanying fluids. The technique of artificial insemination has been responsible for tremendous increases in animal productivity in recent years.

Advantages include the fact that semen that is collected from the male can be used fresh or stored frozen in liquid nitrogen for later use. One **ejaculate,** or the amount of semen produced at one time by a male, can be diluted and used to breed thousands of females. Because semen can be frozen and stored for long periods of time, the use of an outstanding male can be greatly extended. Artificial insemination also greatly reduces the need to keep male animals and reduces the danger of having males around. Other advantages include less chance of injury to breeders and their animals, reduced spread of reproductive diseases, and improved record keeping.

There are some disadvantages to artificial insemination as a method of mating animals. A person trained to inseminate females must be available when the animal is in heat. Semen collection from the male can be a dangerous activity requiring the services of trained technicians; special equipment and facilities; and excellent management. Finally, genetic defects of the male can be spread faster.

In vitro mating occurs outside of the animal's body. Mature eggs are flushed from the female and fertilized by sperm cells collected from the male. The fertilized eggs are then placed in host females for development into offspring. Although this process is very exacting in its requirement and facilities, there are times when this is the only way that a viable fetus can be obtained. This means of mating is often used as part of the new technology of genetic engineering.

SELECTION OF ANIMALS

There are a number of methods by which animals may be selected. They fall into two major types—selection based on physical appearance and selection based on performance or production of either the individual or its progeny.

Selection based on physical appearance is generally used when choosing purebred animals. Often the sole criterion for selection of an animal is how well he or she performs in the show ring. This is an acceptable means of selection when animal breeders are raising animals for show and to sell to others for the same purpose. However, this method often leaves much to be desired when the animal so chosen is expected to produce a product or perform a desired activity. Animals fitted to perform or look their best in the show ring often fall sadly short of expectations in the milking parlor or in other performance settings.

The use of comparative judging helps individuals to develop skills in evaluating animal appearance. Here, one animal is compared with another and frequently judged in groups of four (Figure 28-11).

Selection based on production or performance is usually a more reliable means of choosing animals. If you are selecting dairy cows to produce milk, it makes more sense to select cows with high-production records and high-production relatives. In meat animals, progeny testing, or the testing of performance of the offspring, is often the only way to predict the breeding value of the parents. Other measures of production on which various animals may be selected include rate of gain, feed efficiency, butterfat production, back-fat thickness, loin-eye area, yearly egg production, and pounds of wool produced.

Sometimes, selection of animals is based on pedigree. A **pedigree** is a record of an animal's ancestry and is included on registration papers for purebred animals (Figure 28-12). Although the consideration of pedigree can be important in the selection process, it should always be used in combination with other methods of selecting animals.

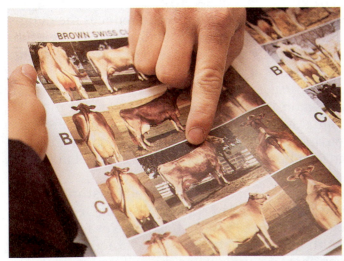

FIGURE 28-11 Comparative judging is a good technique for studying the details of the animal, such as feet, legs, udder, head, body, and overall appearance. *(Courtesy of FFA)*

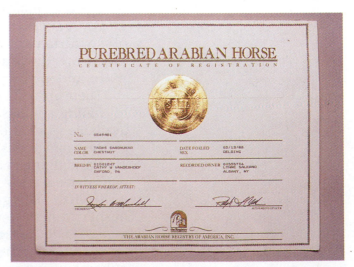

FIGURE 28-12 Pedigree and registration papers are valuable documents when buying purebred animals. *(Courtesy of Michael Dzaman)*

In summary, tremendous gains have occurred in the productivity of animals in the past one hundred years. Fewer and fewer animals are providing the needs of an ever-growing human population. Genetics and gains in animal breeding have been responsible for much of the increase in productivity. This will continue as technology in the field of animal agriscience continues to advance.

STUDENT ACTIVITIES

1. Write the Terms to Know and their meanings in your notebook.
2. Label the parts of the male and female reproductive systems.
3. Sketch and label the various stages in mitosis and meiosis.
4. Study the genetic principles that determine eye color in humans. Compare the factors that control how eye color is passed from parent to child with those that control the tendency toward baldness.
5. Suppose you mate a Hereford bull having horns to six females that were born without horns (polled). If the females were homozygous for the polled characteristic, how many would probably bear calves having no horns? Why?
6. Select a species of animal and determine the origin of the popular breeds in that species.
7. Clip pictures of animals from magazines and newspapers and make a collage of popular breeds for the bulletin board.
8. Develop a bulletin board showing the popular breeds of sheep, cattle, horses, swine, dogs, rabbits, and cats in your community.
9. Suppose that you mated a black male rabbit with a white female rabbit and that the female gave birth to eight bunnies. If white is dominant to black, and the female is heterozygous for hair color, how many of the bunnies are likely to be black?
10. If a red pig with floppy ears were crossed with a white pig with erect ears and all possible characteristics were homozygous, what is the probability that the offspring will be red with erect ears? In pigs, white hair color and erect ears are dominant.
11. Dissect the reproductive organs of a female animal and identify the major parts.
12. Visit a farm or other animal facility and observe artificial insemination or other reproductive techniques being conducted.

SELF EVALUATION

A. Multiple Choice

1. The male sex hormone is called
   a. estrogen.
   b. testosterone.
   c. progesterone.
   d. oxytocin.

2. The gestation period is the
   a.   length of pregnancy.
   b.   time during which an animal is in heat.
   c.   period when an animal is fertile.
   d.   time it takes the egg to mature.

3. Single cell division is called
   a.   meiosis.
   b.   parturition.
   c.   mitosis.
   d.   estrus.

4. A chromosome
   a.   is composed of DNA.
   b.   is a carrier of genes.
   c.   occurs in pairs.
   d.   is all of the above.

5. A pair of genes for characteristics that are alike is said to be
   a.   heterozygous.
   b.   homozygous.
   c.   genotype.
   d.   recessive.

6. Mating of an animal of one breed to an animal of another breed is
   a.   purebreeding.
   b.   grading up.
   c.   crossbreeding.
   d.   outcrossing.

7. The phenotype of an individual is
   a.   what the genes look like.
   b.   the physical appearance.
   c.   the type of animal.
   d.   the expected production.

8. When genes are transferred from one individual to another other than through mating, it is referred to as
   a.   artificial insemination.
   b.   in vitro fertilization.
   c.   genetic engineering.
   d.   hybrid vigor.

9. The inability to reproduce is
   a.   ovulation.
   b.   meiosis.
   c.   biotechnology.
   d.   sterility.

10. Mating a sire to his daughter is
   a.   inbreeding.
   b.   crossbreeding.
   c.   purebreeding.
   d.   grading up.

B. Matching

_____ **1.** Gamete
_____ **2.** Genotype
_____ **3.** Fetus
_____ **4.** Progeny
_____ **5.** Dam

   a.   Developing embryo
   b.   Mother
   c.   Sex cell
   d.   Offspring
   e.   Configuration of genes

C. Completion

1. The _____ manufacture male sex cells.

2. A _____ studies genetics.

3. The passing of traits from parents to offspring is _____.

4. The crossing of half brother to half sister is _____.

5. _____ is the coding mechanism for heredity.

Small Animal Care and Management

COMPETENCIES TO BE DEVELOPED

After studying this unit, you should be able to:

- describe the domestication and history of small animals.
- determine the economic importance of the various classes of small animals.
- list the types and uses of the various classes of small animals.
- describe the approved practices in feeding and caring for small animals.

MATERIALS LIST

- ✓ an observation beehive
- ✓ bulletin board materials
- ✓ magazines and other materials with pictures of poultry and rabbits
- ✓ labels from poultry and rabbit feed bags

Domesticate	Bantams	Hutch
Jungle fowl	Toms	Litters
Waterfowl	Poults	Tattoo
Poultry	Drakes	Honey
Broilers	Ducklings	Hive
Layers	Ganders	Comb
Chicks	Goose	Apiculture
Roasters	Goslings	Swarm
Capons	Squabs	Queen
Castration	Hatchery	Super
Cockerels	Flock	Worker
Cocks	Pelt	Drones
Roosters	Angora	Apiary
Pullets	Buck	
Hens	Does	

As our world population continues to expand and there is less and less room for humans and large animals to coexist, the importance of small animals continues to increase. They are more efficient than are larger animals in converting feed eaten into usable food and other products for humans. They are less intrusive on the lives of people, and more people are keeping animals as pets (Figure 29-1). There are many important species of small animals. This unit explores poultry, rabbits, pets, and honeybees, and their contributions to our daily lives.

FIGURE 29-1 Pet ownership and the animal-care industry have grown extensively in the last two decades. *(Courtesy of Sandy Clark)*

POULTRY

History and Domestication

The domestication of chickens occurred about 4000 B.C. in Southeast Asia. To **domesticate** means to tame for the use of people. The association of **jungle fowl,** ancestors of our present-day chickens, with humans benefitted both. Humans made small clearings in the jungle that attracted insects and other food for the jungle fowl. The jungle fowl provided some eggs and meat for humans. This association over centuries gradually led to the domesticated chicken of today. Chickens came to the New World with the earliest settlers, and the people of Jamestown settlement had their pens of chickens.

Turkeys are the only domesticated animals of agriscience importance to have originated and been domesticated in the New World. When early explorers arrived in the New World, they found that the Native Americans of Central America had highly domesticated turkeys that were being grown for food for animals and humans. While present-day turkeys are direct descendants of the wild turkey of the United States, they have been domesticated to the point where they are totally dependent on humans and cannot survive in the wild.

The various types and breeds of duck and goose have originated from places all over the world. Ducks and geese are also known as **waterfowl.**

Economic Importance

The consumption of red meat has declined slightly in recent years. One reason for this may be negative publicity regarding fat and cholesterol. The cost of red meat such as beef and pork may also have caused a decrease in its demand. As a result, consumption of poultry and poultry products, with the exception of eggs, has been increasing. **Poultry** is a group name given to all domesticated birds. United States consumption of chicken meat increased from 43 lb to 72 lb per person from 1976 to 1990, and turkey meat consumption doubled from 9 lb to 18 lb in the same period. Americans also eat about 235 eggs per person each year. As a result, the poultry industry is and will continue to be an important part of the American agriscience industry. Currently, poultry production ranks third behind beef and swine production in dollar sales of meat. Some of the largest farms in the United States are poultry operations.

Major centers of production in the United States for the nearly 6 billion **broilers** (young chickens grown for meat) produced each year are Arkansas, Georgia, Alabama, North Carolina, Mississippi, Texas, and Maryland. California, Indiana, and Pennsylvania are the leading producers of eggs in the United States, followed by Ohio and Georgia.

The production of turkeys is spread over a wide area, with North Carolina, Minnesota, California, Arkansas, Missouri, and Virginia being leading states. Nearly 60 percent of the more than 10 million ducks produced in the United States each year comes from Long Island, New York. New York, Missouri, Iowa, South Dakota, and Minnesota are major producers of geese.

Types and Uses of Poultry

The types of poultry can be divided into the following general groups: chickens, turkeys, ducks, geese, and captive game birds.

Chickens are usually classified as either layers or broilers, depending on their intended use. **Layers** are chickens developed to produce large numbers of eggs (Figure 29-2). They may produce either white or brown eggs, depending on the breed. Laying chickens are also maintained to produce eggs to be hatched for the production of broiler chicks. **Chicks** are newborn chickens.

Chickens produced for meat are usually classified according to age. Broilers are young-meat chickens usually not more than eight weeks old (Figure 29-3). **Roasters** are mature chickens used for meat. **Capons** are castrated male chickens that are 14 to 17 weeks old when they are marketed. **Castration** is the removal of the male sex organs. This can

FIGURE 29-2 Layers are bred to convert feed into eggs rather than body flesh. *(Courtesy of Bill Muir, Purdue University)*

FIGURE 29-3 Broilers are the most common meat-type chickens. *(Courtesy of USDA)*

be accomplished either surgically or chemically. Game or Cornish chickens are also breeds of chicken raised for meat.

Young male chickens are called **cockerels,** whereas adult males are called **cocks** or **roosters.** These terms also apply to male pheasants. Young female chickens are called **pullets,** and adult female chickens are called **hens.** Adult female turkeys, ducks, and pheasants are also called hens.

Other classes of chicken include **bantams,** or miniature chickens, and ornamental chickens, which are of value strictly for show.

There are more than two hundred recognized breeds of chicken in the United States. However, nearly all of the layer and broiler types used are the result of crossbreeding to maximize production. The foundation breed of most laying-type chickens is the White Leghorn. Most broilers can trace their ancestors back to Cornish or game chickens.

There is really only one type of turkey used commercially in the United States—the Broad-Breasted White (Figure 29-4). This breed accounts for more than 90 percent of the more than 283 million turkeys produced in the United States each year. Other turkey breeds include Broad-Breasted Bronze, Bourbon Red, Holland White, and Beltsville Small White. Male turkeys are called **toms,** and young turkeys are **poults.**

FIGURE 29-4 The most popular breed of turkey is the Broad-Breasted White. *(Courtesy of USDA)*

FIGURE 29-5 Pekin ducks are the most popular breed of duck used for meat. *(Courtesy of Jurgielewicz Duck Farm)*

FIGURE 29-6 Khaki Campbell ducks are the champion egg layers of the bird world. *(Courtesy of John Metzer, Metzer Farms)*

Ducks can be classified as meat producers or egg producers. The primary meat breed is the Pekin (Figure 29-5). They reach a market weight of about 7 lb in eight weeks. This makes them faster growing than broilers, which reach 4 lb in the same period of time. Other breeds of duck used for meat production are Aylesbury, Muscovy, Rouen, and Call.

Egg-laying ducks are generally either Khaki Campbell or Indian Runner. The Khaki Campbell is the champion egg layer of the bird world, often averaging more than three hundred fifty eggs per year (Figure 29-6). This compares to an average of about two hundred fifty eggs laid per year for laying chickens. Male ducks are called **drakes,** and young ducks are **ducklings.**

Geese are raised primarily for meat. There is also a limited market for geese used for weeding certain crops. The Chinese breed is popular for this use. Other breeds of goose are Toulouse, Emden, Pilgrim, and African. Male geese are called **ganders,** a female is a **goose,** and young geese are **goslings.**

Captive game birds include pheasants, quail, chukor partridge, and pigeons (Figure 29-7). The uses of game birds include meat and eggs. Some game birds are also raised to release to the wild or on game preserves for hunting. Pigeons may be raised for sport or for meat. The young pigeons, called **squabs,** are used for meat before they learn to fly.

Some Approved Practices for Poultry Production

Up-to-date information on poultry and other small animals is available from suppliers of animals, equipment, medicine, and building materials. Recommended or approved practices are available from your state university. In the most general terms, approved practices for the production of poultry include the following:

1. Purchase young poultry with a specific use in mind.
2. Purchase young poultry or eggs for hatching from reputable hatcheries or breeders only. A **hatchery** is a business that hatches young poultry from eggs.

FIGURE 29-7 Two types of captive game birds are pictured here: (A) Ruffed Grouse and (B) Chinese Ringnecked Pheasant. *(Courtesy of U.S. Fish and Wildlife Services)*

3. Purchase chicks, pullets, or poults that are immune to and free from disease.

4. Purchase young poultry at the proper time for hitting a target market date. Broilers should be 7 to 8 weeks old before marketing them; ducks, 7 to 8 weeks; turkeys, 12 to 14 weeks; and geese, 12 to 14 weeks old. Layer chicks should be purchased 20 to 22 weeks before you expect them to produce eggs.

5. Ensure that proper housing is available for the type and number of poultry you are planning to raise. Housing considerations include size; ventilation; ease of cleaning; lighting; heating and cooling; feed storage; and maintenance required.

6. Secure and maintain the proper equipment for the type of poultry operation planned. Consider feeder and waterer space, and brooder size.

7. Feed a commercial, balanced ration designed especially for the type of poultry being grown.

8. Plan and follow a flock health program. A **flock** is a group of birds.

9. Plan for marketing at the optimum time.

10. Properly clean and disinfect facilities between flocks of poultry.

RABBITS

History and Domestication

Much of the early history of the rabbit is obscure. It is believed that the Phoenicians brought rabbits to Spain about 1100 B.C. They are also given credit for having introduced rabbits to most of the then-known world.

Romans kept rabbits in special enclosures. Roman women were known to have eaten large quantities of rabbit meat. They felt that it enhanced their beauty.

Early monasteries produced large amounts of rabbit meat and fur. These religious institutions are given credit for having domesticated the rabbit. It is known that great pride was taken in producing good-quality rabbits and that much rabbit trading existed between monasteries.

Rabbit meat has long been an important component of the diets of people in densely populated countries of Europe. Rabbits are efficient converters of feed to meat. They take up relatively little space and reproduce rapidly.

Some rabbits have been raised in the United States since the time of early settlers, but serious rabbit production did not begin until the turn of this century. An intense advertising campaign was conducted for "Belgian Hare" at that time to promote commercial production of rabbits.

Rabbit production also got a boost in the United States during the two world wars. At a time when shortages and rationing of food products occurred, rabbits became an inexpensive source of lean, red meat.

Economic Importance

Rabbit production is an important agriscience enterprise in the United States. Each year, 6 to 8 million rabbits are raised. Americans consume 25 to 30 million lb of rabbit meat each year. Biomedical teaching and research use another six hundred thousand rabbits yearly.

Rabbit production is an ideal enterprise for a young person because it can be started with limited capital (Figure 29-8). With only a small investment in housing and equipment, a person with one pair of rabbits can produce 60 to 80 rabbits each year to eat or to sell. Because they are small and generally accepted by people, rabbits are better adapted for production in more urbanized areas than are most other types of animals. Rabbit meat is low in fat (4 percent), sodium, and cholesterol and high in protein (25 percent).

The outlook for rabbit production in the future is variable with a need for good market analysis, good market development, and careful production management. As the competition between humans and

FIGURE 29-8 Rabbit production can be done on a small basis with very little capital and space. *(Courtesy of FFA)*

animals for grain products increases, the rabbit will play an important role in meeting the protein needs of humans in the future.

Types and Uses

Animals commonly known as rabbits come from two families and three genera, with very distinct differences. These include rabbits, cottontails, and hares. Rabbits bear their young in underground burrows in the wild. The young are born blind, hairless, and completely helpless. In contrast, cottontails and hares usually give birth in nests above ground. The young are born with their eyes open and with hair. They are able to fend for themselves shortly after birth. Hares also have larger hind legs and longer ears.

Hares and cottontails include the jackrabbit, arctic hare, and snowshoe rabbit. Because they belong to different genera, cottontails, hares, and rabbits cannot interbreed.

Domestic rabbits can be divided into a number of groups based on use. These groups include meat, fur, pets, show, and laboratory use. Many of the breeds will fall into several of these use groups.

The primary use of rabbits in the United States is for the production of meat, with pelts being a by-product. A **pelt** is an animal skin with the hair attached. Almost 100 million rabbit pelts are used in the United States each year. Most of these are imported from other countries because, in the United States, rabbits are slaughtered at too young an age to have desirable pelts.

Although all of the breeds of rabbit will produce meat, some breeds are far more efficient in producing desirable-quality meat. The New Zealand White is the most popular breed of rabbit in the United States for meat production. This rabbit occurs in three colors: white, red, and black. It is of medium size. It may be grown into a 4-lb rabbit at eight weeks of age, using about 4 lb of feed for each pound of rabbit produced. Californian and Champage D'Argent are also popular breeds used for meat production (Figure 29-9).

FIGURE 29-9 Important meat-producing breeds of rabbit include the (A) New Zealand, (B) Californian, and (C) Champage D'Argent. *(Courtesy of Isabelle Francais)*

FIGURE 29-10 Important fur-producing breeds of rabbit include the (A) the Satin, (B) Rex, and (C) Havana. *(Courtesy of Isabelle Francais)*

Some breeds of rabbit are grown for their lustrous fur used in the manufacture of fur coats and many other rabbit-fur products. The Satin, Rex, and Havana are examples of rabbit breeds grown for their fur (Figure 29-10).

Rabbits have long been important for use in laboratory work. They are used for research in the development of drugs for treating a wide range of diseases. They are also important participants in nutritional studies and in genetic research. Commonly used breeds of rabbit for laboratory work include New Zealand White, Dutch, and Florida White. Producers who breed rabbits for laboratory work should be aware that many labs will use only white rabbits of medium size (Figure 29-11).

FIGURE 29-11 The (A) Dutch and (B) Florida White join the (C) New Zealand White as important players in biomedical research and education. *(Courtesy of Isabelle Francais)*

FIGURE 29-12 Angora rabbits, such as the French Angora pictured here, are raised for the production of angora wool. *(Courtesy of Isabelle Francais)*

The Angora rabbit is used strictly for the production of wool called **angora.** The wool from Angora rabbits is sheared or pulled from the rabbit about every 10 to 12 weeks. Mature Angora bucks may produce 1 to 1½ lb of wool each year. A **buck** is a male rabbit. Female rabbits are called **does.** There are two breeds of Angora rabbit—French and English (Figure 29-12).

All of the forty breeds of rabbit recognized by the American Rabbit Breeders Association can be used for pets and/or show. They range in size from the Flemish Giant, which can weigh nearly 20 lb, to the Netherland Dwarf, which seldom weighs more than 2 lb and makes a popular pet (Figure 29-13).

Personal preference and the availability of breeding stock usually determine what breed or breeds of rabbit to raise for pets or show.

Approved Practices for Rabbit Production

Approved practices for the production of rabbits include the following:

FIGURE 29-13 Netherland Dwarf rabbits, generally weighing in at less than 2 lb, are very popular as pets. *(Courtesy of Isabelle Francais)*

1. Select the correct breed for the intended use.
2. Use purebred stock if you plan to sell breeding stock and to maintain uniformity in your herd.
3. Purchase breeding stock only from reputable breeders with accurate records.
4. Build or choose a hutch of the proper size for the breed of rabbit that you are growing. A **hutch** is a cage or house for a rabbit. For small- and medium-sized breeds, provide hutches, 30 in. wide × 36 in. long × 18 in. high. For large-sized breeds, provide hutches 30 in. wide × 48 in. long × 18 in. high.
5. Place the hutch where the rabbit will have adequate ventilation and be protected from heat, wind, rain, sleet, and snow.
6. Provide adequate feeder and waterer space. Rabbits should have access to fresh, clean water at all times.
7. Provide a separate hutch or cage for each mature rabbit.

8. Breed rabbits when does are 5 to 8 months old, depending on the breed, and bucks are 6 to 7 months old. It may be wise to delay breeding of large rabbits longer, because they are slower in reaching sexual maturity.

9. Take the doe to the buck's cage for breeding and return her to her own cage immediately after breeding.

10. Maintain one mature buck for every 10 to 25 does, depending on the breed and mating management.

11. Place a nesting box in the doe's cage twenty-five days after mating occurs.

12. Keep the handling of rabbits to a minimum to avoid injury. When handling rabbits, hold them by the skin on the back of the neck, with the other hand supporting the weight of the rabbit.

13. Feed a commercial pelleted-feed free choice (feed available at all times) to does and **litters** (a group of young born at one time to the same parents). Feed single bucks and does 3 to 6 oz of feed each day. Rabbits need to be fed only once a day, and they should be fed in the evening if possible.

14. Maintain accurate breeding, production, and health records for all rabbits.

15. Tattoo all breeding rabbits for identification. A **tattoo** is a means of marking rabbits and other animals for identification. Rabbits are tattooed in the ear.

16. Plan for and maintain a strict herd health program.

17. Dispose of sick and dead rabbits promptly.

18. Market rabbits as soon as they reach market size or weight.

HONEYBEES AND APICULTURE

History and Domestication

Honeybees have been a part of history for at least fifteen thousand years. Cavemen drew pictures on cave walls of bees and of collecting honey. **Honey** is a thick, sweet substance made by bees from the nectar of flowers. In Egypt, mummies were embalmed and stored in a liquid based on honey. Jars of honey have been found in many of the Egyptian tombs.

The Bible makes many references to honey and the use of honey for food. During biblical times, honey was not produced in nice, neat combs as it is today. Rather, in most cases the hive was destroyed in the process of removing the honey and comb. A **hive** is a home for honeybees. **Comb** is the wax foundation in which bees store honey.

Greeks and Romans were very familiar with honeybees and honey. Pompey used poisoned honey to defeat his enemies in at least one engagement. Aristotle wrote in great detail about bees and their production of honey.

Most early civilizations considered honey to be the food of gods. Athletes competing in Olympic games often ate honey before their events in order to gain extra strength and endurance.

Early beekeepers kept their bees in hollow logs, straw hives, or even in crude clay cylinders. All of these containers had to be destroyed in order to remove the honey.

BIO•TECH CONNECTION

Bees in Biocontrol Business

An old adage says, "If you want to get a job done, ask a busy person to do it." Noting another saying, "busy as a bee," entomologists have found a way to have busy honeybees spread biocontrol agents to plants as they pollinate billions of dollars worth of crops each year. Scientists are using unsuspecting honeybees to deliver biocontrol agents right where the agents are needed on plants. Two serious pests are targeted for control by the process. One is the fire blight bacteria, which infects pear and apple trees; and the other is the corn earworm, one of the worst insect pests of corn and cotton.

Fire blight is caused by *Erwinia amylovora*, a bacterium that first colonizes a flower's stigma, the part that receives pollen grains during pollination. The bacteria then multiplies and spreads quickly. The disease causes cankers on twigs and branches and causes the leaves to have a burnt appearance; hence, the name "fire blight." The disorder weakens or kills the trees. On the positive side, scientists have found that spraying blossoms with beneficial bacteria helps to prevent the fire blight disease. The beneficial bacteria apparently compete with the harmful bacteria by consuming the nutrient-rich stigma, and the pathogen cannot get a foothold and invade the tree.

Rather than spraying with the beneficial bacterial, entomologists and plant pathologists at Utah State University and Oregon State University are doing research where honeybees deliver the beneficial bacteria to the point-of-entry for the fire blight pathogen—the stigma. Such a strategy maximizes the use of the biocontrol material. The procedure is being field-tested, wherein numerous beehives are fitted with devices called pollen inserts. The pollen insert is a

Twig-blight phase of fire blight. While collecting nectar and pollen for the hive, honeybees spread a fire-blight-fighting bacteria as well as pollinate the apple blossoms. *(Courtesy of Paul C. Pecknold)*

slotted passageway for bees that automatically dusts the bees with the beneficial bacterium, which is the biocontrol in this case, each time they leave the hive. A hive contains from 10,000 to 100,000 bees, each visiting perhaps 100 blossoms per hour. The beneficial bacteria does no apparent harm to the bees.

Meanwhile, entomologists in Tifton, Georgia, are using honeybees to carry a natural virus, *Heliothis*, that destroys the larva stage of the corn earworm. The device used to "load" the bees is one where the bees use a different entrance and exit to the hive. The exit device includes a metal tray that contains the virus material. The exiting bees carry the material on their feet, legs, and undersides and deposit it on their rounds. The technique reduced earworm infestations in clover and should be effective on other flowering plants.

FIGURE 29-14 Biological technician Gary Delatte and graduate student Lilia de Guzman examine a honeycomb and check the health of the honeybees. *(Courtesy of USDA/ARS #K-5064-2)*

With the invention in the 1850s of movable combs with wax foundations to encourage bees to make neat, straight honeycomb, the whole beekeeping industry changed. Honey was finally a commodity to be enjoyed by nearly everyone (Figure 29-14). Soon after the development of movable comb, the discovery that honey could be whirled out of the comb led to the invention of the honey extractor. It was no longer necessary to destroy the comb in order to get to the honey. The comb, after having been emptied of honey, could now be placed back in the hive to be refilled by the bees.

Today, the production of honey in the United States is a large and profitable business. Far more important than the production of honey is the work that honeybees do in the pollination of crops that are important to agriscience. Modern beekeeping is known as **apiculture.**

Economic Importance

It is very difficult to accurately gauge the true economic importance of honeybees. They are responsible for about 80 percent of insect pollination of plants. Without honeybees, many of the crops important to agriscience would simply disappear from Earth.

Pollination of orchard crops such as citrus, peaches, and apples by honeybees is so important that many beekeepers rent their bees to orchardists when these trees are in bloom. With $10 to $30 in rent per hive, and flatbed trailers to move hives, commercial beekeepers make more money from bee rental than from honey. Such beekeepers operate from Florida to Maine and from Texas to Washington State (Figure 29-15).

There are about three hundred thousand beekeepers in the United States, of which about 99 percent are hobby or part-time beekeepers. These three hundred thousand people care for about 6 million hives of bees. In a normal year, a hive of bees will produce 100 to 150 lb of honey in excess of the approximately 150 lb needed by the bees to live. At $1.50 to $2.00 per pound of honey retail, it is easy to see that the production of honey is big business.

FIGURE 29-15 The care and management of beehives for pollination is an essential component in the production of many farm and orchard crops. *(Courtesy of USDA/ARS #K-4715-1)*

FIGURE 29-16 Worker bees tend the queen (marked), the hive, and the brood, as well as make and store honey for the future. *(Courtesy of USDA/ARS #K-5069-22)*

Approved Practices for Beekeeping

The following is a list of approved practices to be used in the keeping of bees.

1. Check local regulations before starting a beekeeping operation.

2. Locate bees out of direct contact with people and neighbors' yards and gardens.

3. Place hives facing away from prevailing winds. They should also be protected from hot summer sun.

4. Thoroughly clean and disinfect hives before allowing new groups of bees to use them.

5. Purchase bees from reputable sources. It is usually far more profitable to purchase a 3-lb package of bees with a purebred queen than to rely on a swarm to populate a new beehive. A swarm is a group of bees complete with queen that leaves an overcrowded hive to find a new home.

6. Replace queens every two years. A queen bee is the only fertile, egg-laying female in each hive (Figure 29-16).

7. Have your bees inspected for contagious diseases by a federal bee inspector each year.

8. Make sure that each hive of bees has a store of at least 75 lb of honey for the winter. Hives that do not have enough surplus honey stored for winter should be fed a sugar-water mixture to supplement their own honey stores.

9. Always be sure that bees have ample room to store the honey being produced.

10. Remove surplus honey as soon as the bees have capped it over with wax.

11. Remove honey in the evening or at night when nearly all of the bees are in the hive. Supers containing surplus honey can be freed from bees by blowing cool smoke over the bees and brushing them off the comb with a bee brush. A super is a box filled with a movable foundation that is used by the bees to store honey. You can also use a bee excluder between the honey to be removed and the hive body.

12. After moving a hive, put a deflector in the entrance of the hive so the bees will notice that they have been moved. Hives must be moved at least five miles to prevent bees from returning to the former site of their hive.

13. Inspect beehives at least monthly to determine the strength of the hive and the queen. Be sure to observe the number of eggs being layed by the queen. Also note whether the worker bees are building drone or queen cells in the hive. Drone and queen cells look like peanuts. Such cells should be destroyed. Worker bees are undeveloped females and constitute all of the working force of the hive. Drones are males whose only purpose in life is to fertilize the queen once in her life.

14. Reduce or prevent swarming of bees by providing ample hive space for the bees and eliminating queen cells as they are found. Overcrowding often causes bees to develop a second queen. A new queen will attract a group of worker bees and leave the hive to start a new colony. This process is called

FIGURE 29-17 Guide dogs serve dual purposes. When at work, they act as "eyes" for the blind. When at rest or play, they provide companionship. *(Courtesy of Guide Dog Foundation for the Blind, Inc.)*

swarming. Bees will not swarm without a queen because she is their only hope of survival.

15. Be aware of pesticides being used in the area that could kill bees or be stored in the honey being produced.

16. Secure the proper equipment before starting an apiary. An **apiary** is an area for the keeping of beehives.

17. Keep honey that has been removed from bees in an area that bees cannot get to. Otherwise, they will steal all of the honey in a short time.

18. Extract honey from the comb as soon as possible after harvesting it. Honey stored for long periods of time in the comb may granulate, which makes it impossible to extract.

19. Develop a market for your honey.

PET CARE AND MANAGEMENT

Animals make excellent companions and can provide specialized functions for children and adults alike. For many animals, the line between pet, companion animal, and work animal is a very fine one. For example, guide dogs are essential "eyes" for the blind and actually must make intelligent choices for their masters when crossing streets (Figure 29-17). Similarly, dogs and other pets provide companionship for the elderly and may warn of intruders. Good dogs may ward off attackers and provide other protection when appropriately trained. Guard dogs are essential for herding and controlling sheep and protecting sheep from predators.

Many kinds of caged birds and reptiles are used for pets. Their needs are rather specialized and advice for their care is usually obtained from pet shops and other suppliers. Similarly, a large assortment of fish and other aquatic creatures make attractive and challenging aquarium projects for all ages (Figure 29-18). Even exotic animals such as Llamas and kangaroos find their ways onto farms and into the hearts of Americans (Figure 29-19).

FIGURE 29-18 Aquatic animal and plant projects create interest and provide excellent opportunities for learning and profit. *(Courtesy of FFA, Photo by Michael Wilson)*

FIGURE 29-19 The care of exotic animals creates new challenges for experienced pet handlers. *(Courtesy of FFA)*

FIGURE 29-20 Kittens are cute, but they require proper grooming, feeding, and health care. *(Courtesy of FFA)*

Dogs and cats are probably the most prevalent companion animals in the United States. While they have different moods and behave in different ways, they have similar needs (Figure 29-20). The following is a list of approved practices for the care of dogs and cats, courtesy of Pennsylvania State University.

1. Select animals that are alert and healthy.
2. Vaccinate for rabies at age-determined intervals.
3. Select a breed adapted to your situation.
4. Prepare a clean, draft-free living area.
5. Provide an adequate number of feed dishes.
6. Provide a diet of high-quality food for the breed and age.
7. Provide toys and/or treats.
8. Clean feed and water dishes daily.
9. Keep clean, fresh bedding in sleeping area.
10. Provide plenty of clean, fresh water.
11. Clean pen or box and exercise area daily.
12. Develop and implement a sound health plan.
13. Consult with your veterinarian to prevent and control internal and external parasites.
14. Vaccinate animals routinely at proper times.
15. Train animals for show.
16. Properly bathe and groom the animal.
17. Use a proper carrier for transport.
18. Use proper restraint procedures.
19. Maintain accurate breeding and production records.
20. Select a male to mate with your female.
21. Use a superior proven male.
22. Watch females for heat.

FIGURE 29-21 Businesses are needed to supply specialized care, food, health supplies, and equipment for pets and small-animal projects. *(Courtesy of Michael Dzaman)*

23. Prepare a clean area for whelping (giving birth).
24. Properly train.
25. Market and sell young animals.
26. Develop a private market for the animals.
27. Complete a registration application if animals are purebred.
28. Neuter animals not involved in breeding.
29. Summarize and analyze records.

What is one person's pet always becomes someone else's business in providing replacement animals, feed, supplies, and equipment for that pet (Figure 29-21).

Raising small animals provides the opportunity for persons with limited capital and facilities to get a start in animal agriscience. Most small animals are better adapted to production in urban and suburban areas than are larger animals. The same experiences in planning for, caring for, managing, and marketing can be learned with small-animal enterprises without the large outlay of cash needed for the production of large animals.

AGRI·PROFILE

CAREER AREAS: Animal Technician/Grower/Bee Keeper/Manager

The career opportunities for small-animal care and management are good in many areas. The extensive use of laboratory animals for research; small animals for pets; fish for home aquariums and garden pools; small animals for fur; and animals for zoological parks ensures attractive jobs in the future.

The operation of animal hospitals, kennels, grooming services, pet stores, training programs, boarding facilities, public aquariums, animal shelters, and humane societies provide opportunities in a number of fields. These include animal nutrition; facilities construction and maintenance; feeds; health services; care and management; production; breeding; and marketing.

Curriculums in small-animal care and management have been added to many high-school agriscience programs. Similarly, programs are available in many technical schools, community colleges, and universities.

Animal trainers teach animals to obey commands and to perform tricks for audiences. *(Courtesy of Tracy and Pedro Navarro)*

STUDENT ACTIVITIES

1. Write the Terms to Know and their meanings in your notebook.

2. Make a bulletin board display of breeds and types of poultry and rabbit.

3. Attend a fair or show and record the names of the breeds of poultry and rabbit shown there.

4. Interview a local beekeeper about beekeeping practices in your area.

5. Compare the label from a bag of poultry feed with one from a bag of rabbit feed. Determine the differences in ingredients, percentage of protein, additives, and fiber.

6. Set up an observation beehive in the school.

7. Make a list of local crops that are of importance to agriscience and that bees pollinate.

8. Develop a crossword puzzle or word search using the Terms to Know.

SELF EVALUATION

A. Multiple Choice

1. A hive of honeybees needs about _____ lb of honey stored in order to live during the winter.
   a. 25
   b. 50
   c. 75
   d. 100

2. A gosling is a baby
   a. chicken.
   b. duck.
   c. pigeon.
   d. goose.

3. A pair of rabbits can produce _____ lb of meat per year.
   a. 10 to 30
   b. 30 to 50
   c. 50 to 80
   d. 200 or more

4. A castrated male chicken is a
   a. capon.
   b. cock.
   c. cockerel.
   d. rooster.

5. The only bee in a hive capable of laying eggs is the
   a. king.
   b. queen.
   c. worker.
   d. drone.

6. Honeybees account for about _____ percent of all insect pollination.
   a. 20
   b. 40
   c. 60
   d. 80

7. The _____ breed of chicken is the foundation of nearly all types of laying hens.
   a. Cornish
   b. Pekin
   c. Leghorn
   d. game

8. Two breeds of rabbit produce wool called
   a. pelt.
   b. fur.
   c. angora.
   d. wool.

9. The champion egg-laying breed of bird is the
   a. Leghorn.
   b. quail.
   c. Khaki Campbell.
   d. Toulouse.

10. More than _____ lb of rabbit meat are produced in the United States each year.
   a. 1 million
   b. 10 million
   c. 25 million
   d. 50 million

B. Matching

_____	**1.** Drone		a.	Male goose
_____	**2.** Cockerel		b.	Young chicken
_____	**3.** Gander		c.	Male bee
_____	**4.** Doe		d.	Female bee
_____	**5.** Drake		e.	Female pheasant
_____	**6.** Chick		f.	Young turkey
_____	**7.** Buck		g.	Male rabbit
_____	**8.** Worker		h.	Female rabbit
_____	**9.** Hen		i.	Male chicken
_____	**10.** Poult		j.	Male duck

C. Completion

1. More than 90 percent of the turkeys produced in the United States are _____.

2. Ducks raised for meat reach a weight of 7 lb in about _____ weeks.

3. The most popular breed of rabbit for meat production is the _____.

4. A _____ is a home for honeybees.

5. The average laying hen produces about _____ eggs each year.

6. Another word for beekeeping is _____.

7. _____ and _____ are probably the most common companion animals in the United States.

8. Whelping means _____.

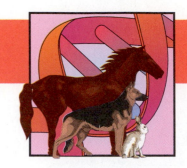

Dairy and Livestock Management

To determine the history, types, uses, care, and management of dairy and livestock.

COMPETENCIES TO BE DEVELOPED

After studying this unit, you should be able to:

- determine the history and economic importance of dairy and livestock.
- recognize major types and classes of livestock.
- list major uses of livestock.
- understand basic approved practices in the care and management of dairy and livestock.

MATERIALS LIST

- ✓ copies of various livestock breed magazines
- ✓ bulletin board materials
- ✓ paper glue
- ✓ scissors
- ✓ paper for notebook covers
- ✓ examples of various animal products or things made from animal products

556

Large animals, including dairy, beef, sheep, goats, and swine, are the backbone of agriscience. Keeping wild animals captured made it necessary for early humans to provide feed for them. It also changed their lives, in that it forced humans to tend the animals. The capturing and domestication of animals changed humans from hunters to farmers. Today, much of agriscience is centered around the production of animals and animal products and the production of feed for those animals.

DAIRY CATTLE

Origin and History

In the wild, mammals normally produce only enough milk to feed their offspring. **Mammals** are animals that produce milk. **Milk** is white or yellowish liquid secreted by the mammary glands of animals for the purpose of feeding young. When early humans realized that milk was good to drink and that some types of animals produced more milk than others, the eventual domestication of milk animals began. Although the cow, buffalo, goat, ewe, mare, and sow have been and are currently being used for the production of milk in various parts of the

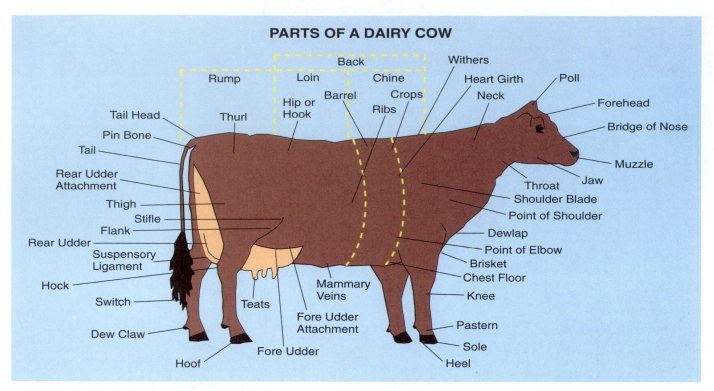

PARTS OF A DAIRY COW

FIGURE 30-1 Dairy type of cattle were developed to convert nutrients into milk rather than meat.

world, this unit will concentrate primarily on cattle in discussing dairy production.

There are records of the use of cattle to produce milk as early as 9000 B.C. There are a number of references in the Bible to milk and the production of milk. Even Hippocrates, the father of modern medicine, recommended milk as a medicine in his writings around 400 B.C.

There were no dairy cows in the New World until Columbus brought them with him on his second voyage in 1493. Cattle also came to the New World in 1611 with the Jamestown colonists. The production of milk was limited to a few cows per family during the colonial period. It was not until the late 1800s that dairying became an important agricultural industry in the United States. Since then efforts have been intensive in breeding and developing the dairy type (Figure 30-1).

Economic Importance

The production of milk is the second most important animal enterprise in the United States, if sales dollars are the criterion for importance. The consumption of milk and other dairy products has remained nearly steady during the past several years, after several decades of steady decline. The average American uses slightly more than 570 lb of milk and dairy products each year. There are about 10 million cows in the United States producing about 148 million lb of milk annually. Dairy is truly big business.

The production of milk is not the only income-generating part of dairy production. Calves not needed as replacement for the dairy herd are sold as **veal** (the meat of young calves). Similarly, cows that are no longer profitable producers of milk are sold for **beef** (meat from cattle). These are replaced by calves that become tomorrow's cows (Figure 30-2).

FIGURE 30-2 Today's calves become tomorrow's dairy cows, so good calf care and management are top priorities on dairy farms. *(Courtesy of FFA)*

BIO•TECH CONNECTION

Of Cows and Cars

Cows and cars are sometimes the recipients of the same technology. Transponders are small devices that can be attached to the ears of cows and other livestock to identify them. As animals come into the electronic fields of certain other devices, their identities can be noted and recorded by computer. The computer can be programmed to simply note and remember what the animal does; or it can direct other devices to provide selected kinds and amounts of feed, water, medicine, pesticide, or other treatment. This technology is common in modern milking parlors and feeding systems.

At the USDA Fort Keogh Livestock and Range Laboratory in southeastern Montana, scientists found transponders to be the answer to a weighty problem! To conduct feeding, grazing, pasture-improvement, and pasture-management experiments, they had to weigh their research subjects, range cattle, at frequent intervals. Animal response to treatment generally shows up in the form of weight gain or loss. At the same time, any disturbance or change in an animal's routine, such a cattle roundup, forced weighing, or catching in a head gate, will influence an animal's weight. Good research required a system that could weigh the animals at frequent intervals without disturbing them—that is, they needed to weigh themselves!

The answer was found in a system utilizing a scale, computer, and transponder. Cattle need water. A continuous supply of fresh water must be readily available if other body functions are to be optimal. Therefore, the researchers set up a scale and guide railing so the animal had to be on the scale to drink. An ear-tag transponder on the animal identifies the animal as it enters the range of

Transponders and other sensing units permit better research and management. The unit being applied here will record motion and other data so scientists can determine how much time the animal spends grazing. *(Courtesy of USDA/ARS #K-4235-8)*

the pickup device and steps on the scale. The electronic scale sends continuous weight readings to the computer, indicating the animal's weight before, during, and after drinking. The computer records the weight data along with time, temperature, weather, and other information monitored at the sight.

As drivers zip across California's Golden Gate Bridge and blissfully bypass the toll booth, they are probably legal motorists simply using the same transponder device that was pioneered by the USDA's Agricultural Research Service. If so, the car-mounted transponder in their car gave the computer the auto's account number; and the computer will have noted the time, date, and auto number for future billing to the motorist. The system is being used by many toll facilities across the country.

Dairy-cattle operations are generally divided into two types—Class A and Class B. The class refers to the intended use for the milk produced. Class A milk is produced under very strict standards and is intended for consumption as fluid milk. Fluid milk includes whole milk, reduced-fat milk, and cream. Class B milk, which can be produced under less strict standards, is intended to be used to make butter, cheese, ice cream, nonfat dry milk, and other manufactured dairy products.

Types and Breeds

More than 90 percent of all dairy cattle in the United States are of the holstein breed. The familiar black and white cattle are the highest average producers of milk of any breed in the country. Because of the large numbers of cattle involved, the breed has also been able to make the most genetic improvement in recent years.

The second most popular breed of dairy cattle is the jersey. Even though they are the smallest of the dairy breeds, they rank number one in butterfat production. **Butterfat** is the fat in milk. Another popular breed of dairy cattle is the Guernsey, which is known for the yellow color of its milk. Ayrshire and Brown Swiss round out the top five breeds of dairy cattle in the United States (Figure 30-3).

Approved Practices

Raising Calves   General approved practices for raising dairy calves include the following:

1. Make sure the newborn **calf** receives colostrum as its first food and for at least the first thirty-six hours of its life. **Colostrum** is the milk that a cow produces for a short time after calving. It contains antibodies that protect the newborn animal from disease until the animal can build up its own natural defenses.

2. Feed milk or milk replacer at 8 to 10 percent of the calf's weight daily until the calf is four weeks old. **Milk replacers** are dry dairy or vegetable products that are mixed with warm water and fed to young calves in place of milk.

3. Start feeding calf starter, a grain mixture, free choice at about 10 days. **Free choice** means making feed available at all times.

4. Wean calves from milk when they are eating 1½ lb of calf starter per day. To **wean** means to remove and keep away from.

5. Feed calves hay and water free choice at age 3 to 9 months. Up to 4 lb of grain and some silage can be fed daily.

6. Make up the bulk of the ration with forages fed free choice after 9 months of age.

7. Remove horns at an early age, preferably as soon as the horns begin to develop.

8. Remove extra teats at an early age.

9. Identify calves with ear tags or tattoos as soon as possible after birth.

10. Prevent calves from sucking each other.

11. Keep hooves properly trimmed.

12. Vaccinate for calfhood diseases at the recommended times.

13. Maintain the calf in clean and sanitary conditions.

FIGURE 30-3 Major breeds of dairy cattle in the United States include (A) Holstein *(Courtesy Holstein-Friesian Association of America)*, (B) Jersey *(Courtesy The American Jersey Cattle Club)*, (C) Guernsey *(Courtesy American Guernsey Association)*, (D) Ayrshire *(Courtesy Ayrshire Breeders' Association)*, and (E) Brown Swiss *(Courtesy Brown Swiss Cattle Breeders)*.

14. Plan for and maintain a disease- and parasite-prevention-and-control program.

15. Breed heifers to calve at 2 to 2½ years of age. A **heifer** is a female that has not given birth to a calf. With cattle, to **calve** means to give birth.

16. Maintain heifers and calves in uniform groups according to size and weight.

Dairy Cows   Some approved practices for dairy cows include the following:

1. Rebreed cows 60 to 90 days after calving. A **cow** is a female of the cattle family that has calved. Cows should be bred to calve once every 12 months.

2. Observe cows for evidence of heat period twice daily (Figure 30-4).

3. Check cows to determine whether they are pregnant. This should be done 45 to 60 days after breeding them.

4. Provide a dry period of about 60 days before calving to allow the cow to rebuild her body. The **dry period** refers to the time when a cow is not producing milk.

5. Feed dairy cows according to their levels of production and stages of pregnancy.

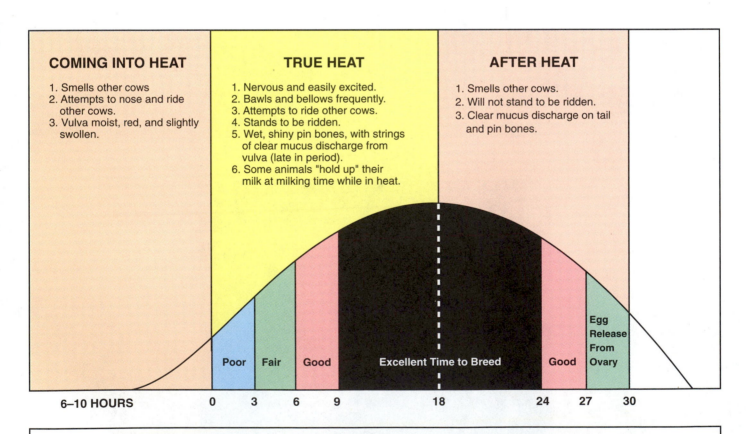

COMING INTO HEAT

1. Smells other cows
2. Attempts to nose and ride other cows.
3. Vulva moist, red, and slightly swollen.

TRUE HEAT

1. Nervous and easily excited.
2. Bawls and bellows frequently.
3. Attempts to ride other cows.
4. Stands to be ridden.
5. Wet, shiny pin bones, with strings of clear mucus discharge from vulva (late in period).
6. Some animals "hold up" their milk at milking time while in heat.

AFTER HEAT

1. Smells other cows.
2. Will not stand to be ridden.
3. Clear mucus discharge on tail and pin bones.

Poor Fair Good   Excellent Time to Breed   Good   Egg Release From Ovary

6–10 HOURS   0   3   6   9   18   24   27   30

NOTE: Suggestions are based on average conditions. Many exceptions are true of normal cows. Cows noticed in "True Heat" at milking time in the morning probably have been in heat for several hours. The numbers 3, 6, 9, etc. indicate hours of heat.

FIGURE 30-4 Breeding at the proper stage of the heat period is essential to maximize pregnancy rates.

Factors Determining Efficiency			
Factor	Results	Goals	State Average
_____ 1. Number of cows in herd			
_____ 2. Total lb of milk produced			
_____ 3. Total lb of butterfat produced			
_____ 4. Average annual milk production/cow (lb)			
_____ 5. Average annual butterfat production/cow (lb)			
_____ 6. Average annual % butterfat for student's herd			
_____ 7. Feed cost/pound of butterfat produced			
_____ 8. Average feed cost/cow			
_____ 9. % of calves sold or kept until 3 months of age			
_____ 10. Profit or loss			
_____ 11. Lb of milk sold/hour of total labor			
_____ 12. Dollar returns/hour of self labor			
_____ 13. Production cost/pound of milk			

FIGURE 30-5 Efficiency factors are good indicators of how the dairy business is operating.

6. Maintain complete health, breeding, and production records for every cow in the herd (Figure 30-5).
7. Establish and maintain a disease- and parasite-prevention-and-control program.
8. Contact a veterinarian anytime that you are unsure of how to treat a dairy-herd problem.
9. Milk dairy cows at regular intervals each day. Two milkings each day, approximately 12 hours apart, is a normal routine. However, milking three times per day is common in many herds.
10. Maintain a regular routine in handling dairy cattle in order to maintain maximum production.
11. Cull unprofitable dairy cows. To **cull** means to remove from the herd.
12. Properly maintain dairy housing and milking equipment.

BEEF CATTLE

Origin and History

It is likely that cattle were domesticated in Europe and Asia at some time during the New Stone Age. Domesticated cattle of today are probably all descended from one of two wild species, *Bos taurus* or *Bos indicus* (Figure 30-6). Some of these wild cattle stood as tall as 7 ft at the shoulders. They were domesticated for meat, milk, and draft. **Draft** means for work.

FIGURE 30-6 *Bos tauras* was one of the wild ancestors of today's cattle.

AGRI·PROFILE

CAREER AREAS: Farm Manager/Herd Manager/Farmer/Rancher/Breed Association Representative

Science and production can no longer be separated. Individuals and teams of scientists visit farms, ranches, and feedlots frequently to gather data, analyze problems, and seek solutions. Farmers, ranchers, and feedlot managers are frequently graduates of agricultural colleges and may have advanced degrees in business, management, or animal sciences.

In addition to scientists and managers, technicians find meaningful careers in the animal production businesses of the nation. Livestock enterprises include beef, dairy, sheep, and swine. There are many nonfarm career opportunities as field representatives for breed associations, feed companies, marketing cooperatives, supply companies, animal-health products, herd-improvement associations, and financial-management firms.

Many individuals utilize their farming and ranching backgrounds in agriscience writing, publishing, and telecasting careers. Further, farm and ranch experience provides excellent background for careers in agriscience teaching and extension work.

Farming, ranching, and feedlot management have become scientific and big business. *(Courtesy of USDA)*

Owning cattle was a symbol of wealth in early times. They were worshipped, used, and often abused. Several early civilizations had "gods" fashioned to look like cattle. The use of cattle in sports such as bullfighting also developed during early times.

Cattle came to the New World with the earliest settlers, who were more interested in animals that could do heavy work than in those that could produce meat. As the European explorers came to the New World, they also brought cattle with them. Spanish settlers brought longhorn-type cattle as a source of food for Christian missions in the Southwest. As demand increased for beef, the cattle industry developed on the frontier, where grass and the large open spaces required for cattle were abundant. Great cattle drives originated in these areas, as cattlemen drove their cattle to human population centers in order to market them.

Today, the beef cattle industry is concentrated in the Midwest and South, where plenty of feed is available and where production of other crops may not be profitable.

Economic Importance

The beef cattle industry is the number one red-meat production industry in the United States. Americans eat about 96 lb of beef per person per year. This is part of a total consumption of about 254 lb of meat and poultry, excluding seafood.

With sales of over $40 billion in cattle and calves, the beef cattle business is one that will likely remain important for many years to come. There are approximately 100 million head of cattle and calves on United States farms today. Leading states in the sales of beef are Texas, Nebraska, Kansas, Colorado, Iowa, and California.

The production of meat is not the only use for beef cattle. Cattle convert inedible grasses into food for people. Further, cattle manure provides fertilizer for crops, and meat by-products are made into many nonfood products that we use every day (Figure 30-7).

By-products from the Production of Meat	
• Bone for bone china. • Horn and bone handles for carving sets. • Hides and skins for leather goods. • Rennet for cheese making. • Gelatin for marshmallows, photographic film, printers' rollers. • Stearin for making chewing gum and candies. • Glycerin for explosives used in mining and blasting. • Lanolin for cosmetics. • Chemicals for tires that run cooler. • Binders for asphalt paving. • Medicines such as various hormones and glandular extracts, insulin, pepsin, epinephrine, ACTH, cortisone, and surgical sutures.	• Drumheads and violin strings. • Animal fats for soap and feed. • Wool for clothing. • Camel's hair (actually from cattle ears) for artists' brushes. • Cutting oils and other special industrial lubricants. • Bone charcoal for high-grade steel, such as ball bearings. • Special glues for marine plywoods, paper, matches, window shades. • Curled hair for upholstery. Leather for covering fine furniture. • High-protein livestock feeds.

FIGURE 30-7 By-products of meat production add greatly to the quality of life of people.

Types and Breeds

The general types of beef-cattle operations include purebred breeders, cow-calf operations, and slaughter-cattle, or feedlot, operations. In a purebred operation, only cattle of a single, pure breed are raised. Operations are geared to produce purebred bulls for cow-calf operations and to produce animals to be sold to other purebred operations. A **bull** is a male of the cattle family. Breeders of purebred cattle have been responsible for much of the genetic improvement in beef cattle in recent years.

Cow-calf operations serve to produce feeder calves for slaughter-cattle producers. They are located mostly in the upper Great Plains states and in the western range states, where grass is in abundance and much of the land is unsuited to produce many other crops. Usually calves are born in the spring, stay with their mothers during the summer, and are weaned in the fall. They are then sold to slaughter-cattle producers. Often these operations use purebred bulls on the grade cows in the herd.

The slaughter-cattle, or feedlot, operator buys calves from cow-calf operators and feeds them until they reach slaughter weight. These operations are generally located in the Midwest, where there is an abundance of corn and other grain available for feed.

Until about 40 years ago, there were basically three breeds of beef cattle in the United States—Hereford, Angus, and Shorthorn. Although these three breeds are still important today, they have been joined by more than fifty other breeds from all over the world (Figure 30-8). These breeds can be divided into many classifications. However, for this study, they are designated as English, exotic, and American.

English breeds originated in the British Isles and include Hereford, Angus, Shorthorn, Galloway, Devon, and Red Poll. In general, English breeds of cattle are of medium size and are noted for the excellent quality of meat that they produce.

Exotic breeds of cattle were first imported to the United States from all over the world when consumers began demanding leaner beef. Beef producers also became aware that calves from these breeds of cattle

FIGURE 30-8 Today's predominant beef breeds in the United States are (A) Hereford *(Courtesy American Hereford Association)* and (B) Angus *(Courtesy American Angus Association)*. **(Continued)**

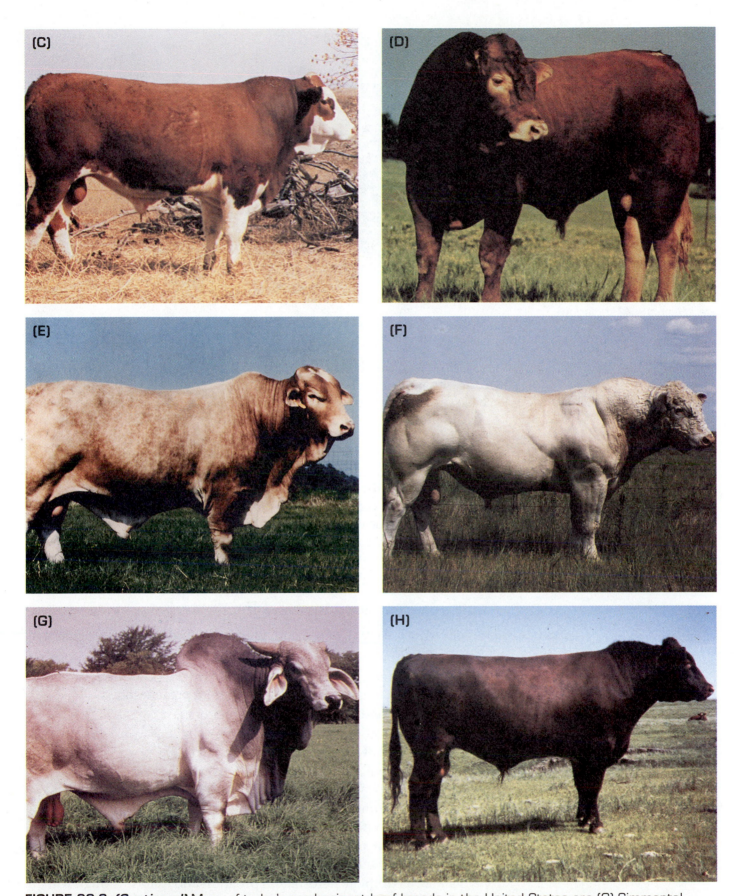

FIGURE 30-8 *(Continued)* More of today's predominant beef breeds in the United States are (C) Simmental *(Courtesy American Simmental Association)*, (D) Limousine *(Courtesy North American Limousine Foundation)*, (E) Beefmaster *(Courtesy Beefmaster Association Universal)*, (F) Charolais *(Courtesy American-International Charolais Association)*, (G) Brahman *(Courtesy American Brahman Breeders' Association)*, and (H) Shorthorn *(Courtesy American Milking Shorthorn Society)*.

FIGURE 30-9 Many commercial producers have crossed and mixed breeds of cattle such as the Brangus bull/cow pictured. *(Courtesy American Red Brangus Association)*

grew faster and much more efficiently than most of the English breeds. Exotic-breed bulls are often used in grade cow-calf operations to increase the weight of calves being produced. Examples of exotic breeds of cattle include Charolais, Limousin, Simmental, Blond D'Aquitaine, and Maine Anjou.

American breeds of beef cattle were developed from necessity. In order to grow beef cattle in the South and Southwest, heat tolerance and disease and parasite resistance were musts. To develop cattle that met these requirements and still had desirable quality meat, breeders crossed Brahman cattle from India with the English breeds. Examples of American breeds of beef cattle are Brangus, Beefmaster, Santa Gertrudis, and Barzona. Many commercial herds are of crossbred, or mixed, breeds (Figure 30-9).

Approved Practices for Beef Production

The following are some of the considerations in the production of beef cattle.

1. Select breeds and individual beef cattle according to intended use, area of the country, and personal preference.
2. Buy cattle from only reputable breeders.
3. Select purebred cattle according to physical appearance, pedigree, and available records (Figure 30-10).
4. Isolate new animals from the herd for at least 30 days to observe for possible diseases and parasites.
5. Provide enough human contact so that beef cattle can be handled when necessary.
6. Break calves to lead as soon as possible, if they are going to be exhibited at fairs and shows.
7. Plan a complete herd-health program and follow through with it.
8. Use the services of a veterinarian for serious health problems.
9. Vaccinate to prevent diseases of local concern.
10. Castrate, dehorn, and permanently identify calves at an early age.

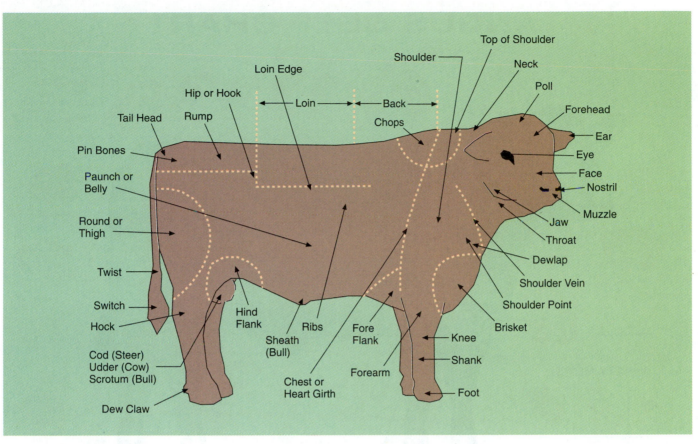

FIGURE 30-10 Learning the proper names of cattle parts is necessary for discussing and evaluating animals.

11. Breed heifers to calve at 2 years of age.

12. Implant steers and heifers grown for slaughter with approved growth hormones. To **implant** is to place a substance under the skin. The substance is released slowly over a long period of time. A **steer** is a castrated male of the cattle family.

13. Wean calves at 205 days of age and at 450 to 500 lb.

14. Provide supplemental nutrition to cattle when natural forages are in short supply.

15. In a pasture-breeding system, allow one mature bull for every 25 to 30 cows. With pen mating, one bull can be used for every 30 to 50 cows.

16. Implement time breeding so that all calves are born in a 30- to 60-day period.

17. Allow cows to calve in clean stalls or pastures.

18. Group cattle being fed for slaughter according to size and sex.

19. Provide shelter from inclement weather.

20. Provide access to clean, fresh water at all times.

21. Feed slaughter cattle to reach market weight at 15 to 24 months of age.

22. Have proper facilities and equipment available for the type of operation planned.

23. Market animals at the optimum time to maximize profits (Figure 30-11).

24. Maintain complete and accurate records.

ANGUS BEEF CHART

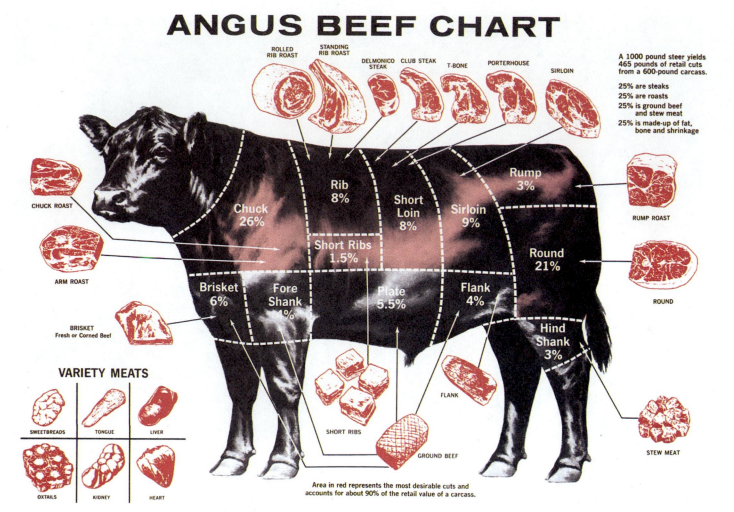

ROLLED RIB ROAST STANDING RIB ROAST DELMONICO STEAK CLUB STEAK T-BONE PORTERHOUSE SIRLOIN

A 1000 pound steer yields 465 pounds of retail cuts from a 600-pound carcass.

25% are steaks
25% are roasts
25% is ground beef and stew meat
25% is made-up of fat, bone and shrinkage

CHUCK ROAST

ARM ROAST

BRISKET
Fresh or Corned Beef

Chuck 26%
Rib 8%
Short Ribs 1.5%
Short Loin 8%
Sirloin 9%
Rump 3%
Round 21%

RUMP ROAST

ROUND

Brisket 6%
Fore Shank 1%
Plate 5.5%
Flank 4%
Hind Shank 3%

FLANK

SHORT RIBS

GROUND BEEF

STEW MEAT

VARIETY MEATS

SWEETBREADS TONGUE LIVER

OXTAILS KIDNEY HEART

Area in red represents the most desirable cuts and accounts for about 90% of the retail value of a carcass.

FIGURE 30-11 Animals should be bred, fed, managed, and marketed in a manner that delivers the best product for the consumer. *(Courtesy of American Angus Association)*

SWINE

Origin and History

Swine were apparently domesticated in China during the Neolithic Age. The first written record of keeping swine appears in 4900 B.C. Mention of swine occurs in the Bible as early as 1500 B.C.

Domestic swine originate from two wild stocks, the European wild boar, *Sus scrofa*, and the East Indian pig, *Sus vittatus*. There are several other wild types of swine that exist today. Even domesticated swine can revert quickly to the wild when the opportunity or need arises.

Swine came to the New World in 1493 on the second voyage of Columbus. As European explorers came to what is now the United States, they brought swine with them as a food supply. The original thirteen head of swine brought to the New World by the explorer de Soto multiplied to over seven hundred head in just three years, and provided food for the explorers.

European settlers also brought swine with them as they settled on the East Coast of the United States. Because swine could find food on their own and reproduce rapidly, they were soon known as "mortgage lifters." Swine in excess of local needs were exported as pork and lard. **Pork** is meat from swine. **Lard** is rendered pork fat. **Render** means to cook and press the oil from.

As westward expansion occurred, swine production followed. In time, the center of swine production settled in the Corn Belt. This area provided large amounts of corn and grain necessary for large-scale pork production.

Economic Importance

Pork production is the number two red-meat industry in the United States. It generates over $10 billion in sales each year. With a population of about 55 million hogs, the United States ranks third in the world in swine production, behind China and Russia. There are about 780 million head of swine in the world today, with more than 45 percent of them in China alone.

On a retail basis, Americans consume about 46 lb of pork per person annually. The consumption of red meat has been slowly decreasing, but the consumption of pork has leveled off.

The modern hog of today is vastly different from its ancestors. It is very lean and trim compared to the lard-type hog in demand fifty years ago and even the meat-type hog of twenty years ago. Hogs are the most efficient converters of feed into meat among the large red-meat animals. It takes about 3½ lb of feed to 1 lb of gain.

Leading states in the United States in the production of swine include Iowa, Illinois, Minnesota, Nebraska, and Indiana. More than 25 percent of the swine production in the country takes place in Iowa alone.

Swine Operations

The basic types of swine operations in the United States are feeder-pig producers and market-hog producers. Operations that produce feeder pigs usually maintain large herds of sows that produce 2 to 2½ litters of piglets each year. **Sows** are females of the swine family that have given birth. A **piglet** is a young member of the swine family. Piglets are usually sold to other producers, who feed them until they reach market weight. In most operations, crossbred sows are bred to purebred boars to produce offspring with more hybrid vigor than would otherwise result (Figure 30-12). A **boar** is a male member of the swine family.

FIGURE 30-12 Feeder pig producers keep brood sows and boars, which provide the young stock for the swine industry. *(Courtesy Coles Swine Farms, Inc.)*

Market-hog operations normally purchase pigs at 5 to 8 weeks of age from feeder-pig producers. They feed the pigs until they reach a market weight of about 220 lb. They are then marketed and sent to slaughter plants for processing.

Another type of swine operation is the purebred producers. They are responsible for producing high-quality boars for feeder-pig operations and purebred stock for other purebred farms. They contribute much to the genetic improvement of swine in general.

Types and Breeds

Until recently, two types of purebred swine predominated in the United States—the lard type and the meat type. With the decreased demand for lard and the demand for lean pork nearly devoid of fat, the lard-type hog has been bred out of existence.

Popular breeds of swine that were developed in the United States include Duroc, Hampshire, Chester White, Poland China, and Spotted Hog. The Berkshire, Yorkshire, and Tamworth breeds were developed in England and the Landrace in Denmark. Duroc, Hampshire, and Yorkshire are the most popular breeds of swine in the United States today (Figure 30-13).

Approved Practices in Swine Production

Some of the approved practices in the production of swine include:

1. Buy pigs from only reputable producers or at certified feeder-pig sales.

2. Observe newly purchased animals for signs of disease and parasites.

3. Group pigs according to size in groups of not more than 20 to 25 animals.

4. Feed a complete, balanced ration based on the age and weight of the animal being fed.

5. Ensure that access to an unlimited supply of fresh water is available at all times.

6. Keep facilities and equipment clean and sanitary.

7. Clean and disinfect all facilities and equipment after each group of animals leaves and before the next group arrives.

8. Select replacement gilts for the breeding herd at an early age and raise them separately from market hogs. A **gilt** is a female pig that has not given birth.

9. Breed gilts at eight months of age and 250 to 300 lb, so they farrow at one year of age. To **farrow** means to give birth to pigs.

10. Use a hand-mating system to breed gilts and sows. In a **hand-mating system,** the boar and sow are kept separate except during mating. Use a boar to check for animals in heat.

11. Put bred gilts or sows in farrowing facilities or farrowing crates three days before they are due to farrow. **Farrowing crates** are specially made cages or pens in which swine give birth. This three-day period gives the female time to adjust to new surroundings before the piglets are born.

FIGURE 30-13 Major breeds of swine in the United States include (A) Duroc, (B) Yorkshire, (C) Hampshire *(photos A, B, and C courtesy Hampshire Swine Registry)*, (D) Spotted *(Courtesy Swine Genetics)*, and (E) Chester White. *(Courtesy Chester White Swine Record)*

12. Perform the following to the piglets at birth:
    a. Clip needle or wolf teeth.
    b. Clip or tie navel cord and dip the end in iodine.
    c. Provide supplemental iron.
    d. Dock tails of pigs to be marketed for meat. To **dock** means to remove all but about 1 in. of the tail.
    e. Weigh all pigs in the litter.

    f. Ear notch all pigs for identification. **Ear notching** is a system of permanently marking animals for identification by cutting notches in their ears at specific locations.

13. Provide creep feed for the baby pigs by the time they are one week old. **Creep feed** is feed provided especially for young animals to supplement milk from their mothers.

14. Castrate males at an early age.

15. Wean pigs at 5 to 8 weeks of age. Weaning at about 6 weeks is normal in most herds of swine.

16. Rebreed sows on the first heat period after weaning the pigs. This usually occurs about three days after the pigs are weaned.

17. Limit the feed for sows to prevent them from getting too fat.

18. Provide protection from heat and cold, especially heat. Swine have no sweat glands, and care must be taken to keep them cool in hot weather.

19. Maintain complete health and production records for each animal in the breeding herd.

20. Set realistic production goals and cull animals not meeting the goals.

SHEEP

Origin and History

The domestication of sheep apparently occurred before the time of recorded history. Fibers of wool have been found in ruins of villages in Switzerland more than twenty thousand years old. Egyptian sculptures showing the importance of sheep date back seven thousand years. Even ancient historical texts are filled with mentions of sheep and **shepherds** (those who take care of sheep).

Sheep were probably domesticated from wild types in Europe and Asia by early humans to use for meat, wool, pelts, and milk. **Wool** is the hair from sheep. Sheep have become so dependent on humans that they can no longer survive on their own in the wild.

As civilization advanced, the production of wool became a priority in sheep production. As a result, specific wool-producing breeds were developed in Europe. Most of today's breeds can be traced back to these breeds developed 500 to 1,000 years ago.

Columbus had sheep with him on his second voyage to the West Indies in 1493. Cortez brought sheep when he explored Mexico in 1519. Spanish missionaries also kept sheep and taught Native Americans of the Southwest how to weave wool into cloth.

English settlers on the East Coast also raised sheep for the production of wool. Lamb and mutton were of secondary concern. **Lamb** refers to young sheep as well as the meat of young sheep. **Mutton** is meat from mature sheep.

Centers of sheep population gradually moved from the Northeast to the West as populations expanded. Areas of open spaces and abundant grasses proved to be ideal for the production of sheep.

Economic Importance

Production of sheep in the United States is relatively unimportant compared to dairy, beef, and swine. Americans eat only about 1.7 lb of lamb and mutton per person each year. This amount is increasing slowly, but is not expected to seriously challenge beef and pork producers. Each person in the United States also uses about 0.8 lb of wool each year.

Sales of sheep and lamb total about $420 million each year. The sale of wool adds about another $70 million to the income of sheep producers.

Because sheep have the ability to survive in areas of limited feed and harsh climates, they are of more economic importance than would be expected. In many such areas, only sheep can survive and constitute a major animal enterprise (Figure 30-14).

Types and Breeds

Sheep operations can be divided into two basic types—farm flocks and range operations. A **flock** is a group of sheep. Farm-flock operations are generally small and are often part of diversified agriscience operations. They may raise either purebred or grade sheep. These flocks usually average less than one hundred fifty animals. They are responsible for about one-third of the sheep and wool produced in the United States.

The other two-thirds of the approximately 11 million sheep in the United States are produced on range operations. Many flocks contain 1,000 to 1,500 head. They are nearly 100 percent grade sheep. Range production is concentrated in the twelve western states. Texas, California, Wyoming, South Dakota, and Colorado are leading states in sheep and wool production.

There are five basic classifications of sheep according to wool type in the United States. They are fine wool, medium wool, long wool, crossbred wool, and fur sheep.

Fine-wool breeds of sheep produce wool that is very fine in texture with a long staple length. It has a wavy texture, is very dense, and is used to make fine-quality garments. Fine-wool sheep often produce as much as 20 lb of wool per sheep per year. Breeds of fine-wool sheep all originated from the Spanish Merino breed. Fine-wool breeds in the United States include the American Merino, Delaine Merino, Debouillet, and Rambouillet.

Medium-wool breeds of sheep were developed for meat, and little emphasis was placed on the production of wool. Popular medium-wool breeds include Suffolk, Shropshire, Dorset, Hampshire, and Southdown.

The long-wool breeds of sheep were developed in England. They tend to be larger than most of the other breeds. The wool produced tends to be long and coarse in texture. Long-wool breeds in the United States include Leicester, Lincoln, Romney, and Cotswold.

Crossbred-wool breeds are the result of crossing fine-wool breeds of sheep with long-wool breeds. They were developed to combine good quality wool with good quality meat. Because they tend to stay together as a group better than other breeds of sheep, they are popular in the western range states. Crossbred-wool breeds include Corriedale, Columbia, Panama, and Targhee.

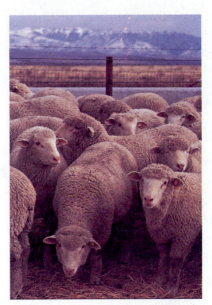

FIGURE 30-14 Sheep can utilize low-quality forage, graze land unsuited for other purposes, survive severe weather conditions, and produce meat and wool. *(Courtesy of USDA/ARS #K-4166-5)*

There is only one breed of fur sheep in the United States. The Karakul is grown for the pelts of its lambs. The pelts are taken from very young lambs and made into expensive Persian lamb coats. The production of wool and meat is of little importance in this breed.

Approved Practices in Sheep Production

Some of the approved practices in the production of sheep and wool include the following:

1. Select lambs that are large and growthy for their age.
2. Select purebred stock based on physical appearance and pedigree.
3. Select breeding stock with a history of multiple births.
4. Provide shelter from severe weather conditions.
5. Provide good-quality forages and unlimited fresh, clean water.
6. Vaccinate for disease problems of local concern.
7. Treat for internal and external parasites.
8. Breed ewes to lamb at no more than two years of age. A **ewe** is a female of the sheep family.
9. Use marking harnesses on rams to tell when ewes have been bred. A **ram** is a male of the sheep family.
10. Use a system of identification to distinguish ewes.
11. Do not disturb ewes during lambing. With sheep, **lambing** means to give birth.
12. Shear ewes at least one month before lambing. **Shearing** is the process of removing wool from sheep.
13. Provide clean, warm, dry stalls for lambing.
14. Make sure that ewes accept their new lambs.
15. Dock lambs' tails at 7 to 10 days of age.
16. Castrate ram lambs to be marketed.
17. Keep hooves properly trimmed.
18. Maintain a complete flock disease- and parasite-prevention-and-control program.
19. Cull ewes that do not lamb or those that have health problems.
20. Maintain a complete and accurate record-keeping system.

GOATS
Origin and History

The domestication of goats probably took place in western Asia during the Neolithic Age, between 7000 and 3000 B.C. Remains of goats have been found in Swiss lake villages of that time period. Mention of the use of mohair from goats is made in the Bible. **Mohair** is hair from angora goats that is used to make a shiny, heavy, wooly fabric.

Goats were imported to the United States from Switzerland for milk production early in the colonial period. Angora goats to be used for mohair production were also imported from Turkey. Nearly 95 percent of the mohair-producing goats in the United States are located in Texas.

Milk or dairy-type goats can be found all over the United States. However, there are concentrations on the East and West coasts.

Economic Importance

Goats are of relatively low economic importance in the United States. Most dairy goats are raised in very small numbers by suburbanites and small farmers to produce milk and meat for their own families. There are few large herds of milk goats. Finding a processor to bottle goat's milk is often difficult, as the number of health standards for food production and handling continue to increase.

However, the United States produces nearly 60 percent of the mohair in the world. Texas provides more than 95 percent of the 17 million pounds produced in the United States annually.

Goats provide little competition for food for cattle and sheep. They prefer to eat twigs and leaves from woody plants rather than grass.

Types and Breeds

There are two types of goats in the United States. They are hair producing and milk producing.

The only hair-producing goat is the Angora. It produces 6 to 7 lb of mohair per year. Mohair ranges in length from 6 to 12 in. Angora goats are best adapted to a dry climate with moderate temperatures. In addition to producing mohair, Angora goats are used for meat and to help control weeds and brush. Goat meat is called **chevon.**

Milk-producing, or dairy, goats are found in every state in the United States. Normal production (per goat) averages 3 to 4 quarts of milk per day during a 10-month lactation period. The **lactation period** is the time during which an animal produces milk. The common breeds of dairy goat are Nubian, Alpine, Saanen, and Toggenburg (Figure 30-15).

Approved Practices in Goat Production

Some approved practices in the production of goats and goat products include the following:

1. Select goats according to intended use.
2. Use physical appearance, pedigree, and records as a basis of selection.

FIGURE 30-15 Dairy goats may utilize high or remote mountain pastures or may be raised in relatively small confinement operations. *(Courtesy American Goat Association)*

3. Purchase replacement animals from reputable breeders.

4. Provide additional feed for hair goats on the range in winter.

5. Feed dairy goats supplemental grains based on amount of milk production.

6. Breed goats in the fall to have **kids** (young goats) in the spring.

7. Breed **does** (female goats) for the first time at 10 to 18 months of age.

8. Use one **buck** (male goat) for every 25 to 50 does.

9. Shear or clip hair goats twice each year.

10. Maintain clean and sanitary conditions for the production of milk.

11. Castrate bucks at an early age that are not to be used for breeding.

12. Maintain a herd-health program.

13. Milk dairy goats twice a day.

14. Dehorn dairy-type goats at an early age.

15. Maintain complete and accurate records of reproduction, production, and health.

The production of dairy and livestock in the United States is big business. Receipts from sales of dairy, beef, swine, sheep, and goat products in the United States total over $70 billion annually. There are types of livestock production that are adapted to almost every locality and situation. Nearly every facet of life is affected in one way or another by animals and animal products. Although food from animals is more expensive than food from crops, animal products add variety and quality to the human diet. Similarly, animals are sources of high-quality fabrics, leather, and many other products in high demand. Therefore, animals and the production of animals will be important enterprises in the future.

STUDENT ACTIVITIES

1. Write the Terms to Know and their meanings in your notebook.

2. Develop a word search or other puzzle using the Terms to Know. Trade your puzzle with someone else in class and solve his or hers.

3. Make a chart showing when and where the various types of animals were domesticated.

4. Participate in a class discussion on how and why certain animals were domesticated and others were not.

5. Take a class survey of all of the breeds of animals owned by students and their families.

6. Invite a breeder of purebred livestock and a breeder of commercial, or grade, livestock to class to discuss advantages and disadvantages of each type of operation.

7. Conduct a survey of your local school district to determine the types of livestock being raised, the breeds, and the numbers of each.

8. Make a bulletin board showing the various types of livestock in your area and the importance of each.

9. Visit a livestock operation to determine the types of jobs that need to be performed there and what training would be necessary to do those jobs.

10. Survey various purebred livestock magazines to determine prices of animals being sold at various purebred sales.

11. Make a notebook-cover collage showing as many breeds as possible of the particular type of livestock in which you are interested.

12. Make a bulletin board showing some items made from animal products.

SELF EVALUATION

A. Multiple Choice

1. The number one livestock industry in the United States is _____ production.
   a. dairy
   b. beef
   c. swine
   d. sheep

2. Milk for the purpose of making cheese is produced in _____ operations.
   a. Grade A
   b. Grade B
   c. Class A
   d. Class B

3. Merino sheep belong to the _____-wool type.
   a. fine
   b. medium
   c. long
   d. crossbred

4. The average Angora goat produces about _____ lb of mohair each year.
   a. 2 to 3
   b. 3 to 5
   c. 6 to 7
   d. 10 to15

5. The most popular breed of dairy cattle in the United States is
   a.   Jersey.
   b.   Ayrshire.
   c.   Brown Swiss.
   d.   Holstein.

6. The leading state in swine production is
   a.   California.
   b.   Texas.
   c.   Iowa.
   d.   Pennsylvania.

7. Docking refers to
   a.   fees charged for selling animals.
   b.   removal of tails.
   c.   backing a cattle trailer to a loading ramp.
   d.   none of the above.

8. Sows normally have _____ litters of piglets per year.
   a.   one
   b.   two
   c.   three
   d.   four

9. The average American eats about _____ lb of beef each year.
   a.   12.1
   b.   55.5
   c.   62.5
   d.   79

10. Dairy heifers should be bred to calve when they are _____ year(s) old.
   a.   1
   b.   2
   c.   3
   d.   4

B. Matching

_____ **1.** Buck	a.	Give birth to cattle
_____ **2.** Mutton	b.	Remove wool from sheep
_____ **3.** Milk	c.	Calf meat
_____ **4.** Calve	d.	Male goat
_____ **5.** Veal	e.	One who takes care of sheep
_____ **6.** Chevon	f.	Meat from mature sheep
_____ **7.** Farrow	g.	Fat in milk
_____ **8.** Shepherd	h.	Give birth to piglets
_____ **9.** Shear	i.	Liquid from mammary glands
_____ **10.** Butterfat	j.	Goat meat

C. Completion

1. _____ were domesticated in China.

2. Goats are of two types—hair and _____.

3. _____ is the most common system of swine identification.

4. _____ is rendered pork fat.

5. Milk-producing animals are called _____.

6. The _____ period is the time when cows are not producing milk.

7. Three classifications of beef-cattle breeds are English, exotic, and _____.

8. Consumption of red meat in the United States has been _____ in recent years.

9. _____ feed is feed provided for young animals to supplement milk from their mothers.

10. Breeds of sheep grown especially for meat belong to the _____ wool class.

Horse Management

OBJECTIVE

To determine the role of horses in our society and how to care for and manage them.

COMPETENCIES TO BE DEVELOPED

After studying this unit, you should be able to:

- understand the origin and history of the horse.
- determine the economic importance of the horse in the United States.
- recognize the various types and breeds of horse.
- understand approved practices for the care and management of horses.
- understand the basics of English and western riding.
- list rules of safety for handling horses.
- understand the vocabulary generally associated with horses.

MATERIALS LIST

- ✓ bulletin board materials
- ✓ horse breed magazines
- ✓ examples of horse tack

TERMS TO KNOW

Light horses	Jennet	Bit
Draft horses	Tack	Horsemanship
Donkey	Gaits	Jog
Coach horses	Equitation	Lope
Horses	Saddle	Filly
Ponies	Stirrups	Foal
Hand	Walk	Colt
Color breeds	Trot	Gelding
Mule	Canter	Shod
Jack	Gallop	Farrier
Mares	Rack	Blemish
Hinny	Pace	Unsoundness
Stallion	Reins	Vice

FIGURE 31-1 Horses are friends, working companions, and pets to America's youth and adults. *(Courtesy of Michael Dzaman)*

Horses are an important part of many industries in the United States. They are often nearly indispensable on working cattle ranches. In sports, they generate millions of dollars in the horse-racing industry. Horses are also important components of rodeos and hunting competitions. They are often status symbols. Horses also serve as food for pets and humans in some cases. But probably most importantly, horses serve as friends, companions, pets, and a form of relaxation to millions of people (Figure 31-1).

ORIGIN AND HISTORY

There are fossil remains of the horse family that have been found in the United States dating back nearly 58 million years. However, by the time of the discovery of America, no horses remained in the New World. They apparently had all died out only a few thousand years before. It should be noted that these early horses were only about the size of a collie dog.

Even though the horse became extinct in the New World a few thousand years ago, it had previously migrated to Europe and Asia when Alaska and Siberia were connected. It is from these horses that humans domesticated the horses of today.

The horse was probably one of the last animals to be domesticated. This occurred first in Central Asia or Persia about five thousand years ago. Horses came to Egypt in about 1680 B.C. from Asia. From there, the Egyptians introduced horses over the known world.

The Arab horse is the ancestor of most modern light breeds of horse. **Light horses** are breeds used for riding. However, horses were not used to any great extent until A.D. 500 to 600.

Draft horses probably originated from the heavy Flanders horse of Europe. **Draft horses** are used for work. At the time of domestication they were often used in religious rites for sacrifice or for food.

Donkeys were domesticated in Egypt some time before 3400 B.C., when they appeared on slates of the First Dynasty. A **donkey** is a member of the horse family with long ears and a short, erect mane. Mention of the use of donkeys appears in many places in the Bible. They were generally used as saddle animals or beasts of burden.

Horses were imported to the New World by Columbus in 1493. When Cortez came to Mexico in 1519, he brought nearly one thousand horses with him. The death of the explorer de Soto in the upper Mississippi region led to the abandonment of many of his horses. These horses, coupled with some of those lost or stolen from Spanish missionaries, probably formed the nucleus of the horses used by the Native Americans of the Plains.

The introduction of the horse to the Native Americans of the Plains completely changed their culture. Horses permitted easier hunting of buffalo in wider areas. This led to competition between various tribes of Plains Indians and a nearly constant state of war in an area where peace had previously prevailed.

The European settlers on the East Coast were far less dependent on the horse than were the explorers. It was of little advantage to have a riding horse with no place to ride. As draft animals, oxen or cattle were more popular because they were stronger and produced meat and milk as well as work.

It was only as Americans prospered that horses became an important part of their lifestyles. Horses have had an illustrious past, and a bright future looms ahead.

ECONOMIC IMPORTANCE

There are approximately 11 million horses in the United States today. Most of these horses are owned by nonfarmers and nonranchers. They are used primarily for pleasure, racing, breeding, and companionship.

Horse owners spend more than $16 billion on horses and their use each year. Horse racing is one of the most popular spectator sports in the United States. Individual racing events may offer as much as $1 million in purses.

Horse shows are also popular events, especially for young people. One breed association alone sponsored more than four thousand shows in a recent year.

There are numerous noneconomic benefits of horses. They help develop a sense of responsibility in young people. They provide physical activity for the young and old alike. They provide opportunities for families to participate in outdoor activities together, and they provide companions when needed (Figure 31-2). Horseback riding is also therapeutic for the healing of certain injuries and disabilities. And, horses are still essential for many cattle roundups.

FIGURE 31-2 Horses provide opportunities for families to participate in outdoor activities together and provide companionship. (Courtesy of Michael Dzaman)

TYPES AND BREEDS

For the purpose of this discussion, members of the horse family are divided into *horses, ponies,* and *donkeys* and *mules.* Horses are further divided into *light horses, coach horses,* and *draft horses.* **Coach horses** were developed to pull heavy coaches and freight wagons.

Members of the horse family that are 14.2 hands or more tall are called **horses.** Animals of the same family that are less than 14.2 hands tall are **ponies.** A **hand** measures 4 in. and is used as a unit of measurement for members of the horse family, determining the size of horses, mules, donkeys, and ponies.

AGRI·PROFILE

CAREER AREAS: Horse Breeder/Trainer/Rancher/Manager/Jockey/Farrier

Careers in horse management are available in a variety of settings. Urban police forces and park services have rediscovered the advantages of patrolling on horseback, and horse-drawn carriages have captured the fancy of sightseers in many areas. Additionally, many families have adopted the horse for teenage animal-production projects in semirural as well as rural settings. These frequently involve the entire family in fairs, shows, or rodeos. Horses are used extensively for herding and handling cattle on ranches and in cattle feedlot operations. Pack mules and riding horses are used extensively for hunting and other recreational pursuits. Finally, the horse-racing industry is huge in the United States, involving many people and extensive career possibilities.

While many raise horses as an avocation, there are rich and varied career opportunities in raising, managing, or caring for horses. Veterinarians, trainers, managers, attendants, jockeys, and farriers are common around racetracks and breeding farms. Tack, feed, and equipment supply centers provide business opportunities for those wishing to interface with the horse industry but preferring not to handle horses directly.

Horse health, soundness, breeding, behavior, and performance command the attention of those in research, veterinarian services, and education. Horse-related jobs may be indoor or outdoor and run the spectrum from laborer to scientist.

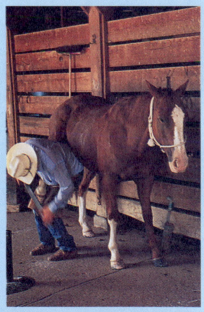

The number of horses is increasing in the United States. The services of the farrier are still in demand, as demonstrated by Mike Woods in Fort Keogh's old cavalry remount barn in Montana. Maintaining working horses in peak condition is vital to the daily operations of many enterprises. *(Courtesy of USDA/ARS K-3906-7)*

The light breeds of horse comprise by far the most popular horses in the United States. Light breeds are further divided into true breeds, or purebreds, and color breeds. True breeds have registration papers and parents of the same breed. **Color breeds** need only to be specific colors or color patterns, although other characteristics may be considered for registration. They need not have purebred parents.

Examples of the nearly fifty true breeds of horse in the United States include Arabian, Morgan, Thoroughbred, Quarter Horse, Standardbred, Tennessee Walking Horse, and American Saddle Horse (Figure 31-3). The most popular of the light breeds of horse in the United States is the Quarter Horse. The Quarter Horse is used for riding, hunting, and sports, and on cattle ranches to work with cattle (Figure 31-4).

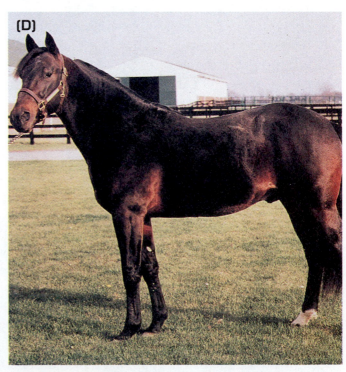

FIGURE 31-3 The most popular breeds of light horses in the United States, based on number registered, are (A) Quarter Horse, (B) Thoroughbred, (C) Arabian, and (D) Standardbred. *(Continued)*

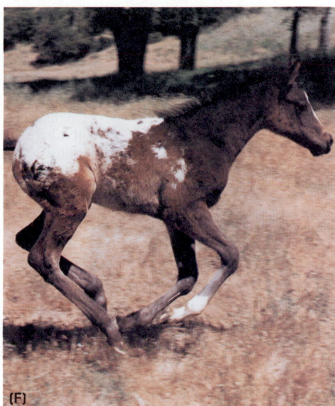

FIGURE 31-3 *(Continued)* The most popular breeds of light horses in the United States, based on number registered, are (E) Paint Horse and (F) Appaloosa.

FIGURE 31-4 Learning the proper names of the parts of a horse aids in discussing, evaluating, and caring for horses.

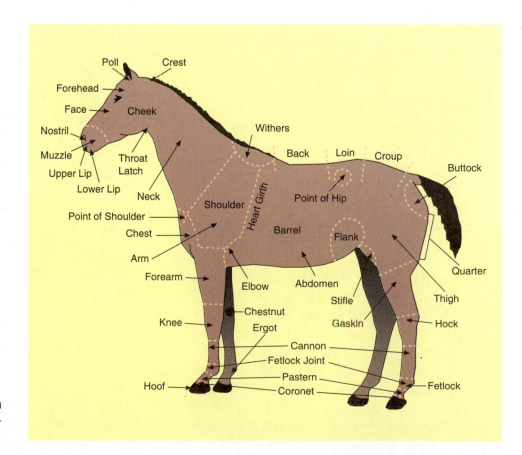

Some of the color breeds of horse include the Palamino, American White, American Creme, and Pinto. They may or may not breed true to color. The uses of the color breeds are the same as those of the true breeds of light horse.

When coaches disappeared from the American scene, coach horses became very rare in the United States because there was little need for them. While they were used, though, they combined the qualities of both the light horse and draft breeds. They were fast (like the light horse) and strong enough (like a draft horse) to pull the heavy stage coaches of their time. The only example of the old coach-style horse in the United States today is the Cleveland Bay.

Draft, or work, horses were once the backbone of American agriculture. Today they are used primarily for show and recreation. Their doom was the invention of the internal combustion engine, which only had to be fed when it was actually working and could do the work of many horses. Breeds of draft horse include Belgian, Clydesdale, Percheron, Shire, and Suffolk (Figure 31-5).

Ponies are smaller versions of the horse. They are used for riding and driving and as pets. Some breeds were originally developed to work in coal mines and in other pursuits where small size is important. The common breeds of pony include Shetland, Welsh, Gotland, Appaloosa, Connemara, and Pony of the Americas. The Shetland is the most popular breed of pony in the United States (Figure 31-6).

BIO•TECH CONNECTION

An International Solution

American veterinarians and other animal health officials are constantly on the alert for a lurking menace that could threaten the horse industry if allowed to get out of hand. A malaria-like disease of horses, equine babesiosis, infects the horses of the world in varying degrees. Horses of Brazil, Argentina, Russia, and Poland have lived with the parasite so consistently that they have developed a level of immunity that protects them from serious consequences of the disease.

Most American horses have not been exposed to the parasite, so they have developed little or no resistance to it. Such horses are very vulnerable, and a bout with equine babesiosis can be fatal. Babesiosis resembles malaria in humans because, in both diseases, insects transmit the infec-

Most American horses can become targets of equine babesiosis. A new test promises to help keep infected horses from entering the United States. *(Courtesy of Barbara Lee Jensen)*

tious agent, which then attacks and destroys red blood cells. Sick animals become feverish and lethargic and refuse to eat.

FIGURE 31-5 Draft horses are kept primarily for historic interest, show, and recreation. *(Courtesy of Michael Dzaman)*

FIGURE 31-6 The Shetland is the most popular breed of pony in the United States. *(Courtesy Multi-World Champion Shetland Harness Pony)*

Microscopic, single-celled parasites, called *Babesia equii,* enter the horse through infected, blood-sucking ticks. The parasites then multiply in the bloodstream and form tiny pear- or ring-shaped bodies, called *merozoites.* The merozoites then invade the red blood cells. The horse's immune system reacts by forming antibodies to attempt to fend off the invaders.

The United States tries to protect its horse population by keeping the parasite out of the country. However, horses coming into the United States from any other part of the world may be carriers of the parasite, even though they appear healthy themselves. There has been a test for the parasite in general use, but the test yields uncertain results.

Donald Knowles of the Agricultural Research Service Animal Disease Research Unit in Pullman, Washington, has found what may be a good test. By studying the blood of infected horses, Dr. Knowles discovered specific proteins on the surface of the merozoites that were present regardless of the origin of the horse. One of the proteins produced an antibody response in two hundred horses from twenty different countries. His colleague, Lance Perryman of Washington State University, made a monoclonal antibody that can accurately detect the presence of this protein. The antibodies are produced by a laboratory cell culture of fused mouse spleen and myeloma cells, called a *hybridoma.* The antibodies are then studied as a component of a diagnostic test.

The National Veterinary Services Laboratory in Ames, Iowa, screens all blood samples from animals attempting entry. However, it is very expensive to ship a horse to an international border only to be required to ship it back home due to rejection for health reasons. If tests are economical and reliable, courtesy tests can be run on blood samples taken before the animal is shipped. If no problems are indicated, the animal can be shipped and then retested at the point of entry. The new procedure developed by Knowles and Perryman should be a step forward in solving the old international dilemma of how to move horses across borders economically without the danger of spreading the dreaded equine babesiosis.

FIGURE 31-7 The mule is a cross between a jack and a mare. *(Mules owned by Jack and Laura Schreiner of West Sand Lake, NY. Courtesy of S & R Photo Acquisitions.)*

Donkeys are used in the United States for work and as pack animals. Miniature donkeys also make excellent pets. The male donkey, called a jack, is used as the male parent of mules. A **mule** is a cross between a jack and a mare. A **jack** is a male donkey or mule. **Mares** are mature female horses or ponies. The opposite cross, a stallion with a jennet, results in a **hinny.** A **stallion** is an adult male horse or pony. A **jennet** is a female donkey or mule (Figure 31-7).

Mules are thought to be more intelligent than horses and can do more work than comparable-sized horses. There are several types and sizes of mule, depending on the breed or type of horse that serves as the female parent. Mules may be used for work, sports, or pleasure.

RIDING HORSES

The styles of riding horses can be divided into two general classifications—English and western. The style of riding and the **tack** (horse equipment) required vary greatly between the two classifications. Even the **gaits** (the ways an animal moves) of the horses involved have different names.

English Equitation

In English equitation, the saddle is smaller and lighter, and has shorter stirrups than the western saddle (Figure 31-8). **Equitation** is the art of riding on horseback. A **saddle** is the padded leather seat for the rider of a horse. **Stirrups** are footrests hung from the saddle. The English riding clothing includes closefitting breeches, jodhpurs, jacket, and hard hat or derby (Figure 31-9, page 592).

The gaits of the English-equitation horse are walk, trot, canter, gallop, rack, and pace. The **walk** is a slow, four-beat gait. The **trot** is a fast, two-beat diagonal gait. The **canter** is a slow, three-beat gait, whereas the **gallop** is a fast, three-beat gait. The **rack** is a fast, four-beat gait, and the **pace** is a side-to-side, two-beat gait (Figure 31-10, page 593).

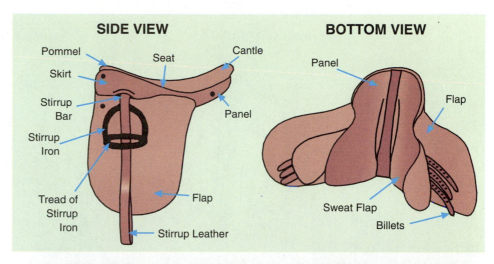

FIGURE 31-8 The English saddle is lighter and structured differently than the western saddle. *(Courtesy of Michael Dzaman)*

The English-equitation horse is controlled by reins that may be held in either one or both hands. **Reins** are leather or rope lines attached to the bit. The **bit** is a metal mouthpiece used to control the horse.

Western Horsemanship

Western horsemanship differs from English equitation in many ways. **Horsemanship** is the art of riding and knowing the needs of the horse. The western saddle is heavy and has a horn that is used when roping livestock. This saddle has longer stirrups than the English saddle and may be plain or have very fancy ornamentation. The western saddle was designed for the comfort of cowboys, who spend much of their time in a saddle (Figure 31-11, page 593).

Western riding attire was also designed for comfort. It consists of jeans and a western shirt and hat. Cowboy boots and chaps complete the typical western outfit.

The gaits of the western horse are walk, jog, lope, and gallop. The **jog** is a slow, smooth, two-beat diagonal gait. The **lope** is a very slow canter.

The western rider controls the horse with the reins held in one hand. The hand holding the reins cannot be changed during competition riding.

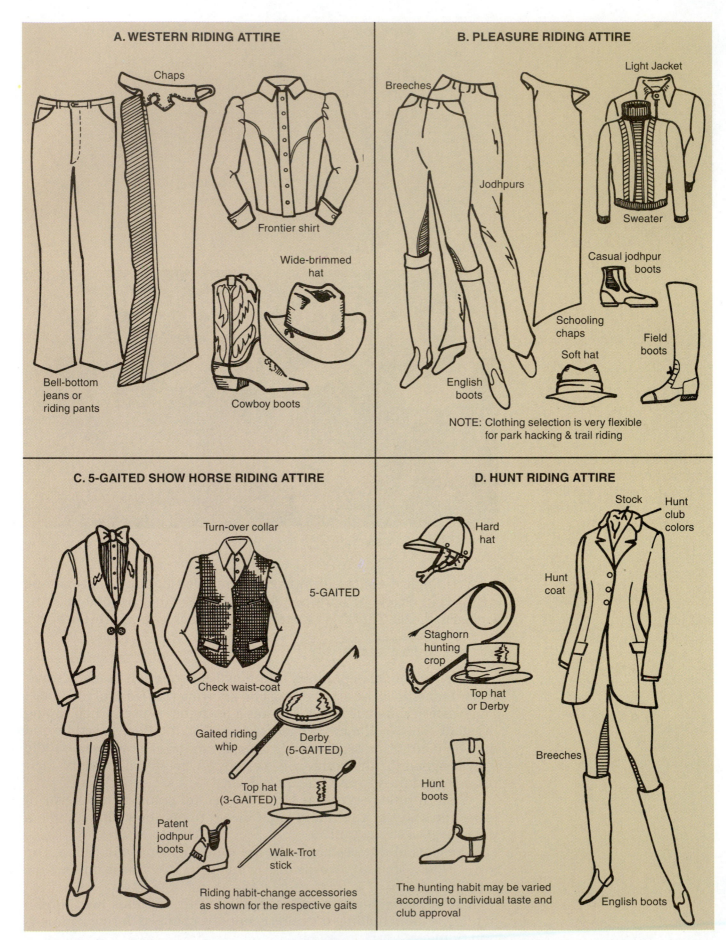

A. WESTERN RIDING ATTIRE

Chaps

Frontier shirt

Wide-brimmed hat

Bell-bottom jeans or riding pants

Cowboy boots

B. PLEASURE RIDING ATTIRE

Breeches

Light Jacket

Jodhpurs

Sweater

Casual jodhpur boots

Schooling chaps

Soft hat

Field boots

English boots

NOTE: Clothing selection is very flexible for park hacking & trail riding

C. 5-GAITED SHOW HORSE RIDING ATTIRE

Turn-over collar

5-GAITED

Check waist-coat

Gaited riding whip

Derby (5-GAITED)

Top hat (3-GAITED)

Patent jodhpur boots

Walk-Trot stick

Riding habit-change accessories as shown for the respective gaits

D. HUNT RIDING ATTIRE

Stock

Hunt club colors

Hard hat

Hunt coat

Staghorn hunting crop

Top hat or Derby

Breeches

Hunt boots

The hunting habit may be varied according to individual taste and club approval

English boots

FIGURE 31-9 Proper riding attire for riders includes western, pleasure, 5-gaited show horse, and hunt.

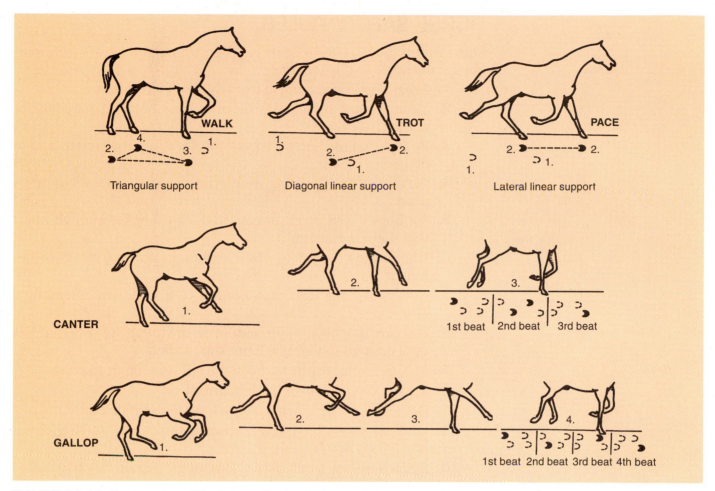

FIGURE 31-10 Basic gaits of English equitation include walk, trot, pace, canter, and gallop.

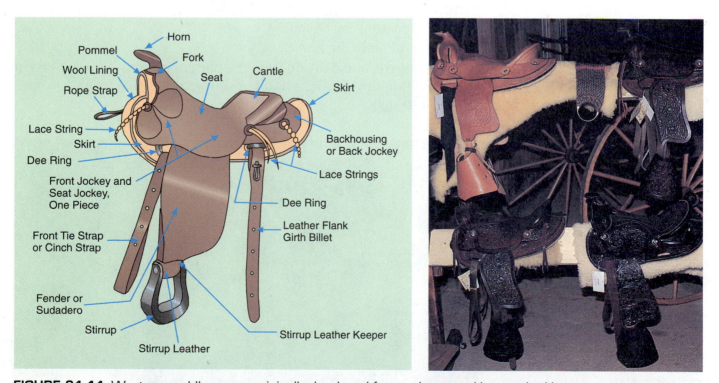

FIGURE 31-11 Western saddles were originally developed for cowboys working cattle. However, present-day designs of these saddles vary from basic and functional to highly ornamental. *(Courtesy of Michael Dzaman)*

HORSE SAFETY RULES

Some of the safety rules to observe when riding and caring for members of the horse family include:

1. Always approach a horse from the front left side.
2. Never do anything to startle or scare a horse. It may kick or rear up.
3. Be sure that the horse always knows exactly what you intend to do to or with it.
4. Pet the horse by putting your hands on its shoulder, not on its nose.
5. Tie horses that are strangers to each other far enough apart that they cannot fight.
6. Walk beside the horse when leading it, not in front of or behind it.
7. Never wrap the lead rope around your hand when leading a horse.
8. Use care in adjusting the saddle on the horse. Be sure that it is tight enough so that it will not slip or slide.
9. Always mount the horse from the left side (Figure 31-12).
10. Always keep the horse under control.
11. Do not allow the horse to misbehave without disciplining it.
12. Walk the horse up and down steep slopes, on rough ground, and across paved roads.
13. Reduce speed when riding on rough terrain or in wooded areas.
14. In groups, ride in single file and on the right side of the road.
15. Be calm and gentle when dealing with your horse.
16. Always wear appropriate riding attire, including hard hats when jumping obstacles.
17. Never tease the horse.
18. Walk the horse to and from the barn or holding area to prevent it from developing a habit of riding to the area when it comes into sight.

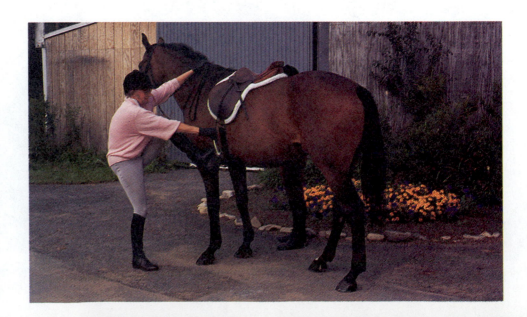

FIGURE 31-12 Always mount a horse from the left side. *(Courtesy of Michael Dzaman)*

19. Know the temperament and vices of the horse.
20. Be especially aware of barking dogs and other things that may frighten the horse.
21. Do not allow other people to ride your horse unsupervised.

APPROVED PRACTICES FOR HORSES
Breeding Horses

Some of the approved practices for breeding horses are:

1. Breed fillies so they will foal for the first time at 3 to 4 years of age. A **filly** is a young female horse or pony. A **foal** is a newborn horse or pony. To foal is to give birth in the horse family.
2. Breed mares in April, May, or June when their fertility is normally highest. Breeding at this time also allows the mare to foal in the spring, when pastures are actively growing.
3. Rebreed mares 25 to 30 days after foaling, if they are in good condition with no reproductive-tract problems.
4. Make sure that the pregnant mares and fillies get plenty of exercise.
5. Allow the mare to foal in a clean pasture free from internal parasites, or in a large, clean pen or stall (Figure 31-13).
6. Put the mare in the facility in which she will foal several days before she is expected to foal. This gives her some time to get accustomed to the facility.
7. Remain out of the mare's sight when she is foaling. Mares prefer to be alone with no interruptions during the foaling process.
8. Contract the services of a veterinarian if the mare shows signs of difficulty in foaling or if the position of the foal is abnormal for delivery.
9. At birth:
   a. Make sure that the foal is breathing. Tickling the foal's nose with a piece of straw will often stimulate it to

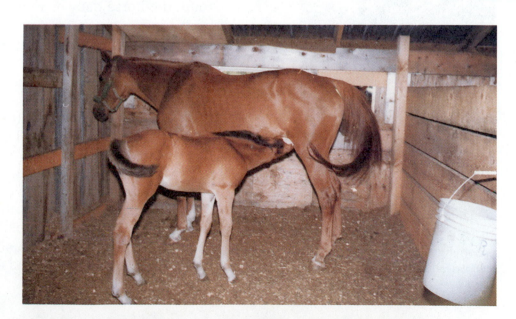

FIGURE 31-13 Mares and foals should be kept in clean, parasite-free surroundings. *(Courtesy of Barbara Lee Jensen)*

start breathing after it is born. More drastic measures, such as artificial respiration, may be needed in more serious situations.

   b. Be sure to remove any mucus from the nose of the foal, and dry the foal off.

   c. Dip the end of the navel cord in iodine after tying it off. This prevents infection from entering the foal through the open navel cord.

   d. Make sure that the foal nurses as soon as it can stand. It should be allowed to stand on its own. Normally, nursing first takes place within one-half hour after birth.

   e. Check to see that the foal has a bowel movement within the first twelve hours of birth. The foal may need an enema if the bowel movement does not occur naturally.

   f. If the foal shows signs of diarrhea, reduce the amount of milk that it is allowed to drink.

10. Provide creep feed for the nursing foal when it reaches 10 days to 2 weeks of age.

11. Begin training the foal as soon as possible. A foal that is trained from an early age seldom needs to be broken.

12. Do not mistreat the mare or foal at any time.

13. Wean the foal at 4 to 6 months of age. The foal and the mare should not be allowed to see each other for several weeks after weaning to break the bond between mother and offspring.

14. Castrate colts that are not to be used for breeding purposes. A **colt** is a young male horse or pony. A castrated male horse is called a **gelding.** Geldings are usually much more docile and easier to handle than stallions.

Care and Management of All Horses

Approved practices for the care and management of all horses and ponies include the following:

1. Groom the horse daily or weekly at a very minimum (Figure 31-14).

2. Always use soft brushes when grooming a horse. Its skin is very sensitive and easily damaged by rough treatment.

3. Shorten the mane and tail when needed by pulling the hairs from the underside.

4. Be especially careful not to get water in the horse's ears when washing. Be sure to dry off the horse quickly so that it does not catch a cold in cool weather.

5. Inspect and clean the horse's hooves daily and always before and after riding.

6. Trim the horse's hooves every 4 to 6 weeks if the horse is not shod. **Shod** means wearing shoes.

7. Replace the shoes of a shod horse every 4 to 6 weeks. A **farrier** (a person who shoes horses and cares for their feet) should be engaged to perform this task.

FIGURE 31-14 Frequent grooming keeps a horse cooperative, content, and healthy. (Courtesy of Michael Dzaman)

8. Shoe a horse for the first time when it is 2 years of age or when it starts being worked.

9. Cool out the horse thoroughly after every exercise period and after riding. This should be done before the horse is allowed to drink large quantities of water.

10. Do not overfeed a horse. The total daily consumption of concentrates and roughages should not total more than 2 to 2.5 percent of the horse's weight.

11. Feed a horse on a regular routine. The number of times that the horse is fed per day makes little difference, as long as it is the same every day.

12. Do not abruptly change the ration being fed. The stomach of the horse is very temperamental and adjusts to changes in feeding practices very slowly.

13. Never feed moldy feed to a horse.

14. Provide unlimited access to fresh, clean water at all times.

15. Maintain a strict health-care program for all horses.

16. Maintain clean and sanitary facilities at all times.

17. When buying a horse, be sure to:

   a. Note blemishes and unsoundnesses of the horse. A **blemish** is an abnormality that does not affect the use of the horse. An **unsoundness** is an abnormality that does affect the use of the horse.

   b. Check the age of the horse by looking at its teeth (Figure 31-15). Examining the teeth can also indicate how long a horse might be usable.

   c. Determine that the horse is not blind or having other problems that may affect its use value and use.

   d. Check for evidence of good health.

   e. Try to determine if the horse has undesirable vices. A **vice** is a bad habit.

   f. Try to determine the personality and spirit of the horse.

   g. Be sure to consider the price of the horse and whether or not you can afford the cost of owning a horse.

   h. Check the pedigree of the horse, if it is a purebred.

Owning and caring for horses and ponies is an excellent way to get outdoor exercise. Horses have given many young people companionship during the trials of youth. Competitors find horses and racing excellent outlets for their excess energy. Finally, cowboys find horses to be an essential part of a team when working with cattle. The horse was, is, and will continue to be an important part of the world of many people in the United States.

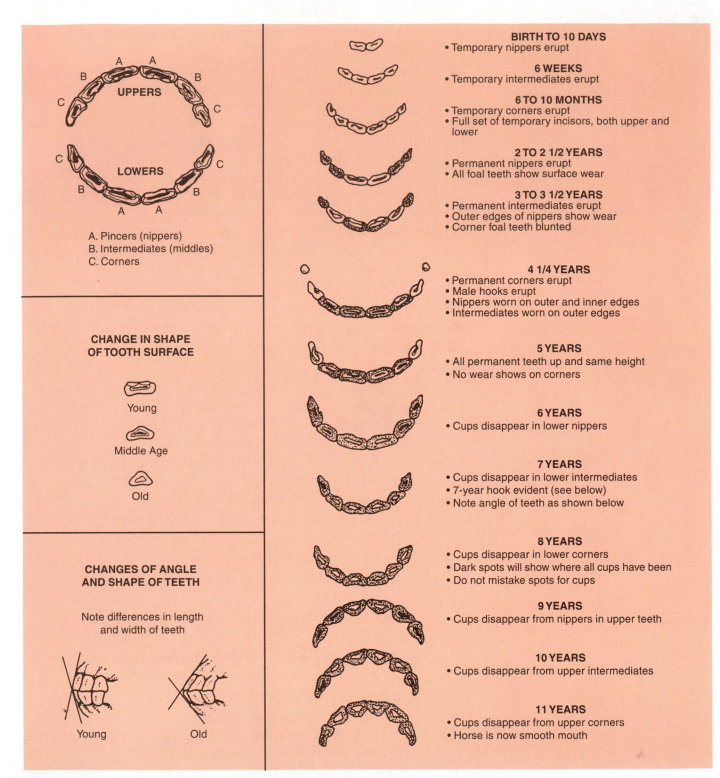

FIGURE 31-15 The age of a horse can be determined by the number and shape of its teeth. *(Courtesy of Gillespie, Modern Livestock & Poultry Production, 5E, Copyright 1997 by Delmar Publishers)*

STUDENT ACTIVITIES

1. Write the Terms To Know and their meanings in your notebook.
2. Survey the students in your school concerning how many have horses and what type and breed of horse that they own.
3. Develop a bulletin board showing as many breeds and types of horse as you can.
4. Attend a horse show and interview participants about why they show horses.
5. Write a report about a breed or type of horse that interests you.
6. Have a horse breeder or owner talk to the class about the care and management of horses.
7. Have a horse veterinarian talk to the class about health-care and first-aid procedures for horses.
8. Develop a word search using the Terms to Know.
9. Visit a tack shop and make a list of the various types of horse equipment sold there.
10. Look at horse-breed-and-care magazines to determine the uses of the various breeds of horse. Also look for the types of job opportunities that may be available in the horse industry.
11. Demonstrate to the class the techniques of grooming a horse.

SELF EVALUATION

A. Multiple Choice

1. Horses were probably domesticated in
   a. Russia.
   b. the United States.
   c. Persia.
   d. China.

2. Light horses are used for
   a. riding.
   b. rodeos.
   c. racing.
   d. all of the above.

3. A mule is a cross between a
   a. jack and a mare.
   b. stallion and a jennet.
   c. jack and a jennet.
   d. stallion and a mare.

4. The number one spectator sport in the United States is
   a. rodeos.
   b. horse racing.
   c. horse shows.
   d. riding.

5. Fillies should be bred to foal at _____ years of age.
   a. 1 to 2
   b. 2 to 3
   c. 3 to 4
   d. 4 to 5

6. Always approach a horse from the
   a. rear.
   b. right side.
   c. left side.
   d. none of the above.

7. Horses should be reshod every _____ weeks.
   - a. 2 to 4
   - b. 4 to 6
   - c. 6 to 8
   - d. 8 to 10

8. The most popular breed of light horse in the United States is the
   - a. Thoroughbred.
   - b. Appaloosa.
   - c. Morgan.
   - d. Quarter Horse.

9. A fast, three-beat gait is the
   - a. gallop.
   - b. canter.
   - c. pace.
   - d. trot.

10. There are approximately _____ horses in the United States.
   - a. 100,000
   - b. 1 million
   - c. 10 million
   - d. 100 million

B. Matching

_____	**1.** Stallion	a.	Way of moving
_____	**2.** Jennet	b.	Young female horse
_____	**3.** Gelding	c.	Female donkey
_____	**4.** Mare	d.	Newborn horse
_____	**5.** Gait	e.	Adult male horse
_____	**6.** Filly	f.	Adult female pony
_____	**7.** Tack	g.	Male donkey
_____	**8.** Jack	h.	Horse equipment
_____	**9.** Foal	i.	Young male horse
_____	**10.** Colt	j.	Castrated male horse

C. Completion

1. A _____ is a person who shoes horses.

2. Horses originated in what is now the _____.

3. _____ horses were developed for work.

4. _____ breeds of horse need only to be certain colors or color patterns in order to be registered.

5. The art of riding is called _____.

6. An abnormality that affects the use of a horse is called a(an) _____.

7. Bad habits in horses are called _____.

8. The result of the cross between a stallion and a jennet is a _____.

9. The most popular breed of pony is the _____.

10. The unit of measurement for horses is the _____.

Food Science and Technology

Wave of the Future

Protein-rich edible-food coatings and biodegradable food packaging are a limited reality now and a wave of the future. Chemist Franklin Shih of the Agricultural Research Service's Southern Research Center in New Orleans is conducting research on soybean film types that promises a variety of new products. These films may have uses ranging from films on citrus and other fruits, to keep them fresh, to fast-food wrappers that can be eaten. Plastics from soybeans have been around since the 1940s, but plastic-like digestible and biodegradable films are new and still developing.

The process developed at the New Orleans facility involves the separation and isolation of proteins of the soybean by freeze-drying protein to remove the water and grinding the protein into fine powder. One product is soy concentrate containing 70 percent protein; another is soy isolate containing 90 percent protein. The proteins can then be mixed with various ingredients and additives before being cast into films for coatings for food products. By using enzymes and other treatments, the protein can be modified for films and coatings with specific uses. Other researchers have developed films and coatings from corn and wheat starches, but soy-protein films promise some additional adaptability.

Protein mixes with oils, and this quality is important in processing products with moisture-resistant characteristics. Further, soy protein can be mixed with starch and gums to form films that keep moisture in but keep oxygen out. This creates an ideal barrier for maintaining food quality. By adjusting other factors, films can be tailored for effective food packaging and preservation. For instance, fat-containing foods, such as meats, can develop off-flavors after being cooked. This creates problems in institutional cooking, such as restaurants and school lunch programs. Coating such foods with the appropriate protein film could solve the problem of off-flavors. The use of protein and other edible films must be approved by the Food and Drug Administration before being put into commercial practice. The development and use of such "agriplastics" will surely continue because they present the advantages over petroleum-based films of being edible, nutritionally valuable, and environmentally safe.

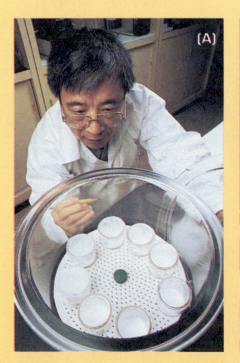

(A) Chemist Fred Shih analyzes the permeability of various soybean film types, (B) treating fruits and vegetables with protective film, and (C) entomologist Jennifer Sharp evaluates effectiveness of insect kills by heat treatment of shrink-wrapped grapefruit. *(Courtesy of USDA/ARS (A) #K-4472-8, (B) #K-3517-5, and (C) #K-3183-10)*

The Food Industry

OBJECTIVE

To explore elements, trends, and career opportunities in the food industry.

COMPETENCIES TO BE DEVELOPED

After studying this unit, you should be able to:

- explain what is meant by the term *food industry*.
- determine the importance of the food industry to the consumer.
- describe the economic scope of the food industry.
- identify government requirements and other assurances of food quality and sanitation.
- compare the major crop and animal commodity production areas in the nation and the world.
- discuss the major food commodity groups and their predominant origins.
- explain the major operations that occur in the food industry.
- describe career opportunities in food science.
- discuss future developments predicted for the food industry.

MATERIALS LIST

✓ bulletin board materials

✓ State Department of Agriculture reports on commodities grown and foods processed in particular states

Gourmet
Food industry
Retailer
Wholesaler
Distributor
Processor
Grader
Packer
Trucker

Harvester
Producer
Grades
Climatic conditions
Technology
Harvesting
Maturity
Underripe
Overripe

Spoiled
Microorganisms
Migratory labor
Processing
Bran
Endosperm
Germ
Edible

Food is all around us—in the school cafeteria, on the dinner table at home, at the fast-food chains that dot our nation's highways, and in our nation's supermarkets. Less visible are the gourmet and specialty food stores and restaurants that satisfy special dietary needs and tastes. **Gourmet** means sensitive and discriminating taste in food preferences. Learning about other cultures is frequently accomplished by tasting their foods and learning how those foods are prepared.

The **food industry** is that industry involved in the production, processing, storage, preparation, and distribution of food for consumption by living things. Pet and animal food, as well as human food, requires a chain of people, places, equipment, regulations, and resources to change farm products into edible foods (Figure 32-1).

FIGURE 32-1 Most modern food products are prepared for the consumer, as seen in ready-to-heat-and-eat grocery items. *(Courtesy of USDA/ARS #K-3550-1)*

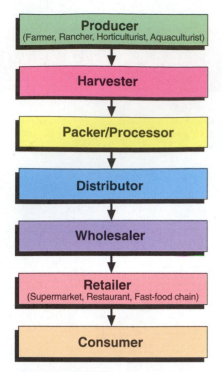

FIGURE 32-2 Producers, such as farmers, ranchers, and aquaculturists who actually produce food, are joined by many others before most food items reach the consumer.

This unit explores the many operations in the food processing industry. Careers are plentiful in this industry. As the different operations are explored, remember that new people are needed to maintain and expand the vital functions that keep the abundant food before us.

THE ECONOMIC SCOPE OF THE FOOD INDUSTRY

When you purchase groceries in the food store or a hamburger at the local fast-food restaurant, does most of your food dollar go directly to the farmer who grew the beef? How about the lettuce, tomato, pickle, bun, and sesame seed that adorns your hamburger? There are many businesses and individuals that join the farmer in dividing your food dollar. The economic chain reaction that begins with your food purchase sends signals to the retailer, wholesaler, distributor, processor, grader, packer, trucker, harvester, producer, and others to replace that food for your next purchase (Figure 32-2).

A **retailer** is the person or store that sells directly to the consumer. The retailer is the end of the marketing chain, whereas a **wholesaler** is a person who sells to the retailer, having purchased fresh or processed food in large quantities. A **distributor** stores the food until a request is received to transport the food to a regional market. A **processor** is anyone involved in cleaning, separating, handling, and preparing a food product before it is ready to be sold to the distributor. A **grader** is the person who inspects the food for freshness, size, and quality, and determines under what criteria it will be sold and consumed. A **packer** is the person or firm that is responsible for putting the food into containers, such as boxes, crates, bags, or bins, for shipment to the processing plant; and a **trucker** is the person responsible for transportation of the product anywhere along the way from farm to consumer. A **harvester** is the person who removes the edible portions from plants in the field. Finally, the **producer** grows the crop and determines its readiness for harvest. It seems like everyone gets a piece of your food dollar (Figure 32-3)! Moreover, the dollars spent on food and fiber in the United States. provide jobs for approximately 20 percent of our working population (Figure 32-4).

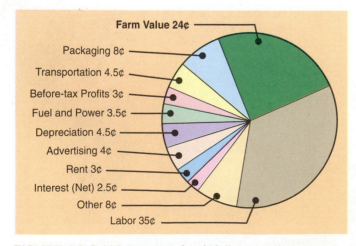

FIGURE 32-3 Where your food dollar goes.

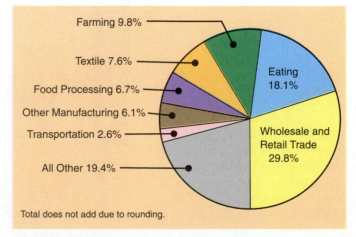

FIGURE 32-4 Career sectors in the food and fiber system.

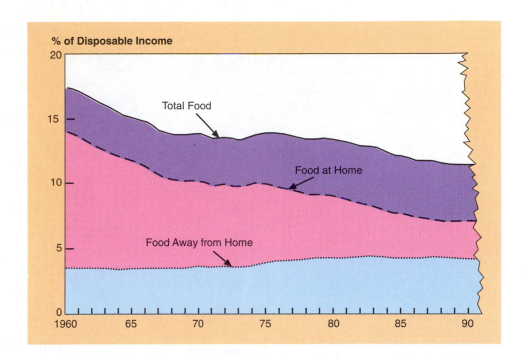

% of Disposable Income

FIGURE 32-5 Proportion of disposable income used for food in the United States.

Where you spend your food dollar also has an influence on who gets how much of your dollar. A meal purchased in a restaurant costs considerably more than a meal prepared from raw food products at home. How has our change in lifestyles and the shift to families with two or more people employed outside the home influenced how and what we eat? More meals are eaten outside the home than was the case a generation ago. Convenience foods for use at home are more in demand today. In 1991, 45 percent of the food dollar in the United States was spent on meals and snacks away from home. This was up from 34 percent in 1970. By 1991 the proportion of income needed for food had dropped below 12 percent of disposable income in the United States. (Figure 32-5).

QUALITY ASSURANCE

Grading and Inspecting

In the United States, we have become accustomed to high-quality food in every state and every store. The grading system established by the U.S. Department of Agriculture (USDA) has provided a uniform set of trading terms known as *grades*. **Grades** are based on quality standards. These improve acceptability of products by the consumer.

Grade standards are established for the following commodities: meat, cattle, wool, poultry, eggs, and dairy products; fresh, frozen, canned, and dried fruits and vegetables; cotton, tobacco, and spirits of turpentine; and rosin. Grades indicate freshness, potential flavor, texture, and uniformity in size and weight, depending on the commodity (Figure 32-6).

Sanitation

Additional quality-assurance programs administered by the USDA include inspection of slaughtering houses and processing plants, and

FOOD	CURRENT GRADE NAMES		WHAT THE GRADES MEAN
MEAT			
Beef	USDA PRIME	*USDA PRIME	Very tender, juicy, flavorful; has abundant marbling (flecks of fat within the lean).
	USDA CHOICE	*USDA CHOICE	Quite tender and juicy, good flavor; slightly less marbling than Prime.
	USDA SELECT	USDA SELECT	Fairly tender; not as juicy and flavorful as Prime and Choice; has least marbling of the three.
Lamb	USDA PRIME	*USDA PRIME	Very tender, juicy, flavorful; has generous marbling.
	USDA CHOICE	*USDA CHOICE	Tender, juicy; flavorful, has less marbling than Prime.
Veal	USDA PRIME	*USDA PRIME	Juicy and flavorful; little marbling.
	USDA CHOICE	*USDA CHOICE	Quite juicy and flavorful; less marbling than Prime.
POULTRY			
Chickens Turkeys Ducks Geese	USDA A GRADE	*U.S. Grade A	Fully fleshed and meaty; uniform fat covering; well formed; good, clean appearance.
		U.S. Grade B	Not quite as meaty as A; may have occasional cut or tear in skin; not as attractive as A.
		U.S. Grade C	May have cuts, tears, or bruises; wings may be removed and moderate amounts of trimming of the breast and

* Indicates grades most often seen at retail.

FOOD	CURRENT GRADE NAMES		WHAT THE GRADES MEAN
EGGS			
	USDA AA GRADE	*U.S. Grade AA	Clean, sound shells; clear and firm whites; yolks practically free of defects; egg covers small area when broken out — yolk is firm and high and white is thick and stands high.
	USDA A GRADE	*U.S. Grade A	The same as AA except egg may cover slightly larger area when broken out and white is not quite as thick.
		*U.S. Grade B	Sound shells, may have some stains or shape may be abnormal; white may be weak and yolk enlarged and flattened; egg spreads when broken out.
DAIRY PRODUCTS			
Instant nonfat dry milk	USDA U.S. EXTRA GRADE U.S. DEPT. OF AGRICULTURE GRADING AND QUALITY CONTROL SERVICE OFFICIALLY GRADED	*U.S. Extra Grade	Sweet, pleasing flavor; natural color; dissolves readily in water.
Butter	USDA AA WHEN GRADE GRADED PACKED UNDER INSPECTION OF THE U. S. DEPT. OF AGRICULTURE OFFICIALLY GRADED	*U.S. Grade AA	Delicate sweet flavor and smooth texture; made from high-quality fresh sweet cream.
	USDA A WHEN GRADE GRADED PACKED UNDER INSPECTION OF THE U. S. DEPT. OF AGRICULTURE OFFICIALLY GRADED	*U.S. Grade A	Pleasing flavor; fairly smooth texture; made from fresh cream.
		U.S. Grade B	May have slightly acid flavor or other flavor or body defects.
Cheddar cheese	USDA AA WHEN GRADE GRADED PACKED UNDER INSPECTION OF THE U. S. DEPT. OF AGRICULTURE OFFICIALLY GRADED	*U.S. Grade AA	Fine, pleasing Cheddar flavor; smooth, compact texture; uniform color.
		U.S. Grade A	Pleasing flavor; more variation in flavor and texture than AA.

* Indicates grades most often seen at retail.

FOOD	CURRENT GRADE NAMES		WHAT THE GRADES MEAN
FISH			
	U S GRADE A	*U.S. Grade A	Uniform in size, practically free of blemishes and defects, in excellent condition, and having good flavor for the species.
		U.S. Grade B	May not be as uniform in size or as free from blemishes or defects as grade A products; general commercial grade.
		U.S. Grade C	Just as wholesome and nutritious as higher grades; a definate value as thrifty buy for use where appearance is not an important factor.
FRESH FRUITS AND VEGETABLES			
The grade is more likely to be found without the shield.	U. S. GRADE NO.1	*U.S. Fancy	Premium quality; only a few fruits and vegetables are packed in this grade.
		*U.S. No. 1	Good quality; chief grade for most fruits and vegetables.
		U.S. No. 2	Intermediate quality between No. 1 and No. 3.
		U.S. No. 3	Lowest marketable quality.
PROCESSED FRUITS AND VEGETABLES AND RELATED PRODUCTS			
Canned and frozen fruits and vegetables	U S GRADE A	*U.S. Grade A	Tender vegetables and well-ripened fruits with excellent flavor, uniform color and size, and few defects.
		U.S. Grade B	Slightly mature vegetables; both fruits and vegetables have good flavor but are slightly less uniform in color and size and may have more defects than A.
		U.S. Grade C	Mature vegetables; both fruits and vegetables vary more in flavor, color, and size and have more defects than B.
Dried or dehydrated fruits. Fruit and vegetable juices, canned and frozen. Jams, jellies, preserves. Peanut butter, honey, catsup, tomato paste.	U S GRADE A	*U.S. Grade A	Very good flavor and color and few defects.
		U.S. Grade B	Good flavor and color but not as uniform as A.
		U.S. Grade C	Less flavor than B, color not as bright, and more defects.

* Indicates grades most often seen at retail.

FIGURE 32-6 Grades of food commodities, as established by USDA. *(Courtesy of USDA)*

oversight of processing operations (Figure 32-7). The USDA oversees food labeling and enforces regulations regarding representation on such labels (Figure 32-8). The National Shellfish Sanitation Program, the U.S. Public Health Service, and the U.S. Food and Drug Administration work with USDA to ensure the safety of food and food products. States, counties, and municipalities also have inspectors. They regulate local conditions to ensure sanitation and safe food handling, especially in restaurants and food-preparation areas.

Nutrition Facts

Serving Size 1 Cup (55g/2.0 oz)
Servings per Container 13

Amount Per Serving	Cereal	Cereal with 1/2 Cup Vitamins A & D Skim Milk
Calories	170	210
Fat Calories	10	10

	% Daily Values**	
Total Fat 1.0g*	**2%**	**2%**
Sat. Fat 0g	**0%**	**0%**
Cholesterol 0mg	**0%**	**0%**
Sodium 300 mg	**13%**	**15%**
Potassium 340mg	**10%**	**16%**
Total Carbohydrate 43g	**14%**	**16%**
Dietary Fiber 7g	**28%**	**28%**
Sugars 17g		
Other Carbohydrate 19g		
Protein 4g		

Vitamin A	15%	20%
Vitamin C	0%	2%
Calcium	2%	15%
Iron	45%	45%
Vitamin D	10%	25%
Thiamin	25%	30%
Riboflavin	25%	35%
Niacin	25%	25%
Vitamin B$_6$	25%	25%
Folate	25%	25%
Vitamin B$_{12}$	25%	35%
Phosphorus	20%	30%
Magnesium	20%	25%
Zinc	25%	25%
Copper	15%	15%

*Amount in cereal. One-half cup skim milk contributes an additional 40 calories, 65mg sodium, 6g total carbohydrate (6g sugars), and 4g protein.

**Percent Daily Values are based on a 2,000-calorie diet. Your daily values may be higher or lower depending on your calorie needs:

		Calories	2,000	2,500
Total Fat	Less than		65g	80g
Sat. Fat	Less than		20g	25g
Cholesterol	Less than		300mg	300mg
Sodium	Less than		2,400mg	2,400mg
Potassium			3,500mg	3,500mg
Total Carbohydrate			300g	375g
Dietary Fiber			25g	30g

Calories per gram:
Fat 9 • Carbohydrate 4 • Protein 4

FIGURE 32-8 The commodity label is the consumer's best assurance of food quality and value.

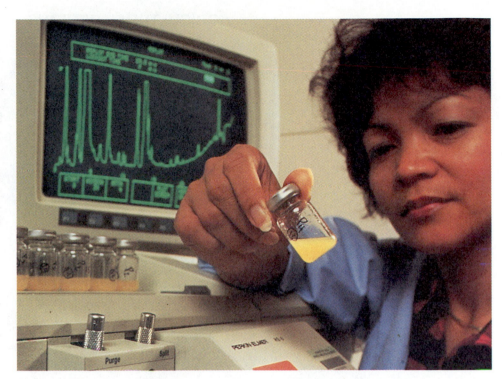

FIGURE 32-7 Quality-control personnel, as well as government inspectors, use USDA standards to monitor foods for cleanliness, wholesomeness, and quality. *(Courtesy of USDA/ARS #K-3512-3)*

COMMODITY GROUPS AND THEIR ORIGINS
What Foods Are Grown Where

Food is grown all over the world. Climatic conditions and available technology dictate that some foods grow better and in greater abundance in certain areas of the world. **Climatic conditions** refer to average temperature, number of days with a certain temperature range, length of growing season, and amount of precipitation for a given geographic area. **Technology** refers to the equipment and scientific expertise available to cultivate, store, process, and transport the crop for consumption in a variety of forms after harvest. Food production in the United States. has always been influenced by geography and climate (Figure 32-9).

Since early times, when humans traveled and traded, foods have been introduced outside the areas where they are grown naturally. The origins of the soybean, for example, can be traced back three thousand years to China, where it is still produced and consumed today. However, major growing areas for the soybean today include the United States, Brazil, and western Europe.

Modern technology has allowed producers to raise crops somewhat artificially with irrigation, and in greenhouses where the conditions of temperature and moisture are controlled. Different varieties of food have been developed to grow under different climatic conditions, such as extreme heat or cold. Similarly, in aquaculture, seafood and fish are produced under controlled conditions.

AGRICULTURAL REGIONS

1 Humid Subtropical Belt	
2 Cotton Belt	
3 Middle-Atlantic Truck Crop Belt	
4 Corn and Winter Wheat Belt	
5 Hard Winter Wheat Belt	**9** Hay Region
6 Corn Belt	**10** Grazing and Irrigated Crop Region
7 Hay and Dairy Region	**11** Western Forest and Hay Region
8 Spring Wheat Region	**12** Columbia Plateau Wheat Region
	13 North Pacific, Hay, and Pasture Region
	14 Pacific Subtropical Crop Region

FIGURE 32-9 Major agricultural regions of the United States.

In the United States, we are accustomed to having almost every food available fresh at any time of the year; but all foods are not grown in all parts of the country. Citrus fruit, including oranges and grapefruit, require warm climates, such as those found in Texas, California, and Florida. Further, over 70 percent of the fresh vegetables that are consumed in the United States are grown in California, Florida, and Arizona. However, people in North Dakota and Maine enjoy the nutrients and good taste of fresh or processed citrus and vegetable products, such as orange juice, daily. Our country is not only "America the Beautiful," it is America the bountiful.

The many operations of the food industry that are discussed later in this unit will explain how we can enjoy foods that are not grown naturally in our particular regions of the country and world. Realizing where some of our food products originate also makes you appreciate the size and scope of the food industry.

Crop Commodities

Grains Various grains have different growing requirements. Therefore, they are produced in different parts of the United States and world. Wheat, which originated in Asia, is grown in the cooler climates of the United States. Different varieties have been developed to accommodate different growing seasons and climatic conditions around the world. Corn is a warm-weather crop, but the many varieties and types permit its growth in every state in the United States. Rice, however, has special moisture requirements; so its production is limited to specific areas of the country.

Oil Crops   These crops are sometimes thought of as the invisible food product. Soybeans, corn, cotton, flax, sunflowers, coconut, peppermint, and spearmint are all significant oil crops in the United States. The United States is a leader in the production of soybeans, corn, cotton, and peanut oils. Soybean products are referred to in ancient Chinese literature, and the origin of the peanut may be traced to Brazil and Paraguay. Sunflowers are native to the United States and are growing in importance. Sunflowers are also grown in Spain, China, Russia, and Mediterranean areas. Safflower oil is a relatively minor oil in terms of proportion to the total oil crop worldwide. It originated in northern India, North Africa, and the Middle East. Its origin is indicative of its drought tolerance.

Sugar Crops   Sugar beets and sugar cane are the principal sugar crops in the United States. Corn is a secondary source of sugar. Sugar beets are grown in temperate areas such as Minnesota, Idaho and California, whereas sugar cane is grown in tropical and subtropical locations around the world. Florida, Hawaii, and Louisiana are the three largest producers of cane sugar in the United States.

Citrus   Oranges, limes, lemons, and grapefruit all require warm temperatures to survive. Consequently, the southern states with warmer climates, such as Florida, Texas, Arizona, and California, are the major producers of citrus.

Tree Fruits   The many varieties of fruit that grow on trees require different weather conditions. Therefore, various fruits are adapted to different parts of the country. Apples and pears require cooler temperatures and do well in mountainous areas. Bananas require very warm conditions and grow best in tropical areas.

Vegetables and Berries   Vegetables and berries are consumed shortly after harvest or are processed by canning, drying, or freezing for future consumption. Some vegetables require warm climates and some require cool climates. Vegetables that require cooler climates include cabbage, broccoli, potatoes, and cauliflower. Vegetables requiring warmer environments include beans, tomatoes, and sweet corn.

Meat Commodities

Animals, like crops, are typically raised in locations with some regard to climatic conditions. Artificial cooling or heating of livestock is costly. Further, large amounts of water must be available for livestock. Where fewer artificial conditions are introduced, the cost of production is minimized. The type, cost, and availability of livestock feed is an important factor influencing where animals are raised in the United States and around the world.

Beef   Most beef is raised near corn, the main feed source of cattle. Over half of the corn in the United States is grown in Nebraska, Iowa, Illinois, and Minnesota. Therefore, beef is raised extensively in the Midwest. The open ranges in the western part of the United States provide other important areas where beef is raised.

Pork   Corn is also the primary food of hogs. Much like beef, the primary area where hogs are raised is the midwestern part of the United States. Grazing on open range is not required for hogs, so the mid-Atlantic and southern states are also important hog-producing areas.

Lamb   Sheep are animals that require large amounts of grazing pasture area. Therefore, they are raised extensively in the range states of the Far West. However, as is true with beef and pork, lamb products are produced in other states.

AGRI·PROFILE

CAREER AREAS: Scientist/Inspector/Quality Controller/Buyer/Seller/Processor/Trucker/Wholesaler/Retailer

The food industry is massive and includes both plant and animal products. It includes the producers, processors, distributors, wholesalers, retailers, fast-food establishments, restaurants, and the home kitchens where food is prepared.

Career opportunities include many that have been discussed previously and some new ones, too. The meat-processing-and-packaging industry is massive and employs a large number of individuals in the United States. Field supervisors and coordinators direct the work of crews to harvest crops at the peak of their quality and transport them to processing or packing plants. Similarly, crews take fish, oysters, clams, lobsters, and other seafood from production habitat to processing centers. In many cases, huge packing and/or processing machines are used in the field, orchard, or boat. Quality-control personnel collect food specimens, label them, test them, and maintain records to ensure quality control on each batch of food coming out of the plant.

Careers in food science; store management; produce management; meat cutting; laboratory testing; field supervision; research; diet and nutrition; health and fitness; and promotion are all possibilities in the food industry.

A poultry research scientist might investigate ways of improving the hatchability of eggs. *(Courtesy of James Strawser, Cooperative Extension Service, The University of Georgia)*

Dairy Products Wisconsin has long been called the "Dairy State." Indeed, many of our dairy products are produced in Wisconsin and other northern-tier states. However, California has a very large dairy industry with many cooperatives and processing plants. These provide the dairy products to consumers nationwide. Dairy animals prefer cooler environments, so the industry is extensive in the northern part of the United States.

Game Each state has native game. Whether or not it is harvested as an agricultural product depends on the demand for the product. Venison is the tasty and popular meat of deer. In some states deer are being raised in captivity to help meet the market demand for venison.

Seafood States that border the Atlantic and Pacific oceans and the Gulf of Mexico are considered the primary suppliers of seafood in the United States. However, the science of aquaculture permits the production of fish in interior states.

Poultry Poultry can be raised in a variety of settings. Typically, chickens and turkeys are raised indoors, where ventilation and temperature are carefully controlled. In the East, most poultry is raised in the mid-Atlantic and southern states. However, important poultry-producing areas are also found in California and other states.

OPERATIONS WITHIN THE FOOD INDUSTRY

The food industry begins with the process of photosynthesis in plants. It progresses through plant and animal growth and on to commodity processing and distribution. The discussion in this unit will focus primarily on the food industry after the crops and animals have been grown.

Harvesting

Harvesting means taking a product from the plant where it was grown or produced. This may involve taking potatoes out of the ground, picking oranges off a tree, removing bean pods from bean plants, or removing and threshing grain from stalks (Figure 32-10). It is most important

FIGURE 32-10 Harvesting means removing the crop from the plant at the proper time. *(Courtesy of Michael Dzaman)*

that the crop be harvested in a timely and careful fashion. The plant should be harvested at the correct stage of maturity. **Maturity** means the state or quality of being fully grown. Harvesting a crop when it is of proper maturity means that it is not underripe, overripe, or spoiled. **Underripe** means that it has not reached maturity. **Overripe** means that the plant is past the optimum maturity; stalks or limbs can easily break or shatter; or fruit can drop. **Spoiled** means that chemical changes have taken place in the food or food product that either reduce its nutritional value or render it unfit to eat.

Spoilage is usually caused by microorganisms. **Microorganisms** that contribute to food spoilage are bacteria, fungi, and nematodes. Bacteria are a group of one-celled plants. Fungi are plants that lack chlorophyll and obtain their nourishment from other plants, thus causing rot, mold, and plant diseases. Nematodes are small worms that feed on or in plants or animals. They live in moist soil, water, or decaying matter and pierce cells of plants and suck the juices. The moisture content of a product determines many changes that the product can undergo. It also dictates the types of handling procedures and storage facilities that are required for certain foods. Certain health considerations are influenced by the various levels of crop maturity. Some fruits, such as bananas and tomatoes, continue to ripen after they have been picked from the tree and vine. Other foods, such as beans and oranges, do not continue to ripen once picked and must be handled accordingly. Knowledge of the complete growth process of a commodity is essential for the producer and harvester.

Harvesting involves the use of equipment and labor. Because timing is critical, migratory labor is often used at harvest time. **Migratory labor** refers to workers who move to the places where harvesting is occurring throughout the year. As crops ripen, laborers migrate to new geographic locations to harvest the crops. Workers are not always available. They may be poorly trained and may not harvest crops as efficiently and economically as do modern machines.

Therefore, engineers and business people have developed and now sell machines to perform many harvesting operations. Such machines make harvesting easier than it was in the past (Figure 32-11). However, machines may not be as gentle as human hands. Consequently, plant breeders have developed new crop varieties that better lend themselves

FIGURE 32-11 Mechanical harvesters replace the back-breaking and difficult work of hand harvesting—here, a cranberry harvester in action. *(Courtesy of USDA/ARS #K-4416-14)*

FIGURE 32-12 Processing is one of many intermediate steps between the producer and the consumer. *(Courtesy of FFA)*

FIGURE 32-13 Tomatoes can be processed into a variety of products. *(Courtesy W. Altee Burpee Company)*

to mechanical harvesting. Such varieties may not be as good to the human touch or taste. For instance, tomato varieties that have skins tough enough for mechanical harvesting are harder to slice and may not be as juicy as varieties suitable for the home garden.

Processing and Handling

The steps involved in turning the raw agricultural product into an attractive and consumable food are collectively known as **processing.** Processing factories or plants clean, dry, weigh, refrigerate, preserve, store, and turn the commodity into a variety of other products (Figure 32-12).

Wheat is cleaned, dried, weighed, and graded for quality. It is then ground into flour. However, before this occurs, it may be separated into bran and germ. The skin or covering of a wheat kernel is known as **bran.** Inside the bran is the **endosperm,** which will become flour, and the **germ,** which is a new wheat plant inside the kernel. Wheat flour is used for breads, cereals, cakes, and pasta. Other grains have similar parts and are used to make similar products.

The processing of tomatoes results in a variety of products (Figure 32-13). Some people claim that the fresh tomato defines summer. The peak of the North American growing season occurs when the tomato is ripe. To the gardener, this means picking tomatoes directly from the vine in a backyard garden and consuming them immediately. However, tomatoes are harvested year-round somewhere in the world, and the food industry can get them to us in edible form. **Edible** means fit to eat or consume by mouth. Processing permits tomatoes to be kept for future use. This begins after the tomato is harvested in the field or greenhouse and arrives at the processing plant.

After tomatoes are cleaned and separated for size and quality, they may be canned whole; chopped, cooked, and strained for juice; or made into other products. Such products include spaghetti and hamburger sauces, relishes, and many other foods.

Transporting

FIGURE 32-14 Food commodities move around the world by every mode of transportation.

Trucks, planes, boats, cars, trains, carts, and bicycles are vehicles used by the food industry in various parts of the world (Figure 32-14). The

FIGURE 32-15 Refrigerated trucks permit supplying of fresh fruits, vegetables, milk, eggs, and meats to every part of the United States, year-round. *(Courtesy of Michael Dzaman)*

transporting of fresh and processed food products composes 5.5 percent of the marketing cost within the food industry in the United States. Timing and the distance foods must travel contribute to the ultimate cost of the foods. The efficiency of transportation can influence food quality in terms of freshness and spoilage. Insulated and refrigerated trucks enable food products to move in fresh form to most parts of the country year-round (Figure 32-15). This luxury is not available to most people of the world.

Approximately 90 percent of our perishable food is shipped by truck. Much of the less-perishable foods, such as wheat, potatoes, and beets, are shipped by rail. Air transportation allows us to enjoy perishable foods from distant regions and different countries. For example, pineapples and papayas from Hawaii are enjoyed all over the world because of air transportation. Because of modern transportation, fresh seafood is enjoyed in many areas where fish are not grown.

The consumer who drives to and from the grocery store is a member of the food industry and provides the final link from farm to table. How far the food is shipped, how the food was wrapped for transportation, how long the food was in transit, and how warm the food became during transport all influence ultimate food quality. From the milk truck driver taking milk from farm to processing center, to the trucker delivering products to the local store, transportation is a key component in the food industry. Knowledgeable and competent employees are great assets when the cash value of loads on their trucks or rail cars is considered. A delay could be costly for both business and the consumer.

Even if food is in perfect condition when you buy it, the quality can decrease substantially before you get it home if precautions are not taken. Food should be packaged correctly in the store, kept cool in the car or truck, and refrigerated or frozen upon arrival home.

Marketing

Wholesalers purchase food products from packing houses, processors, fish markets, and produce terminals. They, in turn, sell to retailers and

BIO•TECH CONNECTION

Keeping the Lid on Drugs

It is generally agreed by authorities that the food supply in the United States is among the safest in the world. Federal and state food inspectors work to ensure that our nation's food supply is wholesome and not contaminated beyond acceptable limits. Avoiding contamination by common bacteria and other living organisms is impossible in any place except a sterile environment. Therefore, we produce our food in sanitary, but not sterile, environments and remove any residual external biological contamination by washing or peeling. We then heat the food to sufficient temperatures to kill internal biological organisms.

Contamination by chemicals is quite a different matter. Plants grown in soils; and fish grown in water containing poisonous chemicals, heavy metals, and other pollutants can result in chemical contamination of fruits, vegetables, nuts, and fish products. Similarly, animals grazing on contaminated pastures or eating contaminated feeds can have chemical contaminants in their meat and milk. Further, drugs used for medicines to keep animals free from diseases, parasites, and insects can contaminate meat and milk if not used according to label instructions. Unfortunately, heating does not remove or neutralize most chemical contaminants. Therefore, food inspectors and other health inspectors keep a vigilant watch with all the tools of modern science to monitor foods from source-to-table and divert unacceptable products from the food stream. In actual practice, reputable food producers, processors, and handlers protect the food supply via quality-control measures in order to protect their businesses and avoid penalties by the government monitoring authorities.

Inspectors must monitor the quality of milk and dairy products to ensure that the products are safe and wholesome. (Courtesy of James Strawser, Cooperative Extension Service, The University of Georgia)

Residues of pesticides and medicines in meat and milk receive continuous scrutiny. Laboratory and other scientific procedures are being improved constantly to help. A case in point is the development of new, high-tech probes to speed up tests for benzimidazoles on meats. These are medicines used to protect cattle, sheep, pigs, chickens, and goats from internal parasites. If the drugs are administered in incorrect doses or with inappropriate timing, some can remain in the meat.

Meat specimens are routinely scrutinized using a high-performance liquid chromatograph. However, future chemists may complement this approach with monoclonal antibody assays. Such assays hold the promise of dramatically speeding up testing without compromising accuracy. Already, monoclonal antibodies have been developed that will inexpensively detect benzimidazoles in meat at concentrations as low as one part per billion—that is the equivalent of one second in thirty-two years!

FIGURE 32-16 Wholesale terminals provide the facilities for trucks and trains to bring food commodities together, and buyers to obtain commodities for their retail outlets. *(Courtesy of FFA, Photo by Bill Stagg)*

FIGURE 32-17 Superstores may stock as many as fifteen thousand items or more. *(Courtesy of FFA)*

institutions such as hospitals, schools, restaurants, and retail stores (Figure 32-16). Grocery stores and fast-food chains are important links in the food chain before the food is purchased by the consumer.

There are many types of retail stores from which consumers can purchase their food items. Such marketing sources meet the needs of consumers in different locations and situations. Superstores carrying fifteen thousand or more items; conventional supermarkets; limited-assortment and box stores; convenience stores; nonconventional food stores; small stores; corner stores; food cooperatives; farmers markets; roadside stands; pick-your-own businesses; and other farm outlets are the most common places that consumers purchase their food items (Figures 32-17 and 32-18). The major differences between the various types of stores are in the number of items stocked and the physical size of the facilities.

FIGURE 32-18 Farmers' markets and other fresh-food markets are on the increase in many areas. *(Courtesy of USDA)*

CAREER OPPORTUNITIES IN THE FOOD INDUSTRY

Each of the areas discussed in this unit and the following unit require people to manage, operate, and carry out the many and varied elements of the food industry. As with all careers in agriscience, those in the food industry present many challenges and rewards. When you consider the chain-reaction nature of the food industry, career opportunities await individuals at the local, county, state, national, and international levels. Careers in the food industry can be divided into seven, often overlapping, categories. Career opportunities in each area are numerous (Figure 32-19).

THE FOOD INDUSTRY OF THE FUTURE

The food industry is ever changing, with new developments each day. Some areas that may attract the food researcher include new food

Some Careers in Food Science and the Food Industry

Business

Accountant
Buyer
Distributor
Financial analyst
Loan officer
Marketing specialist
Salesperson
Statistician

Communications

Advertising specialist
Broadcaster
Media specialist
TV Producer/Demonstrator
Writer

Education

College professor
Extension specialist
Industry educator
Dietician
Teacher

Processing

Butcher
Efficiency expert
Engineer
Plant line worker
Plant supervisor
Refrigeration specialist
Safety expert

Quality Assurance

Food analyst
Grader
Inspector
Lab technician
Quality-control supervisor
Quarantine officer

Research and Development

Distribution analyst
Biochemist
Microbiologist
Packaging specialist
Process engineer

Retailing/Food Service

Baker
Cook/Pizza maker
Counter salesperson
Deli operator
Meat cutter
Nutritionist
Produce specialist
Restaurant owner/operator
Waiter/waitress

Transportation

Dispatcher
Trucker
Rail operator
Merchant marine

FIGURE 32-19 Career opportunities in food science and the food industry are many and varied.

products, new processing and preserving techniques, and new equipment for harvesting labor-intensive crops.

Aquaculture is meeting the increasing demand for fish and will continue to supplement the catches of commercial fishermen. The use of extreme heat and cold in processing has permitted many food items that meet the demand for convenience foods. The convenience food store will continue to play a larger role in the food chain. Economic efficiency in convenience foods and convenience stores is under constant review. Additionally, the USDA and other agencies continue their vigilance regarding safety and nutritional standards at all steps of the food chain.

Improved harvesting equipment for products such as grapes is being tested to lower the labor cost of such crops. Fuel alternatives for cost-effective transportation, and refrigeration with carbon dioxide snow instead of conventional diesel-powered mechanical refrigeration are some of the many developments under constant review in the food industry. This industry must continue to meet the demands for high-quality food in the United States and the world through effective research and qualified employees.

STUDENT ACTIVITIES

1. Write the Terms to Know and their meanings in your notebook

2. Keep a food-dollar diary to document where your food dollars are spent. Record the cost of meals and snacks eaten in and outside of the home.

3. Do a cost comparison of meals prepared at home and similar meals consumed at fast-food places and restaurants.

4. Trace the activities that occur in transforming wheat in the field to a hamburger roll consumed in your home.

5. Draw a diagram tracing the individual food components of a deluxe hamburger back to the places where the components were produced. Label each component, process, and commodity name along the way.

6. Ask your instructor to arrange a field trip to a butcher shop or supermarket to observe demonstrations on meat cutting and packaging.

7. Invite the manager of a food-processing plant in your community to your class to speak about the food-processing industry.

8. Make a collage illustrating the various activities of the food industry.

9. Using resources from your state, identify on a state map the major food commodities produced in different regions of your state.

SELF EVALUATION

A. Multiple Choice

1. Approximately what percentage of the American food dollar is spent on meals away from home?
   - a. 15 percent
   - b. 25 percent
   - c. 35 percent
   - d. 45 percent

2. Which of the following products is native to North America?
   - a. soybeans
   - b. wheat
   - c. sunflowers
   - d. peanuts

3. In the United States, more than one-half of the fresh fruits and vegetables are grown in which states?
   - a. Montana, Oregon, and Washington
   - b. California, Florida, and Texas
   - c. New Jersey, North Carolina, and Georgia
   - d. Arizona, Nebraska, and Ohio

4. Migratory workers would harvest wheat last in which state?
   - a. Arizona
   - b. Nebraska
   - c. Montana
   - d. Ohio

5. When you spend one dollar for food, approximately how much goes into the labor required to harvest that food and process that food after it leaves the farm?

a. $0.94   c. $0.34

b. $0.64   d. $0.04

6. Approximately what percentage of all the jobs in the food and fiber system is related to wholesale and retail sales?

a. 60 percent   c. 40 percent

b. 50 percent   d. 30 percent

7. Which product is consumed away from home the most?

a. fruits   c. vegetables

b. beverages other than milk   d. meat

8. Superstores are likely to carry how many items?

a. 15,000   c. 150

b. 1,500   d. 15

B. Matching

_____ **1.** Harvester   a. Purchases food in large quantities

_____ **2.** Grader   b. Stores food until requested

_____ **3.** Retailer   c. Follows crop harvesting geographically

_____ **4.** Wholesaler   d. Involved in the transportation of food

_____ **5.** Migrant worker   e. Takes the crop from the field

_____ **6.** Trucker   f. Those involved in cleaning, sorting, and preparing a product

_____ **7.** Processor   g. Inspects food and determines how it will be sold

_____ **8.** Distributor   h. Sells directly to the public

C. Completion

1. Three careers that you could pursue in the quality-assurance area of the food industry include _____, _____, and _____.

2. Differences in how grocery stores are categorized are primarily determined by _____ and _____.

3. When buying food at a pick-your-own farm, the producer is also the _____.

4. A beef grade of _____ would indicate very tender, juicy, and flavorful, with abundant marbling.

5. Grocers purchase their supplies through _____.

Food Science

Food

Nutrients

Nutrition

Fermentation

Controlled atmosphere

Refrigeration

Hydrocooling

Blanching

Canning

Dehydration

Freeze-drying

Oxidative deterioration

Dehydrofrozen

Humidity

Retortable pouches

Irradiation

Food additive

Shelf life

Dry-heat cooking

Moist-heat cooking

Conventional ovens

Microwave oven

Convection ovens

Dehydrators

Smokers

Casein

Lactose

Butterfat

Cream

Cheese

Cottage cheese

Condensed milk

Evaporated milk

Vacuum pan

Slaughter

Rendering insensible

Shackles

Hoist

Stuck

Bleeding out

Hide

Viscera

Carcass

Offal

Split carcass

Shrouded

Age (ripen)

Block beef

Disassembly process

Fabrication and
    boxing

Processed meats

Pelt

Evisceration

Singe

Leaf fat

Lard

Giblets

Kosher

Dressing
    percentage

Sweetbreads

Tripe

Rumen

Tankage

Collagen

Y ou are what you eat. Have you ever stopped to consider what that statement means? **Food** is defined by Webster as a material containing or consisting of carbohydrates, fats, proteins, and supplementary substances, such as minerals, used in the body of an organism to sustain growth, repair, and vital processes and to furnish energy, especially parts of the bodies of animals and plants consumed by humans and animals. This unit explores the foods that humans need to maintain health and sustain growth (Figure 33-1). Additionally, it explores how those foods reach our tables from their beginning as raw products.

NUTRITIONAL NEEDS

The body is a complex system that has many nutritional demands. **Nutrients** are substances necessary for the functioning of an organism.

FIGURE 33-1 The future well-being of infants is largely dependent upon the nutrition they get. Good nutrition coupled with good health habits will keep a body productive and vibrant for a lifetime. *(Courtesy of USDA)*

There are more than fifty specific nutrients required for bodily functions. **Nutrition** involves the combination of processes by which all body parts receive and utilize materials necessary for function, growth, and renewal. Nutrition includes the release of energy, the building up of body tissues (both hard and soft), and the regulating of body processes.

After food is digested and in the blood system, nutrients are able to do their work. Nutrients are classified into six major groups. The different groups represent different purposes in the body (Figure 33-2).

Carbohydrates

Carbohydrates serve as the main source of energy for the body. There are three different types of carbohydrates: sugars, starches, and fiber (Figure 33-3). Sugars are simple carbohydrates and are found naturally in many foods, such as fruit, milk, and peas. Refined sugar, or the substance used in many households, comes from sugar beets and sugar cane. Starch is a complex carbohydrate that is found in foods such as bread, potatoes, rice, and vegetables. Starches and sugars are converted to glucose in the body and serve as the major body fuel. Some of the fuel that is generated is stored by the body for later use. However, when the glucose is not used by the body, it is changed to fat. Fiber is also a complex carbohydrate and is found on the walls of plant cells. Humans are unable to digest fiber, yet it plays an important role in moving food through the body and expelling waste after digestion.

Fats

Fats are another source of energy for the body. They are considered to be a more compact source of energy because they have 2 ¼ times the number of calories as the other two energy sources—carbohydrates and proteins. Some vitamins require fats to carry them to the parts of the body where they are needed. Although fat is necessary in the body, too much fat results in obesity and serious diseases, such as heart problems and high blood pressure. Fats are present in differing amounts in most foods. Foods that are known to be high in fat content include cheese, poultry skin, and avocados. Some foods that we are

Carbohydrates

Minerals

Proteins

Water

Fats

Vitamins

FIGURE 33-2 Nutrients are divided into six categories.

FIGURE 33-3 Fruits, vegetables, and grains are healthful sources of carbohydrates. *(Courtesy of Price Chopper Supermarkets)*

accustomed to require fats to prepare them. Baked goods, such as cakes and cookies, salad dressing, and fried foods acquire fats through preparation (Figure 33-4).

Proteins

The body needs food with proteins to build and rebuild its cells. Hair, skin, teeth, and bones are all parts of your body that require protein. Proteins are in a continuous cycle of building up and breaking down. Approximately 3 to 5 percent of your body's protein is rebuilt each day. Beans, peanut butter, meats, eggs, and cheese are high-protein foods.

Vitamins

Vitamins are also essential to the functions of the body. Some vitamins are dissolved in body fat and are stored in the body. Fat-soluble vitamins are not required in the diet each day, because they can be stored in the body. These include vitamins A, D, E, and K. Nine other vitamins—vitamin C and eight B vitamins—are water soluble and must be replenished daily. Specific vitamins have specific jobs in the body, and some foods are known to be richer in specific vitamins (Figure 33-5).

Minerals

There are more than twenty minerals that are needed by the body. The amounts needed may be small, but they are required nonetheless. The twenty minerals are divided into four major groups. Some minerals are parts of bones, others regulate bodily functions, some are needed to make special materials for cells, and others trigger chemical reactions in the body (Figure 33-6).

Water

The human body is more than 50 percent water. Water carries nutrients to cells, removes waste, and maintains the body's proper tem-

FIGURE 33-4 Animal products are good for their protein; but their fatty parts, such as skin and fatty layers, should be removed to avoid excess fat in the diet. *(Courtesy of USDA/ARS #K-4285-2)*

Functions and Sources of Vitamins					
A	**B**	**C**	**D**	**E**	**K**
Functions					
vision bones skin healing wound	using protein, carbohydrates, and fats to keep eyes, skin, and mouth healthy brain nervous system	wound healing blood vessels bones teeth other tissues works with minerals	needed for using calcium and phosphorus bones teeth	preserve cell tissue	blood clotting
Sources					
yellow, orange, and green vegetables	whole-grain and enriched cereals, breads, meats, beans	citrus fruits, melons, berries, leafy green vegetables, broccoli, cabbage, spinach	fatty fish, liver, eggs, butter, added to most milk	vegetable oils, whole-grain cereals	leafy green vegetables, peas, cauliflower, whole grains

FIGURE 33-5 Vitamins have specific functions, and it is important to choose foods for their vitamin content.

Functions of Minerals			
Bone Development	**Fluid Regulation**	**Materials for Cells**	**Trigger Other Reactions**
Sources of Minerals			

calcium
  milk products
 	
magnesium
  nuts, seeds, dark-green vegetables, whole-grain products
 	
phosphorus
  no specific food group
 	
fluorine
  some seafood, some plants, added to drinking water
	sodium
	  salt
	
	potassium
	  bananas
	
	chlorine
	  salt
		iron
		  meats, liver, beans, leafy green vegetables, grains
		
		  works with vitamin C
		
		iodine
		  iodized salt
		
		  added to salt
			zinc
			  whole-grain breads and cereals, beans, meats, shellfish, eggs
			
			copper
			  fish, meats, nuts, raisins, oils, grains

FIGURE 33-6 The careful selection of foods can ensure a correct balance of minerals in our diets.

perature. Fluid foods, such as milk and juice, obviously help to supply the body with water. However, foods such as meat and bread also provide water.

Food Groups that Meet Needs

Each food is different in the types of nutrients it contains and ultimately provides to the body. Foods are divided into five major food groups, which represent the nutritional needs of the body. The five food groups are: bread, cereal, rice and pasta group; vegetable group;

BIO•TECH CONNECTION

New Kid on the Block

After a lengthy period of inquiry and research, the USDA approved the irradiation of raw packaged poultry as a safe and effective procedure for protection against food-borne illness. Irradiation of food is being done in over thirty-five countries.

Proper cooking is still the last point in food handling to eliminate micro-organisms that can cause food poisoning or other illness to those who consume the food. Still, the use of water for washing and cooling fruits, vegetables, and meats during and after processing can spread harmful micro-organisms throughout the batch and leave every item with low concentrations of the micro-organisms. In other words, all of the food has low level contamination; and the organisms will multiply and increase their hazard if conditions are right to do so. Therefore, the handling of raw meat having pathogens on the surface carries the threat of contamination of hands, counter tops, cooking areas, and ultimately, cooked food.

Science has long sought to find a way to eliminate food-borne pathogens without altering or damaging the product or leaving poisonous chemicals on the food. Irradiation seems to be the answer. Irradiation provides the same benefits as when food is processed by heat, refrigeration, or freezing or is treated with chemicals—to destroy insects, fungi, or bacteria that cause food to spoil or cause human disease. Irradiated foods are wholesome and nutritious. The treatment process involves passing food through an irradiation field, but the food itself never contacts a radioactive substance. The process has been subjected to extensive study and declared safe by special scientific committees in Denmark, Sweden, the United

As early as January 1992, irradiated Florida strawberries were sold at a northern Miami supermarket. *(Courtesy of USDA)*

Kingdom, and Canada, and has received official endorsement from the World Health Organization and the International Atomic Energy Agency. No known risks exist from consuming food that is irradiated by legal procedure.

It should be noted that the approved level of irradiation does not kill all micro-organisms—in other words, it does not sterilize food. Therefore, other food preserving measures must be followed when storing irradiated food and preparing it for consumption. Nor does it alter the food in any perceptible way for the consumer. The Cordex Alimentarius, an international committee on food safety, has developed a green irradiation logo. Irradiated foods from the United States will bear this green logo along with the words, "Treated with Radiation" or "Treated by Irradiation."

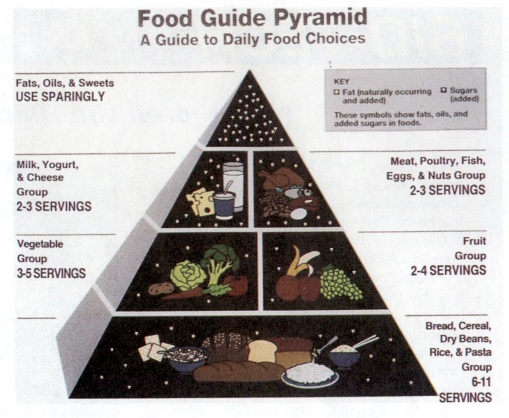

Food Guide Pyramid
A Guide to Daily Food Choices

KEY
☐ Fat (naturally occurring and added)   ☐ Sugars (added)

These symbols show fats, oils, and added sugars in foods.

Fats, Oils, & Sweets
USE SPARINGLY

Milk, Yogurt, & Cheese Group
2-3 SERVINGS

Meat, Poultry, Fish, Eggs, & Nuts Group
2-3 SERVINGS

Vegetable Group
3-5 SERVINGS

Fruit Group
2-4 SERVINGS

Bread, Cereal, Dry Beans, Rice, & Pasta Group
6-11 SERVINGS

FIGURE 33-7 The nutrition pyramid indicates the range in number of daily servings for each major food group as required for a healthful diet. *(Courtesy of USDA)*

fruit group; milk, yogurt, and cheese group; and meat, poultry, fish, dry beans, eggs, and nut group. The USDA represents these groups in a pyramid format to help us remember the relative proportions that each group should make up in a daily diet (Figure 33-7).

You should generally not seek foods in the fats, oils, and sweets group at the top of the pyramid, because plenty of these are used to prepare foods in the other groups, and they show up in adequate amounts in most people's diets in the United States. The names of the groups suggest some of the typical foods that they include. However, the quality of diet can be increased by selecting the most nutritious items from each group (Figure 33-8).

Eating from each of the major food groups daily will ensure a well-balanced diet and provide the essential nutrients needed for growth and development. The number of portions consumed per day differs with each group. At different stages in life, requirements within each group may vary to some degree, but no food group should dominate or be eliminated from the diet. More information on nutrition is provided in the unit on animal nutrition in this text.

FOOD CUSTOMS OF MAJOR WORLD POPULATIONS

What is eaten from each of these food groups varies around the world. Food habits reflect what is most readily available.

In hot and wet climates such as Southeast Asia, a great deal of rice is consumed. In areas of the world where corn grows well, many food items contain corn in some form. For example, cornmeal may be used to make tortillas or pancakes and be mixed as a cereal in countries such as Mexico. Availability of food and technology to prepare that food has dictated eating habits over the years. For instance, introducing

POULTRY, FISH, MEAT, AND EGGS

Lower-scoring foods tend to be high in calories, cholesterol (eggs), fat (red meat), or sodium (processed meats). The foods near the top are relatively low in fat. Most of the foods are rich in protein and iron. (All servings are 4 ounces broiled, baked, or roasted, unless noted otherwise.)

	NUTRITION SCORE
clams, steamed	19
turkey breast, skinless	10
tuna canned in water (3 oz.)	6
cod	1
egg white (1 large)	1
salmon, canned (3 oz.)	1
scallops, steamed	1
flounder	-1
lobster meat, boiled	-4
salmon fillet	-5
blue crab meat, steamed	-6
chicken breast, skinless	-8
turkey breast luncheon meat, 3 slices (2 oz.)	-13
tuna canned in oil (3 oz.)	-18
shrimp, steamed	-21
chicken breast with skin	-23
Canadian bacon, fried, 2 slices (2 oz.)	-24
veal cutlet, breaded, pan-fried	-25
round steak, trimmed (5 oz.)	-29
ham, luncheon meat, 2 slices (2 oz.)	-32
pork chops	-48
bacon, fried, 4½ slices (1 oz.)	-54
egg (1 large)	-59
shrimp, fried	-63
bologna, 2 slices (2 oz.)	-70
leg of lamb	-76
salami, luncheon meat (2 oz.)	-80
hamburger, lean	-81
sirloin steak (5 oz.)	-86
hamburger, regular	-92
round steak, untrimmed (5 oz.)	-97
chicken thighs, fried, home recipe (2)	-103
sausage links, 2 (2 oz.)	-112
pot roast	-114

GRAIN FOODS

Contrary to myth, starchy grain foods are not fattening. Most people would do well to eat more grain foods in place of meat. Grains, especially whole grains, are a nicely balanced, low-fat source of carbohydrate, vitamins, minerals, and protein. (All serving sizes are 1 cup cooked, unless noted otherwise.)

	NUTRITION SCORE
bulgur (cracked wheat)	69
wheat germ (¼ cup)	61
pearled barley	60
brown rice	45
spaghetti or macaroni	45
oatmeal	38
hominy grits	35
whole-wheat bread (2 slices)	31
hamburger or hotdog roll (1)	18
corn muffin (1)	1

FRUITS

Fruits can give you, naturally, all the sweetness you want, plus fiber, vitamins A and C, and other nutrients. Go easy on the dried fruits! Their sugars are sticky and promote tooth decay. (All servings are one medium piece, unless noted otherwise.)

	NUTRITION SCORE
papaya (½ medium)	74
cantaloupe (¼ medium)	67
strawberries (1 cup)	65
orange	62
prunes, uncooked (5)	51
dried apricots (5)	49
tangerine	41
watermelon (2 cups cubed)	36
apple	36
pear	36
blueberries (1 cup)	36
pink grapefruit (½)	35
pineapple, fresh (1 cup)	34
banana	32
cherries (1 cup)	32
honeydew melon (¹/₁₀ melon)	31
raisins (1 oz.)	28
plums (2)	27
applesauce, unsweetened (½ cup)	18
peach	17
grapes (30)	16
peaches in heavy syrup (½ cup)	2

DAIRY

While most dairy foods are rich in protein and calcium, the lower-scoring foods are high in saturated fat, cholesterol, and sodium.

	NUTRITION SCORE
yogurt, nonfat (1 cup)	58
milk, skim (8 oz.)	40
yogurt, plain lowfat (1 cup)	36
milk, 1% lowfat (8 oz.)	28
milk, 2% lowfat (8 oz.)	16
yogurt, fruit-flavored lowfat (1 cup)	13
chocolate milk, 2% lowfat (8 oz.)	6
cottage cheese, 1% fat (½ cup)	3
sour cream, lowfat (2 Tbsp.)	-2
ricotta cheese, part skim (1 oz.)	-3
yogurt, plain (1 cup)	-5
mozzarella cheese, part skim (1 oz.)	-5
milk, whole (8 oz.)	-7
cheddar cheese, reduced fat (1 oz.)	-7
nondairy powder coffee creamer (2 tsp.)	-12
half and half cream (2 Tbsp.)	-15
Swiss cheese (1 oz.)	-15
cottage cheese, 4% fat (½ cup)	-16
mozzarella cheese (1 oz.)	-19
sour cream (2 Tbsp.)	-26
cheddar cheese (1 oz.)	-32
American cheese (1 oz.)	-34
whipped cream (2 Tbsp.)	-59

VEGETABLES

Most vegetables are great sources of vitamins—especially A and C—and minerals. Try a new vegetable today! (All serving sizes are ½ cup cooked, unless noted otherwise.)

	NUTRITION SCORE
sweet potato, baked (1 medium)	184
potato, baked (1 medium)	83
spinach	76
kale	55
mixed vegetables, frozen	52
broccoli	52
winter squash (acorn, butternut), baked	44
Brussels sprouts	37
cabbage, chopped, raw (1 cup)	34
green peas	33
carrot (1)	30
okra	30
corn on the cob (1 ear)	27
tomato (1 medium)	27
green pepper (½)	26
cauliflower, raw	25
artichoke (½)	24
romaine lettuce, raw (1 cup)	24
collard greens	23
asparagus	22
celery (four 5" pieces)	19
green beans	18
turnips	16
sauerkraut	15
summer squash (zucchini)	12
green beans, canned	10
iceberg lettuce, raw (1 cup)	8
bean sprouts (¼ cup)	7
onion, chopped, raw (¼ cup)	7
eggplant	6
cucumber slices, raw	4
mushrooms, raw (¼ cup)	2
dill pickle (½ large)	-3
avocado (½ medium)	-25

LEGUMES

Beans are excellent sources of dietary fiber, protein, vitamins, and minerals. They are also very low in fat. (All serving sizes are ¾ cup cooked, unless noted otherwise.)

	NUTRITION SCORE
kidney beans	91
navy beans	82
black beans, black-eyed peas, or lima beans	78
lentils	74
chickpeas	68
split peas	56
tofu/bean curd (4 oz.)	33

DESSERTS

Most desserts are high in fat, sugar, and calories. Next time, try fresh fruit or nonfat frozen yogurt for a change. (Serving sizes are 1 cup, unless noted otherwise.)

	NUTRITION SCORE
angelfood cake (2 oz.)	1
chocolate pudding (½ cup)	-2
Jell-O (½ cup)	-7
brownie with nuts (1¾" square)	-23
sherbet	-37
vanilla ice cream	-73
cheesecake (4½ oz.)	-161

FIGURE 33-8 Estimated relative nutritional values of selected food items.

dairy products in countries where dairy cows are not raised presents educational as well as transportation and processing challenges.

METHODS OF PROCESSING, PRESERVING, AND STORING FOODS

One of the oldest ways to preserve food for delayed use is fermenting and pickling. **Fermentation** is a chemical change that involves foaming as gas is released. Long ago, it was determined that some foods did not spoil when allowed to ferment naturally or when fermented liquids were added to the foods. Controlled fermentation is now used to produce cheeses, wines, beers, vinegars, pickles, and sauerkraut (Figure 33-9).

Today, the primary objective of processing and preserving is to change raw commodities into stable forms. With refrigeration and various processing techniques, we now expect almost all foods at any time during the year and in any part of the world.

Slowing deterioration is the primary goal in food preservation. Tomatoes and cucumbers that will be sold raw are waxed to retard shriveling while they are in the grocery store. Apples may be treated with a decay inhibitor. Table grapes are fumigated with sulfur dioxide to control mold. Similarly, silos where grains are stored are purged with 60 percent carbon dioxide to control insects.

Carbon dioxide inhibits the growth of bacteria. **Controlled atmosphere** (CA) is the process whereby oxygen and carbon dioxide are adjusted to preserve or enhance particular foods. An example of food preservation by controlled atmosphere is the transporting of cut lettuce in controlled atmosphere to prevent the edges from turning brown.

There are many other processing and preservation techniques. These techniques slow deterioration and allow one to enjoy foods in a variety of forms around the year and around the globe.

Refrigeration is an important key to many processing and preservation techniques. **Refrigeration** is the process of chilling or keeping cool. Low temperatures reduce or stop processes that contribute to the deterioration of products. Refrigeration retards respiration; aging; ripening; textural and color change; moisture loss and shriveling; insect activity; and spoilage from bacteria, fungi, and yeasts. Refrigeration is costly and used widely only in well-developed countries.

When crops are harvested in the field, their temperatures are between 70° and 80°F. The goal of refrigeration is to quickly reduce that temperature to near 32°F. How quickly the food is cooled varies depending on the type of cooling technique used. Some vegetables are precooled in the field by cold air blast, hydrocooling, or vacuum. **Hydrocooling** means cooling with water. Milk is cooled in refrigerated lines and tanks on the farm (Figure 33-10).

The logical step after a product is cooled is to continue cooling it until it is frozen. When foods are kept at 0°F or lower, very little deterioration occurs. However, even frozen foods have storage limits. Fruits and vegetables should be consumed within one year after freezing. Meats should be consumed within 3 to 6 months. Vegetables that are to be frozen often require blanching prior to freezing. **Blanching** is the scalding of food for a brief time before freezing it. This process inactivates enzymes that cause undesirable changes when plant cells are frozen.

FIGURE 33-9 Fermentation is a food preservation process whereby bacteria converts sugars to acids or alcohol that protects the food from spoilage. *(Courtesy of USDA)*

FIGURE 33-10 Food storage and preservation via ice, refrigeration, and freezing is the cornerstone of milk, meat, fish, fruit, and vegetable handling today. *(Courtesy of USDA)*

AGRI·PROFILE

CAREER AREAS: Food Scientist/Food Technician/ Nutritionist/Dietician

Food science programs are available in high schools, colleges, technical institutes, and universities. Careers in food science may be in specialty areas, such as meats, fruits, vegetables, baked goods, dairy products, or wine or other beverages.

Food companies and consumers rely on food scientists to develop new products to meet the ever-changing needs of a busy world. Convenience products, such as instant and freeze-dried coffee; dried and vacuumed-packed fruits and meats; processed chicken tenders; boneless rolled meat; yogurt; ice cream; pasteurized and homogenized milk; shelf-safe milk; low-calorie and low-fat food; space-age food; unrefrigerated foods; and many other products are the handiworks of food science and technology.

Food scientists work for food companies, universities, research centers, food chain stores, dairies, radio and television stations, and government agencies. Some become authors and publish journal and magazine articles, cookbooks, and recipe books.

Are these young visionaries contemplating careers in food science? Food scientists develop better ways to process, handle, package, prepare, and market food products. *(Courtesy of USDA/ARS #K-48191)*

Another popular preservation technique is canning. **Canning** involves putting food in airtight containers and sterilizing the food to kill all living organisms that could cause spoilage. Temperatures of 212° to 250°F are required to kill microorganisms that could cause spoilage. Metal cans are coated to reduce chemical reactions between the can and its contents. A two-year shelf life for canned food is considered normal (Figure 33-11).

Another popular way to process and preserve food is dehydration. **Dehydration** means lowering the moisture content to inhibit growth of microorganisms. Moisture can be removed by the sun, by indoor tunnel or cabinet dehydrators, or by freeze-drying. **Freeze-drying** is the newest method of dehydration. It involves the removal of moisture by rapid freezing at very low temperatures. When foods are dehydrated, they have a 2- to 10-percent moisture content. The shelf life of dehydrated foods is as long as two years. **Oxidative deterioration** is loss of quality due to a reaction with oxygen. This can occur when air reaches dried foods. Glass or metal containers are more airtight than plastic ones. Dried soup mixes, packaged salad dressing, spices, and dried fruits are foods that have been dehydrated. Dehydrated foods are lighter in weight and lower in volume than the same whole foods.

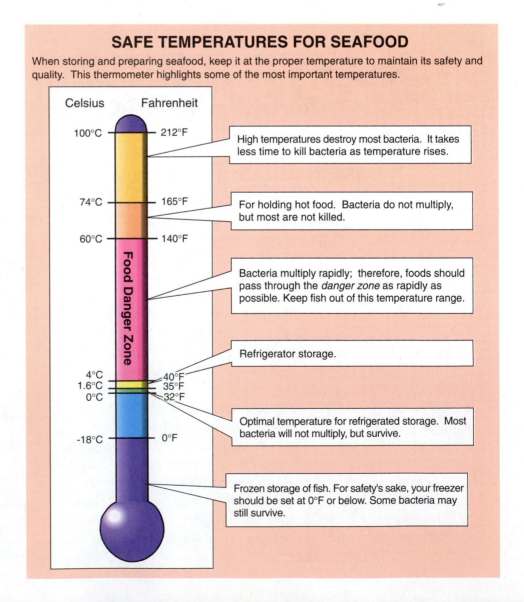

FIGURE 33-11 It is important to choose the appropriate heat intensity and duration for the intended purpose.

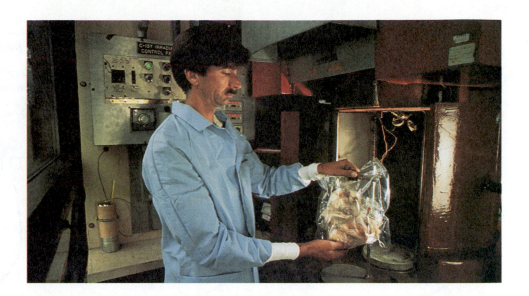

FIGURE 33-12 Irradiation is now an approved method for controlling samonella bacteria in chicken meat. *(Courtesy of USDA/ARS #K-3783-20)*

Another processing technique is dehydrofrozen. **Dehydrofrozen** involves precooking, water evaporation, and freezing. Potatoes are being used in the development of this technique. They are precooked as cubes or slices, and water is evaporated to reduce the weight by 50 percent. The potatoes are then frozen.

Another factor to consider in the processing and handling of foods is humidity during storage and transportation. **Humidity** is the amount of moisture in the air. A humidity level of 90 to 95 percent is required for high-moisture products such as meats and vegetables. Dry onions require only 75 percent humidity for optimum storage.

There is a variety of wrappings used in the processing, preservation, and transportation of foods. Cardboard boxes, wood boxes, molded pulp trays to reduce bruising, and plastic wraps in a variety of thicknesses or plies all meet different needs. Newer retortable pouches provide protection from light, heat, moisture, and oxygen transfer, all in one wrapping. **Retortable pouches** are flexible packages consisting of two layers of film (or plastic) with a layer of foil between them. Although the cost is now high, the benefits of a one- to two-year shelf life are appealing.

Irradiation is a relatively new procedure using gamma rays to kill insects, bacteria, fungi, and other organisms in food products. Gamma rays pass through food without heating or cooking. Therefore, no heat-sensitive nutrients are lost in the process (Figure 33-12).

The cost of each of the preceding processes and techniques is different and must be taken into account by processing plants. Where energy costs are lower, more sophisticated techniques will not cost the consumer as much as they would if energy costs were higher. Research is always looking at ways of preserving food more economically in relation to energy costs.

Food Additives to Enhance Sales of Food Commodities

The processing of some foods may reduce their natural nutritional value. To compensate for that loss, vitamins and minerals are added back into foods to restore their nutritional value. Bread, noodles, and rice have vitamins and minerals added to them before they are

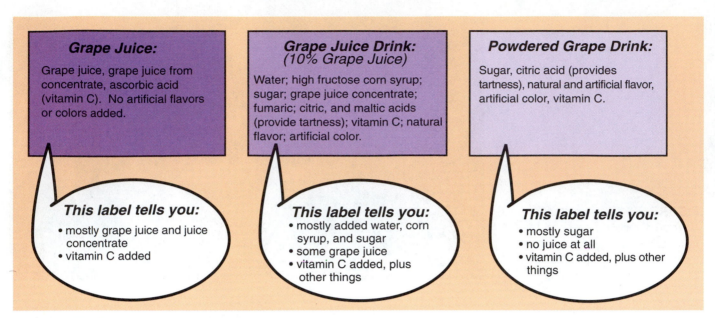

Grape Juice:

Grape juice, grape juice from concentrate, ascorbic acid (vitamin C). No artificial flavors or colors added.

This label tells you:
- mostly grape juice and juice concentrate
- vitamin C added

Grape Juice Drink:
(10% Grape Juice)

Water; high fructose corn syrup; sugar; grape juice concentrate; fumaric; citric, and maltic acids (provide tartness); vitamin C; natural flavor; artificial color.

This label tells you:
- mostly added water, corn syrup, and sugar
- some grape juice
- vitamin C added, plus other things

Powdered Grape Drink:

Sugar, citric acid (provides tartness), natural and artificial flavor, artificial color, vitamin C.

This label tells you:
- mostly sugar
- no juice at all
- vitamin C added, plus other things

FIGURE 33-13 Products may look the same, but read the labels for the real differences. Ingredients are listed in order from the most to the least amount found in the product.

packaged. Vitamins A and D may be added to fluid milk prior to its sale to the public.

A **food additive** is anything that is added to a food during processing and before it goes into a package. In addition to food additives that restore nutritional value, preservatives are also added to food to extend its shelf life. **Shelf life** refers to the amount of time before spoilage begins. Food additives also enhance the color or appearance of foods. Other food additives reduce the cooking time of foods such as oatmeal.

Sugar is probably one of the most widely used food additives. It is found on the labels of many cereals and beverages. Food labels identify the contents of food products. The order of ingredients on a food label indicates the proportions of each, in descending order (Figure 33-13).

The government is continually testing and evaluating the positive and negative effects of food additives. It has been determined that some food additives may cause harm. Such additives are usually banned by the government.

FOOD PREPARATION TECHNIQUES

Some foods can be eaten as they come from the package. Crackers and raisins are examples of foods that do not require additional processing at home to ensure safe eating. However, cooking is often required for other foods, while heating may only improve flavor in others. Raw meat should be cooked before it is eaten to ensure safety. How food is cooked is determined by the type of food and the appliances available.

There are two basic methods of cooking foods. The appliances used to accomplish these methods vary from household to household. **Dry-heat cooking** involves surrounding the food with dry air in the oven or under the broiler. This method is usually used for tender cuts of meat having little connective tissue and for vegetables having a high-moisture content, such as potatoes. **Moist-heat cooking** involves surrounding the food with hot liquid or by steaming, braising, boiling, or stewing the food. The warm moisture breaks down the connective tissues. Moist-heat cooking is a popular method used for less tender cuts of meat and for vegetables with a low-moisture content.

Appliances have been designed to accommodate different types of food, as well as different time and energy demands. The goal of preparing good food in a short period of time, without expending excessive energy, has resulted in the development of several appliances that supplement the traditional gas or electric range and oven. Pressure cookers concentrate moisture by sealing it in, thus reducing cooking time. Crockpots and oven bags allow slow, moist cooking without the operator continually being near the cooking process.

Table-top **conventional ovens** are used to cook small amounts of foods by dry heat. This requires less energy than do large, traditional ovens. The **microwave oven** utilizes electromagnetic waves to heat and cook food. It offers a more energy-efficient way to cook foods that require both dry and moist heat. Approximately 50 percent of the energy goes to the food in the microwave, whereas only 6 to 14 percent of the energy used by a conventional range actually goes to the food.

Convection ovens heat food with the forced movement of hot air. Ovens that combine convection and microwave functions are now on the market. Again, energy efficiency, as well as convenience for the food preparer, is a constant goal.

Dehydrators dry food; **smokers** preserve food by keeping smoke in contact with the food for prolonged periods. The drying or smoking of food for family use is not very popular in the United States. Often, refrigeration is still required after foods are dried or smoked.

FOOD PRODUCTS FROM CROPS

Food from plant sources helps meet body requirements for food in four of the five food groups. Fruits, vegetables, breads, cereals, beans, and oils all come from crops (Figure 33-14).

Fruits, Vegetables, and Nuts

Fruits, vegetables, and nuts are nearly ready to eat when they are harvested. This can be as simple as pick, wash, and eat for items like leafy vegetables, berries, and fruits. Where meals are concerned, the process could include pickling, cutting, peeling, shelling, washing, trimming, cooking the food, or eating it raw. Further, these products may be processed for storage for as short as a few hours to as long as many years. Food processing may be done at home for the family or commercially for the billions of people in the world market.

For some food products the journey from field to table may involve many processes. The first stage may be to simply cool the product and hold it in a temperature- and humidity-controlled environment until the next stage, or it may be to process the food without first precooling. Where processing occurs immediately, the product typically passes through washing equipment designed to avoid injury to the product. Washing may involve flotation, rotary, water-jet, or other cleaning procedures. Skins and hulls may be removed by hot water bath, steam, flame, or cutters. The product may then be trimmed,

FIGURE 33-14 Baked goods from grains form the basic diet in many cultures. *(Courtesy of Price Chopper Supermarkets)*

halved, quartered, sectioned, sliced, diced, or crushed for drying, canning, freezing, or processing into ready-to-cook or precooked products.

Processed foods must be packaged and correctly labeled. All along the way, nutritionists, chemists, inspectors, public health officials, and others monitor the process and product. The law requires that food be safe for consumption and of the weight, quality, and grade that is specified on the label.

Cereal Grains

Cereal grains compose the major diet of most of the world's people. Rice, wheat, corn, barley, oats, and other grains are consumed as whole grains, cracked or rolled grains, flour, bran, and many other products. Grains are economical to process because they can be left on the plant until nearly dry enough to prevent spoilage. If stored in a cool dry place, properly dried grains can remain edible for many years. However, much of the world's food grain is lost every year due to rotting in the field and in storage, or to being consumed or ruined by insects, birds, and rodents.

Processing grain for human consumption generally means separating or milling the grain into its basic components—hulls, bran, flour, and germ. These components are then used to make breakfast cereals, breads, pastries, pasta, and thousands of products that adorn the grocery shelves.

Oil Crops

Soybeans, cottonseed, peanuts, rape, palm nuts, coconuts, olives, and corn are examples of crops that are rich in vegetable oils. These are used for cooking, frying, baking, and other food processes, as well as for many food products such as dressings, coffee creamers, and shortening. These oils are also used in the manufacture of paints, lacquers, plastics, and many products of industry. The seeds or nuts and other oil-rich plant parts are crushed or ground and heated. The oil is then extracted by solvents and purified for food and industrial uses. The meal is dried and ground mostly for livestock feed (Figure 33-15).

Those who do not eat meat products can still meet the requirements for food from all five food groups. People with allergies to milk products can meet their nutritional needs for food from soybeans and other plant sources.

FOOD PRODUCTS FROM ANIMALS

Meats, fish, poultry, dairy products, and the proteins group are nutritious foods from animal sources. (Figure 33-16).

Dairy Products

Milk is used and consumed in a variety of ways. Approximately 37 percent of all milk consumed in the United States is in fluid form. The remainder is used to make many products, including cheese, butter, frozen foods, dried whole milk, cottage cheese, evaporated milk, and

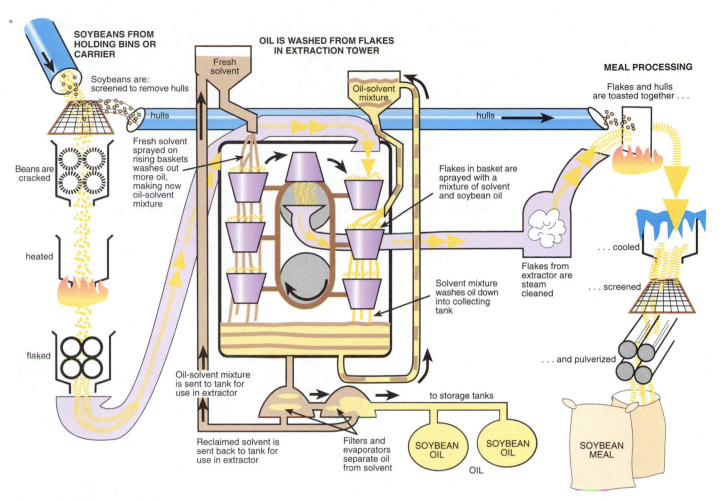

FIGURE 33-15 Processing of grains and oil-crop seeds releases the basic building blocks for many products.

FIGURE 33-16 Meats, fish, poultry, and dairy products are nutritious foods from animal sources.*(Courtesy of Price Chopper Supermarkets)*

condensed milk. The processing of milk in different ways results in these many different products.

Milk and Milk Products

While recommendations for milk consumption are greatly influenced by calcium needs, milk contains some of all the essential food nutrients needed by the body. The major components of milk are water, fat, protein, sugar, and minerals. However, there are numerous other highly important components in milk.

Water   Although cow's milk is a fluid product containing about 88 percent water, it also contains 12 to 13 percent total solids. This is comparable to the solids content of many other foods. Because milk is a food specifically prepared by nature for the nourishment of the very young, it provides the water necessary for life. The water also acts as a carrier for dissolved, suspended, and emulsified components.

Protein   Milk provides a substantial proportion of the total protein in our food supply. **Casein** is the predominate protein in milk. It is found only in milk and makes up about 82 percent of the total milk protein. Casein exists in suspended form and is easily coagulated by the action of acids and enzymes. Milk proteins are high-quality proteins that contain all of the essential amino acids in proper balance for good nutrition.

Milk Fat   Fats are a concentrated source of energy. There are many different fatty acids in milk that give it the distinctive, pleasing flavor that compliments many prepared foods. It also contains those vitamins that are fat soluble—vitamins A, D, E, and K. Milk fat exists in a highly emulsified state, which facilitates its digestion.

Lactose   **Lactose** is milk's major carbohydrate and accounts for about half of the nonfat solids in milk. The relative sweetening power of lactose is about one-sixth that of sucrose (common table sugar).

Minerals   Milk contains seven minerals as major constituents and many more in minor or trace amounts. Calcium and phosphorus are essential in human nutrition for building bony structures and for certain metabolic processes. Milk is the chief source of food calcium in the diet of people in the United States. It has the added advantage of containing phosphorus in the same biological relationship to calcium as occurs in the growing skeleton. It is difficult to provide the recommended daily amounts of dietary calcium without using milk or milk products because calcium is poorly distributed among other foods.

Vitamins   All of the vitamins known to be required by humans are found in milk. Some are fat-soluble and are associated with butterfat, while those that are water-soluble are found in the nonfat portion of milk. Vitamin A and carotene are present in high concentration in milk fat. Carotene, from which vitamin A is formed in the body, gives milk fat its characteristic color. The vitamin D content of fresh milk is low. However, most commercially pasteurized milk is fortified with vitamin D to balance the product for best nutrition.

Milk is an abundant source of riboflavin (vitamin B_2) and an important source of niacin. Although the niacin content of milk is low, it is in a fully available form. Further, milk contains significant amounts of of thiamine (vitamin B$_1$). Other vitamins of the vitamin B complex occurring in milk include pantothenic acid, pyridoxine (vitamin B$_6$), biotin, vitamin B$_{12}$, folic acid, and choline.

Processed Milk Products   The nutritional completeness of milk has led to its reputation as nature's most nearly perfect food. Where it falls short in minor ways, science and the food-processing industry have intervened to improve the options for the consumer.

Fluid milk sold in the United States is pasteurized to ensure safety from disease-causing organisms. Further, it is homogenized to keep the milk fat in suspension so it does not require stirring before each use. Whole milk is marketed with 3.5 percent milk fat. However, for those desiring less fat intake, milk can also be purchased with 2 percent, 1 percent, or no milk fat. The latter is called skim milk. Milk may also be purchased as fortified milk with vitamins A and D added.

The fat in milk is called **butterfat. Cream** is a component of milk that contains up to 40 percent butterfat. Butter, which is about 80 percent fat, is made from cream. Cream is made by concentrating the fat portion of the milk. This is accomplished by passing milk through a cream separator. Whipping cream contains about 40 percent fat, table cream 18 to 20 percent, and "half-and-half" approximately 12 percent.

Ice cream, ice milk, and sherbet are also dairy products. They account for most of the frozen desserts in the United States.

Nonfat dried milk is used both as human food and animal feed. It is frequently used as an ingredient in dairy and other food products.

Cheese is made by exposing milk to certain bacterial fermentation or by treating it with enzymes. Both methods are designed to coagulate some of the proteins found in milk. There are many types and varieties of cheese. **Cottage cheese** is made from skimmed milk.

Condensed and evaporated milks are canned milk products. **Condensed milk** and **evaporated milk** are both produced by removing large portions of water from the whole milk via a machine called a **vacuum pan.** Condensed milk is further treated by adding large amounts of sugar. The sugar content makes condensed milk an important ingredient in the baking and ice cream industries. The many components of milk have resulted in the evolution of an extensive dairy industry. Consumers now can choose from a great array of dairy products (Figure 33-17).

FIGURE 33-17   Milk is the basic ingredient of many food items. *(Courtesy of USDA)*

Meat Products

Meat products, like milk, are enjoyed in a variety of forms. Animal species vary somewhat, but the procedures for slaughtering and processing each are similar. There are common by-products from the processing of animals, as well as the familiar meats that are important to American eating habits.

Beef

Meats of all kinds come from slaughtered animals. **Slaughter** means to kill and process or dress animals for market purposes. The procedure is as follows for beef.

The first step is to render insensible. **Rendering insensible** means making the animal unable to sense pain. There are several methods of accomplishing this that comply with the Humane Slaughter Act of 1958. Packers that do not comply with provisions of this act are unable to sell meat to the federal government. The approved methods must be rapid and effective. Methods used in rendering insensible include a single blow or gunshot, electrical current, or use of carbon dioxide gas. Additionally, rendering insensible may be accomplished by following the ritual requirements of religious faiths.

When the animal is insensible to pain, it is shackled, hoisted, and stuck to permit bleeding. **Shackles** are mechanical devices that confine the legs and prevent movement; **hoist** means to raise into position. *To stick* means to cut a major artery to permit the blood to drain from the body. A large artery is **stuck** for efficient **bleeding out.** The head of the animal is removed during or following the bleeding-out process.

The removal of the **hide** or skin is the next step in the process. The hide is cut open at the median, or midline, of the belly of the animal, and hide pullers are used to remove the hide in one piece. The breast and rump bones are split at this time by sawing.

The term **viscera** refers to organs located in the cavity of the animal. Organs that are removed include the heart, liver, and intestines. The kidneys are not removed at this time. Plants that are regulated by the U.S. Department of Agriculture must have the carcass and viscera inspected to confirm good animal health. **Carcass** refers to the body meat of the animal—the part that is left after the offal has been removed. **Offal** means the nonmeat material that is converted to by-products. It includes the blood, head, shanks, tail, viscera, hide, and loose fat.

The next step is to split the carcass by cutting through the center of the backbone and removing the tail. The **split carcass** (sides of the animal) is then washed with warm water under pressure.

To provide a smooth appearance following cooling, high-quality carcasses are **shrouded,** or wrapped tightly with a cloth. Sides are cooled for a minimum of twenty-four hours prior to ribbing and further processing (Figure 33-18). The meat is kept at 34°F until it is sold and consumed.

Fresh beef must be aged, or ripened. To **age,** or **ripen,** means to leave undisturbed for a period of time so that minor biological changes can take place while the beef cools. Fresh beef is not in its most tender

FIGURE 33-18 After slaughter, the carcass is inspected and cooled before being cut into block or retail cuts. *(Courtesy of USDA/ARS #K-4284-12)*

state immediately after slaughter. While the beef is aging, evaporation (the loss of moisture) and discoloration are kept to a minimum. A fairly thick covering of fat on the carcass helps the aging process. There are three methods used to age beef—traditional aging, fast aging, and vacuum packaging. Time, temperature, and technology help define these three methods.

Beef carcasses are generally disposed of in three ways: as block beef; fabricated, boxed beef; and processed meats. Traditionally, meat is shipped in exposed halves, quarters, or wholesale cuts to be cut into retail cuts in supermarkets. In this condition, it is referred to as **block beef.** It is ready for sale "over the block" or counter. This traditional method creates concern over sanitation, shrinkage, spoilage, and discoloration. Therefore, more packers are using the disassembly process. The **disassembly process** means that the carcass is divided into smaller cuts, vacuum sealed, boxed, moved into storage, and shipped to retailers. This process is also known as **fabrication and boxing. Processed meats** are made from scraps of meat that are not in suitable form for sale over the block. Such meats have the bones removed and are sold as boneless cuts. They can also be canned, made into sausage, dried, or smoked.

Sheep

Sheep are rendered insensible and bled. Next, the front feet are removed, the **pelt** or furlike covering is removed, and the hind feet and head are removed. The opening of the carcass and removal of the viscera, called **evisceration,** are similar to the procedures outlined for beef animals. In view of the small size of a lamb, the forelegs are folded at the knees and are held in place by a skewer after evisceration. Washing and cooling procedures are similar to those used for beef.

Hogs

Hogs are rendered insensible, shackled, hoisted, and bled. The carcasses are then plunged into water at 150°F for about four minutes. This process is required to loosen the hair and dry skin.

The hair of hogs is removed by mechanical scraping. A dehairing machine can remove the hair from about five hundred hogs per hour. After the hair is removed, the hog is returned to overhead racks and processing continues.

The hog is washed and singed prior to the removal of the head. **Singe** means to burn lightly to remove hair. Next, the carcass is opened and eviscerated, prior to being split or halved with a cleaver or electric saw. The leaf fat is removed. **Leaf fat** is layers of fat inside the body cavity. Before the carcass is washed, the kidneys and facing hams are inspected. After washing, the carcass is sent to coolers at 34°F.

Unlike the fat from beef cattle, lard is considered a product along with the meat. **Lard** is the fat from hogs. It is used for a variety of cooking and baking products. Lard is often combined with other animal fats and with vegetable oils, such as cottonseed, soybean, peanut, and coconut. Such mixes are used extensively for baking, cooking, frying, and other food preparations, and for commercial products.

FIGURE 33-19 Broilers nearing the end of the processing line. *(Courtesy of USDA)*

Poultry

The steps in poultry processing are similar in many ways to those required for other animals. The feathers that cover a bird, like the hair that covers the hog, must be removed prior to evisceration.

The process begins by securing the bird on a conveyor belt and bleeding out. Next, the bird is scalded prior to feather removal or picking. Singeing is required to remove the fine hairs that cover a bird under its feathers. After the feathers and hair have been removed, the bird is washed and eviscerated and the giblets cleaned. **Giblets** are the heart, liver, and gizzard of a bird. The bird is then cooled to 40°F, usually with ice (Figure 33-19).

Fish

After fish are caught or harvested, they too must be prepared for processing and consumption. The procedure depends upon the type of fish. Evisceration is usually done after the scales and head are removed. Washing and cooling follow.

Fish, like other foods, are consumed in a variety of forms. Whether a tuna is to be consumed whole or processed to be eaten later will determine whether the fish is left whole or is cut up. Some shellfish, like crab and lobster, are kept alive until they are cooked. The heat of cooking kills them. The meat can then be removed and consumed or processed.

Game

Game processing generally begins in the field. The game is typically killed by either gunshot or bow and arrow. The basic slaughtering procedures previously discussed are followed for game, too. Most game are processed by hunters rather than in processing plants. Large animals such as deer, moose, and elk would be slaughtered in a fashion similar to cattle. Fowl such as geese, quail, and duck would be dressed following procedures similar to those for commercial poultry.

Kosher Slaughter

Kosher means right and proper. This type of animal processing is based on the religious ritual of the Jewish faith. It requires that animals be killed by a rabbi or a specially trained representative. The methods, as well as the time by which the meat of the animal must be sold, are based on ritual that relates to concerns for sanitation. Meat must be consumed quickly. Neither packers nor retailers are permitted to hold kosher meat for more than 216 hours (9 days). Washing is required every seventy-two hours.

Major Cuts of Meat

After the meat is slaughtered, it is further prepared for use. Different areas of the animal are useful for different purposes. Cuts of beef, lamb, and pork, with wholesale and retail terms, are shown in Figures 33-20, 33-21, and 33-22 (see pages 644-646).

By-products—How Waste Products are Used

Although most of the animal is consumed by humans, all of the animal is not edible. The **dressing percentage** is a term used to indicate the percentage or yield of hot carcass weight to the weight of the animal on foot. The offal is removed from the live animal to arrive at the dressing percentage. The formula is: hot carcass weight divided by the live weight times 100. The offal may be 40 percent of the live weight of the animal.

What happens to the offal accounts for many products that are used daily (Figure 33-23, page 647). By-products can generally be divided into twelve categories.

1. Hides—Leather from animal hides is used to make a variety of consumer products, such as shoes, harnesses, saddles, belting, clothing, sports equipment, hats, and gloves (Figure 33-24, page 647).

2. Fats—These are used to make products such as oleomargarine, soaps, animal feeds, lubricants, leather dressing, candles, and fertilizers.

3. Variety meats—The heart, liver, brains, kidneys, tongue, cheek meat, tail, feet, **sweetbreads** (thymus and pancreatic glands), and **tripe** (pickled **rumen,** or stomach, of cattle and sheep) are sold over the counter as variety or fancy meats.

4. Hair—Brushes for artists are made from the fine hairs on the inside of the ears of cattle. Other hair from cattle and hogs is used for toothbrushes, paintbrushes, mattresses, upholstery for residential and commercial furniture, air filters, and baseball mitts.

5. Horns and hoofs—These items are used as a carving medium and are fashioned into decorative knife and umbrella handles, goblets, combs, and buttons.

6. Blood—The blood from animals is used in the refining of sugar, and also in making stock feeds and shoe polish.

7. Meat scraps and muscle tissue—After separation from the fat, meat scraps and muscle tissue are most often made into meat-meal or tankage. **Tankage** is the dried animal residue used as fertilizer and feed.

8. Bones—Some bones are put to the same uses mentioned for horns and hoofs. In addition, bones are converted into stock feed, fertilizers, and glue.

Beef Chart

Retail Cuts of Beef—Where They Come from and How to Cook Them

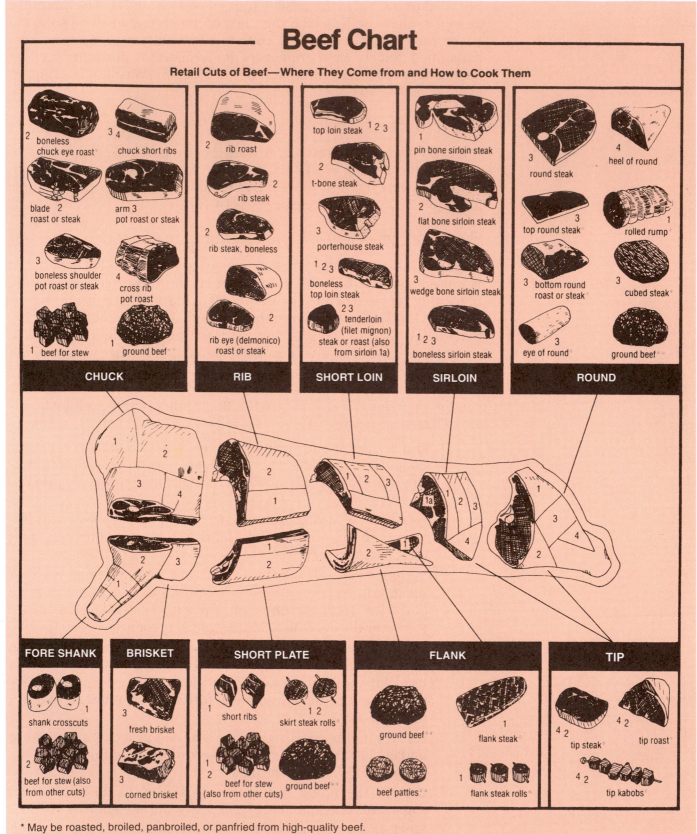

CHUCK
- 2 boneless chuck eye roast
- 3 4 chuck short ribs
- blade 2 roast or steak
- arm 3 pot roast or steak
- 3 boneless shoulder pot roast or steak
- 4 cross rib pot roast
- 1 beef for stew
- ground beef

RIB
- 2 rib roast
- 2 rib steak
- 2 rib steak, boneless
- 2 rib eye (delmonico) roast or steak

SHORT LOIN
- top loin steak 1 2 3
- 2 t-bone steak
- 3 porterhouse steak
- 1 2 3 boneless top loin steak
- 2 3 tenderloin (filet mignon) steak or roast (also from sirloin 1a)

SIRLOIN
- 1 pin bone sirloin steak
- 2 flat bone sirloin steak
- 3 wedge bone sirloin steak
- 1 2 3 boneless sirloin steak

ROUND
- 3 round steak
- 4 heel of round
- 3 top round steak
- 1 rolled rump
- 3 bottom round roast or steak
- 3 cubed steak
- eye of round
- ground beef

FORE SHANK
- 1 shank crosscuts
- 2 beef for stew (also from other cuts)

BRISKET
- 3 fresh brisket
- 3 corned brisket

SHORT PLATE
- 1 short ribs
- 1 2 skirt steak rolls
- 1 2 beef for stew (also from other cuts)
- ground beef

FLANK
- ground beef
- 1 flank steak
- beef patties
- 1 flank steak rolls

TIP
- 4 2 tip steak
- 4 2 tip roast
- 4 2 tip kabobs

* May be roasted, broiled, panbroiled, or panfried from high-quality beef.
** May be roasted, baked, broiled, panbroiled, or panfried.

FIGURE 33-20   Retail cuts of beef—where they come from and how to cook them.

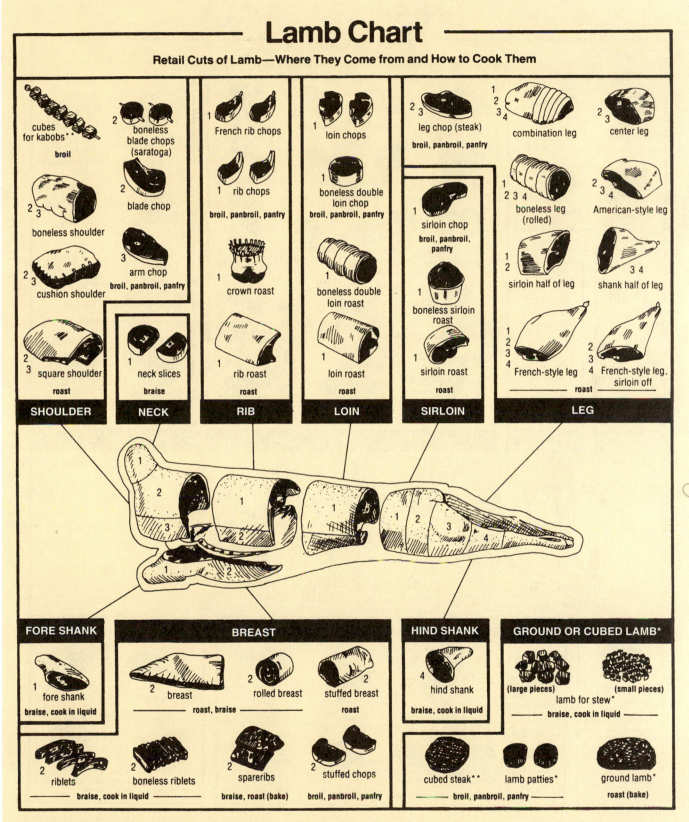

FIGURE 33-21 Retail cuts of lamb—where they come from and how to cook them.

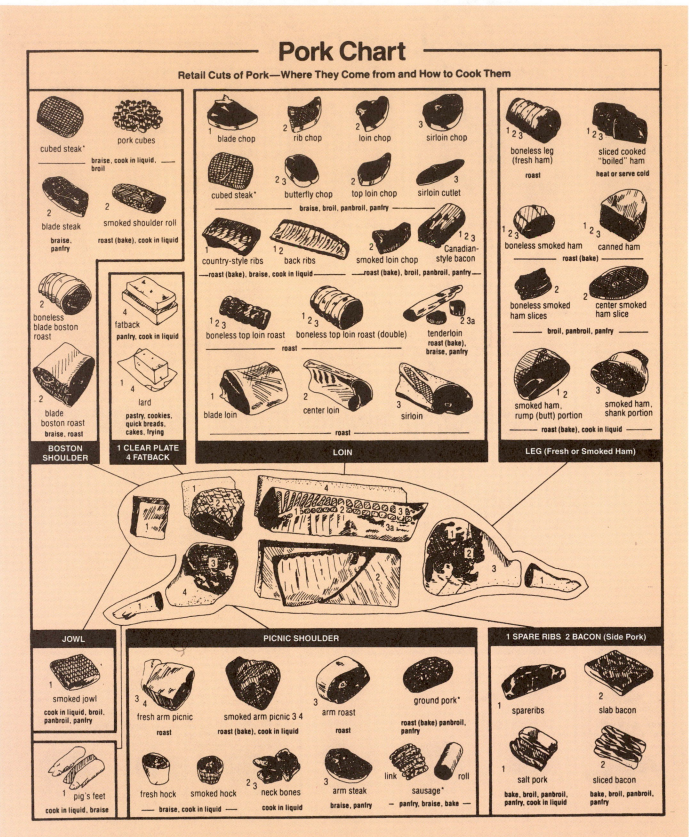

*May be made from Boston shoulder, picnic shoulder, loin, or leg.

FIGURE 33-22 Retail cuts of pork—where they come from and how to cook them.

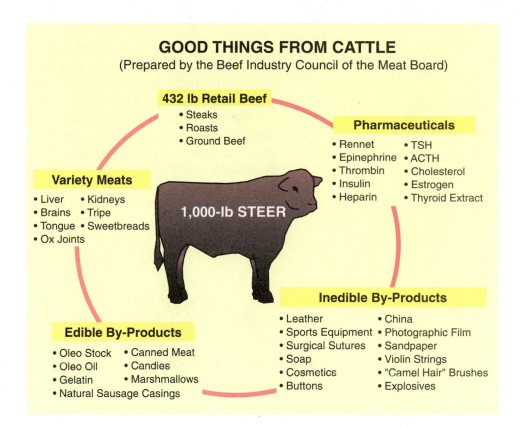

GOOD THINGS FROM CATTLE
(Prepared by the Beef Industry Council of the Meat Board)

432 lb Retail Beef
- Steaks
- Roasts
- Ground Beef

Pharmaceuticals
- Rennet
- TSH
- Epinephrine
- ACTH
- Thrombin
- Cholesterol
- Insulin
- Estrogen
- Heparin
- Thyroid Extract

Variety Meats
- Liver
- Kidneys
- Brains
- Tripe
- Tongue
- Sweetbreads
- Ox Joints

1,000-lb STEER

Inedible By-Products
- Leather
- China
- Sports Equipment
- Photographic Film
- Surgical Sutures
- Sandpaper
- Soap
- Violin Strings
- Cosmetics
- "Camel Hair" Brushes
- Buttons
- Explosives

Edible By-Products
- Oleo Stock
- Canned Meat
- Oleo Oil
- Candies
- Gelatin
- Marshmallows
- Natural Sausage Casings

FIGURE 33-23 There are many useful by-products from animals.

FIGURE 33-24 Leather products are all around us. *(Courtesy of Michael Dzaman)*

9. Intestines and bladders—Sausage, lard, cheese, and snuff all use the intestines and bladders from cattle. Strings for musical instruments and tennis rackets are also made from these by-products.

10. Glands—The pharmaceutical industry relies heavily on animal glands for many drugs that are used today.

11. Collagen—**Collagen** is the chief constituent of the connective tissues. Glue and gelatin are made from collagen. These products, in various forms, are used in the furniture, photography, medical, and baking industries.

12. Contents of the stomach—Stomach contents of slaughtered animals are used primarily in the production of feed and fertilizer.

NEW FOOD PRODUCTS ON THE HORIZON

The foods that we eat and how they reach us is an exciting area of agriscience. New foods, and new versions of familiar foods, are arriving on the market daily. Research is being conducted on improving the nutritional values of foods and keeping our food costs at a minimum.

The fruit industry uses an electronic, fruit-shaped beeswax sensor to log the bumps and bruises sustained by fruit during shipment. Improvements in handling equipment are resulting in better apples and are reducing loss for producers.

Eggs are used in the preparation of other foods, as well as being eaten by themselves. Recently, eggs have been taken out of many diets because of their high cholesterol content, which leads to heart problems. Researchers are investigating ways of reducing the cholesterol level in the egg without negating its nutritional value. Additionally, the poultry industry is looking at factors that make an animal fat and at ways to reduce the high-cholesterol component of poultry products.

Evaluation of different processing techniques with regard to cost, digestibility, nutrient retention, and flavor is constant. For example, nutrient loss in poultry has been found to be lower with freeze-drying than with dehydration.

The conveniences that everyone enjoys will continue to be improved, as there continue to be new developments in the food science field of agriscience.

STUDENT ACTIVITIES

1. Write the Terms to Know and their meanings in your notebook.

2. Keep a diary of what you eat for a week. Note which foods represent each of the five food groups.

3. With your teacher, arrange a visit to a slaughterhouse or meat department of a local grocery store to observe slaughter or meat processing.

4. Compare the end result when a food has been processed in a variety of ways—for example, fresh, canned, frozen, dehydrated, and freeze-dried potatoes.

5. Make a collage of items that exist because of animal by-products.

6. Using an outline of an animal, identify the major cuts of meat and where they come from on the animal.

SELF EVALUATION

A. Multiple Choice

1. Which of the following is *not* a carbohydrate?
   a. sugar
   b. starch
   c. fiber
   d. meat

2. The number of minerals needed by the body is approximately
   a. 2.
   b. 12.
   c. 20.
   d. 200.

3. The suggested number of daily portions from the meat, fish, poultry, eggs, and bean group is
   a. two, 2- to 3-oz servings.
   b. four, 2- to 3-oz servings.
   c. six, 2- to 3-oz servings.
   d. eight, 2- to 3-oz servings.

4. The butterfat content of table cream is approximately
  - a. 10 to 12 percent.
  - b. 18 to 20 percent.
  - c. 30 to 32 percent.
  - d. 38 to 40 percent.

5. To which milk product is sugar added during processing?
  - a. evaporated milk
  - b. condensed milk
  - c. skimmed milk
  - d. dried milk

6. Which mineral is added to most drinking water in the United States to assist in tooth development?
  - a. zinc
  - b. iron
  - c. fluorine
  - d. chlorine

7. Iron is important in the diet for the development of
  - a. bones.
  - b. vision.
  - c. blood.
  - d. hair.

8. Traditional aging of meat takes approximately
  - a. 1 to 6 hours.
  - b. 1 to 6 days.
  - c. 1 to 6 weeks.
  - d. 1 to 6 months.

B. Matching

_____ **1.** Canning	a.	Thymus and pancreatic glands
_____ **2.** Shrouding	b.	Internal organs
_____ **3.** Dehydration	c.	Added prior to packaging
_____ **4.** Viscera	d.	Stored in airtight containers
_____ **5.** Sweetbreads	e.	Reduced moisture content
_____ **6.** Food additive	f.	Wrap in cloth

C. Completion

1. Vitamins _____, _____, _____, and _____ are fat soluble.

2. Minerals that assist in fluid regulation are _____, _____, and _____.

3. Meat from animals that have been slaughtered following Jewish ritual is called _____.

4. The method of precooking food, removing moisture, and then freezing is known as _____.

Communications and Management in Agriscience

Putting it all together

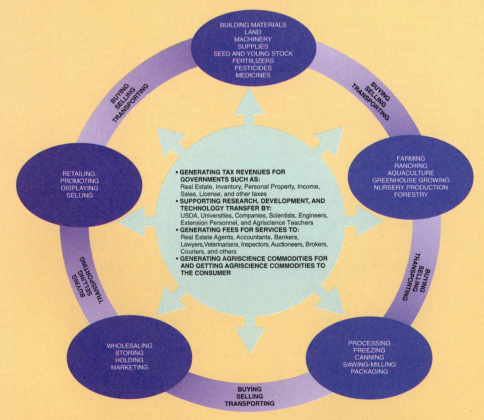

BUILDING MATERIALS
LAND
MACHINERY
SUPPLIES
SEED AND YOUNG STOCK
FERTIILIZERS
PESTICIDES
MEDICINES

BUYING
SELLING
TRANSPORTING

FARMING
RANCHING
AQUACULTURE
GREENHOUSE GROWING
NURSERY PRODUCTION
FORESTRY

BUYING
SELLING
TRANSPORTING

• **GENERATING TAX REVENUES FOR GOVERNMENTS SUCH AS:** Real Estate, Inventory, Personal Property, Income, Sales, License, and other taxes
• **SUPPORTING RESEARCH, DEVELOPMENT, AND TECHNOLOGY TRANSFER BY:** USDA, Universities, Companies, Scientists, Engineers, Extension Personnel, and Agriscience Teachers
• **GENERATING FEES FOR SERVICES TO:** Real Estate Agents, Accountants, Bankers, Lawyers, Veterinarians, Inspectors, Auctioneers, Brokers, Couriers, and others
• **GENERATING AGRISCIENCE COMMODITIES FOR AND GETTING AGRISCIENCE COMMODITIES TO THE CONSUMER**

RETAILING:
PROMOTING
DISPLAYING
SELLING

BUYING
SELLING
TRANSPORTING

BUYING
SELLING
TRANSPORTING

PROCESSING
FREEZING
CANNING
SAWING-MILLING
PACKAGING

WHOLESALING
STORING
HOLDING
MARKETING

BUYING
SELLING
TRANSPORTING

Agriscience has many facets, and communications and management transcends the entire industry.

L ooking at the broad picture, agriscience is the industry that feeds, clothes, shelters, and keeps the nation healthy. These benefits spill over to many other nations of the world. The United States shares the results and products of research in agriscience and donates or sells things such food, fiber, crops, livestock, plant oils, lumber, medicines, breeding stock, germ plasm, seed, agricultural chemicals, machines, and an endless array of agriscience materials for producing food and fiber, as well as the end products themselves. Further, our land grant system of colleges and universities has provided a system of research in agriscience that is the envy of the world. In many countries of the world, the greatest prize in education is to be admitted to and graduate from a land grant college or university in the United States.

Further, the agriscience industry in the United States provides the nation with its playgrounds, outdoor fitness centers, and a source of renewal for a nation boxed in by modern skyscrapers, large industrial facilities, and clogged highways. These take the form of open spaces for softball, baseball, soccer, hockey, horseback riding, golf, hiking, hunting, biking, walking, fishing, bird watching, recreational farming, and gardening. Further, a significant number make their livings as farmers, ranchers, greenhouse operators, aquaculturists, landscapers, nursery operators, zoo workers, and others in the outdoor segment of the food, fiber, and ornamental plant and exotic animal segments.

Agriscience employs about 20 percent of the nation's workers and provides the income for these workers to enjoy the fruits of our free-enterprise system. In contrast to most nations, where most of the income is consumed by the purchase of food, we spend only 11 percent of our income on food and have the remaining 79 percent for clothing, housing, transportation, recreation, health, automobiles, appliances, communications, education, leisure pursuits, and gadgets. No other nation has the abundance of opportunities that the United States offers under an agriscience industry that produces such affordable food and frees so much income for other needs and preferences.

The machinery for this abundance is predominantly planning and management, and the engine for the machinery is the profit incentive of the free-enterprise system. In a system where one can receive and use or accumulate the proceeds from work or innovation, the human spirit and initiative accelerates; the

results are demonstrated in greater productivity. The old adage that "where there's a will, there's a way" is proven correct every day as people apply their education, inventiveness, innovation, and energy in a work place where a substantial amount of the proceeds come back to the individual. Whether one is the owner or a stockholder in line for the profits of the organization; a salesperson or manager in line for bonuses or commissions; a piece-worker or assembly-line worker whose income is tied directly to output; or a per-hour worker with the option to seek a better paying position, the free-enterprise system rewards productivity.

This section addresses the application of communication and management skills in agriscience. These skills transcend agriscience and culminate in the marketplace where the fruits of labor are sold. The marketplace yields the cash to buy labor and other inputs for the business, and profits for the inventors, entrepreneurs, and other risk-takers in the system.

Marketing in Agriscience

OBJECTIVE

To determine the strategies and procedures for marketing agriscience commodities to maximize profits.

MATERIALS LIST

✓ various newspaper and marketing reports

COMPETENCIES TO BE DEVELOPED

After studying this unit, you should be able to:

- describe the marketing strategies that maximize profits.
- describe various pricing strategies.
- distinguish between wholesale and retail marketing.
- describe some methods of marketing at farms, roadside stands, and farmers' markets.
- discuss advantages and disadvantages of terminal markets, auctions, and direct marketing.
- recognize fees, commissions, and other costs of marketing.
- list the procedures for handling livestock to minimize losses during marketing.
- understand the grades of some popular agriscience commodities.
- recognize marketing trends and cycles.
- describe the use of futures in agriscience marketing.

TERMS TO KNOW

Supply	Yardage fee	Pigs
Demand	Commission	Hogs
Consumer demographics	Auction markets	Gilts
Product advertising	Auctioneer	Barrows
Institutional advertising	Direct sales	Lambs
Psychological pricing	Cooperatives	Rams
Penetration pricing	Vertical integration	Ewes
Skimming	Pencil shrink	Wethers
Loss-leader pricing	Veal	Commodity exchange
Prestige pricing	Feeder calves	Futures market
Retail marketing	Yearlings	Futures
Consumers	Slaughter cattle	Opening a position
Middlemen	Quality grades	Offsetting a position
Wholesale marketing	Finish	Out of the market
Terminal market	Yield grades	Profit

I n our free-enterprise system, goods and services may be produced, modified, or improved. The expectation is to sell or market the commodity or service as value is added along the way to the final user or consumer. Many people receive financial benefits in the form of salaries, profits, or "in kind" benefits. Governments receive tax revenues from the various transactions. This unit explores the basics of marketing agriscience commodities to maximize profits.

There are a number of factors that need to be considered when deciding on how to market agriscience products. Some of these factors include the following:

1. Demand for the product to be produced.
2. Supply of the product already available.
3. Types and availability of markets in the area.
4. Competition from similar products.
5. Buying power of intended consumers.
6. Seasonal variations in demand.
7. Government price supports available.

Supply and demand have long been the factors that determine whether or not the production of a product can be profitable. **Supply** is the amount of a product available at a specific time and price. Some of the things that may determine supply are how many people there are

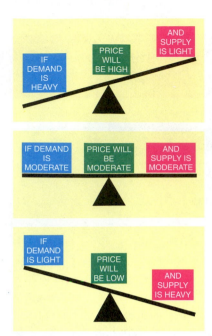

FIGURE 34-1 In a free-market system, the relationship between supply and demand will determine price.

in the business of producing that product in the market area, how much of the product is coming into the market area from other areas, and what is the past history of profitability for that product in your area.

Demand for a product is also determined by numerous factors. **Demand** is the amount of a product wanted at a specific time and price. It is often determined largely by price. The less expensive a product is, the more of it that is wanted. However, there are other factors that influence demand for a product. The amount of money available to consumers to buy that product is a factor often overlooked. Competition from similar products may reduce demand. Seasonal variations in demand also need to be considered. It is easier to sell ice cream in the summer than in the winter. Advertising, personal contacts, and product samples to create demand are also factors to be taken into consideration (Figure 34-1).

FOCUS ON CONSUMERS

Consumers are ultimately the determinant of marketing success. Commodities or services may be on the doorsteps of consumers, but if the consumer chooses not to buy and does not submit to persuasion or coercion, then there is no sale. In our democratic, capitalistic system, we have the right to produce and market goods and services for personal profit within the state and federal legal frameworks. Similarly, the right of refusal to buy goods and services is a basic right of consumers. The marketing system involves the interplay between producers and consumers, and all the people, processes, and jobs that come into play between them.

Consumer Demographics

One of the first marketing functions is to decide who will be the target population. Target population is the people who will be given deliberate exposure to the product. Some factors to consider when determining target populations are: who are the consumers, how many consumers are there, what are their preferences, and where are they located? Categories of information about consumers or potential consumers is known as **consumer demographics.** A knowledge of consumer demographics is helpful in determining target populations and marketing strategies (Figure 34-2).

Some useful categories of consumer demographics for the marketing of agriscience products are population density, ethnic makeup of the population, family income, discretionary money held by individuals within the family, family size, eating preferences, who makes buying decisions in the household, occupation and work locations, tendencies to eat at home or away, clothing preferences, styles, and recreational preferences. People generally satisfy their needs for basic food, clothing, and shelter first, then move on to purchases that reflect group and individual tastes. Because the purchase of food in the United States requires less than 12 percent of the average family income, there is considerable room for choice in the purchase of additional goods and services.

Important Questions to Ask When Determining Who the Consumer Is

- Who are the consumers?
- How large is this consumer population?
- What are their preferences?
- Where are they located?

FIGURE 34-2 Knowing the needs and preferences of prospective customers is essential to effective marketing.

ADVERTISING AND PROMOTION

Advertising is any form of nonpersonal presentation of a product or service. For maximum effectiveness, advertising must be coordinated with other marketing techniques, such as public relations, promotional programs, pricing, personal selling, product description and characteristics, sales promotions, and warranties. Effective advertising performs a variety of functions. These include creating an awareness of the product, motivating a customer to seek out the product, educating the potential customer, and reinforcing the value of purchases already made.

Product advertising focuses on the product itself. This might include the product's usefulness, durability, price, value, and the customer's need for the item (Figure 34-3). On the other hand, institutional advertising is designed to create a favorable image of the firm or institution offering the products or service. Such advertising is likely to claim financial stability, longevity, reliability, and integrity of the company or institution and its employees.

Advertising efforts that seek to find new users for a given product, such as motor boats, are trying to create primary demand. Similarly, advertising intended to attract buyers to a specific brand of the product is attempting to create selective demand. The production of any one agriscience commodity from farms, ranches, ponds, nurseries, or greenhouses is expensive and difficult because there are so many small-volume producers. This is in contrast to other industries, such as automobiles, where there are relatively few manufacturers, and nationwide advertising is practical and affordable.

To address the problem of many small-volume producers of individual agriscience commodities, the United States government has instituted a system that generates a pool of money from producers of a given commodity such as wheat, rice, cotton, tobacco, peanuts, beef, pork, or milk. Here the seller contributes a small amount for each unit sold to a money pool to be used to purchase advertising for the specific commodity. The program is referred to as the *check-off program* for the

FIGURE 34-3 Product advertising focuses on some aspect of the product itself.

given commodity. It is estimated that about 90 percent of all United States farmers contribute to some three hundred federal and state generic promotion programs covering approximately eighty commodities. The spending for research and promotion by these programs is reportedly near the half-billion-dollar mark annually. Media advertising for agriscience goods and services typically includes the use of newspapers, radio, television, outdoor billboards, signs, pamphlets, transportation advertising, direct mail, telephone book classified section, breed journals, product catalogs, and magazines.

The promotion of commodities may be done in numerous ways. Product displays are common in stores and other retail outlets. Similarly, displays and sales of machinery, animals, food, flowers, ornaments, landscape designs, and other commodities and services are common at fairs, shows, open houses, and professional meetings. Free samples may even be offered in stores, at sidewalk promotionals, fairs, shows, public auctions, and wherever prospective customers may gather.

COMMODITY PRICING

Final determination of prices for agriscience commodities is not always under the control of the producer or representative of the producer. This is due to a large extent to the perishable nature of non-processed foods; the relationship between peak quality and price of fruits and vegetables; and the lack of control over production of crops, eggs, milk, and meat after the production cycle is started. Therefore, the producer may not have much control over the supply that is being collectively generated and may have to accept lower prices or loss of the product due to spoilage.

However, the growth of producer-owned marketing cooperatives has enabled producers to negotiate prices on a large-volume basis. Another benefit of cooperatives is that of owning processing plants so fresh commodities can be processed and withheld if necessary until better prices are offered. Cooperatives, privately owned businesses, and individuals all utilize a variety of strategies when setting prices. These strategies are applied according to the benefits the price setter sees in effecting more sales and greater profits (Figure 34-4).

Psychological pricing is a strategy designed to make the price seem lower or less significant than it is. For instance, the pricing of an item at $1.99 seems to attract many buyers that would reject the same item priced at $2.00. Needless to say, the profit from selling more items will easily offset the loss of one cent per item. Another technique is the offering of multiple items for a given price, such as three for twenty-three cents. Here the customer is lured into buying three to save one cent, when one item may be all the customer needs. Discount pricing and bulk selling are also used as techniques of psychological pricing. Sometimes customers are lured into buying more than they would normally buy because of the discount attraction, and then consume more than they would otherwise consume. Some consumers believe that larger packages are automatically priced at less cost per unit. Occasionally, sellers exploit this perception by pricing commodities in larger containers at higher costs per unit than in the standard package (Figure 34-5).

FIGURE 34-4 Modern marketing cooperatives have given small producers much more power to influence the prices they receive for their commodities. (Courtesy of Michael Dzaman)

FIGURE 34-5 Modern computers and scanning devices enable retailers to change prices quickly in accordance with promotional activities or changing market conditions. *(Courtesy of USDA/ARS #K-2656-2)*

Penetration pricing is a strategy wherein price is set below that of competitors to entice customers to try an item. While profits will be low temporarily, they can be regained if new customers continue to buy and the price is gradually raised to new levels. **Skimming** is the opposite of the penetration pricing strategy. Here, the price of a new product is set for unusually high profits at first, when affluent and willing customers are available. This enables the producer to recoup development expenses more quickly and gradually lower the price as sales volume increases or other competing products enter the market. Skimming is also known as "sliding down the demand curve."

Loss-leader pricing is a procedure wherein a popular commodity is offered for sale in a special sales promotion at prices below the cost level—in other words, sold at a loss. The purpose is to attract customers into the business establishment with the hopes that they will buy other items priced at profitable levels. **Prestige pricing** is a procedure used to target buyers with special desires for quality, fashion, or image. The customer's notion or perception that a certain brand is better may be further reinforced by a higher price tag. Prestige pricing seems to work well in the garment and cosmetic industries, and may be very profitable.

In general, in a nonregulated free-enterprise system, there is a tendency to price commodities according to what the market will bear, rather than what it costs to produce the items. This pricing strategy is generally regarded as fair, because our system rewards risk-taking with the potential for profit. If one must be prepared to shoulder losses when offering a commodity that does not reach sales expectations, then it only seems reasonable to permit that individual or company to benefit from the occasional high-profit opportunities.

MARKETING STRATEGIES FOR MAXIMIZING PROFITS

In order to make the most money from a production enterprise, successful marketing is a must. Strategies that can be used to market agriscience products most profitably include:

1. Determine what types of markets are available to you.
2. Determine the costs of various types of marketing.

FIGURE 34-6 Using well-thoughtout marketing strategies pays good dividends.

FIGURE 34-7 Retailing by the producer adds extra dimensions to the business. *(Courtesy of USDA)*

3. Determine transportation costs to market at each of the markets available to you, and sell where transportation costs are favorable.
4. Determine the most profitable form in which to market your product (age, size, weight, degree of preparation).
5. Advertise to create markets where none existed before.
6. In seasonal markets, market your product at the peak of demand (Figure 34-6).

RETAIL MARKETING

Retail marketing is selling a product directly to **consumers** (people who use a product). This may take place on the farm, at roadside markets, or at the farmers' markets that have developed around many centers of population (Figure 34-7).

Retail marketing is generally used for ready-to-eat products. For example, ham, sausage, and fresh fruits and vegetables, rather than live pigs, are usually sold at a retail outlet. This creates many other problems that need to be dealt with and that have an effect on the profitability of a production enterprise. When an animal product is processed for retail sales, various standards and state and federal regulations must be met. Also, changes in facilities are often necessary. Sanitary storage of processed products and the increased labor costs associated with retail customers must also be considered.

Retailing at the Farm

On-the-farm retail sales present special problems. If retail marketing is to take place on the farm where the product is produced, the farm must be well maintained and attractive to the customers buying the product. Facilities for parking are necessary. Animal and sales facilities must be clean, and the animals present must be well cared for and contented. An educational program to make consumers aware of good management techniques is often necessary so that they understand what is happening on the farm. Salespersons or the farm family itself must be available to wait on customers. Privacy is often difficult to achieve, because retail customers often feel that because products are sold at the farm, it is all right to drop by at any time.

AGRI·PROFILE

CAREER AREAS: Auctioneer/Coop Manager/Dealer/Grower/Packer/ Grader/Meat Cutter/Produce Manager/Communications Specialist

The marketing of agriscience products occurs in all segments of our society on a continuing basis. The year-round continuance of bulging displays of produce, dairy, meat, seafood, bread, frozen foods, canned goods, deli foods, and ready-to-eat foods in supermarkets is a testimony to the agriscience industry that precedes this "tip of the iceberg." The producers and breeders of vegetable and fruit crops, grains, flowers, ornamental shrubs, turfgrass, trees, pets, laboratory animals, livestock, horses, dairy animals, poultry, fish, wildlife, milk, eggs, wool, and fur, all market the products of their businesses. This is only the beginning in the process of agriscience marketing. From the farm, ranch, pond, fish tank, feedlot, orchard, nursery, greenhouse, turf farm, or other production unit, there exists a whole system of professional and scientific workers.

Marketing cooperatives, dealers, auction markets, auctioneers, livestock handlers, jobbers, brokers, clerks, truckers, railroad personnel, commodity market managers, futures traders, writers, and broadcasters all get in the marketing act. Government, as well as company inspectors, monitor the products as they make the numerous transitions from production site to finishing site, pet shop, by-product, or dinner table. Teachers and extension workers provide instructional and professional development programs for personnel.

Marketing careers may be launched through programs in technical schools, colleges, or universities in the plant and animal sciences, food science, agricultural economics, marketing, or communications. Additionally, studies in the biological sciences are appropriate for laboratory work associated with agriscience products and marketing.

New products from agriscience are rapidly solving environmental problems. Chemist Sevim Erhan evaluates inks from soybeans that are biodegradable and quickly replacing inks of the past. *(Courtesy of USDA/ARS #K-3998-5)*

On the positive side, profits from farm sales are often higher because retail prices are higher. Also, middlemen are eliminated in the marketing procedure. **Middlemen** are people who handle an agricultural product between the farm and the consumer. Examples of middlemen include buyers, processors, and salespeople. Higher prices can often be charged when a superior product is produced. For farm families who like to meet new people, on-the-farm sales is an excellent opportunity to do so. Also, these types of sales provide the opportunity for urban and suburban people to see the agriscience way of life firsthand.

Roadside Markets

Roadside marketing retains nearly all of the advantages of on-the-farm retail sales while eliminating some of the disadvantages. The sales unit is usually somewhat removed from the actual farm operation when roadside marketing is employed. By removing the customers from the production area, more efficient use of labor can take place. Less perfect care and maintenance of facilities can be tolerated.

On the less positive side, separate facilities for the retail sales unit must be maintained. More labor may be required because it may be inconvenient for the farmer to staff the roadside stand. Furthermore, the positive effects of the consumer seeing the actual farm in operation are diminished.

Farmers' Markets

Farmers' markets have appeared in many large metropolitan areas to cater to the demands of urban consumers. They are normally operated one or two days per week and give urban and suburban consumers access to fresh agriscience products directly from the producers. Farmers' markets give the producer access to markets that would seldom be available otherwise. The producer also has the opportunity to educate the consumer concerning the value of good agriscience products (Figure 34-8).

Of course, there are several costs and inconveniences associated with marketing agriscience products at farmers' markets. There are often fees to be paid for the privilege of participating in farmers' markets. Competition may be higher, especially if several producers are selling the same type of product. Vehicles with heating and/or cooling may be necessary to get products to the market as fresh as possible. Products may be subject to certain food regulations and packaging requirements to meet state and federal standards. Facilities to hold and display products must also be available at the market.

In summary, the decision whether or not to market products of agriscience by retail means requires much consideration. Although the returns are generally higher when products are marketed retail, the costs are also higher. Consideration must be given to the product being produced, availability of markets, labor availability, and personal desires before a decision concerning marketing method can be made.

FIGURE 34-8 Farmers' markets provide opportunities for producers to sell at retail without opening their production facilities to the general public. *(Courtesy of FFA)*

WHOLESALE MARKETING

Wholesale marketing is the marketing of a product through a middleman, who eventually forwards the product to the consumer. Most agriscience products produced in the United States are marketed in this way. Wholesale markets allow for marketing large volumes of products with comparatively little labor. Types of wholesale markets include terminal markets, auction markets, and direct sales.

Terminal Markets

A **terminal market** is usually a stockyard that acts as a place to hold animals until they are sold to another party. The terminal market never actually owns the animals. The animals that are delivered to the terminal market are consigned to a selling agent, who does the actual selling of the animals. The terminal market charges the seller a fee for caring for the animals until they are sold. This is called a **yardage fee.** The selling agent also receives a fee, called a **commission,** for work in selling the animals.

The use of terminal markets to market animals has always been mostly confined to the midwestern and western states. In recent years, the use of terminal markets to market animals has greatly decreased, and most livestock are currently being marketed by other methods.

Auction Markets

At **auction markets,** animals are sold by public bidding on individual lots of animals. An **auctioneer** generally conducts the sales at an auction market. These markets are very widespread and are convenient to most local communities. They have grown in popularity and now represent the most common means of marketing animals of importance to agriscience. Auction markets are usually most practical for the smaller livestock producer.

As is true when selling in terminal markets, commissions are charged for selling animals at auction markets. The amount of commission charged varies with the type and size of the animal. Because many auction markets are small, there may not be the competition that is present at terminal markets to buy smaller numbers of animals (Figure 34-9).

FIGURE 34-9 Auction markets are popular for selling livestock in many communities. *(Courtesy of Bill Angell)*

Direct Sales

Direct sales refers to the producer selling crops and animals directly to processors. This method of marketing has several distinct advantages. There are no commission fees to be paid to selling agents, and there are no yardage fees. Transportation costs are kept to a minimum because the buyer generally comes to the farm to make purchases. The animals look their best for the buyer because they have not been exposed to the stresses of hauling and contact with strange facilities and animals.

There are a few negative aspects to direct marketing of commodities. A producer must have fairly large numbers of animals to be marketed at one time before the purchaser comes to the farm or ranch to buy. The producer is generally at the mercy of the buyer concerning the price received for the animals sold. Further, the producer often is not paid for the animals until after they have been processed.

Direct sales of animals, especially beef cattle, have increased significantly in recent years.

Other Wholesale Marketing Techniques

Approximately 75 percent of the milk marketed in the United States is sold through farmer milk-marketing cooperatives. **Cooperatives** are groups of producers who join together to market a commodity. The cooperatives then either process the milk and sell it directly to consumers or sell it to other large processing plants. Marketing cooperatives have the ability to maintain product quality, arrange for transportation of products from farm to market, balance supply and demand of agriscience products, and plan advertising to increase sales of agriscience products.

In the poultry market, almost 99 percent of the chickens produced for meat are grown under a system called **vertical integration** (several steps in the production, marketing, and processing of animals are joined together). The use of vertical integration in the production and marketing of agriscience products allows for extremely large systems of production that can be very efficient. Only the anticipated number of animals that will be demanded by consumers are produced. There is less competition from other producers, and all phases of production can be controlled (Figure 34-10).

There are a number of methods of marketing commodities of importance to agriscience. The one or more methods chosen by an individual producer are often a matter of what is available and what the producer prefers. Care should be taken to carefully choose the means of marketing. Intelligent marketing practices may well make the difference between profit and loss in a very competitive business.

FIGURE 34-10 Vertical integration in the broiler industry takes the dressed bird directly to the retailer and eliminates the need for live-bird markets. *(Courtesy of USDA)*

MARKETING FEES AND COMMISSIONS

Because there are numerous methods of marketing animals, the fees and commissions vary fairly widely. Livestock are usually sold by the head, and a fee is charged for each animal sold. This fee varies from $1.50 to $3.00 per animal. If animals are kept at a terminal market, yardage fees are also charged for the feed and care of the animals. Charges for insurance may also be deducted from the seller's check.

BIO·TECH CONNECTION

Rainbow Cottons, High-tech Leather, and Biodegradable Detergents

Agriscience research is turning out new colors of no-wrinkle fabrics made with cotton; leather tanned by less-polluting methods for new products; and a detergent made from animal fats that does not pollute our waterways. Cotton has always been a preferred fabric; but its tendency to wrinkle has limited its use. In the past, cotton fabric was dyed before a no-wrinkle finish was applied, because the chemical bond created before the refinishing process would repel dye if applied after the process. The cotton fabric must swell to accommodate the molecules of dye; but this cannot normally occur after the fabric has been heated and treated with a no-wrinkle finish. The fabric industry needs to be able to dye fabric just before clothing manufacture in order to keep up with the rapid changes in clothing fashions. After all, who wants to get stuck with a warehouse full of fabric dyed in the hues and colors that are out of fashion?

To solve the problem, scientists at the ARS Textile Finishing Chemistry Research Unit in New Orleans have developed techniques that allow industry to apply a no-wrinkle finish to cotton fabric before dying. By adding a variety of quaternary ammonium salts to the no-wrinkle treatment solution and a positive charge to the fabric, they have developed a procedure that produces a no-wrinkle fabric that will accept dye. The procedure also broadens the choice of dyes that can be used and gives industry more options for supplies and a wider range of shades with deeper, more vibrant colors.

On another front, scientists at the ARS Eastern Regional Research Center in Phil-

Frank Scholnick and James Chen of the USDA examine the results of experimental treatments for leather. *(Courtesy of USDA/ARS #K-4388-1)*

adelphia are looking for more environmentally friendly ways to decrease bacterial decay in hides and thus preserve leather. One of the most promising techniques is the use of electron beam irradiation. The technique promises to replace the salt-and-brine method of curing and extend the qualities of strength and elasticity for some new leather products.

On yet another front, scientists have long sought laundry detergents that will not pollute ground water when they are discharged with waste water. Tallow is beef fat and a by-product of the beef industry. Agriscientists have developed a tallow-laced soap that is very environmentally friendly. The soap contains no phosphates; will not harm humans, domestic animals, or wildlife; and will usually biodegrade in twenty-four hours. The product is as effective as phosphate detergents and is economical, too.

For animals that are purchased on the farm for slaughter, a fee called *pencil shrink* is sometimes deducted from the selling price of the animals. **Pencil shrink** is the estimated amount of weight that an animal will lose when it is transported to market. This usually varies from 1 to 5 percent.

Fees may also be charged for the sale of commodities based on a percentage of the gross amount received for the product.

MARKETING PROCEDURES

How produce is packaged and how animals are prepared for and transported to markets can often mean the difference between profit and loss. Some procedures to follow when shipping animals are:

1. Always handle live animals quietly and carefully. Animals that are calm and quiet lose less weight during transportation.
2. Move animals when temperatures are moderate, if possible. Usually this means at night during the summer.
3. Do not overcrowd animals on trailers or in lots while they are waiting to go to market.
4. Do not overfeed animals just before hauling them to market. Animals do not travel well on full stomachs. Do provide ample fresh, clean water.
5. Avoid injuring or bruising animals when loading and unloading them. Dead or injured animals are worth very little at the market.
6. Sort animals according to sex and size before shipping them to market.
7. Precondition animals for several days before marketing them.

GRADES AND MARKET CLASSES OF ANIMALS
Cattle

Beef animals are usually classified as either calves or cattle. Calves are less than one year old, whereas cattle are more than one year of age. Calves are further divided into veal calves, feeder calves, and slaughter calves. **Veal** calves are less than three months old and are slaughtered for meat that is also called *veal*. They usually weigh less than 200 lb.

Calves that are between three months and one year old and are marketed for meat are called *slaughter calves*. They have usually been fed at least some grain. **Feeder calves** are six months to one year of age and are sold to people who feed them to market weight as slaughter cattle. The sex classes for feeder and slaughter calves are steers, heifers, and bulls. Steers are castrated males, heifers are young females, and bulls are unaltered males.

Cattle are divided into feeder cattle and slaughter cattle. Feeder cattle are further categorized into age groups of yearlings and two-year-olds and over. **Yearlings** are between one and two years of age, and two-year-olds are two or more years old. These two classes of cattle can also be divided into five sex classes—steer, heifer, bull, cow, and stag. A cow is a female that has had a calf, and a stag is a male that was castrated after reaching sexual maturity.

Slaughter cattle are marketed for the purpose of being slaughtered for meat. They are divided into the same age and sex classes as are feeder cattle. They are also divided into quality and yield grades. **Quality grades** refer to the amount and distribution of **finish** (fat) on the animal. The quality grades for cattle are prime, choice, select, standard, commercial, utility, canner, and cutter (Figure 34-11). **Yield grades** are based on the amount of lean meat an animal will yield in relation to fat and bone (Figure 34-12). The yield grades are 1 through 5, with yield grade 1 producing the largest amount of lean meat.

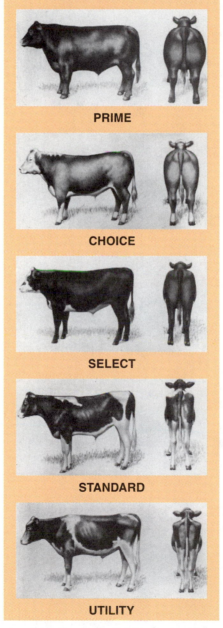

FIGURE 34-11 Quality grades of slaughter cattle are based on the amount and distribution of finish. *(From Gillespie/Modern Livestock and Poultry Production, 5E, copyright 1997 by Delmar Publishers)*

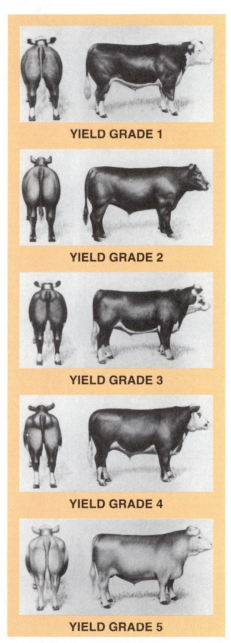

FIGURE 34-12 Yield grades of slaughter cattle are based on the yield of lean meat in relation to the amount of fat and bone. *(From Gillespie/Modern Livestock and Poultry Production, 5E, copyright 1997 by Delmar Publishers)*

Swine

The use classes of swine are feeder pigs, stocker hogs, slaughter pigs, and slaughter hogs. In general, **pigs** are swine less than four months of age, whereas **hogs** are swine more than four months old. Feeder pigs and stocker hogs are sold to be fed to higher weights before being slaughtered. Slaughter pigs and slaughter hogs are sold for immediate slaughter.

The sex classes of swine are gilt, barrow, boar, sow, and stag. **Gilts** are young females. **Barrows** are castrated males. Boars are unaltered males, and sows are mature females.

FIGURE 34-13 Quality grades of feeder pigs. *(From Gillespie/Modern Livestock and Poultry Production, 5E, copyright 1997 by Delmar Publishers)*

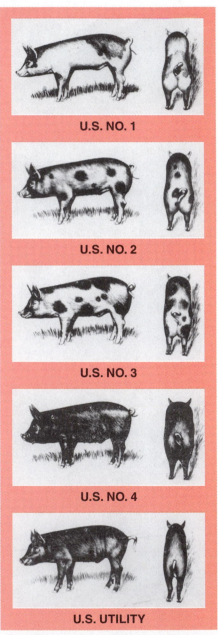

FIGURE 34-14 Quality grades of slaughter hogs. *(From Gillespie/Modern Livestock and Poultry Production, 5E, copyright 1997 by Delmar Publishers)*

Swine are also graded according to quality, with the official USDA grades being U.S. No. 1 through U.S. No. 4. Animals with lean meat of an unacceptable quality are graded U.S. Utility. The highest grade of swine is U.S. No. 1 (Figures 34-13 and 34-14).

Sheep

Sheep that are marketed are classified according to age, use, sex, and weight. The age classes are lambs, yearlings, and sheep. **Lambs** are young sheep and can be divided into hothouse lambs, spring lambs, and lambs. Hothouse lambs are marketed under three months of age, usually for the Christmas or Easter holidays. Spring lambs are three to seven months of age, and lambs are seven to twelve months old. Lambs can also be classified as feeder lambs or slaughter lambs, depending on whether they are to be fed to heavier weights or slaughtered immediately.

Yearlings are between one and two years of age, and sheep are more than two years of age. Sheep are divided into three sex classes—ram, ewe, and wether. **Rams** are unaltered males, **ewes** are females, and **wethers** are castrated males.

Sheep are also graded according to yield and quality. The yield grades are 1 through 5, with 1 being the most desirable. Quality grades are prime, choice, good, utility, and cull (Figure 34-15).

A review of useful terms regarding cattle, swine, and sheep is provided in Figure 34-16.

MARKETING TRENDS AND CYCLES

Agriscience commodity prices previously tended to rise and fall on a fairly well-defined cycle. The cause of general price cycling is supply and demand or government intervention.

Trends and Cycles in Animal Markets

When prices are high, producers increase production of animals and animal products. At first, this action pushes the prices even higher, because females that normally would have been marketed are retained in the breeding herd. When the results of the increased production reach the market and supply exceeds demand, the prices begin to fall. When prices fall to the point where production is not profitable, livestock producers reduce production by selling some of the breeding herd. Of course, this pushes prices even lower. With decreased supplies, the demand increases, and the cycle begins again.

FIGURE 34-15 Quality grades of slaughter lambs.

FIGURE 34-16 Terms given to animals of various sexes and maturities.

	Cattle	Swine/Hogs	Sheep
Young, immature	Calf	Pig	Lamb
Young, but maturing	Feeder or yearling	Feeder or stocker	Yearling
Castrated male	Steer or stag	Barrow	Wether
Mature breeding male	Bull	Boar	Ram
Mature breeding female	Cow	Gilt or sow	Ewe
Meat of the young	Veal	Pig or pork	Lamb
Meat of the mature	Beef	Pork	Mutton

Today, there are many other factors that influence prices. As a result, the traditional livestock price cycles have less variation than in the past. Some factors that currently influence livestock prices are:

1. Importation and exportation of livestock products.
2. Development of new uses for livestock products.
3. Increased advertising of livestock products.
4. Government support prices.
5. World weather conditions.
6. General economic conditions in the world.
7. Changes in consumer demands. In recent years, demand for beef and pork have eased up and down, while demand for poultry has been increasing dramatically. Egg consumption is down, and milk consumption is increasing slightly.

Trends and Cycles in Crop Marketing

The availability of a given crop is stimulated by price, which is influenced by demand. Many grain crops have so many uses that the market for the items is very complex. For instance, corn is the number one choice for most farm-animal and poultry feeds; is frequently used in pet foods; is a favorite breakfast cereal for humans; is the source of corn oil for frying, cooking and manufacturing plastics; is a primary source of alcohol used in motor fuels; and provides ingredients for many baked goods.

However, in many of these commodities, corn can be replaced by substitutes. For instance, barley and other grains will be substituted if the price of corn rises above a certain level. Similarly, when the price of crude oil drops below a certain level, the price of corn may no longer be competitive. Gasoline refiners will then use more petroleum and less grain alcohol in the fuel. Further, weather and other production factors will influence the size of the world grain supply and influence prices here and abroad. Trade policies greatly influence prices of a commodity, because the flow of commodities into the world market or the policy of permitting world supplies to flow into the United States greatly complicates the formulas that determine prices.

A classic example of how United States government policy can affect prices, supplies, and farm conditions occurred in the 1970s. The United States government struck a deal with the former Soviet Union to export greatly increased volumes of United States grain to the former Soviet Union. This new market created a rise in grain prices. United States farmers responded by planting more grain to obtain the new potential income. Anticipating the new grain market and the higher grain prices to continue in the long run, many farmers purchased or leased more land and purchased new and additional equipment and buildings over the next five years. This resulted in additional farm debt, drove up land prices, and helped fuel the increase in interest rates.

About five years after the grain export deal was implemented with the former Soviet Union and during a different presidential administration, the United States implemented an export ban on grain to the Soviet Union in protest of Soviet military activities. This large grain market evaporated quickly, and a damaging domino effect rippled through the nation's agricultural economy. Grain surpluses mounted, grain prices dropped; grain farmers' incomes dropped; farmers scrambled

FIGURE 34-17 Large trucks and modern highways move agriscience commodities quickly from farm to processor or market. *(Courtesy of FFA)*

to change farm enterprises from grain to something else more profitable; some farmers had to default on their loans; some farms were sold because of default on mortgage payments; and banks and agribusinesses came under financial strain.

Many factors make agriscience marketing very unpredictable. The use of rapid transportation on a modern system of interstate highways (Figure 34-17); air shipping of highly perishable foods; refrigerated shipping; food processing procedures that lock in freshness; extensive, specialized, high-technology storage facilities; and an excellent marketing system all contribute to the availability of food and fiber commodities to most any part of the nation on a year-round basis. These factors help to maintain markets and prices that are fairly stable and predictable.

GLOBAL MARKETING IN AGRISCIENCE

Historically, agriscience producers have focused on producing commodities and managing their businesses in ways that held the best promise for acceptable profits. While producers always dream of favorable prices and fair profits, these do not always materialize. Telephones, radios, televisions, fax machines, computers, and other modern communication devices have permitted many changes in agriscience marketing the last decade or two. There have always been many middlemen in agriscience processing and marketing. The number and types are increasing (Figure 34-18).

The Agricultural Commodity Futures Market

FIGURE 34-18 Telecommunications and computers have made agriscience marketing a global enterprise. *(From Schneeman/Paralegals in American Law, copyright 1995 by Delmar Publishers)*

Relatively new on the scene is the use of futures contracts in the agriculture commodity marketing complex. The Chicago Board of Trade, Chicago Mercantile Exchange, Winnipeg Commodity Exchange, and London International Financial Futures Exchange are all commodity exchanges. A **commodity exchange** is an organization licensed to manage the process of buying and selling commodities under specific laws using a system of licensed brokers. Commodity exchanges also manage futures markets. A **futures market** is a procedure conducted by commodity exchanges to provide networks and legal frameworks for sellers and buyers to work through brokers in making contracts called *futures contracts,* or simply *futures.*

Buying and Selling Futures Futures are defined as legally binding agreements, made on the trading floor of a futures exchange, to buy or sell something in the future. Futures prices are determined by competitive bidding of brokers for prospective buyers and sellers from all over the world. Anyone with appropriate assets can work through a broker and buy futures or sell futures on the commodity market. The futures market floats up and down by the moment while the futures exchange is "open," or in session. Telephone and computer networks around the world permit agriscience managers, buyers, sellers, and brokers to communicate almost instantly to place their bids to buy or sell; and the commodity exchange handles the legal work to complete the transactions. Many agriscience commodities, as well as other commodities, are bought and sold in this manner (Figure 34-19).

Farmers, ranchers, cattle feeders, crop growers, and other agriscience owners and managers use the futures market as a tool to stabilize their businesses. The wise buying and selling of futures can protect them from losses due to excessive price fluctuations. For instance, before deciding how much corn to grow, how much to spend on fertilizer and other inputs, or how much corn to keep for feeding out cattle or hogs on the farm, the farmer should know the price per bushel for which the corn could be sold after it is grown. The other decisions would then be easier to make.

By looking in the commodity-price sections of major business newspapers, such as *The Wall Street Journal,* the seller can determine the futures price for corn delivered in a given month. If the farmer regards that price as high enough to commit a certain number of bushels for sale, then he will inform his broker to "sell" that many bushels at the rate of the futures price for the month selected. The broker will negotiate a contract that obligates the farmer to deliver that many bushels of corn on the specified date for the amount of the futures price. Now the farmer is guaranteed the price for the corn, and other decisions about the farming operation can be based on this selling price. If the price of corn is lower in December than the $1.80 futures contract that was made, the payment upon delivery will still be $1.80

FIGURE 34-19 Commodity exchanges make possible the use of futures trading as valuable marketing and management tools. *(Courtesy of Chicago Board of Trade)*

per bushel; and the farmer will have successfully protected the business against a price drop. However, if the price of corn is higher than $1.80 per bushel in December, the farmer might wonder if the purchase of a futures contract was a wise decision.

Here's how a futures contract might be made for 5,000 bushels of corn. Suppose the settlement price on the futures market for December is $1.80 per bushel. The value of the contract for 5,000 bushels is $1.80 \times 5,000$, or $9,000. The farmer is obligated to deliver the corn for $9,000 regardless of the market price of corn at the expiration of the contract.

Opening and Offsetting a Position The purchase of a futures contract does not have to be a final transaction that plays through to the actual buying or selling of the commodity. In fact, in most cases, the goods are not delivered because the participant makes another contract that offsets the first one. The initial step in the futures market is called **opening a position.** At a later date, before the futures contract expires, the participant can take a second position to offset the first. Such action is known as **offsetting a position.** When the position is offset, the participant is said to be **out of the market** and has no obligation to deliver nor take delivery. To offset the position described previously, the farmer would need to buy 5,000 bushels of corn. This would offset the contract to sell 5,000 bushels. If the farmer were fortunate enough to find a day when the futures price was lower than $1.80, and buy futures for 5,000 bushels for the specified date of delivery at that lower price, then the farmer could pocket the difference as income, or **profit,** from the business transaction.

EXPORT MARKETING

As transportation and communications improve, the world seems to get smaller. More people are familiar with more places and people around the world, so more business is being conducted between and among people and nations of different continents. Technology and computer languages are spreading rapidly, so different cultures are able to communicate in business, commerce, engineering, and other fields. Further, fiber optics; lasers; satellites; computers; nationwide and worldwide telephone, television, and computer networks; international airlines and air routes; trade agreements; and improved education all contribute to the increase in international investment, business, trade, and commerce.

The United States now exports an estimated 40 percent of its agriscience products. The livestock, meats, hides, dairy products, grain, lumber, fruits, vegetables, ornamentals, and processed plant and animal products that we export help to offset our trade deficit. This trade activity in agriscience has great implications for marketing agriscience products. In recent years there has been a substantial increase in travel abroad. State delegations, including governors and staff officials, legislators, educators, farmers, ranchers, horticulturists, and consultants of all kinds, are traveling abroad. The purposes of such delegations typically are to develop good will; promote professional linkages, partnerships, and exchanges; and develop markets for commodities. When preparing for careers in agriscience or developing strategies for marketing agriscience products, it is important to consider the international and global opportunities.

The marketing of products is a complex operation if maximum profits are to be realized. There are many different markets for plants and animals and many ways of marketing them and their products. Planning for marketing should take place before entering into production, and it should be kept in mind during the entire production process.

STUDENT ACTIVITIES

1. Write the Terms to Know and their meanings in your notebook.

2. Obtain several copies of marketing reports and newspapers that have marketing reports. Compare the prices received for the various classes and grades of animals listed. Also compare the prices received from various markets.

3. Interview a livestock buyer. Ask how prices are determined and where animals are bought and sold. Also ask how to go about getting a job as a livestock buyer. Report your findings to the class.

4. Go to a wholesale market and talk to a livestock grader about how to determine the various grades and classes of livestock. Try to grade some animals yourself.

5. Invite the operator of a retail market to class to talk about the advantages and disadvantages of marketing animals and animal products retail.

6. List the types of markets for agriscience commodities in your area.

7. Write a brief paper on the "best philosophy for pricing."

8. Develop an ad or other promotional device for an agriscience commodity.

9. Build a display for retailing an agriscience commodity.

10. Research the procedures used to buy and sell in the futures market.

11. Do research on one or more major international trade agreements that influence the marketing of agriscience products.

SELF EVALUATION

A. Multiple Choice

1. Consumer demographics include:
   a. populations, preferences, and product cost.
   b. family size, eating preferences, and discretionary money.
   c. customer location and tax rates.
   d. transportation costs.

2. The grading system based on the amount and distribution of finish on an animal is called
   a.   prime.
   b.   quality.
   c.   yield.
   d.   commercial.

3. The amount of a product that is available at a specific time and price is the
   a.   supply.
   b.   demand.
   c.   commission.
   d.   production.

4. The estimated amount of weight that an animal loses during transportation to market is
   a.   yardage.
   b.   stress.
   c.   loss.
   d.   pencil shrink.

5. Yardage is a fee paid to terminal markets for
   a.   feed.
   b.   insurance.
   c.   selling.
   d.   none of the above.

6. Lambs that are slaughtered when they are less than three months old are called
   a.   mutton.
   b.   chevon.
   c.   feeder lambs.
   d.   hothouse lambs.

7. The least desirable yield grade for cattle is
   a.   U.S. No.1.
   b.   Utility.
   c.   U.S. No. 4.
   d.   5.

8. The sex classes for sheep are
   a.   stag, boar, and sow.
   b.   lamb, sheep, and mutton.
   c.   ram, ewe, and wether.
   d.   none of the above.

9. Roadside stands are
   a.   wholesale markets.
   b.   terminal markets.
   c.   retail markets.
   d.   direct markets.

10. A male castrated after reaching sexual maturity is a
a. wether.
b. steer.
c. stag.
d. barrow.

11. An example of a wholesale market is a/an
a. terminal market.
b. roadside stand.
c. on-the-farm market.
d. farmers' market.

12. A definite trend in agriscience marketing is
a. cost per unit keeps going up.
b. pricing is always based on cost.
c. number of commodities is decreasing.
d. marketing is more global.

B. Matching

_____ **1.** Prime
_____ **2.** Ewe
_____ **3.** U.S. No. 1
_____ **4.** Barrow
_____ **5.** Veal
_____ **6.** Futures purchase
_____ **7.** Offset position

a. Castrated male pig
b. Female sheep
c. Quality grade of cattle
d. Quality grade of swine
e. Calf meat
f. Out of the market
g. Opening a position

C. Completion

1. The lowest quality grade of slaughter hogs is _____.

2. _____ is the amount of a product desired at a specific place and time.

3. A professional livestock seller receives a _____ for selling animals for producers.

4. _____ markets are those where animals are sold by competitive bidding.

5. Futures for agriscience commodities are purchased through a _____.

6. Rather than delivering goods sold on the futures market, most people take an _____ position.

Planning Agribusinesses

OBJECTIVE

To define management, determine management performance, determine how decisions are made, and describe economic principles that affect management.

COMPETENCIES TO BE DEVELOPED

After studying this unit, you should be able to:

- define management.
- describe the importance of management.
- describe kinds of agriscience management decisions.
- list eight steps in decision making.
- describe the economic principles of supply and demand, diminishing returns, comparative advantage, and resource substitutions.
- use capital and credit wisely in business management.

MATERIALS LIST

✓ writing materials

✓ calculator

✓ income statement

✓ newspapers

AGRIBUSINESS MANAGEMENT

Agribusiness management is the human element that carries out a plan to meet goals and objectives in an agriscience business or **enterprise,** generally referred to as an agribusiness. Management decides the types of business or production activities included in the agribusiness, such as horticulture, aquaculture, and farm supplies. Agribusinesses include farming, ranching, nursery operations, landscaping, retail stores, service enterprises, marketing businesses, lending institutions, veterinarian services, and consulting activities.

According to Dun and Bradstreet, 88 percent of all businesses fail because of poor management. Poor management of an agribusiness frequently results from no set objectives or goals. Management is usually considered to be good when maximum profits are achieved from the available resources. Generally, good management has established goals and objectives as guidelines, which translate into good action plans (Figure 35-1).

Influences on Agribusiness Management

Agribusiness management is influenced by the several members who make up the board of directors in corporations and cooperatives. Land, labor, and capital, being in limited supply, also influence management decisions. For instance, limited **capital,** which is money or property, may prevent the agribusiness manager from buying a new piece of equipment that would make the operation more efficient or protect the operation from excessive loses (Figure 35-2). Similarly, limited land will influence the agriscience manager's selection of

FIGURE 35-1 Well-defined goals and objectives should translate into good action plans. *(Courtesy of FFA)*

FIGURE 35-2 Modern irrigation equipment permits profitable farming in areas that would otherwise be inefficient or too risky due to crop loss by drought. *(Courtesy of FFA, Photo by Ken Whitmore Association)*

enterprises. For example, a cow-calf beef enterprise will need more land than will a feedlot operation. Further, limited labor will influence the manager's selection of enterprises. For instance, if labor is limited, it will prevent the use of crop enterprises with high labor requirements.

Estimating a Manager's Performance

The performance or ability of the agribusiness manager can be estimated in several ways. Dollar income is the measurement most often used. According to a study conducted in Ohio, the managerial ability of an individual can be determined by the manager's economic orientation, decisiveness, ability to look at alternatives, and extent of social activities (Figure 35-3).

Agribusiness is constantly changing, and it is expected to change even faster in the future. As agribusiness becomes more complex, errors in management will be more costly. Therefore, the need for better-trained people in agribusiness management is greatly increasing.

- Income goal (Economic orientation)

- Willingness or tendency to make decisions (Decisiveness)

- Manager's ability to recognize alternatives and opportunities

- The extent of social activities or distractions from business goals

FIGURE 35-3 Factors for predicting the managerial performance of agribusiness managers.

CHARACTERISTICS OF DECISIONS IN AGRIBUSINESS

Although there are various types of decisions in agribusiness, decisions of managers are usually organizational or operational (Figure 35-4). Both types of decisions affect the success of the business. The degree of influence on the business is determined in part by the characteristic of the decision. The following discussion covers four characteristics of a decision as well as steps in decision making.

Importance

The importance of a decision may be determined by measuring the potential loss or gain of the decision involved. For example, the selection of an agribusiness enterprise or business venture is likely to be more important than the selection of a brand of a given commodity.

Agribusiness Decisions	
Organizational	**Operational**
• Should I rent more land or should I borrow money and purchase land?	• Should cattle be sold this week or next?
• Should the business be operated as a partnership or corporation?	• Should I use high-magnesium lime or regular lime?
• What lender is the best source of borrowed capital?	• When should I start planting corn? Soybeans? Small grain?
	• When should I change the photoperiod for my poinsettias?

FIGURE 35-4 Kinds of agribusiness decisions.

AGRI·PROFILE

CAREER AREAS: Owner/Manager/Consultant

Agribusiness activities cut across the food and nonfood spectrums of agriscience. This includes the supply side as well as the output side of production. Career opportunities exist in every sector where goods are bought or sold.

Agribusinesses sell products such as feed, seed, pesticide, fertilizer, tools, equipment, plants, or animals. Or, they may sell services such as animal care, crop spraying, or recreational fishing privileges. Careful planning is very important before starting an agribusiness, because many new businesses end in failure. Careful planning increases the chances of success.

Agribusinesses are the backbones of many communities. Here, a consultant utilizes computer and communication resources to predict future business outcomes. *(Courtesy National FFA)*

Career opportunities in agribusiness planning include financial services such as banking; accounting services; management services; teaching; university extension service; marketing; market analysis; product specialization; product engineering; sales; ownership; and management. Agricultural economics, business management, accounting, finance, and personnel management are college programs that can lead to careers in agribusiness planning.

The manager must spend a considerable amount of time selecting a type of business, because this decision cannot be changed easily after it is made. On the other hand, if a certain brand of wire or species of plant being sold is not successful, a different brand or species can be added or substituted at little additional cost.

Frequency

The frequency, or how often a decision is made, varies greatly. Determining the cost per day to rent a truck may be used as an example. The cost of renting the truck for only one day, for a one-time job, might not be an important decision. However, the cost of renting the truck every several weeks as an ongoing expense makes the decision more important.

Urgency

It is important that certain decisions be made immediately. Some decisions can be delayed. The urgency of a decision depends on the cost of waiting. Two examples of decisions with different degrees of urgency are hiring a new employee and buying a new office radio. If it is the busy time of year, an additional person may be needed immediately and should be hired. If you delay, you may lose sales. However, the delay in buying an office radio is not likely to affect the overall management and profit of the agribusiness.

Available Choices

Some situations offer several choices. If several choices are available, the manager should delete the least likely courses of action and focus on the more promising ones. The manager should take time to gather important information regarding the options that is helpful in making the correct decision. If no choices are available, there is no decision to be made.

Steps in Decision Making

Making decisions based on facts will increase your management ability. The agribusiness manager must determine the process to be used for making better decisions. The following eight steps should be helpful in developing a management process (Figure 35-5).

1. Start the management process with a situation or established goal.
2. Gather all available facts and information for accurate analysis of the problem.
3. Analyze the available resources. Human resources are labor, time, skills, and interest. Material resources are land, equipment, and capital. Reevaluate your goal and adjust it if appropriate.
4. Determine the possible ways or established routes of accomplishing the goal or solving the problem.
5. Make a decision concerning the route(s) that will be taken.
6. Follow through with a plan of action.
7. Assume responsibility for implementing the plan.

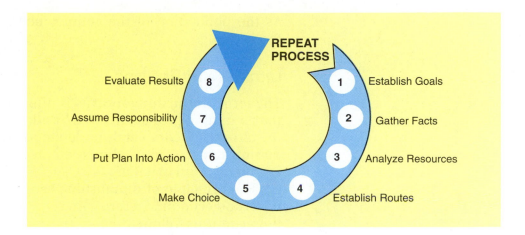

REPEAT PROCESS

Evaluate Results — 8

Assume Responsibility — 7

Put Plan Into Action — 6

Make Choice — 5

4 — Establish Routes

3 — Analyze Resources

2 — Gather Facts

1 — Establish Goals

FIGURE 35-5 Eight steps in the decision-making process.

8. Evaluate the results to determine whether the goals were accomplished. If not, could a better route have been selected? Should the goals be modified?

After going through these eight steps, the manager should determine whether the process worked well. If not, where were the weak links? What changes should be made to improve the process? These and other questions will enable the manager to fine-tune the process until it fits his or her style, the personalities of the staff, and the structure of the organization.

FUNDAMENTAL PRINCIPLES OF ECONOMICS

Price, Supply, and Demand

Price is the amount received for an item that is sold. The price is determined by three price-making factors: (1) the supply of the item, (2) the demand for the item, and (3) the general price level. No one factor can be used to explain all price changes.

The general price level of a commodity is influenced by the supply or availability of the item, the demand for it, and wars, depressions, and many other factors. When supply and demand are in balance, a general price is established. An increase or decrease in either supply or demand is likely to influence the prevailing price.

The quantity of a product that is available to buyers at a given time is called **supply.** A producer of vegetables has control over some factors that influence the vegetable supply. However, one factor that the manager cannot control is the weather. If the rains do not come, the supply of vegetables will be reduced. The ability to buy items needed to produce the vegetables can also increase or reduce the supply. The ability to buy these items is also influenced by their prices. Cost and availability of credit may also influence supply. **Credit** is money loaned. It permits the agribusiness manager to buy items needed for production or for business operation.

Demand may be defined as the quantity of a product that buyers will purchase at varying prices at a given time. The quantity that a buyer is willing to purchase depends upon the quantity available and the price. Both the desire and the ability to purchase influence the extent of demand. The population growth may also influence demand.

As the population of the country changes, the demand for various foods changes.

Diminishing Returns

The term **diminishing returns** is often used in economics. It refers to the amount of profit generated by additional inputs. Most of the time, the term is not completely understood. An understanding of the law of diminishing returns can be extremely helpful to the agribusiness manager in decision making.

The principle of diminishing returns has two parts—physical returns and economic returns. An explanation of both physical and economic returns follows.

Satisfaction from eating may be used to illustrate diminishing physical returns (Figure 35-6). When eating, each bite of food represents an additional unit of input. Each bite helps satisfy a part of the hunger. However, the added amount of satisfaction of hunger diminishes as you eat more. Each mouthful results in less satisfaction. This tendency is known as *diminishing physical returns*. At a certain point, the amount of hunger satisfied becomes negative with each bite taken. This means that each additional bite takes away from the satisfaction gained by previous bites.

An example of diminishing economic returns is demonstrated with the addition of units of nitrogen fertilizer as inputs and the effect on corn yields as outputs (Figure 35-7). Each additional input requires an additional cost. It may be observed that adding the first unit of nitrogen produced the greatest increase in yield. However, the rate of increase in yield diminished as more units of input were added. The decision that the manager needs to make is how many units of nitrogen should be applied to obtain the greatest profit from growing the corn. If the number of inputs is too low or too high, the highest potential income will not be achieved.

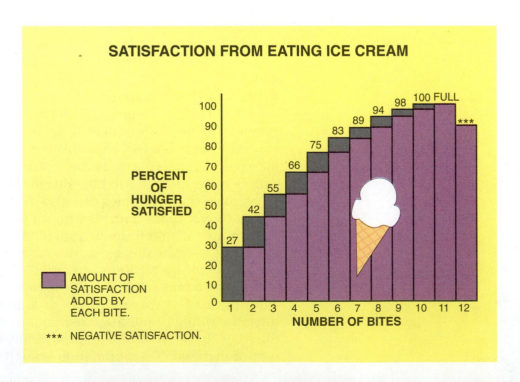

FIGURE 35-6 Diminishing returns.

Effect of Nitrogen on Corn Yields*		
Nitrogen Applied (lb)	Yield (bu.)	Yield Increase (bu.)
0	67	—
60	103	+36
120	133	+30
180	154	+21
240	173	+19
300	169	-4

*Hypothetical and not based on research

FIGURE 35-7 Effect of nitrogen on corn yields.

Comparative Advantage

The United States has nine major farming areas that have developed over a period of years. These areas have developed because of changes in demand or other factors. Within each area, different commodities are produced. Most operations in an area produce similar commodities and have similar systems of production. The reason for the similarity in operations is the comparative advantage found in following the programs that have evolved in that area.

Comparative advantage is the emphasis in a given area where the most returns can be achieved. Comparative advantage may be illustrated in several ways. A recent example has been the change in livestock production in New York State. At one time, most of the farms in New York raised sheep. Today these farms are producing milk, because of the high demand and favorable prices for milk in New York. It has become more advantageous to produce milk than to produce lambs and wool.

Resource Substitution

Resource substitution means the use of a resource or item for another, when the results are the same. It is often possible to substitute a less expensive item for a more expensive one.

For example, it may be possible to substitute barley for corn in making a cheaper dairy feed. This substitution can be done without affecting total production. Barley is frequently less expensive than corn. Because barley and corn have slightly different feeding values for dairy cattle, barley must sell for $2.25 or less per cwt to be a better buy than corn selling for $2.50 per cwt.

The application of the principle of resource substitution is not limited to feeding livestock. This principle can be used in many management situations. Can you think of other examples of resource substitutions?

AGRIBUSINESS FINANCE
Importance and Uses of Credit

The management of finances is the most important single function of the agribusiness manager. Possessing a great deal of technical knowledge and know-how is very important. However, this know-how will not make one a successful agribusiness manager unless you are a competent money manager. Careful planning of finances is probably more important than planning other aspects of the agribusiness. Some very good agribusinesses have failed because the manager was not a good money manager.

Credit is borrowed money. At one time, the manager who sought credit was considered to be a poor manager. Today this is no longer true. The manager approaching a lending institution to negotiate for credit should not do so in an apologetic manner. The lending institution must sell its commodity, money, in order to stay in business. The use of credit is a privilege that must not be abused, however. Once labeled as a poor credit risk, the reputation is difficult to overcome. Credit used wisely, however, is a valuable tool in business.

Classifications of Credit		
Long-Term Loans	**Intermediate-Term Loans**	**Short-Term Loans**
Used to purchase land and buildings, and are for a period of eight to forty years.	Extend for a period of one to seven years for the purchase of breeding livestock, farm equipment, tractors, and similar items.	Usually written for a period of one year or less to cover the cost of feeder livestock, feed, fertilizer, seed, fuel, etc.

FIGURE 35-8 Classifications of credit.

Classifications of Credit

Credit is classified according to its period of use. Loans may be classified as long-term, intermediate-term, or short-term (Figure 35-8).

Long-term loans are used to purchase land and buildings. The loan period ranges from 8 to 40 years. Interest rates for this type of loan are usually lower than they are for other types. These loans are made by federal land banks, the Farmers Home Administration, insurance companies, local banks, and individuals.

Intermediate-term loans are for periods of 1 to 7 years. Capital investments are usually made with the money from this type of loan. **Capital investment** is money spent on commodities that are kept six months or longer. Examples are breeding stock, tractors, store equipment, and warehouse equipment. Production credit associations, the Farmers Home Administration, finance companies, and local banks are sources of this type of credit.

Short-term loans are for a period of one year or less. They are often referred to as *production loans*. Some managers are able to borrow short-term money without security. This is particularly true of managers and businesses with good reputations. Money borrowed without security is referred to as a *signature loan*. The borrower signs a promissory note that the loan will be paid on or before a specified date. Production credit associations and local banks are sources of short-term loans. It is not uncommon for local agribusinesses to permit their reliable customers to make purchases on a short-term credit basis by simply signing a sales slip.

Types of Credit

Two types of credit are productive credit and consumption credit (Figure 35-9). Productive credit is used to increase production or income. This is justifiable when the estimated increase in production will increase profits. This type of credit is used to purchase supplies, plants, flowers, livestock, land, equipment, storage facilities, seed, fertilizer, labor, and other materials.

Consumption credit is used to purchase consumable items used by the individual; it does not contribute to the business income. It is relatively easy for a family to abuse the use of consumption credit. This type of credit can also limit the amount of productive credit available to a family business.

EXAMPLES

PRODUCTIVE CREDIT

Livestock   Plant/Flowers   Equipment

CONSUMPTIVE CREDIT

Vehicles   Clothing/Shoes   Entertainment

FIGURE 35-9 Types of credit based on use.

A form of consumer credit growing in popularity is credit card purchases. Most credit card purchases paid in thirty days or less are interest free. However, the cost of consumer credit is very high if not paid off upon the first billing. Federal statutes require firms charging interest on credit card purchases to publish their interest rates.

Sources of Credit

There are many sources of credit available and differences in lending policies. Because of this, a knowledge and understanding of credit and credit practices should be helpful in securing credit. Lenders differ in the interest rates they charge, the length of the loan period, and the purposes for which money is loaned.

When seeking a loan, remember that the interest rate is not the only important factor. Of substantial importance is the lender's willingness to extend a line of credit should an unexpected event occur. Another important factor is the lender's knowledge of the agribusiness.

Some agencies and institutions make only certain types of loans (Figure 35-10). Commercial banks are the most important source of credit. About 25 percent of the total agricultural debt is owed to banks. This percentage represents about 13.5 percent of the real estate loans and 45 percent of other loans. Commercial banks lead in loan volume because they make all types of loans.

Individuals are the next most important source of agribusiness credit. About 36 percent of agricultural real estate mortgages are held

Sources of Agricultural Credit and Typical Rates				
Source	Length of Loan	Annual Interest Rate*	Percent of Appraisal Loan Value	Purpose of Loan
Commercial Banks & Trust Companies	6-12 Months	8-13%	To 100%	Farm Production Items
	2-3 Years	9-12 ½%	70-80%	Machinery, Equipment, and Livestock
	10-20 Years	9-11%	60-75%	Real Estate
Federal Land Banks	20-35 Years	8 ½% Variable	Up to 85%	Real Estate
Production Credit Association	1 Year	9% Variable	To 100%	Production Items
	3-7 Years	9%	Varies	Machinery, Equipment, and Livestock
Farmers Home Administration	40 Years	5%	To 100%	Real Estate
	7 Years	7-9%	To 100%	Machinery, Equipment, and Livestock
Insurance Companies	20-35 Years	Varies	To 75%	Real Estate
Individuals	15-30 Years	6-7%	To 75%	Real Estate
	2-5 Years	6-8%	To 80%	Machinery, Equipment, and Livestock
Equipment Manufacturers	1 Month-5 Years	17.5%	100%	Machinery and Equipment

*Annual interest rates fluctuate based on the current prime rate. These percentages are for example only and should not be considered current.

FIGURE 35-10 Sources of agribusiness credit.

by individuals. Interest rates, tax deferments and reductions, and payment options are some of the reasons that individuals are willing to finance real estate.

Retail merchants also supply non-real-estate credit. Their credit is usually in the form of an open account. Sometimes cash loans are made. The interest rates are usually higher from these sources than they are from commercial banks. The popularity of this type of credit is based upon its convenience and availability.

Federal land banks were created by Congress in 1916 to provide long-term credit for agriculture. These banks are excellent sources of credit for the purchase of real estate because of the favorable interest rates.

Production Credit Associations (PCA) were established by an Act of Congress in 1933. The purpose was to provide favorable, short-term credit for agriculture. These funds are secured from the federal intermediate credit banks, who in turn secure their money from private lenders.

Life insurance companies have been excellent sources of credit for financing long-term real estate loans. Loans are made through brokers, correspondents, and company representatives. Insurance companies hold about 15 percent of agricultural real estate mortgages.

The **Commodity Credit Corporation (CCC)** was administered as an agency of the U. S. Department of Agriculture (USDA). The CCC is a government-owned corporation. Loans are made on eligible commodities, such as grain, cotton, peanuts, and tobacco. Farmers use the commodities as security for loans. Payment of loans is made by purchasing the loans or delivering the commodities to the CCC.

Farmers Home Administration (FHA) was created by the government during the depression. The FHA's original purpose was to assist tenant farmers in becoming landowners. The FHA provides financing to farmers who are unable to secure credit from any other sources. The advantages of borrowing money from the FHA are as follows:

1. A large percentage of the total cost of the property can be borrowed.

2. The repayment plan is based on the borrower's ability to repay.

3. Supervision of the loan and assistance with planning are provided.

The **Small Business Administration (SBA)** also provides loans to agribusinesses. In 1976 Congress passed legislation that permits the SBA to make agricultural production loans. These loans have a favorable interest rate.

Cost of Credit

Interest is the greatest expense in borrowing money. However, borrowers are also required to pay other fees. These fees include commissions, recording fees, title certification charges, insurance, and service charges. There are many formulas used for calculating interest (Figure 35-11).

A **simple-interest loan** refers to one whereby the full amount of a loan is received by the borrower and is paid back with interest after a short period of time. The borrower generally signs a **promissory note** agreeing to the terms of the loan. If payments are made several times throughout the duration of the loan, the interest is paid to date and interest is charged on the remaining balance of the principal.

In the case of a **discount loan,** interest is subtracted from the principal at the time the loan is made. For example on a $1,000 loan, the

A Comparison of the Different Methods of Calculating Interest		
Example: A Loan of $600.00 for 12 Months at 10%. Repaid in 12 Equal Payments.		
	FORMULAS FOR CALCULATING INTEREST	INTEREST CALCULATED
SIMPLE INTEREST	Interest (I) = Principal X Rate X Time (I = P X R X T)	$600.00 X .10 X 1 = $60.00 = 10% True Interest
DISCOUNTED LOAN	Rate (R) = $\dfrac{\text{Interest (dollar cost)}}{\text{Principal X Time}}$ $\left(R = \dfrac{I}{P \times T}\right)$	$\dfrac{60.00}{540.00}$ X 1 = .11, or 11% True Interest
ADD-ON LOAN	Rate (R) = $\dfrac{2 \text{ X No. of Payments X Interest Charged in \$}}{\text{Beginning Principal X Years X (No. of Payments + 1)}}$	$600.00 X .10 = $60.00 660.00 = Total Payments $\dfrac{2 \text{ X } 12 \text{ X } 60}{600 \text{ X } 1 \text{ X } 13}$ = .184, or 18.4% True Interest

FIGURE 35-11 Formulas for calculating interest.

borrower would receive only $900 of the $1,000 if the interest were 10 percent. Because the full amount of the principal is not received, the resulting true interest rate is 11 percent.

The **add-on loan** method is used for calculating interest on consumer loans. The interest is charged for the entire amount of the principal for the entire length of time. The total principal plus total interest is divided into combined equal installments. This results in a very high rate of true interest. For example, a $1,000 loan at 10 percent interest would require repayment of $1,100 and be repaid in monthly installments of $91.67 ($1,000 principal + $100 interest = $1,100/12 = $91.67).

The **amortized loan** is generally used for the purchase of land, buildings, and other expensive items. The payments are made monthly and are computed so the interest owed plus the payment on the principal is equal throughout the repayment time. By this method, almost all of the payment at the beginning of the repayment period is for interest, while most of the payment near the end is for principal. In other words, the amount being paid on the principal increases proportionately as the amount due for interest decreases. A careful analysis of amortization schedules for a loan over various lengths of time will help the manager determine the cost of borrowing and make wise decisions on when, how much, at what rate, and for how long to borrow capital (Figure 35-12).

Seeking a Loan

Most agribusinesses must borrow money to expand their operations or increase income. The decision to incur debt is not easily reached. A careful examination of available alternatives must be made. Then the best alternative must be chosen.

When seeking a loan, the agribusiness manager must be prepared to fully explain the benefits and risks to potential lenders. Many loan applications are not approved because the borrower does not present the details of the agribusiness. Other loan applications are not approved because the borrower does not effectively present the advantages of expanding. To be effective, the presentation must be organized, and the details of the operation should be put in writing. Lenders are concerned about the benefits and risks associated with providing a loan. Lenders are also concerned about a logical presentation of facts and the desires concerning specific agreements in the loan contract.

Information that should be included in the presentation includes (1) an agribusiness plan; (2) business records (income statements, expense records, net worth statement, and financial history); (3) terms of the loan; and (4) method of repayment.

A borrower should investigate several sources of credit before selecting a lender. The final selection should be one that best meets the needs of the borrower.

Selecting a Lender

Factors that should be considered in selecting a lender (Figure 35-13, page 690) include the:

1. lending institution representative's knowledge of agribusiness problems and practices.

A. 48 PAYMENTS OF $368.67 EACH								
Pmt	Principal	Interest	Balance	Total Interest	Pmt	Principal	Interest	Balance	Total Interest
		12.000%	14,000.00				12.000%	14,000.00	
1	228.67	140.00	13,771.33	140.00	25	290.35	78.32	7,541.63	2,758.38
2	230.96	137.71	13,540.37	277.71	26	293.25	75.42	7,248.38	2,833.80
3	233.27	135.40	13,307.10	413.11	27	296.19	72.48	6,952.19	2,906.28
4	235.60	133.07	13,071.50	546.18	28	299.15	69.52	6,653.04	2,975.80
5	237.95	130.72	12,833.55	676.90	29	302.14	66.53	6,350.90	3,042.33
6	240.33	128.34	12,593.22	805.24	30	305.16	63.51	6,045.74	3,105.84
7	242.74	125.93	12,350.48	931.17	31	308.21	60.46	5,737.53	3,166.30
8	245.17	123.50	12,105.31	1,054.67	32	311.29	57.38	5,426.24	3,223.68
9	247.62	121.05	11,857.69	1,175.72	33	314.41	54.26	5,111.83	3,227.94
10	250.09	118.58	11,607.60	1,294.30	34	317.55	51.12	4,794.28	3,329.06
11	252.59	116.08	11,355.01	1,410.38	35	320.73	47.94	4,473.55	3,377.00
12	255.12	113.55	11,099.89	1,523.93	36	323.93	44.74	4,149.62	3,421.74
13	257.67	111.00	10,842.22	1,634.93	37	327.17	41.50	3,822.45	3,463.24
14	260.25	108.42	10,581.97	1,743.35	38	330.45	38.22	3,492.00	3,501.46
15	262.85	105.82	10,319.12	1,849.17	39	333.75	34.92	3,158.25	3,536.38
16	265.48	103.19	10,053.64	1,952.36	40	337.09	31.58	2,821.16	3,567.96
17	268.13	100.54	9,785.51	2,052.90	41	340.46	28.21	2,480.70	3,596.17
18	270.81	97.86	9,514.70	2,150.76	42	343.86	24.81	2,136.84	3,620.98
19	273.52	95.15	9,241.18	2,245.91	43	347.30	21.37	1,789.54	3,642.35
20	276.26	92.41	8,964.92	2,338.32	44	350.77	17.90	1,438.77	3,660.25
21	279.02	89.65	8,685.90	2,427.97	45	354.28	14.39	1,084.49	3,674.64
22	281.81	86.86	8,404.09	2,514.83	46	357.83	10.84	726.66	3,685.48
23	284.63	84.04	8,119.46	2,598.87	47	361.40	7.27	365.26	3,692.75
24	287.48	81.19	7,831.98	2,680.06	48	365.26	3.65	0.00	3,696.40

B. 24 PAYMENTS OF $659.03 EACH								
Pmt	Principal	Interest	Balance	Total Interest	Pmt	Principal	Interest	Balance	Total Interest
		12.000%	14,000.00				12.000%	14,000.00	
1	519.03	140.00	13,480.97	140.00	13	584.86	74.17	6,832.53	1,399.92
2	524.22	134.81	12,956.75	274.81	14	590.70	68.33	6,241.83	1,468.25
3	529.46	129.57	12,427.29	404.38	15	596.61	62.42	5,645.22	1,530.67
4	534.76	124.27	11,892.53	528.65	16	602.58	56.45	5,042.64	1,587.12
5	540.10	118.93	11,352.43	647.58	17	608.60	50.43	4,434.04	1,637.55
6	545.51	113.52	10,806.92	761.10	18	614.69	44.34	3,819.35	1,681.89
7	550.96	108.07	10,225.96	869.17	19	620.84	38.19	3,198.51	1,720.08
8	556.47	102.56	9,699.49	971.73	20	627.04	31.99	2,571.47	1,752.07
9	562.04	96.99	9,137.45	1,068.72	21	633.32	25.71	1,938.15	1,777.78
10	567.66	91.37	8,569.79	1,160.09	22	639.65	19.38	1,298.50	1,797.16
11	573.33	85.70	7,996.46	1,245.79	23	646.04	12.99	652.46	1,810.15
12	579.07	79.96	7,417.39	1,325.75	24	652.46	6.52	0.00	1,816.67

FIGURE 35-12 Comparison of a loan for $14,000 at 12 percent interest paid over four years versus the same loan repaid in two years.

Factors to Consider in Selecting a Lender

- Lending institution representative's knowledge of agricultural practices and problems.
- Lending institution representative's experience in handling agricultural credit of a similar nature.
- Reputation of lending institution.
- Loan policies (interest rate, repayment schedule, closing cost, penalty clause, optional repayment clause, policy regarding late payments).
- Date loan would be advanced.
- Possibility of increasing loan.
- Availability of credit for other purposes.

FIGURE 35-13 Factors to consider when selecting a lender.

2. lending institution representative's experience in handling agricultural credit of a similar nature.

3. reputation of the lending institution.

4. loan policies (interest rate, repayment schedule, closing costs, penalty clause, optional prepayment clause, and policy regarding failure to meet payment because of circumstances beyond borrower's control).

5. date the loan would be advanced.

6. possibility of increasing the loan.

7. availability of credit for other purposes.

BIO•TECH CONNECTION

Workhorses of Modern Business Planning

The microcomputer has become the workhorse of modern agribusiness planning as well as the workhorse of business operations. Capable of split-second computations and tremendous storage capacity, the microcomputer is found in nearly every business and generally serves as the nerve center for the business. After a small investment in software and a modest investment in accessories, the computer can be used to evaluate marketing alternatives, predict financial outcomes for proposed business ventures, and accumulate data on prospective customers. Further, by using a good computer system, the manager and staff can handle complex mailing lists, individualize communications to large numbers of patrons, track sales and inventories, keep appropriate financial

Computer programs and computer models help the business planner to test alternative practices without risking time and capital. *(Courtesy of Michael Dzaman)*

Ways Whereby Borrower Can Minimize Risk

- Use production loans to increase income
- Limit amount borrowed on new or unfamiliar enterprises
- Keep debt as low as possible while still maintaining efficiency
- Keep abreast of markets and trends
- Maintain proper debt-net worth ratio
- Maintain proper debt-income relationship
- Select lender on dependability and terms
- Have a definite repayment schedule
- Be businesslike, fair, and frank
- Reduce risk by carrying adequate insurance

FIGURE 35-14 Ways whereby borrower can minimize risk.

Using Credit Wisely

There is a chance that the borrower will fail to meet the financial obligations of a loan. In this case, the borrower may lose part or all that is owned. Because of this, the borrower takes a much greater risk than does the lender. The borrower, however, can minimize risk by applying certain rules (Figure 35-14) as follows:

1. Production loans should be used to increase income.

2. The borrower should limit the amount borrowed on new or unfamiliar business ventures.

accounts, and develop presentations for the owners, board of directors, or patrons.

The very small business may have just one computer with a high-quality printer and monitor. However, such businesses soon graduate to multiple computers or one large microcomputer and numerous smaller computers on the same system. Color monitors, printers, and scanners are replacing black and white systems, as the price for computer hardware continues to drop even as memory capacity increases. The software for a business may be an integrated software package, typically with word processing, data-base, spreadsheet, and communications software. Communications software permits the computer to use telephone lines to access outside networks and download data and programs from outside sources. This permits the business access to market quotations, continuous status reports on inventories, up-to-date marketing information, and databases on world business and commerce.

Data for tax computation, annual inventory, business reports, and other government obligations are readily available from an effective computer system. Further, specialized computer programs can be purchased to calculate taxes and prepare tax returns, thereby cutting costs for services and consultants for the business.

Salespeople, buyers, company executives, and other on-the-road personnel use portable computers to gather and synthesize information from sales contacts and conferences, and to phone the information to the home computer for staff to act upon before the traveler returns to the office. With the arrival of voice-input capacity and computers that talk back, typical agribusinesses will continue to find new ways to utilize the computer, and its role as business "workhorse" will undoubtedly expand.

3. The borrower should keep debt as low as possible while still maintaining efficiency.

4. The borrower should keep abreast of markets and trends.

5. A debt-to-net-worth ratio of 1:1 or less should be maintained.

6. A proper debt-to-income relationship should be maintained. The income must be greater than the principal and interest payments.

7. Dependability and terms of the loan should be considered when selecting a lender.

8. The borrower should have a definite repayment plan and schedule.

9. The borrower should be businesslike, fair, and frank.

10. The borrower should have adequate insurance to reduce the lender's risk. Property, liability, and life insurance provide partial protection against risk.

The use of management in agriscience is extensive. Management is a vital part of the business and commerce associated with supplies, services, production, processing, distribution, and selling of plant, animal, and natural-resource commodities. Careful planning and the application of sound agribusiness principles and procedures are necessary for success in many agriscience careers.

STUDENT ACTIVITIES

1. Write the Terms to Know and their meanings in your notebook.

2. Explain how the eight steps in decision making can be used in planning an FFA or other school money-raising activity.

3. Explain how the principal of resource substitution may be used in making an agricultural mechanics project.

4. Discuss with a banker the procedures used in applying for an agribusiness loan.

5. Arrange to have a banker discuss agribusiness capital and credit in a class presentation.

6. Conduct an FFA or other organization fund-raising activity using business-type planning and management strategies.

7. Arrange for an accountant to talk to the class about taxes to consider when planning an agribusiness.

SELF EVALUATION

1. Most businesses fail because of
   a.   death of the manager.
   b.   lack of capital.
   c.   labor problems.
   d.   poor management.

2. Management is considered good if
   a.   labor is adequate at all times.
   b.   maximum profits are achieved.
   c.   the business survives for five years.
   d.   the business survives for ten years.

3. The importance of a decision may be measured by the
   a.   availability of the manager.
   b.   inventory.
   c.   potential for gain or loss.
   d.   time of year.

4. The amount of profit generated by additional inputs is known as
   a.   capital.
   b.   diminishing returns.
   c.   margin.
   d.   profit.

5. Using a resource in the place of another is known as
   a.   bailing out.
   b.   integration.
   c.   resource management.
   d.   resource substitution.

6. Loans for 8 to 10 years are called
   a.   amortized loans.
   b.   intermediate-term loans.
   c.   long-term loans.
   d.   short-term loans.

7. The most important source of credit is
   a.   commercial banks.
   b.   individuals.
   c.   insurance companies.
   d.   land banks.

8. The formula for calculating simple interest is
   a. $I = P \times R \times T$.
   b. $I = P/R \times T$.
   c. $I = R/P \times T$.
   d. $I = P \times T$.

9. A debt-to-net-worth ratio should not exceed
   a.   5:1.
   b.   3:1.
   c.   2:1.
   d.   1:1.

10. The type of insurance that a borrower should have is
   a.   liability.
   b.   life.
   c.   property.
   d.   all of the above.

B. Matching

_____ **1.** Capital

_____ **2.** Price

_____ **3.** Supply

_____ **4.** Credit

_____ **5.** Demand

_____ **6.** Short-term loan

_____ **7.** Intermediate-term loan

_____ **8.** Generated income

_____ **9.** Used for the individual

_____ **10.** Add-on loan

a.   Money loaned

b.   Quantity desired

c.   One year or less

d.   Quantity available

e.   1 to 7 years

f.   Money property

g.   Pay interest in the beginning

h.   Consumptive credit

i.   Amount received

j.   Productive credit

C. Completion

1. List the eight steps in decision making, in their correct order.

2. Study Figure 35-12, page 689, and answer the following:

   a.   What is the total interest paid on the loan if repaid in four years (fourty-eight payments)? $_____

   b.   What is the total interest paid on the loan if repaid in two years (twenty-four payments)? $_____

   c.   How much more does the loan cost if repaid in four years rather than two years? $_____

   d.   What are your observations about stretching a loan out over more time?

Entrepreneurship in Agriscience

OBJECTIVE

To define entrepreneurship and determine considerations for planning and operating an agribusiness.

MATERIALS LIST

✓ variety of publications with pictures

✓ materials to make a collage or bulletin board

COMPETENCIES TO BE DEVELOPED

After studying this unit, you should be able to:

- define and describe entrepreneurship.
- describe steps in planning a business venture.
- state five basic functions performed in the operation of a small business.
- select a product or service for a personal or group enterprise.
- determine the basic functions performed by small-business managers.
- analyze the outcome of a business venture.
- use small-business financial records.
- analyze the benefits of self-employment versus other types of employment.

TERMS TO KNOW

Entrepreneur
Entrepreneurship
Inventor
Buying function
Selling function
Promoting function

Distribution function
Financing function
Short-term plans
Long-term plans
Actuating
Balance sheet

Assets
Liabilities
Net worth
Profit and loss
  statement
Inventory report

Entrepreneurship in agriscience provides extensive career possibilities at all levels of endeavor. It provides opportunities for the inventor, risk taker, profit seeker, owner, manager, and employee. The products and services of entrepreneurs include fresh food, processing plants, equipment sales, fishing, guide services, veterinarian work, forestry work, farming, ranching, teaching, research, mechanics, environmental efforts, law, insurance, real estate, and finance. Opportunities in agriscience exist worldwide (Figure 36-1).

THE ENTREPRENEUR

The **entrepreneur** is the person who organizes a business or trade, or improves an idea. The word is taken from the French word, *entreprendre*, which means to "undertake." **Entrepreneurship** is the process of planning and organizing a small-business venture. It also involves managing people and resources to create, develop, and implement solutions to problems to meet people's needs. The **inventor** is the person responsible for devising something new or making an improvement to an existing idea or product.

The entrepreneur can be the inventor as well as the small-business manager. However, in many cases, they are different people, each with distinct talents. The entrepreneur functions as the liaison between the inventor and the manager or management team. The entrepreneur brings these two groups together for the purpose of getting the invention to the persons it will serve.

It is the entrepreneur who visualizes the venture strategy and is willing to take the risk to get the venture off the ground. Inventors or business managers are not entrepreneurs unless they organize the venture. An understanding of the differences as well as the similarities among these roles is important to comprehending entrepreneurship as a career option.

FIGURE 36-1 FFA members and other youth preparing for life careers in agriscience can expect entrepreneur opportunities around the world. *(Courtesy National FFA)*

697

FIGURE 36-2 Many businesses start with high-school occupational-experience programs in agriscience. *(Courtesy of FFA. Photo by Bill Stagg)*

FIGURE 36-3 Successful businesses generate job opportunities for managers and other employees. *(Courtesy of FFA)*

ENTREPRENEURSHIP

Various types of business enterprises are part of the entrepreneurship system. The element of individual, partner, or corporate ownership means private control as opposed to government ownership control. The profits (or losses) from entrepreneurship go to the owners. However, owners hire managers and other employees; therefore, entrepreneurships provide jobs for everyone (Figure 36-2 and Figure 36-3).

A sales project conducted by a school organization such as the FFA is a type of entrepreneurship. However, it differs from a typical small-business venture in the following ways:

1. The sales project should be a valuable learning experience as well as a money-making activity.
2. The sales project is usually of a limited duration. It is completed in a short period of time. A small business typically is designed to operate indefinitely.
3. The sales project involves the voluntary participation of the class. The small business must hire and pay its employees.
4. The sales project usually involves very little risk of financial loss to the individual. The small-business owner (entrepreneur) could lose his or her investment.
5. The sales project may need to be approved by the school administration, but usually does not need to be licensed by the government, pay taxes, or file employee reports.

With these distinctions in mind, you can plan, organize, and implement a sales project and gain insight into the world of entrepreneurship at the same time (Figure 36-4).

Operating Businesses

Five basic functions are performed in the operation of both a small business and a sales project. These functions represent the basic steps in moving a product or service from the supplier to the consumer. They are listed as follows:

Buying Function The buying function involves selecting a product or service to be marketed or sold for profit. The selection of a product

FIGURE 36-4 Group projects in school can help students develop valuable entrepreneurial skills. *(Courtesy of FFA. Photo by Michael Wilson)*

FIGURE 36-5 Effective selling and promotion is absolutely essential to the success of most businesses. How a product is displayed is very important. The flowers at this greenhouse are eye-catching and visually appealing. This type of presentation is sure to attract customers. *(Courtesy of FFA)*

FIGURE 36-6 Product distribution may be made to retail outlets or to other businesses, homes, farms, and many other destinations. *(Courtesy of Michael Dzaman)*

or service is based on thorough marketing research, which determines consumer (customer) needs and wants. It also determines who, if anyone, is already promoting the product or service.

Selling Function The selling function includes studying the product or service to determine the reasons that customers will want and need the product or service. This function also involves developing suitable customer approaches. Planning sales presentations, determining methods of overcoming objections, and planning for the close of a sale are all part of this function (Figure 36-5).

Promoting Function The promoting function involves developing a plan to identify ways to make potential customers aware of the product or service to be offered. Examples of promotional activities include newspaper, TV, radio, and outdoor advertising.

Distribution Function The distribution function involves physically organizing and delivering the selected product or service. For products, this includes receiving, storing, and distributing the merchandise. Many of the same activities are associated with services, particularly distributing to the customer (Figure 36-6).

Financing Function The financing function includes obtaining capital for the initial inventory, recording sales, maintaining inventory, computing profit or loss, and reporting the results of the venture. A venture is an activity with some risk involved.

SELECTING A PRODUCT OR SERVICE

The first function in the process of establishing a new business or developing a fund-raising project is the selection of a product or service (buying function). This involves analyzing potential products/services in terms of the potential demand (number of customers). The analysis of potential products or services should answer the following questions:

1. Who are the potential purchasers of the proposed product or service? Customers can often be viewed in groups that have similar interests.

2. On what basis does each group make decisions on product or services? Typical responses to this question include price, quality, and service.

3. What are the competing products or services? How do they compare to your product or service relative to price, quality, and service?

4. What is the total estimated demand for your product or service?

5. How much profit would you make on the sale of this product or service? (Multiply the number of units sold by your markup per unit and subtract any expenses you incur in the purchase, sale, and delivery of the product or service.)

AGRI·PROFILE

CAREER AREAS: Owner/Manager/Assistant Manager/ Management Trainee

Agribusiness management is seen by many as an ideal career. This career has the advantage of continuing in the work of the family where one gains experience as they grow up. Or, one may start a business and work in a locality where there may not be jobs in specialized areas.

Many high school and college students are well established in agribusinesses before they finish their educations. Some popular agribusinesses include lawn services, logging, lumber business, greenhouse or nursery operations, machinery repair, hunting and trapping, agricultural supplies, home and garden centers, florist shops, retail flower sales, livestock sales, farming, and ranching.

Preparation for careers in agribusiness management includes both formal education and on-the-job training. High school agribusiness and agriscience programs provide excellent training for business. Such programs should include classroom, laboratory, supervised agriscience experience, and leadership development. Advanced agribusiness programs may be taken at technical schools, colleges, or universities to obtain appropriate training in economics, finance, and management.

The entrepreneur is constantly seeking ways to provide better goods and services to the clientele. *(Courtesy of Michael Dzaman)*

ORGANIZATION AND MANAGEMENT

The role of a venture's organization and management is to make things happen so the venture can achieve its goals. To accomplish this, managers work mainly with data and people. Managing involves getting all the parts of the business—including personnel, marketing strategies, finances, and records—to function together to achieve the venture's goals.

No two managers have jobs that are the same. The jobs are shaped by the type of venture and the personality of the individual manager. However, managers perform the same functions. They are:

Planning work
Organizing people and resources for work
Actuating work
Controlling and evaluating work

Planning

When managers make plans, they set objectives or goals. They recommend policies to achieve the goals, and they develop procedures, methods, or programs to implement those policies. Plans must be constantly reviewed and updated. No matter how thorough, the plans themselves do not guarantee success. In planning, managers must make short-term and long-term plans. **Short-term plans** are accomplished in several days or weeks. **Long-term plans** are accomplished over several months or years.

Organizing

After a plan is developed, the work must be organized. The manager must identify the people needed to carry out the plan and arrange for the necessary equipment and supplies. Staff development may also be required.

Actuating

Actuating simply means putting the plan into action. The manager needs to inform employees of the plan. All persons involved must understand their roles. Actuation also includes motivating people to work efficiently and effectively and to want to get the job done.

Controlling and Evaluating

Managers must carefully control implementation of plans after the work begins. The quality and quantity of all results must be evaluated. If the results are satisfactory, work can continue. If problems arise, changes must be made and alternate plans may need to be developed. Managers must make adjustments in personnel, equipment, policies, or procedures whenever necessary (Figure 36-7).

In a very small venture, all management functions may be carried out by the same person. As a venture grows and its goals and objectives become more involved, additional managers may be required. At such time, organizational lines of responsibility must be developed and job descriptions written to identify the functions that are performed by each manager.

FIGURE 36-7 The manager must be familiar with all the functions and be able to work with all the employees of the business. *(From Schneeman/ Paralegals in American Law, copyright 1995 by Delmar Publishers)*

SMALL-BUSINESS FINANCIAL RECORDS

Financial records are invaluable tools to the entrepreneur and manager. Good financial management will allow the manager to maintain control of the business venture and to increase profits or reduce losses.

Financial records can reveal which items are selling and which are not. These records can also show the extent of success of each person on the sales force. Financial records should indicate the amount of inventory on hand and how much has been sold. How much profit is made and the venture's total value at any time can also be computed via appropriate financial records.

BIO•TECH CONNECTION

Golden Opportunities Waiting

It is widely known that research in agriscience plays a continuous role in keeping America's food, fiber, and natural resources sectors productive and efficient. What is not so well known is the impact of agriscience research on the businesses and industries that are not seen by the general public as being agribusinesses. In a recent tally, the USDA Agricultural Research Service documented nearly seventy-five companies that were manufacturing products or utilizing processes from fifty-three new technologies developed by that agency. Some of these companies were formed to take advantage of new opportunities to meet apparent needs with emerging technologies. Others were established companies that were able to expand their product lines as a result of the research.

The United States government is looking for new ways to transfer newly developed technologies to private firms. Any decrease in time required for technology transfer from laboratory to product development and distribution reduces the pay-back time for the cost of research. New technology releases provide new opportunities for prospective entrepreneurs. Inquiries should be sent to: USDA Alternative Agricultural and Commercialization Center, 12th and C Street SW, Washington, DC 20250 or telephone (202) 401-4860.

PRODUCT	COMPANY	LOCATION	PRODUCT	COMPANY	LOCATION
No-calorie, high-fiber flour	Mt. Pulaski Products Canadian Harvest	Mt. Pulaski, IL Cambridge, MN	Controlled bulk vegetable fermentation	Trumark, Inc.	Roseville, NJ
Plant virus test kit	Agdia, Inc.	Elkhart, IN	Reducing water content of emulsions, suspensions, and dispersions with highly absorbent starch-containing polymeric compositions	Super Absorbent Co. Grain Processing Co. Worne Biotech Promar, Inc.	Lumberton, NC Muscatine, IA Medford, NJ Milwaukie, OR
Super-Slurper (starch-derived absorbent material)	Super Absorbent Co. Grain Processing Co. Henkel Corporation	Lumberton, NC Muscatine, IA Kankakee, IL			
Traps for stable flies	Tiger Farm Products	Hopedale, MA	Removal of heavy metal ions from waste water	Tetrahedron, Inc.	Stanhope, NJ
Turkey hemorrhagic enteritis virus vaccine	Arko Laboratories Oxford Laboratories Willmar Poultry Co.	Jewell, IA Worthington, MN Willmar, MN	In-ovo vaccination	Embrex, Inc.	Morrisville, NC
Dietary supplement	Monarch Nutri. Labs	Ogden, UT	Super Slurper dewatering cartridge	Central Illinois Manufacturing Co.	Bement, IL
Southwestern corn borer pheromone trap	Great Lakes IPM	Vestaburg, MI	Turkey semen extender	Continental Plastic Corp.	Delavan, WI
Microbial insecticides	Reuter Labs	Haymarket, VA	Cotton and soybean seed	Delta and Pineland Seed Co.	Scott, MI

Some new technologies developed by the Agricultural Research Service of USDA, and the companies that have marketed products using those technologies. *(Adapted from USDA/ARS)*

The small-business owner or entrepreneur needs and uses more detailed financial records than are needed to maintain the typical sales project for a club or organization. Financial records for items such as buildings, fixtures and equipment, credit, debt, and return on investments are examples of records maintained by small businesses. However, balance sheets, profit and loss statements, and inventory and sales reports may be very useful for a group project, as well as for a small business. Some typical financial records are discussed in the following paragraphs (Figure 36-8).

PRODUCT	COMPANY	LOCATION	PRODUCT	COMPANY	LOCATION
Improved fire ant insecticide	Griffin Corp.	Valdosta, GA	Recirculating wiper for agricultural chemicals	Apple Machine Co.	Fort Pierce, FL
Japanese beetle trap	Consep Membranes	Bend, OR	Biopesticide for stone fruit	Fermenta	Painesville, OH
Direct marketing of fruits and vegetables	Three Rivers Produce	Southeastern, OK	Marek's disease vaccine	Select Laboratories, Inc. Tri Bio Laboratories	Gainesville, GA State College, PA
Biochemical and physiological factors influencing turkey egg hatchability	Agrimatic Corp.	Paramount, CA	Biodegradable plastic	Agri-Tech	Peoria, IL
Improve flavor, texture, and juiciness of processed poultry meat	Continental Grain Co.	Pendergrass, GA	Vaccination of chicken embryos for protection against avian coccidiosis	Embrex, Inc.	Morrisville, NC
Discover and develop mycoparasites for biocontrol of selected soilborne plant pathogens	Agracetus, Inc.	Middleton, WI	Methods that affect molecular genetic transfer and transformation in crop plants	Pioneer Hi-Bred, Inc.	Johnston, IA
Bee breeding and genetics for bee stock improvement	Weaver Apiaries, Inc.	Navasota, TX	Physiology of agriculturally important spiroplasmas and mycoplasmas	Agdia, Inc.	Elkhart, IN
Modification of starch to provide controlled delivery of agricultural chemicals	Illinois Cereal Mills, Inc.	Paris, IL	High-solids tomatoes and low-sugar potatoes using tissue culture	Northrup King Co.	Gilroy, CA
Cotton gin system design and evaluation to maximize product quality and minimize processing costs	Lummus Industries, Inc.	Columbus, GA	Construction of expression vectors from Marek's disease virus	Select Laboratories, Inc.	Gainesville, GA
Systems to apply irrigation water efficiently, control intake, and reduce nitrate leaching	Elvert Edgar Oest and Co.	Fruita, CO	Computerized control and management of center pivot irrigation systems	Valmont Industries	Valley, NV
Biological control of weeds using exotic and endemic plant pathogens	Confederated Tribes, Coleville Reservation	Nespelem, WA	Colonization factors expressed by *Campylobacter jejuni* in chicken	Southeastern Poultry and Egg Association	Decatur, GA
New engineering concepts for deciduous fruit production, harvesting, sorting	Agri-Tech	Woodstock, VA	New technologies in cotton ginning	Moisture Systems Corp.	Hopkinton, MA
Molecular biology of orbiviruses to diagnose and characterize bluetongue virus	Veterinary Diagnostic Technology, Inc.	Wheat Ridge, CO	Integrated strategies for managing filth-breeding flies on dairy farms	Olson Products, Inc.	Medina, OH
Apparatus and method for rapid analyses of multiple samples	Lachat Chemicals Alpkem Corp.	Mequon, WI Clackamas, OR	Monoclonal antibodies for diagnosis of Marek's disease	Vineland Laboratories Intervet, Inc.	Vineland, NH Millsboro, DE
Novel durable press finishing of textiles	Duro Finishing Corp.	Fall River, MI	Development of avian expression vectors using serotypes 1, 2, and 3 of Marek's disease virus	Solvay Animal Health, Inc.	Charles City, IA
Milk-like products from peanuts	Seabrook Blanching	Edenton, NC	Techniques for automated nondestructive quality evaluation of horticultural crops	Process Equipment Co.	Tipp City, OH
Highly absorbent polymeric compositions derived from flour	Illinois Cereal Mills Industrial Services Venture Chemicals Polysorb	Paris, IL Bradenton, FL Lafayette, LA Smelterville, ID	Development of methods to protect packaged agricultural products from insect infestation	Trécé	Salinas, CA
Rope wick applicator	Rear's Mfg. Co. Brothers Equip. Co. BCP Manufacturing Co. Agman, Inc. Rodgers Sales Stanley Resser	Eugene, OR Friend, NV Winters, TX Riverside, MO Clarksdale, MI Weldon, IL	Effects of residue decomposition in low-input sustainable agricultural systems	Crops Genetics Intl.	Hanover, MD
Starch-based semi-permeable films	Uni-Star Industries	Cuba, IL	Retention of quality of lightly processed fruits and vegetables	Tree Top, Inc.	Selah, WA

FIGURE 36-8 Financial records are fundamental management tools. *(Courtesy of FFA)*

Balance Sheet

A **balance sheet** is like a photograph of the business at a given time. It shows the assets, liabilities, and owner's investment on a particular date. The equation for the balance sheet is: assets = liabilities + net worth. **Assets** are anything the venture owns, including cash on hand, equipment, and inventory. **Liabilities** are both current and long-term debts. **Net worth** is the owner's investment in the business, including profits as they occur. A sample balance sheet is shown in (Figure 36-9).

Profit and Loss Statement

The **profit and loss statement** projects costs and other expenses against sales and revenue over a period of time. The five basic sections in a profit and loss statement are (Figure 36-10):

1. Total sales
2. Cost of goods sold
3. Gross profit
4. Expenses
5. Net profit

Inventory Report

The **inventory report** includes how many units of each product were received, how many were sold, each item's cost, total sales, and profit. It can be used to make decisions such as reducing the price of slow-selling items, reordering other items, and learning which items are yielding the best profits (Figure 36-11).

SELF-EMPLOYMENT VERSUS OTHER FORMS OF EMPLOYMENT

The decision to continue working for someone else or to open a business is a difficult one to make. A way to help make that decision is to look at the advantages and disadvantages of both working for someone else and working for yourself.

Balance Sheet

Date: _____

ASSETS

Current: Cash   _____
         Merchandise   _____
         Accounts Receivable   _____
Fixed: Land   _____
       Building   _____
       Machinery   _____
       Equipment   _____
Other:
       1.   _____
       2.   _____
       3.   _____

TOTAL ASSETS: _____

LIABILITIES

Current: Notes Payable   _____
         Accounts Payable   _____
Noncurrent:
   Debts more than one-
   year maturity   _____

TOTAL:

NET WORTH: _____

TOTAL LIABILITIES
AND NET WORTH: _____

FIGURE 36-9 Elements of a balance sheet.

Profit and Loss Statement

For period ending _____

TOTAL SALES $75,000
COST OF GOODS SOLD
   Beginning Inventory 55,000
   (Plus) Purchases 10,000
   (Less) Ending Inventory................. 15,000

TOTAL COST OF GOODS SOLD 50,000
   (Beginning Inventory plus Purchases
   minus Ending Inventory)............... 25,000

GROSS PROFIT.............................. 25,000
   (Total Sales minus Total Cost of Goods Sold)

EXPENSES
   Salaries 18,000
   Payroll Taxes............................. 2,500
   Rent.. 4,500
   Advertising................................ 1,000

TOTAL EXPENSES.......................... 26,000

NET PROFIT (LOSS) (before taxes) ..... (1,000)
   (Gross Profit minus Total Expenses)

FIGURE 36-10 Sample profit and loss statement.

Inventory Report

Item No.	Description	Beginning Inventory	Items Sold	Ending Inventory	Total Cost Beg./Inv.	Total Cost End/Inv.	Total Sales	Total Profit
000	Tick Tack Tote	45	45	0	33.75	.00	45.00	11.25
001	Brown Tray	32	32	0	25.60	.00	48.00	22.40
002	Fish Tray	20	20	0	15.00	.00	30.00	15.00
003	Lion Tray	6	6	0	4.80	.00	9.00	4.20
004	Guitar Tray	8	8	0	6.40	.00	12.00	5.60
005	Asst. Animal Tray	34	34	0	25.50	.00	51.00	25.50
006	Elephant Tray	10	10	0	5.00	.00	15.00	10.00
007	Screwdriver Set	22	22	0	11.00	.00	33.00	22.00
008	Brush & Shoehorn	36	36	0	32.40	.00	81.00	48.60
009	Fully Auto. Umbrella	9	9	0	20.70	.00	31.50	10.80
010	Auto. Umbrella	8	8	0	16.00	.00	24.00	8.00
011	Dad's No. 1 Keychain	200	190	10	90.00	4.50	190.00	104.50
012	Grandpa Keychain	100	79	31	45.00	13.95	79.00	47.95
013	Dad's Pad	48	30	18	28.80	10.80	45.00	27.00
014	Mini Screwdriver Set	75	75	0	60.00	.00	150.00	90.00
015	Dad's Plaque	72	72	0	28.80	.00	108.00	79.20
016	Tic Tac Toe	36	36	0	21.60	.00	36.00	14.40
017	Dad's Pen	1	1	0	.80	.00	1.50	.70

FIGURE 36-11 Sample inventory report.

Being an Employee

The advantages of being an employee working for someone else center on security. The salaried employee has no personal financial risk or responsibility to the company for which he or she works. Employees generally put in regular hours. If they work additional hours, they may be paid overtime for those hours. In addition, employees are guaranteed vacation time and fringe benefits such as life and health insurance. In many cases, they are offered retirement plans. An employee can count on a somewhat stable lifestyle, with a fairly accurate idea of what the income will be from year to year. Additionally, the employee may move up the career ladder within the company.

However, there are some disadvantages to working for someone else. The company does not have any financial responsibility to the employee should there be a recession resulting in a cutback in personnel. Some companies have a predetermined salary scale, so the employee could remain at a particular salary level until the right combination of years or experience is met. This is also true of promotions that could be based on years accumulated rather than on merit. The work pattern could become fairly routine. If no positions open within the company, the employee must wait until a position opens or look for a position with another company. The company management controls these decisions, not the employee.

Being Self-Employed

One of the major reasons generally cited for opening a business is that it gives the owner control over his or her own destiny. The owner has the opportunity to set personal goals and recruit a team to help carry out those goals. Successfully meeting those goals results in a sense of achievement and, hopefully, financial reward.

Owning the business and being responsible for the decisions related to the business gives the entrepreneur a sense of independence. The owner's ability to make money is not restricted to a particular level.

The disadvantages of owning a business are not as obvious as are the advantages. For instance, the necessary capital outlay may jeopardize family savings or even the family home. Also, the number of hours required to run the business will mean a definite commitment from the owner as well as the family members involved. Should the company have difficulty, the responsibility for both the management and the financial problems rests with the owner.

CONTRIBUTIONS OF SMALL BUSINESSES

Entrepreneurs have been credited with being the cornerstone of the American enterprise system. Furthermore, many see entrepreneurs as the self-renewing agents of the economic environment in the United States. In recent years, numerous socialistic countries of the world, such as China, Russia, and eastern European countries, have encouraged entrepreneurship after forbidding capitalism for decades.

The role of entrepreneurs in the United States is highlighted by the following:

1. Most businesses in the United States (95 percent) are classified as small by the Small Business Administration.

FIGURE 36-12 The small businesses that dot the landscape of the United States are highly productive and responsive to changing needs of communities. *(Courtesy of Michael Dzaman)*

2. New businesses are formed at a rapid rate (about six hundred thousand per year).
3. Small businesses generate almost half (48 percent) of the United States gross national product (GNP).
4. Small businesses employ almost half (48 percent) of all United States workers.
5. Small businesses created 60 percent of the new jobs in our economy in the last two decades.
6. Small businesses produce 2 ½ times as many innovations (products, services, techniques) as do large businesses (Figure 36-12).

As the American economy continues toward an emphasis on services, the role and importance of small business and entrepreneurship should increase. This should occur because small businesses are especially dominant in the service sector of the economy.

New businesses can be the center of innovation because they are not generally tied to existing ways of doing things. They have a sense of energy and vitality that comes from the entrepreneurial spirit.

The contribution of small businesses to the American private-enterprise system is important. It is likely to remain so in the future.

STUDENT ACTIVITIES

1. Write the Terms to Know and their meanings in your notebook.

2. Develop a collage on the bulletin board illustrating entrepreneurship opportunities in agriscience.

3. Make a table showing the advantages and disadvantages of being (1) an employee and (2) an entrepreneur. The following format is suggested:

Being an Employee	Being Self-Employed
Advantages	*Advantages*
1.	1.
2.	2.
etc.	etc.
Disadvantages	*Disadvantages*
1.	1.
2.	2.
etc.	etc.

4. Participate in a group sales project in the FFA, some other school group, 4-H, or some other community group. Encourage the group to operate the project to help all participants obtain useful business skills.

5. Organize a cooperative business within your class or FFA for buying or marketing a product of interest to the group.

6. Become familiar with the record book and record-keeping system, used by members of your local FFA chapter or 4-H club, for keeping records on individual projects.

7. Become an entrepreneur—own and operate your own business venture.

SELF EVALUATION

A. Multiple Choice

1. The selection of a product or service to be sold for profit is called the
   a. buying function.
   b. distribution function.
   c. promoting function.
   d. selling function.

2. Determining reasons that customers may wish to buy a product or service is called the
   a. buying function.
   b. distribution function.
   c. promoting function.
   d. selling function.

3. Development of a plan to identify ways to make potential customers aware of a product or service is called the
   a. buying function.
   b. distribution function.
   c. promoting function.
   d. selling function.

4. When analyzing competing products or services, which is *not* a factor?

    a. price
    c. service
    b. quality
    d. supply

5. Which is *not* a function of managers?

    a. actuating
    c. organizing people and resources
    b. controlling and evaluating
    d. setting policy

6. A disadvantage of being an employee is

    a. overtime pay.
    c. little or no control over future job.
    b. regular hours.
    d. fringe benefits.

7. A major advantage of owning a business is

    a. control over your destiny.
    b. financial responsibility.
    c. the relationship of business income and family finances.
    d. a shorter work week.

B. MATCHING

_____	**1.** Entrepreneur	a.	Business venture
_____	**2.** Entrepreneurship	b.	Organizer; risk taker
_____	**3.** Balance sheet	c.	Expenses versus revenues
_____	**4.** Profit and loss statement	d.	One who devises
_____	**5.** Inventor	e.	Photograph of a business

Glossary

A horizon – layer near the soil surface consisting of mineral and organic matter.

   horizonte A – capa cerca de la superficie que consta de materia mineral y orgánica.

Abiotic – nonliving.

   abiótico – no viviente.

Abortion – loss of a fetus before it is viable.

   aborto – la muerte de un feto antes de que esté viable.

Accent – distinctive feature or quality.

   acento – rasgo o calidad distintiva.

Accent color – attention-getting color.

   color de enfasis – color llamativo.

Acid – pH of less than 6.9.

   ácido – que tiene valor pH de menos que 6,9.

Acidity – sourness.

   ácidez – de sabor agrio.

Active ingredient – a component that achieves one or more purposes of the mixture.

   principo activo – componente que alcanza uno o más de los objetivos de la mixtura.

Actuating – putting a plan into action.

   accionar – ponerse en acción de un programa.

Acute toxicity – a measurement of the immediate effects of a single exposure to a chemical.

   toxidad grave – una medida de los efectos inmediatos de una sola exposición a una sustancia química.

Add-on-loan – the method used for calculating interest on consumer loans. The loan is repaid in installments of equal payments. Interest is added on at the beginning of the payments.

   préstamo adicional – el método que se usa para calcular el interés en los préstamos de consumidor. El préstamo se paga a plazos de pagos iguales. El interés se agrega al comienzo de los pagos.

Adenine – a base in genes designated by the letter *A*.

   adenino – una base en los genes nominado por la letra «A».

Adjourn – a motion used to close a meeting.

   suspender – una moción que se usa para levantar una junta.

Adventitious roots – root other than the primary root or a branch of a primary root.

   raíz adventicia – una raíz diferente de la raíz fundamental o de una rama de la raíz fundamental.

Aeration – the mixing of air with water or soil to improve the oxygen supply of plants and other organisms.

   aireación – el mezclar del aire con agua o tierra para hacer mejoras en la provisión de oxígeno de las plantas y de otros organismos.

Aeroponics – the plant roots hang in the air and are misted regularly with a nutrient solution.

   cultivo aeropónico – las raíces de las plantas flotan en el aire y son rociados regularmente.

Aerosol – a can with contents under pressure.

   aerosol – lata con el contenido bajo presión.

Age or ripen – leave undisturbed for a period of time.

   madurarse – dejar sin tocar por un período.

Aggregate culture – a material such as sand, gravel, or marbles supports the plant roots.

   cultivo árido – una materia como la arena, el cascajo, o las canicas que sostiene las raíces vegetales.

Aggregates – soil units containing mostly clay, silt, and sand particles held together by a gel-type substance formed by organic matter.

   áridos – elementos de tierra que incluyen principalmente partículas de tierra arcillosa, légamo y arena juntados por una sustancia de tipo de gel formada por materia orgánica.

Agribusiness – commercial firms that have developed with or stem out of agriculture.

agroindustria – las firmas comerciales que han desarrollado con o se derivan de la agricultura.

Agribusiness management – the human element that carries out a plan to meet goals and objectives in an agriscience business.

Administración de agroindustria – la parte humana que lleva a cabo un programa para cumplir con las metas y los objetivos de un negocio de agriciencia.

Agricultural – business, employment, or trade in agriculture, agribusiness, or renewable natural resources.

agrícola – negocio, empleo, o industria de la agricultura, de la agroindustria, o de los recursos naturales renovables.

Agricultural economics – management of agricultural resources, including farms and agribusinesses.

economía agropecuaria – administración de recursos agrícolos, inclusive las fincas y las agroindustrias.

Agricultural education – teaching and program management in agriculture.

educación agrícola – enseñanza y dirección de programa en la agricultura.

Agricultural engineering – application of engineering principles in agricultural settings.

ingeniería agrícola – aplicación de los principios de ingeniería dentro del ambiente agrícola.

Agriculture – activities concerned with the production of plants and animals, and the related supplies, services, mechanics, products, processing, and marketing.

agricultura – actividades que tratan de la producción de plantas y de animales, y los artículos, servicios, mecanismo, productos, transformación, y comercio relativos.

Agriculture/agribusiness and renewable natural resources – broad range of activities in agriculture.

agricultura/agroindustria y los recursos naturales renovables – una gran variedad de actividades de la agricultura.

Agriscience – all jobs relating in some way to plants, animals, and renewable natural resources. Also, the application of scientific principles and new technologies to agriculture.

agriciencia – todos los puestos que, de alguna manera, tienen que ver con las plantas, animales, y recursos naturales renovables. Además, la aplicación de los principios científicos y la tecnología reciente a la agricultura.

Agriscience mechanics – design, operation, maintenance, service, selling, and use of power units, machinery, equipment, structures, and utilities in agriscience.

mecanía de agrociencia – la industria que acarrea, clasifica, transforma, envasa, y comercializa los productos desde los orígenes de producción.

Agriscience processing, products, and distribution – industry that hauls, grades, processes, packages, and markets commodities from production sources.

transformación, productos, y distribución de agrociencia – diseño, manejo, mantenimiento, reviso, venta, y uso de los motores, maquinaria, herramientas, estructuras y utilidades de la agrociencia.

Agriscience professions – professional jobs dealing with agriscience situations.

profesiones de agrociencia – puestos profesionales que se ocupan de los asuntos de agrociencia.

Agriscience supplies and services – businesses that sell supplies and agencies that provide services for people in agriscience.

provisiones y servicios de agrociencia – los negocios que venden provisiones, y las agencias que prestan servicios para las personas de agrociencia.

Agronomy – science of soils and field crops.

agronomía – ciencia de las tierras y del fruto del campo.

Air – colorless, odorless, and tasteless mixture of gases.

aire – mezcla de gases incolora, inodora e insípida.

Air layering – plant propagation by girdling a plant stem, wrapping with sphagnum peat, and protecting with plastic.

acodo al aire – propagación de las plantas por anular el tallo de una planta, envolverlo con turba de esfagno, y protegerlo con plástico.

Alkaline – pH of more than 7.1.

alcalino – que tiene valor pH de más que 7,1.

Alkalinity – sweetness.

alcalinidad – lo dulce.

Alluvial – soils transported by streams.

aluvial – tierra que queda después de retirarse las aguas.

Amend – a type of motion used to add to, subtract from, or strike out words in a main motion.

enmendar – un tipo de moción que se usa para aumentar, substraer, o quitar las palabras de una moción principal.

Amendment – in addition to; change in.

enmienda – además de; rectificación en o cambio de.

Ammonia/nitrite/nitrate – the chemical components generated during biological breakdown of animal wastes.

amoníaco/nitrito/nitrato – los componentes químicos que se producen durante la descomposición biológica de los desechos animales.

Amortized loan – loan repaid in equal installments of interest.

préstamo amoritzado – préstamo reembolsado a plazos iguales de interés.

Amphibian – organisms that complete part of their life cycle in water and part on land.

anfibios – organismos que cumplen parte del ciclo de la vida en el agua y otra parte en la tierra.

Anatomy – various parts of the body.

anatomía – varias partes del cuerpo.

Anemia – low red blood cell count often caused by a deficiency of iron, copper, or pyridoxine.

anemia – recuento bajo de glóbulos rojos que muchas veces resulta de una carencia de hierro, de cobre o de piridoxino.

Angiosperm – a plant with its seeds enclosed in a pod or seed case.

angiospermas – plantas cuya semilla está envuelta por una vaina o una cápsula.

Angora – wool from wool-producing rabbits.

angora – lana de los conejos que dan la lana.

Animal science technology – use of modern principles and practices in animal growth and management.

tecnología de las ciencias de animales – uso de principios y prácticas modernos en la cría y el manejo de animales.

Animal sciences – animal growth, care, and management.

ciencias de animales – cría, cuido, y manejo de animales.

Anions – ions that are negatively charged.

aniones – iones que son cargados negativamente.

Annual ring – ring in a cross section of a tree root, trunk, or limb representing one year's growth.

capa cortical – capa de una sección transversal de raíz, tronco, o rama de un árbol que representa el crecimiento de un año.

Annual weed – a weed that completes its life cycle within one year.

mala hierba anual – una mala hierba que cumple el ciclo de vida dentro de un año.

Anorexia – the result of too little nutrition.

anorexia – el resultado de la falta de nutrición.

Anther – portion of the male part that contains the pollen.

antera – parte de lo macho que contiene el polen.

Antibiotic – substance used to help prevent or control certain diseases of animals.

antibiótico – sustancia que se usa para evitar o controlar ciertas enfermedades de los animales.

Apiary – area where beehives are kept.

apiario – sitio en donde se mantienen las colmenas.

Apiculture – beekeeping.

apicultura – arte de criar abejas.

Appressorium – the swollen tip of a fungal hypha which allows the fungus to attach itself to a plant.

apresorio – la punta hinchada de la hifa fungosa que permite que el fongo se una a una planta.

Aquaculturalist – trained professional involved in the production of aquatic plants and animals.

piscicultor – profesional educado, dedicado a la producción de plantas y animales acuáticos.

Aquaculture – raising of finfish, shellfish, and other aquatic animals under controlled conditions. Also, the management of the aquatic environment for production of plants and animals.

piscicultura – la crianza de peces, mariscos, crustáceos, y otros animales acuáticos bajo condiciones controladas. Adicionalmente, la dirección del ambiente acuático para la producción de plantas y animales.

Aquifer – water-bearing rock formation.

acuífero – formación terciaria que embebe agua.

Arboriculture – care and management of trees for ornamental purposes.

arboricultura – cuidado y manejo de árboles para fines de adorno.

Area of cell division – area of the root tip where new cells are formed.

área de división celular – área de la punta de raiz en donde se forman nuevas células.

Area of cell elongation – portion of the new root where the cells start to become specialized and begin their function.

área de estiramiento celular – parte de la nueva raíz en donde las células se ponen a diferenciarse y desempeñar su función.

Area of cell maturation – area where cells mature.

área de maduración celular – área en donde se maduran las células.

Arid – an area deficient in rainfall; dry.

árido – un área que carece de precipitación.

Arthropod – eight-legged animals such as spiders and mites.

artrópodo – animales de cuatro patas como arañas y acáridos.

Artificial insemination – the placing of sperm cells in contact with female reproductive cells by a method other than natural mating.

Inseminación artificial – el meter de las células de esperma en contacto con células reproductivas de la hembra mediante método de otra manera que el apareamiento natural.

Asbestos – heat- and friction-resistant material.

asbesto – materia que es resistente a calor y fricción.

Asexual – propagation utilizing a part or parts of one parent plant.

asexual – reproducción que utiliza una o más partes de una sóla planta madre.

Assets – anything the business owns.

activos – total de lo que posee una empresa.

Asymmetrical – not equal on both sides of center.

asimétrico – que ambos lados no son iguales relativos al centro.

Auction markets – markets where products are sold by public bidding.

mercados de subasta – mercados donde se venden los bienes por hacer pujas.

Auctioneer – person who conducts the sales at auction markets.

subastador – persona que dirige las ventas en los mercados de subasta.

Axillary bud – bud that occurs in the axil of the leaf.

botón axilar – botón que se produce en la axila de la hoja.

B

B & B – an abbreviation for balled and burlapped.

«B y B»– una abreviatura por «balled y burlapped».

B horizon – soil below the A horizon or topsoil and generally referred to as subsoil.

horizonte B – la tierra debajo del horizonte A o de la capa superficial del suelo, y generalmente es conocido a nombre de subsuelo.

Bacteria – one-celled, microscopic plants.

bacterias – microorganismos vegetales unicelulares.

Balance – state of quality and calm between items.

equilibrio – estado de bienestar y tranquilidad entre los cuerpos.

Balled and burlapped – plants that have been dug in the field, wrapped in burlap, and laced with a heavy twine.

«balled y burlapped» – plantas que han sido arrancados del campo, envueltas en arpillera, y atados con bramante grueso.

Balling gun – a device used to place a pill in an animal's throat.

pistola de «balling» – aparato que se usa para meter una pastilla en la garganta de un animal.

Band application – placing fertilizer about 2 in. to one side and slightly below the seed.

abonado en fajas – meter el abono más o menos a os pulgadas de un lado y un poco debajo de la semilla.

Bantum – miniature chicken.

gallina bántum – gallina miniatura.

Barbed wire – wire with sharp points used to discourage livestock from touching fences.

alambre de espino – alambre con puntas agudas que se usa para que el ganado se haga desistir de tocar las cercas.

Bare rooted – plants that are dug and the soil is shaken or washed from the roots.

deraices desnudas – plantas que son arrancadas y la tierra es sacudida o lavada de las raíces.

Barrow – castrated male of the swine family.

cerdo castrado jóven – macho castrado de la familia cochina.

Bases – genetic material that connects strands of DNA.

bases – nombre original por materia genética que conecta fibras de ADN.

Basic color – background color.

color básico – color de fondo.

Bays – natural open water areas along coastlines where freshwater and seawater mix.

bahías – áreas abiertas naturales de agua a lo largo de las costas donde se mezclan el agua dulce y el agua de mar.

Bedrock – the area below horizon C consisting of large soil particles.

roca de fondo – el área debajo del horizonte C que consiste en partículas de tierra más grandes.

Beef – meat from cattle.

carne del ganado – carne del ganado.

Note: In Spanish, there is no equivalent to "beef"—it is just called "cattle meat."

BelRus – superior baking potato bred to grow well in the Northeast.

«BelRus» – patata al horno que es producida para cultivarse bien en el noroeste de los EE.UU.

Beltsville Small White – breed of turkey that weighs only 8 to 12 lb at maturity.

«Beltsville Small White» – raza de pavo que pesa solamente 8 a 12 libras al madurarse.

Biennial weed – a weed that will live for two years.

mala hierba bienal – una mala hierba que vive por dos años.

Binomial – having two names.

binomio – que tiene dos nombres.

Bio – life or living.

bio – vida o viviente.

Biochemistry – chemistry as it applies to living matter.

bioquímica – química que se aplica a la materia viviente.

Biological control – pest control that uses natural control agents.

control biológica – control de los insectos o animales nocivos que utiliza los agentes de control natural.

Biology – basic science of the plant and the animal kingdoms.

biología – la ciencia básica de los reinos vegetal y animal.

Biotechnology – use of cells or components of cells to produce products or processes.

biotecnología – uso de células o componentes de células para producir productos o procesos.

Biotic diseases – diseases caused by living organisms.

enfermedades bióticas – enfermedades causadas por organismos vivientes.

Bit – metal mouthpiece used to control the horse.

bocado – aparato que se mete sobre la boca del caballo para frenarlo.

Blade – the upper portion of the turfgrass leaf.

brizna – la parte superior de la hoja de césped.

Blanching – the brief scalding of food prior to freezing.

escaldar – el sumergir de los alimentos por un tiempo breve en agua hirviendo antes de congelación.

Bleeding out – draining blood from an animal.

hacer sangrías – drenar la sangre de un animal.

Blemish – in horses, any abnormality that does not affect the use of the horse.

defecto – de los caballos, cualquiera abnormalidad que no afecta el uso del caballo.

Block beef – meat sold over the counter to consumers.

carne descuartizada – carne de res que se vende al por menor a los consumidores.

Boar – male of the swine family.

verraco – el macho de la familia cochina.

Board foot – a unit of measurement for lumber that equals $1 \times 12 \times 12$ in.

«board foot» (b.ft.) – unidad de medida de los maderos que es igual a 1" por 1" por 12".

Bogs – water-logged areas.

ciénaga – área de aguas estancadas.

Bone – the main component of the skeletal system.

hueso – el componente principal del esqueleto.

Bone marrow – material inside of bones that makes blood cells.

médula de hueso – materia dentro de los huesos que fabrica los glóbulos.

Border plant – a planting that is used to separate some part of the landscape from another. It might be used as a fence or a windbreak.

arriate – un cuadro de plantas que se usa para separar alguna parte del paisaje de la otra. Se puede usar como cerca o también protección contra el viento.

Bovine somatotropin (BST) – hormone that stimulates increased milk production in cows.

somatotropino bovino – una hormona que estimula un aumento de secreción láctea en las vacas.

Brackish water – waters influenced by tide and river flow with intermediate salinity of 3 to 22 percent.

agua salobre – aguas influidas por la marea y el flujo de río, que tienen salinidad de 3 a 22 por ciento.

Bracts – modified leaf that is often brightly colored and showy.

brácteras – hoja modificada que muchas veces es colorada brillantemente y es llamativa.

Bran – skin or covering of a wheat kernel.

salvado – cascarilla del grano de los cereales.

Breed – a group of animals having similar physical characteristics that are passed on to their offspring.

raza – grupo de animales que tienen características físicas semejantes que se pasan para adelante a su descendencia.

Broadcast planter – scatters seed in a random pattern.

sembrador al voleo – éste esparce las semillas en una pauta hecha al azar.

Broadcasting – scattering seeds rather than sowing them in rows.

sembrar al voleo – esparciendo al aire las semillas en vez de sembrarlas en hileras.

Broiler – young chicken grown for meat.

pollo tomatero – gallina jóven que se cría para comer.

Buck – male of the goat or rabbit family.

cabrón – macho de la cabra.
conejo – macho de la coneja.

Bulbs – short underground stem surrounded by many overlapping, fleshy leaves.

bulbos – tallo corto, subterráneo envuelto por muchas hojas carnosas y traslapadas.

Bull – male of the cattle family.

toro – macho del ganado vacuno.

Bunch-type grass – a turfgrass that grows in clumps.

césped de manojo – un tipo de césped que crece en manojos.

Business meeting – a gathering of people working together to make decisions.

sesión (de negocio) – asamblea de personas trabajando juntas para hacer decisiones.

Butterfat – fat found in milk.

grasa de la leche – grasa que se encuentra en la leche.

Buying function – selection of a product or service to be marketed or sold for a profit.

función adquisitivo – selección de un producto o servicio para que se ponga en venta o se venda algo con ganancia.

By-product – a secondary product left from the production of a primary commodity.

subproducto – un producto solbrante que queda después de la producción de un mercancía primaria.

C horizon – soil below the B horizon; it is important for storing and releasing water to the upper layers of the soil.

horizonte C – la tierra debajo del horizonte B; es importante para la guarda y suministración del agua a las capas superiores de la tierra.

Calf – young member of the cattle family.

becerro – un jóven del ganado vacuno.

Calve – to give birth in cattle.

parir – en el ganado, dar a luz a la cría.

Calyx – group of sepals of a flower.

cáliz – grupo de los sépalos de la flor.

Cambium – growth layer in a tree root, trunk, or limb.

cambium – capa de crecimiento de raíz, tronco, o ramo del árbol.

Cane cutting – stems are canelike and cuttings are cut into sections that have one or two eyes or nodes.

esqueje a la caña – los tallos son parecidos a la caña y se cortan las talas en secciones que tienen uno o dos yemas o nudos.

Canning – storing food in airtight containers.

enlatado – guardar los comestibles en recipientes herméticos.

Cannula – blunt needle.

cánula – aguja desafilada.

Canopy – the top of the plant that has the framework and leaves.

partes aéreas – la parte superior de la planta que tiene la estructura y las hojas.

Canter – in horses, a fast three-beat gait.

medio galope – de caballos, un paso rápido de tres tiempos.

Capability class – soil classification indicating the most intensive but safe land use.

clase de capacidad – clasificación de suelos indicando el uso más intensivo pero seguro de la tierra.

Capability subclass – soil group within a class designated by a small letter.

subclase de capacidad – grupo de suelos dentro de una clase a que se denomina por letra minúscula.

Capability unit – soil group within a subclass.

unidad de capacidad – grupo de suelos dentro de una subclase.

Capillary water – water held by soil particles and available for plant use.

agua capilar – agua sujetada por partículas de tierra y disponible para el uso de las plantas.

Capital – money or property.

capital – dinero o propiedad.

Capital investment – money spent on commodities that are kept six months or longer.

inversión de capital – dinero gastado en mercancía que se guarda por seis meses o más.

Capon – castrated male chicken used for meat.

capón – gallino macho castrado que se usa para comer.

Carbohydrate – starches and sugars that provide energy in the diet.

carbohidrato – féculas y azúcares que proporcionan energía en la dieta.

Carbon dioxide – by-product of respiration.

bióxido de carbono, dióxido de carbono – subproducto de respiración.

Carbon monoxide – colorless, odorless, and highly poisonous gas; carbon dioxide and water combine to make plant food and release oxygen.

monóxido de carbono – gas incoloro, inodoro, y sumamente tóxico; bióxido de carbono y agua se combinan para hacer alimento vegetal y soltar oxígeno.

Carcass – body of meat after the animal has been eviscerated.

res abierta en canal – el cuerpo de carne después de haber sido destripado el animal.

Carcinogen – a chemical capable of producing a tumor.

agente cancerigeno – una sustancia química capaz de producir un tumor.

Carrying capacity – the number of animals that a pasture will provide feed for.

capacidad de carga – la cantidad de animales a las que el pasto proporcionará pienso.

Casein – predominant protein in milk.

caseína – proteina prevalente de la leche.

Cash crop – a crop grown for cash sale.

cultivo comercial – una cosecha cultivada para vender por pago al contado.

Castration – removal of the male sex organs.

castración – ablación de los órganos del macho necesarios a la generación.

Cations – ions that are positively charged.

cationes – iones que son cargados positivamente.

Cattle – members of the bovine family over one year old.

ganado – miembros de la familia vacuna que tienen más de un año de edad.

Causal agent – an organism that produces a disease.

agente causal – un organismo que produce una enfermedad.

Cell – a unit of protoplasmic material with a nucleus and cell walls.

célula – unidad de materia protoplásmica con núcleo y paredes.

Cellulose – woody fiber parts that make up plant cell walls.

celulosa – partes leñosas de fibra que constituyen las paredes de célula vegetal.

Central nervous system – the brain and the spinal cord.

sistema nervioso central – el cerebro y la médula espinal.

Cereal crop – grasses grown for their edible seeds.

cultivo de cereal – hierbas que son cultivadas por sus semillas comestibles.

Cheese – milk that is exposed to bacterial fermentation.

queso – leche que está expuesta a la fermentación bacterial.

Chemical control – the use of pesticides for pest control.

control químico – el uso de pesticidas para control de los insectos y animales nocivos.

Chemistry – science dealing with the characteristics of elements or simple substances.

química – la ciencia que trata de las características de los elementos o sustancias sencillas.

Chevon – meat from goats.

carne de chivo – carne de ganado caprino.

Chick – newborn chicken or pheasant.

polluelo – gallina or faisán recién nacido.

Chlorofluorocarbons – group of compounds consisting of chlorine, fluorine, carbon, and hydrogen used as aerosol propellants and refrigeration gas.

clorofluorocarbonos – grupo de compuestos que se compone de cloro, flúor, carbono, e hidrógeno usados como propulsores de aerosol y de gas de refrigeración.

Chlorophyll – green pigment in leaves.

clorofila – pigmento verde de las hojas.

Chloroplast – membrane-bound body inside a cell containing chlorophyll pigment; necessary for photosynthesis.

cloroplasto – dentro de la célula, cuerpo tapado con membrana que contiene la clorofila; necesario para la fotosíntesis.

Chlorosis – yellowing of the leaf.

clorosis – el volverse amarillo de la hoja.

Chromosome – the rodlike carrier for genes.

cromosoma – cuerpo en forma de bastoncillo que lleva los genes.

Chronic toxicity – a measurement of the effect of a chemical over a long period of time and under lower exposure doses.

toxidad crónica – una medida del efecto de una sustancia química durante un período largo de tiempo y bajo dosises más pequeños de exposición.

Circulation – activity in a plantscaped area.

circulación – actividad en un área ajardinado.

Circulatory system – the system that provides food and oxygen to the cells of the body and filters waste materials from the body.

aparato circulatorio – el conjunto de órganos que proporciona los alimentos y oxígeno a las células del cuerpo, y filtra los residuos desde el cuerpo.

Clay – smallest of soil particles; less than .002 mm.

arcilla – las partículas más pequeñas del suelo; menos de ,002 mm.

Clean culture – any practice that removes breeding or over-wintering sites of a pest.

cultivo limpio – cualquiera práctica que elimina los criaderos o sitios de hibernación de los animales o insectos nocivos.

Clearcut – removal of all marketable trees from an area.

corte raso – el cortar y traslado de todos los árboles comerciales de un área.

Climate – the weather conditions of a specific region.

clima – las condiciones atmosféricas de una región específica.

Climatic conditions – temperature, temperature range, and precipitation.

condiciones climáticas – temperatura, variedad de temperatura, y precipitación.

Clods – a lump or mass of earth.

terrón – masa de tierra.

Clone – exact duplicate.

clon – una duplicación exacta.

Cloning – genetically generating offspring from non-sexual tissue.

clonar – la generación genética de progenie de tejido no sexual.

Closebreeding – mating of father to daughter, mother to son, or brother to sister.

procreación en consanguinidad – acoplamiento de padre a hija, madre a hijo, o hermano a hermana.

Coach horse – type of horse developed to pull stagecoaches.

caballo de coche – tipo de caballo desarrollado para arrastrar diligencias.

Coarse texture – a turfgrass with a wide leaf blade.

grano grueso – césped que tiene una brizna ancha.

Coccidiosis – disease of poultry which costs growers nearly $300 million a year.

coccidiosis – enfermedad de los aves de corral que les cuesta a los agricultores casi 300 milliones de dólares al año.

Cock – adult male chicken or pheasant.

gallo – macho adulto de los aves de corral.

Cockerel – young male chicken or pheasant.

gallo jóven – gallo jóven.
faisán jóven – faisán jóven.

Note: Spanish does not have an equivalent to "cockerel"—a cockerel is referred to as either a "young male chicken" or "young male pheasant."

Collagen – chief component of connective tissue.

colagen – componente principal de tejido conjuntivo.

Collar – light-green or white banded area on the outside of a leaf blade.

cuello – parte alargada y estrecha de color blanco o verde ligero en el exterior del limbo de una hoja.

Colluvial – soils deposited by gravity.

coluvial – suelos depositados por la gravedad.

Color breed – breed of horses based on color.

cría selectiva basado en su color – cría selectiva basado en su color.

Note: The translation for «color breed» is the same as that of the English definition provided.

Colostrum – first milk produced by mammals; high in antibodies.

calostro – primera leche producida por mamíferas; tiene niveles elevados de anticuerpos.

Colt – young male horse or pony.

potro – caballo macho jóven.

Comb – wax material on and in which bees store honey.

panal – material de cera en y sobre la cual las abejas depositan la miel.

Combine – machine that cuts and threshes grain in the field.

segadora trilladora – máquina que siega y trilla el grano en el campo.

Commensalism – one type of wildlife living in, on, or with another without either harming or helping it.

comensalismo – una clase de fauna que vive en, encima de, o por, otra clase sin dañarla ni tampoco ayudarla.

Commission – fee for selling a product.

comisión – honorario por la venta de un producto.

Commodity Credit Corporation – lends money for production of farm commodities.

Sociedad Anómina de Crédito para Mercancía – presta dinero para la producción de productos agrícolos.

Commodity exchange – organization licensed to manage the buying and selling of commodities.

la bolsa – organización autorizada para administrar la compra y la venta de productos.

Common name – name given to a pesticide by a recognized committee on pesticide nomenclature.

nombre genérico – nombre dado a un pesticida por un cómite reconocido de nomenclatura de pesticida.

Comparative advantage – putting the emphasis in the area where the most returns will be received.

ventaja relativa – poner la importancia en el área donde se recibirán la mayoría de los rendimientos.

Competition – two types of wildlife eating the same source of food.

competición – dos clases de fauna comiendo de la misma fuente de alimentación.

Complete fertilizer – a fertilizer having nitrogen, phosphorus, and potassium.

abono completo – abono que tiene nitrógeno, fósforo, y potasio.

Compost – mixture of partially decayed organic matter.

estiércol vegetal – mixtura de materia orgánica descompuesta en parte.

Compound – a chemical substance that is composed of more than one element.

compuesto – una sustancia química que se compone de más de un elemento.

Compound leaf – two or more leaves arising from the same part of the stem.

hoja digitada – dos o más hojas brotando del mismo rabo.

Concentrate – feeds high in total digestible nutrients and low in fiber.

concentrado – pienso con alto nivel de alimentos nutritivos digestibles totales y bajo en fibra vegetal.

Condensed milk – milk that has had water removed and sugar added.

leche condensada – leche de la que se ha quitado el agua y añadido el azúcar.

Condominium – apartment building or unit in which the apartments are individually owned.

condominio – casa de pisos o unidad en la cual tiene su propio dueño cada apartamento.

Conifer – evergreen trees with needlelike leaves.

conífera – árbol de hoja perenne con hojas como agujas.

Conservation tillage – techniques of soil preparation, planting, and cultivation that disturbs the soil the least and leaves the maximum amount of plant residue on the surface.

labranza de conservación – técnicas de preparación del terreno, siembra y cultivación que el menos perturba la tierra y que deja la cantidad máxima de residuos vegetales en la superficie.

Consumers – people who use a product.

consumidores – personas que usan un producto.

Contact herbicide – a herbicide that will not move or translocate within the plant.

herbicida de contacto – un herbicida que no se mueve ni se desplaza dentro de la planta.

Contagious – diseases that can be spread by contact.

contagiosas – enfermedades que se pueden transmitir por contacto.

Container grown – plants that are grown in pots or other type of container and are shipped in the container.

crecido de contenedor – plantas que se crecen en macetas u otros tipos de contenedor y son transportadas en el contenedor.

Contaminate – to add material that will change the purity or usefulness of a substance.

contaminar – añadir material que cambiará la pureza o utilidad de una sustancia.

Continuous flow systems – the nutrient solution flows constantly over the plant roots.

sistemas de flujo continuo – la solución nutritiva fluye constantemente sobre las raíces de las plantas.

Contour – level line around a hill.

curva de nivel – línea a nivel alrededor de un cerro.

Contour practice – operations such as plowing, discing, planting, cultivating, and harvesting across the slope and on the level.

trabajos de cosecha a lo largo del declive – actividades así como arar, abrir los surcos con discos, sembrar, cultivar, y cosechar a lo largo del declive y sobre la llanura.

Controlled atmosphere – adjusting oxygen and carbon dioxide where food is stored and while it is being transported.

atmósfera controlada – que ajusta los oxígeno y dióxido de carbono donde se almacena la comida, y mientras está transportada.

Convection oven – an oven that heats food by circulating hot air.

horno de convección – un horno que calenta la comida por circular aire caliente.

Conventional tillage – land is plowed, turning over all crop residues.

labranza clásica – se ara la tierra, volteando todos los residuos de la cosecha.

Cool-season turfgrass – turfgrass adapted to the northern regions of the U.S. which grow best at 60° to 75° F.

tepe de temporada fresca – tepe adaptado a las regiones norteñas de los Estados Unidos, las cuales crecen mejor a los 60° hasta 65° F.

Cooperative Extension Service – an educational agency of the U.S. Department of Agriculture and an arm of land grant state universities.

Servicio Cooperativo de Extensión – una agencia educativa del Departamento de Agricultura de los Estados Unidos, y un ramo de las universidades estatales de concesión de terrenos.

Cooperatives – groups of producers who join together to market a commodity.

cooperativas – grupos de productores que se juntan para poner en venta un producto.

Corms – short, flattened underground stem surrounded by scaly leaves.

bulbos – tallo corto aplanado, subterráneo envuelto por hojas escamosas.

Corn picker – machine that removes ears of corn from stalks.

recolectora de maíz – máquina que recoge las espigas de maíz de los tallos.

Corolla – collectively, all of the petals of the flower.

corola – colectivamente, todos los pétalos de una flor.

Cottage cheese – a product made of skimmed milk.

requesón – un producto hecho de leche desnatada.

Cotton gin – machine that removes cotton seed from cotton fiber.

desmotadora – máquina que saca la semilla a la fibra del algodón.

Courage – willingness to proceed under difficult conditions.

valor – la voluntad de seguir en adelante a pesar de condiciones duras.

Course-textured (sandy) soil – loose and single grained soil.

tierra gruesa – tierra desmontable y de grano único.

Cover crop – close-growing crop planted to protect the soil and prevent erosion.

planta protectora – cultivo que crece muy junto y se planta para proteger las tierras de la erosión.

Cow – female of the cattle family that has given birth.

vaca – hembra de la familia vacuna que ha parido.

Cream – milk containing 40 percent butterfat.

nata – leche que tiene 40 por ciento de grasa de leche.

Credit – money borrowed.

crédito – dinero prestado.

Creep feed – feed provided to young animals to supplement their mothers' milk.

alimentación suplementaria – pienso suministrado a la cría de animales para suplementar la leche de amamantamiento.

Crop rotation – planting of different crops in a given field every year or every several years.

rotación de cultivos – el sembrar de cultivos diferentes en un campo determinado cada año o cada dos o tres años.

Crop science – use of modern principles in growing and managing crops.

ciencia de cultivos – el uso de principios modernos en el cultivar y manejar de cultivos.

Crossbreed – animal with parents of two different breeds.

híbrido – animal que procede de dos individuos de distinta raza.

Crown – an unelongated stem of major meristematic tissue of turfgrass.

corona – un tallo no alargado de tejido meristemático mayor de césped.

Crumbs – aggregates.

gránulos – áridos.

Crustaceans – aquatic organisms with exoskeletons that molt during growth.

crustáceos – organismos aquáticos con dermatoesqueleto que mudan durante el crecimiento.

Cull – remove from the herd.

entresacar – sacar de entre el rebaño.

Cultivar – a group of plants with a particular species that has been cultivated and is distinguished by one or more characteristics; through sexual or asexual propagation, it will keep these characteristics.

variedad obtenida por selección – un grupo de plantas con una especie particular que ha sido cultivada y que se distingüe por una o más características; por la propagación sexual o asexual mantendrá estas características.

Cultivation – the act of preparing and working soil.

cultivo – la acción de preparar y labrar la tierra.

Cultural control – pest control that adapts farming practices to better control pests.

compaña contra los animales dañinos – control de los animales o insectos nocivos que adapta las prácticas agrícolas para mejorar el control de ellos.

Cuticle – top-most layer of the leaf; waxy protective covering of the leaf.

cutícula – la superficie de la hoja; capa protectora cerosa de la hoja.

Cutting – vegetative part removed from the parent plant and managed so it will regenerate itself.

tala – la parte vegetativa que se quita de la planta progenitora para que se regenere.

Cytosine – a base in genes designated by the letter *C*.

citosino – una base en los genes nominado por la letra «C».

Dam – mother.

progenitora – madre.

Deciduous – plants that lose their leaves every year.

caduco – plantas cuyas hojas se caen cada año.

Decompose – decay into soil-building materials and nutrients.

descomponerse – marchitarse y corromperse, convirtiéndose en materiales y alimentos nutritivos que fortalecen la tierra.

Deficiency – less available than needed for optimum growth.

carencia – cuando no se obtiene lo necesario para el crecimiento óptimo.

Deficiency diseases – conditions resulting from improper levels or balances of nutrients.

enfermedades por carencia – estados de salud resultando de niveles bajos o equilibrios malos de alimentos nutritivos.

Defoliate – to strip a plant of its leaves.

deshojar – quitar las hojas a una planta.

Dehydration – removing moisture with heat.

deshidratación – quitar el agua a algo, usando calefacción.

Dehydrator – a device for drying food.

deshidratador – un aparato para secar alimentos.

Dehydrofrozen – removing moisture after partial cooking and then freezing.

deshidrocongelado – quitar el agua a algo después de guisarlo parcialmente, y entonces congelarlo.

Delta – land created when soil is deposited by water at the mouth of a river.

delta – terreno formado cuando tierra es depositada por las aguas en la desembocadura.

Demand – the amount of a product wanted at a specific time and price.

demanda – conjunto de productos pedidos en un tiempo específico y por un precio específico.

Deoxyribonucleic acid (DNA) – coded genetic material in a cell.

ácido desoxirribonucleico (ADN) – la materia genética codificada de una célula.

Dermatitis – a skin disorder caused by the deficiency of riboflavin, niacin, or biotin.

dermatitis – un trastorno de la piel resultando de la carencia de riboflavina, ácido nicotínico, o biotina.

Development limitations – limitations imposed by resources or use in a plantscape.

límites de explotación – límites impuestos a un paisaje por recursos o utilidad.

Dicotyledon (dicot) – plant with seed leaves.

dicotiledónea – plantas con dos hojas de semilla.

Digestive system – system that provides food for the body of the animal and for all of its systems.

aparato digestivo – el conjunto de órganos que proporciona los alimentos al cuerpo del animal y para todos sus sistemas.

Diminishing returns – point where each additional unit of input decreases the output or returns.

rendimientos decrecientes – el punto en donde cada unidad adicional de insumo disminuye el rendimiento o el reembolso.

Dipping – the process of treating animals for external parasites by walking or swimming them through a medicated bath.

desinfección por inmersión – el proceso de curar animales de los parásitos exteriores por recorrerles por un balneario medicinal.

Direct sales – refers to the selling of animals directly to processors by the producer.

venta sin intermediario – se refiere a la venta de animales directamente a los transformadores por el productor.

Disassembly process – dividing the carcass into smaller cuts.

carnear – tomar la res abierta en canal y descuartizarla.

Discount loan – a loan where interest is subtracted from the principal at the time the loan is made.

empréstito de descuento – un empréstito en que se substraye el interés del principal en el momento en que se presta.

Disease triangle – the term applied to the relationship of the host, pathogen, and the environment in disease development.

triángulo de enfermedad – aplícase al desarrollo de una enfermedad: la relación entre el huésped, el agente patógeno, y el medio ambiente.

Diseases – infective agents that result in lowered health in living things.

enfermedades – agentes infecciosos que resultan en el mal estado de salud en los cuerpos vivientes.

Disinfect – to rid of infective agents.

desinfectar – eliminar agentes infecciosos.

Disinfectants – materials that destroy infective agents such as bacteria and viruses.

desinfectantes – materiales que matan a los agentes infecciosos como bacterias y viruses.

Disinfest – to rid of vermin.

desinfestar – eliminar los bichos.

Dissolved oxygen – oxygen that is dissolved in water.

oxígeno disuelto – oxígeno que es disuelto en agua.

Distribution function – physical organization and delivery of product or service.

función de distribución – organización física y expedición de un producto o servicio.

Distributor – person or business storing food for transport to regional markets.

distribuidor – persona o empresa almacenando alimentos para transportarlos a mercados regionales.

DNA – deoxyribonucleic acid.

ADN – ácido desoxirribonucleico.

Dock – remove or shorten tails of certain animals.

cortar la cola – quitar o acortar las colas de ciertos animales.

Doe – female of the goat or rabbit family.

cabra – hembra del cabrón.
coneja – hembra del conejo.

Domestic – of or pertaining to the household.

doméstico – del hogar o perteneciendo al hogar.

Domesticate – the act of taming an animal for use of man.

domesticar – la obra de domar al animal para que lo puede usarlo el hombre.

Dominant – gene that expresses itself to the exclusion of other genes.

dominante – el gene que se manifiesta con exclusión de otros genes.

Donkey – member of the horse family with long ears and a short, erect mane.

burro – miembro de la familia de caballos con orejas grandes y crines rapados.

Dormant – resting stage, no active growth.

dormición – estado de descanso; no hay crecimiento activo.

Double-eye cutting – used when the plant has leaves that are opposite.

esqueje de yema doble – empleado cuando la planta tiene hojas simétricas.

Draft – animals used for work.

tiro – animales que se usan para cargar.

Draft horse – type of horse bred for work.

caballo de tiro – tipo de caballo criado para cargar.

Drake – male duck.

pato – el macho de los patos.

Drenching – a process of administering drugs orally to animals.

administrar una pócima – el proceso de administrar las drogas oralmente a los animales.

Dressing percentage – the proportion of the live weight of an animal and the weight of the carcass prior to cooling.

rendimiento a la canal – la proporción del peso vivo del animal con el peso de la canal antes de enfriarse.

Drift – movement of a pesticide through the air to nontarget sites.

arrastre – movimiento de un pesticida por el aire hasta sitios intencionales.

Drill planter – plants seed in narrow rows at a high population.

sembradora a chorillo – siembra la semilla en surcos muy juntos a población alta.

Drip line – the edge of the tree where the branches stop.

lindero para viento – el límite del árbol donde terminan las ramas.

Drone – male bee.

zángano – macho de la abeja.

Drupe – a stone fruit.

drupa – fruta que contiene un hueso.

Dry heat cooking – surrounding food with dry air while cooking.

cocinar en estufa de secado – rodeando la comida con aire seco mientras cocinarla.

Dry matter – material left after all water is removed from a feed material.

materia seca – materia que queda después de quitar toda el agua de material de pienso.

Dry period – period of time when a cow is not producing milk.

período seco – período de tiempo cuando la vaca no está secretando leche.

Duckling – young duck.

patito – cría del pato.

Dwarf – tree that has rootstock that limits growth to 10 ft or less.

enano – árbol que tiene el rizoma, lo cual que limita el crecimiento a 10 pies o menos.

Ear notch – system of permanently marking animals by cutting notches from their ears.

poner una señal – práctica de señalar animales permanentemente por hacer una muesca en las orejas.

Ease of working – refers to the difficulty in cutting, shaping, nailing, and finishing wood.

facilidad de tallar – se refiere a la dificultad en cortar, labrar, clavar, y acabar madera.

Economic threshold – the level of pest damage to justify the cost of a control measure.

«precio de umbral» – el nivel de daño a causa de animales o insectos nocivos que justifica el costo de una medida de control.

Edible – fit to eat or consume by mouth.

comestible – apropiado para comerse o tomar por la boca.

Egg – female reproductive cells.

óvulo – células reproductores de la hembra.

Ejaculate – the amount of semen produced at one time by a male.

ejaculación – la cantidad de semen expulsada a una vez por el macho.

Element – a uniform substance that cannot be further decomposed by ordinary means.

elemento – una sustancia uniforme que no puede descomponerse por medios ordinarios.

Embryo – fertilized egg.

embrión – óvulo fecundado.

Embryo transfer – a process that removes fertilized eggs from a female and places them in another female who carries them until birth.

traslado de embrión – un proceso que quita óvulos fecundados de la hembra y los coloca en otra hembra, que los lleva hasta el parto.

Endocrine or hormone system – a group of ductless glands that release hormones into the body.

sistema endocrino u hormonal – conjunto de glándulas de secreción interna las cuales secretan hormonas por el cuerpo.

Endosperm – interior of a wheat kernel that will become wheat flour.

endosperma – parte interior de un grano de trigo que llegará a ser harina de trigo.

Enterprise – commercial establishment to generate profits.

empresa – establecimiento comercial que produce ganancias.

Enthusiasm – energy to do a job and inspiration to encourage others.

entusiasmo – energía para hacer una obra y inspiración para animar a otras personas.

Entomology – science of insect life.

entomología – ciencia de la vida de insectos.

Entomophagous insects – insects that feed on other insects.

insectos entomófagos – insectos que se alimentan con otros insectos.

Entrepreneur – person organizing a business, trade, or entertainment.

empresario – persona que organiza un negocio, comercio, o espectáculo.

Entrepreneurship – process of planning and organizing a business.

  empresiales – el proceso de planear y organizar un negocio.

Environment – space and mass around us.

  medio ambiente – espacio y masa que nos rodea.

Epidermis – surface layer on the lower and upper side of the leaf.

  epidermis – capa de las partes superior e inferior de la hoja.

Equitation – the art of riding on horseback.

  equitación – el arte de montar caballo.

Eradicant fungicide – a fungicide applied after disease infection has occurred.

  fungicida desarraigante – un fungicida aplicado después de que se ha ocurrido la infección.

Eradication – complete control or removal of a pest from a given area.

  desarraigo – control completo o extirpación de un animal o insecto dañino de un área determinada.

Erosion – wearing away.

  erosión – desgasto o deterioro lento.

Estrogen – hormone that regulates the heat period.

  estrógeno – hormona que regula el período en que el animal está en celo.

Estrus – heat period or time when female is receptive to the male.

  celo – período o tiempo fecundo cuando la hembra está receptiva al macho.

Estuaries – ecological systems including bays, streams, and tidal areas influenced by brackish water.

  estuarios – sistemas ecológicos inclusive bahías, arroyos, y tierras bajas del litoral que son afectados por agua salobre.

Evaporated milk – milk that has had water removed.

  leche deshidratada – leche a que el agua se ha quitado.

Evaporation – changing from a liquid to a vapor or gas.

  evaporación – transformación de un líquido o vapor a gas.

Evergreen – plants that do not lose their leaves on a yearly basis.

  planta de hoja perenne – plantas a que no se caen las hojas cada año.

Evisceration – removal of the viscera.

  destripación – extirpación de las vísceras.

Ewe – female of the sheep family.

  oveja – hembra del carnero.

Executive meeting – meeting of the officers to conduct business of the organization between regular meetings.

  reunión del consejo de dirección – reunión de los directores para dirigir el negocio de la organización en medio de reuniones corrientes.

Exoskeleton – the external body wall of an insect.

  dermatoesqueleto – caparazón o pared exterior del cuerpo de un insecto.

Experience – anything and everything observed, done, or lived through.

  experiencia – cada y toda cosa observada, hecha, o vivida.

Extemporaneous – speech delivered with little or no preparation.

  extemporáneo – dícese de un discurso pronunciado con poca preparación.

Extravaginal growth – turfgrass growth in which growth originates from an axillary bud on the crown.

  crecimiento extravaginal – crecimiento de césped en que brota el crecimiento del botón axilar en la corona.

Fabricated, boxed beef – vacuum sealing smaller cuts of meats prior to shipment.

  carne de res fabricada – envasar al vacío a los cortes más pequeños de carne antes de transportar.

Famine – widespread starvation.

  escasez – inanición o hambre general.

Farmers Home Administration – assists farmers to become landowners.

  Administración de Hogares de Agricultores – ésta presta apoyo a los agricultores para que se hagan propietarios.

Farrier – a person who shoes horses.

  herrador de caballos – persona que clava y ajusta las herraduras a un caballo.

Farrow – giving birth in the swine family.

  parir a cerdos – parir a cerdos.

Note: In Spanish, the word "parir" means "to give birth in the nonhuman mammal." Thus, there is no other word that differentiates the birth process of a particular species of animal.

Farrowing crate – special cage or pen used for swine to give birth in.

box de parición – jaula o pocilga especial dentro de la cual paren las cerdas.

Fat – nutrients that have 2.25 times as much energy as carbohydrates.

grasa – alimentos nutritivos que tienen 2,25 veces más la energía que los carbohidratos.

Feces – solid body waste.

heces fecales – excrementos sólidos del cuerpo.

Federal Land Banks – provide long-term credit for agriculture.

Bancos Federales de Tierras – éstos dan crédito a plazo largo para la agricultura.

Feed – animal food.

pienso – alimentos para animales.

Feed additive – a nonnutritive substance added to feed to improve growth, increase feed efficiency, or to maintain health.

aditivos a pienso – una sustancia no nutritiva añadida al pienso para mejorar el crecimiento, para aumentar la eficacia del pienso, o para mantener la salud.

Feeder calves – calves intended to be fed to heavier weights before slaughter.

becerros de engorde – becerros a que les da tanto para comer para que se engorden mucho antes de ser matados.

Feedlots – areas in which large numbers of animals are grown for food.

«feedlotes» – áreas en que muchos animales son cultivados para que sirvan como carne.

Feedstuff – any edible material used for animal feed.

forraje – cualquiera material comestible que sirve para pienso.

Fermentation – a chemical change that results in gas release.

fermentación – un cambio químico que resulta en desprendimiento de gases.

Fertility – amount and type of nutrients in the soil.

feracidad – cantidad y clase de alimentos nutritivos en la tierra.

Fertilizer – material that supplies nutrients for plants.

abono – material que proporciona los alimentos nutritivos a las plantas.

Fertilizer analysis – the percentage of nutrients by weight in the fertilizer.

análisis de abono – por peso, es el porcentaje de alimentos nutritivos que hay en el abono.

Fertilizer grade – percentages of primary nutrients in fertilizer.

grado del abono – proporción de alimentos nutritivos en el abono.

Fetus – embryo from the attachment of the uterine wall until birth.

feto – embrión desde el tiempo de fijación a la pared uterina hasta el nacimiento.

FFA – an intra-curricular organization for students enrolled in agriscience programs.

AFF – una organización dentro del programa para estudiantes matriculados en los programas de agriciencia.

Fibrous root – one of the two major root systems, consisting of many fine, hair-like roots.

raíz fasfciculada – uno de dos sistemas principales de raíz, que consiste de muchas raíces finas como el pelo.

Filament – structure that supports the anther.

filamento – base que sostiene la antera.

Filly – young female horse or pony.

potra – hembra jóven del caballo o poney.

Financing function – obtaining capital for a business.

función de financiar – obtenier capital para un negocio.

Fine texture – a turfgrass with narrow leaves.

textura fina – césped con hoja delgada.

Fine-textured (clay) soils – usually forms very hard lumps or clods when dry; plastic when wet.

tierra fina – usualmente forma terrones duros cuando seca; como plástico cuando mojada.

Finish – fat covering on a market animal.

acabado de engorde – envultura de gordo en el animal de mercado.

Flock – group of birds or sheep.

bandada – grupo de aves juntas.
rebaño – grupo de ovejas juntas.

Floriculture – production and distribution of cut flowers, potted plants, greenery, and flowering herbaceous plants.

floricultura – producción y distribución de flores cortadas, de plantas en maceta, ramas y hojas verdes, y plantas herbáceas florecientes.

Flower – reproductive part of the plant.

flor – parte reproductiva de la planta.

Flukes – very small, flat worms that are parasites.

trematodos – gusanos de cuerpo plano muy pequeños que son parásitos.

Foal – newborn horse or pony. Also, to give birth in horses.

potro – caballo o poney recién nacido.
parir – dar a luz a los animales.

See note under Farrow.

Foliage – stems and leaves.

follaje – ramas y hojas.

Foliar sprays – liquid fertilizer sprayed directly to the leaves of plants.

abonado foliar – abono líquido que se pulveriza directamente a las hojas de las plantas.

Food – material needed by the body to sustain life.

alimento – material que necesita el cuerpo para sostener la vida.

Food additive – anything added to food prior to packaging.

suplemento aditivo – cualquiera cosa añadida a la alimentación antes de envasarla.

Food chain – interdependence of plants or animals on each other for food.

cadena trófica – dependencia reciproca de plantas o animales unos a otros para alimentarse.

Food industry – production, processing, storage, preparation, and distribution of food.

industrias alimenticias – producción, tratamiento, almacenaje, preparación, y distribución de alimentos.

Foot-candle – amount of light found one foot from a standard candle or candela.

candela por pie cuadrado – cantidad de luz que se encuentra de un pie de una candela o vela estándar.

Forage – crop plants grown for their vegetative growth and fed to animals.

forraje – plantas cultivadas por su vegetación y dadas de comer a los animales.

Forest – large group of trees and shrubs.

bosque – gran conjunto de árboles y arbustos.

Forest land – land at least 10 percent stocked by forest trees.

tierra forestal – tierra poblada de por lo menos 10 por ciento de árboles forestales.

Forester – a person who studies and manages forests.

silvicultor – una persona que estudia y maneja los bosques.

Forestry – industry that grows, manages, and harvests trees for lumber, poles, posts, panels, and many other commodities.

silvicultura – industria que cultiva, maneja, y cosecha los árboles para maderos, varas, palos, tableros, y muchos otros productos.

Formal – equal in size and number.

formal – igual en cantidad, tamaño y textura.

Formulation – the physical properties of the pesticide and its inert ingredients.

formulación – las propiedades físicas del pesticida y sus ingredientes inertes.

4-H – network of clubs directed by Cooperative Extension Service personnel to enhance personal development and provide skill development in many areas.

4-H – una red de asociaciones dirigida por el personal del Servicio Cooperativo de Extensión para fomentar el desarrollo personal y enseñar técnicas en muchas áreas.

Free choice – making feed available at all times.

alimentación a elección – dejando que está disponible todo el pienso que quiera el animal todo el tiempo.

Free water – water that drains out of soil after it has been wetted.

agua libre – agua que avena del suelo después de haber sido mojado el suelo.

Freemartin – sexually imperfect female calf (usually sterile) born twin to a male in cattle.

ternera estéril – ternera hembra que es sexualmente defectuosa, nacida gemela a un macho del ganado.

Note: The Spanish term for "freemartin" implies sterility (*estéril*), so the "(usually sterile)" part of the definition was deleted in Spanish.

Freeze-drying – removing moisture with cold.

deshidratar por congelación – quitar a algo el agua por el frío.

Fresh water – water that flows from the land to oceans and contains little or no salt.

agua dulce – agua que corre desde la tierra hasta los mares y que tiene poco o no sal.

Fructose – simple fruit sugar.

fructosa – azúcar simple de frutas.

Fruit – mature ovary; seed.

fruto – óvulo maduro, semilla.

Fry – small, newly hatched fish.

Alevines – peces recién salidos del huevo.

Fungal endophytes – microscopic plants growing within a plant.

endofitos fungosos – plantas microscópicas creciendo dentro de una planta.

Fungi – plants that lack chlorophyll.

hongos – plantas que faltan de clorofila.

Fungicide – a material used to destroy fungi or protect plants against their attack.

fungicida – una sustancia que se usa para desarraigar hongos o proteger las plantas de su ataque.

Furrows – grooves made in the soil.

surcos – hendiduras hechas en la tierra.

Futures – legally binding agreements made on the trading floor of a futures exchange.

futuros – convenios obligatorios legalmente, hechos en el parquet de un cambio a futuro.

Futures market – legal frameworks for sellers and buyers to buy and sell futures contracts.

mercado afuturo – estructuras legales para vendedores y compradores para comprar y vender contratos a futuro.

Gait – way of moving.

paso – manera de pasar.

Galactose – simple milk sugar.

galactosa – azúcar simple de leche.

Gallop – fast, three-beat gait.

galope – paso rápido de tres tiempos.

Gamete – a reproductive cell.

gameto – una célula reproductiva.

Gander – male goose.

ganso – macho de la gansa.

Gelding – castrated male of the horse family.

caballo castrado – caballo cuyos órganos necesarios a la generación han sido extirpados.

Note: Because there is no term for "gelding" in Spanish other than "castrated horse," the Spanish definition reads, "castrated horse – horse whose organs necessary for procreation have been removed."

Gene – a unit of hereditary material located on a chromosome.

gene – unidad de materia hereditaria ubicada en un cromosoma.

Gene mapping – finding and recording the locations of genes in a cell.

trazar un mapa de genes – localizando y recordando las posiciones de genes en una célula.

Gene splicing – the process of removing and inserting genes in cells.

empalme de genes – el proceso de extirpar y meter las genes en las células.

General-use pesticide – a pesticide that poses a minimal amount of risk when applied according to label directions.

pesticida para uso general – un pesticida que plantea un riesgo mínimo cuando se aplica según las instrucciones de etiqueta.

Generation – the collective offspring of common parents.

generación – la descendencia colectiva de padres comunes.

Genes – components of cells that determine the characteristics of living things.

genes – componentes de las células que determinan las características de los seres vivientes.

Genetic control – pest control using resistant varieties of crops.

control genético – control de animales o insectos dañinos usando variedades resistentes de cultivos.

Genetic engineering – movement of genes from one cell to another.

ingeniería genética – traslado de genes de una célula a otra.

Geneticist – a person who studies genetics.

genetista – persona que especializa en genética.

Genetics – the biology of heredity.

genética – la biología de la herencia.

Genotype – what the genes look like.

genotipo – cómo se manifiestan los genes.

Genus – taxonomic category between family and species.

género – grupo de clasificación taxonómica que cae entre familia y especie.

Genus – (plural is genera) a closely related and definable group of plants comprising one or more species.

genus – un grupo definible y estrechamente relacionado de plantas que consta de úna o más especies.

Germ – new wheat plant inside the kernel.

germen – progenie de trigo dentro del grano.

Germinate – a seed sprouting or starting to grow.

germinar – una semilla brotando o empezando a crecer.

Gestation – length of pregnancy.

gestación – tiempo que dura en el estado de embarazo.

Giblets – heart, liver, and gizzard of poultry.

menudillos – corazón, higadillo, y molleja de las aves.

Gills – organ of aquatic animal that absorbs oxygen from the water.

agalla – órgano de animal acuático que absorbe el oxígeno del agua.

Gilt – female of the swine family that has not given birth.

cerda jóven – hembra de la familia cochina que no ha parido.

Ginning – the process of removing the seeds from cotton.

desmotar – el proceso de sacar la semilla al algodón.

Girl Scouts and Boy Scouts – organizations that provide opportunities for leadership development and skill development.

Niñas Exploradores y Niños Exploradores – asociaciónes que da oportunidades del desarrollo de liderazgo y de técnica.

Glacial – soils deposited by ice.

glacial – tierras depositadas por hielo.

Glucose – simple sugar and the building blocks for other nutrients.

glucosa – azúcar simple y los fundamentos para otros alimentos nutritivos.

Goiter – enlarged thyroid gland caused by a deficiency of iodine.

bocío – glándula tiroides dilatada causada por la carencia de yodo.

Goose – female of the goose family.

gansa – hembra del ganso.

Gosling – young goose.

ansarino – ganso jóven.

Gourmet – sensitive and discriminating taste in food preferences.

gastronómico – gusto sensible y muy bueno en cuanto a las preferencias de comida.

Grade – animal that is not purebred or of unknown ancestry.

cruce – animal que no es de pura sangre or raza; o de linaje desconocido.

Grader – person who inspects the food for freshness, size, and quality.

clasificadora – persona que inspecciona los alimentos en cuanto a su frescura, tamaño, y calidad.

Grades – quality standards.

grados – estándares de calidad.

Gradient – a measurable change in an amount over time or distance.

índice – un cambio mensurable de una cantidad por un tiempo o una distancia.

Grading up – crossing a purebred male with a grade female.

procreación por sustitución – Acoplar un macho de pura raza y una hembra de cruce.

Grafting – joining two parts of a plant together so that they will grow as one.

injertar – juntando dos partes de planta para que crezcan así como úna.

Grain crop – crop grown for its edible seeds.

cultivo de cereales – plantas cultivadas por sus semillas comestibles.

Grass waterway – strip of grass growing in the low area of a field where water can gather and cause erosion.

estepa de gramíneas – franja de hierba creciendo en la parte baja del campo donde el agua puede acumularse y resultar en erosión.

Gravitational water – water that drains out of soil after it has been wetted.

agua de gravitación – agua que avena del suelo después de haber sido mojado el suelo.

Green manure crop – crop grown to be plowed under to provide organic matter to the soil.

plantas para abono verde – plantas cultivadas para ser enterradas con el objetivo de proporcionar materia orgánica al suelo.

Green revolution – process in which many countries became self-sufficient in food production.

revolución verde – proceso en donde muchos países logran a ser autosuficientes en la producción de alimentos.

Green roughage – plant materials with high moisture content such as pastures and root plants.

forraje grosero verde – materias vegetales con alto contenido de humedad así como pastos y tubérculos.

Group planting – trees or shrubs planted together so as to point out some special feature or to provide privacy or a small garden area.

siembra por agrupación – árboles o arbustos están plantados juntos para señalar alguna característica especial o para proveer intimidad o un jardín pequeño.

Guanine – a base in genes designated by the letter *G*.

guanino – una base en los genes nominado por la letra «G».

Guard cells – cells that surround the stoma.

células guardias – células que rodean el estoma.

Gully erosion – removal of soil to form relatively narrow and deep trenches known as gullies.

erosión en cárcavas – extirpación de tierra para formar acequias relativamente hondas y estrechas que son conocidas por el nombre de cárcavas.

H

Habitat – area or type of environment in which an organism or biological population normally lives.

hábitat – área o tipo de medio ambiente en que normalmente vive un organismo o una población biológica.

Hand – unit of measurement for horses equal to 4 inches.

palmo – unidad de medir a los caballos que es igual a cuatro pulgadas.

Hand mating – system of breeding in which the male and female are kept apart except during mating.

cubrición a mano – el sistema de acoplamiento en donde el macho y la hembra son separados excepto durante la época de celo.

Hardness – wood's resistance to compression.

dureza de la madera – la resistencia de madera a la compresión.

Hardwood – wood from deciduous trees.

madera dura – madera de los árboles caducos.

Hardy – the ability of a plant to survive and grow in a given environment.

resistente – la capacidad de una planta de sobrevivir y crecer en un ambiente determinado.

Harvester – person responsible for taking products from plants in the field.

cosechadora – persona responsable por recoger el fruto del campo.

Harvesting – taking a product from the plant where it was grown or produced.

cosechar – recoger un producto de la planta donde fue cultivado o producido.

Hatchery – place where eggs are hatched.

criadero – sitio donde sale de los huevos la cría.

Hay – forages that have been cut and dried to a low level of moisture, used for animal feed.

heno – forrajes que han sido cortados y secados a un nivel bajo de humedad, usado para pienso.

Haylage – silage made from forages dried to 40 to 55 percent moisture.

hierba presecada y ensilada – ensilaje hecho de forajes secados a 40 o 55 por ciento de humedad.

Heartwood – inactive core of a tree trunk or limb.

duramen – núcleo inactivo de un tronco o rama de árbol.

Heel cutting – a shield-shaped cut made about halfway through the wood around the leaf and the axial bud.

escudete – un corte en forma de escudo hecho a medio camino a través de la madera alrededor de la hoja y el botón axilar.

Heifer – female of the cattle family that has not given birth.

vaquilla – hembra del vacuno que no ha parido.

Hen – adult female chicken, duck, turkey, and pheasant.

gallina – adulta hembra del gallo.

Herb – plant kept for aroma, medicine, or seasoning.

hierba – planta guardada para aroma, medicina, o condimiento.

Herbaceous – a plant that has a stem that does not turn woody; it is more or less soft and succulent, and not winter hardy.

herbáceo – una planta que tiene el tallo que no vuelve a ser leñoso; es más o menos blando y carnoso, tampoco es resistente al invierno.

Herbicide – a substance for killing weeds.

herbicida – una sustancia empleada para matar las malas hierbas.

Heredity – transmission of characteristics from parent to offspring.

herencia – la transmisión de rasgos de los progenitores a la progenie.

Heterozygous – pairs of genes that are different.

hetercigótico – pares de genes que son diferentes.

Hide – skin of an animal.

cuero – piel de los animales.

High technology – use of electronics and ultramodern equipment to perform tasks and control machinery and processes.

alta tecnología – el uso de electrónica y equipo ultra-moderno para desempeñar funciones y controlar maquinería y procesos.

Hinny – cross between a stallion and a jennet.

burdégano – un cruce entre un semental y una burra.

Hive – place or box in which bees are kept.

colmena – sitio o caja en donde se guardan las abejas.

Hog – swine more than four months old.

cerdo, marrano – cerdo que tiene más de cuatro meses de edad.

Hoist – lift into position.

subir – llevar a un lugar o posición más alta.

Homozygous – pairs of genes that are alike.

homocigótico – pares de genes que son semejantes.

Honey – a thick, sweet substance made by bees from the nectar of flowers.

miel – una sustancia viscosa y dulce hecha por las abejas con el néctar de las flores.

Horizon – layer.

horizonte – capa.

Horizon A – soil located near the surface that is made up of desirable proportions of mineral and organic matter.

horizonte A – tierra ubicada cerca de la superficie que es compuesta de proporciones convenientes de materia mineral y orgánica.

Horizon O – natural layer of organic matter on the surface of soils.

horizonte O – capa natural de materia orgánica en la superficie de suelos.

Hormones – chemicals that regulate many of the activities of the body.

hormonas – sustancias químicas que regulan muchas de las actividades del cuerpo.

Horse – member of the horse family 14.2 or more hands tall.

caballo – miembro de la familia de los caballos que mide 14,2 o más de palmos de alto.

Horsemanship – the art of riding and knowing the needs of the horse.

equitación – el arte de montar a caballo y conocerlo en cuanto a sus necesidades.

Horticulture – the science of producing, processing, and marketing fruits, vegetables, and ornamental plants.

horticultura – la ciencia de producir, procesar, y poner en venta las frutas, verduras, y plantas de adorno.

Host animal – a species of animal in or on which diseases or parasites can live.

huésped animal – una especie de animal en o sobre el cual viven parásitos o enfermedades.

Hothouse lamb – lamb less than three months old.

cordero lechal – cordero que tiene menos de tres meses de edad.

Humidity – moisture in the air.

humedad – impregnación de agua en el aire.

Hutch – a rabbit cage or house.

conejera – jaula o morada para conejos.

Hybrid – plant or animal offspring from crossing two different species or varieties.

híbrido – progenitura vegetal o animal que procede del cruzar dos distintas especies o razas.

Hybrid vigor – offspring of greater strength and potential resulting when two different breeds or varieties are crossed.

vigor híbrido – progenitura de más fuerza y potencial resultando cuando se cruzan dos distintas razas o cruces.

Hydrocarbons – organic compounds containing hydrogen and carbon.

hidrocarburos – compuestos orgánicos que contienen hidrógeno y carbono.

Hydrocooling – removal of heat by immersing in cold water.

enfriamiento del agua – quitarse del calor por inmersión en agua fría.

Hydroponics – the practice of growing plants without soil.

cultivo hidropónico – la práctica de cultivar plantas sin tierra.

Hygroscopic water – water that is held too tightly by soil particles for plant roots to absorb.

agua higroscópico – agua que está sujetada con demasiado rigor por las partículas de tierra para que las raíces de planta la absorban.

Hyphae – the threadlike vegetative structure of fungi.

hifas – la estructura vegetal filiforme de los hongos y mohos.

Ice-minus – bacteria that retards frost formation on plant leaves.

«ice-minus» – bacterias que retrasan la formación de escarcha en las hojas vegetales.

Imbibition – the absorption of water into the cell causing it to swell.

embebición – la absorción de agua en la célula, resultando en que se hincha.

Immune – not affected by.

inmune – que no se afecta.

Impatiens – popular, easy-to-grow, summer flowering plant.

balsamina nicaragua – planta de floración del verano que es popular y fácil para cultivar.

Imperfect flower – flower that is missing one or more of the following parts: stamen, pistil, petals, or sepals.

flor imperfecta – flor que falta de úna o más de las partes siguientes: estambre, pistilo, pétalos, o sépalos.

Implant – to place a substance under the skin to improve growth of animals.

implantar – meter una sustancia bajo la piel para mejorar el crecimiento de animales.

Improvement by selection – picking the best parents for the next generation.

cultivo mejoramiento – seleccionando los mejores padres para la generación siguiente.

Improvement projects – projects that improve beauty, convenience, safety, value, or efficiency learned outside of the regularly scheduled classroom or laboratory classes.

proyectos de hacer mejoras – proyectos que hacen mejoras en la belleza, conveniencia, seguridad, valor, o eficacia y que los aprende uno fuera de las clases de escuela o laboratorios.

In vitro – fertilization of eggs outside of the body.

in vitro – fertilización de óvulos fuera del cuerpo.

Inbreeding – mating of animals that are related.

procrear en consanguinidad – apareamiento de animales del mismo parentesco.

Incomplete dominance – neither gene expresses itself to the exclusion of the other.

dominancia incompleta – ninguno de los genes se manifiesta con exclusión al otro.

Incorporated – mixed into the soil by spading, plowing, or discing.

incorporación del abono al suelo – metido en la tierra y mezclado por arar, usar la pala, o usar los discos abresurcos.

Induction – a specific set of conditions must occur to cause flower development.

inducción – tiene que suceder unas condiciones específicas para causar el desarrollo de la flor.

Inert – inactive.

inerte – inactivo.

Inflorescence – the flowers and ultimately the seed area of the plant.

florescencia – son las flores y ultimamente el área de las semillas de la planta.

Informal – when speaking of landscaping it refers to planting areas that contain elements that are not equal in number, size, or texture.

informal – cuando se refiere al ajardinar, se refiere a las áreas de plantar que contienen elementos que no son iguales en cantidad, tamaño, ni textura.

Infusion – the process for treating udder problems through the teat canal.

infusión – el proceso usado para tratar problemas de la ubre por el canal de la teta.

Inheritability – the capacity to be passed down from parent to offspring.

capacidad para heredar – la capacidad para ser transmitido del progenitor a la progenitura.

Injection – the process of administering drugs by needle and syringe.

inyección – el proceso de administrar medicamento por medio de jeringuilla hipodérmica.

Inner bark – phloem; transport food downward in a tree.

corteza interior – líber; transporte de alimentación hacia abajo en un árbol.

Inorganic compound – a compound that does not contain carbon.

compuesto inorgánico – un compuesto que no contiene el carbono.

Insect – a six-legged animal with three body segments.

insecto – un animal de seis patas con tres sementos de cuerpo.

Insecticide – a material used to kill insects or protect against their attack.

insecticida – un material que se usa para matar insectos o proteger contra su ataque.

Insemination – the placement of semen in the female reproductive tract.

inseminación – la introducción de semen en las vías genitales de la hembra.

Instar – the state of the insect during the period between molts.

etapas de crecimiento – la fase del insecto durante el período en medio de las mudas.

Institutional advertising – advertising designed to create a favorable image of the firm or institution.

publicidad institucional – publicidad creada para establecer un concepto favorable de la empresa o institución.

Insulin – chemical used to control blood sugar levels.

insulina – sustancia química que se emplea en la regulación de niveles de la cantidad de glucosa.

Integrated pest management – pest control program based on multiple-control practices.

lucha integrada – programa de control de los animales o insectos nocivos basado en ejercicios de varios controles.

Integrity – honesty.

integridad – honradez.

Intermediate loans – loans with payment periods that range from one to seven years.

préstamos intermedios – préstamos que tienen plazos de pagos que oscilan entre uno y siete años.

Internode – area between two nodes.

entrenudos – área entre dos nudos.

Intradermal – between layers of skin.

intradérmico – entre las capas de piel.

Intramuscular – in a muscle.

intramuscular – en el interior de los músculos.

Intraperitoneal – in the abdominal cavity.

intraperitoneal – en el interior de la cavidad abdominal.

Intraruminal – in the rumen.

Intraherbarial – en el interior del herbario.

Intravaginal growth – a type of growth in which shoots develop within the lower leaf sheath at a crown's axillary bud.

crecimiento intravaginal – un tipo de creimiento en que desarrollan los brotes en el interior de la vaina de la hoja exterior, en la corona del botón axilar.

Intravenous – in a vein.

intravenoso – en el interior de la vena.

Inventor – person devising something new or improving an existing idea or product.

inventor – persona que crea algo nuevo o mejora una idea o producto existente.

Inventory report – units received and sold, unit cost, total sales, and profit.

inventario – unidades recibidas y vendidas, coste por unidad, ventas totales, y ganancia.

Involuntary muscles – muscles that operate in the body without control by the will of the animal.

músculos involuntarios – músculos que funcionan en el cuerpo sin control por la voluntad del animal.

Ions – atom or a group of atoms that have an electrical charge.

iones – átomo o grupo de átomos que tienen una carga eléctrica.

IPM – integrated pest management.

lucha integrada – control integrado de animales e insectos nocivos.

Irradiation – treating food with gamma rays.

irradiación – tratamiento de comestibles con rayos gamma.

Irrigation – addition of water to plants to supplement that provided by rain or snow.

irrigación – adición de agua a los cultivos para suplementar la proporcionada por la lluvia o la nieve.

Jack – male donkey or mule.

burro – macho de la burra o mula.

Jennet – female donkey or mule.

burra – hembra del burro o mula.

Jog – in horses, a slow, smooth, two-beat, diagonal gait.

trote corto – de caballos, paso diagonal, lento y suave, de dos tiempos.

Jungle fowl – wild ancestor of the chicken.

ave de corral de la selva – antepasado salvaje de la gallina.

Katahdin – popular potato variety of the 1930s.

«**Katahdin**» – variedad popular de patata durante los 1930.

Key pest – a pest that occurs on a regular basis for a given crop.

animal o insecto dañino regular – animal o insecto dañino que ocurre en una basis regular para un cultivo determinado.

Kid – young goat.

chivo – cabrío jóven.

Knife application – injection of fertilizer into soil in gaseous form.

fertilización por inyección – inyección del abono al suelo en forma gaseosa.

Knowledge – familiarity, awareness, and understanding.

conocimiento – familiaridad, conciencia, y entendimiento.

Kosher – prepared in accordance with Jewish dietary laws.

«**Kosher**» – preparado según las leyes dietéticas judaicas.

Lactation – milk production.

lactancia – secreción de leche.

Lactation period – period of time when mammals are producing milk.

período de amamantamiento – época en que los mamíferos están secretando leche.

Lactose – compound milk sugar.

lactosa – compuesto de azúcar de leche.

Lacustrine – soils deposited by lakes.

lacustre – los suelos depositados por lagos.

Lamb – member of the sheep family less than one year old; also, meat from young sheep.

cordero – miembro de la familia de ovejas que tiene menos de un año de edad; también la carne de la oveja jóven.

Lambing – act of giving birth in sheep.

parir a corderos – parir a corderos.

See note under Farrow.

Lard – rendered pork fat.

manteca de cerdo – grasa rendida de cerdo.

Larvae – mobile aquatic organisms that become fixed and grow into nonmobile adults.

larvas – organismos acuáticos movibles que logran a ser fijados y crecen a ser adultos no movibles.

Lay on the table – a motion used to stop discussion on a motion until the next meeting. The way to table a motion is to say, "I move to table the motion."

presentar una moción – una moción usada para terminar discusión sobre otra moción hasta la próxima reunión. La manera de presentar una moción es decir, "Yo propongo que se presente la moción."

Layer – chicken developed for the purpose of laying eggs.

gallina ponedora – una gallina criada para el objetivo de poner huevos.

Layering – a practice inducing formation of roots for new plants while maintaining the health and vigor of the parent plant.

acodar – una práctica que induce la formación de raíces para plantas nuevas mientras manteniendo la salud y vigor de la planta madre.

LC$_{50}$ – lethal concentration of a pesticide in the air required to kill 50 percent of the test population.

CM$_{50}$ – concentración mortal de un pesticida en el aire necesaria para matar 50 por ciento de una población prueba.

LD$_{50}$ – lethal dose of a pesticide required to kill 50 percent of a test population.

DM$_{50}$ – dosis mortal de un pesticida necesaria para matar 50 por ciento de una población prueba.

Leached – certain elements have been washed out of the soil.

lixiviado – ciertos elementos han sido derrubiados del suelo.

Lead – to show the way by going in advance or to guide the action or opinion of.

guiar – mostrar por medio de ir delante o dirigir la acción o la opinión de una.

Leadership – the capacity or ability to lead.

dirección – la capacidad de guiar.

Leaf – plant part consisting of a stipule, petiole, and blade.

hoja – parte vegetal consistiendo de estípulo, pecíolo, y brizna.

Leaf fat – loose fat on hogs.

gordo suelto – grasa flácida en el cuerpo de los cerdos.

Leaf mold – partially decomposed plant leaves.

mantillo – hojas vegetales parcialmente descompuestas.

Legume – plant in which certain nitrogen-fixing bacteria utilize nitrogen gas from the air and convert it to nitrates that the plant can use as food.

legumbre – planta en la cual ciertas bacterias fijadores del nitrógeno utilizan el gas nitrogenado del aire y lo convierten a nitratos para que los puede usar la planta como alimentación.

Lenticels – pores in the stem that allow the passage of gases in and out of the plant.

lenticelas – poros en el tallo o el pedúnculo que permiten que los gases entren a y salgan de la planta.

Lethal – genetic characteristic that causes an animal to be born dead or to die shortly after birth.

mortal – característica genética que causa que un animal nazca muerto o muera poco depués del nacimiento.

Liabilities – current and long-term debts.

pasivos – deudas actuales y a largo plazo.

Light horse – type of horse developed for riding.

caballo de silla – tipo de caballo criado para montar.

Light intensity – brightness of light.

intensidad luminosa – el brillo de la luz.

Ligule – a membranous or hairy structure located on the inside of the leaf.

lígula – está ubicada en la vaina foliar interior y es una estructura membranosa y pelosa.

Lime – material that reduces the acid content of soil and supplies nutrients such as calcium and magnesium to improve plant growth.

cal – material que reduce el contenido de ácido del suelo y proporciona a minerales así como calcio y magnesio para mejorar el crecimiento de las plantas.

Linebreeding – mating of individuals with a common ancestor.

cría por líneas puras – apareamiento de animales que tienen antepasado común.

Linen – fabric produced from fibers in flax plants.

lino – tejido producido de las fibras de linácea.

Linseed oil – oil produced from flax seed.

aceite de linaza – aceite producido de semilla de lino.

Litter – group of young born at the same time with the same parents.

cama de paja – la cría nacida a la misma vez de los mismos padres.

Loamy – a granular soil containing a good balance of sand, silt, and clay.

marga – un suelo granular que contiene un buen equilibrio de arena, légamo, y arcilla.

Loess – soils deposited by wind.

loess – tierras depositadas por el viento.

Long-term loans – loans with payment periods that range from eight to forty years.

préstamos a largo plazo – préstamos cuyos plazos de pago oscilan entre ocho y cuarenta años.

Long-term plans – plans accomplished over months or years.

planes a largo plazo – planes realizados por meses o años.

Lope – in horses, a very slow canter.

paso largo – de caballos, medio galope muy lento.

Loss-leader pricing – commodity offered for sale at prices below the cost level.

valoración del artículo barato para atraer clientes – se ofrece en venta la mercancía a precios por debajos del nivel de coste.

Loyalty – reliable support for an individual, group, or cause.

lealtad – apoyo seguro para un individuo, un grupo, o una causa.

Lumber – boards cut from trees.

maderos – trozos cortados de la sustancia dura de los árboles.

Lymph glands – glands that secrete disease-fighting materials for the body.

ganglios linfáticos – glándulas que secretan materiales que combaten las enfermedades del cuerpo.

Macronutrients – elements used in relatively large quantities.

elementos nutritivos esenciales – elementos usados en cantidades relativamente grandes.

Main motion – a basic motion used to present a proposal for the first time. The way to state it is to get recognized and then say, "I move . . . "

moción mayor – la moción básica para presentar una propuesta por la primera vez. La manera de declararla es para ser reconocido y luego decir, «Yo propongo . . . »

Malting – process of preparing grain for the production of beer and alcohol.

malteado – proceso de preparar el grano para la producción de cerveza y alcohol.

Maltose – compound milk sugar.

maltosa – azúcar de malta compuesto.

Mammal – milk-producing animal.

mamífero – animal que secrete la leche.

Mammary system – milk-production system in the female.

aparato mamario – aparato de producción de leche de la hembra.

Manage – to use people, resources, and processes to reach a goal.

administrar – utilizar a personas, recursos, y procesos para lograr una meta.

Mange – crusty skin condition caused by mites.

sarna – enfermedad de la piel que es causada por los acáridos.

Mare – adult female horse or pony.

yegua – hembra adulta del caballo o poney.

Margins – edge of the leaf.

linde – el límite de la hoja.

Mastitis – infection of the milk-secreting glands of cattle, goats, and other milk-production animals.

mastitis – infección de las glándulas que secretan la leche del ganado, cabras y otros mamíferos.

Maturity – the state or quality of being fully grown.

madurez – estado o cualidad del desarrollo o crecimiento completo.

Mechanical control – pest control that affects the pest's environment or the pest itself.

control mecánico – control de los animales o insectos nocivos que afecta el animal o insecto nocivo mismo o el ambiente de éso.

Media – material that provides plants with nourishment and support through their root systems.

medio – material que proporciona alimentación y sostenimiento a las plantas por sus raíces.

Medium – surrounding environment in which something functions and thrives.

medio ambiente – los alrededores en los cuales funciona y crece alguna cosa.

Medium-textured (loamy) soil – a relatively even mixture of sand, silt, and clay.

suelo limoso – una mixtura relativamente igual de arena, légamo, y arcilla.

Meiosis – cell division that results in the formation of gametes.

meiosis – división celular que resulta en la formación de los gametos.

Meristematic tissue – plant tissue responsible for plant growth.

tejido meristemático – tejido vegetal que causa el crecimiento de la planta.

Mesophyll – tissue of the leaf where photosynthesis occurs.

mesofilo – tejido de la hoja en donde ocurre el fotosíntesis.

Metamorphosis – the change in growth stages of an insect.

metamorfosis – la mudanza de forma en las etapas de crecimiento del insecto.

Microbe – living organism that requires the aid of a microscope to be seen.

microbio – organismo viviente que requiere la ayuda del microscopio para que se vea.

Micronutrients – elements used in very small quantities.

micronutrientes – elementos usados en cantidades pequeñas.

Microorganisms – tiny plants or animals that may contribute to food spoilage.

microorganismos – microbios animales o vegetales que pueden contribuir a la putrefacción de los comestibles.

Microwave oven – an oven that heats food by using an electromagnetic wave.

horno de microonda – un horno de cocina que calenta la comida por medio de usar ondas electromagnéticas.

Middlemen – people who handle an agricultural product between the farm and the consumer.

intermediarios – personas que manejan un producto agrícolo entre la finca y el consumidor.

Migratory labor – workers who move to places where harvesting is occurring.

obreros migratorios – trabajadores del campo que se mudan a lugares donde se está ocurriendo el cosechar.

Milk – white or yellowing liquid produced by the mammary glands for the purpose of feeding young.

leche – líquido blanco o amarillento que es segregado por las glándulas mamarias para el objetivo de alimentar la cría.

Milk replacer – dried dairy or vegetable products that are mixed in warm water and fed to calves in place of milk.

sustitivo de leche – productos secados lácteos o vegetales que son mezclados con agua tibia y dado de comer a los becerros en lugar de leche.

Milking machine – machine that milks cows and goats.

máquina de ordeñar – aparato o máquina que extrae la leche de la ubre de las vacas y cabras.

Mineral – element essential for normal body functions.

mineral – elemento esencial para las funciones normales de cuerpo.

Mineral matter – nonliving items such as rocks.

materia mineral – cosas no vivientes como rocas.

Minimum tillage – soil is worked only enough so that seed will germinate.

labranza mínima – la tierra es labrada justo lo suficiente para que germinen las semillas.

Minutes – is the name of the official written record of a business meeting.

actas – el nombre del record escrito oficial de una junta directiva.

Misuse statement – pesticide label information which reminds the user to follow the label directions when using the products.

aviso de mal manejo – información en la etiqueta que avisa al usuario que haya de seguir las direcciones de la etiqueta mientras usando los productos.

Mitosis – simple cell division for growth.

mitosis – división celular simple que resulta en el crecimiento.

Mode of action – how a pesticide affects its target.

manera de afectar – cómo el pesticida afecta su objetivo.

Mohair – hair from Angora goats used to make a shiny, heavy, woolly fabric.

Mohair – pelo de la cabra de Angora usado para hacer un tejido lanoso, grueso, y con brillo.

Moist heat cooking – surrounding the food with liquid while cooking.

cocinar por humectación – rodeando a la comida con líquido mientras cocinando.

Moldboard plow – plow with a curved bottom that will turn prairie soils.

arado de vertedera – arado con orejera la cual puede voltear el suelo de llanura.

Molt – softening and cracking of the exoskeleton so that crustaceans may escape and grow a larger exoskeleton.

mudar – ablandarse y agrietarse del dermatoesquéleto para que los crustáceos puedan escaparse y dejarse crecer un dermatoesquéleto más grande.

Monitor – to check or observe pest numbers and damage.

determinar – ver u observar la cantidad de animales o insectos nocivos y el daño resultando de ellos.

Monocotyoledon (monocot) – plant with one seed leaf.

monocotiledóneo – planta con solamente una hoja de semilla.

Monogastric – animal with a single-compartment stomach.

monogástrico – animal con solamente una cavidad en el estómago.

Mosaic – a virus disease of plants in which a leaf shows a symptom of light and dark green mottling of the foliage.

mosaico – una enfermedad viral de las plantas en donde la hoja manifiesta un síntoma de vetas manchas de verde claro y oscuro en el follaje.

Mulch – material placed on soil to break the fall of raindrops (preventing erosion), prevent weeds from growing, or improve the appearance of the area.

pajote – material puesto encima del suelo para cortar la caída de las gotas de lluvia (para prevenir la erosión), para prevenir el crecimiento de malas hierbas, o para hacer mejoras del aspecto del área.

Mule – cross between a jack and a mare.

mulo – una cruce entre el burro macho y la yegua.

Muscular system – the lean meat of the animal.

sistema muscular – la carne sin grasa del animal.

Mutation – change in genes.

mutación – cambios en los genes.

Mutton – meat from mature sheep.

cordero – carne de cordero mayor.

Mutualism – two types of wildlife living together for the mutual benefit of both.

mutualismo – dos clases de fauna que viven juntas para el beneficio mutuo de las dos.

Mycelium – a collection of fungal hyphae.

micelio – un grupo de hifa fungosa.

Natural fisheries – existing breeding ground of wild fishes, i.e., oceans, continental shelves, lakes, and bays.

criaderos naturales de peces – criaderos existentes de la cría de peces en estado salvaje; por ejemplo, mares, el plataformo continental, lagos, y bahías.

Natural service – male mates directly with the female.

servicio natural – cuando el macho acopla directamente con la hembra.

Nematode – a type of roundworm.

nematodos – un tipo de ascáride.

Net contents – the amount of pesticide in the container.

contenidos netos – la cantidad de pesticida en el contenido.

Net worth – owner's investment in the business.

fortuna neta – las inversiones del dueño en su negocio.

Neutral – neither acid nor alkaline.

neutro – ni ácido ni siquiera alcalino.

Nitrates – a form of nitrogen used by plants.

nitratos – una forma de nitrógeno usado por plantas.

Nitrogen fixation – conversion of nitrogen gas to nitrate by bacteria.

fijación del nitrógeno – la conversión de gas de nitrógeno en nitrato por bacteria.

Nitrogen-gas laser – device used to determine wavelengths given off by plants.

láser de gas nitrógeno – aparato que se usa para determinar las longitudas de onda emitidas por las plantas.

Nitrous oxides – compounds containing nitrogen and oxygen; they make up about 5 percent of the pollutants in automobile exhaust.

óxido nitroso – compuestos que contienen nitrógeno y oxígeno; constituyen aproximadamente 5 por ciento de los agentes contaminadores en el gas de escape de los automóviles.

Node – portion of the stem that is swollen or slightly enlarged that gives rise to buds.

nudo – la parte del tallo o pedúnculo que es hinchada o una poca expandida que echa renuevos.

Nomenclature – a systematic method of naming plants or animals.

nomenclatura – un método sistemático de dar nombres a las plantas o los animales.

Noncontagious – diseases that cannot be spread to other animals.

no contagiosas – enfermedades que se transmiten a otros animales por contacto.

Nonselective herbicide – a chemical that kills all plants it comes in contact with.

herbicida no-selectivo – una sustancia química que mata todas las plantas a las que toca.

No-till – seed is planted directly into the residue of the previous crop, without exposing the soil surface.

sin labranza – dícese de cuando se siembra la semilla directamente en el residuo de la cosecha previa, sin exponer la superficie del suelo.

Noxious weed – plant that is prohibited by state law.

mala hierba perjudicial – planta que es prohibida por la ley del Estado.

Nucleic acid – original name for deoxyribonucleic acid.

ácido nucleico – nombre original por ácido desoxirribonucleico.

Nurse crop – a crop used to protect another crop until it can get established.

planta protectora – un cultivo que se usa para proteger a otro cultivo hasta que éste pueda establecerse.

Nursery – a place where young trees, shrubs, and other plants are grown.

vivero – un lugar donde se recrían árboles, arbustos, y otras plantas jóvenes.

Nut – a fruit or seed contained within a removable outer cover.

nuez – fruto o semilla envuelto en una cáscara o cubierta exterior a que se quita.

Nutriculture – water culture.

nutricultura – cultivo de agua.

Nutrients – substances necessary for the functioning of an organism.

alimentos nutritivos – sustancias necesarias para el funcionar del organismo.

Nutrition – the process where all body parts receive materials needed for function, growth, and renewal.

nutrición – el proceso en que todas las partes del cuerpo reciben materiales necesarios para función, crecimiento, y renovación.

O horizon – surface soil layer consisting mostly of organic material.

horizonte O – capa de la superficie de suelo que consta mayormente de materia orgánica.

Obesity – the result of too much or improper types of food being eaten.

obesidad – el resultado de comer demasiada comida o de comer demasiado de alimentos de tipo inadecuado.

Offal – materials thrown away during slaughter, including the head, blood, viscera, loose fat, etc.

asaduras – materias desechadas durante el sacrificio, incluso la cabeza, la sangre, las vísceras, el gordo suelto, etc.

Offsetting a position – taking a second position on the futures market to offset a first.

compensar una posición – tomar una posición secundaria en el mercado a futuro para compensar una primera.

Oil meal – by-product of the production of vegetable oil.

harina de aceite – subproducto de aceite vegetal.

Oilseed crop – crop produced for the oil content of their seeds.

cultivo de semilla oleaginosa – cultivo cultivado para el contenido de aceite de sus semillas.

Olericulturalist – one who studies the cultivation of vegetables.

oleicultor – uno que estudia la cultivación de verduras.

Olericulture – the cultivation of vegetables.

oleicultura – la cultivación de verduras.

On-the-job training – experience obtained while working in an actual job setting.

aprendizaje – experiencia obtenida mientras trabajando en un empleo verdadero.

Opening position – initial step in the futures market.

posición inicial – paso inicial en el mercado a futuro.

Order of business – refers to the items and sequence of activities conducted at a meeting.

orden de negocio – se refiere a los puntos y la sucesión de actividades conducidos en una reunión.

Organic compound – a compound that contains carbon.

compuesto orgánico – un compuesto que contiene carbono.

Organic food – food that has been grown without the use of certain chemical pesticides.

alimentos orgánicos – alimentos que han sido cultivados sin el uso de ciertos pesticidas químicos.

Organic matter – dead plant and animal tissue that originates from living sources such as plants, animals, insects, and microbes.

materia orgánica – tejido muerto animal o vegetal que origina de orígenes vivientes como plantas, animales, insectos, y microbios.

Ornamental plant – used to improve the appearance of a structure or an area.

planta de adorno – usada para hacer mejoras al aspecto o estructura de un área.

Ornamentals – science of fruits, vegetables, and ornamental plants.

cultivo de plantas de adorno – ciencia de frutas, verduras, y plantas de adorno.

Osmosis – the flow of a fluid through a semipermeable membrane separating two solutions, which permits the passage of the solvent but not the dissolved substance. The liquid will flow from a weaker to a stronger solution, thus tending to equalize concentrations.

Osteomalacia – weak or brittle bones in mature animals caused by a deficiency of calcium, phosphorus, or vitamin D.

osteomalacia – huesos débiles o quebradizos en los animales maduros causados por una carencia de calcio, fósforo, o vitamina D.

Out of the market – when a futures position is offset by a second position.

fuera del mercado – cuando una posición a futuro es compensado por una posición segunda.

Outcrossing – mating of unrelated animal families within one breed.

«outbreeding» – acoplamiento de familias animales no relacionadas dentro de la misma raza.

Ovary – female organ that produces eggs or female sex cells; also, that portion of the flower that contains the ovules or seeds.

ovario – órgano de la hembra que produce los óvulos o células sexuales femeninas; además aquella parte de la flor que consta de los óvulos o semillas.

Overgrazing – damage to plants or soil due to animals eating too much of the plants at one time or reducing the plants' ability to hold soil or recover after grazing.

sobrepastorear – daño a las plantas o al suelo causado por el comer demasiado de parte de los animales a las plantas de una vez o el reducir de la capacidad de la planta de guardar tierra o recuperar después del apacentamiento.

Overripe – beyond maturity.

demasiado maduro – más allá de madurez.

Overseeding – seeding a second crop into one that is already growing.

sobresembrar – sembrar un cultivo segundo encima de úno que ya está creciendo.

Ovulation – process of releasing mature eggs from the ovary.

ovulación – el proceso de desprender óvulos maduros del ovario.

Ovule – unfertilized seed.

óvulo (vegetal) – semilla ya no fecundada.

Oxidative deterioration – decay resulting from exposure to air.

deterioro oxidativo – deterioro resultando de la exposición al aire.

Ozone – compound that exists in limited quantities about fifteen miles above the Earth's surface.

ozono – compuesto que existe en cantidades limitadas hacia quince millas encima de la superficie de la tierra.

Pace – two-beat side to side gait.

amblar – paso de dos tiempos, un lado tras otro lado.

Packer – person or firm responsible for preparing commodities for shipment.

empaquetador – persona o firme responsable de preparar productos para envío.

Palisade cells – elongated, vertical cells that give the leaf strength and manufacture food.

células empalizadas – células verticales, alaragadas que dan fuerza a la hoja y fabrican alimentación.

Parakeratosis – a skin disease similar in appearance to mange caused by a deficiency of zinc in the diet.

parakeratosis – una enfermedad de la piel semejante a un aspecto a la sarna, causada por una carencia de zinc en la dieta.

Parasite – organism that lives in or on another organism with no benefit to the host.

parásito – organismo que vive en o encima de otro organismo sin beneficio al huésped.

Parasitism – one type of wildlife living and feeding on another without killing it.

parasitismo – una clase de fauna viviendo asido y alimentándose con otra sin matarla.

Parent material – the horizon of unconsolidated material from which soil develops.

materiales madres del suelo – el horizonte de material descomprimido desde lo cual desarrolla el suelo.

Parliamentary procedure – a system of guidelines or rules for making group decisions in business meetings.

procedimiento parliamentario – un sistema de principios o normas para hacer decisiones de grupo en reuniones de negocio.

Parturition – birth.

parto – nacimiento.

Pasture – forages that are harvested by the livestock itself.

pasto – forrajes que se cosechan por el ganado mismo.

Pasture mating – male allowed to roam freely with the herd and mate at random.

servicio a campo – cuando se permite que el macho ronde libremente por el rebaño y acople al azar.

Pathogen – organisms that produce disease.

agente patógeno – organismos que producen las enfermedades.

Peat moss – a type of organic matter made from spagnum moss.

turba – un tipo de materia orgánica hecha de esfagno.

Pedigree – record of ancestry.

carta de origen – registro de linaje.

Pelt – skin of an animal with the hair attached.

pellejo – la piel de un animal, inclusive el pelo.

Pencil shrink – the estimated amount of loss that an animal will sustain during the trip to market.

contracción – la cantidad estimada de pérdida que sostendrá el animal durante el transporte al mercado.

Penetration pricing – a strategy where price is set below that of competitors.

valoración psicológica – una estrategia donde se fija el precio más bajo que lo de los competidores.

Percolation – movement of water through the soil.

filtración – paso del agua a través de la tierra.

Perennial weed – a weed that lives for more than two years.

mala hierba perenne – una mala hierba que vive más de dos años.

Perfect flower – flower containing all of the parts: stamen, pistil, petals, and sepals.

flor perfecta – flor que consta de todas las partes: estambre, pistilo, pétalos, y sépalos.

Peripheral nervous system – system that controls the functions of the body tissues including the organs.

sistema nervioso periférico – sistema que controla las funciones de los tejidos de cuerpo inclusive los órganos.

Perlite – natural volcanic glass material having water-holding capabilities.

perlita – material vítreo volcánico natural que tiene capacidades de aguantar el agua.

Permeable – permitting movement.

permeable – que permite movimiento a través de si mismo.

Pest – any organism that adversely affects man's activities.

insecto, animal dañino – cualquier organismo que afecta adversamente las actividades del hombre.

Pest resurgence – ability of a pest population to recover.

ensurgimiento renacimiento – la capacidad de una población de insectos o animales dañinos para recuperar.

Pesticide – chemical used to control pests.

pesticida – sustancia química empleada para combatir los animales e insectos dañinos.

Pesticide resistance – the ability of a pest to tolerate a lethal level of a pesticide.

resistencia de pesticida – la capacidad de un animal o insecto dañino para tolerar un nivel mortal de pesticida.

Petals – brightly colored, sometimes fragrant portion of the flower.

pétalos – parte de la flor de color vivo y a veces fragrante.

Petiole – stem of the leaf.

pecíolo – rabillo de la hoja.

pH – measurement of acidity or alkalinity from 1 to 14.

valor pH – medida de la acidez o alcalinidad desde 1 a 14.

Phenotype – physical appearance of an individual.

fenotipo – características externas, o aspecto de un individuo.

Pheromone – a chemical secreted by an organism to cause a specific reaction by another organism of the same species.

feromona – una sustancia química secretada por un organismo para causar una reacción específica por otro organismo de la misma especie.

Phloem – cells that carry manufactured food to areas of the plant where it is stored or used.

líber – células que llevan alimentación fabricada a las áreas de la planta donde se conserva o se usa.

Phloem tissue – conductive tissue responsible for transport of food downward in a tree.

líber – tejido conductivo que es responsable del transporte de los alimentos nutritivos hacia abajo en un árbol.

Photodecomposition – chemical breakdown caused by exposure to light.

fotodescomposición – la descomposición química causada por la exposición a la luz.

Photosynthesis – process in which chlorophyll in green plants enables those plants to utilize light to manufacture sugar from carbon dioxide and water.

fotosíntesis – proceso en donde el clorafilo de las plantas verdes las capacita utilizar la luz, fabricar azúcar de dióxido de carbono y agua.

Physiology – study of the functions and vital processes of living creatures and their organs.

fisiología – el estudio de las funciones y procesos vitales de criaturas vivientes y sus órganos.

Pig – swine less than four months of age.

gorrino – cerdo que tiene menos de cuatro meses de edad.

Piglet – newborn of the swine family.

lechón – el recién nacido del cerdo.

Pistil – female part of the flowers consisting of stigma, style, ovary, and ovule.

pistilo – parte femenina de las flores que consta de estigma, estilo, ovario, y óvulo.

Plan – to think through.

planear – pensar muy bien en cómo se va a hacer algo.

Plant disease – any abnormal plant growth.

enfermedad de planta – cualquier crecimiento vegetal anormal.

Plant fertilization – addition of elements to a plant-growing environment.

fertilización de planta – adición de elementos a un ambiente en que se crecen las plantas.

Plant Hardiness Zone map – a map developed by the U.S. Department of Agriculture, dividing the country into ten zones based on average winter temperatures.

Mapa de Zonas de Plantas Resistentes al Frío – mapa desarrollado por el Departamento de Agricultura de los EE.UU., que divide el país en diez zonas basado en las temperaturas medias de invierno.

Plant nutrition – provision of elements to plants.

alimentación nutritiva vegetal – el suministro de elementos a las plantas.

Playability level – the suitability for the intended recreational or sporting use of the turf.

nivel de resistencia – la conveniencia del uso intencional recreativo o deportivo del césped.

Plugging – establishment of turf by using small pieces of existing turfgrass.

cubrir de trasplantes de tepe – iniciando el césped por plantear tepes ya existentes.

Plywood – construction material made of thin layers of wood glued together.

contrachapado – material de construcción hecho de hojas delgadas de madera que son pegadas juntas.

Point of order – a procedure used to object to some item in or about the meeting that is not being done properly. The procedure to use is to say, "Point of order." The presiding officer should then recognize the member by saying, "State your point."

cuestión de orden – un procedimiento que sirve para oponerse a algún punto de o sobre la reunión que no está haciéndose adecuadamente. El procedimiento de usar es decir, «¡Cuestión de orden!» Entonces el director ha de reconocer al miembro por decir, «Exprese su cuestión.»

Polled – hornless, either naturally or otherwise.

mocho – sin cuernos, naturalmente o de otro modo.

Pollen – small male sperm or grains that are necessary for fertilization in the flower.

polen – esperma pequeño o polvillo macho que es necesario para la fecundación de la flor.

Pollination – transfer of pollen from anther to stigma.

polinización – transporte del polen del antero hasta el estigma.

Polluted – containing harmful chemicals or organisms.

contaminado – que contiene sustancias químicas u organismos dañosos.

Pome – fleshy fruits with embedded core and seeds.

pomo – frutos carnosos con corazón fijado y semillas.

Pomologist – a fruit grower or scientist.

pomólogo – cientista o cultivador de frutos.

Pomology – the science and practices of fruit growing.

pomología – la ciencia y el ejercicio de cultivar frutos.

Pony – member of the horse family less than 14.2 hands tall.

jaca – miembro de la familia de los caballos menos de 14,2 palmos de alto.

Porcine somatotropin (PST) – hormone that increases meat production in swine.

somatotropino porcino – una hormona que aumento la producción de carne en los cerdos.

Pork – meat from swine.

puerco – carne del cerdo.

Port – town having a harbor for ships to take on cargo.

puerto – ciudad o pueblo que tiene un lugar en la costa o el río que es dispuesto para seguridad y las carga y descarga de las naves.

Postemergence herbicide – a herbicide applied after the weed or crop is present.

herbicida postemergente – un herbicida aplicado después de que sea presente la mala hierba o el cultivo.

Potable – drinkable, that is, free of harmful chemicals and organisms.

potable – que puede beberse, o sea que es libre de organismos y sustancias químicas dañosos.

Poult – young turkey.

pavipollo – pavo jóven.

Poultry – group name for domesticated birds.

aves de corral – nombre de grupo para los aves domesticados.

PPM – parts per million.

partes por millón – partes por millón.

PPT – parts per thousand.

partes por mil – partes por mil.

Note: In Spanish, abbreviations are not used for "PPM" and "PPT," as they would both be "PPM," translated verbatim.

Precipitate – a chemical action that results in the dropout of solids in a solution.

precipitado – una reacción química que resulta en el depositarse de sedimento en una solución.

Precipitation – formation of rain and snow.

precipitación – formación de lluvia y nieve.

Precooling – rapid removal of heat before storage or shipment.

prerefrigeración – el quitarse del calor rápidamente antes de almacenamiento o transporte.

Predation – a way of life where one type of wildlife eats another type.

rapacidad – una manera de vivir en que un tipo de fauna se alimenta con otro tipo.

Predator – an animal that feeds on a small or weaker animal.

animal de rapiña – un animal que se alimenta con otro animal más pequeño o más débil.

Preemergence herbicide – a herbicide applied prior to weed or crop germination.

herbicida pre-emergente – un herbicida aplicado antes de la germinación de malas hierbas o cultivos.

Presiding officer – a president, vice president, or chairperson who is designated to lead a business meeting.

presidente – el presidente, vicepresidente, o presidente interino designado para dirigir una sesión de negocio.

Prestige pricing – pricing to buyers with special desires for quality, fashion, or image.

valoración de prestigio – valoración a los compradores que tienen deseos especiales para calidad, moda, or reputación.

Prey – animal eaten by another animal.

presa – animal comido por otro animal.

Price – amount received for an item that is sold.

precio – cantidad recibido por un artículo que es vendido.

Primary nutrients – in agriculture, nitrogen, phosphorus, and potash.

alimentos nutritivos mayores – en la agricultura, nitrógeno, fósforo, y potasa.

Processed meats – meat that is not suitable for sale over the counter.

carne elaborada – carne que no es adecuado para venderse al por menor.

Processing – turning raw agricultural products into consumable food.

elaboración – el convertirse de los productos agrícolas en bruto en alimentos comestibles.

Processor – person or business cleaning, separating, handling, and preparing a product before it is sold to the distributor.

elaborador – persona o empresa que limpia, separa, maneja, y prepara un producto antes de que se venda al distribuidor.

Producer – person who grows the crop.

productor – persona que cultiva el cultivo.

Product advertising – advertising that focuses on the product itself.

publicidad de producto – publicidad que fija en el producto si mismo.

Production agriscience – farming and ranching.

agriciencia de producción – labrar y llevar un rancho.

Production Credit Association – provides short-term credit for agriculture.

Asociación de Crédito de Producción – proveer de crédito a corto plazo.

Production or productive project – enterprise conducted for wages or profit.

producción o proyecto productivo – empresa conducida por ganancia o lucro.

Profession – occupation requiring an education, especially law, medicine, teaching, or the ministry.

profesión – vocación que exige una enseñanza superior, especialmente el derecho, la medicina, la enseñanza, o el ministerio.

Profile – a cross-section view of soil.

pérfil – una vista del suelo por corte transversal.

Profit – proceeds from a business transaction.

beneficio – ganancias de una transacción comercial.

Profit and loss statement – projection of costs and expenses against sales and revenue over time.

estado de ganancias y pérdidas – la previsión de los costos y gastos en contra a las ventas e ingresos por un tiempo.

Progeny – offspring.

progenie – descendecia.

Progesterone – hormone that prevents estrus during pregnancy and causes development of the mammary system.

progesterona – hormona que impide el estro durante el embarazo y causa el desarrollamiento del aparato mamario.

Program – total plans, activities, experiences, and records.

programa – planes, actividades, experiencias, y registros totales.

Project – a series of activities related to a single objective or enterprise.

proyecto – una serie de actividades relativas a un sólo objetivo o empresa.

Promissory note – note signed by borrower agreeing to the terms of the loan.

pagaré – obligación escrita firmada por el prestatario, consintiendo a los plazos del préstamo.

Promoting function – plan to make potential customers aware of the product or service.

función de promocionar – plan para hacer que los clientes se den cuenta del producto o servicio.

Propagated – the increase or multiplication of animals and plants by sexual or asexual methods.

propagar – la multiplicación de animales y plantas por métodos sexuales o asexuales.

Propagation – process of increasing the numbers of a species or perpetuating a species.

propagación – el proceso de aumentar la población de una especie o de perpetuar una especie.

Protectant fungicide – a fungicide applied prior to disease infection.

fungicida preventivo – un fungicida aplicado antes de infeccionarse.

Protein – nutrient made up of amino acids and essential for maintenance, growth, and reproduction.

proteina – alimento nutritivo que consta de aminoácidos y que es esencial para el mantenimiento, el crecimiento, y la reproducción.

Protozoa – microscopic, one-celled animals that are parasites of animals.

protozario – microorganismos unicelulares que son parásitos de otros animales.

Pruning – the removal of dead, broken, unwanted, diseased, or insect-infested wood.

podar – el quitar de madera muerta, trozada, no deseada, enferma o plagada por los insectos.

Psychological pricing – a strategy designed to make a price seem lower or less significant.

valoración psicológica – una estrategia creada para hacer que un precio parezca más bajo o menos significante.

Pullet – young female chicken.

polla – hembra jóven del pollo.

Pulpwood – wood used for making fiber for paper and other products.

madera para pasta de papel – madera usada para hacer fibras para papel y otros productos.

Purebred – animal with both parents of the same breed and with registration papers.

animal de pura raza – animal cuyos dos padres son de la misma raza.

Purify – removal of all foreign material.

purificar – extirpación de toda materia extranjera.

Putting green – golf playing area with turfgrass mowed very short.

campo más pequeño que es utilizado para jugar al golf – área para jugar al golf que tiene el césped muy segado.

Quality grade – grade based on amount and distribution of finish on an animal.

grado de calidad – grado basado en la cantidad y distribución del acabado de engorde de un animal.

Quarantine – isolation of pest-infested material.

cuarentena – aislamiento de material plagado.

Queen – only fertile, egg-laying female bee in the hive.

reina – la sóla abeja hembra que es fecunda y pone huevos.

Rack – in horses, a fast, four-beat gait.

pasitrote – de los caballos, un paso rápido de cuatro tiempos.

Radioactive material – material that is emitting radiation.

materia radioactivo – materia que emite la radiación.

Radon – colorless, radioactive gas formed by disintegration of radium.

radón – gas radioactivo incoloro, formado por la desintegración de radium.

Ram – male member of the sheep family.

carnero – macho de la oveja.

Ratio – proportion.

razón – proporción.

Ration – the amount of feed fed in one day.

ración – la porción de pienso que se da de comer en un día.

Real-world experience – an activity conducted in the daily routine of our society.

experiencia de la vida diaria – una actividad conducida dentro de la rutina diaria de la sociedad.

Reaper – machine that cuts grain.

segador – máquina que corta mieses.

Recombinant DNA technology – gene splicing.

tecnología de ADN preparado en el laboratorio – empalme de genes.

Rectum – the last organ in the digestive tract.

recto – el último órgano del aparato digestivo.

Recuperative potential – the ability of a plant to recover after being damaged.

potencialidad de recuperar – la capacidad de la planta para reestablecerse después de ser dañada.

Refer – a motion used to refer some other motion to a committee or person for finding more information and/or taking action on the motion on behalf of the members. The way to state a referral is to say, "I move to refer this motion to . . . "

referir – una moción usada para referir a otra moción a una comisión para averiguar más información y/o para tomar acción sobre la moción en nombre de los socios. La manera de declarar una referencia es para decir, «Yo propongo referir esta moción a . . . »

Refrigeration – keeping cool.

refrigeración – mantenerse congelado.

Registration papers – records proving that an animal is purebred.

cartas de registro – registros probando que un animal determinado es de pura raza.

Reins – leather or rope lines attached to the bit for control of the horse.

riendas – cuerdas de cuero o soga fijadas al bocado y empleada para controlar el caballo.

Relative humidity – the relationship between the actual amount of water vapor in the air to the greatest amount possible, at a given temperature.

humedad relativa – relación entre la cantidad verdadera de vapor de agua en el aire a la cantidad más grande que sea posible, a una temperatura determinada.

Render – cook and press the oil from.

derretir – cocinar y extraer el óleo de algo.

Rendering insensible – making the animal unable to feel pain.

rendir insensible – hacer que el animal no pueda sentir el dolor.

Renewable natural resources – resources provided by nature that can replace themselves.

recursos naturales renovables – recursos suministrados por la naturaleza que pueden volver a llenarse de nuevo.

Reproduction – the making of a new plant or animal.

reproducción – el hacerse de planta o animal nuevo.

Residual soils – parent materials formed in place.

suelos residuos – materiales madres del suelo formados en lugar.

Resource substitution – the use of a resource or item for another when the results are the same.

sustitución de recursos – el uso de un recurso o cuerpo en vez de otro cuando los resultados salen semejantes.

Resources inventory – summary of the resources that may be available for conducting a SAEP.

inventario de recursos – resumen de los recursos que pueden ser disponibles para conducir un PEAS.

Respiration – a process in which living cells take in oxygen and give off carbon dioxide.

respiración – un proceso en lo cual las células vivientes absorben el oxígeno y expulsan el bióxido de carbono.

Respiratory system – system that provides oxygen to the blood of the animal.

aparato respiratorio – conjunto de órganos que proporciona el oxígeno a la sangre del animal.

Restricted-use pesticide – a pesticide which poses a greater risk to man and the environment than one that is not so labeled.

pesticida de uso restringido – un pesticida que representa una amenaza más grande al hombre y al medio ambiente que otro que no se califica así.

Retail marketing – the selling of a product directly to consumers.

mercadeo al por menor – la venta de un producto directamente al consumidor.

Retailer – person or store that sells directly to the consumer.

minorista – persona o tienda que vende directamente al consumidor.

Retortable pouches – packages with foil between two layers of plastic.

embalaje a devolver – paquetes que tienen hoja de aluminio entre dos hojas de plástico.

Rhizome – horizontal, underground stems.

rizoma – tallo subterráneo, horizontal.

Rickets – deformity of bones in young animals caused by a deficiency of calcium, phosphorus, or vitamin D.

raquitismo – deformidad de los huesos en animales jóvenes causada por una carencia de calcio, fosforo, o vitamina D.

Roaster – old chicken used for meat.

pollo para asar – pollo viejo usado para carne.

Rollover – rapid change in water quality.

derribo – cambio rápido de la calidad del agua.

Rooster – adult male chicken or pheasant.

gallo – adulto macho del pollo.
faisán – adulto macho de la faisana.

Root crop – crop grown for the thick, fleshy storage root that it produces.

raíz planta – cultivo cultivado por el tubérculo grueso carnoso que produce.

Root cutting – section of a root cut and used for propagation purposes.

estaca de raíz – cuando se corte parte de la raíz y se usa para fines de propagación.

Root pruning – systematic cutting of the roots by hand or machine to encourage the roots to develop within the root ball range.

poda de restauración – el podar sistematicamente de las raíces a mano o a máquina para apoyar que las raíces desarrollen dentro del área del cepillón.

Rootbound – roots restricted by a container.

raíces restringidas por contenidor – raíces restringidas por un contenidor.

Note: The translation of "rootbound" is the same as that of the English definition provided.

Rooting hormone – chemical used to stimulate root formation on a cutting.

hormona radicular – sustancia química usada para estimular la formación de raíz en una tala.

Rootstock – root system and stem of a plant on which another plant is grafted.

portainjerto – sistema radicular y tallo de una planta sobre la cual se injerta otra planta.

Roughage – grass, hay, or silage and other feeds high in fiber and low in TDN.

forraje – hierba, heno, o ensilaje y otros piensos altos en forraje vegetal y bajos en NDT.

Roundworms – slender worms that are tapered on both ends.

ascárides – lombrices delgados que tienen aspecto afilado en dos extremos.

Row crop planter – plants seed in precise rows and with even spacing within the row.

sembradora a hilera – sembra la semilla en hileras precisas y de distancia uniforme dentro de las hileras.

Rumen – multicompartment stomach of cattle and sheep.

herbario – estómago de varias cavidades del ganado o de la oveja.

Ruminant – animal that has a stomach with more than one compartment.

rumiante – animal que tiene estómago con más de una cavidad.

Saddle – padded leather seat for the rider of a horse.

silla – aparejo de asiento rellenado y de cuero que usa el jinete.

SAEP – supervised agricultural experience program.

PEAS – programa de experiencia agrícola supervisado.

Safe water – water that is free of harmful chemicals and disease-causing organisms.

agua potable – agua que es libre de las químicas dañosas y de los organismos que causan las enfermedades.

Salinity – a measure of total dissolved mineral solids in water.

salinidad – una medida de minerales disueltos totales que hay en el agua.

Salmonids – any of the family of soft-finned fish such as salmon or trout that have the last vertebrae upturned.

salmonides – cualquieres de la familia de peces de alas suaves como salmón o trucha que tienen el vértebra última respingona.

Saltwater marsh – the ecological systems of plants and animals influenced by tidal waters with salinity of 15 to 34 PPT.

salina – sistemas ecológicos de plantas y animales influidos por mareas con salinidades de 15 a 34 partes por mil.

Sand – largest soil particles; 1 to .05 mm.

arena – las partículas de tierra más grandes; de 1 a ,05 mm.

Sapwood – transports water and minerals upward in tree roots, trunks, and stems.

albura – traslada los agua y minerales hacia arriba en raices, troncos, y tallos de árbol.

Saturated – water added until all the spaces or pores are filled.

saturado – el añadido del agua hasta que se llenen todos espacios o poros.

Scale – size of items.

escala – tamaño de un cuerpo.

Scarify – to soak, keep moist, or mechanically scrape a seed coat to aid germination.

escarificar – empapar, mantener húmedo, o raspar a máquina de la capa de una semilla para apoyar la germinación.

Scurvy – vitamin C deficiency disease.

escrobuto – una enfermedad por la carencia de vitamina C.

Seasonal – pertaining to a particular season of the year.

estacional – que pertenece a una estación determinada del año.

Secondary host – a plant or animal which carries a disease or parasite during part of the life cycle.

huésped secundario – una planta o un animal que lleva una enfermedad o parásito durante parte del ciclo de la vida.

Secretary – a person elected or appointed to take notes and prepare minutes of the meeting.

secretaria – una persona elegida o apuntada para recordar y preparar las actas de la sesión.

Seed blend – combination of different cultivars of the same species.

mezcla de semilla – combinación de cultivos diferentes de la misma especie.

Seed certification – program administered by state governments to insure seed quality.

certificación de semillas – programa administrado por los gobiernos de los estados para asegurar la calidad de semilla.

Seed culm – stem that supports the inflorescence of the plant.

caña de semilla – tallo que sostiene la florescencia de la planta.

Seed legume – crop that is nitrogen-fixing and produces edible seeds.

legumbre – cultivo que es fijador del nitrógeno y produce semillas comestibles.

Seed mixture – combination of two or more species.

mixtura de semilla – combinación de dos o más especies.

Seed pieces – vegetative means of propagating sugar cane.

multiplicación por estacas – medidas vegetativas de propagar la caña de azúcar.

Seedling – young plant grown from seed.

planta de semillero – plantón cultivado de semilla.

Seining – physical netting of fish in open water.

pescar con jábega – el coger de los peces con red en las aguas abiertas.

Selective breeding – mating adults who have characteristics desired in the offspring.

crianza selectiva – selección de padres para apareamiento que tienen los rasgos que se quiere que tenga la descendencia.

Selective herbicide – a chemical that kills or affects certain types of plants.

herbicida selectivo – una sustancia química que mata o afecta ciertos tipos de plantas.

Selling function – market research, sales plans, and sales closures.

función de vender – las investigaciones del mercado, las ventas, los planes, y cierres de venta.

Semen – the sperm cells and accompanying fluids.

semen – las espermatazoides y los líquidos que los acompañan.

Semi-arid – an area partially deficient in rainfall; dry.

semi-árido – un área que parcialmente carece de la precipitación.

Semi-dwarf – tree that has rootstock that limits growth to 15 ft or less.

semi-enano – árbol que tiene el rizoma, lo cual que límite el crecimiento a lo de 15 pies o menos.

Seminal roots – roots formed at the time of seed germination to anchor the seed in soil.

raices seminales – raices formadas en momento de la germinación de semilla para fijar la semilla en el suelo.

Semi-permeable membrane – membrane that permits a solution to move through it.

membrana semipermeable – membrana que permite que la penetre una solución.

Sepals – small, green, leaflike structures found at the base of the flower.

sépalos – cuerpos verdes pequeños y de aspecto de hoja que se encuentran al base de la flor.

Sequence – related or continuous series.

sucesión – series continuas o relacionadas.

Sewage system – receives and treats human waste.

alcantarillado – recibe y trata los desechos humanos.

Sex-linked – genes carried on chromosomes that determine sex.

(herencia) ligada al sexo – genes que se transmiten por los cromosomas que determinan el sexo.

Sexual reproduction – union of an egg and sperm to produce a seed or fertilized egg.

reproducción sexual – unión de un óvulo y un esperma para producir una semilla u óvulo fecundado.

Shackles – mechanical devices that restrict movement.

trabas – dispositivos mecánicos que restringen el movimiento.

Shear – remove wool from sheep.

esquilar – quitar la lana a la oveja.

Sheath – lower portion of the turfgrass leaf.

vaina foliar – parte a la base de la hoja de césped.

Sheet erosion – removal of soil from broad areas of the land.

erosión laminar – eliminación del suelo de áreas amplias del terreno.

Shelf life – time between packaging and spoilage.

tiempo de conservación de un producto en venta – el tiempo entremedias de envasar y de putrefacción.

Shellfish – any aquatic animal having a shell or shell-like exoskeleton.

mariscos – cualquier animal acuático que tiene cáscara o dermatoesqueleto parecido a la cáscara.

Shepherd – one who cares for sheep.

pastor – persona que guarda las ovejas.

Shod – wearing shoes.

herrado – dícese del caballo llevando herraduras.

Short-term loans – loans with a payment period of one year or less.

préstamos a corto plazo – préstamos con plazos de pago de un año o menos.

Short-term plans – plans accomplished in days or weeks.

planes a corto plazo – planes realizados dentro de días o semanas.

Shrinkage – changes in dimensions of wood as it reacts to changes in humidity and temperature.

contracción – disminución de volumen de la madera mientras reacciona a los cambios de humedad y temperatura.

Shroud – wrap tightly with a cloth.

amortajar – envolver ajustadamente con un paño.

Shrubs – woody perennial plants that normally grow low, produce many stems or shoots from the base, and do not reach more than 15 ft high.

arbustos – Plantas perennes leñosos que normalmente crecen bajo, producen muchas ramas o brotes desde la base, y no alcanzan a más que 15 pies de alto.

Side dressing – placing fertilizer in bands about 8 in. from the sides of growing plants.

abonado en fajasa lado – meter el abono más o menos a ocho pulgadas de los lados de los cultivos crecientes.

Signal word – the required word on the label which denotes the relative toxicity of the product.

palabra señalada – la palabra requerido en la etiqueta que indica la toxicidad relativa del producto.

Silage – feed resulting from the storage and fermentation of green crops in the absence of air.

ensilaje – pienso formado a causa del ensilar y fermentación de las plantas verdes debido a la ausencia del aire.

Silo – air-tight storage facility for silage.

silo – almacenamiento estanca al aire para el ensilaje.

Silt – intermediate soil particles; .05 to .002 mm.

légamo – partículas de tierra intermedias; ,05 a ,002 mm.

Silviculture – the scientific management of forests.

silvicultura – el manejo científico de los bosques.

Simple interest – interest calculated when the full amount of a loan is received by the borrower and paid back with interest after a short period of time.

interés simple – interés calculado cuando la cantidad total del préstamo es recibido por el prestatario y pagado con interés después de un período corto.

Simple layering – stem is bent to the ground, held in place, and covered with soil.

acodo simple – el tallo es metido debajo de la tierra, sujetado y cubierto con tierra.

Simulate – to look or act like.

simular – dar la apariencia de algo o imitar algo.

Singe – burn lightly.

chamuscar – quemar superficialmente.

Single eye – cutting made with the node of a plant with alternate leaves.

esqueje de yema – tala hecha con la yema de una planta que tiene hojas alternativas.

Sire – father.

semental – padre.

Skeletal system – bones joined together by cartilage and ligaments.

esquelético – huesos juntados por el cartílago y los ligamentos.

Skimming – setting the price of a new product for unusually high profits at first when affluent and willing customers are available.

importe de los derechos – fijar el precio de un producto nuevo para que se saquen ganancias extraordinariamente altas al principio, cuando hay clientes opulentos y dispuestos.

Slaughter – the killing of animals for market.

sacrificar – la matanza de animales para el mercadeo.

Slaughter cattle – cattle intended for slaughter.

ganado de sacrificio – ganado para el matadero.

Small Business Administration – provides loans to agribusinesses.

Administración de Comercios Pequeños – hace empréstitos a las agroindustrias.

Smoker – device used to add smoke flavor and taste to food.

secador al humo – dispositivo que se usa para añadir el sabor ahumado a los comestibles.

Sodding – removal of a portion of the soil and turfgrass plant for vegetative establishment purposes.

cubrir con césped – excavación de una parte del suelo y césped para fines de establecimiento vegetal.

Softwood – wood from conifers.

madera blanda – la madera de las coníferas.

Soil – top layer of the earth's surface suitable for the growth of plant life.

tierra, suelo – capa de la superficie de la tierra que es conveniente para el crecimiento de seres vegetales.

Soil science – study of the properties and management of soil to grow plants.

ciencia de estudio de los suelos – el estudio de las propiedades y administración de los suelos para cultivar plantas.

Solution culture – same as water culture.

cultivo de solución – el mismo que cultivo hidropónico.

Sow – female of the swine family that has given birth.

cerda – hembra del cerdo que ha parido.

Spat – the young shellfish just after they attach to underwater structures.

freza – los mariscos jóvenes, inmediatamente después de ligarse a estructuras subterráneas.

Spawn – egg-laying process of fish.

frezar – de los peces, el proceso de poner huevos.

Species – the basic unit in the classification system whose members have similar structure, common ancestors, and maintain their characteristics; subgroup of genus.

especies – la unidad básica del sistema de clasificación cuyos miembros tienen estructura semejante, antepasados comunes, y mantienen sus características; sub-grupo de genus.

Specimen plant – plant that is used as a single plant to highlight it or some other special feature of the landscape.

planta maestra – planta que se usa como planta singular para demostrar a ella o a otro rasgo especial del paisaje.

Sperm cells – male reproductive units.

células de esperma – unidades reproductivas del macho.

Sphagnum – pale and ashy mosses used to condition soil.

esfagno – musgo pálido y ceniciento usado para acondicionar el suelo.

Split carcass – halves of the animal after it is killed.

res abierta en canal – las mitades del animal después del sacrificio.

Split vein cutting – cutting made by slitting a large leaf on the veins before placing in a rooting medium.

esqueje del nervio partido – una tala hecha por hender una gran hoja en los nervios antes de meterla en un medio para echar raíces.

Spoiled – chemical changes that reduce nutritional value or render food unfit to eat.

estropeado – cambios químicos que reducen el valor nutritivo o render el comestible inadecuado para comer.

Spongy layer – lower, irregular layer in the leaf that allows the veins, or vascular bundle, to extend into the leaf.

capa esponjosa – capa del vaina exterior desigual, que permite que los nervios, o el haz vascular, se extienda a la hoja.

Sprigging – planting a section of a rhizome or stolon (sprig) in the soil.

propagación por estacas de raíz – plantar una parte de un rizoma o estolón (ramito) en el suelo.

Spring lamb – lamb between three and seven months old.

lechazo – cordero entre tres y siete meses de edad.

Sprinklers – devices that spray water into the air to water crops; its effect is much like rain.

regaderas – dispositivos que espuman agua al aire para regar los cultivos; tiene casi el mismo efecto como la lluvia.

Square foot – a unit of area measuring one foot on each side of a square.

pie cuadrado – una unidad de medir un pie en cada lado del cuadro.

Stag – male hog or cattle castrated after reaching sexual maturity.

verraco castrado – macho de la familia cochina castrado después de madurarse sexualmente.

Stallion – adult male horse or pony.

semental – adulto macho del caballo o poney.

Stamen – male part of the flower that contains the pollen, anther, and filament.

estambre – órgano sexual masculino de la flor que consiste de los polen, antero, y filamento.

Starch – major energy source in livestock feeds.

fécula – gran fuente de energía en los piensos.

Starter solutions – dilute mixtures of single or complete fertilizers used when plants are transplanted.

abono de material inicial – mixturas diluidas de abonos individuos o completos que se usan cuando se trasplantan las plantas.

Steer – castrated male member of the cattle family.

bueyezuelo – macho castrado de la familia vacuna.

Stem – supports the leaves and conducts the flow of water and nutrients.

tallo – sostiene las hojas y conduce el flujo de agua y alimentos nutritivos.

Stem tip cutting – cutting taken from the end of a stem or branch, normally including the terminal bud.

esqueje de punta de tallo – una tala sacada del extremo de un tallo o ramo, normalmente inclusive el botón terminal.

Sterility – inability to reproduce.

esterilidad – incapacidad para reproducir.

Stiff lamb disease – muscle degeneration in lambs caused by a deficiency of vitamin E.

distrofia muscular de los corderos por falta de nutrición – degeneración de los músculos de los corderos resultando de una carencia de vitamina E.

Stigma – part of the pistil that receives the pollen.

estigma – parte del pistilo que recibe el polen.

Stimulant crop – crop that stimulates the senses of users.

cosecha estimulante – cosecha que estimula los sentidos de los usuarios.

Stipule – a small appendage that is located at the base of the petiole on some species of plants.

estípula – un apéndix que está ubicado en la base del pecíolo de algunas especies de planta.

Stirrup – attached to the saddle by the stirrup leather, this device is a footrest used when riding horses.

estribo – atado a la silla por la barriguera, este aparato es un descanso para el pie que se usa al montar caballos.

Stolon – a stem that grows aboveground.

estolón – un tallo que crece encima de la tierra.

Stolonizing – type of vegetative establishment in which sprigs are broadcast onto the soil surface.

sembrar estolones al voleo – tipo de establecimiento vegetal en que los ramitos son sembrados al voleo a la superficie del suelo.

Stoma – small openings, usually on the lower side of the leaf, that control movement of gases.

estoma – aberturas pequeñas, usualmente en la vaina exterior de la hoja, que controlan el movimiento de gases.

Strip cropping – alternating strips of row crops with strips of close-growing crops.

cultivo en fajas – fajas alternas de cultivos a hilera con fajas de cultivos que crecen muy juntos.

Structure – refers to the tendency of soil particles to cluster together and function as soil units.

estructura – refiere a la tendencia de las partículas de tierra de agrupar y funcionar como unidades de tierra.

Stuck – cutting of a large blood vessel.

«stuck» – el cortar de un gran vaso sanguíneo.

Style – enlarged terminal part of the pistil.

estilo – parte terminal engrandizado del pistilo.

Subcutaneous – under the skin.

subcutáneo – por debajo de la piel.

Subsurface irrigation – water is supplied to the crop from underground pipes.

riego subterráneo – agua es proporcionada al cultivo desde tubería subterránea.

Successive – following one right after the other.

sucesivo – en seguida.

Succulent – in horticulture, thick, fleshy leaves or stems that store moisture.

planta carnosa – de la horticultura: hojas gruesas y carnosas que almacena agua.

Sucrose – compound cane sugar.

sucrosa – azúcar de caña compuesto.

Sugar crop – crop grown for the sugar content of its stems or roots.

cultivo de azúcar – un cultivo cultivado por el contenido de azúcar en los tallos o raíces.

Sulfur – pale-yellow element occurring widely in nature.

azufre – elemento amarillo blanco que aparece mucho en la naturaleza.

Super – box in which bees store excess honey.

super – parte superior del conjunto de la colmena.

Supervised – looked after and directed.

supervisado – guiado y dirigido.

Supervised Agricultural Experience Program – SAEP; all supervised agriscience experiences.

programa de experiencia agrícola supervisado – PEAS; todas experiencias agrícolas supervisadas.

Supplement – extra nutrient added to animal feed.

suplemento – alimento nutritivo añadido al pienso.

Supplementary agriscience skills – agriscience skills learned anywhere except in the instructional settings at school or in production enterprises or improvement projects in the SAEP.

técnicas suplementarias de agriciencia – técnicas aprendidas en cualquiera parte excepto en las clases de instrucción en la escuela o en las empresas o proyectos de mejoramientos del PEAS.

Supply – the quantity of a product that is available to buyers at a given time.

oferta – la cantidad de un producto que está disponible a los consumidores.

Surface irrigation – water flows over the soil surface to the crop.

riego de superficie – el agua fluye sobre la superficie del suelo al cultivo.

Susceptible – subject to some influence.

susceptible – dispuesto a alguna influencia.

Swarm – group of bees with a queen that leaves an overcrowded hive to find a new home.

enjambre – grupo de abejas con una reina que sale de una colmena atestada para hallar una morada nueva.

Sweetbreads – thymus and pancreatic glands of animals.

mollejas – glándulas de timo y páncreas de los animales.

Symmetrical – in landscaping, a planting that is equal in number, size, and texture on both sides.

simétrico – en labrar el paisaje, cuando hay plantas iguales en número, tamaño, y textura en ambos lados.

Symptom – the visible change to the host caused by a pest.

síntoma – el cambio visible al huésped causado por un animal o insecto nocivo.

Synthetic nutrients – nutrients manufactured chemically.

alimentos nutritivos sintéticos – alimentos nutritivos que son fabricados quimicamente.

Syringe – an instrument used to give injections of medicine or to draw body fluids from animals.

jeringuilla – un instrumento usado para administrar inyecciones de medicina o para sacar fluidos corporales de los animales.

Syringing – a light application of water to a turfgrass.

aplicación con mango de riego – una aplicación ligera de agua a un césped.

Systemic herbicide – a herbicide that is absorbed by the roots of the plant and translocated throughout the plant.

herbicida sistemático – un herbicida que es absorbido por las raíces de la planta y desplazado por la planta.

Tack – equipment or gear for horses.

arreo – guarniciones o aparejo para caballos.

Tact – skill of encouraging others in positive ways.

tacto – característica o facilidad de animar a los demás en maneras positivas.

TAN – measurement of total ammonial nitrogen.

NAT – medida de nitrógeno amoniacal total.

Tankage – dried animal residue after slaughter.

fertilizante orgánico – los desechos secados del animal después del sacrificio.

Taproot – large main root of the system; usually has little or no branch roots.

raíz eje – gran raíz primaria del sistema radicular; usualmente no tiene pocos o ningunos otros raices secundarias.

Targeted pest – identified pest that, if introduced, poses a major economic threat.

animal o insecto nocivo objetivo – animal o insecto nocivo que, si se introduce, representa una amenaza económica mayor.

Tattoo – means of marking animals for identification.

tatuaje – medios para marcar los animales para fines de identificación.

Taxonomy – systematic classification of plants and animals.

taxonomía – clasificación sistemática de plantas y animales.

TDN – total digestible nutrients; the measure of digestibility of feed.

NDT – nutrientes digestibles totales.

Teats – the appendages of an udder.

tetas – los apéndices del ubre.

Technology – application of science to an industrial or commercial objective; also, the equipment and expertise to cultivate, harvest, store, process, and transport crops for consumption.

tecnología – la aplicación de ciencia a un objetivo industrial o comercial; también, el equipo y pericia para cultivar, cosechar, almacenar, procesar, y transportar las cosechas para consumpción.

Terminal – growing at the end of a branch or stem.

terminal – dícese de algo que está creciendo en el extremo de una rama o un rabo.

Terminal bud – bud at the end of a twig or branch.

botón terminal – yema que está en el extremo de una rama o ramita.

Terminal market – a stockyard that acts as a place to hold animals until they are sold to another party.

mercado terminal – un corral que sirve para guardar los animales hasta que se vendan a otro individuo.

Terrace – soil or wall structure built across the slope to capture water and move it safely to areas where it will not cause erosion.

bancal – estructura de tierra o pared construida a lo largo del declive para coger el agua y transportarla seguramente a los áreas en donde no causarán la erosión.

Terrestrial – land organism.

terrícola – organismo que vive en la tierra.

Testes – organs that produce sperm cells.

testes – órganos que producen los espermatozoides.

Testosterone – male sex hormone.

testosterona – hormona sexual masculino.

Tetraethyl lead – colorless, poisonous, and oily liquid.

plomo tetraetílico – líquido incoloro, tóxico, y aceitoso.

Texture – visual or surface quality.

textura – cualidad visual o del superficie de algo.

Thatch – building of organic matter on the soil and around turfgrass plants.

fertilización del suelo – la fomentación de materia orgánica en el suelo y alrededor de las plantas de césped.

Thermo – classification of plants according to growing season required.

termo – clasificación de las plantas según la temporada de cultivación que se exigen.

Threshing – removal of grain from the plant.

trilla – acción de separar del grano de la paja de la planta.

Thymine – a base in genes designated by the letter *T*.

timino – una base en los genes nominado por la letra «T».

Tidewater – water that flows up the mouth of a river as the ocean tide rises and comes in.

agua de marea – agua que fluye a la boca de un río mientras la marea sube y la entra.

Tillable – soil that is workable with tools and equipment.

arable – tierra que se puede labrar con herramientas y equipo.

Tillers – new shoots of a grass plant that develop at the axillary bud of the crown and form within the lower leaf sheath.

retoños – vástagos o brotes que brotan del botón axilar de la corona y desarrollan dentro de la vaina exterior de hoja.

Timberland – forest land capable of producing over 20 ft³ of industrial wood per year.

bosque maderable – tierra forestal capaz de producir más de 20 ft³ de maderos industriales por año.

Tip layering – tip of a shoot place in media and covered.

acodo de punta – la punta del vástago es metido en la media y tapada.

Tissue culture – plant reproduction using very small, actively growing plant parts under sterile conditions and medium.

cultivo de tejidos – reproducción vegetal usando partes pequeñas y activamente crecientes de la planta bajo condiciones y medios estériles.

Tofu – food made by boiling and crushing soybeans and letting it coagulate into curds.

tofu – comestible hecho por hervir y triturar la soja y dejándo que se coagule para formar requesón.

Tom – male turkey.

pavo – macho de la pava.

Topdressing – a procedure in which fertilizer is broadcast lightly over close-growing plants.

abonado de cobertera – un procedimiento en lo cual se siembra al voleo el abono ligeramente sobre los cultivos que crecen muy juntos.

Topsoil – desirable proportions of plant nutrients, chemicals, and living organisms located near the surface which support good plant growth.

superficie – proporciones deseosas de alimentos nutritivos, sustancias químicas vegetales y organismos ubicados cerca de la superficie, que sostienen buen crecimiento vegetal.

Townhouse – one of a row of houses connected by common side walls.

casa de ciudad de tres o más pisos – una de una filera de casas conectadas por paredes laterales comunes.

Toxicity – a measurement of how poisonous a chemical is.

toxicidad – una medida de la noxidad de una sustancia química.

Toxin – a poisonous substance, causing injury to animals or plants.

toxina – una sustancia capaz de producir efectos tóxicos, resultando en daño a animales o plantas.

Tractor – source of power for belt-driven machines as well as for pulling.

locomotora de tracción – fuente de potencia para máquinas propulsadas por correa, tanta como para arrastrar.

Trade name – the manufacturer's name for its pesticide product.

marca registrada – el nombre del fabricante por su producto de pesticida.

Transpiration – process by which a plant loses water vapor.

transpiración – proceso por lo cual la planta se pierde de su vapor.

Transplants – plants grown from seed in special structures.

trasplantes – plantas cultivadas de semilla en estructuras especiales.

Trap crop – a susceptible crop planted to attract a pest into a localized area.

planta-tampa – un cultivo susceptible sembrado para atraer a un animal o insecto dañino a un área localizada.

Tree – woody plant that produces a main trunk and has a more or less distinct and elevated head (a height of 15 ft or more).

árbol – planta leñosa que produce un tronco principal y que tiene una copa más o menos distinta y elevada (de 15 pies o más de alto).

Tree spade – a piece of equipment, operated by hydraulics, that will dig a tree in a matter of a few minutes to a very specific size ball.

pala mecánica – un aparato, operado por la hidráulica, que arrancará a un árbol en meros minutos a un cepillón muy específico.

Tripe – the pickled rumen of cattle and sheep.

callos – el herbario adobado de ganado y corderos.

Trot – in horses, fast two-beat diagonal gait.

trote – de los caballos, un paso rápido diagonal de dos tiempos.

Trucker – person transporting commodities from farm to consumer.

camionero – persona que transporta la mercancía de la finca al consumidor.

Tuber crop – crop grown for its fleshy, underground storage stem.

tubérculo – cultivo cultivado por su raíz carnosa de almacenamiento.

Tubers – specialized food-storage stem that grows underground.

tubérculos radiculares – raíz especializada de almacenamiento de alimentación que crece subterráneamente.

Turbidity – measure of suspended solids.

turbiedad – medida de sólidos suspendidos.

Turf – the turfgrass plant and soil immediately below it; also, grass used for decorative or soil-holding purposes.

tepe – la planta de césped y la tierra inmediatamente por debajo de éste; también la hierba usada para fines de adorno o de sujetar el suelo.

Turfgrass – grasses that are mowed frequently for a short and even appearance.

césped – hierbas que se siegan frecuentemente para obtener un aspecto corto y plano.

Turgor – swollen or still condition as a result of being filled with liquid.

turgencia – condición hinchada o rígida como resulto de llenarse con líquido.

Two-year-old – cattle that are two or more years old.

de dos años – ganado que tienen dos años o más.

Udder – the milk-secreting glands of the animal.

ubre – las glándulas del animal que secretan la leche.

Underripe – as applied to vegetation, any that is not mature.

insuficientemente madura – como se aplica a la vegetación, cualquiera que no es madura.

Universal solvent – material that dissolves or otherwise changes most other materials.

solvente universal – material que disuelve o de otro modo cambia la mayoría de los otros materiales.

Unselfishness – placing the desires and welfare of others above oneself.

desinterés – poniendo los deseos y el bienestar de los demás mas allá de los de si mismo.

Unsoundness – in horses, any abnormality that affects the use of the horse.

defectuoso – de los caballos, cualquiera abnormalidad que afecta el uso del caballo.

Urea – synthetic source of nitrogen made from air, water, and carbon.

urea – fuente sintética de nitrógeno hecho de aire, agua, y carbono.

Urinary system – system that removes waste materials from the blood.

aparato urinario – conjunto de órganos que elimina los desechos de la sangre.

Urine – liquid body waste.

orina – desechos líquidos corporales.

USDA – United States Department of Agriculture.

DAEU – Departamento de Agricultura de los Estados Unidos.

Utility-type grass – a turfgrass adapted to lower maintenance levels.

césped de utilidad – un césped adaptado a niveles bajos de mantenimiento.

Vaccination – the injection of an agent into an animal to prevent disease.

vacunación – la inyección de un agente a un animal para evitar la enfermedad.

Vacuum pan – a device used to remove water from milk.

tacho de vacío – un dispositivo usado para quitar el agua de la leche.

Vapor drift – movement of pesticide vapors due to chemical volatilization of the product.

arrastre de vapor – el traslado de vapores de pesticidas debido a la volatilización química del producto.

Variegated – a leaf having streaks, marks, or patches of color.

jaspeado – una hoja que tiene rayas, marcas, o manchas de color.

Variety – a subdivision of a species, it has various inheritable characteristics of form and structure that are continued through both sexual and asexual propagation.

variedad – una subdivisión de una especie, tiene varias características capaces de transmitir por herencia, de forma y estructura que se transmiten por la propagación sexual y asexual.

Vascular bundle – food, water, and nutrient-carrying tissue extending from roots to tips of plants.

haz vascular – tejido que conduce la alimentación, el agua y los nutrientes, que extiende desde las raíces hasta las puntas de las plantas.

Veal – young calves or the meat from young calves.

ternura – terneros jóvenes o el carne de ellos.

Vector – a living organism that transmits a disease.

portador – un organismo viviente que transmite una enfermedad.

Vegetable – any herbaceous plant whose fruit, seeds, roots, tubers, bulbs, stems, leaves, or flower parts are used as food.

verdura – cualquiera planta herbácea cuyos frutos, semillas, raíces, bulbos, tallos, hojas, o partes de flor, son usados como comestibles.

Vegetative – plant parts such as stems, leaves, roots, and buds, but not flowers, fruits, and seeds.

vegetativo – partes de la planta como los tallos, las hojas, las raíces, y los botones, pero no las flores, los frutos, y las semillas.

Vegetative spreading – reproduction in plants other than by seed.

propagación vegetativa – reproducción de las plantas de otro modo de por semilla.

Veneer – thin sheet of wood used in paneling and furniture.

chapa – hoja delgada de madera que se usa en revestimiento de madera y en mueblería.

Vermiculite – mineral matter used for starting plant seeds and cuttings.

vermiculita – materia mineral usada para empezar semillas y talas de plantas.

Vertebrates – animals with backbones.

vertebrados – animales que tienen columnas vertebrales.

Vertical integration – several steps in the production, marketing, and processing of animals are joined together.

integración vertical – varias medidas en la producción, mercadeo, y elaborar de animales son unidos.

Veterinarian – animal doctor.

veterinario – doctor de animales.

Viable – alive and capable of functioning.

viable – vivo y capaz de funcionar.

Vice – bad habit in horses.

resabio – mala costumbre de un caballo.

Virgin – a forest that has never been harvested.

selva virgen – un bosque que nunca jamás ha sido cosechado.

Virus – pathogenic entities consisting of nucleic acid and a protein sheath.

virus – cuerpos patogénicos que consisten de ácido nucléico y una vaina de proteina.

Viscera – internal organs of an animal including heart, liver, and intestines.

víscera – órganos internos del animal inclusive los corazón, higado e intestinos.

Vitamin – complex chemicals essential for normal body functions.

vitamina – sustancias químicas esenciales para funciones normales del cuerpo.

Volatilization – process in which a liquid or solid changes to the gaseous phase.

volatilización – proceso en que un líquido o sólido transforma en el estado gaseoso.

Voluntary muscles – muscles that can be controlled by animals to do things such as walk and eat food.

músculos voluntarios – músculos que pueden ser controlados por los animales para hacer cosas como pasar y comer alimentos.

Walk – in horses, a slow four-beat gait.

ponerse al paso – de los caballos, un paso lento de cuatro tiempos.

Warm-blooded animal – animal with the ability to regulate its body temperature.

animal de sangre caliente – animal con la capacidad de regular su temperatura de cuerpo.

Warm-season turfgrass – turfgrass adapted to the southern region of the United States which grow best at 80° to 95° F.

césped de temporada de verano – césped adaptado a la región sureña de los Estados Unidos que crece mejor a temperaturas de 80° a 95° F.

Warp – the tendency of wood to bend permanently due to moisture change.

alabearse – la tendencia de la madera de torcer permanentemente debido al cambio de la humedad.

Water – clear, colorless, tasteless, and nearly odorless liquid.

agua – líquido claro, incoloro, insípido, y casi inodoro.

Water culture – growing of plants with roots in a nutrient solution.

cultivo acuático – la cultivación de plantas con raíces en una solución nutritiva.

Water cycle – movement of water surface, to atmosphere, to surface.

ciclo de agua – movimiento del agua superficial al atmósfera, y de nuevo a las superficies.

Water hardness – measured as the amount of calcium (ppm) in the solution.

dureza de agua – medida como la cantidad de calcio (partes por mil) en la solución.

Water quality – measurement of factors that affect the utilization of water.

cualidad de agua – medida de los factores que afectan la utilización de agua.

Water resources – all aspects of water conservation and management.

recursos de agua – todos aspectos de conservación y administración de agua.

Water table – level below which soil is saturated or filled with water.

nivel hidrostático – nivel debajo de lo cual el suelo se llena o se empapa de tierra.

Waterfowl – ducks and geese.

aves acuáticos – patos y gansos.

Waterlogged – soaked or saturated with water.

anegado – mojado o empapado.

Wean – to accustom an animal to take food other than by nursing.

destetar – hacer que el animal se acostumbre alimentarse de otro modo de mamar.

Weed – a plant that is not wanted.

mala hierba – una planta no deseable.

Wether – castrated male member of the sheep family.

carnero castrado – carnero castrado.

Wetlands – a lowland area often associated with ponds or creeks that is saturated with freshwater.

tierras mojadas – una región de tierra baja que muchas veces es asociada con charcos o arroyos, que es saturada con agua dulce.

White muscle disease – muscle degeneration in calves caused by a deficiency of vitamin E.

distrofia muscular por falta de nutrición – degeneración de los músculos de los becerros causada por la carencia de vitamina E.

Wholesale marketing – the marketing of a product through a middleman.

mercado al por mayor – el mercadeo de un producto por un intermediario.

Wholesaler – person who sells to the retailer.

mayorista – persona que vende al minorista.

Wildlife – animals that are adapted to live in a natural environment without the help of humans.

fauna – animales que se han adaptado a vivir en un ambiente natural sin la ayuda de los seres humanos.

Wood grain – hard and soft patterns in wood caused by growth of the tree as it adds successive annual growth rings.

hilo de madera – diseños duros y blandos en la madera causados por el crecimiento del árbol mientras crece la capa cortical anualmente.

Wood lot – small, privately owned forest.

rodal pequeño – bosque pequeño que tiene propietario individual.

Woody – hard, stiff, dark-colored growth of plants; they are winter hardy.

leñoso – crecimiento duro, resistente, y de color oscuro de las plantas; son resistentes al invierno.

Worker – female bee that does the work in the hive.

obrera – abeja hembra que hace el trabajo de la colmena.

X-Gal – compound that causes marked bacteria to turn blue.

«X-Gal» – una compuesta que causa que se vuelva azúl la bacteria marcada.

Xylem – vessels of the vascular bundle that carry the water and nutrients from roots to leaves.

xilema – vasos del haz vascular que llevan el agua y los alimentos nutritivos desde las raíces a las hojas.

Xylem tissue – conductive tissue responsible for transport of water and nutrients within the plant.

tejido de xilema – tejido conductivo responsable de la transportación de agua y alimentos nutritivos dentro de la planta.

Yardage fee – a fee for caring for animals until they are sold.

gastos del encierro – un honrario por cuidar los animales hasta que sean vendidos.

Yearling – cattle between one and two years of age.

añal – ganado entre uno y dos años de edad.

Yield grade – grade based on the amount of lean meat in relation to the amount of fat and bone in cattle and sheep.

proporción carne/grasa – grados basados en la cantidad de carne magra en proporción con la cantidad de grasa y hueso en el ganado vacuno y las ovejas.

Zygote – fertilized egg.

cigoto – óvulo fecundado.

INDEX